T0390628

PANGOLINS

Biodiversity of the World: Conservation From Genes to Landscapes

Series Editor: Philip Nyhus

Snow Leopards, 9780128022139
Edited by Tom McCarthy and David Mallon

Cheetahs: Biology and Conservation, 9780128040881
Edited by Laurie Marker, Lorraine K. Boast, and Anne Schmidt-Kuentzel

Whooping Cranes: Biology and Conservation, 9780128035559
Edited by John B. French, Jr., Sarah J. Converse, and Jane E. Austin

Biodiversity of the World: Conservation From Genes to Landscapes

PANGOLINS

SCIENCE, SOCIETY AND CONSERVATION

Series Editor

PHILIP J. NYHUS

Environmental Studies Program, Colby College, Waterville, ME, United States

Volume Editors

DANIEL W.S. CHALLENDER

Department of Zoology and Oxford Martin School, University of Oxford, Oxford, United Kingdom

IUCN SSC Pangolin Specialist Group, ℅ Zoological Society of London, Regent's Park, London, United Kingdom

HELEN C. NASH

Department of Biological Sciences, National University of Singapore, Singapore

IUCN SSC Pangolin Specialist Group, ℅ Zoological Society of London, Regent's Park, London, United Kingdom

CARLY WATERMAN

Conservation and Policy, Zoological Society of London, Regent's Park, London, United Kingdom

IUCN SSC Pangolin Specialist Group, ℅ Zoological Society of London, Regent's Park, London, United Kingdom

Academic Press is an imprint of Elsevier
125 London Wall, London EC2Y 5AS, United Kingdom
525 B Street, Suite 1650, San Diego, CA 92101, United States
50 Hampshire Street, 5th Floor, Cambridge, MA 02139, United States
The Boulevard, Langford Lane, Kidlington, Oxford OX5 1GB, United Kingdom

British Library Cataloguing-in-Publication Data
A catalogue record for this book is available from the British Library

Library of Congress Cataloging-in-Publication Data
A catalog record for this book is available from the Library of Congress

ISBN: 978-0-12-815507-3

For Information on all Academic Press publications
visit our website at https://www.elsevier.com/books-and-journals

Publisher: Charlotte Cockle
Acquisition Editor: Anna Valutkevich
Editorial Project Manager: Devlin Person
Production Project Manager: Punithavathy Govindaradjane
Cover Designer: Matthew Limbert

Typeset by MPS Limited, Chennai, India

Map tiles in Chapters 4–11 by Stamen Design, under CC BY 3.0. Data by OpenStreetMap, under ODbL.

Contents

Section One

WHAT IS A PANGOLIN?

Part 1: Evolution, Phylogeny and Taxonomy

Part 2: Biology, Ecology and Status

Section Three
CONSERVATION SOLUTIONS

Part 1: Law Enforcement and Regulation

17. Conserving pangolins through international and national regulation and effective law enforcement

STUART R. HARROP

18. Combating the illegal pangolin trade - a law enforcement practitioner's perspective

CHRISTIAN PLOWMAN

19. Addressing trade threats to pangolins in the Convention on International Trade in Endangered Species of Wild Fauna and Flora (CITES)

DANIEL W.S. CHALLENDER AND COLMAN O'CRIODAIN

20. Understanding illegal trade in pangolins through forensics: applications in law enforcement

ANTOINETTE KOTZE, ROB OGDEN, PHILIPPE GAUBERT, NICK AHLERS, GARY ADES, HELEN C. NASH AND DESIRE LEE DALTON

Part 2: Awareness Raising and Behavior Change

21. No longer a forgotten species: history, key events, and lessons learnt from the rise of pangolin awareness

PAUL THOMSON AND LOUISE FLETCHER

22. Changing consumer behavior for pangolin products

GAYLE BURGESS, ALEGRIA OLMEDO, DIOGO VERÍSSIMO AND CARLY WATERMAN

Part 3: Site-Based Protection and Local Community Engagement

23. Engaging local communities in responses to illegal trade in pangolins: who, why and how?

ROSIE COONEY AND DANIEL W.S. CHALLENDER

24. Exploring community beliefs to reduce illegal wildlife trade using a theory of change approach

DIANE SKINNER, HOLLY DUBLIN, LEO NISKANEN, DILYS ROE AND AKSHAY VISHWANATH

25. Community conservation in Nepal – opportunities and challenges for pangolin conservation

AMBIKA P. KHATIWADA, TULSHI LAXMI SUWAL, WENDY WRIGHT, DILYS ROE, PRATIVA KASPAL, SANJAN THAPA AND KUMAR PAUDEL

26. The Sunda pangolin in Singapore: a multi-stakeholder approach to research and conservation

HELEN C. NASH, PAIGE B. LEE, NORMAN T-L LIM, SONJA LUZ, CHENNY LI, YI FEI CHUNG, ANNETTE OLSSON, ANBARASI BOOPAL, BEE CHOO NG STRANGE AND MADHU RAO

27. Holistic approaches to protecting a pangolin stronghold in Central Africa

ANDREW FOWLER

Part 4: Ex Situ Conservation

28. Captive husbandry of pangolins: lessons and challenges

LEANNE VIVIAN WICKER, FRANCIS CABANA, JASON SHIH-CHIEN CHIN, JESSICA JIMERSON, FLORA HSUAN-YI LO, KARIN LOURENS, RAJESH KUMAR MOHAPATRA, AMY ROBERTS AND SHIBAO WU

36. Conservation planning and PHVAs in Taiwan

JIM KAO, JUNG-TAI CHAO, JASON SHIH-CHIEN CHIN, NIAN-HONG JANG-LIAW, JOCY YU-WEN LI, CAROLINE LEES, KATHY TRAYLOR-HOLZER, TINA TING-YU CHEN AND FLORA HSUAN-YI LO

37. Leveraging support for pangolin conservation and the potential of innovative finance

OLIVER WITHERS AND TENKE ZOLTANI

38. Supporting pangolin conservation through tourism

ENRICO DI MININ AND ANNA HAUSMANN

Section Four
THE FUTURE

39. Taking pangolin conservation to scale

DANIEL W.S. CHALLENDER, HELEN C. NASH, CARLY WATERMAN AND RACHEL HOFFMANN

List of contributors

Levita Acosta-Lagrada Palawan Council for Sustainable Development Staff (PCSDS), Puerto Princesa City, Philippines; IUCN SSC Pangolin Specialist Group, ℅ Zoological Society of London, Regent's Park, London, United Kingdom

Gary Ades Fauna Conservation Department, Kadoorie Farm & Botanic Garden, Hong Kong SAR, P.R. China

Nick Ahlers TRAFFIC, ℅ IUCN, Hatfield Gables, Pretoria, South Africa

Faraz Akrim Department of Wildlife Management, PMAS-Arid Agriculture University, Rawalpindi, Pakistan

Daniel Alempijevic Integrative Biology, Florida Atlantic University, Boca Raton, FL, United States

Dexter Alvarado Katala Foundation Inc., El Rancho, Puerto Princesa City, Philippines

Shaista Andleeb Department of Wildlife Management, PMAS-Arid Agriculture University, Rawalpindi, Pakistan

Brou Guy-Mathieu Assovi Université Felix Houphouët-Boigny, Abidjan, Côte d'Ivoire

Sam Ayebare Wildlife Conservation Society, Bronx, NY, United States

Guillaume Billet Centre de Recherche en Paléontologie-Paris (CR2P), UMR CNRS 7207, Sorbonne Université, Muséum National d'Histoire Naturelle, Paris, France

Timothy C. Bonebrake School of Biological Sciences, The University of Hong Kong, Hong Kong SAR, P.R. China

Anbarasi Boopal ACRES Wildlife Rescue Centre (AWRC), Singapore

Tom Bruce Conservation and Policy, Zoological Society of London, Regent's Park, London, United Kingdom

Gayle Burgess Institution of Environmental Sciences, London, United Kingdom; Society for the Environment, Coventry, United Kingdom

Francis Cabana Wildlife Reserves Singapore, Singapore

Rod Cassidy Sangha Pangolin Project, Dzanga-Sangha, Central African Republic

Tamar Cassidy Sangha Pangolin Project, Dzanga-Sangha, Central African Republic

Daniel W.S. Challender Department of Zoology and Oxford Martin School, University of Oxford, Oxford, United Kingdom; IUCN SSC Pangolin Specialist Group, ℅ Zoological Society of London, Regent's Park, London, United Kingdom

Jung-Tai Chao Taiwan Forestry Research Institute, Taipei, Taiwan

Shavez Cheema 1StopBorneo Wildlife, Kota Kinabalu, Malaysia; IUCN SSC Pangolin Specialist Group, ℅ Zoological Society of London, Regent's Park, London, United Kingdom

Tina Ting-Yu Chen Taipei Zoo, Taipei, Taiwan

Wenda Cheng School of Biological Sciences, The University of Hong Kong, Hong Kong SAR, P.R. China

Jason Shih-Chien Chin Taipei Zoo, Taipei, Taiwan

Ju Lian Chong School of Science and Marine Environment & Institute of Tropical Biodiversity and Sustainable Development, Universiti Malaysia Terengganu, Kuala Nerus, Malaysia; IUCN SSC Pangolin Specialist Group, ℅ Zoological Society of London, Regent's Park, London, United Kingdom

Yi Fei Chung Conservation Division, National Parks Board, Singapore

Rosie Cooney IUCN CEESP/SSC Sustainable Use and Livelihoods Specialist Group, Gland, Switzerland; Fenner School of Environment and Society, Australian National University, Canberra, ACT, Australia

Desire Lee Dalton National Zoological Garden, South African National Biodiversity Institute, Pretoria, South Africa; University of Venda, Thohoyandou, South Africa

Tim R.B. Davenport Tanzania Program, Wildlife Conservation Society (WCS), Zanzibar, Tanzania

Enrico Di Minin Department of Geosciences and Geography, University of Helsinki, Helsinki, Finland; Helsinki Institute of Sustainability Science (HELSUS), University of Helsinki, Helsinki, Finland; School of Life Sciences, University of KwaZulu-Natal, Westville, South Africa

Holly Dublin IUCN CEESP/SSC Sustainable Use and Livelihoods Specialist Group, Nairobi, Kenya

Sérgio Ferreira-Cardoso Institut des Sciences de l'Evolution, UMR 5554, CNRS, IRD, EPHE, Université de Montpellier, Montpellier, France

Louise Fletcher IUCN SSC Pangolin Specialist Group, ℅ Zoological Society of London, Regent's Park, London, United Kingdom

Andrew Fowler Conservation and Policy, Zoological Society of London, Regent's Park, London, United Kingdom

Philippe Gaubert Laboratoire Évolution & Diversité Biologique (EDB), Université de Toulouse Midi-Pyrénées, CNRS, IRD, UPS, Toulouse, France; CIIMAR, University of Porto, Matosinhos, Portugal

Timothy J. Gaudin Department of Biology, Geology and Environmental Science, University of Tennessee at Chattanooga, Chattanooga, TN, United States

Maja Gudehus Sangha Pangolin Project, Dzanga-Sangha, Central African Republic

Lam Kim Hai Save Vietnam's Wildlife, Cuc Phuong National Park, Ninh Binh Province, Vietnam

Stuart R. Harrop Kingston University, London, United Kingdom

John Hart Lukuru Foundation, Kinshasa, Democratic Republic of the Congo

Terese Hart Lukuru Foundation, Kinshasa, Democratic Republic of the Congo

Anna Hausmann Department of Geosciences and Geography, University of Helsinki, Helsinki, Finland; Helsinki Institute of Sustainability Science (HELSUS), University of Helsinki, Helsinki, Finland

Lionel Hautier Institut des Sciences de l'Evolution, UMR 5554, CNRS, IRD, EPHE, Université de Montpellier, Montpellier, France

Martin Hega Gabon Program, Wildlife Conservation Society (WCS), Libreville, Gabon

Sean P. Heighton Laboratoire Évolution & Diversité Biologique (EDB), Université de Toulouse Midi-Pyrénées, CNRS, IRD, UPS, Toulouse, France; Department of Zoology and Entomology, University of Pretoria, Pretoria, South Africa

Sarah Heinrich School of Biological Sciences, University of Adelaide, Adelaide, SA, Australia; Monitor Conservation Research Society, Big Lake Ranch, BC, Canada; IUCN SSC Pangolin Specialist Group, ℅ Zoological Society of London, Regent's Park, London, United Kingdom

Michael Hoffmann Conservation and Policy, Zoological Society of London, Regent's Park, London, United Kingdom

Rachel Hoffmann IUCN Species Survival Commission, IUCN, Cambridge, United Kingdom

Ichu Godwill Ichu Pangolin Conservation Network, ℅ Central Africa Bushmeat Action Group (CABAG), Yaoundé, Cameroon; IUCN SSC Pangolin Specialist Group, ℅ Zoological Society of London, Regent's Park, London, United Kingdom

Daniel J. Ingram African Forest Ecology Group, Biological and Environmental Sciences, University of Stirling, Stirling, United Kingdom

Nausheen Irshad Department of Zoology, University of Poonch, Rawalakot, Pakistan

Nian-Hong Jang-Liaw Taipei Zoo, Taipei, Taiwan

Raymond Jansen Department of Environmental, Water and Earth Sciences, Tshwane University of Technology, Pretoria, South Africa; African Pangolin Working Group, Pretoria, South Africa

Zhigang Jiang Institute of Zoology, Chinese Academy of Sciences, Beijing, P.R. China

Jessica Jimerson Save Vietnam's Wildlife, Cuc Phuong National Park, Ninh Binh Province, Vietnam

Jim Kao Taipei Zoo, Taipei, Taiwan

Prativa Kaspal Women for Conservation/ Bhaktapur Multiple Campus, Tribhuvan University, Kirtipur, Nepal

Lydia K.D. Katsis IUCN SSC Pangolin Specialist Group, ℅ Zoological Society of London, Regent's Park, London, United Kingdom

Ambika P. Khatiwada National Trust for Nature Conservation, Lalitpur, Nepal

Antoinette Kotze National Zoological Garden, South African National Biodiversity Institute, Pretoria, South Africa; Genetics Department, University of the Free Sate, Bloemfontein, South Africa

Ade Kurniawan Wildlife Reserves Singapore, Singapore; IUCN SSC Pangolin Specialist Group, ℅ Zoological Society of London, Regent's Park, London, United Kingdom

Juan Lapuente Comoé Chimpanzee Conservation Project, Comoé Research Station, Côte d'Ivoire & Animal Ecology and Tropical Biology, Biozentrum, Universität Würzburg Tierökologie und Tropenbiologie (Zoologie III), Würzburg, Germany

Natalie Lawrence Department of History and Philosophy of Science, University of Cambridge, Cambridge, United Kingdom

Paige B. Lee Department of Conservation, Research and Veterinary Services, Wildlife Reserves Singapore, Singapore

Caroline Lees IUCN SSC Conservation Planning Specialist Group, Apple Valley, MN, United States

Chenny Li Conservation Division, National Parks Board, Singapore

Hou-Feng Li Department of Entomology, National Chung-Hsing University, Taichung, Taiwan

Jocy Yu-Wen Li Taipei Zoo, Taipei, Taiwan

Norman T-L Lim National Institute of Education, Nanyang Technological University, Singapore; IUCN SSC Pangolin Specialist Group, ℅ Zoological Society of London, Regent's Park, London, United Kingdom

Chung-Chi Lin Department of Biology, National Changhua University of Education, Changhua, Taiwan

Flora Hsuan-Yi Lo Taipei Zoo, Taipei, Taiwan

Karin Lourens Johannesburg Wildlife Veterinary Hospital, Johannesburg, South Africa

Sonja Luz Department of Conservation, Research and Veterinary Services, Wildlife Reserves Singapore, Singapore; IUCN SSC Pangolin Specialist Group, ℅ Zoological Society of London, Regent's Park, London, United Kingdom; IUCN SSC Conservation Planning Specialist Group, Apple Valley, MN, United States

Tariq Mahmood Department of Wildlife Management, PMAS-Arid Agriculture University, Rawalpindi, Pakistan

Fiona Maisels Wildlife Conservation Society, Bronx, NY, United States; Biological and Environmental Sciences, University of Stirling, Stirling, United Kingdom

David Mills School of Life Sciences, University of Kwazulu-Natal, Durban, South Africa; Panthera, New York, NY, United States

Rajesh Kumar Mohapatra Nandankanan Zoological Park, Bhubaneswar, India

Dana J. Morin Department of Wildlife, Fisheries and Aquaculture, Mississippi State University, Starkville, MS, United States; Cooperative Wildlife Research Laboratory, Southern Illinois University, Carbondale, IL, United States; IUCN SSC Pangolin Specialist Group, ℅ Zoological Society of London, Regent's Park, London, United Kingdom

Helen C. Nash Department of Biological Sciences, National University of Singapore, Singapore; IUCN SSC Pangolin Specialist Group, ℅ Zoological Society of London, Regent's Park, London, United Kingdom

Constant Ndjassi Liberia Programme, Fauna and Flora International, Monrovia, Liberia

Bruno Nebe Pangolin Research Mundulea, Swakopmund, Namibia

Leo Niskanen IUCN Eastern and Southern Africa Regional Office, Nairobi, Kenya

Stuart Nixon Field Programmes, North of England Zoological Society, Chester Zoo, Chester, United Kingdom

Rob Ogden TRACE Wildlife Forensics Network, Edinburgh, United Kingdom; Royal (Dick) School of Veterinary Studies and the Roslin Institute, University of Edinburgh, United Kingdom

Alegria Olmedo Department of Zoology, University of Oxford, Oxford, United Kingdom; People for Pangolins, London, United Kingdom

Annette Olsson Conservation International, Singapore

Colman O'Criodain WWF International, The Mvuli, Nairobi, Kenya

Wendy Panaino Brain Function Research Group, School of Physiology and Centre for African Ecology, School of Animal, Plant and Environmental Sciences, University of the Witwatersrand, Johannesburg, South Africa

Sudarsan Panda Satkosia Tiger Reserve, Angul, India

Elisa Panjang Danau Girang Field Centre, Sabah Wildlife Department, Kota Kinabalu, Malaysia; Organisms and Environment Division, Cardiff School of Biosciences, Cardiff University, Cardiff, United Kingdom; IUCN SSC Pangolin Specialist Group, ℅ Zoological Society of London, Regent's Park, London, United Kingdom

Keri Parker Save Pangolins, ℅ Wildlife Conservation Network, San Francisco, CA, United States; IUCN SSC Pangolin Specialist Group, ℅ Zoological Society of London, Regent's Park, London, United Kingdom

Kumar Paudel Greenhood Nepal, New Baneshwor, Kathmandu, Nepal

Priyan Perera Department of Forestry and Environmental Science, University of Sri Jayewardenepura, Nugegoda, Sri Lanka

Darren W. Pietersen Mammal Research Institute, Department of Zoology and Entomology, University of Pretoria, Hatfield, South Africa; IUCN SSC Pangolin Specialist Group, ℅ Zoological Society of London, Regent's Park, London, United Kingdom

Christian Plowman Wildlife Conservation Society, Brazzaville, Republic of Congo

Paul Rankin Deceased

Madhu Rao Wildlife Conservation Society, Singapore

Amy Roberts Chicago Zoological Society/ Brookfield Zoo, Brookfield, IL, United States

Dilys Roe International Institute for Environment and Development, London, United Kingdom

Sabine Schoppe Katala Foundation Inc., El Rancho, Puerto Princesa City, Philippines; IUCN SSC Pangolin Specialist Group, ℅ Zoological Society of London, Regent's Park, London, United Kingdom

Gono Semiadi Pusat Penelitian Biologi Lembaga Ilmu Pengetahuan Indonesia, Cibinong Science Center, Bogor, Indonesia; IUCN SSC Pangolin Specialist Group, ℅ Zoological Society of London, Regent's Park, London, United Kingdom

Sandhya Sharma Conservation Biologist, Sindhupalchowk, Nepal

Debbie Shaw IUCN SSC Pangolin Specialist Group, ℅ Zoological Society of London, Regents Park, London, United Kingdom

Chris R. Shepherd Monitor Conservation Research Society, Big Lake Ranch, BC, Canada; IUCN SSC Pangolin Specialist Group, ℅ Zoological Society of London, Regent's Park, London, United Kingdom

Matthew H. Shirley Tropical Conservation Institute, Florida International University, North Miami, FL, United States

Diane Skinner IUCN CEESP/SSC Sustainable Use and Livelihoods Specialist Group, Harare, Zimbabwe

Olufemi Sodeinde Department of Biological Sciences, New York City College of Technology, City University of New York, Brooklyn, NY, United States

Withoon Sodsai Nottingham Trent University, Nottingham, United Kingdom; IUCN SSC Pangolin Specialist Group, ℅ Zoological Society of London, Regent's Park, London, United Kingdom

Durojaye Soewu Department of Fisheries and Wildlife Management, College of Agriculture, Ejigbo Campus, Osun State University, Osogbo, Nigeria

Bee Choo Ng Strange Vertebrate Study Group, Nature Society Singapore, Singapore

Nick Ching-Min Sun Graduate Institute of Bioresources, National Pingtung University of Science and Technology, Pingtung, Taiwan

Tulshi Laxmi Suwal Small Mammals Conservation and Research Foundation (SMCRF), Kathmandu, Nepal; Department of Tropical Agriculture and International Cooperation, National Pingtung University of Science and Technology, Pingtung, Taiwan

Jonathan Swart Welgevonden Game Reserve, Vaalwater, South Africa

Sanjan Thapa Small Mammals Conservation and Research Foundation (SMCRF), Kathmandu, Nepal

Paul Thomson Save Pangolins, ℅ Wildlife Conservation Network, San Francisco, CA, United States

Kathy Traylor-Holzer IUCN SSC Conservation Planning Specialist Group, Apple Valley, MN, United States

Michael 't Sas-Rolfes School of Geography and the Environment and Oxford Martin School, University of Oxford, Oxford, United Kingdom

Diogo Veríssimo Department of Zoology and Oxford Martin School, University of Oxford, Oxford, United Kingdom; Institute for Conservation Research - San Diego Zoo, Escondido, CA, United States

Akshay Vishwanath IUCN Eastern and Southern Africa Regional Office, Nairobi, Kenya

Martin T. Walsh Wolfson College, University of Cambridge, Cambridge, United Kingdom

Muhammad Waseem WWF-Pakistan, Islamabad, Pakistan

Carly Waterman Conservation and Policy, Zoological Society of London, Regent's Park, London, United Kingdom; IUCN SSC Pangolin Specialist Group, ℅ Zoological Society of London, Regent's Park, London, United Kingdom

John R. Wible Section of Mammals, Carnegie Museum of Natural History, Pittsburgh, PA, United States

Leanne Vivian Wicker Australian Wildlife Health Centre, Healesville Sanctuary, Zoos Victoria, Healesville, VIC, Australia

Daniel Willcox Save Vietnam's Wildlife, Cuc Phuong National Park, Ninh Binh Province, Vietnam; IUCN SSC Pangolin Specialist Group, ℅ Zoological Society of London, Regent's Park, London, United Kingdom

Oliver Withers Conservation and Policy, Zoological Society of London, Regent's Park, London, United Kingdom

Nicci Wright Humane Society International – Africa, Johannesburg, South Africa; African Pangolin Working Group, Johannesburg, South Africa

Wendy Wright School of Health and Life Sciences, Federation University Australia, Gippsland, VIC, Australia

Shibao Wu School of Life Science, South China Normal University, Guangzhou, P.R. China

Shuang Xing School of Biological Sciences, The University of Hong Kong, Hong Kong SAR, P.R. China

Yishuang Yu School of Life Science, South China Normal University, Guangzhou, P.R. China

Fuhua Zhang School of Life Science, South China Normal University, Guangzhou, P.R. China

Mingxia Zhang Xishuangbanna Tropical Botanical Garden, Chinese Academy of Sciences, Mengla, P.R. China

Youlong Zhou Henan University of Chinese Medicine, Zhengzhou, P.R. China

Tenke Zoltani Better Finance, Geneva, Switzerland

Foreword

> "Night miniature artist engineer... Leonardo da Vinci's replica — impressive animal and toiler of whom we seldom hear."
>
> *"The Pangolin," by poet Marianne Moore.*

One of nature's most fascinating and otherworldly animals, the pangolin, is so uniquely constructed and evolutionarily distinct, it seems like it must be a product of the imagination — too exotic and extraordinary to be true.

And then you see one in real life, and you realize that no human mind, even da Vinci's, could conjure up such a magical creature.

I remember the first time I laid eyes on one more than two decades ago, in Lao PDR while I was exploring the region. Think of a small dog covered in scales with a dinosaur tail, a long snout, serpentine tongue and no teeth, and then imagine the little animal sit up and look you in the eye so that you have a brief but unforgettable moment of connection.

My curiosity about pangolins was piqued months before I first met one. I spent the earlier part of 1996 helping produce the Red List of Threatened Species for the International Union for Conservation of Nature (IUCN), the first assessment of all mammals using a set of quantitative criteria to evaluate extinction risk.

As part of that massive undertaking, conducted with a global network of scientists, I realized how little was known about pangolins. Research was limited, including everything from their population status and ecology to poaching data. In fact, at that time, pangolins were not even classified as "threatened," but I had read about how they were increasingly being trafficked and thought they might be more imperiled than previously thought.

Soon, the illegal pangolin trade was estimated to be in the tens of thousands. In 2012, along with Dan Challender, my longtime friend and co-editor of this book, we re-formed IUCN's Species Survival Commission Pangolin Specialist Group (PSG) in an attempt to fill the knowledge gaps and inform the long-term protection of these fascinating creatures. We convened specialists and conducted an intensive status assessment of the eight extant species, and it turned out that the threat level was much higher than previously understood.

In fact, not only were all eight species of pangolin, split between Asia and Africa, considered at risk of extinction, they were soon to become regarded as the most trafficked wild mammals in the world. We made it our mission to bring the plight of pangolins to the public's attention.

Why have pangolins become such a prime target for exploitation? They have been considered culturally significant in Africa and Asia throughout human history because of their unique appearance and habits. They are the world's only scaly mammals; their armor-like scales comprised of keratin, which is the same protein in human nails and hair. Pangolins have evolved over 80 million years. Their closest relatives are the carnivores, but unlike carnivores, they subsist primarily on ants and termites, which they root out with their powerful claws, snout and long sticky tongue.

Pangolins are small but intensely powerful creatures. Several years after meeting my first pangolin in Lao PDR, I came across several while studying gorillas in a rainforest in Gabon. One evening, I saw a giant pangolin, which can weigh more than 30 kg. I snuck up behind it to watch as it clambered away. Then, I reached out and tried to hold the animal to get a better look, and it just kept moving and dragging me forward. At that moment, face down on the forest floor, I realized how amazingly strong these animals were. I also realized that they are better left alone.

Yet, pangolins can also be heartbreakingly innocent and vulnerable. Predominantly nocturnal and solitary by nature, they are perhaps nature's biggest introverts, quietly going about their business in the small hours of the night. When threatened, they roll themselves up into a ball, their scales protecting them like a shield. While this can fend off predators in the wild, it makes them easy targets for human poachers and traffickers who can simply pick them up and carry them away.

The pangolins' independent streak also makes it difficult to keep them in captivity. Their survival rate in zoos and other wildlife collections has historically been poor, although this is now starting to improve.

This brings us to the most important part of the pangolins' story: why are they being captured and trafficked in epic proportions? Some estimates suggest that more than one million pangolins have been trafficked globally since the year 2000. They are sought after mainly for their scales for use in traditional medicines as well as for their meat, which is considered a delicacy in Asia, especially in China and Vietnam.

International trade in the four species of Asian pangolin has been prohibited since 2000. And in 2017, a ban on international commercial trade of all eight species took effect. Despite these actions, nearly 70 countries and territories have been involved in the illegal pangolin trade in the last decade.

In China, wildlife protection laws prohibit the sale and consumption of pangolin meat and the sale of illegally-sourced scales. However, some 200 pharmaceutical companies produce traditional medicines containing pangolin scales, which are sold by about 700 hospitals. This is happening even though there are other pathways and multiple viable alternatives in Traditional Chinese Medicine and western medicine has not determined any physiological benefits to humans. However, news came to light in August 2019 that traditional medicines containing pangolin scales would no longer be covered by China's state insurance funds. The change, which begins in 2020, is a vital step toward reducing demand for pangolin scales.

This leads us to public awareness, where much of the battle to save the pangolin is now being waged. Here we are seeing some signs of hope, as efforts to increase understanding and engagement of different audiences are beginning to gain traction.

One method of changing sentiments and, hopefully, habits is to look to key influencers, such as celebrities, to embrace the cause and help raise awareness of pangolins and their threats. Such individuals may potentially sway their followers to support pangolin conservation and/or discourage use and consumption of pangolin products. In 2014, Dan and I worked with the Royal Foundation's United for Wildlife coalition and Rovio Entertainment (creators of Angry Birds) to develop an app called "Roll with the Pangolins." After Prince William, the Duke of Cambridge, played it in a high-profile video, people around the world learned about the pangolins and their plight. The hope was to inspire further action for pangolin conservation.

While this effort primarily focused on saving pangolins, we also learned lessons that have implications for much broader conservation work. We created a body of knowledge about how to take a species from obscurity to being relatively well-known among key audiences — and then how to scale this effort to reach the broader public. In 2016, the pangolin even made a splash in Hollywood with a cameo in Disney's "The Jungle Book," a mainstream milestone for all those who care about the conservation of these amazing creatures. I am also heartened to say that at recent meetings of the Convention on International Trade in Endangered Species of

Wild Fauna and Flora (CITES), talk of pangolins has been widespread and the animals have started to receive the attention they deserve.

But much more needs to be done to secure a future for pangolins. One of the keys will be to find effective ways to curb human consumption and use of pangolins. With the global conservation movement increasingly embracing behavioral science, there are more opportunities than ever before to identify and target root causes of demand. At the same time, local non-governmental organizations are playing a larger role and also becoming interested in behavior change.

Another way to drive awareness and change is through compelling thought leadership like this book, which compiles the latest information on pangolins and the dangers they face, outlines solutions that need to be implemented, and details the levers that can be pulled to address the escalating threats — from local community engagement, to demand reduction, to effective law enforcement.

There is simply no other mammal like the pangolin, which makes it special and important to save. Pangolins are the most innocent, shy, captivating creatures, and yet they are being hunted and poached into extinction. The injustice in that should compel us all to action.

For me, the decision to fight for the future of pangolins has also become deeply personal. In many ways, I see pangolins as more than a group of species in peril, they are a metaphor for all wildlife in our modern age, and as a conservationist this goes to my own sense of purpose. If we want to maintain the great diversity of life, we want to save species like the pangolin.

In the face of humankind's relentless invasion and pillaging of natural habitats and insatiable consumption of animals, the planet's wild creatures are becoming as defenseless as a small, solitary pangolin, curled in a ball and carried away by a poacher to a painful and certain death. In the same way, the fate of all wild animals now lies in human hands. Only we can save them.

When people ask me what can they do, I always say choose one or two things in your life and stick with it. For me, pangolins are a common cause in my life that I will keep focusing on, no matter what I am doing. We all can have similar focus points and make sure we dedicate our time and energy to them. That is how you drive change. It does not happen over one or two years; it is a lifetime commitment.

In my office at the National Geographic Society headquarters in Washington, DC, I keep a small pangolin sculpture on my desk as a reminder of this personal charge. Across from my desk is an iconic pangolin image by renowned photographer Joel Sartore, a National Geographic Fellow. When I walk out of the door, I pass the pangolin portrait with its intense, almost human stare.

"Impressive animal and toiler of whom we seldom hear." Times have changed dramatically since poet Marianne Moore wrote that line in the 1930s. Today, with the future of pangolins at dire risk, we have to continue to lift them out of obscurity and continue to spotlight them to the world. We have to make sure that more people hear and know about them. We have to create a paradigm shift.

Fortunately, there are individuals around the world who are already working to realize this goal. Dedicated scientists, researchers, conservationists — and countless other champions for pangolins and the natural world — have made, and will continue to make, vital contributions in driving awareness, engagement and progress for the pangolin.

To all those who read this book, and your choice to carry forward its message, thank you for taking one more important step forward in this quest. It is one more reason to have hope that we can and *will* save these most exquisite, enchanting and precious species.

Dr. Jonathan E.M. Baillie

Executive Vice President and Chief Scientist, National Geographic Society, Washington, DC, United States

Preface

The Intergovernmental Science-Policy Platform on Biodiversity and Ecosystem Services recently published a landmark assessment of the state of the world's biological diversity. The sobering report concludes there is overwhelming evidence that humans are contributing to a significant decline in the world's life support system through activities such as deforestation, overharvesting, climate change, pollution, and invasive alien species.

Recognition that the world's ecosystems are under threat and species extinctions are increasing at unprecedented rates is, of course, not news to conservation professionals. Scholars, practitioners, activists, and officials spanning nearly every discipline are engaged in extraordinary efforts to understand these complex threats and to develop and support innovative conservation and restoration solutions.

In recognition of these conservation challenges and opportunities, in 2016 Academic Press, an imprint of Elsevier, launched *Biodiversity of the World: Conservation from Genes to Landscapes*. Books in this series are edited and written by prominent scholars and practitioners to illuminate and advance biodiversity science and conservation. Each volume focuses on one species or taxonomic group of conservation concern and covers topics spanning behavior, conservation, ecology, evolution, genetics, management, physiology, policy, restoration, and sustainability. Contributions cross disciplinary boundaries and spatial scales—from genes to landscapes.

Pangolins: Science, Society and Conservation, edited by Daniel W.S. Challender, Helen C. Nash, and Carly Waterman, is the fourth title in the series. *Pangolins* includes contributions from the world's leading authorities on the science and conservation of the species. This landmark volume presents for the first time the best available information about the evolution and taxonomy, natural history, culture, threats, and conservation of the world's pangolins. This was a herculean effort by the editors to pull together more than three dozen cutting-edge chapters from the world's leading pangolin experts.

Pangolins was completed in 2019 just as the 18th meeting of the Conference of the Parties to the Convention on International Trade in Endangered Species of Wild Fauna and Flora (CITES) was convening. CITES is the global treaty established almost half a century ago to prevent species from becoming endangered or extinct because of international trade. The timing of this book is notable because it was just three short years ago at the 17th meeting of the Conference of the Parties in South Africa that this book began its journey following a meeting I had with Dr. Daniel Challender, Chair of the Pangolin Specialist Group of the International Union for Conservation of Nature (IUCN) Species Survival Commission (SSC). After representatives of 182 countries unanimously agreed to a ban on commercial, international trade of all pangolin species, one of the signature events of the conference, delegates and attendees exploded in cheers and applause. Minutes later I talked with Dr. Challender about contributing a book on pangolins for the series. I was thrilled he was already thinking about developing such a book and that he agreed to do this. He has leveraged his infectious enthusiasm and boundless energy to bring together a world-class group of co-editors and contributors to complete this truly comprehensive and authoritative book.

This is the fourth volume in the series. The inaugural volume in 2016 was *Snow Leopards* , edited by Thomas McCarthy and David Mallon, winner of the Wildlife Society's 2017 Wildlife Publication Award in the Book Category. This was followed in 2017 by the publication of *Cheetahs: Biology and Conservation*, edited by Laurie Marker, Lorraine Boast, and Anne Schmidt-Kuentzel, and in 2018 by the publication of *Whooping Cranes: Biology and Conservation*, edited by John French, Sarah Converse, and Jane Austin. One goal of the series is to publish timely books that will be recognized internationally for comprehensive, multi-disciplinary, and authoritative treatment of species and taxonomic groups of high conservation concern. In completing this remarkable synthesis of the science and conservation of the unique and endearing pangolins, Daniel, Helen, and Carly have raised the bar for all future books in the series. It was an honor to have these exceptional editors and the chapter authors contribute their expertise to this series, and this book will undoubtedly have an enormous influence on pangolin research and conservation for years to come.

Books of this complexity are not possible without a supportive publisher. This series was possible because of the encouragement and enthusiasm of Kristi Gomez, former Senior Acquisition Editor for Life Sciences at Elsevier, who championed the original idea. Anna Valutkevich, Acquisitions Editor for Animal Science, Organismal, and Evolutionary Biology has expertly and actively continued to support the growth of the series. Devlin Person, Editorial Project Manager, and the rest of the Elsevier team skillfully maneuvered this volume through production.

In a time when threats to the world's biological diversity are increasing, it is heartening to know there are countless individuals and organizations working to develop and apply the best available science to study and conserve the world's most threatened species. The story of the pangolin is at once a tragic example of how humans can do great damage to valuable and enigmatic species and their habitats—but importantly it is also an optimistic example illustrating how dedicated scientists and practitioners can organize and address a seemingly insurmountable conservation challenge.

Additional books in the series are underway. I hope you will enjoy and value *Pangolins: Science, Society and Conservation*, and I hope you will share my enthusiasm and excitement for the upcoming volumes on similarly important species written and edited by experts exploring the science and conservation of the world's biological diversity from genes to landscapes.

Philip Nyhus, Series Editor
Biodiversity of the World: Conservation from Genes to Landscapes
Director, Environmental Studies Program, Colby College, Waterville, ME, United States

Acknowledgments

Pangolins, historically overlooked in conservation, received unprecedented levels of attention in October 2016 when they were included in Appendix I of the Convention on International Trade in Endangered Species of Wild Fauna and Flora (CITES), thus establishing a ban on international commercial trade in the species. Amid the international media attention, the concept for this book was born. Recognizing that there existed no "go to" resource on the science and conservation of pangolins, Philip Nyhus, suggested writing such a volume as part of Elsevier's *Biodiversity of the World: Conservation from Genes to Landscapes* series.

With little hesitation we accepted the challenge. Before long, we had drafted an outline and the process had begun. However, we couldn't do it without contributors and so we turned to our professional and social networks, built up over the last decade of research and conservation work on pangolins and similarly imperilled species. We were delighted and heartened by the enthusiastic and virtually unanimous "yes" we received from prospective contributors who shared our vision for a book that compiled the latest knowledge on pangolins, their threats and conservation. Encouragingly, the concept sailed through the international peer-review process, with the reviewers also seeing great value in the volume.

To our knowledge, this is the first comprehensive book on pangolins and their science and conservation. The strength of the volume truly lies in the contribution of the 134 authors from around the globe; their disciplines and areas of expertise range from conservation science and policy to veterinary health and innovative finance. We are enormously grateful to each and every one of them for their investment in the book – it simply would not have been possible to produce it without them.

It would have been easy for the book to focus on the biological conservation of pangolins, but that would have arguably achieved little. Contemporary understanding of pangolins is informed in part by the knowledge of local people and indigenous communities who live among or close to them. The threats facing pangolins are virtually all anthropogenic, and the solutions to conserving them, be they local or global, will certainly be devised and implemented by varied stakeholders working together. In short, pangolins need multifaceted conservation interventions incorporating a wide variety of actors across society, an approach explicitly recognized in this book as you will read in the following pages.

This book would not have been possible without the support of a broad range of people. First, we are grateful to Philip Nyhus for the opportunity to edit this book and for his support and guidance in making it a reality. At Elsevier, we thank Kavitha Balasundaram, Billie Jean Fernandez, Kristi Gomez, Punitha Govindaradjane, Praveen Kumar, Sandhya Narayanan, Anna Valutkevich, Andre Wolff, and most of all our editorial project manager, Devlin Person, for all their help and support.

We are thankful to David Mallon for his generous advice on taking on a book of this size in spite of our lack of experience and to Chris Shepherd, who provided substantial input into the initial concept and outline. We are indebted to Michael Hoffmann for his advice and support throughout the process, and indeed for access to the "Hoffmann library." We extend our gratitude to friends and colleagues who reviewed chapters of this book and provided technical input and advice when we reached out for it. We especially thank Gary Ades, Rosie Cooney, Animesh Ghose, Amy Hinsley, Rachel Hoffmann, Daniel Ingram, Jessica Jimerson, Helen O'Neill, Matthew Shirley, Nick Ching-Min Sun, Michael 't Sas-Rolfes, Scott Trageser, Tessa Ullmann, and Diogo Veríssimo. We owe a big thank you to Thomas Starnes for producing the geographic range maps for the volume. Thank you also to Rajesh Mohapatra for allowing us to use his image of Indian pangolins that adorns the front cover.

Pangolins are the most enchanting of creatures. There is much to discover to inform their conservation, and much to do to secure a future for these endearing and increasingly iconic species. If this book serves in any way as a catalyst for positive outcomes for pangolins in the future then our collective effort will have been well worth it.

Daniel W.S. Challender, Helen C. Nash and Carly Waterman

SECTION ONE

What is a Pangolin?

Overview

Section One, Part 1 of this volume discusses the evolution, morphology, phylogeny, and taxonomy of pangolins. It details the latest scientific thinking on when, where and how these unique mammals likely evolved, drawing on evidence from the fossil record and DNA to explain tens of millions of years of natural history. While evaluating this evolutionary history, Chapter 1 discusses the unique morphology of pangolins, including their epidermal scales and anatomical adaptations to a myrmecophagous diet. Chapter 2 provides an authoritative account of the phylogeny and taxonomy of fossil and extant pangolins. It also presents an updated, synthetic classification of the extant species, and discusses contemporary research revealing cryptic diversity – an avenue for future research.

Part 2 explores the natural history and status of pangolins. Chapter 3 provides an overview of the ecosystem roles of pangolins, including as predators of social insects, as prey, as parasite hosts, and providers of other ecosystem services. Chapters 4–11 present authoritative accounts of the knowledge of each species of pangolin. This includes what is known about their taxonomy, and morphology, and while all pangolins share a basic morphology, species-specific characteristics and adaptations are discussed and differences between the species highlighted. Morphometric data is also presented for each species. These chapters include up to date information on pangolin distribution and fully updated geographic range maps. Knowledge of habitats, ecology and behavior, ontogeny and reproduction, populations, status, and threats is also presented. Overexploitation, the main threat to most species, is expanded on in Section Two.

PART 1

Evolution, Phylogeny and Taxonomy

CHAPTER

1

Evolution and morphology

Timothy J. Gaudin[1], *Philippe Gaubert*[2,3], *Guillaume Billet*[4], *Lionel Hautier*[5], *Sérgio Ferreira-Cardoso*[5] *and John R. Wible*[6]

[1]Department of Biology, Geology and Environmental Science, University of Tennessee at Chattanooga, Chattanooga, TN, United States [2]Laboratoire Évolution & Diversité Biologique (EDB), Université de Toulouse Midi-Pyrénées, CNRS, IRD, UPS, Toulouse, France [3]CIIMAR, University of Porto, Matosinhos, Portugal [4]Centre de Recherche en Paléontologie-Paris (CR2P), UMR CNRS 7207, Sorbonne Université, Muséum National d'Histoire Naturelle, Paris, France [5]Institut des Sciences de l'Evolution, UMR 5554, CNRS, IRD, EPHE, Université de Montpellier, Montpellier, France [6]Section of Mammals, Carnegie Museum of Natural History, Pittsburgh, PA, United States

OUTLINE

Introduction

Pangolins, or scaly anteaters, comprise the placental mammal order Pholidota, and represent one of the most unusual orders of mammals, morphologically speaking. As implied by the name "scaly anteater," their most noteworthy feature is an external "armor" of overlapping epidermal scales that often earns them colorful nicknames like "walking pinecones" or "perambulating artichokes." They are the only mammals to possess such an armor (although xenarthran armadillos also possess a covering of epidermal scales, these scales are closely associated with underlying bony osteoderms, and as such are organized quite differently; Grassé, 1955a; Vickaryous and Hall, 2006), and pangolins will generally employ this defensive armor by curling up into a ball so that only the scales are exposed (hence the

Pangolins
DOI: https://doi.org/10.1016/B978-0-12-815507-3.00001-0

name "pangolin," derived from the Malay word "peng-goling" meaning "one that rolls up"; Kingdon, 1997). However, in addition to this feature, pangolins are unusual among mammals in their loss not only of teeth, but also of functional enamel genes (Meredith et al., 2009), as well as their extraordinarily elongate tongue (Chan, 1995; Kingdon, 1974), their myrmecophagous diet (Heath, 2013), their pungent anal glands (Kingdon, 1974), and their specializations for digging and climbing (Gaudin et al., 2009, 2016).

The origin and early evolution of Pholidota has been difficult to elucidate. Lacking teeth, the most durable and readily preserved part of the mammalian skeleton, their fossil record is sparse, although, as noted by Gaudin et al. (2016; see also Gaudin et al., 2006, 2009), it includes a number of well-preserved taxa. The scarcity of fossil pangolins is almost certainly exacerbated by their preference for forested habitats and the low population densities at which some species exist in the wild (Gaudin, 2010; Gaudin et al., 2016). Moreover, their relationships to other groups of placental mammals have historically been controversial, although a consensus appears to be emerging (see subsequent discussion). Here too their lack of teeth is a hindrance, because dental morphology has figured importantly in elucidating relationships among mammals in general (Emry, 2004; Ungar, 2010).

The goal of this chapter is to outline the phylogenetic relationships of pangolins as a whole to other groups of mammals, to summarize what is known of their fossil record, and to discuss the historical biogeography of the group. It then explores in more detail the unusual morphological features that characterize this highly distinctive order of mammals.

Pangolin supraordinal relationships

Linnaeus (1758) named *Manis pentadactyla* (Chinese pangolin) and placed it in the order Bruta along with elephants, manatees, sloths, and anteaters, to which he added armadillos in 1766. Storr (1780) removed the elephants and sloths and named the order Mutici for pangolins, anteaters, sloths, and armadillos. The name Mutici forgotten or ignored, the animals included by Storr formed the core of the order Edentati Vicq d'Azyr, 1792, a clade that with various additions and deletions was supported into the 20th century as Edentata Cuvier, 1798. Although not widely recognized, this clade was even raised to the level of sub-class named Paratheria by Thomas (1887) to be counterpart to Eutheria and Metatheria. Undoubtedly the reason Thomas's (1887) hypothesis received little support is the fact that pangolins show clear affinities with other placental mammals, possessing most of the synapomorphic traits of Placentalia, among them a chorioallantoic placenta with a prolonged gestational period, along with the loss of the epipubic bones, a brain with distinct olfactory bulbs, a corpus callosum and a gyrencephalic cerebral cortex, and a bicornuate uterus (Elliot Smith, 1899; Grassé, 1955b; O'Leary et al., 2013).

Weber (1904) gave the pangolins ordinal status as Pholidota, within a supraordinal Edentata, which also included Xenarthra (armadillos, sloths, and anteaters) and Tubulidentata (aardvarks). Xenarthrans and pangolins continued to be linked in supraordinal groupings by various authors in the early 20th century (Rose et al., 2005), but Simpson's (1945) separation of the two in his influential classification of mammals largely ended such claims. For example, in the first cladistic study of placental mammal phylogeny, McKenna (1975) made Pholidota *incertae sedis* within a broad clade including all extant placental orders except Xenarthra, Lagomorpha, and Macroscelidea. However, the association of pangolins and xenarthrans was revived by Novacek and colleagues in the late 1980s and early 1990s (e.g., Novacek, 1986, 1992;

Novacek and Wyss, 1986). They proposed a number of derived resemblances between Pholidota and Xenarthra, including features traditionally used to link the "edentate" mammals to one another, i.e., reduction of the dentition and fossorial specializations (in particular, extensive sacro-innominate fusion), along with several novel cranial features (e.g., reduction of the subarcuate fossa, restriction of the palatine to the ventral orbital wall). These putative synapomorphies have been criticized on morphological grounds (Rose et al., 2005), but perhaps more significantly, not a single molecular analysis (or combined morphological and molecular study) of placental mammal phylogeny published in the last two decades has supported a sister-group relationship between Xenarthra and Pholidota.

Pangolins were first grouped with carnivorans based on DNA sequence data by Shoshani et al. (1985), a result essentially universally accepted since then by additional sequence data (e.g., Murphy et al., 2001a,b; Meredith et al., 2011) and the total evidence analysis of O'Leary et al. (2013). However, it should be noted that the morphology-only results of the latter study yield an "Edentata" grouping of Pholidota+Xenarthra+Tubulidentata. There are several unusual derived features shared by living pangolins and carnivorans, including an ossified tentorium cerebelli and fusion of the scaphoid and lunate bones in the wrist, but these are reduced or absent in the earliest members of each group (Rose et al., 2005), so as yet there are not any identified morphological synapomorphies linking the two taxa.

Several extinct clades have been purported to be closely related to pangolins. The hypothesized relationship that has endured the longest is with palaeanodonts, a small Laurasian group known from the early Paleocene to early Oligocene (Rose, 2006). Palaeanodonts have a reduced dentition, presumably sharing the myrmecophagous diet of pangolins and many xenarthrans, and show adaptations for fossoriality (Rose, 2006). Matthew (1918) was the first to propose close affinities between palaeanodonts, pangolins, and xenarthrans, stating "On the whole, I can find no very conclusive evidence against deriving *Manis* as well as Loricata [=Cingulata] (and through them the remaining Xenarthra) from the primitive type represented by *Palaeanodon*. Just how direct the ancestors may be in each case is a highly speculative matter." Matthew (1918) foretold two schools of thought on the affinities of palaeanodonts: some (e.g., Patterson et al., 1992; Simpson, 1945; Szalay, 1977) supporting closer ties with xenarthrans and others (e.g., Emry, 1970; McKenna and Bell, 1997; Rose et al., 2005) supporting closer ties with pangolins. The latter group has received the most support, with Gaudin et al. (2009) naming a new superorder Pholidotamorpha including Palaeanodonta and Pholidota, which has been supported by the total evidence analysis of O'Leary et al. (2013). Curiously, the Pholidotamorpha of Gaudin et al. (2009) also includes the middle Eocene taxon *Eurotamandua joresi*. The latter, known mostly from a single skeleton from the famed Messel lagerstätten (Storch, 1981), was initially described as a relative of xenarthran anteaters (i.e., vermilinguans), which would make it the only Old World xenarthran. However, more recent studies have disputed this allocation (Rose, 1999; Szalay and Schrenk, 1998), and Gaudin et al. (2009) actually placed *Eurotamandua* within Pholidota proper, in between two other Messel "edentates," *Euromanis krebsi* and *Eomanis waldi*, as a sister taxon to a clade including the latter plus all other undoubted living and fossil pangolins.

Pangolin evolutionary history

It is likely that the Pholidota has its origin in Laurasia, given that the oldest fossil pangolins derive from Europe, and the two possible closest relatives of Pholidota, Palaeanodonta and Carnivora, originate in North America (Flynn and Wesley-Hunt, 2005; Gaudin et al., 2016; Rose et al., 2005). This pattern of supraordinal relationships implies a substantial ghost lineage for Pholidota at its origin, as both Palaeanodonta and Carnivora date to the early Paleocene (Gaudin et al., 2016; O'Leary et al., 2013), whereas the oldest undoubted fossil pangolins, *Euromanis* and *Eomanis*, are derived from the lower middle Eocene Messel deposits (~45 Ma, Lutetian Age; Gaudin et al., 2009; Rose et al., 2005; Fig. 1.1). Molecular phylogenies imply an even older ghost lineage, extending back into the Cretaceous (Emerling et al., 2018; Meredith et al., 2011).

The pangolins from Messel are clearly fossorial, and share strong resemblances to palaeanodonts. For example, they possess an elongated scapular acromion, a humeral supinator crest with a free-standing proximal extension and an elongated deltopectoral crest that is canted medially at its distal end, short and broad metapodials, and an enlarged third manual ungual (Fig. 1.1; Gaudin et al., 2009; Rose et al., 2005; Storch, 2003). The oldest fossil pangolins are also the smallest fossil

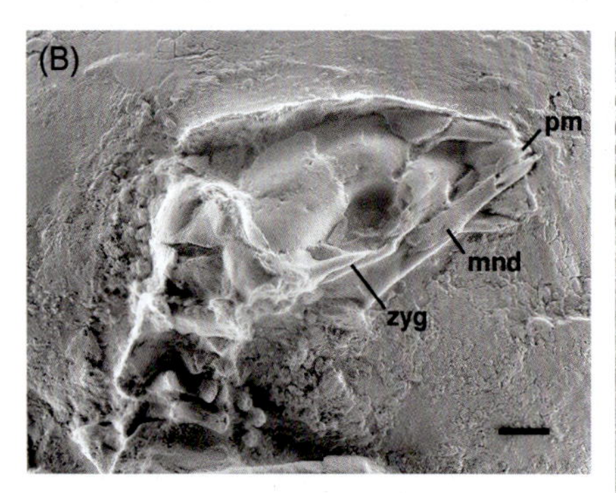

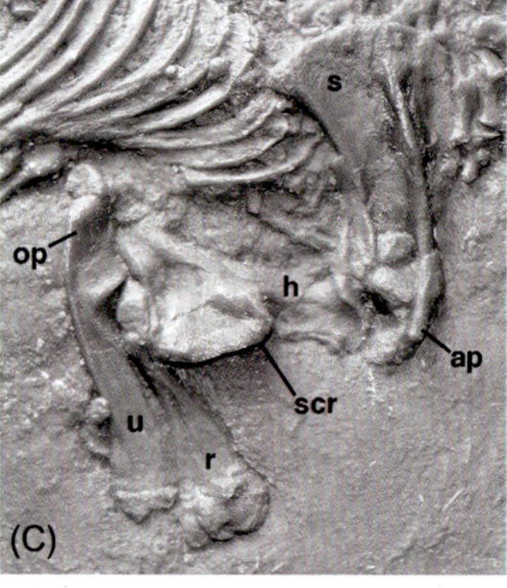

FIGURE 1.1 (A) Skeleton of *Eomanis waldi* in right lateral view. (B) Close-up of the skull of *E. waldi* in right lateral view (SMF MEA 263 cast). (C) Close-up of the right scapula, humerus, radius and ulna of *Eomanis waldi* (SMF MEA 263 cast) in lateral view. Abbreviations: ap, acromion process; h, humerus; mnd, mandible; op, olecranon process; pm, premaxilla; r, radius; s, scapula; scr, supinator crest; u, ulna; zyg, zygomatic arch. *Photos in (A & B) courtesy of Gerhard Storch. Photo in (C) reprinted by permission from Springer/J. Mammal. Evol. Gaudin, T.J., Emry, R.J., Wible, J.R., 2009. The phylogeny of living and extinct pangolins (Mammalia: Pholidota) and associated taxa: a morphology based analysis. J. Mammal. Evol. 16 (4), 235–305. (https://doi.org/10.1007/s10914-009-9119-9).*

pangolins. *Euromanis krebsi* is known from a single incomplete specimen lacking a skull (Gaudin et al., 2009; Storch and Martin, 1994). It is a little larger than *Eomanis waldi*, which is known from multiple specimens, including several nearly complete skeletons (Gaudin et al., 2009; Storch, 1978, 2003), and is smaller than the arboreal African pangolins, the smallest living species (total length of nearly complete skeleton of *E. waldi*, SMF MEA 263, ~ 47 cm; Gaudin et al., 2016). *Eomanis waldi* lacked any clear arboreal adaptations, possessing a rather short tail and short digits. It was edentulous, like living pangolins and other extant myrmecophagous mammals, and thus presumably consumed an ant and termite based diet. It also preserved epidermal scales, though the extent of its scale cover is unknown (Koenigswald et al., 1981). In most respects, however, its skeleton was considerably more primitive than that of other known pangolins, lacking such classic pangolin features as fissured unguals and enrolled lumbar zygapophyses, and retaining a complete zygomatic arch in the skull with a sizable jugal contribution (Gaudin et al., 2009; Rose et al., 2005; the jugal is lost in all extant pangolins except *Manis pentadacytla*, Emry, 2004).

The Paleogene fossil record for pangolins is sparse, but includes several well-known taxa, and is confined to portions of Laurasia outside the range of extant pangolins (Gaudin et al., 2006, 2016). Subsequent to the Messel taxa, only three Paleogene fossil pangolin genera are known. The genus *Cryptomanis* is based on a single partial skeleton lacking a skull, from the late middle Eocene of northern China (*C. gobiensis*; ~40 Ma, Bartonian Age; Gaudin et al., 2006; Fig. 1.2). This taxon had the most robust skeleton of all the Paleogene fossil pangolins, and was likely scansorial like the living Asian pangolins, with elongate digits but lacking a prehensile tail (Gaudin et al., 2006). The genus *Patriomanis* is known from several nearly complete skeletons and some partial remains from two localities in western North America, both latest Eocene in age (~35–37 Ma, Priabonian Age; Gaudin et al., 2016; Fig. 1.2). All the *Patriomanis* material is assigned to a single species, *P. americana*. This species was less robust than *Cryptomanis*, but still showed distinct fossorial adaptations, presumably for excavating ant and termite mounds, as it lacked teeth like *Eomanis*. In comparison to *Cryptomanis*, it had shorter digits but a longer, likely prehensile tail. This tail differed in form but was functionally equivalent to that of the extant prehensile tailed (and arboreal or semi-arboreal) smaller African pangolins (see Chapters 8 and 9), suggesting *Patriomanis* was arboreal (Gaudin et al., 2016). The two genera together comprise the extinct family Patriomanidae (Gaudin et al., 2009, 2016).

The final Paleogene fossil pangolin, the genus *Necromanis*, has the widest stratigraphic and geographic range of any extinct pholidotan, ranging from the middle Oligocene to the middle Miocene (i.e., extending from the late Paleogene into the early Neogene; ~28–14 Ma; Alba et al., 2018; Gaudin et al., 2009; Koenigswald, 1999; Fig. 1.2), and is known from multiple localities in Germany, France and Spain (Alba et al., 2018; Crochet et al., 2015; Koenigswald, 1999). Despite this, the three species in this genus, *N. franconica*, *N. quercyi*, and *N. parva*, remain incompletely known in terms of their skeletal anatomy, though an undescribed skeleton of *N. franconica* from Germany may yet rectify this situation (Hoffmann et al., 2009). The taxonomic affinity of *Necromanis* is also unsettled, with recent phylogenetic analyses allying it either with Patriomanidae (Hoffmann and Martin, 2011), or as the sister taxon to Patriomanidae or to the clade containing all the extant taxa, Manidae (Gaudin et al., 2009).

There is one additional putative Paleogene pangolin record, in the form of several isolated unguals of questionable taxonomic value, from the Oligocene of North Africa

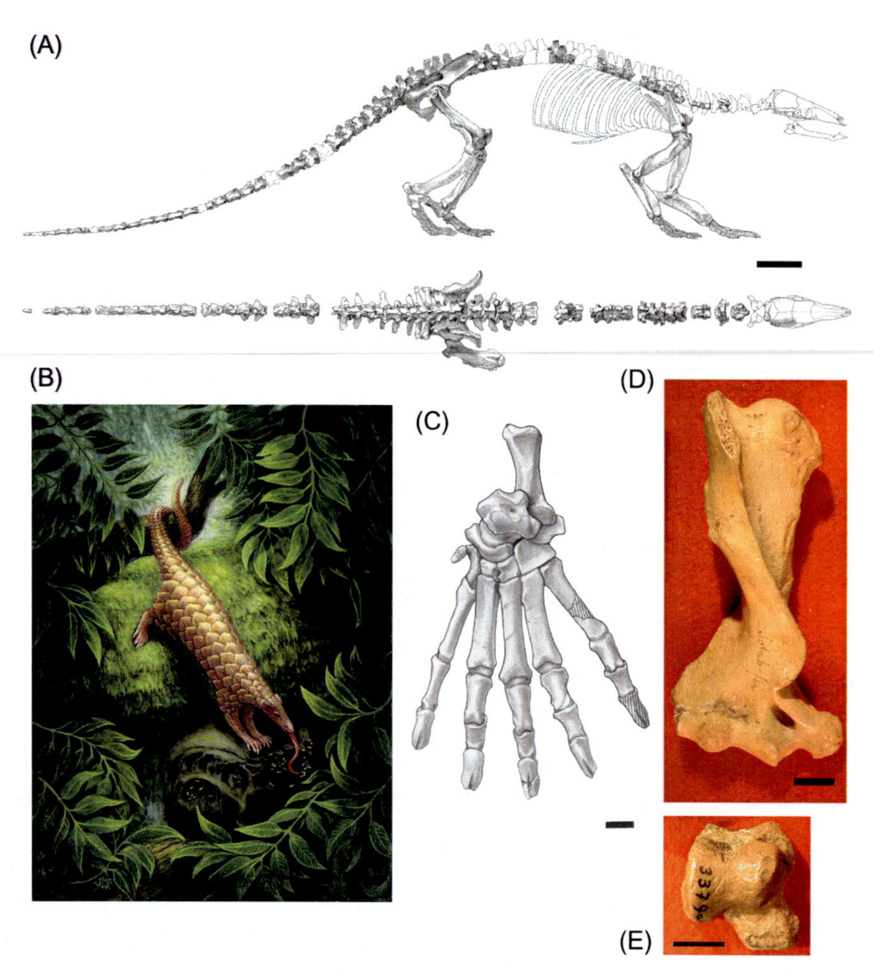

FIGURE 1.2 (A) Skeleton of *Patriomanis* in left lateral (above) and dorsal views (below). (B) Life reconstruction of *Patriomanis*. (C) Left pes of *Cryptomanis* (AMNH 26140) in dorsal view. (D) Right humerus of *Necromanis franconica* in anterior view. (E) Right astragalus of *Necromanis franconica* (SMF M3379a) in dorsal view. Scale bar=1 cm. *(A and B) modified from Gaudin, T.J., Emry, R.J., Morris, J., 2016. Description of the skeletal anatomy of the North American pangolin* Patriomanis americana *(Mammalia, Pholidota) from the latest Eocene of Wyoming (USA). Smithson. Contrib. Paleobiol. 98, 1–102, (C) modified from Gaudin, T.J., Emry, R.J., Pogue, B., 2006. A new genus and species of pangolin (Mammalia, Pholidota) from the late Eocene of Inner Mongolia, China. J. Vertebr. Paleontol. 26 (1), 146–159 (modified by permission from Taylor and Francis Ltd, www.tandfonline.com). Photos in (D and E) by T. Gaudin.*

(Gaudin, 2010; Gaudin et al., 2009; Gebo and Rasmussen, 1985).

The Neogene record of pangolins is little better than the Paleogene record. The group persisted in Europe at least until the middle Miocene (Koenigswald, 1999), and perhaps until the late Pliocene, based on fragmentary material from Hungary (Kormos, 1934), although, as noted above, the early Neogene material of *Necromanis* is not very complete. Undoubted fossil pangolins are known from Africa in the Pliocene and Pleistocene, the Pliocene forms being assigned to the extant giant pangolin species *Smutsia gigantea*, though their remains are somewhat smaller than those of extant members, and the late Pleistocene record assigned to the living Temminck's pangolin (*Smutsia* cf. *S. temminckii*; Gaudin, 2010). Two of the three records are from South Africa, outside the geographic range of the extant African species, and the Pliocene *S. gigantea* from South Africa is the only African fossil pangolin in which a substantial portion of the skeleton is preserved (Botha and Gaudin, 2007; Gaudin, 2010). Fossil pangolins do not appear in South Asia and the East Indies until the Pleistocene (Emry, 1970). All remains are assigned to extant taxa except for *Manis lydekkeri*, known only from an isolated phalanx, and *M. palaeojavanica*, the giant pangolin from the Pleistocene of Java (Emry, 1970; Gaudin et al., 2009). The latter is known from relatively complete skeletal remains, and is the largest pangolin ever to have lived, with an estimated total length of 2.5 m (Dubois, 1926), roughly one and a half times the length of the largest living form, *S. gigantea*.

Morphological specializations of extant pangolins

Relative to other lineages of mammals, pangolins can be considered a uniform group despite their ancient origin, the adaptive cost of myrmecophagy possibly acting as a strong evolutionary constraint. In fact, a cohort of traits associated with feeding on ants and termites, also observed in South American anteaters, are present in all eight species of extant pangolins (Manidae), including the lack of teeth, weakly developed masticatory muscles, sharp claws, forelegs with strong flexor muscles, thick integument, reduced pinnae and valvular nostrils.

As noted above, the most characteristic feature of pangolins is their scaled armor. It covers the dorsal part of the body, the whole tail and the outer side of the legs (Fig. 1.3). Scales are cornified extrusions of the epidermis, composed of flattened, solid and keratinized cells (Tong et al., 1995; Wang et al., 2016). They do not consist of compressed hairs like in rhino (Rhinocerotidae) horns or echidna (Tachyglossidae) spines, but instead are homologous with primate nails (Spearman, 1967). Pangolins could thus be the only living mammals that have developed nail-like structures on their body.

Wearing a scaled armor may be evolutionarily costly. Indeed, scales are metabolically inactive and represent, together with the thick skin, anywhere from 1/10 to 1/4 or 1/3 of the total body weight of pangolins. Moreover, because scales are made of a significant proportion of scleroprotein (Mitra, 1998), they

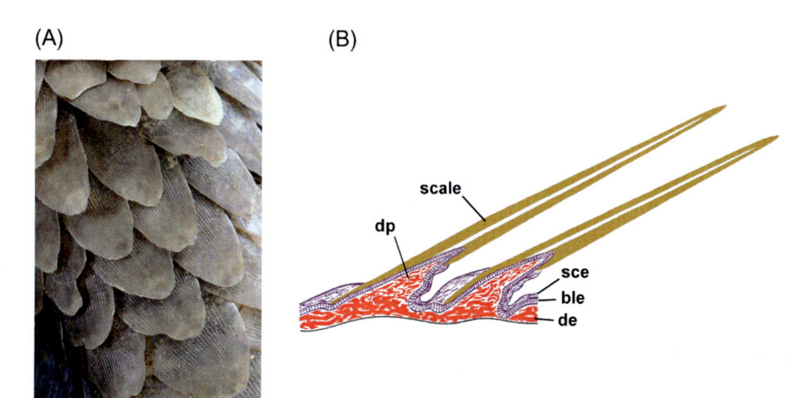

FIGURE 1.3 (A) Close-up photograph of scales at the base of the tail in *Phataginus tricuspis*, in dorsal view. (B) Schematic drawing of pangolin scale histology in *P. tricuspis*. Abbreviations: ble: basal layers of epidermis; de: dermis; dp: dermal papilla; sce, stratum corneum of epidermis. *(B) modified from Grassé, P.-P., 1955b. Ordre des Pholidotes. In: Grassé P.-P. (Ed.), Traité de Zoologie, vol. 17 Mammifères. Masson et Cie, Paris, pp. 1267–1282. Photograph of pangolin scales by T. Gaudin.*

likely require a large consumption of protein for their production (Gaubert, 2011).

Contrary to popular belief, scales contribute little to thermal insulation or protection against ants and termites or dermal parasites (e.g., Heath and Hammel, 1986). Rather, they seem more useful for protecting against injuries from large predators and fossorial activities (e.g., Tong et al., 1995). Scales are sensitive to touch and their orientation is adjustable by dermal muscles. When pangolins roll up into their typical defensive position, the large, salient scales on the sides of the tail are protruded backwards as a threatening series of sharp asperities.

The skull of extant pangolins is very distinctive relative to that of most extant mammals, as it bears many unusual morphological features related to their shift to an almost exclusively myrmecophagous diet. The skull is characterized by its elongated and toothless rostrum, and the incomplete or very slender zygomatic arch. The narrow snout confers a characteristic triangular shape to the skull in dorsal and ventral view (Fig. 1.4). Though lacking teeth, the maxillary bone of pangolins presents sharpened edges that border the palate laterally. The mandible is equally toothless but bears an osseous pseudo-tooth anteriorly. The lower jaw

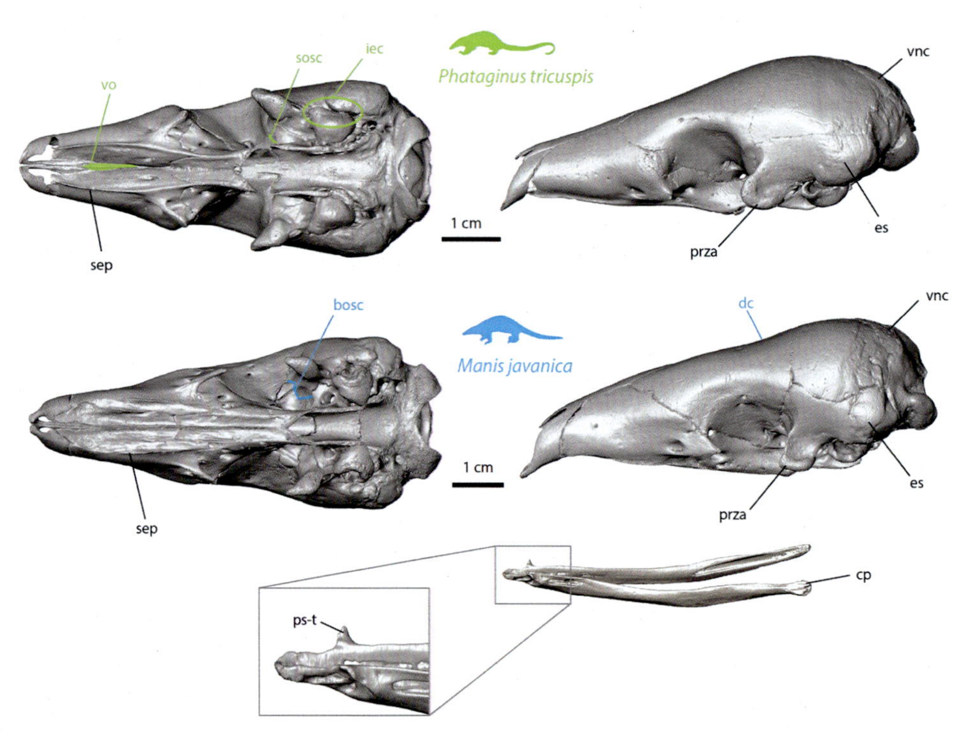

FIGURE 1.4 Cranial characters in extant pangolins. The left part of the figure shows ventral views and the right part lateral views of skulls of *Phataginus tricuspis* (BMNH 12-12-3-3) and *Manis javanica* (BMNH 9-1-5–858). Features labeled in black are shared by all extant pangolins, those labeled in green are typical of African pangolins (*Phataginus* and *Smutsia* spp.), while those labeled in blue characterize Asian pangolins (*Manis* spp.). Abbreviations: bosc: broad orbitosphenoid-squamosal contact; cp: condylar process of the mandible, flat and low-positioned; dc: dorsal constriction; es: inflated epitympanic sinus; iec: inflation of ectotympanic; prza: posterior root of (incomplete) zygomatic arch; ps-t: osseous pseudo-tooth; sep: sharpened edges of the palate; sosc: short orbitosphenoid-squamosal contact; vnc: vestigial nuchal crest; vo: vomer visible in palatal view.

is long and mediolaterally compressed with an ascending ramus consisting solely of a flat and low-positioned condylar process, lacking coronoid and angular processes (Fig. 1.4). Besides these general features, some of which were already present in Paleogene pholidotans (Gaudin et al., 2016), extant pangolins may be further diagnosed by a number of more subtle cranial specializations. The skulls of extant pangolins (Manidae) are unusual and highly diagnostic, characterized by multiple unambiguous cranial synapomorphies, some illustrated in Fig. 1.4 and the remainder listed in Gaudin et al. (2009). As a complement to detailed observation of discrete anatomical characters, recent geometric morphometric analyses highlighted the main patterns of cranial shape variation in seven of the eight extant pangolin species (Fig. 1.5; Ferreira-Cardoso et al., in press). These analyses showed that allometry plays a significant role in shaping intrageneric skull variation, especially among African pangolins, with larger species showing a longer rostrum. The Asian species (*Manis* spp.) generally exhibit more robust (higher and wider) rostra, long zygomatic processes (complete zygomatic arches are rare, but present in some old individuals) and varying degrees of orbital constrictions (Fig. 1.5A, lower left). In contrast, the small African species (*Phataginus* spp.) present narrow and slender rostra, lack orbital and dorsal constrictions, and have relatively wide braincases (Figs. 1.4 and 1.5A (lower right)). Gaudin et al. (2009) stressed similarities between African ground pangolins and their Asian relatives, which partly explains the weak morphological support for the monophyly of the African clade. Ferreira-Cardoso et al. (in press)

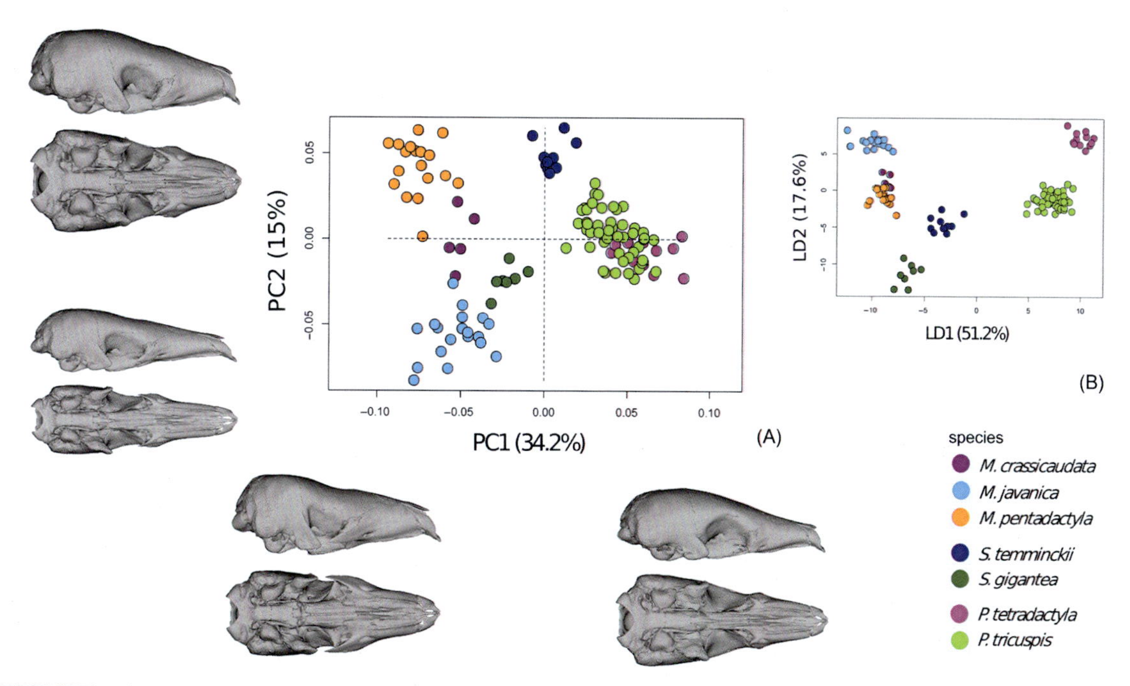

FIGURE 1.5 Morphological variation of the skull of extant pangolins using 3D geometric morphometric methods on a set of 75 anatomical landmarks (Ferreira-Cardoso et al., in press). (A) Principal component analysis and associated patterns of morphological transformation for crania. (B) Linear Discriminant Analysis (LDA) performed on the first principal components explaining 90% of the variance.

confirmed that African ground pangolins (*Smutsia* spp.) display intermediate skull shapes, with the Temminck's pangolin (*S. temminckii*) slightly closer to white-bellied pangolins (*P. tricuspis*), whereas the giant pangolin (*S. gigantea*) resembles the Sunda pangolin (*Manis javanica*). In general, no species represents a clear outlier in the morphospace and the interspecific variation pattern of cranial shape seems to reflect the different geographical distributions (Maninae – Asian; Smutsiinae – African *sensu* Gaudin et al., 2009). Functional correlates of cranial morphology remain to be better assessed for many structures whose shape is unique to pangolins. In this regard, the functional implications of the very thick semicircular canals and the low cochlear spiral of the bony labyrinth in *P. tricuspis* (Ekdale, 2013), as well as the variable morphologies of the epitympanic sinus, posterior root of the zygomatic arch, and glenoid surface (Fig. 1.4) deserve further scrutiny.

Gaudin et al. (2009) noted few postcranial apomorphies that define Pholidota as a whole, among them a triangular subungual process, a prominent ischial spine, and a small obturator foramen. The postcranial skeleton of pangolins has long been considered highly distinctive, due in large part to adaptations for fossoriality and arboreality or scansoriality (Fig. 1.6), and to a lesser degree, myrmecophagy. It should be noted, however, that many of the postcranial skeletal features typically associated with this group are not evident in its earliest members, i.e., in the middle Eocene Messel taxa. However, the Manoidea, the clade including only the extinct Patriomanidae and the Manidae (encompassing all the extant taxa), is diagnosed by a number of distinctive postcranial skeletal features, many linked to fossoriality. These would include fissured ungual phalanges and enrolled or embracing lumbar zygapophyses (for a more complete listing see Figs. 1.7 and 1.8 and Gaudin et al., 2009).

The living pangolins show an additional suite of distinctive postcranial features, including strongly keeled metapodials, a concave astragalar head, and reduction of the crests on the long bones of the hindlimbs (again, for a more complete listing see Figs. 1.7 and 1.8 and Gaudin et al., 2009). According to Gaudin et al. (2006, 2016), many of these features are related to a reduction in fossorial features in the proximal limb elements, and an enhancement in the distal elements, perhaps related to a modified digging style in extant pangolins relative to the extinct forms.

Forefoot claws and powerful forelimbs are used to open termite mounds and ant nests. In arboreal species, forefoot claws are curved and hindfoot claws are longer for moving along tree branches. In terrestrial pangolins, forefoot

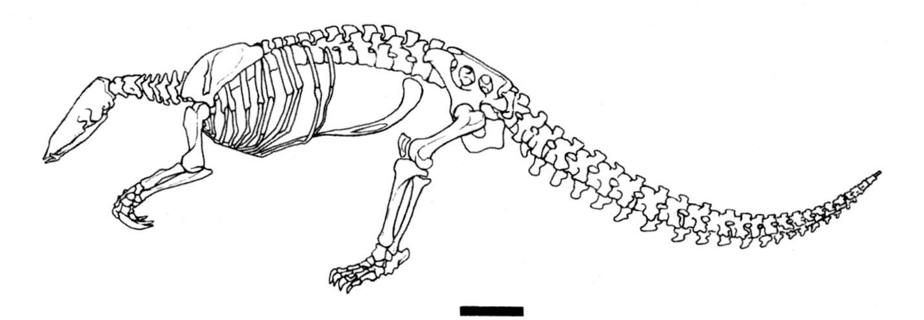

FIGURE 1.6 Skeleton of *Smutsia gigantea* in left lateral view. Scale bar=10 cm. *From Gaudin, T.J., 2010. Pholidota. In: Werdelin, L., Sanders W.J. (Eds.), Cenozoic Mammals of Africa. University of California Press, Berkeley, pp. 599–602.*

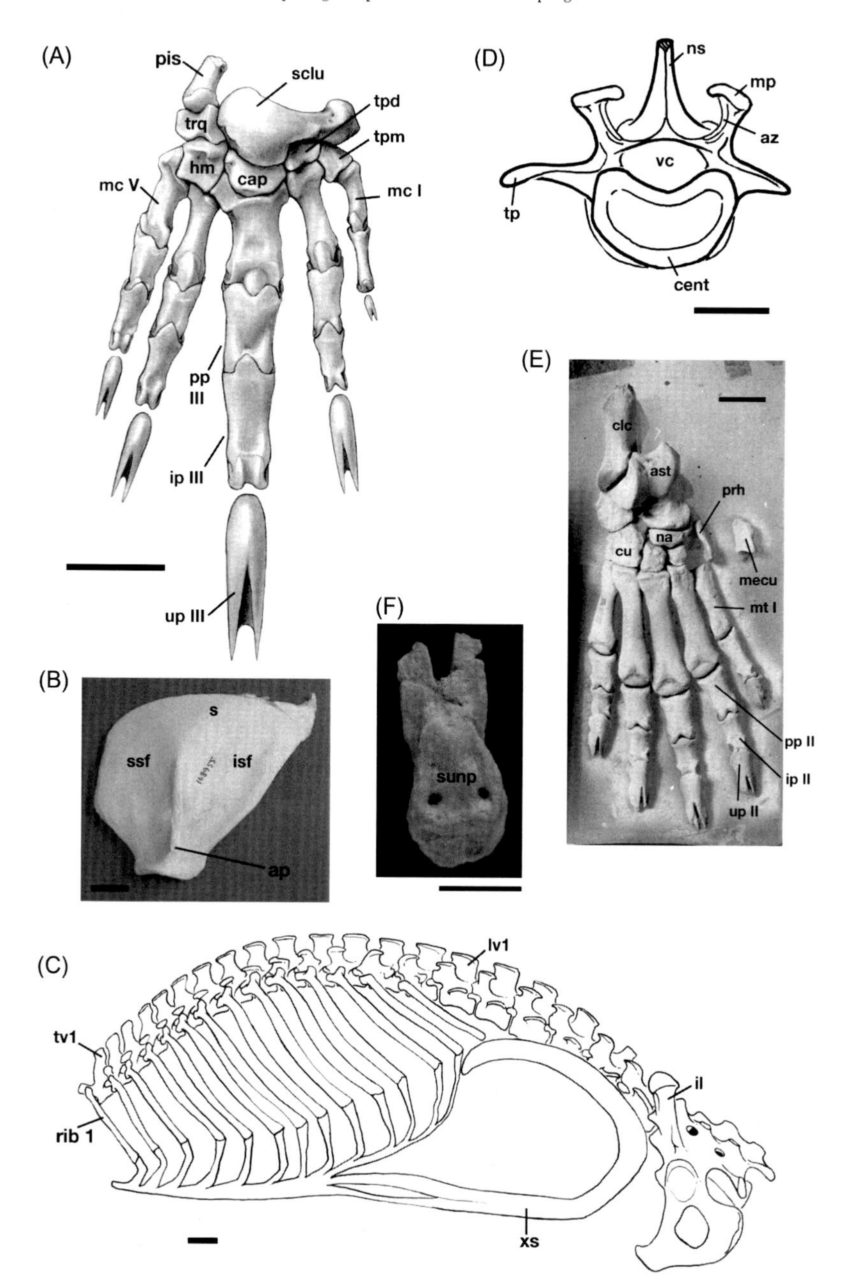

FIGURE 1.7 Pangolin postcranial skeletal features. (A) Right manus of *Phataginus tricuspis* (CM 16206) in dorsal view. (B) Left scapula of *Smutsia temminckii* (AMNH 168955) in lateral view. (C) Dorsal vertebrae, ribcage, sternum, sacrum and pelvis of *Phataginus tricuspis* (CM 16206) in left lateral view. (D) Lumbar vertebra of *Patriomanis americana* (USNM-P 299960) in anterior view. (E) Right pes of *Patriomanis americana* (USNM-P 299960) in dorsal view (the medial cuneiform illustrated is from the left side, as the right one is missing). (F) Ungual phalanx II of right pes of *Patriomanis americana* (USNM-P 299960) in ventral view. The following unambiguous synapomorphies of Manoidea (following Gaudin et al., 2009)

claws are proportionally longer and less curved than in arboreal species, and show greater wear due to their role in supporting the weight of the pangolins while walking and digging in the soil, whereas in arboreal species claws are held perpendicular to the ground and the weight is taken on the folded up wrists. Hindfoot claws are kept short in the plantigrade, terrestrial species by soil abrasion. The forelimbs of terrestrial pangolins, which are predominantly used for burrowing and digging into ant nests, show osteological features reflecting specifically enhanced capacity for shoulder protraction and retraction, strong elbow extension, flexion of the carpus and digits, as well as pronation and supination of the antebrachium (Gaubert, 2011; Kingdon, 1974; Steyn et al., 2018). The clavicle is variably present in extant species regardless of their locomotory mode (Kawashima et al., 2015), which makes its absence difficult to interpret functionally.

The tail of pangolins is very muscular and can be used by members of the larger species as a club to strike predators. It is also used to aid in balance when walking bipedally or to assist climbing in arboreal species. In the semi-arboreal and arboreal African pangolins, the tail is highly prehensile and bears at its ventral tip a touch-sensitive pad impregnated with numerous Pacinian corpuscles (vibration/pressure receptors; Doran and Allbrook, 1973). A similar, but narrower and perhaps less sensitive pad, is also present in the Sunda, Philippine and Chinese pangolins (*Manis javanica, M. culionensis*, and *M. pentadactyla*), but not in other species (Pocock, 1924; see Chapters 4, 6–7).

The semi-arboreal and arboreal African pangolins are particularly distinctive in their postcranial anatomy, showing the greatest number of postcranial synapomorphies of any of the pangolin genera in Gaudin et al.'s (2009) phylogenetic analysis, including "more uniform, slender digits ...[and] more gracile limb elements" (Gaudin et al., 2009). Both species in this genus possess a greatly elongated tail, the black-bellied pangolin (*Phataginus tetradacytla*) having 47–50 caudals, the most of any known extant mammal (Flower, 1885; Gaubert, 2011; Gaudin et al., 2016).

The skeletal musculature in pangolins has been described by numerous authors. Much of the older literature has been summarized admirably by Windle and Parsons (1899). The epaxial musculature has been described by Slijper (1946) for "*Manis* spec.," whereas Jouffroy (1966) and Jouffroy et al. (1975) have published detailed descriptions of the muscles of the distal forelimb, manus, distal hindlimb and pes in the giant pangolin. All the living pangolin taxa show an enlargement of the cartilaginous xiphisternum, where the sternoglossus muscle of the tongue takes its origin

◀ are illustrated: fissured ungual phalanges; fusion of scaphoid and lunate bones; acromion process of scapula rudimentary; neural spines of anterior thoracic vertebrae not dramatically elongated relative to those of more posterior thoracics; embracing lumbar zygapophyses; prehallux present; astragalus/cuboid contact present; triangular subungual process. Abbreviations: ap, acromion process; ast, astragalus; az, anterior zygapophysis; cap, capitate; cent, vertebral centrum; clc, calcaneus; cu, cuboid; hm, hamate; il, ilium; ip, intermediate phalanx; isf, infraspinous fossa; lv, lumbar vertebra; mc, metacarpal; mecu, medial cuneiform; mp, metapophysis; mt, metatarsal; na, navicular; ns, neural spine; pis, pisiform; pp, proximal phalanx; prh, prehallux; s, scapula; sclu, scapholunate; ssf, supraspinous fossa; sunp, subungual process; tp, transverse process; tpd, trapezoid; tpm, trapezium; trq, triquetrum; tv, thoracic vertebra; up, ungual phalanx; vc, vertebral canal; xs, xiphisternum. Scale bars=1 cm. *(A–E) modified from Gaudin, T.J., Emry, R.J., Wible, J.R., 2009. The phylogeny of living and extinct pangolins (Mammalia, Pholidota) and associated taxa: a morphology based analysis. J. Mammal. Evol. 16 (4), 235–305; (F) modified from Gaudin, T.J., Emry, R.J., Morris, J., 2016. Description of the skeletal anatomy of the North American pangolin* Patriomanis americana *(Mammalia, Pholidota) from the latest Eocene of Wyoming (USA). Smithson. Contrib. Paleobiol. 98, 1–102.*

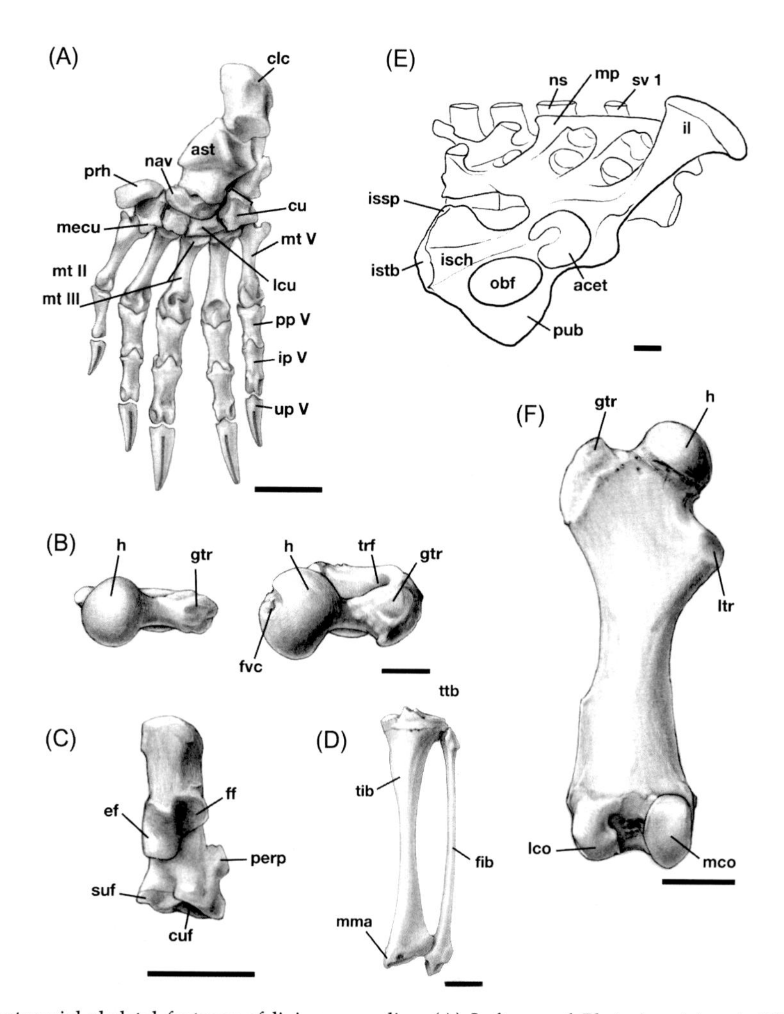

FIGURE 1.8 Postcranial skeletal features of living pangolins. (A) Left pes of *Phataginus tricuspis* (CM 16206) in dorsal view. (B) Left femur of *Phataginus tricuspis* (CM 16206, on left) and *Patriomanis americana* (USNM-P 299960, on right) in proximal view. (C) Left calcaneus of *Phataginus tricuspis* (CM 16206) in dorsal view. (D) Left tibia and fibula of *Phataginus tricuspis* (CM 16206) in anterior view. (E) Pelvis and sacrum of *Smutsia gigantea* (AMNH 53858) in right lateral view. (F) Left femur of *Phataginus tricuspis* (CM 16206) in posterior view. The following unambiguous synapomorphies of Manidae (following Gaudin et al., 2009) are illustrated: distal keel on metatarsals and metacarpals extends along entire dorsoventral length of condyle; proximal end of metatarsal II expanded transversely; proximal articular facet of metatarsal III overlaps dorsal surface of shaft; lateral cuneiform widened transversely, ratio of width to height ≥ 1.4; width of astragalar neck $\geq 60\%$ maximum width of astragalus; astragalar head of navicular with concave distal surface; greater trochanter of femur compressed anteroposteriorly, anteroposterior depth $\leq$ transverse width; fovea capitis absent; sustentacular facet on calcaneus situated well distal to astragalar and fibular facets, contacting distal margin of calcaneus; femoral trochanteric fossa and intertrochanteric ridge rudimentary or absent; sacroiliac attachment fused; metapophyses of sacral vertebrae elongated, $>2/3$ neural spine height; gluteal fossa poorly demarcated, iliac crest rounded, weak, dorsal flange absent, caudal dorsal iliac spine incorporated in sacroiliac junction; ischial spine situated close to ischial tuberosity, dorsal to posterior portion of obturator foramen; dorsal edge of ischium ventral to transverse processes of sacral vertebrae; cnemial crest of tibia weak, rounded, lacking lateral excavation. Abbreviations: acet, acetabulum; ast, astragalus; clc, calcaneus; cu, cuboid; cuf, cuboid facet; ef, ectal facet; ff, fibular facet; fib, fibula; fvc, fovea capitis; gtr, greater trochanter; h, head; il, ilium; ip, intermediate phalanx; isch, ischium; issp, ischial spine; istb, ischial tuberosity; lco, lateral condyle; lcu, lateral cuneiform; ltr, lesser trochanter; mco, medial condyle; mecu, medial cuneiform; mma, medial malleolus; mp, metapophysis; mt, metatarsal; nav, navicular; ns, neural spine; obf, obturator foramen; perp, peroneal process; pp, proximal phalanx; prh, prehallux; pub, pubis; suf, sustentacular facet; sv, sacral vertebra; tib, tibia; trf, trochanteric fossa; ttb, tibial tuberosity; up, ungual phalanx. Scale bars=1 cm. *Adapted by permission from Springer/J. Mammal. Evol. Gaudin, T.J., Emry, R.J., Wible, J.R., 2009. The phylogeny of living and extinct pangolins (Mammalia: Pholidota) and associated taxa: a morphology based analysis. J. Mammal. Evol. 16 (4), 235–305. (https://doi.org/10.1007/s10914-009-9119-9).*

(Grassé, 1955b), but this takes on extraordinary proportions in the African pangolins, where it extends into the abdominal cavity, actually turning dorsally in front of the pelvis toward the vertebral column (Doran and Allbrook, 1973; Kingdon, 1974; Figs. 1.6 and 1.7). Heath (2013) reviewed various published descriptions of the elongated tongue musculature in African pangolins and the Sunda pangolin. The tongue is enormously elongated (Chan, 1995; Doran and Allbrook, 1973; Heath, 2013; Kingdon, 1974; Fig. 1.9A), and capable of extending nearly half its resting length outside the mouth (Heath, 2013). It extends posteriorly through the neck and thorax to the ventral abdominal wall and beyond. In the thorax and throat region, the tongue is housed in a glossal tube lined by mucous membrane that serves as a guide and facilitates movement. The tongue is stored in a pocket in the throat when retracted (Chan, 1995; Doran and Allbrook, 1973; Heath, 2013). Among the noteworthy features of pangolin musculature are the thickened panniculus and the dramatic reduction of the masticatory muscle (Grassé, 1955b; Windle and Parsons, 1899). Windle and Parsons (1899) listed a variety of distinctive features of the skeletal musculature in pangolins, including the absence of muscles commonly present in other mammals (e.g., cleidomastoid), the enlargement of other muscles (e.g., flexor fibularis, popliteus, gracilis, and supinator brevis, the last of which houses a proximal sesamoid element), modified attachments (e.g., the biceps brachii lacks a long head or radial attachment, whereas the plantaris acquires an attachment to the greater trochanter of the femur), and a variety of muscle fusions

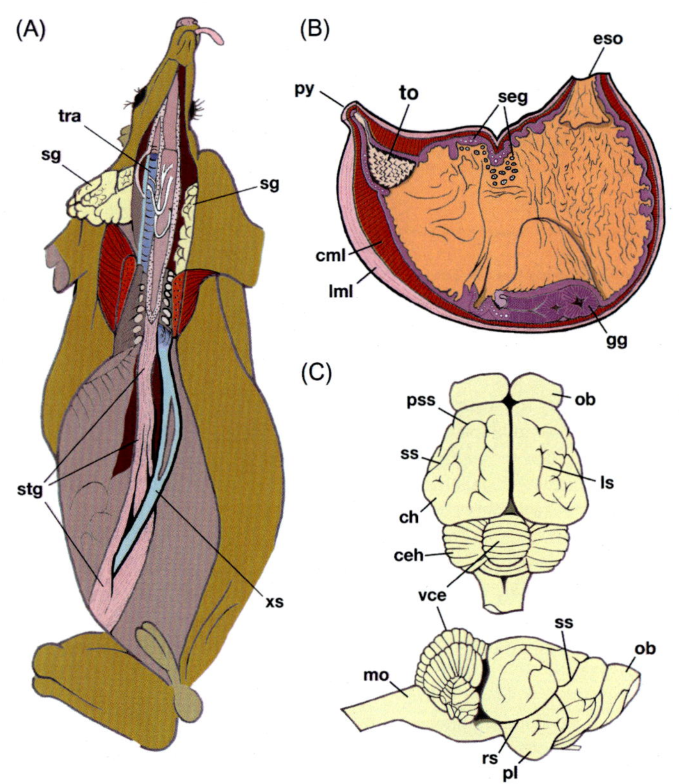

FIGURE 1.9 (A) Dissected *Smutsia gigantea* in ventral view, showing salivary glands, xiphisternum and elongated tongue musculature. (B) Frontal section of stomach of *Manis javanica* shown in ventral view. (C) Brain of *Manis javanica* in dorsal (above) and right lateral (below) views. Abbreviations: ceh, cerebellar hemisphere; ch, cerebral hemisphere; cml, circular layer of smooth muscle fibers; eso, esophagus; gg, gastric glands; lml, longitudinal layer of smooth muscle fibers; ls, lateral sulcus; mo, medulla oblongata; ob, olfactory bulb; pl, piriform lobe; pss, presylvian sulcus; py, pylorus; rs, rhinal sulcus; seg, serous glands; sg, salivary glands; ss, Sylvian sulcus; stg, sternoglossus muscle; to, triturating organ, lined with cornified denticles; tra, trachea; vce, vermis of cerebellum; xs, xiphisternum. *(A) modified from Kingdon, J., 1974. East African Mammals. vol. 1. University of Chicago Press, Chicago. (B, C) Modified from Grassé, P.-P., 1955b. Ordre des Pholidotes. In: Grassé P.-P. (Ed.), Traité de Zoologie, vol. 17. Mammifères. Masson et Cie, Paris, pp. 1267–1282.*

(e.g., spinodeltoid/acromiodeltoid, caudofemoralis/gluteus superficialis, gluteus medius/gluteus minimus/pyriformis, lateral head of gastrocnemius/plantaris). Additionally, Jouffroy (1966) noted that the distal gastrocnemius is fused to the soleus, and the brachioradialis and deltoid abut one another. Pangolins exhibit a number of peculiar myological resemblances to xenarthrans, including the presence of a pterygo-tympanic muscle (Grassé, 1955b), the presence of a rectus thoracis lateralis muscle, a femoral head to the biceps femoris, a strong development of all muscles from the shoulder region (Shrivastava, 1962), and the absence of the sphincter colli muscle (Windle and Parsons, 1899), as well as the presence of a superficial and deep layer in the extensor digitorum breves muscles (Jouffroy, 1966; Jouffroy et al., 1975). Kawashima et al. (2015) proposed that pangolins evolved a unique shoulder girdle musculature in response to the loss of mobility linked to the presence of a continuous hard armor along the body. They showed that the attachments of some shoulder girdle muscles have moved to fully cover the scapula, and suggested that opposite head swinging contributes to the rotation of the shoulder and the protraction of the forelimb.

Like other myrmecophagous mammals (see Gaudin et al., 2018), the salivary glands of pangolins are tremendously enlarged, covering much of the pharyngeal and throat region almost to the shoulder, and secreting a very sticky, alkaline mucus (Fang, 1981; Heath, 2013; Fig. 1.9A). The stomach is comprised of either a single or double chamber (Fang, 1981; Grassé, 1955b; Heath, 2013), with a mass of cornified denticles lining the area near the pylorus to grind up the ants and termites consumed (Krause and Leeson, 1974; Nisa et al., 2010; Fig. 1.9B). Pangolins will also swallow small stones and soil during feeding to facilitate this grinding process in the stomach (Grassé, 1955b). Pangolins lack a cecum, and possess greatly enlarged anal glands that form a projecting perianal ring around the anal opening (Grassé, 1955b; Heath, 2013; Kingdon, 1974).

Female pangolins typically possess two axillary nipples, with a bicornuate uterus and an indeciduate placenta covered by diffuse, villous tufts (Grassé, 1955b; Heath, 2013). The penis in males is described as "petit" by Grassé (1955b), but "well-developed" by Heath (2013), with the corpora cavernosa fused anteriorly (Grassé, 1955b). The testes are housed in a fold of skin in the inguinal area, but there is technically no scrotum, nor do pangolins possess a bulbourethral gland (Grassé, 1955b; Heath, 2013).

The vascular system of the black-bellied pangolin (*Phataginus tetradactyla* [=*Manis macroura*]) was studied by Hyrtl (1854) who found extensive arteriovenous retia mirabilia in the forelimb, hindlimb, and within the vertebral canal in the sacrocaudal region. Along with the black-bellied pangolin, the cranial arteries have been studied in the Sunda pangolin by Bugge (1979) and Wible (1984) and in the Chinese pangolin and white-bellied pangolin by du Boulay and Verity (1973). The pattern in the head was similar in all taxa: the vertebral-basilar arteries are the primary supplier of blood to the brain with supplement from the internal carotids, and most of the stapedial system is obliterated with its end branches annexed to the external carotid system. du Boulay and Verity (1973) reported a rete mirabile on the vertebral artery in white-bellied pangolins, but this was not noted for black-bellied pangolins by Hyrtl (1854). Regarding the arterial supply to the tongue, in white-bellied pangolins and giant pangolins, the left and right lingual arteries join at the base of the tongue to form a central artery that passes through the lyssa (lytta), a supporting fibromuscular cylinder in the tongue apex (Doran and Allbrook, 1973). Union of the lingual arteries has been reported in other mammals with protrusive tongues (e.g., *Tachyglossus* and *Tarsipes*, Doran and Badgett, 1971).

Historically, the most notable contribution on the pangolin brain was by Weber (1894) who illustrated and provided an overview for the Sunda pangolin. A more comprehensive snapshot has been provided recently by Imam et al. (2017) for the white-bellied pangolin. The average encephalization quotient (n = 5) for *P. tricuspis* was 0.844 (average mammalian EQ is 1); however, if the roughly 1/5 of the mass of scales is removed from the calculations, the EQ average is 0.997 (Imam et al., 2017). In general, the gyral and sulcal patterns of the cerebral cortex resemble those in carnivorans, although simpler (Fig. 1.9C). An unusual feature, also reported for the white-bellied pangolin (Chang, 1944), is the position of the pyramidal tract decussation, far rostral to that in other mammals, with apparent association with the hypoglossal nucleus presumably related to the highly modified tongue. Another unusual feature noted by Imam et al. (2017) is a very short spinal cord ending in the thoracic region with an extremely long cauda equina.

Conclusion

Much progress has been made in understanding the relationships of pangolins to other placental mammals, and in reconstructing the morphological evolution and biogeographic history of the Pholidota. A poor fossil record for the group nevertheless remains a significant hindrance in understanding all aspects of pangolin evolutionary history. A better understanding of the anatomy of the oldest fossil pangolins, from the middle Eocene Messel fauna of Germany, would represent an important step forward. Only *Eurotamandua* has received very detailed published treatments (Storch and Habersetzer, 1991; Szalay and Schrenk, 1998), and the only specimen of *Eomanis waldi* that has been described in any detail is the type (Storch, 1978), which is the poorest among multiple known skeletons (T. Gaudin, pers. obs.). Additional finds of new Paleogene fossil pangolin taxa would undoubtedly add much to our knowledge of pholidotan morphological evolution and systematic interrelationships, whereas additional Neogene fossils could clarify the biogeographic origins of the extant forms, which remain poorly understood and reconstructed on fairly scant evidence (Gaudin et al., 2006, 2016). Regarding the morphology of extant forms, interesting studies have been and continue to be published (e.g., Gaubert and Antunes, 2005; Gaudin et al., 2009; Imam et al., 2017; Nisa et al., 2010; Ofusori et al., 2008; Wang et al., 2016), but the group remains significantly understudied. This is no doubt attributable to both the small number of species, and their obscure nature relative to western science, existing as they do largely in Old World tropical forests, at low population densities in some cases, and exhibiting secretive, primarily nocturnal habits. Yet given the distinctiveness of pangolins and the imperative created by their threatened conservation status, additional studies are clearly warranted. It is hoped that this chapter, and this book, will help to spur new investigations of this fascinating group of mammals.

References

Alba, D.M., Hammond, A.S., Vinuesa, V., Casanovas-Vilar, I., 2018. First record of a Miocene pangolin (Pholidota, Manoidea) from the Iberian Peninsula. J. Vertebr. Paleontol. 38 (1), e1424716.

Botha, J., Gaudin, T., 2007. An early pliocene pangolin (Mammalia; Pholidota) from Langebaanweg, South Africa. J. Vertebr. Paleontol. 27 (2), 484–491.

Bugge, J., 1979. Cephalic arterial pattern in New World edentates and Old World pangolins with special reference to their phylogenetic relationships and taxonomy. Acta Anat. (Basel) 105 (1), 37–46.

Chan, L.-K., 1995. Extrinsic lingual musculature of two pangolins (Pholidota: Manidae). J. Mammal. 76 (2), 472–480.

Chang, H.T., 1944. High level decussation of the pyramids in the pangolin (*Manis pentadactyla dalmanni*). J. Comp. Neurol. 81 (3), 333–338.

Crochet, J.-Y., Hautier, L., Lehmann, T., 2015. A pangolin (Manidae, Pholidota, Mammalia) from the French Quercy phosphorites (Pech du Fraysse, Saint-Projet, Tarn-et-Garonne, late Oligocene, MP 28). Palaeovertebrata 39 (2), e4.

Cuvier, G., 1798. Tableau Élémentaire de l'Histoire Naturelle des Animaux. J. B. Baillière, Paris.

du Boulay, G.H., Verity, P.M., 1973. The Cranial Arteries of Mammals. William Heinemann Medical Books Limited, London.

Doran, G.A., Allbrook, D.B., 1973. The tongue and associated structures in two species of African pangolins, *Manis gigantea* and *Manis tricuspis*. J. Mammal. 54 (4), 887–899.

Doran, G.A., Badgett, H., 1971. A structural and functional classification of mammalian tongues. J. Mammal. 52 (2), 427–429.

Dubois, E., 1926. *Manis palaeojavanica*, the giant pangolin of the Kendeng fauna. Proceedings of the Koninklijke Nederlandsche Akademie van Wetenschappen, Amsterdam 29, 1233–1243.

Ekdale, E.G., 2013. Comparative anatomy of the bony labyrinth (inner ear) of placental mammals. PLoS One 8 (6), e66624.

Elliot Smith, G., 1899. The brain in the Edentata. Transactions of the Linnean Society of London. Zoology 7 (7), 277–394.

Emerling, C.A., Delsuc, F., Nachman, M.W., 2018. Chitinase genes (CHIAs) provide genomic footprints of a post-Cretaceous dietary radiation in placental mammals. Sci. Adv. 4 (5), eaar6478.

Emry, R.J., 1970. A North American Oligocene pangolin and other additions to the Pholidota. Bull. Am. Museum Nat. Hist. 142, 457–510.

Emry, R.J., 2004. The edentulous skull of the North American pangolin, *Patriomanis americanus*. Bull. Am. Museum Nat. Hist. 285, 130–138.

Fang, L.-X., 1981. Investigation on pangolins by following their trace and observing their cave. Nat., Beijing Nat. Hist. Museum 3, 64–66. [In Chinese].

Ferreira-Cardoso, S., Billet, G., Gaubert, P., Delsuc, F., Hautier, L., in press. Skull shape variation in extant pangolins (Manidae, Pholidota): allometric patterns and systematic implications. Zool. J. Linn. Soc.

Flower, W.H., 1885. An Introduction to the Osteology of the Mammalia. Macmillan, London.

Flynn, J.J., Wesley-Hunt, G.D., 2005. Carnivora. In: Rose, K. D., Archibald, J.D. (Eds.), The Rise of Placental Mammals. Origins and Relationships of the Major Extant Clades. Johns Hopkins University Press, Baltimore, pp. 175–198.

Gaubert, P., 2011. Family Manidae. In: Wilson, D.E., Mittermeier, R.A. (Eds.), Handbook of the Mammals of the World, vol. 2. Hoofed Mammals. Lynx Edicions, Barcelona, pp. 82–103.

Gaubert, P., Antunes, A., 2005. Assessing the taxonomic status of the Palawan pangolin *Manis culionensis* (Pholidota) using discrete morphological characters. J. Mammal. 86 (6), 1068–1074.

Gaudin, T.J., 2010. Pholidota. In: Werdelin, L., Sanders, W. J. (Eds.), Cenozoic Mammals of Africa. University of California Press, Berkeley, pp. 599–602.

Gaudin, T.J., Emry, R.J., Pogue, B., 2006. A new genus and species of pangolin (Mammalia, Pholidota) from the late Eocene of Inner Mongolia, China. J. Vertebr. Paleontol. 26 (1), 146–159.

Gaudin, T.J., Emry, R.J., Wible, J.R., 2009. The phylogeny of living and extinct pangolins (Mammalia, Pholidota) and associated taxa: a morphology based analysis. J. Mammal. Evol. 16 (4), 235–305.

Gaudin, T.J., Emry, R.J., Morris, J., 2016. Skeletal anatomy of the North American pangolin *Patriomanis americana* (Mammalia, Pholidota) from the latest Eocene of Wyoming (USA). Smithson. Contrib. Paleobiol. 98, 1–102.

Gaudin, T.J., Hicks, P., Di Blanco, Y., 2018. *Myrmecophaga tridactyla* (Pilosa: Myrmecophagidae). Mammal. Sp. 50 (956), 1–13.

Gebo, D.L., Rasmussen, D.T., 1985. The earliest fossil pangolin (Pholidota: Manidae) from Africa. J. Mammal. 66 (3), 538–540.

Grassé, P.-P., 1955a. Ordre des Édentés. In: Grassé, P.-P. (Ed.), Traité de Zoologie, vol. 17, Mammifères. Masson et Cie, Paris, pp. 1182–1266.

Grassé, P.-P., 1955b. Ordre des Pholidotes. In: Grassé, P.-P. (Ed.), Traité de Zoologie, vol. 17, Mammifères. Masson et Cie, Paris, pp. 1267–1282.

Heath, M., 2013. Order Pholidota – Pangolins. In: Kingdon, J., Hoffmann, M. (Eds.), Mammals of Africa, vol. V, Carnivores, Pangolins, Equids and Rhinoceroses. Bloomsbury Publishing, London, pp. 384–386.

Heath, M.E., Hammel, H.T., 1986. Body temperature and rate of O_2 consumption in Chinese pangolins. Am. J. Physiol.-Regul., Integr. Comp. Physiol. 250 (3), R377–R382.

Hoffmann, S., Martin, T., 2011. Revised phylogeny of Pholidota: implications for Ferae. J. Vertebr. Paleontol. 31 (Suppl. 2), 126A–127A.

Hoffmann, S., Martin, T., Storch, G., Rummel, M., 2009. Skeletal reconstruction of a Miocene pangolin from southern Germany. J. Vertebr. Paleontol. 29, 115A–116A.

Hyrtl, J., 1854. Beiträge zur vergleichenden Angiologie. V. Das arterielle Gefäss-system der Edentaten. Denkschriften Akademie der Wissenschaft, Wien, mathematisch-naturwissenschaftliche Klasse 6, 21–65.

Imam, A., Ajao, M.S., Bhagwandin, A., Ihunwo, A.O., Manger, P.R., 2017. The brain of the tree pangolin (*Manis tricuspis*). I. General appearance of the central nervous system. J. Comp. Neurol. 525 (11), 2571–2582.

Jouffroy, F.K., 1966. Musculature de l'avant-bras et de la main, de la jambe et du pied chez *Manis gigantea*, III. Biol. Gabon. 2, 251–286.

Jouffroy, F.K., Lessertisseur, J., Renous, S., 1975. Le problème des muscles extensores breves profundi (manus et pedis) chez les Mammifères (Xenarthra et Pholidota). Mammalia 39 (1), 133–145.

Kawashima, T., Thorington, R.W., Bohaska, P.W., Chen, Y. J., Sato, F., 2015. Anatomy of shoulder girdle muscle modifications and walking adaptation in the scaly Chinese pangolin (*Manis pentadactyla pentadactyla*: Pholidota) compared with the partially osteoderm-clad armadillos (Dasypodidae). Anat. Rec. 298 (7), 1217–1236.

Kingdon, J., 1974. East African Mammals, vol. 1. University of Chicago Press, Chicago.

Kingdon, J., 1997. The Kingdon Field Guide to African Mammals. Academic Press, London.

Koenigswald, W. von, 1999. Order Pholidota. In: Rössner, G.E., Heissig, K. (Eds.), The Miocene Land Mammals of Europe. Verlag Dr. Friedrich Pfeil, Munich, pp. 75–80.

Koenigswald, W. von, Richter, G., Storch, G., 1981. Nachweis von Hornschuppen bei *Eomanis waldi* aus der "Grube Messel" bei Darmstadt (Mammalia, Pholidota). Senckenbergiana lethaea 61, 291–298.

Kormos, T., 1934. *Manis hungarica* n. s., das erste Schuppentier aus dem europäischen Oberpliozän. Folia Zoologica et Hydrobiologica 6, 87–94.

Krause, W.J., Leeson, C.R., 1974. Stomach of pangolin (*Manis pentadactyla*) with emphasis on pyloric teeth. Acta Anat. 88 (1), 1–10.

Linnaeus, C., 1758. Systema Naturæ Per Regna Tria Naturæ, Secundum Classes, Ordines, Genera, Species, Cum Characteribus, Differentiis, Synonymis, Locis. Tomus I. Editio decima, reformata. Salvius, Stockholm.

Matthew, W.D., 1918. Edentata. A revision of the lower Eocene Wasatch and Wind River faunas. Part V—Insectivora (continued), Glires, Edentata. Bull. Am. Museum Nat. Hist. 38, 565–657.

McKenna, M.C., 1975. Toward a phylogenetic classification of the Mammalia. In: Luckett, W.P., Szalay, F.S. (Eds.), Phylogeny of the Primates. Plenum Press, New York and London, pp. 21–46.

McKenna, M.C., Bell, S.K., 1997. Classification of Mammals Above the Species Level. Columbia University Press, New York.

Meredith, R.W., Gatesy, J., Murphy, W.J., Ryder, O.A., Springer, M.S., 2009. Molecular decay of the tooth gene Enamelin (ENAM) mirrors the loss of enamel in the fossil record of placental mammals. PLoS Genet. 5 (9), e1000634.

Meredith, R.W., Janěcka, J.E., Gatesy, J., Ryder, O.A., Fisher, C.A., Teeling, E.C., et al., 2011. Impacts of the Cretaceous terrestrial revolution and KPg extinction on mammal diversification. Science 334 (6055), 521–524.

Mitra, S., 1998. On the scales of the scaly anteater *Manis crassicaudata*. J. Bombay Nat. Hist. Soc. 95 (3), 495–498.

Murphy, W.J., Eizirik, E., Johnson, W.E., Zhang, Y.P., Ryder, O.A., O'Brien, S.J., 2001a. Molecular phylogenetics and the origins of placental mammals. Nature 409 (6820), 614–618.

Murphy, W.J., Eizirik, E., O'Brien, S.J., Madsen, O., Scally, M., Douady, C.J., et al., 2001b. Resolution of the early placental mammal radiation using Bayesian phylogenetics. Science 294 (5550), 2348–2351.

Nisa, C., Agungpriyono, S., Kitamura, N., Sasaki, M., Yamada, J., Sigit, K., 2010. Morphological features of the stomach of Malayan pangolin, *Manis javanica*. Anat. Histol. Embryol. 39 (5), 432–439.

Novacek, M.J., 1986. The skull of leptictid insectivorans and the higher-level classification of eutherian mammals. Bull. Am. Museum Nat. Hist. 183, 1–111.

Novacek, M.J., 1992. Mammalian phylogeny: shaking the tree. Nature 356, 121–125.

Novacek, M.J., Wyss, A.R., 1986. Higher-level relationships of the recent eutherian orders: morphological evidence. Cladistics 2 (4), 257–287.

Ofusori, D.A., Caxton-Martens, E.A., Keji, S.T., Oluwayinka, P.O., Abayomi, T.A., Ajayi, S.A., 2008. Microarchitectural adaptation in the stomach of the African tree pangolin (*Manis tricuspis*). Int. J. Morphol. 26 (3), 701–705.

O'Leary, M.A., Bloch, J.I., Flynn, J.J., Gaudin, T.J., Giallombardo, A., Giannini, N.P., et al., 2013. The placental mammal ancestor and the post-KPg radiation of placentals. Science 339 (6120), 662–667.

Patterson, B., Segall, W., Turnbull, W.D., Gaudin, T.J., 1992. The ear region in xenarthrans (=Edentata, Mammalia). Part II. Sloths, anteaters, palaeanodonts, and a miscellany. Fieldiana, Geology n.s. 24, 1–79.

Pocock, R.I., 1924. The external characters of the pangolins (Manidae). Proc. Zool. Soc. Lond. 94 (3), 707–723.

Rose, K.D., 1999. *Eurotamandua* and Palaeanodonta: Convergent or related? Paläontologische Zeitschrift 73 (3-4), 395–401.

Rose, K.D., Emry, R.J., Gaudin, T.J., Storch, G., 2005. Chapter 8. Xenarthra and Pholidota. In: Rose, K.D., Archibald, J.D. (Eds.), The Rise of Placental Mammals. Origins and Relationships of the Major Extant Clades. Johns Hopkins University Press, Baltimore, pp. 106–126.

Rose, K.D., 2006. The Beginning of the Age of Mammals. Johns Hopkins University Press, Baltimore.

Shoshani, J., Goodman, M., Czelusniak, J., Braunitzer, G., 1985. A phylogeny of Rodentia and other eutherian orders: parsimony analysis utilizing amino acid sequences of alpha and beta hemoglobin chains. In: Luckett, W.P., Hartenberger, J.-L. (Eds.), Evolutionary Relationships Among Rodents: A Multidisciplinary Approach. Plenum, New York, pp. 191–210.

Simpson, G.G., 1945. The principles of classification and a classification of mammals. Bull. Am. Museum Nat. Hist. 85, 1–350.

Slijper, E.J., 1946. Comparative biologic-anatomical investigations on the vertebral column and spinal musculature of mammals. Verhandelingen der Koninklijke Nederlandsche Akademie van Wetenschappen, Afdeeling Natuurkunde, Tweede Sectie 17, 1–128.

Spearman, R.I.C., 1967. On the nature of the horny scales of the pangolin. Zool. J. Linn. Soc. 46 (310), 267–273.

Shrivastava, R.K., 1962. The deltoid musculature of the Edentata, Pholidota and Tubulidentata. Okajimas Folia. Anat. Jpn. 38 (1), 25–38.

Steyn, C., Soley, J.T., Crole, M.R., 2018. Osteology and radiological anatomy of the thoracic limbs of Temminck's ground pangolin (*Smutsia temminckii*). Anat. Rec. 301 (4), 624–635.

Storch, G., 1978. *Eomanis waldi*, ein Schuppentier aus dem Mittel-Eozän der "Grube Messel" bei Darmstadt (Mammalia: Pholidota). Senckenbergiana lethaea 59, 503–529.

Storch, G., 1981. *Eurotamandua joresi*, ein Myrmecophagidae aus dem Eozän der "Grube Messel" bei Darmstadt (Mammalia, Xenarthra). Senckenbergiana lethaea 61, 247–289.

Storch, G., 2003. Fossil Old World "edentates." In: Fariña, R.A., Vizcaíno, S.F., Storch, G. (Eds.), Morphological studies in fossil and extant Xenarthra (Mammalia). Senckenbergiana Biologica 83, 51-60.

Storch, G., Habersetzer, J., 1991. Rückverlagerte Choanen und akzessorische Bulla tympanica bei rezenten Vermilingua und *Eurotamandua* aus dem Eozän von Messel (Mammalia: Xenarthra). Zeitschrift für Säugetierkunde 56, 257–271.

Storch, G., Martin, T., 1994. *Euromanis krebsi*, ein neues Schuppentier aus dem Mittel-Eozän der Grube Messel bei Darmstadt (Mammalia: Pholidota). Berliner geowissenschaftliche Abhandlungen E13, 83–97.

Storr, G.C.C., 1780. Prodromus methodi mammalium. Respondente F. Wolffer, Tubingae.

Szalay, F.S., 1977. Phylogenetic relationships and a classification of the eutherian mammals. In: Hecht, M.K., Goody, P.C., Hecht, B.M. (Eds.), Major Patterns in Vertebrate Evolution. Plenum Press, New York, pp. 315–374.

Szalay, F.S., Schrenk, F., 1998. The middle Eocene *Eurotamandua* and a Darwinian phylogenetic analysis of "edentates". Kaupia: Darmstädter Beiträge zur Naturgeschichte 7, 97–186.

Thomas, O., 1887. On the homologies and succession of the teeth in the Dasyuridae, with an attempt to trace the history of the evolution of mammalian teeth in general. Philos. Trans. R. Soc. Lond. 1887 (B), 443–462.

Tong, J., Ren, L.Q., Chen, B.C., 1995. Chemical constitution and abrasive wear behaviour of pangolin scales. J. Mater. Sci. Lett. 14 (20), 1468–1470.

Ungar, P.S., 2010. Mammal Teeth: Origin, Evolution, and Diversity. Johns Hopkins University Press, Baltimore.

Vickaryous, M.K., Hall, B.K., 2006. Osteoderm morphology and development in the nine-banded armadillo, *Dasypus novemcinctus* (Mammalia, Xenarthra, Cingulata). J. Morphol. 267 (11), 1273–1283.

Vicq d'Azyr, F., 1792. Système anatomique des Quadrupèdes. Encyclopédie méthodique. Vve. Agasse, Paris.

Wang, B., Yang, W., Sherman, V.R., Meyers, M.A., 2016. Pangolin armor: overlapping, structure, and mechanical properties of the keratinous scales. Acta Biomater. 41, 60–74.

Weber, M., 1894. Beiträge zur Anatomie und Entwickelung des Genus *Manis*. Zoologische Ergebnisse einer Reise in Niederländisch Ost-Indien 2, 1–116.

Weber, M., 1904. Die Säugetiere. Einführung in die Anatomie und Systematic der recenten und fossilen Mammalia. Verlag von Gustav Fischer, Jena.

Wible, J.R., 1984. The Ontogeny and Phylogeny of the Mammalian Cranial Arterial Pattern. Ph.D. Dissertation, Duke University, Durham, United States.

Windle, B.G., Parsons, F.G., 1899. Myology of the Edentata. Proc. Zool. Soc. Lond. 314–338, 990–1017.

2

Phylogeny and systematics

Philippe Gaubert[1,2], John R. Wible[3], Sean P. Heighton[1,4] and Timothy J. Gaudin[5]

[1]Laboratoire Évolution & Diversité Biologique (EDB), Université de Toulouse Midi-Pyrénées, CNRS, IRD, UPS, Toulouse, France [2]CIIMAR, University of Porto, Matosinhos, Portugal [3]Section of Mammals, Carnegie Museum of Natural History, Pittsburgh, PA, United States [4]Department of Zoology and Entomology, University of Pretoria, Pretoria, South Africa [5]Department of Biology, Geology and Environmental Science, University of Tennessee at Chattanooga, Chattanooga, TN, United States

O U T L I N E

Introduction

Pangolins (Order Pholidota) have a chaotic taxonomic history. As discussed in Chapter 1, earlier studies proposed that Pholidota were closely related to other ant-eating mammals (e.g., Xenarthra), whereas with the advent of molecular phylogenetics, the edentate pangolins became strongly associated (sister-group) with the Carnivora, a dentally-diversified group of mammals. This association, together with the unresolved relationships of some

Pangolins
DOI: https://doi.org/10.1016/B978-0-12-815507-3.00002-2

fossil lineages (e.g., palaeanodonts), continues to feed the debate around the evolution of Pholidota. This chapter briefly reviews pangolin classification from traditional ecological knowledge, and the relationships between fossil and extant forms within the Pholidota and the extant family Manidae. On the basis of developments in the molecular phylogenetics and phylogeography of extant pangolins, this chapter elaborates on a biogeographic scenario regarding the diversification of the group in the tropics since its origin, with extant pangolins arising probably in Eurasia between the late Eocene and Oligocene (c. 45–31 Ma; Gaubert et al., 2018). On the grounds of such a multidisciplinary reassessment, the chapter provides an updated taxonomy of this group threatened with extinction.

Traditional ecological knowledge

Phylogeny and systematics are often seen as matters for specialists, rarely involving the realm of traditional knowledge (Freeman, 1992). Yet, the diversity of vernacular names attributed to pangolins across their geographic ranges remain a remarkable evocation of their peculiar appearance and way of life. The word "Pangolin" derives from the Malay language (see Chapter 1), and the genus name *Manis* comes from the Latin *"manes"* for ghost, or spirit of the dead in Roman religion, echoing its nocturnal and secretive way of life (Gotch, 1979). In China, the pangolin is called "Chuan Shan Jia", which literally means "the animal with full scales that can go through the mountain" (S-J. Luo, in litt. July 2018). In South Africa, the Afrikaans name for the pangolin is "Ietermagog", which is said to refer to the fact that these animals eat creeping insects ("goggas"; Lynch, 1980). In Venda, the name "Khwara" refers to a shortage of rain stemming from the belief that if pangolin blood is spilt on a piece of land, that land will receive no rain (Netshisaulu, 2012). In Benin, the pangolin is called "Agnika" in Nago (Kwa), which means –as in Malay– the one that rolls up (C.A.S. Djagoun, in litt. July 2018).

Although it remains unclear whether local populations in range states, and notably hunters, attribute specific names to different species of pangolin, and how this relates to the identification of divergent morphotypes in local knowledge, the latter sometimes mirrors the western scientific standards of taxonomy. In Cameroon, the three species of pangolin sold in the bushmeat markets of Yaoundé are clearly differentiated from each other. The black-bellied pangolin (*Phataginus tetradactyla*) is called "Pangolin Sorcier" (sorcerer pangolin; Gaubert, 2011), probably as a reflection of its secretive habits in the "bas-fonds" (seasonally flooded bottomlands; Pagès, 1970), whereas the giant pangolin (*Smutsia gigantea*) is named as such ("Pangolin Géant") and the white-bellied pangolin (*P. tricuspis*) receives the generic name "Pangolin" (P. Gaubert and F. Njiokou, pers. obs. 2007). In southern Democratic Republic of the Congo (DRC), the Kisakata language makes the distinction between the giant pangolin, named "Ikonfre" for the animal that is above all the others, and the small (most probably white-bellied) pangolin called "Nkoo", which means the animal that pulls its tail (C. Keboy, in litt. August 2018). In Benin, pangolins are named "Lihui" in Fon, the main language in the southern part of the country. Although it is unclear to what extent different species of pangolins co-occur in the country (Neuenschwander et al., 2011), "Lihui" seems to be applied to different morphotypes, including one that resembles the giant pangolin (S. Zanvo and P. Gaubert, pers. obs.).

Ordinal taxonomy

The proper contents of the Order Pholidota have been a matter of some controversy, particularly when fossil taxa have been taken into account. Although it is clear that the extant pangolins pertain to a monophyletic clade exclusive of other living placentals (see Chapter 1), there are a number of fossil taxa whose membership in Pholidota has not been universally accepted, especially members of the clade Palaeanodonta (Emry, 1970), and the enigmatic Eocene taxon *Eurotamandua* (Gaudin et al., 2009). As discussed in Chapter 1, most recent studies have preferred to link palaeanodonts with pangolins (Gaudin et al., 2009; O'Leary et al., 2013; Rose et al., 2005), in contrast to earlier research that suggested the group might be more closely related to Xenarthra (the armadillos, sloths and vermilinguans; e.g., Patterson et al., 1992). Gaudin et al. (2009) asserted that the alliance of palaeanodonts and pangolins raises nomenclatural problems. If one conceives of the Order Pholidota in a broad sense, including palaeanodonts, as in Emry (1970), then there is no widely accepted term for the monophyletic clade including extant pangolins and their close, non-palaeanodont allies (i.e., stem pangolins). If, however, one restricts the order to living pangolins and their stem fossil taxa, there is no previously used term available for the palaeanodont/pangolin clade. It does not help that there exists no real definition of what ought to constitute a mammalian "Order" (Cantino and de Queiroz, 2000). Nonetheless, Gaudin et al. (2009) offered a suggested nomenclature that maximized both taxonomic stability and consistency with past use. They restricted the Order Pholidota to extant pangolins and their stem fossil taxa by erecting a stem-based clade defined as "the most inclusive clade including the common ancestor of [*Euromanis*] *krebsi* ...and *Manis pentadactyla* and its [sic] descendants, plus all taxa more closely related to this common ancestor than to [the palaeanodont] *Metacheiromys dasypus*" (Gaudin et al., 2009). This definition excluded palaeanodonts from Pholidota. These authors then applied the new name "Pholidotamorpha" to the clade including both pholidotans and palaeanodonts by erecting a node-based clade defined as "the least inclusive clade including the common ancestor of *Metacheiromys dasypus* and *M. pentadactyla* and its [sic] descendants" (Gaudin et al., 2009).

Family-level phylogeny

Except for a somewhat inexplicable taxonomic arrangement suggested by McKenna and Bell (1997), there has been little debate that the three genera encompassing all eight extant pangolin species comprise a monophyletic family, the Family Manidae, exclusive of other extinct pangolin genera (Gaubert et al., 2018; Gaudin and Wible, 1999; Gaudin et al., 2009). Gaudin et al. (2009) defined this family as a stem-based clade, "the most inclusive clade including the common ancestor of *Phataginus tricuspis* and *M. pentadactyla* and its [sic] descendants, plus all taxa more closely related to this common ancestor than to [the late Eocene North American pangolin] *Patriomanis americana*" (including *Manis palaeojavanica*, and more questionably, *M. lydekkeri* and *M. hungarica*, extinct taxa from the Plio-Pleistocene of Europe and Asia) (Emry, 1970; Kormos, 1934). The other families within Pholidota are comprised entirely of extinct taxa. The Patriomanidae includes the two extinct species *Patriomanis americana* and the late Eocene Chinese *Cryptomanis gobiensis* (Gaudin et al., 2006, 2009, 2016). The Oligocene-Miocene European genus *Necromanis* may represent a third member of this family, but its relationships are not completely resolved – in some analyses it forms the immediate sister taxon to Manidae exclusive of the other two patriomanids, whereas in others the sister taxon to *Patriomanis* + *Cryptomanis* within Patriomanidae (Gaudin et al., 2009; Hoffmann and Martin, 2011; Fig. 2.1). In either event, Patriomanidae is the next closest relative to Manidae, the two comprising together a clade called "Manoidea" by Gaudin et al. (2009). The Manoidea is one of the strongest nodes on the tree, and is diagnosed by 27 unambiguous skeletal synapomorphies, six of them entirely unique to this clade. The family Eomanidae was originally erected by Storch (2003) to contain the European taxa *Eomanis waldi* and *"Eomanis" krebsi* (now *Euromanis krebsi*; see Gaudin et al., 2009), but the latter two did not cluster together in the phylogenetic analysis of Gaudin et al. (2009), so the family was restricted to a monotypic group containing only *Eomanis waldi*. This restricted

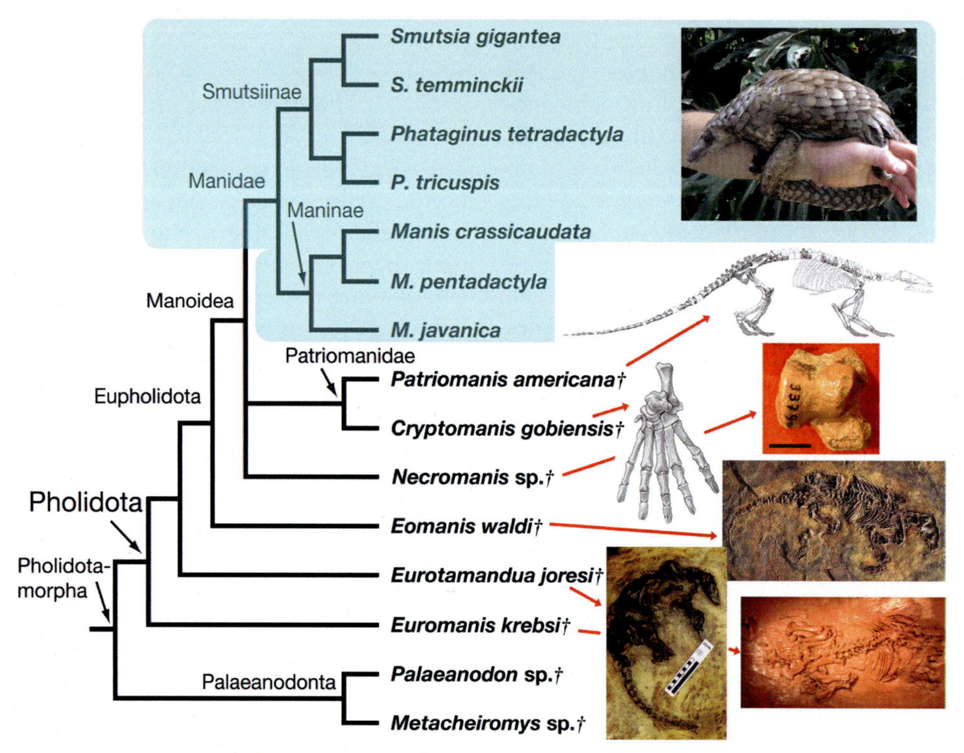

FIGURE 2.1 Morphology-based phylogeny of Pholidota from Gaudin et al. (2009). Strict consensus tree based on a cladistic analysis of 395 osteological characters in 15 ingroup taxa, including seven of the eight extant pangolin species (*Manis culionensis* was not included in their study), five fossil pangolins, *Eurotamandua joresi*, and two metacheiromyid palaeanodont genera. †=extinct taxon. *Adapted by permission from Springer/J. Mammal. Evol. Gaudin, T.J., Emry, R.J., Wible, J. R., 2009. The phylogeny of living and extinct pangolins (Mammalia: Pholidota) and associated taxa: a morphology based analysis. J. Mammal. Evol. 16 (4), 235–305. (https://doi.org/10.1007/s10914-009-9119-9).* Photographs of living *Phataginus tricuspis*, right astragalus of *Necromanis franconica* (dorsal view, scale bar=1 cm) and skeletons of *Euromanis* and *Eurotamandua* by T.J. Gaudin. Photograph of *Eomanis* by G. Storch. *Drawings of skeleton of* Patriomanis *modified from Gaudin, T.J., Emry, R.J., Morris, J., 2016. Skeletal anatomy of the North American pangolin* Patriomanis americana *(Mammalia, Pholidota) from the latest eocene of Wyoming (USA). Smithson. Contrib. Paleobiol. 98, 1–102; left pes of* Cryptomanis *(dorsal view) from Gaudin, T.J., Emry, R.J., Pogue, B., 2006. A new genus and species of pangolin (Mammalia, Pholidota) from the late Eocene of inner Mongolia, China. J. Vertebr. Paleontol. 26 (1), 146–159, modified by permission from Taylor and Francis Ltd, www.tandfonline.com.*

Eomanidae was identified as the sister taxon of Manoidea, together comprising a clade that Gaudin et al. (2009) termed the "Eupholidota." Two extinct species of pangolins from the middle Eocene Messel deposits (the same deposits from which *Eomanis waldi* derives), *Eurotamandua joresi* and *Euromanis krebsi*, formed successive sister taxa to Eupholidota in Gaudin et al.'s (2009) phylogeny. Given the rather weak support for the basal nodes in their tree, Gaudin et al. (2009) declined to make any definitive family level assignments to these two taxa, leaving them essentially as *incertae sedis* (i.e., with uncertain taxonomic status) within Pholidota.

Phylogeny of extant taxa: morphological evidence

The first cladistic study of pangolin systematics based on morphology was by Gaudin and Wible (1999), based solely on cranial skeletal anatomy. It recognized three main clades of living pangolins, including an Asian clade, a clade for the arboreal African pangolins, and a clade for the African ground pangolins. The more extensive analysis of Gaudin et al. (2009), including all the characters from the Gaudin and Wible (1999) study, but adding many more characters derived from the entire skeleton, confirmed these three groupings, elevating each to genus level status, with the genus *Manis* assigned to the Asian species, the genus *Phataginus* assigned to the arboreal African pangolins, and the genus *Smutsia* assigned to the African ground pangolins (as in Koenigswald, 1999) (Fig. 2.1). The first two genera (*Manis* and *Phataginus*) received robust support. Indeed, the node uniting arboreal African pangolins was the strongest of the entire tree, and was diagnosed by 49 unambiguous synapomorphies (in a matrix of 395 characters; Gaudin et al., 2009 – see also Chapter 1). In contrast to the Gaudin and Wible (1999) study, the analysis of Gaudin et al. (2009) also grouped the two African genera into a monophyletic clade, which they designated as the subfamily Smutsiinae. Although the grouping is only weakly supported (as determined by Bremer support and bootstrap values), it is diagnosed by 21 unambiguous morphological synapomorphies, including five that are unique to this clade (Gaudin et al., 2009).

One area where morphology-based phylogenetic studies have so far been unsuccessful is in unambiguously explaining the relationships among the four Asian species. Neither Gaudin and Wible (1999), nor Gaudin et al. (2009) included *M. culionensis* in their analyses (though the latter authors hypothesized its close relationship to the Sunda pangolin [*M. javanica*], an idea confirmed by recent molecular phylogenetics; Gaubert et al., 2018). However, the analysis of Gaudin and Wible (1999) weakly supported an alliance of *M. javanica* with the Chinese pangolin (*M. pentadactyla*), whereas the more extensive analysis of Gaudin et al. (2009) somewhat more robustly supported linking *M. pentadactyla* to the Indian pangolin (*M. crassicaudata*). This "continental" Asian pangolin clade was diagnosed by 23 unambiguous skeletal synapomorphies, including five unique features. Unfortunately, it is not consistent with results from the most recent, comprehensive molecular study, which united *M. javanica* and *M. crassicaudata* as sister taxa (Gaubert et al., 2018), a result that differs from both morphological studies.

Phylogeny of extant taxa: molecular evidence

Molecular phylogenies have been hampered by incomplete taxonomic sampling and/or weak locus representation (du Toit et al., 2014; Gaubert and Antunes, 2005; Hassanin et al., 2015; Zhang et al., 2015). Nevertheless, they all confirmed the deep divergence between African and Asian pangolins as suggested by their morphological differentiation (Gaudin et al., 2009). Gaubert et al. (2018) presented the most comprehensive phylogeny of pangolins to date, based on the assessment of variation in complete mitogenomes and nine nuclear genes among the eight extant species (Fig. 2.2). This study confirmed the deep split between Asian and African pangolins, as previously suggested from morphological observations (Patterson, 1978), morphological phylogenetics (Gaudin et al., 2009) and preliminary molecular phylogenies (du Toit et al., 2014; Gaubert and Antunes, 2005; Hassanin et al., 2015). Large mitogenomic distances (ca. 18–23%) supported the delineation of three genera within extant pangolins (as proposed by Gaudin et al., 2009 on the basis of morphological disparity), including *Manis* (Asian pangolins), *Smutsia* (large/terrestrial African pangolins), and *Phataginus* (small/arboreal African pangolins) (Gaubert et al., 2018). Whether three genera are sufficient to describe the diversity of extant pangolins remains an open question, which requires more in-depth comparative investigation (see *Taxonomy of extant taxa*).

Large mitogenomic distances coupled with high morphological distinctiveness among the three genera seem to justify the designation of three subfamilies within extant pangolins: Maninae (*Manis* spp.), Smutsiinae (*Smutsia* spp.), and Phataginae subfam. nov. (*Phataginus* spp.). Indeed, Gaudin et al. (2009) described the osteological distinctiveness of *Phataginus*, a clade supported by ten unique synapomorphies (compared with seven and three unique synapomorphies for *Manis* and *Smutsia*, respectively). In addition, Gaubert et al. (2018) described eight external traits unique to *Phataginus*, compared to four and one for *Manis* and *Smutsia*, respectively, totaling 18 diagnostic traits for *Phataginus*, 12 for

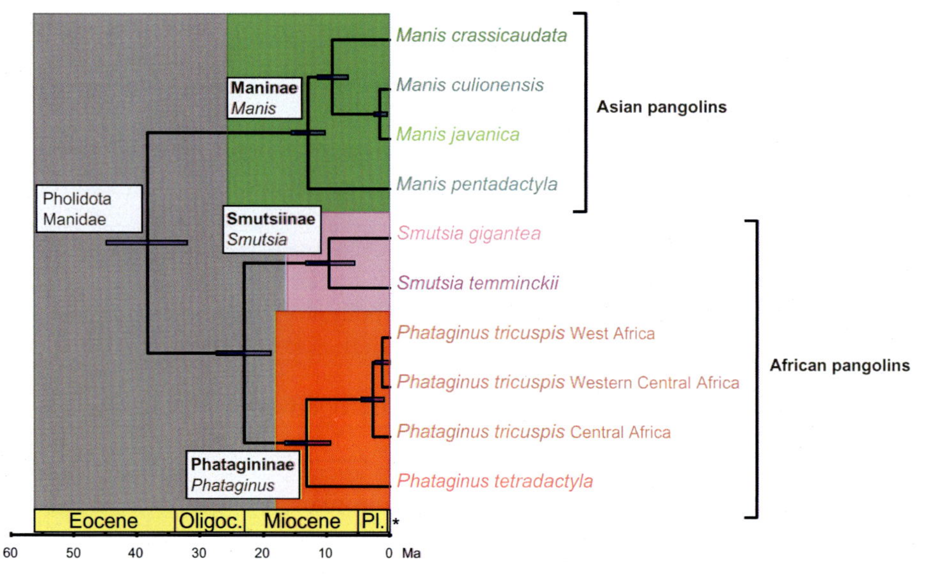

FIGURE 2.2 Time-calibrated tree of extant pangolins derived from the phylogenetic analysis of mitogenomes and nine nuclear genes. Median values and 95% highest posterior densities (HPD; horizontal bars) of time to most recent common ancestors are indicated for the supported nodes. Geological time periods: Oligoc.=Oligocene; Pl.=Pliocene; *=Holocene. *Adapted from Gaubert, P., Antunes, A., Meng, H., Miao, L., Peigné, S., Justy, F., et al., 2018. The complete phylogeny of pangolins: scaling up resources for the molecular tracing of the most trafficked mammals on Earth. J. Hered. 109 (4), 347–359.*

Manis and four for *Smutsia* (see *Taxonomy of extant taxa*).

The combined phylogenetic tree of Gaubert et al. (2018) yielded a fully resolved picture of the extant pangolin species tree (Fig. 2.2). *Smutsia gigantea* and *P. tricuspis* were sister-species, respectively, of *S. temminckii* and *P. tetradactyla*. Within Maninae, the molecular tree was in conflict with the morphological hypothesis supporting the *Manis pentadactyla* – *M. crassicaudata* sister-species relationship (Gaudin et al., 2009). Instead, *M. crassicaudata* was the sister-species of the clade (*M. javanica* – *M. culionensis*), with *M. pentadactyla* positioned as the sister-species of the three other species (Gaubert et al., 2018).

Biogeographic scenario of diversification in extant pangolins

Using molecular clock estimates, du Toit et al. (2014) and Gaubert et al. (2018) proposed a date of origin for the extant Manidae. These dates are older than those proposed in previous mammal-level phylogenies (e.g., Bininda-Emonds et al., 2007; Meredith et al., 2011), but correspond to the known age of the patriomanids, the sister-family to Manidae (Gaudin et al., 2009). Furthermore, Gaubert et al. (2018) posited that the original diversification of extant pangolins into *Manis, Smutsia* and *Phataginus* occurred between the late Eocene and late middle Miocene (ca. 38–13 Ma). Divergence between Asian and African pangolins probably occurred before the Oligocene-Miocene boundary (ca. 23 Ma), coinciding with the estimated origin of extant African pangolins, and with hypothesized patterns of diversification in other mammalian taxa of Eurasian origin (e.g., Gaubert and Cordeiro-Estrela, 2006; Steiner and Ryder, 2011). As the fossil record indicates a Laurasian (most likely European), late Paleocene origin (ca. 60 Ma) of Pholidota (Storch, 2003), Gaudin et al. (2006, 2009) hypothesized a dispersal from Europe into Africa and Asia later in the Cenozoic, with extant pangolins restricted to the tropics due to global cooling during the Plio-Pleistocene (starting ca. 5 Ma). Such a scenario is consistent with the divergence time estimates of Gaubert et al. (2018), the authors invoking a possible dispersal through "filter routes" across the Tethys seaway between Africa and Eurasia as early as the late Eocene (Sen, 2013). An alternative, more recent route of dispersal between Eurasia and Africa could have also occurred through the collision between Eurasia and the Arabian microplate – the "Gomphotherium landbridge" (Koufos et al., 2005; Rögl, 1999) – some 16–20 Ma. However, a knowledge gap exists in the fossil record of extant pangolins, as African and Asian fossils have not been described earlier than 5 Ma (Botha and Gaudin, 2007; Emry, 1970), except for an ambiguous record from early Oligocene deposits in the Fayum, Egypt (Gaudin et al., 2009; Gebo and Rasmussen, 1985), further complicating the biogeographic history of pangolins.

Gaubert et al. (2018) estimated that the three extant pangolin genera diversified in the middle-late Miocene (9.8–12.9 Ma), during a period of drastic global climatic degradation – the mid Miocene event or middle Miocene climatic cooling (Costeur et al., 2007). This event involved gradual, worldwide cooling from ca. 14 Ma onwards, which disrupted interregional mammalian fauna connectivity and thus promoted diversification (e.g., Maridet et al., 2007).

Morphologically similar, sister-group species (*M. javanica* and *M. culionensis*) and cryptic lineages within *P. tricuspis* (see *Phylogeography and cryptic diversity in tropical Africa and Asia*) emerged more recently, mostly during the Pleistocene (1.7–2.7 Ma), coinciding with cyclical rainforest contractions (deMenocal, 2004) that may have factored into their diversification (Gaubert et al., 2016, 2018). The phylogeographic assessment of Gaubert et al. (2016) minimized the role of traditional biogeographic barriers and riverine refuges on the diversification of *P. tricuspis* in tropical Africa. Rather, major periods of aridity in the Pleistocene, accompanying rainforest contraction into refugia (deMenocal, 2004), were likely to have shaped lineage diversification within the species, which fits with the species' preference for rainforest habitats (Gaubert, 2011). One outlier is the lineage endemic to the Dahomey Gap (from Togo to southeastern Nigeria), which could be derived from isolated populations in western Africa that adapted to drier conditions at a time when the Dahomey Gap was covered by a mosaic of forest-savanna patches (Dupont and Weinelt, 1996; Maley, 1996), therefore fitting the vanishing refuge model of diversification (Damasceno et al., 2014).

It has also been suggested that the split between the Sunda and Philippine pangolins (*M. javanica* and *M. culionensis*, respectively) occurred because proto-Philippine pangolins coming from Borneo through Pleistocene land bridges were subsequently isolated by the rising sea level (Gaubert and Antunes, 2005). On the other hand, the non-monophyly of *M. crassicaudata* from Sri Lanka (relative to a specimen from India) suggested that the prolonged connections between India and Sri Lanka across the shallow, 20 km-wide Palk Strait, probably as recently as 10,000 ya (Rohling et al., 1998), facilitated multiple dispersals of pangolins during the Pleistocene, as has been observed in other mammalian species (e.g., Gaubert et al., 2017; Vidya et al., 2011).

Overall, it appears that climatic conditions have acted as the main driver of pangolin diversification from the middle Miocene, as global cooling affected the tropics (deMenocal, 2004), to which extant pangolin species are largely restricted.

Phylogeography and cryptic diversity in tropical Africa and Asia

Geographic variation within the Manidae remains largely unexplored. Gaubert et al. (2015), in a study focusing on the molecular tracing of the bushmeat trade in Africa, were the first to show the existence of marked mitochondrial divergences corresponding to geographic patterns within a single pangolin species, *P. tricuspis*. Expanding on these results, Gaubert et al. (2016) assessed genetic variation in two mitochondrial, three nuclear and one Y-borne genes in more than 100 pangolins sampled across the species' range. On this basis, they delineated the existence of six divergent geographic lineages within the species. Those phylogeographic lineages were circumscribed to western Africa (west of Ghana), Ghana, the Dahomey Gap, western Central Africa, Gabon and Central Africa (Fig. 2.3). The fact that those lineages had non-overlapping ranges was remarkable, with the exception of western Central Africa and Gabon, which have uncertain, reciprocal range delimitations.

The six phylogeographic lineages within *P. tricuspis* were identified as Evolutionarily Significant Units (ESUs) that may warrant species or subspecies status, given their genetic and geographic isolation (Gaubert et al., 2016, 2018). This would have as a consequence the re-erection and the new description of several taxa, pending support from more integrative approaches involving nuclear genomics and comparative morphology.

Preliminary phylogeographic assessments of *S. temminckii* in Southern Africa, using mitochondrial and nuclear markers, have suggested weak population structuring and divergence within the southern half of the species' range (du Toit, 2014; S. Heighton and P. Gaubert, unpubl. data), in contrast with the marked phylogeographic patterns in *P. tricuspis*.

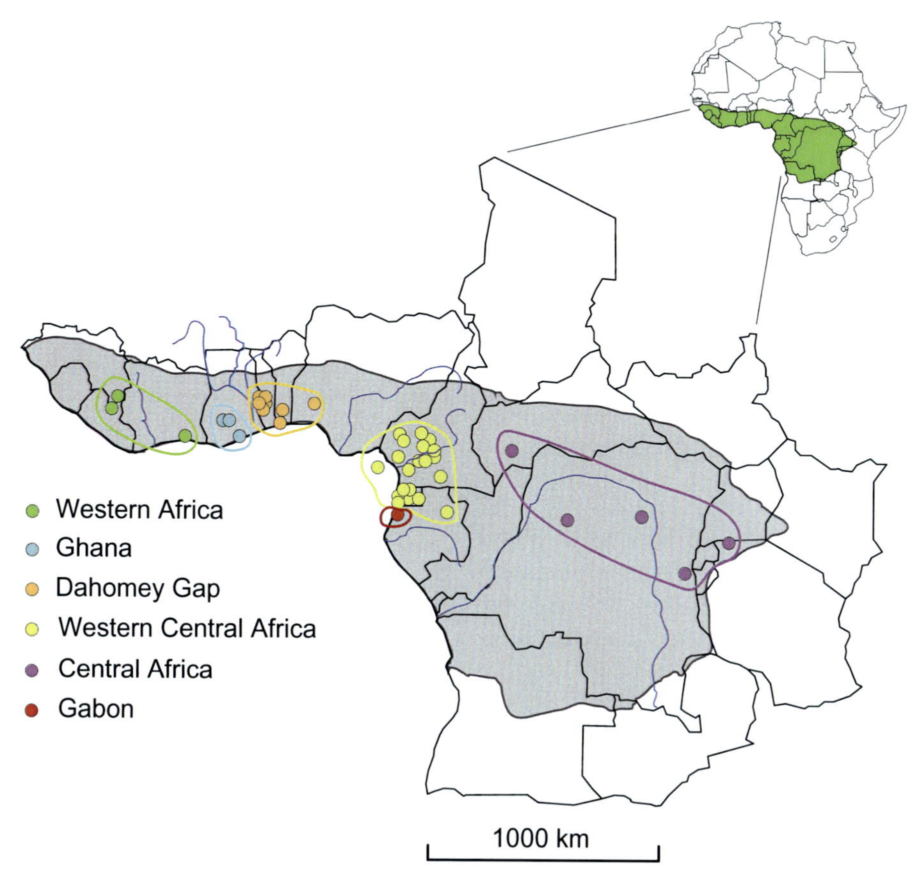

FIGURE 2.3 Map synthesizing the distribution of the six cryptic lineages across the range of *Phataginus tricuspis* on the basis of mitochondrial and nuclear variation. Note the distribution map is from The IUCN Red List of Threatened Species (version 2014-3). *Adapted from Gaubert, P., Njiokou, F., Ngua, G., Afiademanyo, K., Dufour, S., Malekani, J., et al., 2016. Phylogeography of the heavily poached African common pangolin (Pholidota,* Manis tricuspis*) reveals six cryptic lineages as traceable signatures of Pleistocene diversification. Mol. Ecol. 25 (23), 5975–5993.*

The assessment of geographic variation within *M. javanica* has been less thorough, but yet has already revealed some level of cryptic diversity among Asian pangolins. A seminal study based on the mitochondrial DNA (mtDNA) typing of 239 pangolin scales seized in Hong Kong revealed the existence of two deeply divergent genetic clades (ca. 9%), one corresponding to *M. javanica* and the other one likely representing a cryptic pangolin lineage in Asia that did not match with any known Asian pangolin sequences (Zhang et al., 2015). Preliminary analyses based on mtDNA sequences and museum specimens with known geographic origins further confirmed the existence of several cryptic lineages within the species, as divergent from known *M. javanica* as it is from *M. culionensis*. However, these efforts have so far failed to recover the cryptic pangolin lineage identified in Zhang et al. (2015) (P. Gaubert, unpubl. data). A study based on thousands of nuclear markers (SNPs) also echoed a certain level of cryptic diversity within *M. javanica*, with possibly three distinct lineages delineated from Borneo, Java and Singapore-Sumatra, although some level of admixture possibly mediated by humans could jeopardize such lineage-delimitation (Nash et al., 2018).

Taxonomy of extant taxa

Gaudin et al. (2009) discussed in some detail the unsettled history of pangolin systematics, characterized by broad disagreement over species names and generic level assignments. Historically, early classification schemes including genus- to subgenus-level attributions relied heavily on external features of pangolin anatomy, i.e., scale patterns and scale morphology, the presence or absence of hairs between scales and hair color, the presence or absence of ear pinnae, the size of the eyes, the presence and shape of the sensitive pad at the end of the tail, the plantar structure and shape of the claws, and the size and shape of the cartilaginous xiphisternum (Grassé, 1955; Jentink, 1882; Patterson, 1978; Pocock, 1924). These early authors recognized a clear distinction between the Asian and African taxa, a distinction that has been supported in subsequent phylogenetic studies (see *Phylogeny of extant taxa*). Pangolins have been grouped into a single genus *Manis* (Emry, 1970; Jentink, 1882; Schlitter, 2005) or subdivided into a variety of subfamilies, genera, and/or subgenera. Pocock (1924) classified pangolins into six genera, including *Manis* (*M. pentadactyla*), *Phatages* (*Phatages crassicaudata*), *Paramanis* (*Paramanis javanica*), *Smutsia* (*S. temminckii*, *S. gigantea*), *Phataginus* (*P. tricuspis*) and *Uromanis* (*Uromanis tetradactyla*). Seven species have generally been described among extant pangolins. Frequently considered a subspecies of the Sunda pangolin, *Manis culionensis* (Palawan Isl., Philippines) was raised to the species-level on the basis of fixed morphological and molecular divergences with its sister-species *M. javanica* (Feiler, 1998; Gaubert and Antunes, 2005; Gaubert et al., 2018).

Common names for the different species of pangolins have also constituted a non-consensual subject (Gaubert, 2011; IUCN, 2018; Schlitter, 2005). Below, a synthetic classification is provided which summarizes the most updated taxonomic information on the extant Manidae (see Fig. 2.4).

An updated classification of extant pangolins

Family Manidae Gray, 1821

Subfamily Maninae Gray, 1821 [Asian pangolins]

Genus *Manis* Linnaeus, 1758

Manis pentadactyla Linnaeus, 1758: Chinese pangolin, short-tailed pangolin

Manis crassicaudata E. Geoffroy Saint-Hilaire, 1803: Indian pangolin, thick-tailed pangolin

Manis javanica Desmarest, 1822: Sunda pangolin, Malayan pangolin

[consists of several cryptic lineages that warrant further taxonomic investigation]

Manis culionensis (de Elera, 1915): Philippine pangolin, Palawan pangolin

Morphological diagnosis

Seven unique osteological synapomorphies (Gaudin et al., 2009): presence of deep groove for calcaneal-navicular "spring" ligament on ventral margin of astragalar head; deep groove for tendon of m. tibialis posterior on posterior distal surface of tibia, closed over by soft tissue to form a tunnel; transverse foramen of axis visible in anterior view; proximal articulation on capitate very wide, $\geq 85\%$ of maximum dorsoventral depth of capitate; broad orbitosphenoid/squamosal contact; facial nerve travels within closed canal formed by promontorium and crista parotica; body of incus stout and rectangular, crura short.

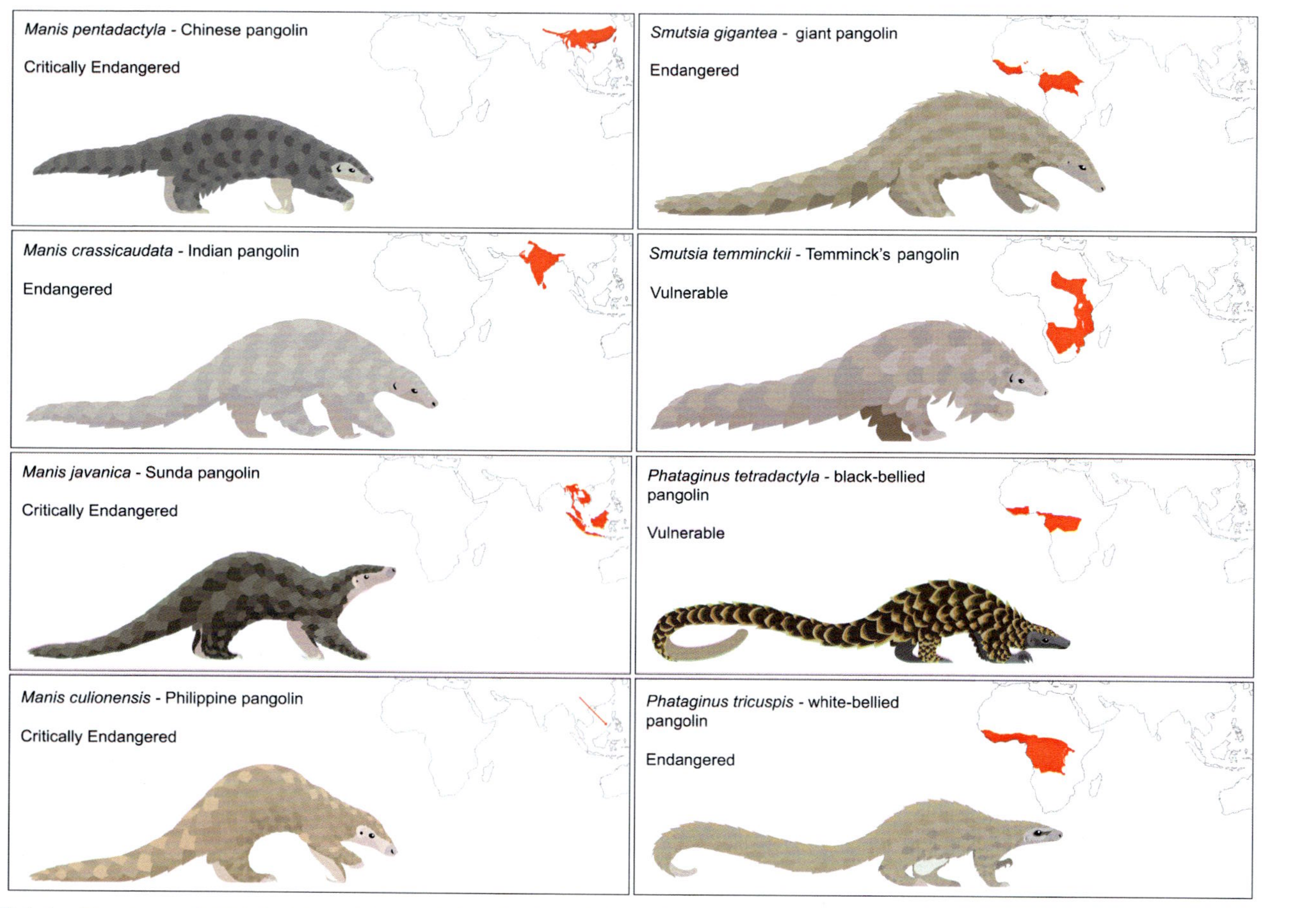

FIGURE 2.4 Taxonomy, IUCN Red List status and distribution of the eight extant species of pangolins. Distribution maps from The IUCN Red List of Threatened Species (version 2019-3). *Illustrations by Sheila McCabe.*

One soft tissue synapomorphy (Gaudin et al., 2009): cartilaginous extension of xiphisternum elongated, length much greater than ossified portion of xiphisternum, shovel shaped at distal end with central perforation.

Five unique external traits (Gaubert et al., 2018): hairs projecting between scales [no hairs in African pangolins] (Pocock, 1924); third claw of hindfoot much longer than fourth claw [slightly longer in African pangolins]; median row of scales on upper side of tail continues to the tip [interrupted before the tip in African pangolins] (Pocock, 1924); posterior margin of scales, median row on dorsal side of tail, smooth/V-shaped [three-cusped in African pangolins]; pinna (ear) on posterior border of auditory orifice [absent in African pangolins].

Subfamily Smutsiinae Gray, 1873 [large African pangolins]

Genus *Smutsia* Gray, 1865

Smutsia gigantea (Illiger, 1815): giant pangolin, giant ground pangolin

Smutsia temminckii (Smuts, 1832): Temminck's pangolin, Cape pangolin, ground pangolin, steppe pangolin

Morphological diagnosis

Three unique osteological synapomorphies (Gaudin et al., 2009): enlarged attachment surface for Achilles' tendon extending forward along plantar surface of calcaneus for more than half its length; wide anconeal process of ulna with a maximum width $>15\%$ of maximum ulnar length; presence of an elongated lateral perforation in mandibular canal.

One unique external trait (Gaubert et al., 2018): ventral parts of body, including feet, with hairs short (<0.5 cm) and sparse [longer and denser hairs in other pangolins].

Subfamily Phataginінae Gaubert, 2018 [small African pangolins]

Type and only genus: *Phataginus* Rafinesque, 1820

Phataginus tetradactyla (Linnaeus, 1766): black-bellied pangolin, long-tailed pangolin

Phataginus tricuspis (Rafinesque, 1820): white-bellied pangolin, African white-bellied pangolin, three-cusped pangolin, tree pangolin, common African pangolin

[may consist of six distinct species/subspecies, corresponding to the lineages from western Africa, Ghana, Dahomey Gap, western Central Africa, Central Africa and Gabon; Gaubert et al., 2016, 2018].

Morphological diagnosis

Ten unique osteological synapomorphies (Gaudin et al., 2009): lateral flange of metatarsal V elongated dorsoventrally, separated from cuboid facet by pit enclosed by dorsal and ventral ridges; cuboid facet of metatarsal V transversely compressed with width $\leq$ depth, but expanded ventrally; navicular facet of lateral cuneiform butterfly-shaped, expanded transversely on dorsal and ventral ends with concave medial and lateral margins; concavity on astragalar facet of navicular restricted to ventral side of convexity; proximal edge (=posterior edge) of astragalar trochlea straight or convex in dorsal view; distal tibia compressed, ratio of maximum width to anteroposterior depth ≥ 2; lesser trochanter directed medially, largely obscured by head but visible medially in proximal view; acetabular fossa opens ventrally; distal edge of trochlea of humerus convex in anterior view; manual ungual phalanx on digit I greatly reduced, $<1/2$ the length of ungual phalanx V.

Eight unique external traits (Gaubert et al., 2018): ratio head+body/tail length <1 [>1 in other pangolins]; upper part of forefoot hairy and not covered with scales [covered with scales in other pangolins]; upper part of hind foot (dorsal view) not covered with scales, with a hairy zone present between base of claws and hind foot scales [completely covered with scales, thus without hairy zone, in other pangolins] (Pocock, 1924); base of first claw on forefoot posterior to that of fifth claw [almost aligned in other pangolins]; base of first claw on hind foot clearly posterior to that of fifth claw [almost aligned in other pangolins]; length of third claw on forefoot ≥ twice longer than second and fourth claws [less than twice in other pangolins]; space between rhinariaum and the last cranial scales (dorsal view; in adults) ≥1 cm [<1 cm in other pangolins]; large terminal pad of tail (ventral side, in adults) due to the absence of two median and two lateral scales [narrow or absent in other pangolins] (Pocock, 1924).

Conclusion

Despite interest in the study of pangolins notably motivated by the critical, media-amplified conservation predicament the animals are in, based on their threats, pangolins remain an evolutionary "black-box." Although the classification of extant pangolins into three genera seems to reach consensus, several issues including fossil affinities with extant taxa and species delimitation within the latter, will require further evidence and additional investigations to be clarified. Given the results from the few, seminal studies on the genetic diversity in pangolin species (white-bellied pangolin, Sunda pangolin), there is strong potential for unraveling additional cryptic diversity within extant pangolins. This in turn could have a direct impact on the traceability of trafficking (see Chapter 20) and on the implementation of efficient conservation measures.

References

Bininda-Emonds, O.R.P., Cardillo, M., Jones, K.E., MacPhee, R.D.E., Beck, R.M.D., Grenyer, R., et al., 2007. The delayed rise of present-day mammals. Nature 446 (7135), 507–512.

Botha, J., Gaudin, T., 2007. An early pliocene pangolin (Mammalia; Pholidota) from Langebaanweg, South Africa. J. Vertebr. Paleontol. 27 (2), 484–491.

Cantino, P.D., de Queiroz, K., 2000. PhyloCode: a phylogenetic code of biological nomenclature, Version 2a. Available from: <http://www.ohio.edu/phylocode/>. [April 9, 2019].

Costeur, L., Legendre, S., Aguilar, J.-P., Lécuyer, C., 2007. Marine and continental synchronous climatic records: towards a revision of the European Mid-Miocene mammalian biochronological framework. Geobios 40 (6), 775–784.

Damasceno, R., Strangas, M.L., Carnaval, A.C., Rodrigues, M.T., Moritz, C., 2014. Revisiting the vanishing refuge model of diversification. Front. Genet. 5, 1–12.

deMenocal, P.B., 2004. African climate change and faunal evolution during the Pliocene-Pleistocene. Earth Planet. Sci. Lett. 220 (1–2), 3–24.

du Toit, Z., 2014. Population Genetic Structure of the Ground Pangolin based on Mitochondrial Genomes. M.Sc. Thesis, University of the Free State, Bloemfontein, South Africa.

du Toit, Z., Grobler, J.P., Kotzé, A., Jansen, R., Brettschneider, H., Dalton, D.L., 2014. The complete mitochondrial genome of Temminck's ground pangolin (*Smutsia temminckii*; Smuts, 1832) and phylogenetic position of the Pholidota (Weber, 1904). Gene 551 (1), 49–54.

Dupont, L.M., Weinelt, M., 1996. Vegetation history of the savanna corridor between the Guinean and the Congolian rain forest during the last 150,000 years. Veg. Hist. Archaeobot. 5 (4), 273–292.

Emry, R.J., 1970. A North American Oligocene pangolin and other additions to the Pholidota. Bull. Am. Museum Nat. Hist. 142, 457–510.

Feiler, A., 1998. Das Philippinen-Schuppentier, *Manis culionensis* Elera, 1915, eine fast vergessene Art (Mammalia: Pholidota: Manidae). Zoologische Abhandlungen - Staatliches Museum Für Tierkunde Dresden 50, 161–164.

Freeman, M.M., 1992. The nature and utility of traditional ecological knowledge. Northern Perspect. 20, 9–12.

Gaubert, P., 2011. Family Manidae. In: Wilson, D.E., Mittermeier, R.A. (Eds.), Handbook of the Mammals of the World, vol. 2. Hoofed Mammals. Lynx Edicions, Barcelona, pp. 82–103.

Gaubert, P., Antunes, A., 2005. Assessing the taxonomic status of the Palawan pangolin *Manis culionensis* (Pholidota) using discrete morphological characters. J. Mammal. 86 (6), 1068–1074.

Gaubert, P., Cordeiro-Estrela, P., 2006. Phylogenetic systematics and tempo of evolution of the Viverrinae (Mammalia, Carnivora, Viverridae) within feliformians: implications for faunal exchanges between Asia and Africa. Mol. Phylogenet. Evol. 41 (2), 266–278.

Gaubert, P., Njiokou, F., Olayemi, A., Pagani, P., Dufour, S., Danquah, E., et al., 2015. Bushmeat genetics: setting up a reference framework for the DNA-typing of African forest bushmeat. Mol. Ecol. Resour. 15 (3), 633–651.

Gaubert, P., Njiokou, F., Ngua, G., Afiademanyo, K., Dufour, S., Malekani, J., et al., 2016. Phylogeography of the heavily poached African common pangolin (Pholidota, *Manis tricuspis*) reveals six cryptic lineages as traceable signatures of Pleistocene diversification. Mol. Ecol. 25 (23), 5975–5993.

Gaubert, P., Patel, R.P., Veron, G., Goodman, S.M., Willsch, M., Vasconcelos, R., et al., 2017. Phylogeography of the small Indian civet and origin of introductions to western Indian Ocean islands. J. Hered. 108 (3), 270–279.

Gaubert, P., Antunes, A., Meng, H., Miao, L., Peigné, S., Justy, F., et al., 2018. The complete phylogeny of pangolins: scaling up resources for the molecular tracing of the most trafficked mammals on Earth. J. Hered. 109 (4), 347–359.

Gaudin, T.J., Wible, J.R., 1999. The entotympanic of pangolins and the phylogeny of the Pholidota (Mammalia). J. Mammal. Evol. 6 (1), 39–65.

Gaudin, T.J., Emry, R.J., Pogue, B., 2006. A new genus and species of pangolin (Mammalia, Pholidota) from the late Eocene of Inner Mongolia, China. J. Vertebr. Paleontol. 26 (1), 146–159.

Gaudin, T., Emry, R., Wible, J., 2009. The phylogeny of living and extinct pangolins (Mammalia, Pholidota) and associated taxa: a morphology based analysis. J. Mammal. Evol. 16 (4), 235–305.

Gaudin, T.J., Emry, R.J., Morris, J., 2016. Skeletal anatomy of the North American pangolin *Patriomanis americana* (Mammalia, Pholidota) from the latest Eocene of Wyoming (USA). Smithson. Contrib. Paleobiol. 98, 1–102.

Gebo, D.L., Rasmussen, D.T., 1985. The earliest fossil pangolin (Pholidota: Manidae) from Africa. J. Mammal. 66 (3), 538–541.

Gotch, A.F., 1979. Mammals – Their Latin Names Explained. A Guide to Animal Classification. Blandford Press, Poole.

Grassé, P.P., 1955. Traité de Zoologie. vol. 17, Mammifères, Masson et Cie, Paris.

Hassanin, A., Hugot, J.-P., van Vuuren, B.J., 2015. Comparison of mitochondrial genome sequences of pangolins (Mammalia, Pholidota). C. R. Biol. 338 (4), 260–265.

Hoffmann, S., Martin, T., 2011. Revised phylogeny of Pholidota: implications for Ferae. J. Vertebr. Paleontol. 31 (Suppl), 126A–127A.

IUCN, 2018. The IUCN Red List of Threatened Species. Version 2018-1. Available from: <http://www.iucnredlist.org>. [August 13, 2018].

Jentink, F.A., 1882. Note XXV. Revision of the Manidae in the Leyden Museum. Notes from the Leyden Museum IV, 193–209.

Koenigswald, W. von., 1999. Order Pholidota. In: Rössner, G.E., Heissig, K. (Eds.), The Miocene Land Mammals of Europe. Verlag Dr. Friedrich Pfeil, Munich, pp. 75–80.

Kormos, T., 1934. *Manis hungarica* n. s., das erste Schuppentier aus dem europäischen Oberpliozän. Folia Zoologica et Hydrobiologica 6, 87–94.

Koufos, G.D., Kostopoulos, D.S., Vlachou, T.D., 2005. Neogene/Quaternary mammalian migrations in Eastern Mediterranean. Belg. J. Zool. 135, 181–190.

Lynch, C., 1980. Mammalian names and scientific names. Culna 19, 14–16.

Maley, J., 1996. The African rain forest - main characteristics of changes in vegetation and climate from the Upper Cretaceous to the Quaternary. Proceedings of the Royal Society of Edinburgh, Section B 104, 31–73.

Maridet, O., Escarguel, G., Costeur, L., Mein, P., Hugueney, M., Legendre, S., 2007. Small mammal (rodents and lagomorphs) European biogeography from the late Oligocene to the mid Pliocene. Glob. Ecol. Biogeogr. 16 (4), 529–544.

McKenna, M.C., Bell, S.K., 1997. Classification of Mammals Above the Species Level. Columbia University Press, New York.

Meredith, R.W., Janečka, J.E., Gatesy, J., Ryder, O.A., Fisher, C.A., Teeling, E.C., et al., 2011. Impacts of the Cretaceous terrestrial revolution and KPg extinction on mammal diversification. Science 334 (6055), 521–524.

Nash, H.C., Wirdateti, Low, G.W., Choo, S.W., Chong, J.L., Semiadi, G., et al., 2018. Conservation genomics reveals possible illegal trade routes and admixture across pangolin lineages in Southeast Asia. Conserv. Genet. 19 (5), 1083–1095.

Netshisaulu, N.C., 2012. Metaphor in Tshivenda. Ph.D. Thesis, Stellenbosch University, Stellenbosch, South Africa.

Neuenschwander, P., Sinsin, B., Goergen, G., 2011. Nature conservation in West Africa: Red List for Benin, International Institute of Tropical Agriculture, Ibadan.

O'Leary, M.A., Bloch, J.I., Flynn, J.J., Gaudin, T.J., Giallombardo, A., Giannini, N.P., et al., 2013. The placental mammal ancestor and the post-K-Pg radiation of placentals. Science 339 (6120), 662–667.

Pagès, E., 1970. Sur l'écologie et les adaptations de l'oryctérope et des pangolins sympatriques du Gabon. Biol. Gabon. 6, 27–92.

Patterson, B., 1978. Pholidota and Tubulidentata. In: Maglio, V.J., Cooke, H.B.S. (Eds.), Evolution of African Mammals. Harvard University Press, Cambridge, pp. 268–278.

Patterson, B., Segall, W., Turnbull, W.D., Gaudin, T.J., 1992. The ear region in xenarthrans (=Edentata, Mammalia). Part II. Sloths, anteaters, palaeanodonts, and a miscellany. Fieldiana, Geology, n.s. 24, 1–79.

Pocock, R.I., 1924. The external characters of the pangolins (Manidae). Proc. Zool. Soc. Lond. 94 (3), 707–723.

Rögl, F., 1999. Mediterranean and Paratethys. Facts and hypotheses of an Oligocene to Miocene paleogeography (short overview). Geol. Carpath. 50, 339–349.

Rohling, E.J., Fenton, M., Jorissen, F.J., Bertrand, G., Ganssen, G., Caulet, J.P., 1998. Magnitudes of sea level lowstands of the last 500,000 years. Nature 394 (6689), 162–165.

Rose, K.D., Emry, R.J., Gaudin, T.J., Storch, G., 2005. Xenarthra and Pholidota. In: Rose, K.D., Archibald, J.D. (Eds.), The Rise of Placental Mammals. Origins and Relationships of the Major Extant Clades. Johns Hopkins University Press, Baltimore, pp. 106–126.

Schlitter, D.A., 2005. Order Pholidota. In: Wilson, D.E., Reeder, D.M. (Eds.), Mammals Species of the World - A Taxonomic and Geographic Reference, third ed. Johns Hopkins University Press, Baltimore, pp. 530–531.

Sen, S., 2013. Dispersal of African mammals in Eurasia during the Cenozoic: ways and whys. Geobios 46 (1–2), 159–172.

Steiner, C.C., Ryder, O.A., 2011. Molecular phylogeny and evolution of the Perissodactyla. Zool. J. Linn. Soc. 163 (4), 1289–1303.

Storch, G., 2003. Fossil Old World "edentates". In: Fariña, R.A., Vizcaíno, S.F., Storch, G. (Eds.), Morphological studies in fossil and extant Xenarthra (Mammalia), vol. 83. Senckenbergiana Biologica, pp. 51–60.

Vidya, T.N.C., Sukumar, R., Melnick, D.J., 2011. Range-wide mtDNA phylogeography yields insights into the origins of Asian elephants. Proc. R. Soc. Lond. B: Biol. Sci. 276 (1706), 893–902.

Zhang, H., Miller, M.P., Yang, F., Chan, H.K., Gaubert, P., Ades, G., et al., 2015. Molecular tracing of confiscated pangolin scales for conservation and illegal trade monitoring in Southeast Asia. Glob. Ecol. Conserv. 4, 414–422.

PART 2

Biology, Ecology and Status

CHAPTER

3

The role of pangolins in ecosystems

Jung-Tai Chao[1], *Hou-Feng Li*[2] *and Chung-Chi Lin*[3]

[1]Taiwan Forestry Research Institute, Taipei, Taiwan [2]Department of Entomology, National Chung-Hsing University, Taichung, Taiwan [3]Department of Biology, National Changhua University of Education, Changhua, Taiwan

OUTLINE

Introduction

Pangolins occur in tropical and subtropical Asia and sub-Saharan Africa (Allen, 1938; Heath, 2013; Chapters 4–11). Macro-determinants of their distribution are likely (1) the distribution of their prey, i.e., ants and termites, (2) ambient temperature (pangolins can quickly perish in inappropriate temperatures; see Chapters 6 and 28), and (3) access to water, though Temminck's (*Smutsia temminckii*) and Indian pangolins (*Manis crassicaudata*) persist in arid and desiccated environments (see Chapters 5 and 11). Pangolins inhabit diverse ecosystems ranging from primary and secondary tropical and subtropical forest, including dipterocarp, broadleaf, coniferous and bamboo forests, to savanna woodland, grasslands, and artificial landscapes including monoculture plantations and gardens (see Chapters 4–11). They are principally nocturnal, with the exception of the black-bellied pangolin (*Phataginus tetradactyla*; see Chapter 8), and are solitary except when mating or females have young. Social interactions are mediated by scent and pangolins rely on a well-developed olfactory system to find prey (Choo et al., 2016; Imam et al., 2018). They exhibit a combination of fossorial (i.e., burrow dwelling), semi-arboreal and arboreal lifestyles and shelter in burrows and tree hollows among other structures. This chapter provides a brief overview of the role of pangolins in ecosystems, specifically as predators of social insects, as creators of burrows,

Pangolins
DOI: https://doi.org/10.1016/B978-0-12-815507-3.00003-4

as prey species, and as hosts of endo- and ectoparasites.

Predators of ants and termites

Predating almost exclusively on ants and termites, pangolins are myrmecophagous and termitophagous. They consume all life stages of their prey, including eggs, larva, pupa and adults, but are prey selective (Irshad et al., 2015; Pietersen et al., 2016). In Sudan, Temminck's pangolin fed on only two out of 22 available ant species (Sweeney, 1956), and Swart et al. (1999) noted that the common pugnacious ant (*Anoplolepis custodiens*) constituted 77% of the overall diet of this species in Sabi Sand Wildtuin, South Africa. In Taiwan, more than 90 ant species have been recorded in pangolin habitat, and in eastern Taiwan >70% of ant species were found in fecal samples of the Chinese pangolin (*M. pentadactyla*; C.-C. Lin, unpubl. data). This involved species from the major ant subfamilies Dolichoderinae, Dorylinae, Formicinae, Ponerinae, and Myrmicinae (C.-C. Lin, unpubl. data), and included the yellow crazy ant (*A. gracilipes*) which is native to Asia but invasive in Taiwan (Fig. 3.1; see Chapter 4).

Ant communities are seasonally dynamic, which appears to affect their availability and abundance as prey for pangolins (Wu et al., 2005). However, this has been subject to little research (though see Chapter 4). The composition of ant communities is influenced by extrinsic factors including temperature, humidity, light, and rainfall, and biological factors such as food availability, as well as predation (Lach et al., 2009). Ants are typically less active in the winter months in the northern hemisphere (Kharbani and Hajong, 2013; Nondillo et al., 2014): in Taiwan, ground-dwelling ants reach peak activity in warmer weather (typically July–October) and are least active in cooler periods (in particular November–December; C.-C. Lin, unpubl. data). Correspondingly, a higher number of prey species and abundance of ants were found in feces of Chinese pangolins in Taiwan in the summer compared to winter (C.-C. Lin, unpubl. data; see Chapter 4). Seasonal prey dynamics can affect pangolins in a number of ways, including body weight fluctuations (e.g., Chinese pangolin; see Chapter 4). This may apply to other species and warrants further research.

Pangolins also exhibit prey selectivity regarding termites. Chinese pangolins appear to feed on termites that inhabit soil and have large colonies (i.e., several million individuals), rather than large bodied termites that nest in wood and have smaller colonies (i.e., several thousand individuals; Li et al., 2011). In Taiwan, fecal

FIGURE 3.1 The Chinese pangolin (*M. pentadactyla*) predates on the yellow crazy ant (*A. gracilipes*) despite it being invasive to Taiwan. *Photo credit: Wei-Ren Liang.*

FIGURE 3.2 The Formosan fungus-growing termite (*O. formosanus*) can comprise a major proportion of the termite prey of the Chinese pangolin (*M. pentadactyla*). *Photo credit: Wei-Ren Liang.*

samples of Chinese pangolins contained the remains of the Formosan fungus-growing termite (*Odontotermes formosanus*), which can comprise a major proportion of termite prey (see Chapter 4; Fig. 3.2). Other species included a wood-feeding species (*Nasutitermes parvonasutus*), soil-feeding species (*Pericapritermes nitobei*), and a subterranean termite (*Reticulitermes flaviceps*; Liang, 2017). Chinese pangolins avoided four dry- or dampwood termite species (Kalotermitidae and Stylotermitidae; Li et al., 2011; Liang et al., 2017). Fungus-growing termites and pangolins are exclusively Old World taxa (Kharbani and Hajong, 2013), and are sympatric, and it is likely that fungus-growing termites (subfamily Macrotermitinae) are important prey species for pangolins (see Chapter 4).

By predating on ants and termites, pangolins act as a control agent for social insects, and although subject to little research, likely influence local ant and termite abundance and community structure, thus affecting local ecosystem function (e.g., trophic interaction, decomposition, nutrient cycling and energy flows; see Del Toro et al., 2012). Ants have diverse ecological niches and feeding habits, but most species feed on small arthropods, plant tissue, or the detritus of animal bodies and, as omnivores or detritivores, act as decomposers in the ecosystem (Lach et al., 2009). Termites are social cockroaches and specialist scavengers of dead plant tissue. They forage from diverse food sources, including wood, leaf litter, humus, algae, and fungi, among others. Synergizing with gut symbiotic microbes and external mutualistic fungi or bacteria, termites are able to digest cellulose and hemi-cellulose, the major component of plant tissue, and they play a critical role in energy and material recycling in ecosystems. However, if uncontrolled, ants can become pests to humans and provide a number of disservices (e.g., cause structural damage to buildings; Del Toro et al., 2012), while termites can, among other things, cause crop damage (Sileshi et al., 2005). Pangolins may therefore perform an important ecosystem service by contributing to the regulation of ant and termite populations. Harrison (1961) estimated the stomach contents of a Sunda pangolin (*M. javanica*) to include >200,000 ants, and Lee et al. (2017) calculated that a Chinese pangolin had more than 26,000 prey items in its stomach, 97% (25,803) of which were ants. Assuming predation of these insects took place in the few days preceding analysis, this suggests that an individual pangolin consumes several million prey items each year. Although myrmecophagous, pangolins do consume other prey, including beetles and sweat bees (see Chapters 4–11).

Burrow creation

With the exception of the black-bellied pangolin (see Chapter 8), all pangolin species excavate burrows, to varying extents, for shelter and in order to locate prey. They dig burrows in varied soil types including but not limited to loam, clay and sandy substrates (Dorji, 2017; Fan, 2005). Burrows are important because they provide shelter with stable temperatures: pangolins are poor at temperature self-regulation (see Chapter 1). By digging burrows, pangolins likely influence a number of soil processes, including turnover of organic matter, aeration, and mineralization rates. They may also act as bioturbators, contributing to soil destratification and mixing, and creating preferential flow paths for soil gas and infiltrating water. The structure of burrows, their depth, length and complexity, which varies by species and season, likely influences the extent of impact on soil processes. Fan (2005) measured the diameter and length of 294 Chinese pangolin burrows as 17.3 ± 3.0 cm and 80.6 ± 48.3 cm (mean $\pm$ SD) respectively, and estimated the mean volume of soil turned over from a typical burrow as 0.019 m^3. With a mean density of 56.7 burrows/ha (Fan, 2005), this suggests the total volume of soil turned over by Chinese pangolins in Taiwan is 1.08 m^3/ha.

A range of commensal taxa use pangolin burrows, which provide them with shelter and thermal refugia. There is little knowledge of how this affects organismal distribution and abundance and if, and to what extent, it influences ecosystem processes and functioning, which is an avenue for further research. Species recorded using pangolin burrows include Travancore tortoise (*Indotestudo travancorica*; Deepak et al., 2016), pouched rats (*Cricetomys* spp.), African bush-tailed porcupine (*Atherurus africanus*; Bruce et al., 2018), crab-eating mongoose (*Herpestes urva*), yellow-throated marten (*Martes flavigula*) and Chinese ferret-badger (*Melogale moschata*) among others (see Chapters 4–11).

Pangolins as prey and host species

When threatened, pangolins either flee or characteristically curl up into a ball, relying on

FIGURE 3.3 Male African lion (*P. leo*) with a Temminck's pangolin (*S. temminckii*). *Photo credit: Foto Mous/Shutterstock.com.*

their scales for protection. The latter is a frequently effective defense against some predators such as African (*Panthera leo*; Fig. 3.3) and Asiatic lions (*P. l. persica*) and tigers (*P. tigris*); other predators include chimpanzees (*Pan troglodytes*), crocodiles (Crocodilia), and pythons (Pythonidae; see Chapters 4–11; Kharbani and Hajong, 2013; Kawanishi and Sunquist, 2004). Shine et al. (1998) observed that of 229 individual reticulated pythons (*Malayopython reticulatus*) with identifiable remains in their stomach, six (0.26%) contained remains of Sunda pangolins. However, there is little knowledge of predation rates on pangolin populations.

Pangolins host a range of endo- and ectoparasites, which have been recorded across the Manidae. This includes myriad species of bacteria, protozoans, viruses, ticks, mites, nematodes, cestodes, acanthocephalans and pentastomes (see Chapter 29). The clinical significance of most of these organisms is unknown, though healthy pangolins frequently carry a tick load, especially *Amblyomma* spp. Hutton (1949) found several ticks beneath nearly every scale of an Indian pangolin (*M. crassicaudata*) found in the Nilgiris, India.

Although limited, evidence indicates that pangolins play important roles in the ecosystems in which they occur. Further research is needed to better understand the precise interactions between pangolins, their prey, and ecosystem function in order to quantify the importance of these roles. This should include the impact of pangolins on ecosystem services valued by humans, and the potential consequences of their extirpation.

References

Allen, G.M., 1938. The Mammals of China and Mongolia. Natural History of Central Asia, vol. XI. Part I. The American Museum of Natural History, New York.

Bruce, T., Kamta, R., Mbobda, R.B.T., Kanto, S.T., Djibrilla, D., Moses, I., et al., 2018. Locating giant ground pangolins (*Smutsia gigantea*) using camera traps on burrows in the Dja Biosphere Reserve, Cameroon. Trop. Conserv. Sci. 11, 1–5.

Choo, S.W., Rayko, M., Tan, T.K., Hari, R., Komissarov, A., Wee, W.Y., et al., 2016. Pangolin genomes and the evolution of mammalian scales and immunity. Genome Res. 26 (10), 1312–1322.

Deepak, V., Noon, B.R., Vasudevan, K., 2016. Fine scale habitat selection in travancore tortoises (*Indotestudo travancorica*) in the Anamalai Hills, Western Ghats. J. Herpetol. 50 (2), 278–283.

Del Toro, I., Ribbons, R.R., Pelini, S.L., 2012. The little things that run the world: a review of ant-mediated ecosystem services and disservices (Hymenoptera: Formicidae). Myrmecological News 17, 133–146.

Dorji, D., 2017. Distribution, habitat use, threats and conservation of the Critically Endangered Chinese pangolin (*Manis pentadactyla*) in Samtse District. Bhutan. Unpublished Report for Rufford Small Grants, UK.

Fan, C.Y., 2005. Burrow Habitat of Formosan Pangolins (*Manis pentadactyla pentadactyla*) at Feitsui Reservoir. M.Sc. Thesis, National Taiwan University, Taipei, Taiwan. [In Chinese].

Harrison, J.L., 1961. The natural food of some Malayan mammals. Bull. Singapore Natl. Museum 30, 5–18.

Heath, M., 2013. Family Manidae. In: Kingdon, J., Hoffmann, M. (Eds.), Mammals of Africa., vol. V, Carnivores, Pangolins, Equids, Rhinoceroses. Bloomsbury Publishing, London, p. 387.

Hutton, A.F., 1949. Notes on the Indian pangolin (*Manis crassicaudata*, Geoffer St. Hilaire). J. Bombay Nat. Hist. Soc. 48, 805–806.

Imam, A., Bhagwandin, A., Ajao, M.S., Spocter, M.A., Ihunwo, A.O., Manger, P.R., 2018. The brain of the tree pangolin (*Manis tricuspis*). II. The olfactory system. J. Comp. Neurol. 526 (16), 2571–2582.

Irshad, N., Mahmood, T., Hussain, R., Nadeem, M.S., 2015. Distribution, abundance and diet of the Indian pangolin (*Manis crassicaudata*). Anim. Biol. 65, 57–71.

Kawanishi, K., Sunquist, M.E., 2004. Conservation status of tigers in a primary rainforest of Peninsular Malaysia. Biol. Conserv. 120 (3), 329–344.

Kharbani, H., Hajong, S.R., 2013. Seasonal patterns in ant (Hymenoptera: Formicidae) activity in a forest habitat of the West Khasi Hills, Meghalaya, India. Asian Myrmecology 5, 103–112.

Lach, L., Parr, C.L., Abbott, K.L. (Eds.), 2009. Ant Ecology. Oxford University Press, Oxford.

Lee, R.H., Cheung, K., Fellowes, J.R., Guénard, B., 2017. Insights into the Chinese pangolin's (*Manis pentadactyla*) diet in a peri-urban habitat. Trop. Conserv. Sci. 10, 1–7.

Li, H.-F., Lin, J.-S., Lan, Y.-C., Pei, K.J.-C., Su, N.-Y., 2011. Survey of the termites (Isoptera: Kalotermitidae, Rhinotermitidae, Termitidae) in a Formosan pangolin habitat. Florida Entomol. 94 (3), 534–538.

Liang, C.-C., 2017. Termite Species Composition in Soil and Feces of Formosan Pangolin (*Manis pentadactyla pentadactyla*) at Luanshan, Taitung. M.Sc. Thesis, National Pingtung University of Science and Technology, Pingtung, Taiwan. [In Chinese].

Liang, W.-R., Wu, C.-C., Li, H.-F., 2017. Discovery of a cryptic termite genus, *Stylotermes* (Isoptera: Stylotermitidae), in Taiwan, with the description of a new species. Ann. Entomol. Soc. Am. 110 (4), 360–373.

Nondillo, A., Ferrari, L., Lerin, S., Bueno, O.C., Bottona, M., 2014. Foraging activity and seasonal food preference of *Linepithema micans* (Hymenoptera: Formicidae), a species associated with the spread of *Eurhizococcus brasiliensis* (Hemiptera: Margarodidae). J. Econ. Entomol. 107 (4), 1385–1391.

Pietersen, D.W., Symes, C.T., Woodborne, S., McKechnie, A.E., Jansen, R., 2016. Diet and prey selectivity of the specialist myrmecophage, Temminck's ground pangolin. J. Zool. 298 (3), 198–208.

Shine, R., Harlow, P.S., Keogh, J.S., Boeadi, 1998. The influence of sex and body size on food habits of a giant tropical snake, *Python reticulatus*. Funct. Ecol. 12 (2), 248–258.

Sileshi, G., Mafongoya, P.L., Kwesiga, F., Nkunika, P., 2005. Termite damage to maize grown in agroforestry systems, traditional fallows and monoculture on nitrogen-limited soils in eastern Zambia. Agric. For. Entomol. 7 (1), 61–69.

Swart, J.M., Richardson, P.R.K., Ferguson, J.W.H., 1999. Ecological factors affecting the feeding behaviour of pangolins (*Manis temminckii*). J. Zool. 247 (3), 281–292.

Sweeney, R.C.H., 1956. Some notes on the feeding habits of the ground pangolin, *Smutsia temminckii* (Smuts). Ann. Mag. Nat. Hist. 9, 893–896.

Wu, S., Liu, N., Li, Y., Sun, R., 2005. Observation on food habits and foraging behavior of Chinese Pangolin (*Manis pentadactyla*). Chin. J. Appl. Environ. Biol. 11 (3), 337–341. [In Chinese].

CHAPTER

4

Chinese pangolin *Manis pentadactyla* (Linnaeus, 1758)

Shibao Wu[1,*], *Nick Ching-Min Sun*[2,*], *Fuhua Zhang*[1], *Yishuang Yu*[1], *Gary Ades*[3], *Tulshi Laxmi Suwal*[4,5] *and Zhigang Jiang*[6]

[1]School of Life Science, South China Normal University, Guangzhou, P.R. China [2]Graduate Institute of Bioresources, National Pingtung University of Science and Technology, Pingtung, Taiwan [3]Fauna Conservation Department, Kadoorie Farm & Botanic Garden, Hong Kong SAR, P.R. China [4]Small Mammals Conservation and Research Foundation (SMCRF), Kathmandu, Nepal [5]Department of Tropical Agriculture and International Cooperation, National Pingtung University of Science and Technology, Pingtung, Taiwan [6]Institute of Zoology, Chinese Academy of Sciences, Beijing, P.R. China

OUTLINE

* These authors contributed equally to this chapter.

DOI: https://doi.org/10.1016/B978-0-12-815507-3.00004-6

Taxonomy

Previously included in the genera *Pholidotus*, *Phatages* (Brisson, 1762; Fitzinger, 1872) and *Pangolinus* (Rafinesque, 1820), the species is here included in the genus *Manis* based on morphological (Gaudin et al., 2009) and genetic evidence (Gaubert et al., 2018). The type specimen *Manis pentadactyla* Linnaeus (1758) is from Taiwan. Three subspecies are reported and recognized by some authors (e.g., Ellerman and Morrison-Scott, 1966) based on morphological characteristics including size, skull length and the shape and length of the nasal bones (Allen, 1906, 1938; Wang, 1975). They are *M. pentadactyla pentadactyla* (Linnaeus, 1758) for populations in Taiwan, nominally referred to as the Formosan pangolin; *M. p. aurita* (Hodgson, 1836) for populations in mainland Asia; and *M. p. pusilla* (Allen, 1906) for populations on Hainan island, China. However, the status of these subspecies has not been confirmed using morphological and molecular methods (though see Wu et al., 2007) and warrants further research.

Synonyms: *Manis brachyura* (Erxleben, 1777), *Manis auritus* (Hodgson, 1836), *Manis dalmanni* (Sundevall, 1842), *Pholidotus assamensis*, *Phatages bengalensis* (Fitzinger, 1872), *Manis pusilla* (Allen, 1906) and *Pholidotus kreyenbergi* (Matschie, 1907).

Etymology: *Manis*, coined from the Latin *"manes"* for spirit of the dead, or ghost in Roman religion, referring to the nocturnal behavior of pangolins and their unusual appearance; *pentadactyla* references the five (*pente-*) digits (*-daktulos*) on the fore- and hindlimbs (Gotch, 1979). The Order name, Pholidota, derives from the Greek *"pholis"* or *"pholidos"* (genitive), a horny scale (Gotch, 1979).

Description

The Chinese pangolin (*Manis pentadactyla*) is a small to medium sized mammal with a body weight of 3–5 kg, but individuals may reach more than 8 kg (Table 4.1). Total body length ranges up to 89 cm. The tail measures up to 40 cm, i.e., less than half the total body length and is notably shorter than in other pangolin species (Table 4.1; Wu et al., 2004a). Heath (1992) reports the tail is <0.42 of overall length. Males are up to 30% heavier than females (Wu et al., 2005a). The body is streamlined, elongate and covered in overlapping keratinous scales (20–50 mm diameter) that grow from the skin in a grid-like formation (Fig. 4.1A); they grow continuously and wear at the edges over time (Heath, 1992). Scales cover the dorsal surface of the head and trunk, the lateral surfaces of the trunk, the outside of the limbs, and the tail. Scales on the trunk are arranged from the mid-dorsal line outward (Heath, 1992). There are 14–18 transversal, 15–19 longitudinal, and 14–20 scale rows along the tail margins, which are folded and sharply pointed, and cover the tail both dorsally and ventrally (Table 4.1; Frechkop, 1931). The medio-dorsal scale row continues to the tail tip (Pocock, 1924), though Thomas (1892) reported that a specimen from Myanmar did not display this characteristic. All scales grow directed to the posterior with the exception of those on the hindlimbs, which point down (Heath, 1992). Post-scapular scales are of similar size to those on the distal part of the body. Total scales number between ~520 and 580 (Table 4.1). Zhou et al. (2012) estimated the dry weight of scales at 573.47 g (mean; n = 35). Scale color is dark brown, dark olive brown, yellow-brown or dark gray (Allen, 1938; Heath, 1992); occasionally scales lack pigmentation. Like other Asian pangolins, thick bristle-like hairs, whitish to pale brown in color, grow at the base of the scales (Allen, 1938). The muzzle, face (except the forehead), ventral surface of the trunk and inside of the limbs lack scales and are naked except for sparse white hair (Allen, 1938; Heath, 1992).

The skull is thick and demonstrates a number of features associated with adaptation to a

TABLE 4.1 Chinese pangolin morphometrics.

	Measurement		Country	Source(s)
Weight	Weight (♂)	4.5 (2.1–8.5) kg, n = 20	China	Wu et al., 2004b; S. Wu, unpubl. data
		5 (3.5–7.6) kg, n = 19	Taiwan	Chin et al., 2015
	Weight (♀)	3.5 (2.2–5.7) kg, n = 20	China	Wu et al., 2004b; S. Wu, unpubl. data
		4.7 (4–6) kg, n = 14	Taiwan	Chin et al., 2015
Body	Total length (♂)	749 (596–890) mm, n = 18	China	Wu et al., 2004b; S. Wu, unpubl. data
	Total length (♀)	699 (598–810) mm, n = 20	China	Wu et al., 2004b; S. Wu, unpubl. data
	Head-body length (♂)	437 (356–590) mm, n = 20	China	Wu et al., 2004b; S. Wu, unpubl. data
	Head-body length (♀)	413 (357–475) mm, n = 18	China	Wu et al., 2004b; S. Wu, unpubl. data
	Tail length (♂)	310 (238–400) mm, n = 18	China	Wu et al., 2004b; S. Wu, unpubl. data
	Tail length (♀)	285 (241–340) mm, n = 20	China	Wu et al., 2004b; S. Wu, unpubl. data
Vertebrae	Total number of vertebrae	59		Jentink, 1882
	Cervical	7		Jentink, 1882
	Thoracic	16		Jentink, 1882; Mohr, 1961
	Lumbar	6		Jentink, 1882; Mohr, 1961
	Sacral	3		Jentink, 1882; Mohr, 1961
	Caudal	27		Jentink, 1882; Mohr, 1961
Skull	Length (♂)	99.4 mm, n = 1	China	Wu et al., 2004b
	Breadth across zygomatic processes (unsexed)	31.9 (26.9–37.6) mm, n = 12	China	Luo et al., 1993
Scales	Total number of scales	554 (527–581), n = 10	China, India, Nepal, Myanmar, Unknown	Ullmann et al., 2019
	No. of scale rows (transversal, body)	14–18		Frechkop, 1931; Wu et al., 2004b
	No. of scale rows (longitudinal, body)	15–19		Frechkop, 1931
	No. of scales on outer margins of tail	14–20		Frechkop, 1931; Wu et al., 2004b
	No. of scales on median row of tail	16–20		Frechkop, 1931
	Scales (wet) as proportion of body weight	No data		
	Scales (dry) as proportion of body weight[a]	11–19%		

[a]*Based on weight of scales in Zhou et al. (2012), and body weight of 3–5 kg.*

FIGURE 4.1 (A) A wild Chinese pangolin in Bangladesh. The tail is shorter compared to other species of pangolin. (B) A rescued Chinese pangolin. This species has very large claws on the forelimbs and prominent ear pinnae. *(A) Photo credit: Scott Trageser. (B) Photo credit: Gary Ades/Kadoorie Farm and Botanic Garden.*

myrmecophagous diet (e.g., absence of certain cranial bones; see Heath, 1992). The jugal bones are present (Emry, 2004), but are absent in other pangolin species. The head is cone-shaped and the snout is short, an adaptation to the species' fossorial lifestyle. Scales cover the forehead and can give the species a helmeted-like appearance, but terminate before the rhinarium. The Chinese pangolin has prominent ear pinnae, which are the largest of all pangolins (the outer rim measures 20–30 mm), and comprise well-developed flaps that extend away from the head, anterior to which is the conspicuous auditory orifice (Fig. 4.1B; Pocock, 1924; Wu et al., 2004a). The facial skin is yellow-pinkish gray. The eyes have dark irises and are protected by thick, heavy eyelids. The nostrils are naked, moist and pink (Pocock, 1924). Like all pangolins, the species is toothless and the oral cavity is small. The tongue is long (16–40 cm) and thin, reaching up to 1 cm at the thickest point, is coated in a tenacious saliva, and is extended (up to 8–10 cm) and retracted quickly to consume prey (Heath, 1992). The base of the tongue is the xiphisternum in the abdomen, which is spade-shaped and formed from the last pair of cartilaginous ribs, and is short compared to the African pangolins; when retracted the tongue folds into a pocket in the throat (Heath, 1992). The stomach is composed of two chambers, the first a storage chamber, comprising ~80% of total volume, and the second serves the function of mastication, i.e., is a grinding compartment (Lin et al., 2015). The thick muscular walls are equipped with rugae and small keratinous spines which grind up prey, aided by small stones and soil that are ingested (Heath, 1992; Krause and Leeson, 1974; Lin et al., 2015).

All four limbs are stocky, with the hindlimbs slightly shorter, and have five digits with claws that are slightly curved (Heath, 1992). The claws on the third (middle) digit of the forelimbs are the largest (up to 66 mm) and most powerful (Pocock, 1924; Wu et al., 2004a). The first and fifth digits are small and essentially vestigial, and the first digit is set slightly advanced of the fifth (Pocock, 1924). The claws on the hindfeet are much shorter than on the forefeet. There are ill-defined granular pads on the forefeet behind the first and fifth digits, and the anterior part of the hindfeet is padded and granular (Pocock, 1924). The proximal end of the tail is stocky and muscular, and narrows towards the distal end – the final section being almost parallel-sided, and there is a granular pad on the ventral surface at the tail tip (Pocock, 1924; Tate, 1947). The dorsal surface of the tail is slightly arched and the ventral surface flat.

A post-anal depression comprising a transversely oblong pit on the anal eminence, which may be glandular, is present, and is unique to the Chinese pangolin (Heath, 1992; Pocock, 1924; Wu et al., 2005a). Anal glands, which produce a musky odor, are also present. Males have a well-developed penis. Like all pangolins, the testes do not descend into a scrotum but pass through the inguinal canal at sexual maturity and reside in a fold of skin in the groin (Heath, 1992). In females, the vulva is anterior to the anus (Pocock, 1924); they have two pectoral nipples.

Body temperature is regulated at 33.5–35 °C and average resting metabolic rate has been estimated at 183.6 ml O_2/kg/h (Heath and Hammel, 1986). Chinese pangolins will increase their metabolic rate when the ambient temperature is <25 °C (Heath and Hammel, 1986). Weber et al. (1986) noted that O_2 affinity of the blood and hemoglobin is higher than in non-burrowing mammals of similar size, suggesting an adaptation to extended periods in hypoxic environments, i.e., burrows, especially if they are sealed (see *Ecology*; Heath, 1992). The species has a subcutaneous fat layer (up to 1 cm) which likely provides considerable insulation (Fang, 1981).

The Chinese pangolin may be confused with other Asian pangolins, including the Sunda (*M. javanica*), Philippine (*M. culionensis*) and Indian pangolin (*M. crassicaudata*). The species is most easily distinguishable by its more thickly set body and noticeably shorter tail, larger ear pinnae, longer claws on the forelimbs (especially the middle claw), and the number of transverse scale rows (14–18) and scales on the tail margins (14–20), which are more or less numerous in congeners (see Chapters 5–7). The Chinese pangolin has a granular pad at the tail tip on the ventral side whereas the Indian pangolin has a terminal scale.

Distribution

Widely distributed in East Asia, northern Southeast Asia, and parts of South Asia (Fig. 4.2). Distributed south of the Yangtze River in China, from the southern provinces of Guangdong, Guangxi, Hainan and Yunnan, to Guizhou and Sichuan (Junlian, Mabian, Xichang and Miyi counties), the Tibet Autonomous Region (Chayu and Mangkang counties) and Chongqing (Xiushan, Nanchuan, Youyang and Fuling counties), across Hunan, Jiangxi and Fujian Provinces, the Xianning region in southeast Hubei Province, southern Anhui, and Zhejiang Province to the east (Allen, 1938; Heath, 1992; Jiang et al., 2016; Tate, 1947; Wu et al., 2002). To the north, the species is distributed in the low mountains of Ningzhen, Maoshan, Laoshan, and Yisu in the south of Jiangsu Province and in Jinshan and Fengxian counties in Shanghai (Wu et al., 2002). Allen and Coolidge (1940) note the species is (or was) distributed on the island of Zhoushan at the mouth of the Yangtze River. Zhai (2000) suggest the species occurs in the Funiu mountain district of Xichuan county, Henan Province, which would be the only locale north of the Yangtze River and requires further investigation. In Hong Kong Special Administrative Region (hereafter "Hong Kong"), the Chinese pangolin has been recorded in the central and northeast New Territories and on Lantau Island where it occurs at low altitudes, but not the smaller outlying islands (Shek et al., 2007). In Taiwan, the species is widely distributed, with the exception of the west of the island (Fig. 4.2), occurring in lowland agricultural

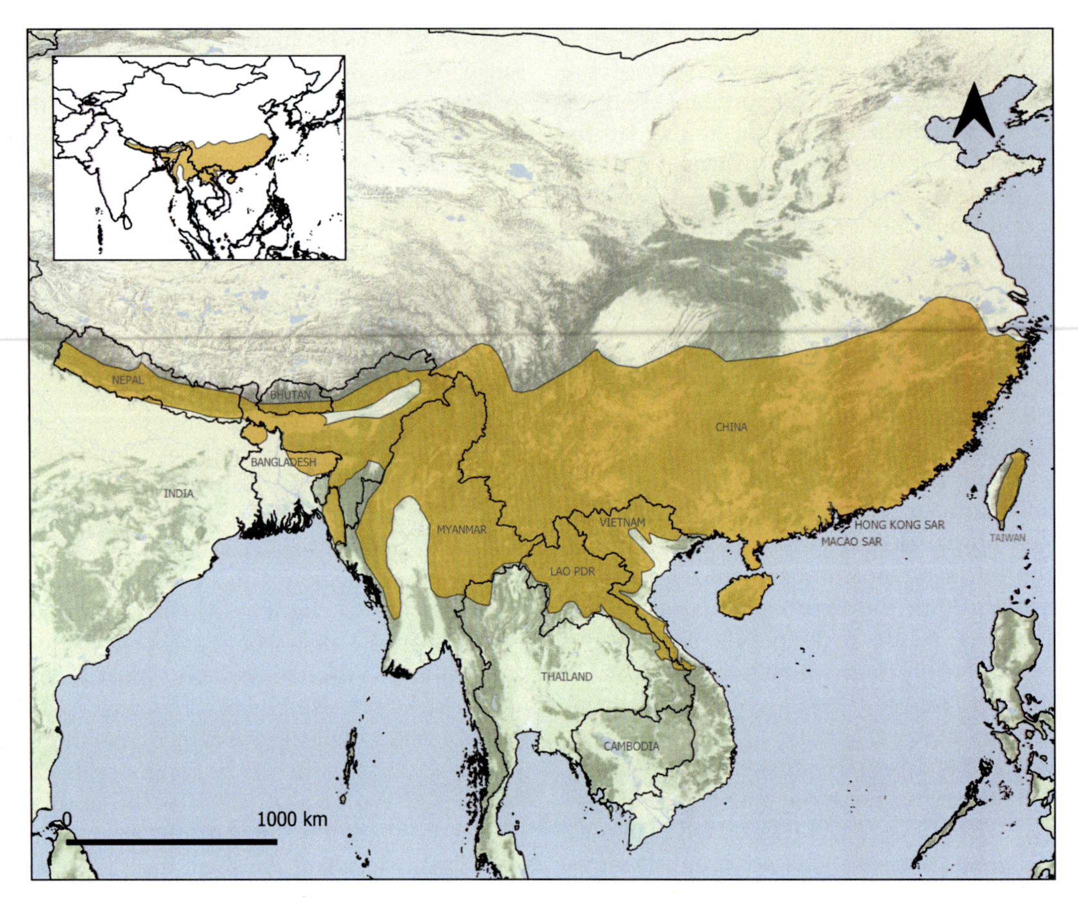

FIGURE 4.2 Chinese pangolin distribution. *Source: Challender et al., 2019.*

fields (up to 1000 m asl) and in the foothills of the Central Mountain Range, Western Foothill Range, Taoyuan Tableland, East Coast Mountain Range, Ouluanpi Tableland, East Coast Mountain Range, Tatun Volcano Group, Taipei Basin, Puli Basin, and the Pingtun Plain (Chao, 1989).

Beyond China, the species is distributed in northern and central Vietnam, from Ha Giang Province as far south as Quang Tri Province, but is absent from the northeast coast (Newton et al., 2008). The species is found in Cuc Phuong National Park, Khe Net Nature Reserve, Ke Go Nature Reserve, and Ba Na National Park (Newton et al., 2008). The southern limit of the species' distribution is the Annamite Mountain range, which straddles Vietnam and Lao Peoples' Democratic Republic (Lao PDR), and the species is distributed in northern and central regions of the latter (Duckworth et al., 1999). In Thailand, a single specimen has been collected from Doi Inthanon National Park and there is a record from Doi Sutep, Chiang Mai Province (Allen and Coolidge, 1940). There are also records from Loei (Chiang Khan) and Mae Hong Son (Lumnam Pai; Nabhitabhata and Chan-ard, 2005). There are few records from Myanmar, but the species has been collected in Pyinmana near Naypyidaw (Hopwood, 1929) and is likely distributed in the north, east and west of the country (Lekagul and McNeely,

1988). Naing et al. (2015) recorded the species in Hukaung Valley Wildlife Sanctuary in northern Myanmar in the early 2000s. Rao et al. (2005) recorded pangolins in Hkakaborazi National Park which likely refer to this species. Pangolins do occur in Karen State (Moo et al., 2017) but observations in the south likely refer to the Sunda pangolin.

West of Myanmar, the distribution extends to North and Northeast India, Bangladesh, Bhutan and Nepal. In Northeast India, the species has been recorded in Arunachal Pradesh north of the Brahmaputra River and to the south in Meghalaya, Nagaland, Manipur, Tripura and Mizoram, and in Assam state (Srinivasulu and Srinivasulu, 2012; Zoological Society of India, 2002). There are also records from northern West Bengal and Sikkim (Misra and Hanfee, 2000; Srinivasulu and Srinivasulu, 2012). In northern India, the species occurs in northern Bihar, to the south of the Nepalese border (Muarya et al., 2018). In Bhutan, there are records from Samtse in the southwest, Sarpang and Tsirang in the central south and Samdrup Jongkhar in the southeast (Dorji, 2017; Kinley et al., 2018; Srinivasulu and Srinivasulu, 2012). Trageser et al. (2017) reviewed the distribution of pangolins in Bangladesh, and confirmed the species' presence in the Chittagong Hill Tracts in the southeast, Lawachara National Park and adjacent protected areas in the northeast and Thakurgao region in the northwest. Choudhury (2004) referred to the species as common in northern Bangladesh but without verified records. The species is widely distributed in Nepal, having been recorded in eastern, central and western regions of the country, both within and outside protected areas, including Bardia, Chitwan, Makalu Barun, Parsa, Sagarmartha and Shivapuri-Nagarjun National Parks and the Annapurna, Gaurishankar and Kanchenjunga Conservation Areas (DNPWC and DoF, 2018; Shrestha, 1981; Srinivasulu and Srinivasulu, 2012).

The Chinese pangolin appears to be sympatric with the Sunda pangolin in parts of Southeast Asia (Lao PDR, Myanmar, Thailand and Vietnam), and the Indian pangolin in southern Nepal (e.g., Chitwan and Suklaphanta National Parks), Northeast India (Manas National Park) and theoretically Bangladesh (Goswami and Ganesh, 2014; Lahkar et al., 2018; Suwal and Verheugt, 1995; H. Baral, pers. comm.). Duckworth et al. (1999) hypothesized that the Chinese pangolin occurs at higher altitudes where it is sympatric with the Sunda pangolin, but habitat and ecological differences between Asian pangolins where they are sympatric have not been elucidated and warrant further research.

Habitat

The Chinese pangolin inhabits primary and secondary tropical and subtropical rainforest, limestone, bamboo, coniferous, mixed coniferous and broadleaf forests, low mountain or hill forest (400–500 m asl), grasslands and agricultural landscapes, and can tolerate some degree of disturbance (Chao, 1989; Gurung, 1996; Wu et al., 2003a,b). The species occurs across a wide altitudinal gradient with burrows recorded from less than 100 m asl in low altitude hill forest in Taiwan (N. Sun, pers. obs.) up to 3000 m in eastern Nepal (A. Khatiwada, unpubl. data). Wu et al. (2003a) reported that the species occurs in mixed coniferous and broadleaf, evergreen broadleaf, coniferous, and shrub forest in Dawuling Natural Reserve in Guangdong Province, China, and has a preference for mixed coniferous and broadleaf forest, followed by shrub forest in winter. A thick shrub layer, which reduces air convection around burrow entrances, is particularly important in the winter months (Wu et al., 2003a). In Bangladesh, the species occurs in the Chittagong Hill Tracts, which comprise primary and secondary evergreen, mixed evergreen and bamboo forest, and degraded habitat. There are also records from replanted mixed evergreen forest in Lawachara National Park, a heavily disturbed site dominated by teak (*Tectona grandis*) and *Xylia dolabriformis* among others (Trageser et al., 2017). In the Royal Nagarjung Forest in Nepal, Gurung (1996) recorded Chinese pangolin burrows, primarily in grassland dominated by cogon grass (*Imperata cyclindrica*) and ferns (*Gleichemia* spp.), which provided vegetative cover, mainly on slopes at the base of *Woodfordia fruiticosa* and chilaune (*Schima wallichi*). Some burrows were close to human settlements. Kaspal (2008) recorded burrows in mixed hardwood, and pine (*Pinus* spp.) forests in Bhaktapur district, Nepal. Gurung (1996) reported that farmers observed the species in croplands (maize, bean, yam and bamboo) when prey were abundant in July and August. Adaptable, the species can tolerate some disturbance, but will avoid habitat close to major highways (Chao, 1989; Lin, 2011; Wu et al., 2005a,b). In southern Taiwan, the species has been observed burrowing in the foundations of an abandoned building (N. Sun, unpubl. data).

Ecology

Most ecological knowledge of the Chinese pangolin stems from research in Taiwan and southern China. The species appears to occupy defined home ranges and, being fossorial, digs burrows for shelter and while foraging for prey, and available research suggests a polygynous social structure. In Fu-san, northern Taiwan, Lu (2005) used telemetry and estimated male home range size to be 69.9 ha (based on 12 months tracking; n = 1), and for females, 24.4 ha (based on six months tracking; n = 6). In Taitung, southeastern Taiwan, Lin (2011) estimated the home range of a single male to be 96 ha (100% Minimum Convex Polygon [MCP]; 66.6 ha [90% Fixed Kernel Density {FKD} estimate]), and female home range to be 14.3–30.3 ha (100% MCP; 15.9–19.6 ha [90% FKD estimate], n = 6). Male home ranges overlapped with those of several females suggesting a polygynous social structure, a finding corroborated by research in the Coastal Mountain range of eastern Taiwan. Microsatellite data revealed females mated with more than one male between, but not within, a season (N. Sun, unpubl. data; see also *Ontogeny and reproduction*).

Chinese pangolins use burrows to access prey, for shelter (including parturition), and to avoid predation (Bao et al., 2013; Wu et al., 2004c). Burrows can broadly be categorized into resident (or resting) and feeding burrows (Lin, 2011) but the species will dig burrows for both purposes, e.g., will dig a resident burrow adjacent to an ant nest and feed in the burrow before resting (Heath, 1992). Feeding burrows are typically much shorter than resident burrows, and may be found in more open areas including grassy hillsides (Fig. 4.3A; N. Sun, unpubl. data). In southeastern Taiwan, Lin (2011) estimated feeding burrow density of 110.8/ha and found no statistically significant difference between burrow frequency and habitat type (forested and more open areas).

Burrow length and depth can reflect prey behavior; in southern China burrows are deeper (1–3 m in length) in winter compared to summer (0.3–1 m; Fang and Wang, 1980; Wu et al., 2004c) and there are similar observations from Taiwan (N. Sun, unpubl. data). In winter, Chinese pangolins consume at least two species of termite including *Macrotermes barneyi*, with nests located up to 2 m beneath the surface, but in summer, the nests are shallower (15–50 cm) and closer to the surface (Fang and Wang, 1980; Heath, 1992).

Resident burrows typically consist of an entrance and an unbranched tunnel that terminates in a chamber and may be used throughout the year (Wu et al., 2004c; N. Sun, unpubl. data). Tunnels are typically dug in a downwards direction, but will then incline to prevent rainwater from entering the burrow (Gurung, 1996). Burrow entrances may be circular (Gurung,

(A) (B)

FIGURE 4.3 (A) Feeding burrow on an open, grassy hillside in southeastern Taiwan. (B) A resident burrow on a steep slope with dense surrounding vegetation (cut away for research purposes). *(A and B) Photo credit: Nick Ching-Min Sun.*

1996; Newton et al., 2008) or semi-circular (SMCRF, 2017) and up to 21 × 21 cm (Gurung, 1996). Burrow chambers have a diameter of 30–70 cm and are typically large enough for 2–3 adults to occupy simultaneously, though chambers up to 2 m in diameter have been reported (Fang and Wang, 1980; Liu and Xu, 1981; Wang, 2005; Wu et al., 2004c). Chambers may be lined with leaves from *Cinnamomum micranthum, Turpinia formosana, Smilax* sp., and *Dicranopteris linearis* (Chao, 1989). Rate of burrow excavation has been estimated at 2–3 m/h (Fang and Wang, 1980; Shi and Wang, 1985a).

Observations indicate that the species has specific preferences on the location of resident burrows. Research conducted during winter at Dawuling Natural Reserve in southern China, demonstrates the species prefers to dig resident burrows on steep slopes (30–60°) that receive direct sunlight, have a thick shrub layer, some canopy cover, and are close (< 1000 m) to water sources but more than 1000 m from human disturbance (Wu et al., 2003a,b). There are similar observations in Nepal (Gurung, 1996). At Dawuling, most burrows were dug between 760 and 1500 m elevation; northern facing slopes and those with bare ground were avoided (Wu et al., 2003a, 2004c, 2005a). This corroborates findings that resident burrows are usually well hidden by shrubs and other vegetation (Fig. 4.3B; Jiang et al., 1988; Katuwal et al., 2017; Liu and Xu, 1981). In Taiwan, resident burrows have been recorded under large boulders (N. Sun, unpubl. data), and Fan (2005) found that resident burrow density at Feitsui Reservoir did not differ significantly between bamboo, coniferous or broadleaf tree habitat, or secondary forest.

Lin (2011) reports that in southern Taiwan a male Chinese pangolin had an estimated 72.5–83.3 resident burrows in its home range, and females an estimated 29.4–39.6 burrows, reflecting the polygynous social structure, i.e., male home ranges overlap with those of several females (Lin, 2011). Resident burrows with a depth of up to 5.05 m have been recorded in Taiwan (Lin, 2011) but in Bangladesh local hunters report that burrows may reach a depth of 10 m (Trageser et al., 2017). Once in a burrow, Chinese pangolins may create an earthen wall in the tunnel, leaving a short gap at the top (~3 cm, presumably for ventilation), and which likely serves as a predator deterrent (Trageser et al., 2017; Wang, 1990). Resident burrows are typically used for 2–3 consecutive days in the dry season (November–April), but for an average of only one day during the wet season (June–October) in Taiwan (Lin, 2011). In winter at Dawuling Natural Reserve, China, the species will use the same burrow for an estimated 9–15 days (mean = 12) before moving to another burrow (Wu et al., 2004c). Different pangolins will also use the same resident burrow, but rarely on the same day (N. Sun, unpubl. data).

Burrows play an important role in providing Chinese pangolins with a stable thermal environment throughout the year (Bao et al., 2013; Fan, 2005). Bao et al. (2013) determined temperature variation in a Chinese pangolin burrow in winter at Luofushan Nature Reserve, China. They found slight fluctuations in diurnal burrow air temperature (0–0.5 °C) compared to diurnal air temperature outside the burrow (0.7–20 °C). Throughout the winter, air temperature in the burrow fluctuated much less (17.8–21°C) compared to outside air temperature (4.6–38.3°C; Bao et al., 2013). Mean temperature inside the burrow was 19.0 °C and was statistically significantly higher than the mean outside temperature, 15.2 °C (Bao et al., 2013). Mean burrow temperature in the summer was <26.3 °C (South China Normal University [SCNU], unpubl. data). Khatri-Chhetri et al. (2015) observed reduced heart rate and body temperature during autumn-winter in individuals in Taiwan, and suggested this to be a metabolic response to conserving energy. The species also requires high and constant humidity: 80–95% in the rainy season (June–August) and 70–85% in the dry season (December–February) in China (SCNU, unpubl. data).

The Chinese pangolin, like other pangolins, is prey selective, but will consume all life stages of prey including eggs, larvae and imagoes, and castes (e.g., worker ants; Fang and Wang, 1980). Published and gray literature indicates that the species predates on 23 species of ant and 12 species of termite (Table 4.2), though Sun et al. (2017) report predation on 70 ant species (5 Families, 25 genera) and four termite species (2 families and 4 genera; see Table 4.3) in Taiwan. Lee et al. (2017) examined the gut contents of a juvenile Chinese pangolin from Hong Kong and found >26,000 prey items, and a tendency for predation on larger (>4 mm) arboreal or semi-arboreal ants (as opposed to termites), including larvae. This included *Camponotus nicobarensis* and *Polyrhachis* spp., in particular *P. tyrannica*, as well as *Crematogaster dohrni* (1.88 mm), while Wu et al. (2005b) note a preference for *P. dives*. In contrast, Sun et al. (2017) found that arboreal ants comprised less than 5% of diet (by prey number), indicating diet adaptability. The presence of *Anoplolepis gracilipes*, an invasive species of ant associated with disturbed habitat, including near human settlements, in the diet analyzed by Lee et al. (2017) suggests that this individual foraged at forest margins and in scrubland. This contrasts with disturbance-avoiding behavior observed in rural populations (e.g., in China; Wu et al., 2003a). Feces in the wild weighs 59 g (mean; n = 123; N. Sun, unpubl. data).

Diet varies by season. In China, based on feeding frequency, ants are the major prey items in summer and termites in winter (Wu et al., 2005b). Sun et al. (2017) observed clear seasonal fluctuations in prey consumption in Taiwan. In the wet season (June–October), pangolins consumed nearly twice the number of insects than in the dry season (November–April), with the proportion of ants ranging from 56% to 98% (Sun et al., 2017). Species diversity and ant abundance were significantly higher in the wet season,

TABLE 4.2 Ant and termite species the Chinese pangolin has predated on based on published and gray literature.

Species	Source
Ants (23 species)	
Aphaenogaster exasperata	Lee et al., 2017
Anoplolepis gracilipes	Lee et al., 2017
Camponotus friedae	Yang et al., 2007
Camponotus mitis	Lee et al., 2017
Camponotus nicobarensis	Lee et al., 2017
Camponotus variegatus	Lee et al., 2017
Camponotus herculeanus	Ke et al., 1999
Camponotus sp.	Wu et al., 2004d
Carebara yanoi	Lee et al., 2017
Crematogaster rogenhoferi	Liu and Xu, 1981; Xu et al., 1983; Yang et al., 2007
Crematogaster macaoensis	Li et al., 2010a
Crematogaster dohrni	Lee et al., 2017; Yang et al., 2007
Dolichoderus affinis	Li et al., 2010a
Tetramorium bicarinatum	Ke et al., 1999
Nylanderia bourbonica	Lee et al., 2017
Oecophylla smaragdina	Li et al., 2010a
Odontomachus monticola	Wu et al., 2004d
Pheidole sp.	Wu et al., 2004d

(Continued)

TABLE 4.2 (Continued)

Species	Source
Pheidologeton yanoi	Yang et al., 2007
Paratrechina bourbonica	Wu et al., 2004d
Polyrhachis demangei	Lee et al., 2017
Polyrhachis tyrannica	Lee et al., 2017
Polyrhachis dives	Liu and Xu, 1981; Xu et al., 1983; Wu et al., 2004d; Yang et al., 2007
Termites (12 species)	
Capritermes nitobi	Wu et al., 2004d
Coptotermes formosanus	Ke et al., 1999; Liu and Xu, 1981; Xu et al., 1983; Wu et al., 2004d; Zhu-Ge and Huang, 1989
Coptotermes hainanensis	Wu et al., 2005b
Macrotermes barneyi	Lee et al., 2017; Liu and Xu, 1981; Shi and Wang, 1985b; Xu et al., 1983; Wu et al., 2004d; Zhu-Ge and Huang, 1989
Odontotermes formosanus	Liu and Xu, 1981; Shi and Wang, 1985b; Wu et al., 2004d; Yang et al., 2007; Zhu-Ge and Huang, 1989
Odontotermes zunyiensis	Wu et al., 2004d
Odontotermes hainanensis	Xu et al., 1983; Wu et al., 2004d
Pericapritermes nitobei	Liang, 2017
Reticulitermes flaviceps	Lee et al., 2017
Reticulitermes chinensis	Li et al., 2010b; Liu and Xu, 1981; Zhu-Ge and Huang, 1989
Reticutitertmes hainanensis	Xu et al., 1983
Reticulitermes speratus	Yang et al., 2007

TABLE 4.3 Composition of ant and termite species to Chinese pangolin diet in Taiwan based on 132 fecal samples collected between 2009 and 2015.

Species	Proportion of diet (%)
Ants	
Pheidologeton yanoi	27.4
Pheidole nodus	18.2
Pheidole fervens	11.6
Anoplolepis gracilipes	9.8
Crematogaster schimmeri	7.1
Camponotus monju	4.8
Pseudolasius binghami	4
Others	17.1
Termites	
Odontotermes formosanus	84.4
Others	15.6

*Source: Sun, N.C.-M., Pei, K.J.-C., Lin, C.-C., Li, H.-F., Liang, C.-C., 2017. The feeding ecology of Formosan pangolin (*Manis pendatactyla pendatactyla*) in the southern Coastal Mountain Range, Taiwan. In: South Asian Conference on Small Mammals Conservation "Small Mammals: Sustaining Ecology and Economy in the Himalaya", 27–29 August 2017, Kathmandu, Nepal.*

corresponding to an increase in adult body weight (Sun et al., 2017); in contrast, the species may lose up to 25–30% of body weight in the dry season (N. Sun, unpubl. data). Shi and Wang (1985a) report that in China the species can go without food for 10 days and 5–7 days in the winter and summer respectively.

There are few reports of predators, but they likely include leopard (*Panthera pardus*) and potentially pythons. The reticulated python (*Malayopython reticulatus*) will predate on the Sunda pangolin (Lim and Ng, 2008), and thereby likely the Chinese pangolin. Other species interactions include crab-eating mongoose (*Herpestes urva*) and masked palm civet (*Paguma larvata*), which will use Chinese pangolin burrows, including when occupied by a pangolin (N. Sun, unpubl. data). Several ecto- and endoparasites have been reported associated with the species (see Chapter 29).

Behavior

The species is nocturnal, spending most of the day resting or sleeping in an underground burrow and emerging at night to forage (Wu et al., 2005b). Fang and Wang (1980) reported active behavior between 1900 and 2200. Liu and Xu (1981) report that the species will forage up to 5–6 km away from a resident burrow each night. In a captive study in the United States, individuals never emerged before 1600, and usually after 1700, and were active intermittently (for between 30 s and 1.5 h) until 0200 (Heath, 1987). Active time comprised 2.9–7.6% (mean = 5.6%) of any 24 hour period (Heath, 1987; Heath and Vanderlip, 1988). Chen et al. (2005) observed captive animals in Taiwan emerging earlier (1600) in winter than in summer (1700), with activity ceasing by 0300, but occasionally animals were active up to 0800. Activity in captivity in China has been recorded between 1700 and 0730 (SCNU, unpubl. data).

The Chinese pangolin is quadrupedal (but is capable of bipedal locomotion) and walks with the head raised and the snout away from the ground, taking the weight on the hindlimbs, which are plantigrade, and forelimbs, which are curled inwards, with the weight taken on the folded up wrists (Wang, 1990; Wu et al., 2005a). The tail may be held out straight, parallel to the ground, or may be dragged, leaving "tail drag" marks on softer substrates (Allen, 1938; Heath, 1992).

When searching for prey, the species will walk more slowly and stop periodically to search, using its snout to sniff the air or ground, and having poor eyesight, relies on its acute olfactory senses to locate prey (Shi and Wang, 1985b; Wu et al., 2005a,b). Foraging includes locating an ant nest or termitaria, but also searching under leaf litter and rotten wood. On finding a nest or termitaria, the strong claws on the forelimbs are used to dig into it, with the hindlimbs used to support the body. If the substrate is hard, the tail is used for additional support enabling the forelimbs to exert greater excavating force (Shi and Wang, 1985b). Accumulated soil is pushed beneath and behind the body. Ant nests can be broken apart quickly, and the species extends it long, viscous tongue into the nest to "lap up" prey into the mouth (Shi and Wang, 1985a; Wu et al., 2005a,b). To consume dispersed ants, a pangolin may lie on the ground and erect its scales allowing ants to climb on the body, including beneath the scales, before quickly closing the scales, trapping the prey. The animal will then enter a body of water and re-erect its scales, releasing the ants into the water, and will consume them (Tao, 480–498; Wang, 1990; Wu et al., 2005a,b). If accessing an ant nest or termitaria necessitates digging a substantial burrow, the burrow may subsequently be used for shelter (Wu et al., 2005a).

Having fed, the Chinese pangolin will dig a hole in which to urinate and defecate and fill it with soil. The hole (5–10 cm deep) is dug using the forelimbs or the head, which is placed between the legs and dragged backwards and forwards (Heath, 1992). In captivity, the species generally defecates in the corner of enclosures or near the base of a wall. The animal will dig a pit and face the wall with the forelimbs slightly off the ground, sometimes standing upright, while drooping the head and contracting the abdominal muscles to discharge the feces, which is then buried.

Although predominantly terrestrial, the species is a capable climber and will do so in order to forage for arboreal prey (Fang, 1981), or for exploration. When climbing, the strong limbs, and foreclaws are used to grasp tree trunks and branches, supported by the flexible, prehensile tail. When descending, the species may slide, tumble or even fall directly from the tree. Chinese pangolins are also able swimmers. The head is held above the surface of the water with the rest of the body submerged, and all four limbs are used to swim. In the Czech Republic, an escaped individual was caught having swum across the 40 m wide Moldau River (Yang et al., 2007).

When threatened or having detected a predator, Chinese pangolins will either stop suddenly and freeze, flee, or roll up into a ball. If the latter, the head is buried in the chest and the forelimbs placed around the head, with the hindlimbs tucked up towards the neck and the tail curled up over the head making an impregnable ball. The species may roll down hillsides to evade predators (Liu and Xu, 1981), and may hiss when threatened (Zhang et al., 2016); the scales may be erected to withstand predators (Wang, 1989). A musky odor secreted from the anal glands may also help to deter predators.

Ontogeny and reproduction

Predominantly solitary, individuals only come together to mate when females are in estrus, and there is a defined breeding season. Otherwise, females and their young remain together until the young are independent. Mating occurs in spring and summer (February–July), and females give birth during autumn or early spring (September–February; Chin et al., 2011; Heath, 1992; Masui, 1967; Wu, 1998; Yang et al., 2001; Zhang et al., 2016). However, research in eastern Taiwan recorded males visiting female resident burrows from December to May, and staying with females in the burrow for periods of 2 min to more than 24 hours (n = 8; N. Sun, unpubl. data). Thus, it is possible there is greater variation in the timing of mating than previously recorded. When more than one male is in the presence of a female in estrus, they will fight violently until ones wins and the other leaves the area (Fang, 1981; Heath, 1992; Wang, 2005). Copulation entails the male mounting the female from the side and entwining tails to ensure the genitals meet.

In eastern Taiwan, 54 pangolins (29 males and 25 females) were captured between 2009 and 2013 in a 1000 ha area and genotyped at 10 microsatellite markers. 186 pairs of kin relationships were established, including 73 full-sibling pairs, 107 half-sibling pairs, and six parent-offspring relationships, suggesting close and complex kin relationships, but notably a low level of genetic diversity and limited gene flow (N. Sun, unpubl. data). This affirms previous research conducted across Taiwan (Wang, 2007) and may be due to geographical isolation or poor dispersal ability. Non-significant subpopulation structure and heterozygote deficiency indicated that inbreeding might exist in the eastern Taiwan population (N. Sun, unpubl. data).

Gestation lasts ~6–7 months (180–225 days), during which period females gradually gain weight and the mammary glands swell (Heath and Vanderlip, 1988; Yang et al., 2001; Zhang et al., 2016). Some reports suggest gestation may last up to eight months (Wang, 1990), or even 372 days (Chin et al., 2011), but this appears extreme and may be the result of delayed implantation which warrants further investigation. The placenta is furcated and the uterus is bipartite. One young is born at parturition (Zhang et al., 2016); twins are rare (Wu, 1998; Wang, 1990). Evidence indicates that females can conceive while still nursing young. In one case, an embryo was found in the uterus of a rescued, lactating female (S. Wu, unpubl. data). In Taiwan, a male Chinese pangolin mated with a female, which was nursing a 4 month old juvenile (Fig. 4.4; N. Sun, unpubl. data). Liu and Xu (1981) observed four Chinese pangolins in one burrow, including a male and female and two young that weighed 1500 g and 250 g respectively, potentially suggesting the species can give birth twice a year.

Females give birth in a burrow. Young are born with heads, limbs, claws and tails well developed. This includes scales, most of which are closely bound to the skin, and are gray or purplish brown and darker at the base: the

FIGURE 4.4 A male and female Chinese pangolin exhibiting mating behavior in Taiwan. The female was nursing a 4-month old juvenile at the time. *Photo credit: Nick Ching-Min Sun.*

non-vascular parts are milky white and 1–2 mm in length (Chao et al., 1993; Wu, 1998; Zhang et al., 2016). White hair between the scales is clearly visible. The claws are curled inwards and wrapped in a soft colloidal membrane, which dries and is shed the day after parturition. This may help prevent the neonate's claws from scratching the female's uterus and vagina during parturition (Zhang et al., 2016). The young's eyes are open and motor coordination is good (Heath and Vanderlip, 1988). At birth, young weigh 80–180 g (Chin et al., 2011; Heath and Vanderlip, 1988; Liu and Xu, 1981), and are 185–265 mm in length (Chao et al., 1993; Zhang et al., 2016). When hungry, young will seek out the mother's nipples to suckle. The mother will wrap the juvenile in her chest tightly, usually exposing only the head, uncurling her body slightly during lactation, and will curl tightly around the juvenile if threatened (Masui, 1967).

Young are left in the burrow while the female forages for the first few weeks, but will accompany the female thereafter, being "taxied" around on the base of the tail (Fig. 4.5; Sun et al., 2018). An infant was first observed being taxied around by its mother after 29 days (Sun et al., 2018). After 11 weeks, the young will start to leave the burrow independently, displaying digging and licking behavior, and the duration of time spent exploring, and distance covered, increases considerably after 15 weeks (Heath, 1992; Sun et al., 2018). These observations suggest that young start to forage independently at ~4 months old, and are carried by the female only when changing burrow (Sun et al., 2018). Young grow at a rate of 1.2 cm in length each week during the nursing period; in one case an individual measured 47 cm immediately prior to independence (Sun et al., 2018). Growth rate is slower in captivity (0.7 cm/per week; Wang et al., 2012). Wang (2007) reports that a captive individual weighed ~1.3 kg when it was 47 cm in length.

Weaning occurs at 5–6 months old (Liu and Xu, 1981; Sun et al., 2018; Wang, 2007) when

FIGURE 4.5 A female Chinese pangolin and its young. Young begin to emerge from the burrow at around 30 days old. *Photo credit: Nick Ching-Min Sun.*

young weigh 2–3 kg (Masui, 1967; SCNU, unpubl. data). Females generally reach sexual maturity at 1–1.5 years old, but potentially as early as 6 months old in some individuals (Chin et al., 2011; Zhang et al., 2016). Females have become pregnant when they weigh 2–3 kg (n = 8; Zhang et al., 2016). Lifespan in the wild is unknown, but in captivity individuals have lived to be more than 20 years old (see Chapter 36).

Population

There is little reliable quantitative data, but populations are suspected to be declining in most parts of the range. The main factor affecting abundance is exploitation by humans, and other influences likely include the availability of prey, including seasonally, and predation. Once common in China, populations declined severely between the 1960s and 1990s because of overexploitation (Wu et al., 2004e, 2005a; Zhang, 2009; see also Chapter 16). Wu et al. (2004e) reported population declines of up to 94% in this period. In 2002, China's population was estimated at 50,000–100,000 individuals based on sampling methods including burrow counts and local ecological knowledge (Wu et al., 2002). China's National Forestry Administration (now National Forestry and Grassland Administration [NFGA]) estimated the national population to comprise 64,000 individuals in the late 1990s (National Forestry Administration, 2008). The population of Guangxi was estimated at 990 individuals with a density of 0.043 individuals/km^2 (National Forestry Administration, 2008). Zhang et al. (2010) estimated the population at 25,100–49,450 individuals in 2008. At some sites (e.g., Dawuling Natural Reserve and Luofushan Natural Reserve in Guangdong Province), new burrows have not been seen in approaching 20 years, and it is suspected that the species may have been extirpated from Jiangsu, Henan and Shanghai Provinces (Wu et al., 2005a; Zhang et al., 2010). Yang et al. (2018) estimated that in eastern China (Fujian, Jiangxi and Zhejiang Provinces) the species' range contracted by 52% between the 1970s and 2000s, and populations are now largely confined to the Wuyi Mountains. On Hainan Island, interviews with local people suggest the species is present but of very low abundance (Nash et al., 2016). However, there have been observations since 2010 in parts of Anhui, Zhejiang, Jiangxi, Fujian, Guangdong and Hainan, including evidence of breeding, i.e., females with young (Zhang et al., 2017; S. Wu, unpubl. data). The species is present in Hong Kong but there are no abundance estimates (Pei et al., 2010). Once common in Taiwan, the species underwent steep declines in the 1950s–70s because of overexploitation (Chao, 1989; see Chapter 16). Populations have recovered in some places since (see Chapter 36) and Taiwan, and perhaps Hong Kong, likely represent the only parts of the range where populations are not in decline. Pei (2010) estimated densities of 12–13 pangolins/km^2 in Taitung, Taiwan. Using *VORTEX* to conduct Population Viability Analysis (PVA), Kao et al. (2019), estimated, very approximately, a metapopulation for Taiwan of ~15,000 individuals divided into four sub-populations on the island: north, central, south and east.

Available evidence indicates that populations in mainland Southeast Asia have declined since the late 20th century because of overexploitation. For instance, in Vietnam, the species is rare, with local hunters in three parts of the country independently reporting that populations declined dramatically in the 1990s and 2000s due to overhunting (Newton et al., 2008; P. Newton, pers. comm.). There are little or no data for Lao PDR, Thailand, Myanmar and Bhutan.

The species was common in the hill forests of Arunachal Pradesh in India in the 1980s but there is a lack of contemporary data. In Bangladesh, Chinese pangolins may have been extirpated from parts of the southeast, but small, isolated populations likely persist in regions within the Chittagong Hill Tracts where there is an absence of high-skilled hunters: however, populations in the country are suspected to be declining (Trageser et al., 2017). The population in Nepal occurs within and outside protected areas (see Chapter 25) and has been estimated at 5000 individuals but is suspected to be declining due to excessive hunting and poaching, and to a lesser degree habitat loss (Jnawali et al., 2011; Thapa, 2013).

Status

The Chinese pangolin is listed as Critically Endangered on The IUCN Red List of Threatened Species (Challender et al., 2019). This is based on past, ongoing and suspected future declines because of hunting and poaching for local and international use, and the species and its derivatives are heavily trafficked internationally (see *Threats*). The species is categorized as Critically Endangered on China's National Red List (Jiang et al., 2016) and Endangered in Nepal (Jnawali et al., 2011). National legislation in range states affords the species protection from exploitation and in 2016 the Chinese pangolin was included in CITES Appendix I.

Threats

The major threat to the Chinese pangolin is unsustainable hunting and poaching, both targeted and untargeted, for local and international use. The severity of the threat is difficult to quantify across the species' range, but evidence indicates that it is pervasive—the only exceptions appear to be Hong Kong and Taiwan—and disentangling local and international use is challenging. In China, the species is hunted and its body parts, especially scales, are used locally for medicinal and cultural purposes (e.g., Hainan Province; Nash et al., 2016; see also Chapter 14). In northeastern India, the meat is consumed locally by tribal communities, the legality of which depends on the local district (see D'Cruze et al., 2018), and the scales and other body parts are used for a wide range of medicinal applications (Mohapatra et al., 2015). However, it appears that in many places local consumption is foregone in favor of selling scales in particular, into illegal, international trade because of the high financial rewards available; this is the case in Bangladesh (Trageser et al., 2017), parts of China (Nash et al., 2016), northeastern India (D'Cruze et al., 2018) and Nepal (Katuwal et al., 2015).

International trade reported to CITES during the 1980s and 1990s involved more than an estimated 50,000 Chinese pangolins, primarily in the form of skins, but this trade has largely declined following the introduction of zero export quotas in the year 2000 (see Chapter 16). However, as indicated, illegal, international trade persists seemingly unabated. Challender et al. (2015) estimated that between the year 2000 and 2013, international trafficking likely involved more than 50,000 Chinese pangolins based on seizure records. This takes place to meet consumer demand for scales, which are used as an ingredient in traditional medicines, and meat which is consumed as a luxury food, primarily in China and Vietnam (Challender and Waterman, 2017). Evidence indicates the trade is in part operated by criminal groups and is organized from the village level up to the point of, and beyond, export (Katuwal et al., 2015; see Chapter 16).

Secondary threats include infrastructure development, habitat loss and fragmentation, pesticide use and, in China, major developments including hydropower stations and mining. In Taiwan and Hong Kong, where populations are not threatened by heavy hunting or poaching, threats comprise feral dogs, roadkill and development, i.e., conversion of land for human use. Research in Taiwan also suggests that Chinese pangolins can become trapped in burrows or tree hollows, despite being proficient burrowers and climbers, and is a cause of mortality; they are also killed by gin traps (Sun et al., 2019).

References

Allen, J.A., 1906. Mammals from the island of Hainan, China. Bull. Am. Museum Nat. Hist. 22, 463–490.

Allen, G.M., 1938. The Mammals of China and Mongolia. Natural History of Central Asia, vol. XI. Part I. The American Museum of Natural History, New York.

Allen, G.M., Coolidge, H.J., 1940. Mammal collections of the Asiatic Primate Expeditions. Bull. Museum Comp. Zool., Harvard 97 (3), 131–166.

Bao, F., Wu, S., Su, C., Yang, L., Zhang, F., Ma, G., 2013. Air temperature changes in a burrow of Chinese pangolin, *Manis pentadactyla*, in winter. Folia Zool. 62 (1), 42–47.

Brisson, M.-J., 1762. Regnum Animale in Classes IX. Apud Theodorum Haak, Lugduni Batavorum.

Challender, D.W.S., Harrop, S.R., MacMillan, D.C., 2015. Understanding markets to conserve trade-threatened species in CITES. Biol. Conserv. 187, 249–259.

Challender, D., Waterman, C., 2017. Implementation of CITES Decisions 17.239 b) and 17.240 on Pangolins (*Manis* spp.), CITES SC69 Doc. 57 Annex. Available from <https://cites.org/sites/default/files/eng/com/sc/69/E-SC69-57-A.pdf>. [April 3, 2018].

Challender, D., Wu, S., Kaspal, P., Khatiwada, A., Ghose, A., Sun, N.C.-M., et al., 2019. *Manis pentadactyla. The IUCN Red List of Threatened Species* 2019: eT12764 A123585318. Available from: <http://dx.doi.org/10.2305/IUCN.UK.2019-3.RLTS.T12764A123585318.en>.

Chao, J.-T., 1989. Studies on the Conservation of the Formosan Pangolin (*Manis pentadactyla pentadactyla*). General Biology and Current Status. Division of Forest Biology, Taiwan Forestry Research Institute. Council of Agriculture, Executive Yuan, Taiwan. [In Chinese].

Chao, J.-T., Chen, Y.-M., Yeh, W.-C., Fang, K.-Y., 1993. Notes on a newborn Formosan pangolin *Manis pentadactyla pentadactyla*. Taiwan Museum 46 (1), 43–46.

Chen, S.-H., Hsi, C.-C., Chen, Y.-M., Chang, M.-H., 2005. Activity pattern of Formosan pangolin (*Manis pentadactyla pentadactyla*) in captivity. Taipei Zoo, Taiwan, Unpublished Report, pp. 1–7. [In Chinese].

Chin, S.-C., Lien, C.-Y., Chan, Y.-T., Chen, C.-L., Yang, Y.-C., Yeh, L.-S., 2011. Monitoring the gestation period of rescued Formosan pangolin (*Manis pentadactyla pentadactyla*) with progesterone radioimmunoassay. Zoo Biol. 31 (4), 479–489.

Chin, S.-C., Lien, C.-Y., Chan, Y., Chen, C.-L., Yang, Y.-C., Yeh, L.-Y., 2015. Hematologic and serum biochemical parameters of apparently healthy rescued Formosan pangolins (*Manis pentadactyla pentadactyla*). J. Zoo Wildlife Med. 46 (1), 68–76.

Choudhury, A., 2004. On the pangolin and porcupine species of Bangladesh. J. Bombay Nat. Hist. Soc. 101 (3), 444–445.

D'Cruze, N., Singh, B., Mookerjee, A., Harrington, L.A., Macdonald, D.W., 2018. A socio-economic survey of pangolin hunting in Assam, Northeast India. Nat. Conserv. 30, 83–105.

DNPWC and DoF (Department of National Parks and Wildlife Conservation and Department of Forests), 2018. Pangolin Conservation Action Plan for Nepal (2018-2022). Department of National Parks and Wildlife Conservation and Department of Forests, Kathmandu, Nepal.

Dorji, D., 2017. Distribution, habitat use, threats and conservation of the Critically Endangered Chinese pangolin (*Manis pentadactyla*) in Samtse District, Bhutan. Unpublished Report for Rufford Small Grants, UK.

Duckworth, J.W., Salter, R.E., Khounboline, K., 1999. Wildlife in Lao PDR: 1999 Status Report. IUCN, Wildlife Conservation Society and Centre for Protected Areas and Watershed Management, Vientiane, Lao PDR.

Ellerman, J.R., Morrison-Scott, T.C.S., 1966. Checklist of Palaearctic and Indian Mammals 1758 to 1946, second ed. British Museum, London.

Emry, R.J., 2004. The edentulous skull of the North American Pangolin, *Patriomanis americanus*. Bull. Am. Museum Nat. Hist. 285, 130–138.

Erxleben, U.C.P., 1777. Systema Regna Animalis, Classis 1, Mammalia. Lipsize, Impensis Weygandianus.

Fan, C.Y., 2005. Burrow Habitat of Formosan Pangolins (*Manis pentadactyla pentadactyla*) at Feitsui Reservoir. M.Sc. Thesis, National Taiwan University, Taipei, Taiwan. [In Chinese].

Fang, L.-X., 1981. Investigation on pangolins by following their trace and observing their cave. Nat., Beijing Nat. Hist. Museum 3, 64–66. [In Chinese].

Fang, L.X., Wang, S., 1980. A preliminary survey on the habits of pangolin. Mem. Beijing Nat. Hist. Museum 7, 1–6. [In Chinese].

Fitzinger, L.J., 1872. Die naturliche familie der schuppenthiere (Manes). Sitzungsberichte der Kaiserlichen Akademie der Wissenschaften. Mathematisch-Naturwissenschaftliche Classe, CI., LXV, Abth. I, 9–83.

Frechkop, S., 1931. Notes sur les mammifères. VI. Quelques observations sur la classification des pangolins (Manidae). Bulletin du Musee royal d'Histoire naturelle de Belgique VII (22), 1–14.

Gaubert, P., Antunes, A., Meng, H., Miao, L., Peigné, S., Justy, F., et al., 2018. The complete phylogeny of pangolins: scaling up resources for the molecular tracing of the most trafficked mammals on Earth. J. Hered. 109 (4), 347–359.

Gaudin, T., Emry, R., Wible, J., 2009. The phylogeny of living and extinct pangolins (Mammalia, Pholidota) and associated taxa: a morphology based analysis. J. Mammal. Evol. 16 (4), 235–305.

Goswami, R., Ganesh, T., 2014. Carnivore and herbivore densities in the immediate aftermath of ethno-political conflict: the case of Manas National Park, India. Trop. Conserv. Sci. 7 (3), 475–487.

Gotch, A.F., 1979. Mammals – Their Latin Names Explained. A Guide to Animal Classification. Blandford Press, Poole.

Gurung, J.B., 1996. A pangolin survey in Royal Nagarjung Forest in Kathmandu, Nepal. Tiger Paper 23 (2), 29–32.

Heath, M.E., 1987. Twenty-four-hour variations in activity, core temperature, metabolic rate, and respiratory quotient in captive Chinese pangolins. Zoo Biol. 6 (1), 1–10.

Heath, M.E., 1992. *Manis pentadactyla*. Mammal. Sp. 414, 1–6.

Heath, M.E., Hammel, H.T., 1986. Body temperature and rate of O_2 consumption in Chinese pangolins. Am. J. Physiol.-Regul., Integr. Comp. Physiol 250 (3), R377–R382.

Heath, M.E., Vanderlip, S.L., 1988. Biology, husbandry, and veterinary care of captive Chinese Pangolins (*Manis pentadactyla*). Zoo Biol. 7 (4), 293–312.

Hodgson, B.H., 1836. Synoptical description of sundry new animals enumerated in the catalogue of Nepalese mammals. J. Asiatic Soc. Bengal 5, 231–238.

Hopwood, S.F., 1929. Some notes on the pangolin (*Manis pentadactyla*) in Burma. J. Bombay Nat. Hist. Soc. XXXIII (1 & 2), 1–471.

Jentink, F.A., 1882. Note XXV. Revision of the Manidae in the Leyden Museum. Notes from the Leyden Museum IV, 193–209.

Jiang, H., Feng, M., Huang, J., 1988. Preliminary observation on pangolin's active habits. Chin. Wildlife 9, 11–13. [In Chinese].

Jiang, Z.G., Jiang, J., Wang, Y., Zhang, E., Zhang, Y., Li, L., et al., 2016. Red list of China's vertebrates. Biodivers. Sci. 24 (5), 500–551. [In Chinese].

Jnawali, S.R., Baral, H.S., Lee, S., Acharya, K.P., Upadhyay, G.P., Pandey, M., et al., 2011. The Status of Nepal's Mammals: The National Red List Series. Department of National Parks and Wildlife Conservation, Kathmandu, Nepal.

Kao, J., Li, J.Y.W., Lees, C., Traylor-Holzer, K., Jang-Liaw, N.H., Chen, T.T.Y., et al., (Eds.), 2019. Population and Habitat Viability Assessment and Conservation Action Plan for the Formosan Pangolin, *Manis p. pentadactyla*. IUCN SSC Conservation Planning Specialist Group, Apple Valley, Minnesota.

Kaspal, P., 2008. Status, Distribution, Habitat Utilization and Conservation of Chinese Pangolin in the Community Forests of Suryabinayak Range Post, Bhaktapur District. M.Sc. Thesis, Khowpa College, Tribhuvan University affiliated, Nepal.

Katuwal, H.B., Naupane, K.R., Adhikari, D., Sharma, M., Thapa, S., 2015. Pangolins in eastern Nepal: trade and ethno-medicinal importance. J. Threat. Taxa 7 (9), 7563–7567.

Katuwal, H.B., Sharma, H.P., Parajuli, K., 2017. Anthropogenic impacts on the occurrence of the critically endangered Chinese pangolin (*Manis pentadactyla*) in Nepal. J. Mammal. 98 (6), 1667–1673.

Ke, Y.Y., Chang, H., Wu, S.B., Liu, Q., Fenf, G.X., 1999. A study of Chinese pangolin's main food nutrition. Zool. Res. 20 (5), 394–395. [In Chinese].

Khatri-Chhetri, R., Sun, C.-M., Wu, H.-Y., Pei, K.J.-C., 2015. Reference intervals for hematology, serum biochemistry, and basic clinical findings in free-ranging Chinese pangolin (*Manis pentadactyla*) from Taiwan. Vet. Clin. Pathol. 44 (3), 380–390.

Kinley, Dorj, C., Thapa, D., 2018. New distribution record of the Critically Endangered Chinese pangolin *Manis pentadactyla* in Bhutan. The Himalayan Naturalist 1 (1), 13–14.

Krause, W.J., Leeson, C.R., 1974. The stomach of the pangolin (*Manis pentadactyla*) with emphasis on the pyloric teeth. Acta Anat. 88 (1), 1–10.

Lahkar, D., Ahmed, M.F., Begum, R.H., Das, S.K., Lahkar, B.P., Sarma, H.K., et al., 2018. Camera-trapping survey to assess diversity, distribution and photographic capture rate of terrestrial mammals in the aftermath of the ethnopolitical conflict in Manas National Park, Assam, India. J. Threat. Taxa 10 (8), 12008–120017.

Lee, R.H., Cheung, K., Fellowes, J.R., Guénard, B., 2017. Insights into the Chinese pangolin's (*Manis pentadactyla*) diet in a peri-urban habitat. Trop. Conserv. Sci. 10, 1–7.

Lekagul, B., McNeely, J.A., 1988. Mammals of Thailand, second ed. Darnsutha Press, Bangkok.

Li, W., Tong, Y., Xiong, Q., Huang, Q., 2010a. Efficacy of three kinds of baits against the subterranean termite *Reticulitermes chinensis* (Isoptera: Rhinotermitidae) in rural houses in China. Sociobiology 56 (1), 209–222.

Li, X., Zhou, J.L., Guo, Z.F., Guo, A.W., Chen, F.-F., 2010b. The analysis on nutrition contents of ants preyed on by *Manis pentadactyla*, Xishuangbanna of China. Sichuan J. Zool. 29 (5), 620–621. [In Chinese].

Liang, C.-C., 2017. Termite Species Composition in Soil and Feces of Formosan Pangolin (*Manis pentadactyla*

pentadactyla) at Luanshan, Taitung. M.Sc. Thesis, National Pingtung University of Science and Technology, Pingtung, Taiwan. [In Chinese].

Lim, N.T.-L., Ng, P., 2008. Predation on *Manis javanica* by *Python Reticulatus* in Singapore. Hamadryad 32 (1), 62–65.

Lin, J.S., 2011. Home Range and Burrow Utilization in Formosan Pangolin (*Manis pentadactyla pentadactyla*) at Luanshan, Taitung. M.Sc. Thesis, National Pingtung University of Science and Technology, Pingtung, Taiwan. [In Chinese].

Lin, M.F., Chang, C.-Y., Yang, C.W., Dierenfeld, E.S., 2015. Aspects of digestive anatomy, feed intake and digestion in the Chinese pangolin (*Manis pentadactyla*) at Taipei Zoo. Zoo Biol. 34 (3), 262–270.

Linnaeus, C., 1758. Systema Naturæ Per Regna Tria Naturæ, Secundum Classes, Ordines, Genera, Species, Cum Characteribus, Differentiis, Synonymis, Locis. Tomus I. Editio decima, reformata. Salvius, Stockholm.

Liu, Z.H., Xu, L.H., 1981. Pangolin's habits and its resource protection. Chin. J. Zool. 16, 40–41. [In Chinese].

Lu, S., 2005. Study on the Distribution, Status and Ecology of Formosan Pangolin in Northern Taiwan (2/2). Taiwan Forestry Research Institute, Taipei, Taiwan. [In Chinese].

Luo, R., Chen, Y., Wei, K., 1993. The Mammalian Fauna of Guizhou. Guizhou Science and Technology Publishing House, Guiyang. [In Chinese].

Masui, M., 1967. Birth of a Chinese pangolin *Manis pentadactyla* at Ueno Zoo, Tokyo. International Zoo Yearbook 7, 147.

Matschie, P., 1907. Über chinesische süagetiere. In: Filchner, W. (Ed.), Wissenschaftliche Ergebinisse der Expedition Filchner. Ernst Siegfied Mittler und Sohn, Berlin, pp. 41–45.

Misra, M., Hanfee, N., 2000. Pangolin distribution and trade in east and northeast India. TRAFFIC Dispatches 14, 4–5.

Mohapatra, R.K., Panda, S., Acharjyo, L.N., Nair, M.V., Challender, D.W.S., 2015. A note on the illegal trade and use of pangolin body parts in India. TRAFFIC Bull. 27 (1), 33–40.

Mohr, E., 1961. Schuppentiere. Neue Brehm-Bucherei. A. Ziemsen Verlag, Wittenburg Lutherstadt.

Moo, S.S.B., Froese, G.Z.L., Gray, T.N.E., 2017. First structured camera-trap surveys in Karen State, Myanmar, reveal high diversity of globally threatened mammals. Oryx 52 (3), 537–543.

Muarya, K.K., Shafi, S., Gupta, M., 2018. Chinese Pangolin: sighting of Chinese pangolin (*Manis pentadactyla*) in Valmiki Tiger Reserve, Bihar, India. Small Mammal Mail 416. In: Zoo's Print 33 (1), 15–18.

Nabhitabhata, J., Chan-ard, T., 2005. Thailand Red Data: Mammals, Reptiles and Amphibians. Office of Natural Resources and Environmental Policy and Planning, Bangkok, Thailand.

Naing, H., Fuller, T.K., Sievert, P.R., Randhir, T.O., Po, S. H.T., Maung, M., et al., 2015. Assessing large mammal and bird richness from camera-trap records in the Hukaung Valley of Northern Myanmar. Raffles Bulletin of Zoology 63, 376–388.

Nash, H.C., Wong, M.H.G., Turvey, S.T., 2016. Using local ecological knowledge to determine status and threats of the Critically Endangered Chinese pangolin (*Manis pentadactyla*) in Hainan, China. Biol. Conserv. 196, 189–195.

National Forestry Administration, 2008. Investigation of Key Terrestrial Wildlife Resources in China. China Forestry Publishing House, Beijing, China. [In Chinese].

Newton, P., Nguyen, V.T., Roberton, S., Bell, D., 2008. Pangolins in peril: using local hunters' knowledge to conserve elusive species in Vietnam. Endanger. Species Res. 6, 41–53.

Pei, K.J.-C., 2010. Ecological Study and Population Monitoring for the Taiwanese Pangolin (*Manis pentadactyla pentadactyla*) in Luanshan Area, Taitung. Taitung Forest District Office Conservation Research, Taitung, Taiwan. [In Chinese].

Pei, K.J.-C., Lai, Y.C., Corlett, R.T., Suen, K.-Y., 2010. The larger mammal fauna of Hong Kong: species survival in a highly degraded landscape. Zool. Stud. 49 (2), 253–264.

Pocock, R.I., 1924. The external characters of the pangolins (Manidae). Proc. Zool. Soc. Lond. 94 (3), 707–723.

Rafinesque, C.S., 1820. Sur le genre *Manis* et description d'une nouvelle espèce: *Manis ceonyx*. Annales Générales des Sciences Physiques 7, 214–215.

Rao, M., Myint, T., Zaw, T., Htun, S., 2005. Hunting patterns in tropical forests adjoining the Hkakaborazi National Park, north Myanmar. Oryx 39 (3), 292–300.

Shek, C.-T., Chan, S.S.M., Wan, Y.-F., 2007. Camera Trap Survey of Hong Kong Terrestrial Mammals in 2002-06. Hong Kong Biodiversity. Agric., Fish. Conserv. Depart. Newsl. 15, 1–15.

Shi, Y., Wang, Y., 1985a. The pangolins' habit of eating ants. Chin. J. Wildlife 28 (6), 42–43. [In Chinese].

Shi, Y., Wang, Y., 1985b. The preliminary study on captive breeding pangolins. For. Sci. Technol. 10, 28–29. [In Chinese].

Shrestha, T.K., 1981. Wildlife of Nepal. A Study of Renewal Resources of Nepal, Himalayas. Curriculum Development Center, Tribhuvan University, Kathmandu, Nepal.

SMCRF (Small Mammals Conservation and Research Foundation), 2017. Pangolin Monitoring Protocol for Nepal. Department of Forests, Government of Nepal.

Srinivasulu, C., Srinivasulu, B., 2012. South Asian Mammals: Their Diversity, Distribution, and Status. Springer, New York.

Sun, N.C.-M., Pei, K.J.-C., Lin, C.-C., Li, H.-F., Liang, C.-C., 2017. The feeding ecology of Formosan pangolin (*Manis pendatactyla pendatactyla*) in the southern Coastal Mountain Range, Taiwan. In: South Asian Conference on Small Mammals Conservation "Small Mammals: Sustaining Ecology and Economy in the Himalaya", 27–29 August 2017, Kathmandu, Nepal.

Sun, N.C.-M., Sompud, J., Pei, K.J.-C., 2018. Nursing period, behavior development, and growth pattern of a newborn Formosan pangolin (*Manis pentadactyla pentadactyla*) in the wild. Trop. Conserv. Sci. 11, 1–6.

Sun, N.C.-M., Arora, B., Lin, J.-S., Lin, W.-C., Chi, M.-J., Chen, C.-C., et al., 2019. Mortality and morbidity in wild Taiwanese pangolin (*Manis pentadactyla pentadactyla*). PLoS One 14 (2), e0212960.

Sundevall, C.J., 1842. Om slägtet Sorex, med några nya arters beskrifning. Kungliga Vetenskapsakademien, Stockholm.

Suwal, R., Verheugt, Y.J.M., 1995. Enumeration of Mammals of Nepal. Biodiversity Profiles Project Publication No. 6. Department of National Parks and Wildlife Conservation, Ministry of Forest and Soil Conservation, Nepal.

Tao, H.J., 480–498. Annotation of Materia Medica - Animals, China. [In Chinese].

Tate, G.H.H., 1947. Mammals of Eastern Asia. Macmillan, New York.

Thapa, P., 2013. An overview of Chinese pangolin (*Manis pentadactyla*): Its general biology, status, distribution and conservation threats in Nepal. Initiation 5, 164–170.

Thomas, O., 1892. On the Mammalia collected by Signor Leonardo Fea in Burma and Tenasserim. In: Doria, G., Gestro, R., (Ed.), Annali del Museo civico do storia naturale di Genova, Series 2a vol. X, pp. 913–949.

Trageser, S.J., Ghose, A., Faisal, M., Mro, P., Mro, P., Rahman, S.C., 2017. Pangolin distribution and conservation status in Bangladesh. PLoS One 12 (4), e0175450.

Ullmann, T., Veríssimo, D., Challender, D.W.S., 2019. Evaluating the application of scale frequency to estimate the size of pangolin scale seizures. Glob. Ecol. Conserv. 20, e00776.

Wang, X.Q., 1975. The taxonomy of aves and mammals. Northeast Forestry University, Heilongjiang. [In Chinese].

Wang, P.L., 1989. The habits and resource protection of pangolins. Environ. Prot. Technol. 4, 27–28. [In Chinese].

Wang, Q.S., 1990. The Mammal Fauna of Anhui. Anhui Publishing House of Science and Technology, Hefei. [In Chinese].

Wang, S., 2005. Preliminary observation on wild life habits of 558 Chinese pangolins. Introd. Consult. 4, 52–53.

Wang, P.J., 2007. Application of Wildlife Rescue System in Conservation of the Formosan Pangolin (*Manis pentadactyla pentadactyla*). M.Sc. Thesis, National Taiwan University, Taipei, Taiwan. [In Chinese].

Wang, L.M., Lin, Y.J., Chan, F.T., 2012. The first record of successfully fostering a young Formosan pangolin (*Manis pentadactyla pentadactyla*). Taipei Zoo Bull 23, 71–76. [In Chinese].

Weber, R., Heath, M.E., White, F.N., 1986. Oxygen binding functions of blood and hemoglobin from the Chinese pangolin, *Manis pentadactyla*: possible implications for burrowing and low body temperature. Respir. Physiol. 64, 103–112.

Wu, S., 1998. Notes on a newborn Chinese pangolin (*Manis pentadactyla aurita*). J. Qinghai Normal Univ. (Nat. Sci.) 1, 40–42. [In Chinese].

Wu, S., Ma, G., Tang, M., Chen, H., Liu, N., 2002. The current situation of the resource on Chinese pangolins and countermeasures for protection. J. Nat. Resour. 17 (2), 174–179. [In Chinese].

Wu, S.B., Liu, N.F., Ma, G.Z., Xu, Z.R., Chen, H., 2003a. Habitat selection by Chinese pangolin (*Manis pentadactyla*) in winter in Dawuling Natural Reserve. Mammalia 67 (4), 493–501.

Wu, S.B., Liu, N.F., Ma, G.Z., Xu, Z.R., Chen, H., 2003b. Studies on habitat selection by Chinese pangolin (*Manis pentadactyla*) in winter in Dawuling Natural Reserve. Acta Ecol. Sin. 23 (6), 1079–1086. [In Chinese].

Wu, S., Liu, N., Zhang, Y., Ma, G., 2004a. Physical measurement and comparison for two species of pangolins. Acta Theriol. Sin. 24 (4), 361–364. [In Chinese].

Wu, S., Liu, N., Zhang, Y., Ou, Z., Chen, H., 2004b. Measurement and comparison for skull variables in Chinese pangolin and Malayan pangolin. Acta Theriol. Sin. 24 (3), 211–214. [In Chinese].

Wu, S., Ma, G., Chen, H., Xu, Z., Li, Y., Liu, N., 2004c. A preliminary study on burrow ecology of *Manis pentadactyla*. Chin. J. Appl. Environ. Biol. 15 (3), 401–407. [In Chinese].

Wu, S.B., Liu, N.F., Ma, G.Z., Tang, M., Chen, H., Xu, Z.R., 2004d. A current situation of ecology study on pangolins. Chin. J. Zool. 39 (2), 46–52. [In Chinese].

Wu, S., Liu, N., Zhang, Y., Ma, G., 2004e. Assessment of threatened status of Chinese pangolin (*Manis pentadactyla*). Chinese J. Appl. Environ. Biol. 10 (4), 456–461. [In Chinese].

Wu, S.B., Ma, G.Z., Liao, Q.X., Lu, K.H., 2005a. Studies of Conservation Biology on Chinese Pangolin. Chinese Forest Press, Beijing. [In Chinese].

Wu, S., Liu, N., Li, Y., Sun, R., 2005b. Observation on food habits and foraging behavior of Chinese Pangolin (*Manis pentadactyla*). Chin. J. Appl. Environ. Biol. 11 (3), 337–341. [In Chinese].

Wu, S.-H., Chen, M., Chin, S.-C., Lee, D.-J., Wen, P.-Y., Chen, L.-W., et al., 2007. Cytogenetic analysis of the Formosan pangolin, *Manis pentadactyla pentadactyla* (Mammalia: Pholidota). Zool. Stud. 46 (4), 389–396.

Xu, L., Liu, Z., Liao, W., 1983. Birds and Animals of Hainan Island. Beijing Science Press, Beijing. [In Chinese].

Yang, C.W., Guo, J.C., Li, Z.W., Yuan, X.W., Cai, Y.L., Fan, Z.Y., 2001. Studies on Taiwan Chinese pangolin. Taipei Zoo, Taipei, Taiwan. [In Chinese].

Yang, C.W., Chen, S., Chang, C.-Y., Lin, M.F., Block, E., Lorentsen, R., et al., 2007. History and dietary husbandry of pangolins in captivity. Zoo Biol. 26 (3), 223–230.

Yang, L., Chen, M., Challender, D.W.S., Waterman, C., Zhang, C., Hou, Z., et al., 2018. Historical data for conservation: reconstructing long-term range changes of Chinese pangolin (*Manis pentadactyla*) in eastern China (1970–2016). Proceedings of the Royal Society B 285 (1885), 20181084.

Zhai, W., 2000. Rare and Endangered Animals in Henan Province. Henan Science and Technology Press, Zhengzhou. [In Chinese].

Zhang, Y., 2009. Conservation and trade control of pangolins in China. In: Pantel, S., Chin, S.-Y. (Eds.), Proceedings of the Workshop on Trade and Conservation of Pangolins Native to South and Southeast Asia, 30 June–2 July 2008, Singapore Zoo, Singapore. TRAFFIC Southeast Asia, Petaling Jaya, Selangor, Malaysia, pp. 66–74.

Zhang, L., Li, Q., Sun, G., Luo, S., 2010. Population status and conservation of pangolins in China. Bull. Biol. 45 (9), 1–4. [In Chinese].

Zhang, F., Wu, S., Zou, C., Wang, Q., Li, S., Sun, R., 2016. A note on captive breeding and reproductive parameters of the Chinese pangolin, *Manis pentadactyla* Linnaeus, 1758. ZooKeys 618, 129–144.

Zhang, S., Zheng, F., Li, J., Bao, Q., Lai, J., Cheng, H., 2017. Monitoring diversity of ground-dwelling birds and mammals in Wuyanling National Nature Reserve using infrared camera traps. Biodivers. Sci. 25 (4), 427–429. [In Chinese].

Zhou, Z.M., Zhao, H., Zhang, Z.X., Wang, Z.H., Wang, H., 2012. Allometry of scales in Chinese pangolins (*Manis pentadactyla*) and Malayan pangolins (*Manis javanica*) and application in judicial expertise. Zool. Res. 33 (3), 271–275.

Zhu-Ge, Y., Huang, M., 1989. Fauna of Zhejiang (Mammalia). Zhejiang Science and Technology Press, Hangzhou. [In Chinese].

Zoological Society of India, 2002. Pangolins (Mammalia: Pholidota) of India. ENVIS Newsl. 9 (1 and 2).

CHAPTER

5

Indian pangolin *Manis crassicaudata* (Geoffroy, 1803)

Tariq Mahmood[1], Rajesh Kumar Mohapatra[2], Priyan Perera[3], Nausheen Irshad[4], Faraz Akrim[1], Shaista Andleeb[1], Muhammad Waseem[5], Sandhya Sharma[6] and Sudarsan Panda[7]

[1]Department of Wildlife Management, PMAS-Arid Agriculture University, Rawalpindi, Pakistan [2]Nandankanan Zoological Park, Bhubaneswar, India [3]Department of Forestry and Environmental Science, University of Sri Jayewardenepura, Nugegoda, Sri Lanka [4]Department of Zoology, University of Poonch, Rawalakot, Pakistan [5]WWF-Pakistan, Islamabad, Pakistan [6]Conservation Biologist, Sindhupalchowk, Nepal [7]Satkosia Tiger Reserve, Angul, India

OUTLINE

Taxonomy

Included in the genus *Manis* based on both morphological (Gaudin et al., 2009) and genetic evidence (Gaubert et al., 2018). Previously included in the genus *Phatages* (Allen, 1938; Pocock, 1924; Sundevall, 1842) and *Pholidotus* (Gray, 1865).

Synonyms: *Manis laticauda* (Illiger, 1815), *Phatages laticauda* (Sundevall, 1842), *Pholidotus*

Pangolins
DOI: https://doi.org/10.1016/B978-0-12-815507-3.00005-8

indicus (Gray, 1865), and *Phatages laticaudatus* (Fitzinger, 1872). The species is monotypic. Chromosome number is not known.

Etymology: *Manis* (see Chapter 4); *crassicaudata* in reference to the thick or heavy (*crassus-*) tail (*cauda*) and provided with (*–atus*) (Gotch, 1979).

Description

The Indian pangolin (*Manis crassicaudata*) is a medium-sized mammal. Adults typically weigh 8–16 kg and reach a total body length of up to ~148 cm, though extremes have been recorded (Table 5.1). An adult male weighing 32.2 kg and measuring 170 cm in length was recorded in Rajasthan, India (Sharma, 2002). The species is sexually dimorphic, with males larger and heavier than females (Roberts, 1977), but the extent of sexual dimorphism requires further research. Based on individuals (n = 12) studied at the Potohar Plateau, Pakistan, Irshad et al. (2016) revealed a positive correlation between body weight and total body length and divided the species into three age classes: juveniles (≤2.5 kg, 40–65 cm total length), sub-adults (2.51–8 kg, 66–120 cm) and adults (≥8 kg, ≥ 120 cm). The tail is thickset, heavy and highly muscular and is as wide as the posterior of the body at the proximal end, tapering gradually to the distal end (Heath, 1995; Roberts, 1977). It is rounded on the dorsal surface and flat on the ventral surface, and is prehensile (Heath, 1995). The tail constitutes 39–54% of total body length (Aiyapann, 1942; Irshad et al., 2016). A terminal scale is present on the ventral surface of the tail (Pocock, 1924).

Overlapping, keratinous and bluntly pointed scales, which grow from the epidermis, cover the body including the dorsal and lateral surfaces, the outside of the limbs and both the dorsal and ventral surfaces of the tail (Jentink, 1882; Heath, 1995). Scales extend on to the forehead but terminate before, and leave a gap to the rhinarium. The scales are massive, the largest among the Asian pangolins, and measure up to 70 mm long and more than 85 mm wide (Mitra, 1998). Growing outward from a medio-dorsal scale row, the largest scales are on the posterior of the body and proximal portion of the tail (Irshad et al., 2016). These scales are up to twice as wide as the first row of post-scapular scales, and weigh 7–10 g (Mitra, 1998). On the tail the medio-dorsal scale row continues to the tail tip. There are 11–14 transversal (i.e., across the body at the midpoint of the back along the anterioposterior axis) and 11–13 longitudinal scale rows on the body, and 14–17 rows of scales on the tail margins (Table 5.1). Scales on the tail margins are folded, covering part of the dorsal and ventral surfaces and are sharply pointed. There is intra-specific variation in total scale number, ranging from ~440 to 530 (Table 5.1). The scales are finely striated at the base and range in color from pale yellow-olive to khaki or dark brown (Fig. 5.1; Mitra, 1998; Mohr, 1961). Several yellowish bristle-like hairs (40–50 mm) protrude from the base of the scales (Prater, 1971; R.K. Mohapatra, unpubl. data). Mohapatra et al. (2015a) estimated that the skin and scales of a 10.3 kg individual weighed 3.5 kg, i.e., 34% of total mass.

The head is small and conical and the muzzle pointed. The face and snout, throat, neck, entire ventral surface, and the inner sides of the limbs lack scales and are covered in sparse, light pink hair (Fig. 5.1; Pocock, 1924). The skin on the face is whitish-pink to light brown, and the rhinarium darker brown. The eyes are small and dark and protected by thick, swollen eyelids (Heath, 1995). There are small crescent-shaped ear pinnae anterior to which are the auditory orifices (Roberts, 1977) but the species has a meager auditory range and relies on highly acute olfactory senses in order to locate prey (Prater, 1971). Lacking teeth, Indian pangolins use their long salivary-coated tongue to "lick up" ants and termites. The tongue measures up to 42.5 cm in length and

TABLE 5.1 Indian pangolin morphometrics.

Measurement			Country	Source(s)
Weight	Weight (♂)	14.25 (11–19.3) kg, n = 4	Pakistan	Irshad et al., 2015; Roberts, 1977
		10.92 (9.88–12.05) kg, n = 3	India (Odisha)	Mohapatra and Panda, 2013
		32.2 kg, n = 1	India (Rajasthan)	Sharma, 2002
	Weight (♀)	14.55 (9.1–20) kg, n = 2	Pakistan	Irshad et al., 2015; Roberts, 1977
		9.8 (9–10.59) kg, n = 3	India (Odisha)	Mohapatra and Panda, 2013
Body	Total length (♂)	1396 (1340–1473) mm, n = 3	Pakistan	Irshad et al., 2015
		957 (920–1020) mm, n = 3	India (Odisha)	Mohapatra and Panda, 2013
		1700 mm, n = 1	India (Rajasthan)	Sharma, 2002
	Total length (♀)	1370 mm, n = 1	Pakistan	Irshad et al., 2015
		883 (780–1000) mm, n = 3	India (Odisha)	Mohapatra and Panda, 2013
	Head-body length (♂)	805 (762–843) mm, n = 3	Pakistan	Irshad et al., 2015
	Head-body length (♀)	833 mm, n = 1	Pakistan	Irshad et al., 2015
	Tail length (♂)	650 (530–710) mm, n = 3	Pakistan	Irshad et al., 2015
		487 (470–510) mm, n = 3	India (Odisha)	Mohapatra and Panda, 2013
	Tail length (♀)	538 mm, n = 1	Pakistan	Irshad et al., 2015
		453 (390–510) mm, n = 3	India (Odisha)	Mohapatra and Panda, 2013
Vertebrae	Total number of vertebrae	57		Jentink, 1882
	Cervical	7		Jentink, 1882
	Thoracic	15		Jentink, 1882
	Lumbar	6		Jentink, 1882
	Sacral	3		Jentink, 1882
	Caudal	26		Jentink, 1882
Skull	Length (♂)	77.5 mm, n = 1		Heath, 1995
	Breadth across zygomatic processes	No data		

(Continued)

TABLE 5.1 (Continued)

	Measurement		Country	Source(s)
Scales	Total number of scales	485 (444–527), n = 17	India, Sri Lanka, unknown	Mohapatra et al., 2015a; Ullmann et al., 2019
	No. of scale rows (transversal, body)	11–14		Frechkop, 1931; Mohr, 1961
	No. of scale rows (longitudinal, body)	11–13		Frechkop, 1931; Jentink, 1882
	No. of scales on outer margins of tail	14–17		Frechkop, 1931; Jentink, 1882
	No. of scales on median row of tail	14–17		Frechkop, 1931
	Scales (wet) as proportion of body weight	No data		
	Scales (dry) as proportion of body weight	No data		

FIGURE 5.1 Indian pangolin. The scales get wider and longer along the anterioposterior axis and are largest on the rump and proximal portion of the tail. Indian pangolins have the largest scales of the Asian pangolins. *Photo credit: Vickey Chauhan/Shutterstock.com.*

constitutes 37% of total body length (Irshad et al., 2016); large salivary glands in the thorax secrete a sticky saliva that coats the tongue enabling prey to be consumed. The tongue originates from its base, the caudal end of the xiphoid process in the abdomen (which is shorter than in African pangolins) and passes through the thorax to the small oral cavity (Doran and Allbrook, 1973; Irshad et al., 2015). Prey are masticated in the stomach, aided by small stones that are ingested when consuming prey (Prater, 1971).

The forelimbs are slightly longer than the hindlimbs, each of which has five digits (Heath, 1995; Irshad et al., 2016). The forelimbs terminate with powerful elongated and slightly curved claws, which are used for digging and to break into ant nests and termite mounds (see *Behavior*; Roberts, 1977). The three middle claws are noticeably larger, and the central claw is the largest and strongest; the first and fifth claws are vestigial (Pocock, 1924; Roberts, 1977). The hindlimbs are short and stout, with short blunt claws, and the hindfeet have a coarsely granular pad extending to the base of the claws (Pocock, 1924).

Anal glands are present which produce a foul smelling yellow liquid (Hutton, 1949). In females, the uterus is bicornuate and the placenta diffuse and non-deciduae (Grassè, 1955). Two nipples (5–8 mm) are present in the ventral thoracic region (Aiyapann, 1942).

Body temperature is regulated at ~33.4 °C and the species has a low metabolic rate; in a 16 kg individual basal metabolic rate was estimated at 78 ml O_2/kg/h (McNab, 1984). Low metabolic rate has been observed in other species of pangolin and is low among mammals, due in part to the proportion of mass that comprises the metabolically inactive scales (McNab, 1984; Heath and Hammel, 1986).

The Indian pangolin may readily be confused with its Asian congeners, the Sunda (*M. javanica*) and Philippine (*M. culionensis*), and Chinese pangolin (*M. pentadactyla*) with which it is sympatric in parts of its range (see *Distribution*). The species is most easily distinguishable from other Asian pangolins by its larger size, larger scales, smaller ear pinnae (especially compared to the Chinese pangolin), the number of transversal scales rows (11–14), which number 15–18 in the Chinese pangolin and are more numerous in the Sunda and Philippine species (see Chapters 6 and 7), and the presence of a terminal scale on the ventral tail tip; other Asian pangolins have fleshy tail pads (Pocock, 1924). The Chinese pangolin also has a distinct anal depression (see Chapter 4).

Distribution

The species is distributed in South Asia, from northern and southeastern Pakistan, throughout the Indian sub-continent south of the Himalayas, to northeastern India, and Sri Lanka (Fig. 5.2). Locally distributed in Pakistan, the species has been recorded from all four provinces. In Khyber Pakhtunkhwa there are records from Nowshera, Maneshra, Mardan, Peshawar and Swabi districts, and in Punjab from Gurjat, Kasur, Kohat, Lahore and Sialkot districts, as well as the Potohar Plateau region from Attock, Chakwal, Jhelum and Rawalpindi (Roberts, 1977; Irshad et al., 2015; T. Mahmood, unpubl. data). The species has also been recorded in the Margalla Hills National Park north of Islamabad (Mahmood et al., 2015a) and there are records, including from 2008 to 2013, from the Indus and Chenab River (which flows into the Indus) floodplains (Bhakar, Mianwali and Jhang districts) in southern Punjab (Roberts, 1977; Mahmood et al., 2018; F. Abbas, pers. comm.). In Azad Jammu and Kashmir, there are records from the districts of Kotli (including Pir Lasura National Park), Mirpur and Bhimber (Akrim et al., 2017; R. Hussain, pers. comm.). In the south, the species occurs in Las Bela and Mekran districts in Balochistan, and in Sind has been recorded in Kirthar National Park north of Karachi, and in Hyderabad, Tharparker, Dadu and Larkana districts, and Kutch to the east (Roberts, 1977).

The species is widely distributed in India from the Himalayan foothills to the south of the country, excluding the far north and northeast (Tikader, 1983) but there is a need for a more precise understanding of present distribution. In the north the species occurs in Uttarakhand, Uttar Pradesh, Bihar and Jharkhand, as well as Delhi, Rajasthan (including Keoladeo National Park and Mukundara Hills Tiger Reserve [Latafat and Sadhu, 2016]), Madhya Pradesh (Gwalior [Saxena, 1985] and Achanakur Wildlife Sanctuaries and Chambal National Park), Gujarat (including Kutch [Himmatsinhji, 1984] and Gir National Park), Chhattisgarh, and West Bengal where there are records from several districts including Cooch Behar, Jalpaiguri and South-24 Parganas districts (see Agrawal et al., 1992; Choudhury, 2001; Srinivasulu and Srinivasulu, 2012). The northeastern limits of distribution are not well understood. Choudhury (2001) reports an individual was caught in Nagaon district, Assam in the 1990s, the eastern most record of the species, and it possibly occurs in Meghalaya (Agrawal et al., 1992). Goswami and Ganesh (2014) report observing an individual in Manas National Park, Assam in 2008. These records imply the species is sympatric with the Chinese pangolin in northeastern India. Further south there are records from Odisha and Mishra and Panda (2012)

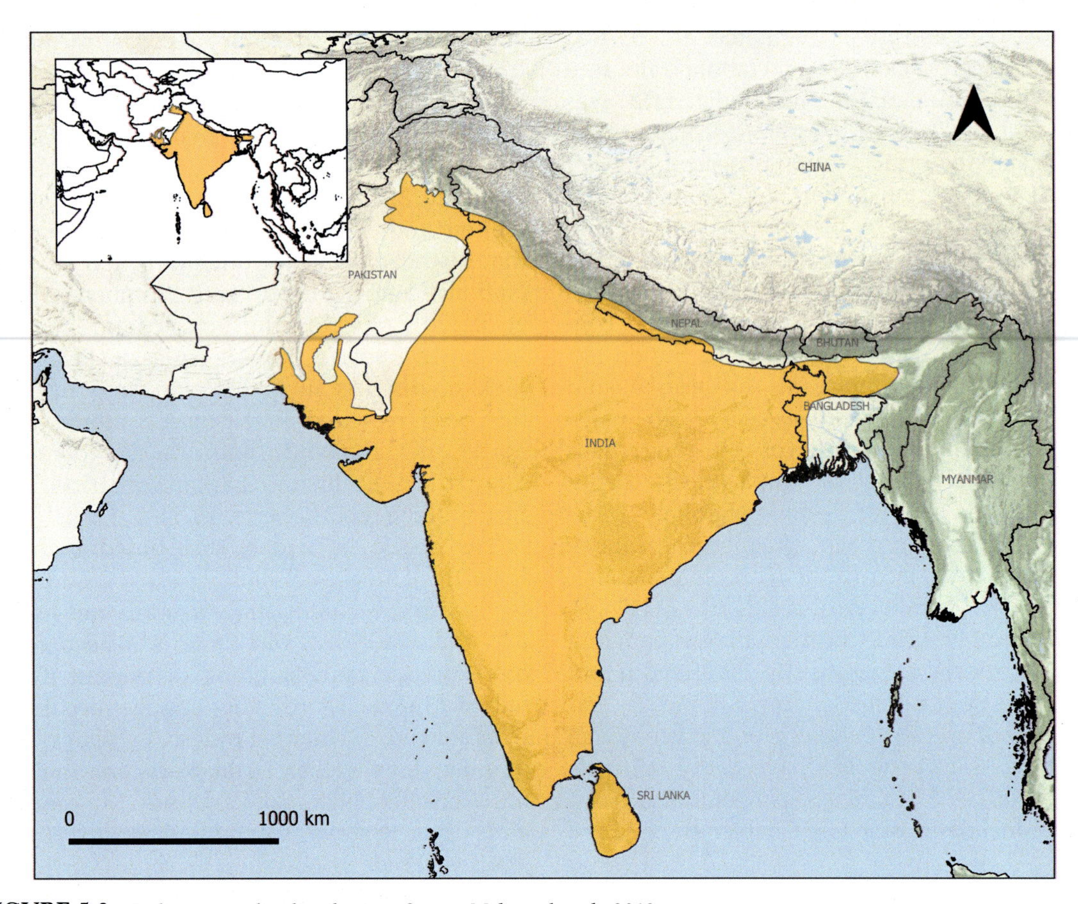

FIGURE 5.2 Indian pangolin distribution. *Source: Mahmood et al., 2019.*

note the presence of the species in 14 out of 30 districts in the state based on animals rescued by Nandankanan Biological Park; there are also records from Kotgarh, Nandankanan, Chandaka, Kuldiha, Satkosia Gorge and Sunabedha Wildlife Sanctuaries and Similipal National Park (Mohapatra, 2016). The Indian pangolin occurs in Maharashtra – two individuals were camera trapped in Pratchidgad Community Reserve in 2017 (Freedman, 2017) – Karnataka (Bandipur, Bhadra, Dalma and Dandeli Wildlife Sanctuaries and Bandipur Tiger Reserve), Goa (Catugao Wildlife Sanctuary) and Andhra Pradesh (Kambalakonda Wildlife Sanctuary; Murthy and Mishra, 2010). There are unverified, older reports from Sriharikota Island (see Manakadan et al., 2013), and historical records from Kanyakumari, and Kerala; the species also occurs in Tamil Nadu (Srinivasulu and Srinivasulu, 2012).

In Nepal, the Indian pangolin is distributed in lowland areas in the south and west of the country (Baral and Shah, 2008; Fig. 5.2), including in Bardia, Banke, Chitwan, Parsa and Shuklaphanta National Parks (DNPWC and DOF, 2018; Suwal and Verheugt, 1995; H. Baral, pers. comm.). In Parsa National Park the species appears to be sympatric with the Chinese pangolin (as in northeastern India) and further research is needed to explicate presumed ecological differences between the species. There are also reports from Makwanpur district, and the species' distribution may extend into the eastern foothills and more widely in the Terai regions, possibly contiguously into northern India (DNPWC and DoF, 2018).

The occurrence of the Indian pangolin in Bangladesh is unclear. Khan (1985) reported that the species occurred throughout the country historically, but has been extirpated from northwest (Rangpur, Dinajpur, Rajshahi) and west-central (Kushtia, Jessore, Pabna, Bogra) regions, and Dhaka and Cumilla; Heath (1995) reported that the species had been extirpated altogether. Trageser et al. (2017) reviewed the status of pangolins in Bangladesh but did not find empirical evidence for the occurrence of the species. However, based on the purported historical distribution, the species' ecology, and records from 2018 in neighboring West Bengal, India, it seems likely that the species occurs in the north- and southwestern part of the country, excluding the coastal areas of Khulna, Satkhira, Bagerhat, Barisal and Patuakhali (A. Ghose and S. Trageser, pers. comm.). Rajshahi University maintains a museum specimen of likely, but unconfirmed, Bangladeshi origin (S. Trageser, pers. comm.) and further research is needed on the occurrence of the species in Bangladesh.

The Indian pangolin is distributed throughout Sri Lanka, occurring locally in lowland areas in all nine provinces, including Nuwara Eliya District in the center of the country where it may occur at higher elevations (Perera and Karawita, 2019; Phillips, 1981). There are records from all dry zone national parks, and in coastal areas in the northeast (Mullaitivu, Trincomalee and Kuchchaweli), northwest (Norochcholai, Ilanthadiya and Kalpitiya) and south (Kalamatiya, Waligama, Dikwella, Bundala and Unawatuna; Perera and Karawita, 2019).

The presence of the species in China warrants discussion. A number of older and contemporary reference works mention occurrence of the species in far southwestern Yunnan Province (Allen, 1938; Heath, 1995; Jiang et al., 2015; Smith and Xie, 2013). Examination of the literature suggests that this was a case of mistaken identity by Allen (1938) who included in the proposed species *"Phatages crassicaudata"* Howell's (1929) description of *M. aurita* (=Chinese pangolin; see Chapter 4) and *M. javanica* as it occurs in the proposed synonymy of *M. aurita* by Anderson (1878). Based on this evidence and the fact that there have been no further records in the country, suggests it is unlikely that the species occurs in China. There are no records from Myanmar (see CITES, 2000) and the most eastern record otherwise is in Assam, northeastern India (Choudhury, 2001).

Habitat

The Indian pangolin inhabits tropical and sub-tropical forest, dry-mixed evergreen, monsoon, sub-mountain and riverine forest (Phillips, 1981; Roberts, 1977), and has been recorded in mangrove forest, grasslands, agricultural land, artificial landscapes (e.g., plantations), home-gardens, scrubland and desiccated areas (Karawita et al., 2018; Pabasara et al., 2015; Roberts, 1977). The species is thought to adapt well to modified habitats provided there is abundant prey and not undue exploitation pressure. Altitudinal limits range from near sea level (e.g., in coastal Sri Lanka) up to 1538 m asl in Nepal (Frick, 1968; Mitchell, 1975; District Forest Office, Surkhet, Nepal, unpubl. data) while Hutton (1949) reported that an individual was found at ~2100 m in the Nilgiris, India. There are records from Nuwara Eliya District in central Sri Lanka where the species may occur at up to 1850 m, but it is thought to be more abundant in drier zones (Karawita et al., 2018; Pabasara et al., 2015).

In Pakistan, the species occurs in, and may have a preference for sub-tropical thorn forests and barren hilly areas (Fig. 5.3; Roberts, 1977). Mahmood et al. (2014) report that in the Potohar Plateau, there is a close association between the occurrence of the species and its

FIGURE 5.3 Indian pangolin in Chakwal District in the Potohar Plateau, Pakistan. *Photo credit: Faraz Akrim/Tariq Mahmood.*

burrows, and dominant tree (Arabic gum [*Acacia nilotica*], Indian plum [*Zizyphus mauritiana*] and phulai [*A. modesta*]) and shrub (*Z. nummularia*, *Calotropis procera*, and mesquite [*Prosopis juliflora*]) species. In Margalla Hills National Park, the species was recorded in areas dominated by phulai, *Z. nummularia*, North Indian rosewood (*Dalbergia sissoo*) and chir pine (*Pinus roxburghii*) while West Indian lantana (*Lantana camara*) and pomegranate (*Punica granitum*) appear important to the species' ecology (see *Ecology*; Mahmood et al., 2015a). In Mukundara Tiger Reserve, Rajasthan, India, the species was recorded on flat terrain in forest dominated by *Anogeissus pendula* and *Acacia catechuoides* (Latafat and Sadhu, 2016). At the tropical lowland Yagirala Forest Reserve in Sri Lanka, Karawita et al. (2018) recorded the species' resident burrows in naturally regenerated secondary forest dominated by *Dipterocarpus zeylanicus*, Sri Lankan ironwood (*Mesua ferrea*), nandu wood (*Pericopsis mooniana*), jack tree (*Artocarpus heterophyllus*) and Ceylon breadfruit (*A. nobilis*) with canopy cover of 75–85%, and a height of 25–40 m. The sub-canopy was dominated by *Chaetocarpus castanocarpus*, *Garcinia hermonii*, *Xylopia championi*, *Horsfieldia iriyaghedhi* and *Myristica dactyloides*. Resident burrows were also observed in managed rubber (*Hevea brasiliensis*) plantations. Feeding burrows were found in *Pinus*-dominated (*P. carribea*) forest with dense undergrowth dominated by *Ochlandra stridula*, and in human-modified habitats, predominantly tea (*Camellia sinensis*) and fruit tree cultivation (Karawita et al., 2018).

Ecology

There is limited knowledge of Indian pangolin ecology. There are no estimates of size but it is assumed that the species has defined home ranges. Principally fossorial, i.e., burrow dwelling, the species excavates burrows both for shelter and in order to find prey (see also *Behavior*: Prater, 1971; Roberts, 1977); shelter may also take the form of chambers or crevices

FIGURE 5.4 Indian pangolins excavate burrows for both shelter and to find prey. (A) A resting or resident burrow. When digging this type of burrow Indian pangolins leave a soil "apron" outside the burrow, which is noticeable in this image. (B) A feeding burrow. *Photo credit: Faraz Akrim/Tariq Mahmood.*

among rocks and boulders (Prater, 1971). In some habitats (e.g., lowland tropical forest in Sri Lanka and presumably tropical India) the species is arboreal (Heath, 1995). Resting or resident burrows are much larger and deeper than feeding burrows and specific characteristics (e.g., length and width) vary with habitat and soil type (Mahmood et al., 2013). Entrances are circular or elliptic in shape and may be up to 46 × 59 cm in width and height respectively (see Fig. 5.4; Karawita et al., 2018). They then consist of a tunnel that slopes downward initially but then gradually inclines at an angle of 20–30°, and may curve slightly, before terminating in a resting chamber (Fig. 5.5). This is presumably to avoid flooding. Once inside a resting burrow, Indian pangolins create a wall of earth thus concealing the resting chamber, perhaps to aid in predator avoidance, but leave a small gap for ventilation (see Fig. 5.5; Karawita et al., 2018). In loose soil, resting chambers may be up to 6 m or more beneath the surface, but shallower in more rocky ground; they vary in size and chambers of 60 cm in diameter have been recorded (Prater, 1971). In Margalla Hills National Park, Pakistan, most burrows were dug under West Indian lantana and pomegranate shrubs (Mahmood et al., 2015a). In contrast, in the Potohar Plateau, the majority of resting burrows were excavated under or within the roots of karira (*Capparis decidua*) and *Salvador aoleoides*, which is potentially related to an association between these shrubs and prey species (Mahmood et al., 2013). In tropical lowland rainforest in Sri Lanka, resting burrows were located at an elevation of 75–100 m, on moderately steep (45–60°) slopes, with canopy cover of >75%, and away from human habitation (Karawita et al., 2018). Karawita et al. (2018) note that in this habitat the species appears to have a tendency to dig resting burrows in rocky substrates, potentially because there is less risk of burrow collapse. There is little knowledge of burrow use (e.g., how frequently they are excavated, duration of use, co-inhabitation) but there are observations of mothers with young (Mahmood et al., 2015b) and a male and female sharing burrows (Roberts, 1977) respectively. Research on burrow use is needed (see Chapter 34).

The Indian pangolin predates on ants and termites, consuming the eggs, young and adults, and also ingests grit, sand and small stones that aid mastication (Prater, 1971). The species uses olfaction to locate prey and

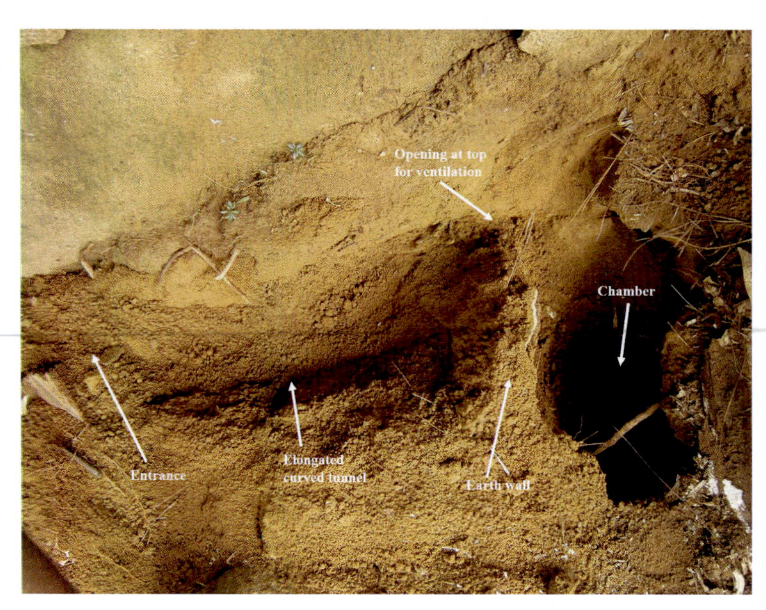

FIGURE 5.5 Cross-section of an Indian pangolin resident burrow in Sri Lanka. *Photo credit: Priyan Perera.*

will dig feeding burrows in order to access subterranean ant nests and termitaria. Feeding burrows are shallower than resting burrows and more numerous. In Pakistan, there were an estimated 18.85 ± 2.63 feeding burrows/km^2 compared to 2.57 ± 0.29 resting burrows/km^2, and the former had a mean depth of 19.46 ± 2.86 cm ($n = 55$; Mahmood et al., 2013). Mean feeding burrow depth in southwestern Sri Lanka was 68.12 cm ($n = 54$; Karawita et al., 2018). Feeding burrows may also have multiple access points leading to a single ant nest or termite colony. An Indian pangolin will repeatedly visit a feeding burrow but will do so periodically to allow its prey to recover predation events (Karawita et al., 2018).

The species is prey selective (Irshad et al., 2015; Prater, 1971) and has been recorded consuming non-ant and termite prey suggesting it is not exclusively myrmecophagous. Based on stomach content analysis, a female Indian pangolin collected in Kerala, India had predated exclusively on the ant genera *Leptogenys* sp., including eggs and adults, and 57% of the stomach content mass was grit (Ashokkumar et al., 2017). In Maharashtra, India, the species has been recorded predating on *Camponotus angusticollis*, *C. compressus*, *C. parius*, *Carebara affinis*, *Polyrhachis menelas*, *Pheidole malinsii*, and the weaver ant (*Oecophylla smaragdina*; B. Katdare, unpubl. data). In Pakistan, two species of black ants, *C. confucii* and *C. compressus* comprised the majority of prey consumed in the Potohar Plateau, with one termite species, *Odontotermis obesus* comprising a much smaller proportion based on fecal analysis (Irshad et al., 2015). An average 58% of fecal volume was clay, and grass and wood fragments had also been ingested (Irshad et al., 2015). As well as finding the remains of black ants in the stomach of an individual from the Nilgiris, India, Hutton (1949) found beetle wing sheaths, cockroach remains and worms, vegetable matter, and 20 stones up to 0.6 cm in diameter. The species has also demonstrated prey selectivity in captivity (Phillips, 1928). Weaver ants (*O. smaragdina*) provided with eggs were consumed at Nandankanan Zoological Park, India (Mohapatra and Panda, 2014a). Indian

FIGURE 5.6 Mugger crocodiles (*Crocodylus palustris*) are one of a number of predators of Indian pangolins. *Photo credit: Charlotte Arthun.*

pangolins are assumed to persist without water in desiccated areas (Prater, 1971).

Major predators include tiger (*Panthera tigris*) and leopard (*P. pardus*) in India (Benatar, 2018; Mohapatra, 2018; Ramakrishnan et al., 1999), and in Gir National Park, Gujarat, Asiatic lion (*P. leo persica;* Anon, 2011), but there is little knowledge of predation rates. In Yala National Park, Sri Lanka, predation by mugger crocodile (*Crocodylus palustris*) has been observed (see Fig. 5.6; Mohapatra, 2018). Sloth bears (*Melursus ursinus*) may also predate on the species (Mohapatra, 2018). Approximately 78% of the diet of sloth bears comprises ants and termites (Bargali et al., 2004) and it may be that the sharing of food resources and competition for prey brings the two species together. Other potential predators include caracal (*Caracal caracal*), spotted hyena (*Crocuta crocuta*), Indian python (*Python molurus*) and the domestic dog (*Canis familiaris*) in Pakistan (Roberts, 1977; M. A. Beg, pers. comm.).

Behavior

Typically solitary, except when mating and rearing young, but there is little knowledge of social structure. It is assumed that social relations are mediated by scent. Males are known to maintain territories by spraying urine and scat at territory boundaries, and may use their anal glands for this purpose. In captivity, both sexes appear to mark their territories, but the behavior is more frequent in males (R.K. Mohapatra, unpubl. data). Roberts (1977) reports a male and female sharing a burrow in Las Bela, southeastern Pakistan in November.

Nocturnal, the species shelters and sleeps curled up in a burrow during the day (Israel et al., 1987; Prater, 1971). Activity patterns of wild individuals are not known but in captivity at Nandankanan Zoo, India, activity peaked between 2000 and 2100 (Mohapatra and Panda, 2013). Captive animals ($n = 6$) were active for 129.02 ± 46.45 min/night and spent ~59% of their time walking in the enclosure and 14% feeding (Mohapatra and Panda, 2013). As with other species of pangolin there appears to be variation in activity; Roberts (1977) notes an observation of an individual active in broad daylight (1630) in September in Punjab, Pakistan.

The species is quadrupedal, bearing weight on the hindlimbs – the hindfeet are plantigrade – and the forelimbs, but the front claws are folded inward and the weight is taken on the folded up wrists. Typically moving slowly, the back is arched and the tail is held off the ground and aids balance when walking (see Fig. 5.3; Israel et al., 1987; Prater, 1971). Indian pangolins may also stand on their hindlimbs with the head elevated in order to survey their surroundings and to detect odors in the air; this likely assists in locating prey, detecting predators, locating conspecifics (e.g., females finding young) and social interactions, including finding a mate (Israel et al., 1987; Acharjyo, 2000). The species can walk a few paces when bipedal (Roberts, 1977). Able climbers, Indian pangolins climb trees in search of arboreal prey (e.g., *O. smaragdina*) in some habitats, using their forelimbs to grasp tree trunks and branches and supported by the hindlimbs and prehensile tail, which serves as an additional limb (Prater, 1971).

The powerful claws are used to excavate both resting and feeding burrows. When digging a deep burrow the forefeet are used to loosen and scoop out soil, and the pangolin will periodically back out of the burrow to remove excavated earth (Underwood, 1945). This material is pushed under the body and behind the animal in a rapid movement of the front part of the body before the pangolin returns down the burrow (Roberts, 1977; Underwood, 1945).

Acute olfactory senses are used to locate prey, and sniffing frequency increases as the source of prey, and the best point of attack of a subterranean or epigeal ant nest or termite mound is identified (Underwood, 1945). The strong claws are used to break into these structures and Underwood (1945) reports that when a large ant nest is found pangolins will dig guided by scent and will excavate the nest by lying on their back or side if needed, and may end up digging upwards with their hindlimbs on the nest ceiling enabling better purchase. Once prey are located digging ceases while prey are consumed. When feeding, the tongue is rapidly extended and retracted into the galleries of ant nests or termitaria, and coated in a sticky saliva, is used to lick up prey. In the process, soil, fragments of wood and small stones are also ingested. Water is drunk freely where available and is lapped up (Prater, 1971).

When threatened the Indian pangolin will quickly roll into a ball, placing the head down towards the chest and folding in all the limbs, curling the tail over the head, thus exposing only the protective scales. The species may secrete a noxious substance from the anal glands when threatened or disturbed (Hutton, 1949; Roberts, 1977), and may hiss loudly (Aiyapann, 1942; Acharjyo, 2000; Mohapatra and Panda, 2014b). In Ripley (1965), W. W. A. Phillips tells the tale of a villager in Sri Lanka who found a pangolin and clubbed it, knocking it unconscious, draped it around his neck and proceeded home intent on eating the animal. Only stunned, the pangolin regained consciousness on the journey and immediately tried to curl into a ball around the man's neck; the villager was found dead some time later with the pangolin still curled around his neck.

Ontogeny and reproduction

There is little knowledge of breeding biology. It is assumed that males locate receptive females using olfaction, and during breeding males and females will occupy the same burrow (Roberts, 1977). There appears to be no defined breeding season, at least not across the species' range. Mahmood et al. (2015b) report that in the Potohar Plateau region, Pakistan the species breeds once a year, with mating occurring between July and October, and young have been observed in January, April and December. Prater (1971) reports that on the Deccan Plateau, India, young are born between January and March, and there is a record of parturition in South India in July; a pregnant female was also found in July in Sri Lanka. In Rajasthan, India, a captive Indian pangolin gave birth in November (Prakash, 1960), and breeding behavior has been observed in captivity in all months except May and June (Mohapatra and Panda, 2014a).

During mating, the male may chase the female and preliminary interaction involves a long period of noso-nasal or noso-genital inspection (Mohapatra and Panda, 2014b). The male mounts the female from the rear or side, using the claws of the forelimbs to hold the female's body; the female raises her tail, allowing the male to align their genitals and copulation to take place; the two pangolin's tails are typically entwined (Mohapatra et al., 2015b). Gestation period is estimated to be 165–251 days (Mohapatra et al., 2018). Older references to a much shorter (65–70 day) gestation period are erroneous (see Roberts, 1977).

Parturition takes place in a resting burrow and one young is typically born, but twins have been reported (Prater, 1971). At Nandankanan Zoo, India, all captive births (n = 20) have been to a single young (Mohapatra et al., 2018; Mohapatra and Panda, 2014a). There are unverified reports by local people of females with two young in Pakistan; eight out of 11 observations in independent locations referred to twins (Mahmood et al., 2015b). Young weigh 200–400 g at birth and measure 300–450 mm (see Ogilvie and Bridgwater, 1967; Prater, 1971). They are precocial with soft scales, which harden with time, reaching the robustness of adult scales after 18 months (R.K. Mohapatra, unpubl. data). Little is known about post-partum care, but the mother nurses the young and will coil around it to protect it (Acharjyo, 2000; Mohapatra and Panda, 2014a). Juveniles will accompany their mother on foraging trips as they grow and are taxied around on the base of her tail, sometimes crosswise, or on her back (Israel et al., 1987; Roberts, 1977). Young become independent at 5–8 months of age (Mohapatra and Panda, 2014a; Mohapatra, 2016) and reach adult size and sexual maturity at about 3 years of age (R.K. Mohapatra, unpubl. data). Longevity in the wild is unknown; an Indian pangolin at Oklahoma Zoo lived to be more than 19 years old (Weigl, 2005).

Population

Indian pangolins occur in a diversity of habitats and major influences on abundance are the availability of prey and predation, including by humans. Historical accounts suggest the species is uncommon, rarely observed and occurs at low densities (Phillips, 1981; Roberts, 1977). There is limited data on abundance, but there are reported declines in parts of the species' range. In the Potohar Plateau, Pakistan, Irshad et al. (2015) estimated population density in four districts based on counts of active burrows between 2010 and 2013 (see Willcox et al., 2019 for discussion on the method used). Estimated density declined by ~79% across all four districts between 2010 and 2012 (Irshad et al., 2015), which has been attributed to illegal capture and killing for international trade in pangolin scales (see Mahmood et al., 2012). In 2012 estimated densities comprised 0.37 ± 0.12 individuals/km^2 in Chakwal, 0.12 ± 0.08 individuals/km^2 in Jhelum, 0.10 ± 0.00 individuals/km^2 in Rawalpindi, and 0.33 ± 0.24 individuals/km^2 in Attock (Irshad et al., 2015). Using the same method, Mahmood et al. (2015a) estimated an average density 0.36 individuals/km^2 in Margalla Hills National Park, and 0.00044 individuals/km^2 in Maneshra district of Khyber Pakhtunkhwa Province, Pakistan (Mahmood et al., 2018).

Pabasara et al. (2015) estimated population density at Yagirala Forest Reserve in southwestern Sri Lanka to be 5.69 individuals/km^2, based on modeling camera trap records without identification of individual animals (see Rowcliffe et al., 2008). This is potentially an underestimate based on known exploitation of populations in the area. These densities are far higher than anywhere else in the species' range and may be due to habitat differences, i.e., tropical lowland rainforest supporting a broader and more abundant prey base. Overexploitation of populations by excessive hunting and poaching may have almost eliminated pangolins from some parts of the country (P. Perera, unpubl. data).

There are no quantitative population data available for India, Nepal or Bangladesh, though there are contemporary records of the species in protected areas in the former two countries (see *Distribution*). Further research is needed to determine the status of populations in all range states, including Bangladesh.

Status

The Indian pangolin is listed as Endangered on The IUCN Red List of Threatened Species due to hunting for its meat and scales for local use, and apparently increasing poaching and international trafficking of its scales, mainly to East Asia, for use in traditional medicines (Mahmood et al., 2019). Using the IUCN Red List Categories and Criteria the species is listed as Endangered in India and Nepal (Jnawali et al., 2011), Vulnerable in Pakistan (Sheikh and Molur, 2005), and Near Threatened in Sri Lanka, and is protected by law in each range state, which provides protection from exploitation. However, threats persist (see *Threats*). The species is included in CITES Appendix I.

Threats

The main threat to the Indian pangolin is overexploitation for local use and consumption and international use, and the scales of the species are trafficked internationally, mainly to China (Mahmood et al., 2012; see Chapter 16). Quantifying exploitation across the species' range is challenging but hunting occurs in many parts of the range, including Odisha, Arunachal Pradesh, Assam and the Western and Eastern Ghats among other states in India (D'Cruze et al., 2018; Gubbi and Linkie, 2012; Mitra, 1998; Mohapatra et al., 2015a; Kanagavel et al., 2016). This includes hunting by tribal communities which use the meat as a protein source and the scales and claws as curios and for medicinal purposes, which is steeped in tradition (see Chapter 14; Mohapatra et al., 2015a; Sharma, 2014). The legality of hunting varies by region in India (see D'Cruze et al., 2018). In Pakistan, the scales continue to be used by "Hakims," practitioners of country medicine (Roberts, 1977; T. Mahmood, unpubl. data). In Sri Lanka, pangolin meat is a delicacy among local hunters and overexploitation is the main threat (Perera et al., 2017). There is also demand from foreign laborers working on major development projects in the country (P. Perera, unpubl. data) and Karawita et al. (2016) note the existence of a niche market for pangolin meat in urban restaurants that cater for foreign visitors. Indigenous communities use powdered scales as an ingredient in medicinal ointment to cure diseases in cattle (P. Perera, unpubl. data).

Little international trade involving the Indian pangolin has been reported to CITES historically (see Chapter 16; Challender and Waterman, 2017). However, the species has been trafficked since at least the early 2000s for consumptive use in China (Wu and Ma, 2007). Contemporary trafficking primarily involves scales, which are sourced in India, Pakistan, Sri Lanka and potentially Nepal (both within and outside protected areas) and largely destined to China, and are either shipped directly or are trafficked along routes through Myanmar and Nepal (Challender and Waterman, 2017; Mahmood et al., 2012; Perera et al., 2017). Seemingly increasing trafficking of this species is perhaps related to declines in populations of the Chinese and Sunda pangolins (see Chapters 4 and 6), and could be related to increasing awareness of the monetary value of scales (D'Cruze et al., 2018). Seizures suggest that the Indian pangolin is under heavy collection pressure. Available data indicate that between 2011 and 2017 scales from an estimated 1724 Indian pangolins were trafficked internationally, though the actual number of animals involved is likely higher, with trade going undetected (see Chapter 16; Challender and Waterman, 2017).

Secondary threats include agricultural expansion and habitat loss, which opens up previously inaccessible areas to hunting and poaching, and pesticide use and roadkill (Karawita et al., 2016; Murthy and Mishra, 2010). Pangolins are considered pests in some oil palm plantations in southwestern Sri Lanka, where they dig around the base of oil palm trees in search of prey, damaging the plants (P. Perera, unpubl. data). Thus, while the species is adaptable to habitat modification, it may inadvertently invite persecution.

References

Acharjyo, L.N., 2000. Management of Indian pangolin in captivity. In: The Managing Committee (Ed.), Souvenir, 125 Years of Calcutta Zoo (1875–2000). Zoological Garden, Alipore, Calcutta, pp. 27–30.

Agrawal, V.C., Das, P.K., Chakraborty, S., Ghose, R.K., Mandal, A.K., Chakraborty, T.K., et al., 1992. Mammalia. In: Director (Ed.), State Fauna Series 3: Fauna of West Bengal, Part 1. Zoological Survey of India, Calcutta, pp. 27–169.

Aiyapann, A., 1942. Notes on the Pangolin (*Manis crassicaudata*). J. Bombay Nat. Hist. Soc. 43, 254–257.

Akrim, F., Mahmood, T., Hussain, R., Qasim, S., Zangi, I., 2017. Distribution pattern, population estimation and threats to the Indian pangolin *Manis crassicaudata* (Mammalia: Pholidota: Manidae) in and around Pir Larusa National Park, Azad Jammu and Kashmir, Pakistan. J. Threat. Taxa 9 (3), 9920–9927.

Allen, G.M., 1938. The Mammals of China and Mongolia. Natural History of Central Asia, vol. XI. Part I. The American Museum of Natural History, New York.

Anderson, J., 1878. An Account of the Zoological Results of the Two Expeditions to Western Yunnan in 1868 and 1875; and a Monograph of the Two Cetacean Genera, Platanista and Orcella. Bernard Quaritch, London.

Anon, 2011. Tenacious pangolin defies pride of lions. The Hindu: National Daily News Paper. Available from <http://www.thehindu.com/news/tenacious-pangolin-defies-pride-of-lions/article2059874.ece>. [May 31, 2019].

Ashokkumar, M., Valsarajan, D., Suresh, M.A., Kaimal, A. R., Chandy, G., 2017. Stomach contents of the Indian pangolin *Manis crassicaudata* (Mammalia: Pholidota: Manidae) in Tropical Forests of Southern India. J. Threat. Taxa 9 (5), 10246–10248.

Baral, H.S., Shah, K.B., 2008. Wild Mammals of Nepal. Himalayan Nature, Kathmandu.

Bargali, H.S., Akhtar, N., Chauhan, N.P.S., 2004. Feeding ecology of sloth bears in a disturbed area in central India. Ursus 15 (2), 212–217.

Benatar, S., 2018. Not so invincible. Sanct. Asia 38 (4), 38–39.

Challender, D., Waterman, C., 2017. Implementation of CITES Decisions 17.239 b) and 17.240 on Pangolins (*Manis* spp.), CITES SC69 Doc. 57 Annex. Available from <https://cites.org/sites/default/files/eng/com/sc/69/E-SC69-57-A.pdf>. [May 31, 2019].

Choudhury, A., 2001. A Systematic Review of Mammals of North-East India With Special Reference to Non-Human Primates. D.Sc. Thesis, Guahati University, Assam, India.

CITES, 2000. Amendments to Appendices I and II of the Convention, Prop. 11.13 Transfer of *Manis crassicaudata, Manis pentadactyla, Manis javanica* from Appendix II to Appendix I. CITES, Geneva, Switzerland.

D'Cruze, N., Singh, B., Mookerjee, A., Harrington, L.A., Macdonald, D.W., 2018. A socio-economic survey of pangolin hunting in Assam, Northeast India. Nat. Conserv. 30, 83–105.

DNPWC and DoF (Department of National Parks and Wildlife Conservation and Department of Forests), 2018. Pangolin Conservation Action Plan for Nepal (2018-2022). Department of National Parks and Wildlife Conservation and Department of Forests, Kathmandu, Nepal.

Doran, G.A., Allbrook, D.B., 1973. The tongue and associated structures in two species of African pangolins, *Manis gigantea and Manis tricuspis*. J. Mammal. 54 (4), 887–899.

Fitzinger, L.J., 1872. Die naturliche familie der schuppenthiere (Manes). Sitzungsberichte der Kaiserlichen Akademie der Wissenschaften. Mathematisch-Naturwissenschaftliche Classe, CI., LXV, Abth. I, 9–83.

Frechkop, S., 1931. Notes sur les mammifères. VI. Quelques observations sur la classification des pangolins (Manidae). Bulletin du Musee royal d'Histoire naturelle de Belgique VII (22), 1–14.

Freedman, E., 2017. Indian Pangolins Spotted in Proposed Reserve. Available from: <https://www.rainforest-trust.org/indian-pangolin-spotted-proposed-reserve/>. [May 31, 2019].

Frick, F., 1968. Die Höhenstufenverteilung der Nepalesischen Säugetiere. Säugetierkundliche Mitteilungen 17, 161–173.

Gaubert, P., Antunes, A., Meng, H., Miao, L., Peigné, S., Justy, F., et al., 2018. The complete phylogeny of pangolins: scaling up resources for the molecular tracing of the most trafficked mammals on Earth. J. Hered. 109 (4), 347–359.

Gaudin, T., Emry, R., Wible, J., 2009. The phylogeny of living and extinct pangolins (Mammalia, Pholidota) and associated taxa: a morphology based analysis. J. Mammal. Evol. 16 (4), 235–305.

Goswami, R., Ganesh, T., 2014. Carnivore and herbivore densities in the immediate aftermath of ethno-political conflict: the case of Manas National Park, India. Trop. Conserv. Sci. 7 (3), 475–487.

Gotch, A.F., 1979. Mammals – Their Latin Names Explained. A Guide to Animal Classification. Blandford Press, Poole.

Grassè, P.P., 1955. Ordre des Pholidotes. In: Grassè, P.P. (Ed.), Traite de Zoologieé. vol. 17, Mammifères, Masson et Cie, Paris, pp. 1267–1284.

Gray, J.E., 1865. 4. Revision of the genera and species of entomophagous Edentata, founded on the examination of the specimens in the British Museum. Proc. Zool. Soc. Lond. 33 (1), 359–386.

Gubbi, S., Linkie, M., 2012. Wildlife hunting patterns, techniques, and profile of hunters in and around Periyar Tiger Reserve. J. Bombay Nat. Hist. Soc. 109 (3), 165–172.

Heath, M.E., 1995. *Manis crassicaudata*. Mammal. Sp. 513, 1–4.

Heath, M.E., Hammel, H.T., 1986. Body temperature and rate of O_2 consumption in Chinese pangolins. Am. J. Physiol.-Regul., Integr. Comp. Physiol. 250 (3), R377–R382.

Himmatsinhji, 1984. On the presence of the pangolin *Manis crassicaudata* Gray and a fox *Vulpes* sp. in Kutch. J. Bombay Nat. Hist. Soc. 81, 686–687.

Howell, A.B., 1929. Mammals from China in the collections of the United States National Museum. Proceedings of the United States National Museum 75 (2772). Smithsonian Institution Press, Washington D.C., pp. 1–82.

Hutton, A.F., 1949. Notes on the Indian pangolin (*Manis crassicaudata*, Geoffer St. Hilaire). J. Bombay Nat. Hist. Soc. 48, 805–806.

Illiger, J.K.W., 1815. Ueberblick der Säugethiere nach ihrer Vertheilung über die Welttheile. Abhandlungen der Physikalischen Klasse der Königlich-Preussischen Akademie der Wissenschaften aus den Jahren 1815, 39–159.

Irshad, N., Mahmood, T., Hussain, R., Nadeem, M.S., 2015. Distribution, abundance and diet of the Indian pangolin (*Manis crassicaudata*). Anim. Biol. 65, 57–71.

Irshad, N., Mahmood, T., Nadeem, M.S., 2016. Morpho-anatomical characteristics of Indian pangolin (*Manis crassicaudata*) from Potohar Plateau, Pakistan. Mammalia 80 (1), 103–110.

Israel, S., Sinclair, T., Grewal, B., Hoofer, H.J., 1987. Indian Wildlife. Apa Production. Hong Kong.

Jentink, F.A., 1882. Note XXV. Revision of the Manidae in the Leyden Museum. Notes from the Leyden Museum IV, 193–209.

Jiang, Z., Ma, Y., Wu, Y., Wang, Y., Zhou, K., Liu, S., et al., 2015. China's Mammal Diversity of Geographic Distribution. Science Press, Beijing. [In Chinese].

Jnawali, S.R., Baral, H.S., Lee, S., Acharya, K.P., Upadhyay, G.P., Pandey, M., et al., 2011. The Status of Nepal's Mammals: The National Red List Series. Department of National Parks and Wildlife Conservation. Kathmandu, Nepal.

Kanagavel, A., Parvathy, S., Nameer, P.O., Raghavan, R., 2016. Conservation implications of wildlife utilization by indigenous communities in the southern Western Ghats of India. J. Asia-Pacific Biodivers. 9 (3), 271–279.

Karawita, K., Perera, P., Pabasara, M., 2016. Indian pangolin (*Manis crassicaudata*) in Yagirala Forest Reserve: Ethnozoology and implications for conservation. Proceedings of the 21st International Forestry and Environment Symposium, 2016, Sri Lanka.

Karawita, H., Perera, P., Pabasara, G., Dayawansa, N., 2018. Habitat preference and den characterization of Indian Pangolin (*Manis crassicaudata*) in a tropical lowland forested landscape of Sri Lanka. PLoS One 13 (11), e0206082.

Khan, M.A.R., 1985. Mammals of Bangladesh. Nazma Reza, Dhaka.

Latafat, K., Sadhu, A., 2016. First photographic evidence of Indian pangolin *Manis crassicaudata* E. Geoffrey, 1803 in Mukundara Hills Tiger Reserve (MHTR), Rajasthan, India. J. Bombay Nat. Hist. Soc. 113, 21–22.

McNab, B.K., 1984. Physiological convergence amongst ant-eating and termite-eating mammals. J. Zool. Soc. Lond. 203 (4), 485–510.

Mahmood, T., Hussain, R., Irshad, N., Akrim, F., Nadeem, M.S., 2012. Illegal mass killing of Indian Pangolin (*Manis crassicaudata*) in Potohar Region, Pakistan. Pak. J. Zool. 44 (5), 1457–1461.

Mahmood, T., Jabeen, K., Hussain, I., Kayani, A.R., 2013. Plant species association, burrow characteristics and the diet of the Indian Pangolin, *Manis crassicaudata*, in the Potohar Plateau, Pakistan. Pak. J. Zool. 45 (6), 1533–1539.

Mahmood, T., Irshad, N., Hussain, R., 2014. Habitat preference and population estimates of Indian Pangolin (*Manis crassicaudata*) in District Chakwal of Potohar Plateau, Pakistan. Russ. J. Ecol. 45 (1), 70–75.

Mahmood, T., Andleeb, S., Anwar, M., Rais, M., Nadeem, M.S., Akrim, F., et al., 2015a. Distribution, abundance and vegetation analysis of the scaly anteater (*Manis crassicaudata*) in Margalla Hills National Park Islamabad, Pakistan. J. Anim. Plant Sci. 25 (5), 1311–1321.

Mahmood, T., Irshad, N., Hussain, R., Akrim, F., Hussain, I., Anwar, M., et al., 2015b. Breeding habits of the Indian pangolin (*Manis crassicaudata*) in Potohar Plateau, Pakistan. Mammalia 80 (2), 231–234.

Mahmood, T., Kanwal, K., Zaman, I.U., 2018. Records of the Indian Pangolin (Mammalia: Pholidota: Manidae: *Manis crassicaudata*) from Mansehra District, Pakistan. J. Threat. Taxa 10 (2), 11254–11261.

Mahmood, T., Challender, D., Khatiwada, A., Andleeb, S., Perera, P., Trageser, S., Ghose, A., Mohapatra, R., 2019. *Manis crassicaudata. The IUCN Red List of Threatened Species* 2019: e.T12761A123583998. Available from: <http://dx.doi.org/10.2305/IUCN.UK.2019-3.RLTS.T12761A123583998.en>.

Manakadan, R., Sivakumar, S., David, P., Murugan, B.S., 2013. The mammals of Sriharikota Island, Southern India, with insights into their status, population and distribution. J. Bombay Nat. Hist. Soc. 110 (2), 114–121.

Mishra, S., Panda, S., 2012. Distribution of Indian Pangolin *Manis crassicaudata* Gray (Pholidota, Manidae) in Orissa: a rescue prospective. Small Mammal Mail – Bi-Annu. Newsl. CCINA RISCINSA 3 (2), 51–53.

Mitchell, R.M., 1975. A checklist of Nepalese mammals. Säugetierkundliche Mitteilungen 23, 152–157.

Mitra, S., 1998. On the scales of the scaly anteater *Manis crassicaudata*. J. Bombay Nat. Hist. Soc. 95 (3), 495–498.

Mohapatra, R.K., Panda, S., 2013. Behavioural sampling techniques and activity pattern of Indian pangolins *Manis crassicaudata* (Mammalia: Manidae) in captivity. J. Threat. Taxa 5 (17), 5247–5255.

Mohapatra, R.K., Panda, S., 2014a. Husbandry, behaviour and conservation breeding of Indian pangolin. Folia Zool. 63 (2), 73–80.

Mohapatra, R.K., Panda, S., 2014b. Behavioural descriptions of Indian pangolins (*Manis crassicaudata*) in captivity. Int. J. Zool. 795062.

Mohapatra, R.K., Panda, S., Acharjyo, L.N., Nair, M.V., Challender, D.W.S., 2015a. A note on the illegal trade and use of pangolin body parts in India. TRAFFIC Bull. 27 (1), 33–40.

Mohapatra, R.K., Panda, S., Nair, M.V., 2015b. On the mating behaviour of captive Indian pangolin (*Manis crassicaudata*). TAPROBANICA: J. Asian Biodivers. 7 (1), 57–59.

Mohapatra, R.K., 2016. Study on Some Biological Aspects of Indian Pangolin (*Manis crassicaudata* Gray, 1827). Ph. D. Thesis, Utkal University, Bhubaneswar, India.

Mohapatra, R.K., 2018. Rare observations of inter-specific interaction of sloth bear and leopard with Indian pangolin at Satpura Tiger Reserve, Central India. Biodivers. Int. J. 2 (4), 331–333.

Mohapatra, R.K., Panda, S., Sahu, S.K., 2018. On the gestation period of Indian pangolins (*Manis crassicaudata*) in captivity. Biodivers. Int. J. 2 (6), 559–560.

Mohr, E., 1961. Schuppentiere. Neue Brehm-Bucherei. A. Ziemsen Verlag, Wittenburg Lutherstadt.

Murthy, K.L.N., Mishra, S., 2010. A note on road killing of Indian pangolin *Manis crassicaudata* Gray at Kambalakonda Wildlife Sanctuary of Eastern Ghat Ranges. Small Mammal Mail – Bi-Annu. Newsl. CCINSA RISCINSA 2 (2), 8–10.

Ogilvie, P.W., Bridgwater, D.D., 1967. Notes on the breeding of an Indian pangolin *Manis crassicaudata*. Int. Zoo Yearbook 7 (1), 116–118.

Pabasara, M.G.T., Perera, P.K.P., Dayawansa, N.P., 2015. Preliminary Investigation of the Habitat Selection of Indian Pangolin (*Manis crassicaudata*) in a Tropical lowland Forest in South-West Sri Lanka. Proc. Int. For. Environ. Symp. 20, 4.

Perera, P.K.P., Karawita, K.V.D.H.R., Pabasara, M.G.T., 2017. Pangolins (*Manis crassicaudata*) in Sri Lanka: a review of current knowledge, threats and research priorities. J. Trop. For. Environ. 7 (1), 1–14.

Perera, P.K.P., Karawita, K.V.D.H.R., 2019. An update of distribution, habitats and conservation status of the Indian pangolin (*Manis crassicaudata*) in Sri Lanka. Glob. Ecol. Conserv. 21, e00799.

Phillips, W.W.A., 1928. A note on the habits of the Indian pangolin (*Manis crassicaudata*). Spoila Zeylan 14, 333.

Phillips, W.W.A., 1981. Manual of the Mammals of Sri Lanka. Wildlife and Nature Protection Society of Sri Lanka, Colombo.

Pocock, R.I., 1924. The external characters of the pangolins (Manidae). Proc. Zool. Soc. Lond. 94 (3), 707–723.

Prakash, I., 1960. Breeding of mammals in Rajasthan desert, India. J. Mammal. 41 (3), 386–389.

Prater, S.H., 1971. The Book of Indian Animals, third ed. Bombay Natural History Society, Bombay.

Ramakrishnan, U., Coss, R.G., Pelkey, N.W., 1999. Tiger decline caused by the reduction of large ungulate prey: evidence from a study of leopard diets in southern India. Biol. Conserv. 89 (2), 113–120.

Ripley, S.D., 1965. The Land and Wildlife of Tropical Asia. Life Nature Library Series. Time-Life International, Nederland.

Roberts, T.J., 1977. The Mammals of Pakistan. Ernest Benn Ltd, London.

Rowcliffe, J.M., Field, J., Turvey, S.T., Carbone, C., 2008. Estimating animal density using camera traps without the need for individual recognition. J. Appl. Ecol. 45 (4), 1228–1236.

Saxena, R., 1985. Instance of an Indian pangolin (*Manis crassicaudata*, Gray) digging into a house. J. Bombay Nat. Hist. Soc. 83, 660.

Sharma, S.K., 2002. Abnormal weight and length of the Indian pangolin *Manis crassicaudata* Gray, 1827, from Sirohi District, Rajasthan. J. Bombay Nat. Hist. Soc. 99 (1), 103–104.

Sharma, B.K., 2014. Pangolins in trouble. Sanct. Asia XXXIV 3, 38–41.

Sheikh, K.M., Molur, S. (Eds.), 2005. Status and Red List of Pakistan's Mammals based on the Pakistan Mammal Conservation Assessment and Management Plan Workshop 18–22 August 2003. IUCN, Islamabad.

Smith, A.T., Xie, Y., 2013. Mammals of China. Princeton University Press, Princeton.

Srinivasulu, C., Srinivasulu, B., 2012. South Asian Mammals: Their Diversity, Distribution, and Status. Springer, New York.

Sundevall, C.J., 1842. Om slägtet Sorex, med nâgra nya arters beskrifning. Kungliga Vetenskapsakademien, Stockholm.

Suwal, R., Verheugt, Y.J.M., 1995. Enumeration of Mammals of Nepal. Biodiversity Profiles Project Publication No. 6. Department of National Parks and Wildlife Conservation, Ministry of Forest and Soil Conservation. Nepal.

Tikader, B.K., 1983. Threatened Animals of India. Zoological Survey of India, Calcutta.

Trageser, S.J., Ghose, A., Faisal, M., Mro, P., Mro, P., Rahman, S.C., 2017. Pangolin distribution and conservation status in Bangladesh. PLoS One 12 (4), e0175450.

Ullmann, T., Veríssimo, D., Challender, D.W.S., 2019. Evaluating the application of scale frequency to estimate the size of pangolin scale seizures. Glob. Ecol. Conserv. 20, e00776.

Underwood, G., 1945. Note on the Indian pangolin (*Manis crassicaudata*). J. Bombay Nat. Hist. Soc. 45, 605–607.

Weigl, R., 2005. Longevity of Mammals in Captivity: From the Living Collections of the World. Kleine Senckenberg-Reihe 48, Stuttgart.

Willcox, D., Nash, H., Trageser, S., Kim, H.J., Hywood, L., Connelly, E., et al., 2019. Evaluating methods for detecting and monitoring pangolin (Pholidota: Manidae) populations. Glob. Ecol. Conserv. 17, e00539.

Wu, S.B., Ma, G.Z., 2007. The status and conservation of pangolins in China. TRAFFIC East Asia Newsl. 4, 1–5. [In Chinese].

CHAPTER

6

Sunda pangolin *Manis javanica* (Desmarest, 1822)

Ju Lian Chong[1,2], *Elisa Panjang*[2,3,4], *Daniel Willcox*[2,5], *Helen C. Nash*[2,6], *Gono Semiadi*[2,7], *Withoon Sodsai*[2,8], *Norman T-L Lim*[2,9], *Louise Fletcher*[2], *Ade Kurniawan*[2,10] and *Shavez Cheema*[2,11]

[1]School of Science and Marine Environment & Institute of Tropical Biodiversity and Sustainable Development, Universiti Malaysia Terengganu, Kuala Nerus, Malaysia [2]IUCN SSC Pangolin Specialist Group, ℅ Zoological Society of London, Regent's Park, London, United Kingdom [3]Organisms and Environment Division, Cardiff School of Biosciences, Cardiff University, Cardiff, United Kingdom [4]Danau Girang Field Centre, Sabah Wildlife Department, Kota Kinabalu, Malaysia [5]Save Vietnam's Wildlife, Cuc Phuong National Park, Ninh Binh Province, Vietnam [6]Department of Biological Sciences, National University of Singapore, Singapore [7]Pusat Penelitian Biologi Lembaga Ilmu Pengetahuan Indonesia, Cibinong Science Center, Bogor, Indonesia [8]Nottingham Trent University, Nottingham, United Kingdom [9]National Institute of Education, Nanyang Technological University, Singapore [10]Wildlife Reserves Singapore, Singapore [11]1StopBorneo Wildlife, Kota Kinabalu, Malaysia

OUTLINE

Pangolins
DOI: https://doi.org/10.1016/B978-0-12-815507-3.00006-X

Taxonomy

Included in the genus *Manis* based on morphological (Gaudin et al., 2009) and genetic evidence (Gaubert et al., 2018; Hassanin et al., 2015). Previously included in the genus *Paramanis* (Pocock, 1924), subgenus *Paramanis* (Ellerman and Morrison-Scott, 1966) and genus *Pholidotus* (Fitzinger, 1872; Gray, 1865).

Synonyms: *Manis leptura* (Blyth, 1842), *Manis aspera* (Sundevall, 1842), *Manis leucura* (Blyth, 1847), *Manis guy* (Focillon, 1850), *Manis sumatrensis* (Ludeking, 1862), *Pholidotus malaccensis* and *Pholidotus labuanensis* (Fitzinger, 1872). The type specimen is from Java, Indonesia (Desmarest, 1822).

Pangolin populations in the Philippines were previously ascribed to this species but morphological research by Feiler (1998) and later Gaubert and Antunes (2005) supported classification of a distinct species, the Philippine pangolin (*Manis culionensis*; see Chapter 7). The latter classification was based on six discrete morphological characters, including total number of transversal scale rows on the back, scale size in nuchal, scapular and postscapular regions, ratio of nasal bone to total skull length, ratio of head and body to tail length, posterior region of the palatine bone, and posterior extension of the zygomatic process. This was supported by genetic analyses (see Gaubert et al., 2018) and estimates indicate that the Philippine pangolin diverged from the Sunda species approximately 500,000–800,000 years ago (Gaubert and Antunes, 2005).

Research using genome-wide markers has revealed highly divergent subpopulations of the Sunda pangolin, with distinct lineages in Borneo, Java, and Singapore/Sumatra (see also Chapter 2; Nash et al., 2018). The species' wide geographic distribution suggests that there could be further, unreported cryptic diversity across the range. Further genetic and morphological research would enable a better understanding of taxonomy. Diploid number 2n = 38.

Etymology: *Manis* (see Chapter 4); the epithet *javanica* references the island of Java, Indonesia.

Description

The Sunda pangolin is a medium-sized mammal typically reaching a weight of 4–7 kg, and a total length of up to 140 cm (Table 6.1), but larger individuals have been recorded. Sulaiman et al. (2017) recorded a male weighing 13.5 kg and records from the logbooks of a trafficking syndicate in Sabah, Malaysia suggest that the species can attain a weight of up to 21 kg; the trimmed mean weight of these individuals was 4.96 kg (n = 20,857; Table 6.1; Pantel and Anak, 2010). Head-body length ranges up to 79 cm, and tail length up to 72 cm, the tail comprising >0.42 of overall length (Heath, 1992). Body mass is positively correlated with overall length (Sulaiman et al., 2017; Yang et al., 2010). The species is sexually dimorphic with males larger and heavier than females, which is likely related to territoriality in males (see *Behavior*; Lim, 2007; Sulaiman et al., 2017).

Hard, keratinous and overlapping, rounded scales grow from the skin (Fig. 6.1), the distal edge of which is sharp. Scales cover the streamlined body on the dorsal and lateral surfaces, and both the dorsal and ventral surfaces of the tail, and the outer margins of the limbs, but are absent from the throat and belly, and inner surfaces of the limbs (Jentink, 1882). On the forehead the scales are small and terminate just before the muzzle; they are absent from the lateral parts of the face (Pocock, 1924). There are 15–19 transversal, and 15–21 longitudinal scale rows on the body (Frechkop, 1931; Jentink, 1882), and the scales on the dorsum are similar in size to the first row of postscapular scales. On the tail there are 20–30 rows of scales on the median scale row and on the margins (Frechkop, 1931; Jentink, 1882). On

TABLE 6.1 Sunda pangolin morphometrics.

Measurement			Country	Source(s)
Weight	Weight (♂)	5.09 (2.8–9.1) kg, n = 21		Save Vietnam's Wildlife, unpubl. data
	Weight (♀)	4.5 (2.9–6.3) kg, n = 21		Save Vietnam's Wildlife, unpubl. data
	Weight (unsexed)	4.96[a] (<1–21) kg, n = 20,857	Sabah, Malaysia	Pantel and Anak, 2010
Body	Total length (♂)	1019 (370–1375) mm, n = 15	Malaysia	Sulaiman et al., 2017
	Total length (♀)	897 (320–1400) mm, n = 16	Malaysia	Sulaiman et al., 2017
	Head-body length (♂)	524 (253–790) mm, n = 15	Malaysia	Sulaiman et al., 2017
	Head-body length (♀)	473 (170–680) mm, n = 16	Malaysia	Sulaiman et al., 2017
	Tail length (♂)	495 (110–680) mm, n = 15	Malaysia	Sulaiman et al., 2017
	Tail length (♀)	422 (150–720) mm, n = 16	Malaysia	Sulaiman et al., 2017
Vertebrae	Total number of vertebrae	61		Frechkop, 1931; Jentink, 1882
	Cervical	7		Jentink, 1882
	Thoracic	15		Jentink, 1882
	Lumbar	5–6		Jentink, 1882; Mohr, 1961
	Sacral	3–4		Jentink, 1882; Mohr, 1961
	Caudal	29–30		Jentink, 1882; Mohr, 1961
Skull	Length (unsexed)	60–100 mm		Gaubert, 2011
	Breadth across zygomatic processes	No data		
Scales	Total number of scales	873 (817–952), n = 12	Indonesia, Malaysia, Singapore, Thailand	Ullmann et al., 2019; D. Challender and C. Shepherd, unpubl. data
	No. of scale rows (transversal, body)	15–19		Gaubert and Antunes, 2005; Frechkop, 1931; Jentink, 1882
	No. of scale rows (longitudinal, body)	15–21		Frechkop, 1931; Jentink, 1882

(*Continued*)

TABLE 6.1 (Continued)

Measurement		Country	Source(s)
No. of scales on outer margins of tail	20–30		Frechkop, 1931; Jentink, 1882
No. of scales on median row of tail	20–30		Frechkop, 1931; Jentink, 1882
Scales (wet) as proportion of body weight	12.3%, n = 1		D. Challender and C. Shepherd, unpubl. data
Scales (dry) as proportion of body weight	10.9%, n = 1		D. Challender and C. Shepherd, unpubl. data

[a]*Trimmed mean.*

the latter, the scales are kite-shaped and folded around the dorsal and ventral surfaces. The medio-dorsal scale row extends to the tail tip. The scales are striated and keeled on the flanks and hindlimbs, and though less pronounced, keeled on the forelimbs (Fig. 6.1; Jentink, 1882; Pocock, 1924). The tail is prehensile, immensely strong, and has an unscaled, cutaneous pad on the ventral side at the tip (Pocock, 1924). Scale coloration ranges from light olive-brown to dark gray-brown and may or may not be uniform. Individuals may possess scales that are yellowish-white or translucent on the dorsum and tail (see Fig. 6.1). A small number (3–4) of thick hairs grow from the base of each scale which are white-gray, brown or even black (Hafiz and Chong, 2016).

The head is short and conoid, tapering to the muzzle which is bluntly truncated (Pocock, 1924), and the skull has adaptations to a myrmecophagous diet (see Chapter 1). The face is gray or pinkish-gray, and may have a blue tinge (M. Hafiz, pers. obs.); the belly and ventral surface are the same color and covered in short light-colored hairs. The rhinarium is dark pink-brown. The eyes are small and have a dark iris and covered by thick swollen eyelids. Ear pinnae are present but are reduced to a thickened subvertical ridge of integument with a slightly convex posterior edge anterior to which is the auditory meatus (Fig. 6.1; Pocock, 1924).

Tarmizi and Sipangkul (2019) noted the presence of a glandular structure in the neck of a male Sunda pangolin, which excreted a white oily substance with a musky odor when squeezed. They suspect that it develops in males after reaching a certain age (or weight), though its function is not known.

The jaw is elongated but lacks teeth and the oral cavity is small. The tongue is elongated and like in other pangolins is not attached to the hyoid bone but to the caudal end of the xiphoid process (xiphisternum), which is spatulate in shape, in the abdomen (see Chan, 1995). Chan (1995) divided the retracted tongue into three sections: the non-protruded part, which is attached to the xiphisternum and extends to the thorax; the glossal tube, which comprises the majority of the tongue and extends to the oral cavity; and the free part, which is coated in a mucous membrane and stems from the base of the glossal tube, through the glossal-tube cavity and into the oral cavity. The tongue can be extended by 25 cm to consume prey (Nowak, 1999). Prey passes down to the pyloric region where it is ground up, and the stomach is lined by thick,

FIGURE 6.1 The Sunda pangolin is semi-arboreal and an excellent climber. The keeled scales on the flank and hindlimbs are visible in this image. *Photo credit: David Tan.*

cornified stratified squamous epithelium, which likely affords the mucosa protection from mechanical abrasion during mastication (Nisa et al., 2010). Like the Chinese pangolin, there are keratinous spines in the stomach which appear to aid digestion (Nisa et al., 2010; see Lin et al., 2015). The kidneys have highly active and pronounced proximal straight tubules, which may be related to saliva production (Pongchairerk et al., 2008).

The species is pentadactylous with five digits on the fore- and hindfeet, and a granular pad is present on both, though is less well defined on the former (Pocock, 1924). The digits terminate with claws, the largest and strongest of which is the middle, and the second and fourth claws are smaller but of approximately equal size, while the first and fifth claws are rudimentary and non-functional; digit length and disposition on the hindfeet approximates the forefeet (Pocock, 1924). The claws on the hindfeet are almost as long as on the forefeet (Jentink, 1882). A large pair of anal glands are present in the circumanal area. The vulva and penis are located anterior to the anus respectively (Pocock, 1924). Females have two pectoral mammae.

Body temperature is regulated at 33–35 °C (Nguyen et al., 2014). Basal metabolic rate has been estimated at 262 ml O_2/kg/h (McNab, 1984).

The Sunda pangolin may be confused with the Philippine and Chinese pangolin. Differences between the Sunda and Philippine species are discussed under *Taxonomy* and in Chapter 7. The Sunda pangolin can be distinguished from the Chinese species by the less thickly set body, longer tail, smaller ear pinnae, and smaller and shorter claws on the forefeet, which are noticeably larger and longer in the Chinese pangolin (see Chapter 4). The Chinese species has fewer transversal (14–18) scale rows and scales on the tail margins (14–20), though there is some overlap in the latter (see Table 6.1).

Distribution

Widely distributed in Southeast Asia, except the Philippines (Fig. 6.2; Corbet and Hill, 1992). The northern and western limits of the distribution are not well defined. The species occurs in central and southern Myanmar and there are records from Karen state in 2014–2015 (Moo et al., 2017) and Tanintharyi region from 2014 to 2016 (Aung et al., 2017). Wu et al. (2005) asserted that the species occurs in Yunnan Province, southwestern China (see Fig. 6.2) based on specimens held at the Kunming Institute of Zoology, but there is uncertainty over the provenance of these specimens, and thereby distribution in the country. Similarly, some sources (e.g., Khan, 1985) refer

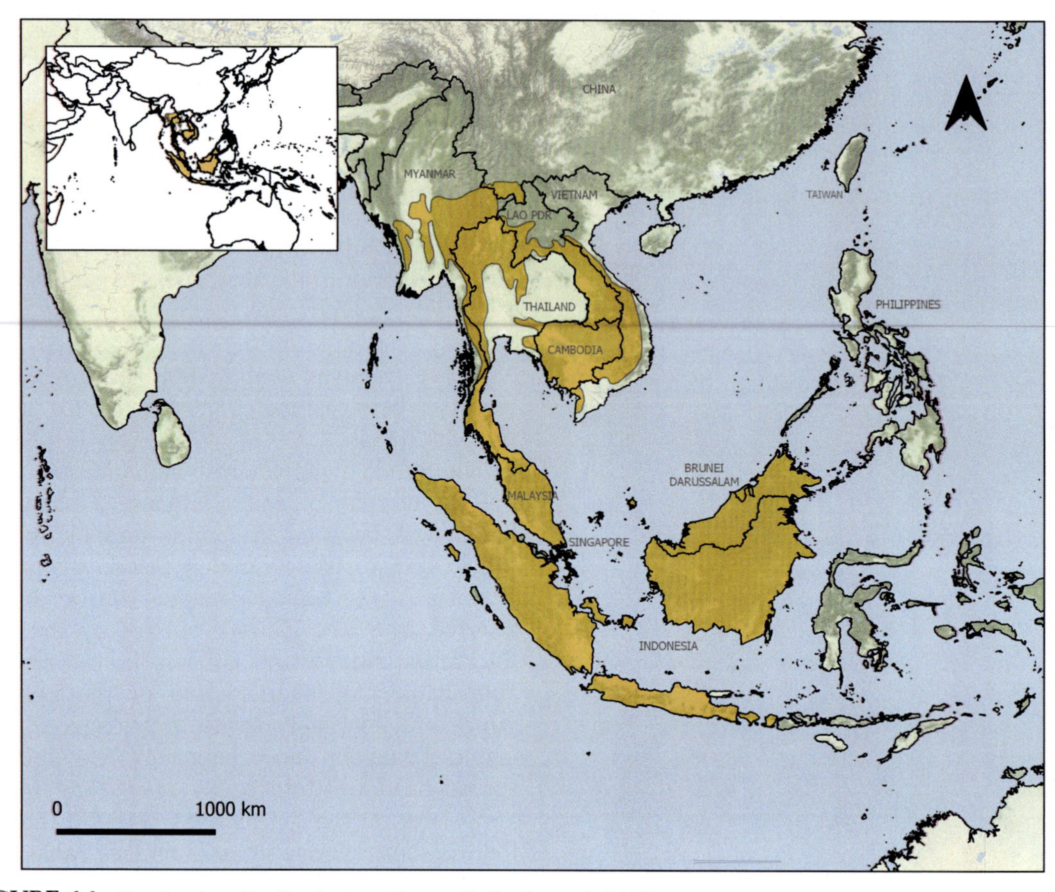

FIGURE 6.2 Sunda pangolin distribution. *Source: Challender et al. (2019).*

to the occurrence of the species in Bangladesh but there are no supporting records (see Trageser et al., 2017).

In Thailand, the species is distributed in the west, south and southeast (Lekagul and McNeely, 1988), and there are records from post-2000 in Khao Ang Ru Nai Wildlife Sanctuary (Jenks et al., 2012), Khao Yai and Kaeng Krachan National Parks (W. Sodsai, unpubl. data), Khlong Nakha Wildlife Sanctuary, and the southeastern western forest complex (sWEFCOM) in Kanchanaburi Province (ZSL, 2017), and Hala-Bala Wildlife Sanctuary in the far south (Kitamura et al., 2010). The species also occurs on surrounding islands, including Ko Ra, Phang Nga Province (GVI, unpubl. data), but distribution on such islands is poorly documented. The species is broadly distributed in Lao People's Democratic Republic (Lao PDR) except the north and north-east (see Fig. 6.2) though there are few recent records. Duckworth et al. (1999) recorded the species from Nam Kading National Biodiversity Conservation Area (NBCA) as far south as Xe Pian NBCA; there are unverified camera trap records from Nakai-Nam Theun NBCA in the early 2010s (Coudrat, 2017). The species is distributed across Cambodia, including the east (Desai and Lic, 1996) and there are records from post-2012 in the north (e.g., Virachey National

Park [McCann and Pawlowski, 2017] and Chhep Wildlife Sanctuary [Suzuki et al., 2017]) and south (e.g., Botum Sakor and Southern Cardamom National Parks [Gray et al., 2017a]), including coastal areas (Peam Krasop Wildlife Sanctuary; Thaung et al., 2017). In Vietnam, the species occurs in central and southern regions (Newton et al., 2008) from Nghe An Province (e.g., Pu Mat National Park; Save Vietnam's Wildlife, unpubl. data) as far south as Ca Mau Province (U Minh Ha National Park; Willcox et al., 2017) and there are records from Kon Tum, Tay Ninh and Quang Nam Provinces (Bourret, 1942) and Ha Tinh (Timmins and Cuong, 1999), Quang Binh (Le et al., 1997a) and Dak Lak (Le et al., 1997b).

The Sunda pangolin is widely distributed in Peninsular Malaysia and Malaysian Borneo. In the former, distribution extends to all states, including Penang (Azhar et al., 2013; Chong et al., 2016; Ickes and Thomas, 2003; Numata et al., 2005). The species occurs throughout Sabah and Sarawak and there are recent records from both protected (e.g., Danum Valley, Sepilok-Kabili Forest Reserve, and Lower Kinabatangan Wildlife Sanctuary) and unprotected areas (e.g., adjacent to Universiti Malaysia Sabah) in Sabah (E. Panjang, unpubl. data). There are also records from surrounding islands including Gaya Island and Labuan (E. Panjang, unpubl. data). In Sarawak, there are camera trap records from post-2000 in Lanjak Entimau Wildlife Sanctuary (Mohd-Azlan and Engkamat, 2013), and from 2015 to 2017 in Kubah National Park (Mohd-Azlan et al., 2018) and the Mt. Penrissen area (Kaicheen and Mohd-Azlan, 2018).

The species occurs across Singapore (see Chapter 26) including adjacent islands e.g., Pulau Tekong and Pulau Ubin (Lim and Ng, 2008a). Little is known about distribution in Brunei Darussalam but the species occurs in all four districts (Brunei Muara, Kuala Belait, Temburong and Tutong; Fletcher, 2016; S. Cheema, unpubl. data).

In Indonesia the species is widely distributed including in Kalimantan, Sumatra, Java, Kiau and the Linngga archipelago, Bangka and Belitung, Nias and the Pagi islands, and Bali and nearby islands (Corbet and Hill, 1992). Lyon (1909) noted that the species occurred on Pulau Bulan in the Northern Riau Islands, and there were observations in 2018 on Pulau Poto, adjacent to Bintan (H.C. Nash, pers. obs.). The species also occurs on the Natuna Islands, off the northwest coast of Borneo (Phillips and Phillips, 2018).

Habitat

The Sunda pangolin occurs in diverse habitats ranging from primary and secondary tropical forest (including dipterocarp), evergreen and hill forest, peat swamp forest, grasslands, and monoculture (e.g., oil palm and rubber) plantations, gardens and urban areas (Azhar et al., 2013; Ketol et al., 2009; Lim and Ng, 2008a; Payne et al., 1985; Wearn, 2015). Exact altitudinal limits are not well defined. Duckworth et al. (1999) noted that in Lao PDR the species was primarily found in the lowlands and lower hills up to 600 m asl and distribution in mainland Southeast Asia may be limited to <1000 m asl. Payne and Francis (2007) report that the species occurs up to 1700 m on Mt. Kinabalu, Sabah and there are records from above 1200 m in the Mt. Penrissen area, Sarawak (Kaicheen and Mohd-Azlan, 2018). Occurrence in artificial landscapes (e.g., oil palm plantations) and degraded forest (e.g., Singapore), including evidence of breeding, suggests that the species is adaptable to modified habitats providing an adequate prey base exists and it is not unduly persecuted. Khwaja et al. (2019) found that the species was more likely to occupy locations outside rather than inside protected areas, though a coarse measure of protected area status was used. Further ecological research is needed to determine whether more frequent sightings in modified habitats reflect greater abundance or ease of detection (Davies and Payne, 1982), and the ability of the species to persist and reproduce in these habitats long-term. Evidence from Singapore suggests that this is the case regarding degraded tropical forest (see Chapter 26). The species may be well adapted to wetland and riverine ecosystems; there are records from the seasonally inundated U Minh wetlands in Vietnam where individuals were observed in *Phragmites* reeds and Melaleuca forest (Willcox et al., 2017), and of individuals crossing the Kinabatangan River in Sabah where it is >100 m wide (E. Panjang, unpubl. data). Khwaja et al. (2019) report that occupancy increased with increasing distance from the nearest river. The Sunda pangolin is intolerant of colder temperatures with individuals kept outside of their natural range in areas that experience moderately cold winters frequently dying rapidly from exposure (Hua et al., 2015; Save Vietnam's Wildlife [SVW], unpubl. data).

Ecology

There is limited knowledge of Sunda pangolin ecology and behavior and most of what is known is based on research in Singapore. The species is semi-arboreal and appears to have stable home ranges. On Pulau Tekong, Singapore, Lim (2007) tracked adult males and estimated a mean 100% MCP (Minimum Convex Polygon) home range of 41 ha (range 8.2–76.6 ha; n = 4). Lim also tracked a female and estimated a home range of 7 ha (100% MCP) but this estimate is confounded as the individual had a juvenile in tow. Lim further observed aggressive behavior and a stand-off between two males, suggesting that the species is territorial and will defend a territory. Wild-caught males in captivity are known to exhibit territorial behavior if housed together or in close proximity (SVW, unpubl. data).

The species utilizes diverse structures for shelter, which range from tree hollows, either in standing or fallen trees (and live or dead), burrows, either freshly dug or modified from existing structures, to tall grasses (e.g., *Imperata cylindrica*), and the branches of oil palm trees in commercial plantations (see Fig. 6.3A–C; Lim and Ng, 2008a; E. Panjang, unpubl. data). These structures may have more than one entrance; tree hollow entrances in Singapore measured 13–24 cm in diameter (Lim and Ng, 2008a). The species is known to dig burrows in captivity (SVW, unpubl. data), and in Singapore will make use of urban structures including large pipes (see Chapter 26). Lim (2007) noted that in Singapore, adult males (n = 4) will use the same resting site for 1–2 consecutive days only, whereas the female with young had a high level of den fidelity (see *Ontogeny and reproduction*).

The Sunda pangolin predates on both ants and termites and is prey selective. In Singapore, adult males spent a higher proportion of time feeding on ants (67%) rather than termites (32.9%; mean, n = 3) and foraging bouts were typically short, lasting only about 2 min (Lim, 2007). Feeding bouts are longer when attacking an ant nest or termitaria. Eleven ant genera were predated upon: *Anoplolepis, Dolichoderus, Dorylus, Odontomachus, Oecophylla, Papyrius, Paratopula, Paratrechina, Pheidole, Philidris* and *Polyrhachis* (Lim, 2007). The species showed a preference for *Polyrhachis* spp. and *A. gracilipes*. Notably, the weaver ant *Oecophylla smaragdina* was also preferred and Sunda pangolins will exert considerable amounts of energy to ascend trees in order to access nests of this species. Willcox et al. (2017) observed predation on *O. smaragdina* in U Minh Ha National Park, Vietnam, and this species is readily consumed in captivity (Nguyen et al., 2014). *Crematogaster* spp. ("heart-shaped" ants), *Monomorium, Rhoptromyrmex, Technomyrmex* were avoided in Singapore (Lim, 2007), but *Crematogaster* spp. are consumed in captivity in Vietnam

FIGURE 6.3 (A) Sunda pangolins shelter in a range of structures including tree hollows, like this one in Singapore. (B–C) A Sunda pangolin den in Singapore. (B) A fallen tree. (C) Entrance to the den at the base of the tree. *Reproduced with permission from Lim, N.T.-L., Ng, P.K.L., 2008a. Home range, activity cycle and natal den usage of a female Sunda pangolin Manis javanica (Mammalia: Pholidota) in Singapore. Endanger. Sp. Res. 4, 233–240.*

(Nguyen et al., 2014). Beebe (1919) examined the stomach contents of a Sunda pangolin and found the species predates on fire ants (presumably *Solenopsis* spp.). Harrison (1961) estimated that the stomach contents of one individual contained >200,000 ants and pupae.

Predators include tiger (*Panthera tigris*), leopard (*P. pardus*), and clouded leopard (*Neofelis nebulosa*; Grassman et al., 2005; Kawanishi and Sunquist, 2004), pythons (e.g., reticulated python [*Malayopython reticulatus*]; Lim and Ng, 2008b; Shine et al., 1998), sun bears (*Helarctos malayanus*; Hedges and Aziz, 2013) and feral dogs. Little is known about predation rates.

Endo- and ecto-parasites include, but are not limited to, protozoans (*Eimeria tenggilingi*; Else and Colley, 1976), helminths (e.g., *Brugia malayi*, *B. pahangi*; Laing et al., 1960), ticks (*Amblyomma* spp.) and bacteria (*Mycoplasma* sp.) respectively (see Chapter 29; Hafiz et al., 2012; Jammah et al., 2014; Mohapatra et al., 2016).

Behavior

Like all pangolins, the species is solitary, but females are sometimes observed with young (Lim and Ng, 2008a); these are the only lasting social bonds, i.e., until young become independent. Social interactions appear to be scent-based and are mediated through scent marks, which likely comprise a combination of urine and secretions from the anal glands, and contain information about sexual and health status.

The Sunda pangolin is principally nocturnal, resting by day in a place of shelter (e.g., a tree hollow or burrow) and active at night. In Singapore, Lim (2007) estimated mean daily active duration of adult males to be 165 ± 14 min (mean ± S.E; n = 4), and for a female with young, 127 ± 13.1 min (Lim and Ng, 2008a). Peak activity occurred between 0300 and 0600 (Lim, 2007). The female was observed active diurnally (0800–1800) in December, which is potentially related to increased activity as the juvenile approached weaning age (Lim, 2007). Nocturnal behavior has been recorded in captivity. Challender et al. (2012) observed seven individuals and peak activity occurred between 1800 and 2100, but intermittent activity was recorded between 1700 and 0500. Nguyen et al. (2014) observed that individuals will emerge from their burrows at 1700 in winter, but later in summer (1800–1900), which is potentially related to daylight levels.

The species is quadrupedal and when walking the head and tail are held below the level of the body, which is arched resulting in a humped appearance. The forelimbs are turned inwards and the weight taken on the folded up wrists (van Strien, 1983). When walking, and when foraging, the species will pause periodically, either on all four limbs or will bear the weight on the hindlimbs and raise the head, body, and forelimbs and will sniff the air in different directions, presumably to detect odors in the air in order to locate prey and conspecifics, and potentially to detect predators. Being semi-arboreal, the species is an excellent climber and has enormous core strength, and uses its long tail as a climbing aid. The species moves its limbs in pairs when climbing thick tree trunks, the strong claws enabling grip (see Fig. 6.1), and is supported by the tail. In the forest canopy, ambling along tree branches is simple and the tail is often curled around branches for support. The species can easily move along branches upside down and the tail is capable of supporting the entire body weight; the species will occasionally hang from a branch by only its tail. Challender et al. (2012) found idiosyncrasies in how individuals (n = 7) partitioned their active time in captivity in Vietnam. In some individuals more than 70% of their active time was spent feeding and in locomotion (walking and/or climbing), and a maximum of 43.5 ± 6.5% (mean ± S.E) and 40.7 ± 3.9% of time was spent on feeding, and walking and climbing, respectively (Challender et al., 2012).

Being macrosmatic the species searches for prey both on the ground and in the forest canopy using its acute olfactory senses. On locating an ant nest the powerful forelimbs are used to strike it and break it apart and the long salivary-coated tongue is protracted and withdrawn into the nest and chambers within, to lick up ants and their eggs (Fig. 6.4). Depending on nest size, individuals may end up inside the nest. This behavior is noisy, and the species can be detected in forest habitat by listening for sounds of individuals breaking into ant nests (Lim, 2007).

The species is an able swimmer and has been observed crossing rivers (see Fig. 6.5; E. Panjang, pers. obs.), swimming within flooded drains and canals in Singapore (H.C. Nash, pers. obs.) and swimming in the sea from one island to another (S. Cheema, unpubl. data). In captivity the species will readily use water pools if the ambient temperature is >30 °C, but will avoid them in winter; individuals either walk or slide into the pools from a tree branch. One individual in Vietnam repeatedly dropped into a pool from tree branches at a height of 2 m (Nguyen et al., 2014). The species will defecate in water pools (Nguyen et al., 2014).

FIGURE 6.4 A Sunda pangolin consuming prey from a rotting tree trunk in Sabah, Malaysia. *Photo credit: Shavez Cheema.*

When threatened individuals freeze. The scales provide a physical defensive barrier and may offer some form of camouflage (N. Lim, pers. obs.). The scales likely evolved to be strong and thick owing to the species' adaptation to an insectivorous diet and form of prey consumption, i.e., spending periods with the head positioned downward and inside ant nests and termitaria (see Stankowich and Campbell, 2016). Depending on the threat, the species will seek refuge in a tree or curl up into a defensive ball. If the latter, the glandular pad on the tip of the tail hooks onto a post-scapular scale, and Sunda pangolins are nearly impossible to uncurl (Lekagul and McNeely, 1988).

Vocalizations are limited to "hisses" or "puffs" but are not thought to play a role in communication.

Ontogeny and reproduction

Breeding is suspected to occur all year round (Lim and Ng, 2008a; Nguyen et al., 2014) and in captivity is aseasonal (Zhang et al., 2015). However, in northern parts of the range, where winters are colder and ant activity is reduced, breeding may be seasonal (SVW, unpubl. data). Little is known about mating behavior but it is assumed that males locate receptive females using olfaction. In captivity, mating occurs many times between a male and a female in estrus (Zhang et al., 2015). Gestation period is 168–188 days (approximately 6 months; Nguyen et al., 2014; Zhang et al., 2015), and one young is typically born at parturition; twins are reported but they are very rare (see Chapter 28). Before giving birth in captivity, females may appear anxious and

FIGURE 6.5 Sunda pangolins are capable swimmers, like this individual in Sabah, Malaysia. Individuals have been observed swimming between islands in the sea. *Photo credit: Shavez Cheema.*

during parturition will face the wall in a corner, and assume the defecating position. The weight is taken on the hindlimbs and tail, with the forelimbs held off the ground and the head lowered toward the belly as the abdominal muscles are contracted. Births can last 145–270 min (mean = 209, n = 4), and at parturition young weigh 110–140 g (Nguyen et al., 2014; Zhang et al., 2015; A. Kurniawan, unpubl. data).

Maternal care lasts for 3–4 months (Lim and Ng, 2008a; Nguyen et al., 2014). In Singapore, a female with young was documented to use three natal dens over this period, which comprised hollows in large trees (>50 cm DBH; Fig. 6.3A–C), suggesting that mature forests are important for reproduction in this species (Lim and Ng, 2008a). The mother will leave the young while embarking on foraging trips and by around 30 days old the young will begin exploring areas around the denning site (Lim and Ng, 2008a). In the second and third months, the juvenile will begin foraging with its mother, and at times will be transported around on the base of the mother's tail (see Fig. 6.6). Observations in captivity suggest that juveniles of 2–3 months old will readily leave the mother and forage independently (Nguyen et al., 2014). The suspected period of maternal care (3–4 months) is corroborated by observations of "wrestling" behavior in captivity between a female and a male offspring of weaning age, which was aggressive in nature (Challender et al., 2012).

There is little knowledge of age at sexual maturity. Based on an extrapolated growth rate of 10.44 g/day (mean, n = 3), Sunda pangolins may reach a weight of 3–5 kg in the first year (Zhang et al., 2015) but growth rates require further research. In males, sexual maturity may occur by the time they are 1.5 years old; viable sperm have been collected from captive males of this age that weighed 4.6 kg

FIGURE 6.6 A female Sunda pangolin with a male offspring approaching weaning age in Vietnam. *Photo credit: Daniel W.S. Challender/Save Vietnam's Wildlife.*

(A. Kurniawan, unpubl. data). Females may reach sexual maturity by 1 year old; in captivity a female was confirmed to be pregnant at 6–7 months old (and weighed 1.75 kg; Zhang et al., 2015), but this may not mimic reproductive behavior in the wild. Ageing pangolins is difficult but Lim (2007) and Yang et al. (2010) considered individuals of <3 kg in weight to be juvenile and those >3 kg to be adults, the latter also based on characteristics including wear to the scales and claws. Yang et al. (2010) categorized individuals weighing 1.5–3 kg to be sub-adult, but age categories and composition, and more broadly demographics, are poorly understood.

Population

There is a lack of quantitative data on populations in most range states, but available evidence suggests that the species is declining in many parts of its range. An exception is Singapore where in 2019 the population was estimated at 1046 individuals (see Chapter 26). Major factors affecting abundance appear to be availability and quantity of suitable prey species, the availability of large (>50 cm DBH) trees or other structures providing suitable denning sites for breeding females (Lim and Ng, 2008a), and predation and exploitation rates.

There are no recent data from Myanmar, but the species has ostensibly been extirpated from lowland areas due to agricultural expansion and hunting (Challender et al., 2019). Seizure data suggest that illegal trade to China from Myanmar involves this species (Nijman et al., 2016) and populations are under threat (see *Threats*). In Thailand, the species has similarly been largely eradicated from lowland areas and is considered rare (see Anon, 1999;

Lekagul and McNeely, 1988). There are records from a number of sites (see *Distribution*) but encounter rates are low; ZSL (2017) recorded detection rates (no. of events/no. of trap nights) of <0.01 at Khlong Nahka Wildlife Sanctuary and the sWEFCOM, and Kitamura et al. (2010) recorded five encounters from 11,106 camera trap days over 3 years at Hala-Bala Wildlife Sanctuary. Based on available evidence, the species is extremely rare in northern parts of its range in mainland Southeast Asia where there have been huge declines historically, especially in Lao PDR and Vietnam. In the former, local communities in three areas independently inferred that populations had declined by up to 99% between the 1960s and 1990s because of overexploitation (Duckworth et al., 1999; see Nooren and Claridge, 2001). In Vietnam, local hunters reported severe declines in the 1990s and 2000s for the same reason, and that the species is extremely rare (Newton et al., 2008). There are corroborating reports from southern Vietnam (e.g., MacMillan and Nguyen, 2014; Nuwer and Bell, 2014). In Cambodia, populations are suspected to be declining and interviews with local hunters suggest that the species has been extirpated from multiple protected areas due to over-hunting (A. Olsson, unpubl. data). The species has been recorded at a number of sites post-2012 (see *Distribution*), but encounter rates (encounters per 100 camera trap days) were low, comprising 0.1, <0.1, and 0.09 at Southern Cardamom and Botum Sakor National Parks, and Peam Krasop Wildlife Sanctuary respectively (Gray et al., 2017a; Thaung et al., 2017).

Variously described as common and uncommon historically in Malaysia (see Davies and Payne, 1982; Ickes and Thomas, 2003; Payne et al., 1985), evidence indicates that populations are declining in places. Interviews with hunters and villagers in Kelantan, Pahang, and Terengganu in 2007 and 2011 indicated that populations declined during the 1980s and 1990s due to hunting pressure for trade (D. Challender, unpubl. data; J.L. Chong, unpubl. data); this reportedly caused the commercial extinction of the species in some localities (see Chapter 16). There is little data for Sabah and Sarawak but populations are under heavy exploitation pressure (see *Threats*) and encounter rates are low. In Mt Penrissen, Sarawak, the species was recorded 32 times from 7382 camera trap days over 24 months (Kaicheen and Mohd-Azlan, 2018), and Mohd-Azlan et al. (2018) estimated naïve occupancy in Kubah National Park of 0.1 based on two encounters from 2161 camera trap days over seven months. There is little data for Indonesia, but the magnitude of illegal trade originating from the country suggests that the species occurs in some number but is under threat (see *Threats*): the species is very rare on Bali (H.C. Nash, unpubl. data).

Status

The Sunda pangolin is listed as Critically Endangered on The IUCN Red List of Threatened Species (Challender et al., 2019) due to past, ongoing and inferred future population declines because of overexploitation. In Malaysia and Singapore the species is categorized as Critically Endangered, and it has been assessed as Endangered in Vietnam. In Cambodia the species is considered "rare." National legislation affords the species protection in all range states, typically meaning that exploitation for personal and commercial purposes is prohibited. In Brunei Darussalam the species is not a protected species *per se* but does receive protection under national wildlife and forestry legislation. In 2016, the species was included in CITES Appendix I. Exceptions to prohibitions on exploitation include the rights of Indigenous peoples to hunt and use the species under customary laws e.g., in Malaysia. In Kelantan and Terengganu, the Orang Asli will burn dried pangolin scales to keep away species including Asian elephants (*Elephas maximus*; J.L. Chong, unpubl. data) while in Malaysian Borneo pangolin meat is consumed by Indigenous groups, and the Kadazan in Sabah use scales as protection from crocodiles (J.L. Chong, unpubl. data). Local taboos afford the species protection in places. Orang Asli around the Krau Wildlife Sanctuary, Pahang, believe pangolins to be an incarnation of the human placenta and will not consume the species (Hafiz and Chong, 2016).

Threats

The principal threat is overexploitation from unsustainable hunting and poaching for both local and international use, and the species is subject to targeted and untargeted exploitation across its range. Sunda pangolins have been consumed for subsistence purposes (i.e., as a source of protein) by indigenous peoples and local communities for centuries (Corlett, 2007), with assumed curative and health benefits (see Anon, 1999), and the scales in particular are used for a variety of applications. In parts of Malaysia it is believed that the scales cure asthma and provide protection from witchcraft (Anon, 1999; Hoi-Sen, 1977) and in Indonesia that they afford protection from harmful magic; they are also attached to fence posts to deter pests such as deer from damaging crops (Puri, 2005; see also Chapter 14). A range of hunting methods are used including snares, nets and dogs, and pangolins are caught opportunistically (Gray et al., 2017b; Newton et al., 2008), including by "hobbyists" who are not reliant on hunting for sustenance or as a primary means of income (Pantel and Anak, 2010). Quantifying the extent of local use and the threat it poses is challenging across the species' wide geographic range, but it is pervasive (Newton et al., 2008; Nijman et al., 2016; Pantel and Chin, 2009). Extricating local from international use is also difficult but evidence indicates that in many places local consumption is foregone in favor of selling the animals into illegal trade due to the high prices they fetch (MacMillan and Nguyen, 2014; G. Semiadi, unpubl. data). In some parts of Southeast Asia, selling one pangolin can provide the equivalent of several months' income (D. Challender, unpubl. data).

Commercial, international trade involving the species occurred throughout the early and mid-20th century, mainly comprising skins and scales, and involved an estimated tens of thousands of individuals annually (see Chapter 16). This continued throughout the 1970s–1990s and the species accounted for the vast majority of trade reported to CITES in this period, with the source of commercial harvesting appearing to shift to the southern portion of the range (Chapter 16). Notably, there was a parallel illicit trade, potentially involving more than 200% of trade in the species (by estimated number of individuals) reported to CITES up to the year 2000, mainly involving live and dead animals and scales (Challender et al., 2015; see Chapter 16).

Despite international policy decisions in CITES (see Chapter 19) illegal trade continued post-2000 and between 2001 and 2019 the Sunda pangolin was the most frequently seized pangolin species globally (D. Challender, unpubl. data), implicating virtually all range states (see Chapter 16). Challender et al. (2015) estimated that potentially over 200,000 Sunda pangolins were trafficked between 2000 and 2013, most of which were destined to China and Vietnam. Evidence suggests that the species continues to be targeted for illicit international trade, especially in the southern parts of its range. Between 2007 and 2009 a criminal syndicate illegally exported >22,000 individuals from Sabah, Malaysia (Pantel and Anak, 2010), and in February 2019 authorities in Sabah seized 30 tonnes of Sunda pangolins involving a combination of live and dead animals and quantities of scales (Anon, 2019). The species is also targeted in northern parts of its range (e.g.,

Myanmar; see Nijman et al., 2016), and expansion of agro-industrial plantations (e.g., rubber and oil palm) and roads, including in protected areas, is increasing accessibility (see Clements et al., 2014). Hunters and traders are now very probably targeting populations that were previously insulated from hunting pressure, which may partly explain why large quantities of individuals are still in illicit trade, several decades after commercial exploitation began. The drivers of trafficking include incentives along international supply chains (see discussion in Chapter 16) and the threat from overexploitation is compounded by poorly resourced and ineffective law enforcement in range countries (Challender and Waterman, 2017), and weak evidence that demand reduction efforts are effective (Veríssimo and Wan, 2019). Brunei Darussalam, Singapore and potentially smaller islands within the species' range appear to be the only places where overexploitation is not a major threat.

Habitat loss is an indirect threat where it results in the opening up of previously inaccessible areas, making the species more susceptible to hunting and poaching. The same applies to mining, the creation of hydropower dams and logging concessions, and there is an increased risk of roadkill and feral dog attacks. Roadkill is the main threat in Singapore (see Chapter 26). There is a need for comparative ecological research to understand densities in artificial and modified landscapes (e.g., oil palm plantations) and natural habitats (Davies and Payne, 1982). This should encompass how the species uses these habitats, including foraging behavior and the use of den sites, and the impact of monoculture crop re-planting (~25 years in the case of palm oil; Woittiez et al., 2017).

References

Anon, 1999. Review of Significant Trade in Animal Species included in CITES Appendix II, Detailed Review of 37 species. *Manis javanica*. World Conservation Monitoring Centre, IUCN Species Survival Commission and TRAFFIC, Cambridge, UK.

Anon, 2019. Malaysia makes record 30-tonne seizure. Available from: <https://phys.org/news/2019-02-malaysia-tonne-pangolin-seizure.html>. [July 15, 2019].

Aung, S.S., Sitwe, N.M., Frechette, J., Grindley, M., Connette, G., 2017. Surveys in southern Myanmar indicate global importance for tigers and biodiversity. Oryx 51 (1), 13.

Azhar, B., Lindenmayer, D., Wood, J., Fischer, J., Manning, A., McElhinny, C., et al., 2013. Contribution of illegal hunting, culling of pest species, road accidents and feral dogs to biodiversity loss in established oil-palm landscapes. Wildlife Res. 40 (1), 1–9.

Beebe, C.W., 1919. The Pangolin or Scaly Anteater. Zool. Soc. Bull. XVII (5), 1141–1145.

Blyth, E., 1842. The Journal of the Asiatic Society of Bengal V, XI, 444–470.

Blyth, E., 1847. Report of the Curator, Zoological Department. The Journal of the Asiatic Society of Bengal V, XVI pt. 2, 1271–1276.

Bourret, R., 1942. Les mammifères de la collection du Laboratoire de Zoologie de l'Ecole Supérieure des Sciences. Notes et Travaux de l'Ecole Supérieure, Université Indochinoise No. 1.

Challender, D.W.S., Nguyen, V.T., Jones, M., May, L., 2012. Time-budgets and activity patterns of captive Sunda pangolins (*Manis javanica*). Zoo Biol. 31 (2), 206–218.

Challender, D.W.S., Harrop, S.R., MacMillan, D.C., 2015. Understanding markets to conserve trade-threatened species in CITES. Biol. Conserv. 187, 249–259.

Challender, D., Waterman, C., 2017. Implementation of CITES Decisions 17.239 b) and 17.240 on Pangolins (*Manis* spp.), CITES SC69 Doc. 57 Annex. Available from <https://cites.org/sites/default/files/eng/com/sc/69/E-SC69-57-A.pdf>. [April 3, 2018].

Challender, D., Willcox, D.H.A., Panjang, E., Lim, N., Nash, H., Heinrich, S., Chong, J., 2019. *Manis javanica. The IUCN Red List of Threatened Species* 2019: e.T12763A123584856. Available from: <http://dx.doi.org/10.2305/IUCN.UK.2019-3.RLTS.T12763A123584856.en>.

Chan, L.-K., 1995. Extrinsic lingual musculature of two pangolins (Pholidota: Manidae). J. Mammal. 76 (2), 472–480.

Chong, J.L., Sulaiman, M.H., Marina, H., 2016. Conservation of the Sunda pangolin (*Manis javanica*) in Peninsular Malaysia: important findings and conclusions. Malayan Nat. J. 68 (4), 161–171.

Clements, G.R., Lynam, A.J., Gaveau, D., Yap, W.L., Lhota, S., Goosem, M., et al., 2014. Where and how are roads endangering mammals in Southeast Asia's forests? PLoS One 9 (12), e115376.

Corbet, G.B., Hill, J.E., 1992. The Mammals of the Indomalayan Region: A Systematic Review. Oxford University Press, Oxford.

Corlett, R., 2007. The Impact of Hunting on the Mammalian Fauna of Tropical Asian Forests. Biotropica 30 (3), 292–303.

Coudrat, C., 2017. Report on Camera Trap Survey in Nakai-Nam Theun National Protected Area. Project Analouk, Lao PDR. Unpublished report.

Davies, G., Payne, J., 1982. A Faunal Survey of Sabah. WWF Malaysia, Kuala Lumpur.

Desai, A.A., Lic, V., 1996. Status and Distribution of Large Mammals in Eastern Cambodia: Results of the First Foot Surveys in Mondulkiri and Rattankiri Provinces. IUCN, FFI, WWF Large Mammal Conservation Project, Phnom Penh, Cambodia.

Desmarest, M.A.G., 1822. Mammalogie ou Description des Espèces de Mammifères. Second partie, contenant Les Ordres des Rongeurs, des Édentatés, des Pachydermes, des Ruminans et des Cétacés. Chez Mme Veuve Agasse, Imprimeur-Libraire, Paris.

Duckworth, J.W., Salter, R.E., Khounboline, K., 1999. Wildlife in Lao PDR: 1999 Status Report. IUCN, Wildlife Conservation Society, Centre for Protected Areas and Watershed Management, Vientiane, Lao PDR.

Ellerman, J.R., Morrison-Scott, T.C.S., 1966. Checklist of Palaearctic and Indian Mammals 1758 to 1946, second ed. British Museum, London.

Else, J.G., Colley, F.C., 1976. *Eimeria tenggilinggi* sp. n. from the scaly anteater *Manis javanica* Desmarest in Malaysia. J. Eukaryotic Microbiol. 23 (4), 587-488.

Feiler, A., 1998. Das Philippinen-Schuppentier, *Manis culionensis* Elera, 1915, eine fast vergessene Art (Mammalia: Pholidota: Manidae). Zoologische Abhandlungen - Staatliches Museum Für Tierkunde Dresden 50, 161–164.

Fitzinger, L.J., 1872. Die naturliche familie der schuppenthiere (Manes). Sitzungsberichte der Kaiserlichen Akademie der Wissenschaften. Mathematisch-Naturwissenschaftliche Classe, CI., LXV, Abth. I, 9–83.

Fletcher, L., 2016. Developing a strategy for pangolin conservation in Brunei: Refining guidelines for the release of confiscated animals and gathering baseline data. Unpublished report for 1st Stop Brunei, Brunei Darussalam.

Focillon, A.D., 1850. Du genre Pangolin (*Manis Linn.*) et de deux nouvelles espèces de ce genre. Revue et Mag. de Zool. 2 Sér., Tome VII, 465–474, 513–534.

Frechkop, S., 1931. Notes sur les mammifères. VI. Quelques observations sur la classification des pangolins (Manidae). Bulletin du Musee royal d'Histoire naturelle de Belgique VII (22), 1–14.

Gaubert, P., 2011. Family Manidae. In: Wilson, D.E., Mittermeier, R.A. (Eds.), Handbook of the Mammals of the World, vol. 2. Hoofed Mammals. Lynx Edicions, Barcelona, pp. 82–103.

Gaubert, P., Antunes, A., 2005. Assessing the taxonomic status of the Palawan pangolin *Manis culionensis* (Pholidota) using discrete morphological characters. J. Mammal. 86 (6), 1068–1074.

Gaubert, P., Antunes, A., Meng, H., Miao, L., Peigne, S., Justy, F., et al., 2018. The complete phylogeny of pangolins: scaling up resources for the molecular tracing of the most trafficked mammals on earth. J. Hered. 109 (4), 347–359.

Gaudin, T.J., Emry, R.J., Wible, J.R., 2009. The phylogeny of living and extinct pangolins (Mammalia, Pholidota) and associated taxa: a morphology based analysis. J. Mammal. Evol. 16 (4), 235–305.

Grassman Jr., L.I., Tewes, M.E., Silvy, N.J., Kreetiyutanont, K., 2005. Ecology of three sympatric felids in a mixed evergreen forest in north-central Thailand. J. Mammal. 86 (1), 29–38.

Gray, J.E., 1865. 4. Revision of the genera and species of entomophagous Edentata, founded on the examination of the specimens in the British Museum. Proc. Zool. Soc. Lond. 33 (1), 359–386.

Gray, T.N.E., Billingsley, A., Crudge, B., Frechette, J.L., Grosu, R., Herranz-Muñoz, V., et al., 2017a. Status and conservation significance of ground-dwelling mammals in the Cardamom Rainforest Landscape, southwestern Cambodia. Cambodian J. Nat. Hist. 2017 (1), 38–48.

Gray, T.N.E., Marx, N., Khem, V., Lague, D., Nijman, V., Gauntlett, S., 2017b. Holistic management of live animals confiscated from illegal wildlife trade. J. Appl. Ecol. 54 (3), 726–730.

Hafiz, M.S., Marina, H., Afzan, A.W., Chong, J.L., 2012. Ectoparasite from confiscated Malayan pangolin (*Manis javanica* Desmarest) in peninsular Malaysia. UMT 11th International Annual Symposium on Sustainability Science and Management, Terengganu, Malaysia, 9–11 July.

Hafiz, S., Chong, J.L., 2016. Tenggiling Sunda Khazanah Alam Malaysia. Penerbit UMT, Kuala Terengganu.

Harrison, J.L., 1961. The natural food of some Malayan mammals. Bull. Singapore Natl. Museum 30, 5–18.

Hassanin, A., Hugot, J.-P., van Vuuren, J.B., 2015. Comparison of mitochondrial genome sequences of pangolins (Mammalia, Pholidota). C. R. Biol. 338 (4), 260–265.

Heath, M.E., 1992. *Manis pentadactyla*. Mammal. Sp. 414, 1–6.

Hedges, L., Aziz, S.A., 2013. A novel interaction between a sun bear and a pangolin in the wild. Int. Bear News 22 (1), 31–32.

Hoi-Sen, Y., 1977. Scaly Anteater. Nat. Malays. 2 (4), 26–31.

Hua, L., Gong, S., Wang, F., Li, W., Ge, Y., Li, X., et al., 2015. Captive breeding of pangolins: current status, problems and future prospects. ZooKeys 507, 99–114.

Ickes, K., Thomas, S.C., 2003. Native, wild pigs (*Sus scrofa*) at Pasoh and their impacts on the plant community. In: Okuda, T., Manokaran, N., Matsumoto, Y., Niiyama, K., Thomas, S.C., Ashton, P.S. (Eds.), Pasoh: Ecology and Natural History of a Southeast Asian Lowland Tropical Rain Forest. Springer, Japan.

Imai, M., Shibata, T., Mineda, T., Suga, Y., Onouchi, T., 1973. Histological and histochemical investigations on the stomach in man, Japanese monkey (*Macaca fuscata yakui*) and some other kinds of animals. Report V. On the stomach of the pangolin (*Manis pentadactyla* Linne). Okajimas Fol. Anat. Jap. 49, 433–454.

Jammah, O., Faizal, H., Chandrawathani, P., Premaalatha, B., Erwanas, A.I., Lily, R., et al., 2014. Eperythrozoonosis (*Mycoplasma* sp.) in Malaysian pangolin. Malays. J. Vet. Res. 5 (1), 65–69.

Jenks, K.E., Songasen, N., Leimgruber P., 2012. Camera trap records of dholes in Khao Khang Rue Nai Wildlife Sanctuary, Thailand. Canid news, 5, 1–5.

Jentink, F.A., 1882. Note XXV. Revision of the Manidae in the Leyden Museum. Notes from the Leyden Museum IV, 193–209.

Kaicheen, S.S., Mohd-Azlan, J., 2018. Camera trapping wildlife on Mount Penrissen area in Western Sarawak. Malays. Appl. Biol. 47 (1), 7–14.

Kawanishi, K., Sunquist, M.E., 2004. Conservation status of tigers in a primary rainforest of Peninsular Malaysia. Biol. Conserv. 120 (3), 329–344.

Ketol, B., Anwarali, F.A., Marni, W., Sait, I., Lakim, M., Yanmun, P.I., et al., 2009. Checklist of mammals from Gunung Silam, Sabah, Malaysia. J. Trop. Biol. Conserv. 5, 61–65.

Khan, M.A.R., 1985. Mammals of Bangladesh. A field guide: Nazma Reza, Dhaka.

Khwaja, H., Buchan, C., Wearn, O.R., Bahaa-el-din, L., Bantlin, D., Bernard, H., et al., 2019. Pangolins in global camera trap data: implications for ecological monitoring. Glob. Ecol. Conserv. 20, e00769.

Kitamura, S., Thon-Aree, S., Madsri, S., Poonswad, P., 2010. Mammal diversity and conservation in a small isolated forest of southern Thailand. Raffles Bull. Zool. 58 (1), 145–156.

Krause, W.J., Leeson, C.R., 1974. The stomach of the pangolin (*Manis pendactyla*) with emphasis on the pyloric teeth. Acta Anat. 88, 1–10.

Laing, A.B.G., Edeson, J.F.B., Wharton, R.H., 1960. Studies on filariasis in Malaya: the vertebrate hosts of *Brugia malayi* and *Brugia pahangi*. Ann. Trop. Med. Parasitol. 53 (4), 92–99.

Le, C.X., Truong, L.V., Dang, D.T., Ho, C.T., Ngo, D.A., Nguyen, N.C., et al., 1997a. A report of field surveys on biodiversity in Phong Nha Ke Bang forest (Quang Binh Province, central Vietnam). IEBR, FIPI, Forestry College, University of Vinh, WWF Indochina Programme, Hanoi, Vietnam.

Le, C.X., Pham, T.A., Duckworth, J.W., Vu, N.T., Lic, V., 1997b. A survey of large mammals in Dak Lak Province, Viet Nam. Unpublished report to IUCN and WWF. Hanoi, Viet Nam.

Lekagul, B., McNeely, J.A., 1988. Mammals of Thailand, second ed. Darnsutha Press, Bangkok.

Lim, N.T.-L., 2007. Autoecology of the Sunda Pangolin (*Manis javanica*) Singapore. M.Sc. Thesis, National University of Singapore, Singapore.

Lim, N.T.-L., Ng, P.K.L., 2008a. Home range, activity cycle and natal den usage of a female Sunda pangolin *Manis javanica* (Mammalia: Pholidota) in Singapore. Endanger. Sp. Res. 4, 233–240.

Lim, N.T.-L., Ng, P.K.L., 2008b. Predation on *Manis javanica* by *Python reticulatus* in Singapore. Hamadryad 32 (1), 62–65.

Lin, M.F., Chang, C.-Y., Yang, C.W., Dierenfeld, E.S., 2015. Aspects of digestive anatomy, feed intake and digestion in the Chinese pangolin (*Manis pentadactyla*) at Taipei zoo. Zoo Biol. 34 (3), 262–270.

Ludeking, E.W.A., 1862. Natuur- en Geneeskundige Topographische der Schets der Residentie Agam. In: Wassink, G. (Ed.), Geneeskundig Tijdschrift voor Nederlandsch Indie, Uitgegeven door de Vereeniging Tot Bevordering der Geneeskundige Wetenschappen in Nederlandsch Indie. Lange and Co, Batavia, pp. 1–153.

Lyon Jr., M.W., 1909. Additional notes on mammals of the Rhio Lingga archipelago, with descriptions of new species and a revised list. Proc. US Natl. Museum XXXVI 1684, 479–493.

McCann, G., Pawlowski, K., 2017. Small carnivores' records from Virachey National Park, north-east Cambodia. Small Carnivore Conserv. 55, 26–41.

MacMillan, D.C., Nguyen, Q.A., 2014. Factors influencing the illegal harvest of wildlife by trapping and snaring among the Katu ethnic group in Vietnam. Oryx 48 (2), 304–312.

McNab, B.K., 1984. Physiological convergence amongst ant-eating and termite-eating mammals. J. Zool. Soc. Lond. 203 (4), 485–510.

Mohd-Azlan, J., Engkamat, L., 2013. Camera trapping and conservation in Lanjak Entimau wildlife sanctuary, Sarawak, Borneo. Raffles Bull. Zool. 61 (1), 397–405.

Mohd-Azlan, J., Kaicheen, S.S., Yoong, W.C., 2018. Distribution, relative abundance and occupancy of selected mammals along paved road in Kubah National Park, Sarawak, Borneo. Nat. Conserv. Res. 3 (2), 36–46.

Mohapatra, R.K., Panda, S., Nair, M.V., Acharjyo, L.N., 2016. Check list of parasites and bacteria recorded from pangolins (*Manis* sp.). J. Parasit. Dis. 40 (4), 1109–1115.

Mohr, E., 1961. Schuppentiere. Neue Brehm-Bucherei. A. Ziemsen Verlag, Wittenberg Lutherstadt.

Moo, S.S.B., Froese, G.Z.L., Gray, T.N.E., 2017. First structured camera-trap surveys in Karen State, Myanmar, reveal high diversity of globally threatened mammals. Oryx 52 (3), 1–7.

Nash, H.C., Wirdateti, Low, G., Choo, S.W., Chong, J.L., Semiadi, G., et al., 2018. Conservation genomics reveals possible illegal trade routes and admixture across pangolin lineages in Southeast Asia. Conserv. Genet. 19 (5), 1083–1095.

Newton, P., Nguyen, V.T., Roberton, S., Bell, D., 2008. Pangolins in peril: using local hunters' knowledge to conserve elusive species in Vietnam. Endanger. Sp. Res. 6, 41–53.

Nguyen, V.T., Clark, V.L., Tran, Q.P., 2014. Sunda Pangolin (*Manis javanica*) Husbandry Guidelines. Carnivore and Pangolin Conservation Program – Save Vietnam's Wildlife, Vietnam.

Nijman, V., Zhang, M.X., Shepherd, C.R., 2016. Pangolin trade in the Mong La Wildlife market and the role of Myanmar in the smuggling of pangolins into China. Glob. Ecol. Conserv. 5, 118–126.

Nisa, C., Agungpriyono, S., Kitamura, N., Sasaki, M., Yamada, J., Sigit, K., 2010. Morphological features of the stomach of Malayan pangolin, *Manis javanica*. Anat. Histol. Embryol. 39 (5), 432–439.

Nooren, H., Claridge, G., 2001. Wildlife trade in Laos: the End of the Game. Netherlands Committee for IUCN, Amsterdam.

Nowak, R.M., 1999. Walker's Mammals of the World. Johns Hopkins University Press, Baltimore.

Numata, S., Okuda, T., Sugimoto, T., Nishimura, S., Yoshida, K., Quah, E.S., et al., 2005. Camera trapping: a non-invasive approach as an additional tool in the study of mammals in Pasoh Forest Reserve and adjacent fragmented areas in peninsular Malaysia. Malay. Nat. J. 57 (1), 29–45.

Nuwer, R., Bell, D., 2014. Identifying and quantifying the threats to biodiversity in the U Minh peat swamp forests of the Mekong Delta, Vietnam. Oryx 48 (1), 88–94.

Pantel, S., Chin, S.-Y., (Eds.), 2009. Proceedings of the Workshop on Trade and Conservation of Pangolins Native to South and Southeast Asia, 30 June–2 July 2008, Singapore Zoo, Singapore. TRAFFIC Southeast Asia, Petaling Jaya, Selangor, Malaysia.

Pantel, S., Anak, A.N., 2010. A preliminary assessment of Sunda pangolin trade in Sabah. TRAFFIC Southeast Asia, Petaling Jaya, Selangor, Malaysia.

Payne, J., Francis, C.M., Phillipps, K., 1985. A Field Guide to the Mammals of Borneo. The Sabah Society and WWF Malaysia, Kota Kinabalu and Kuala Lumpur.

Payne, J., Francis, C.M., 2007. A Field Guide to the Mammals of Borneo. The Sabah Society, Kota Kinabalu.

Phillips, Q., Phillips, K., 2018. Phillips's Field Guide to the Mammals of Borneo and their Ecology: Sabah, Sarawak, Brunei and Kalimantan. John Beaufoy Books, Oxford.

Pocock, R.I., 1924. The external characters of the pangolins (Manidae). Proc. Zool. Soc. Lond. 94 (3), 707–723.

Pongchairerk, U., Kasorndorkbua, C., Pongket, P., Liumsiricharoen, M., 2008. Comparative histology of the Malayan Pangolin kidneys in normal and dehydrated condition. Kasetsart J. (Nat. Sci.) 42, 83–87.

Puri, P.K., 2005. Deadly dances in the Bornean rainforest: hunting knowledge of the Penan Benalui. Royal Netherlands Institute of Southeast Asian and Caribbean Studies Monograph Series. KITLV Press, Leiden.

Shine, R., Harlow, P.S., Keogh Boeadi, J.S., 1998. The influence of sex and body size on food habits of a giant tropical snake, *Python reticulatus*. Funct. Ecol. 12, 248–258.

Stankowich, T., Campbell, L.A., 2016. Living in the danger zone: exposure to predators and the evolution of spines and body armor in mammals. Evolution 70 (7), 1501–1511.

Sulaiman, M.H., Azmi, W.A., Hassan, M., Chong, J.L., 2017. Current updates on the morphological measurements of the Malayan pangolin (*Manis javanica*). Folia Zool. 66 (4), 262–266.

Sundevall, C.J., 1842. Om slägtet Sorex, med nâgra nya arters beskrifning. Kungliga Vetenskapsakademien, Stockholm.

Suzuki, A., Thong, S., Tan, S., Iwata, A., 2017. Camera trapping of large mammals in Chhep Wildlife Sanctuary, northern Cambodia. Cambodian J. Nat. Hist. 1, 63–75.

Tarmizi, M.R., Sipangkul, S., 2019. Assisted reproduction technology: Anaesthesia and sperm morphology of the Sunda pangolin (*Manis javanica*). My Wildlife Vets 2019 (2), 5–6.

Thaung, R., Muñoz, V.H., Holden, J., Willcox, D., Souter, N.J., 2017. The vulnerable fishing cat *Prionailurus viverrinus* and other globally threatened species in Cambodia's coastal mangroves. Oryx 52 (4), 636–640.

Timmins, R.J., Cuong, T.V., 1999. An Assessment of the Conservation Importance of the Huong Son (Annamite) Forest, Ha Tinh Province, Vietnam, Based on the Results of a Field Survey for Large Mammals and Birds. Center for Biodiversity and Conservation and American Museum of Natural History, New York, USA.

Trageser, S.J., Ghose, A., Faisal, M., Mro, P., Mro, P., Rahman, S.C., 2017. Pangolin distribution and conservation status in Bangladesh. PLoS One 12 (4), e0175450.

Ullmann, T., Veríssimo, D., Challender, D.W.S., 2019. Evaluating the application of scale frequency to estimate the size of pangolin scale seizures. Glob. Ecol. Conserv. 20, e00776.

van Strien, N.J., 1983. Guide to the tracks of mammals of Western Indonesia. School of Environmental Conservation Management, Ciawi, School of Environmental Conservation Management, Indonesia.

Veríssimo, D., Wan, A.K.Y., 2019. Characterizing efforts to reduce consumer demand for wildlife products. Conserv. Biol. 33 (3), 623–633.

Wearn, O.R., 2015. Mammalian Community Responses to a Gradient of Land-Use Intensity on the Island of Borneo. M.Sc. Thesis, Imperial College London, UK.

Willcox, D., Bull, R., Nhuan, N.V., Tran, Q.P., Nguyen, V. T., 2017. Small carnivore records from the U Minh Wetlands, Vietnam. Small Carnivore Conserv. 55, 4–25.

Woittiez, L.S., van Wijk, M.T., Slingerland, M., van Noordwijk, M., Giller, K.E., 2017. Yield gaps in oil palm: a quantitative review of contributing factors. Eur. J. Agron. 83, 57–77.

Wu, S.B., Wang, Y.X., Feng, Q., 2005. A new record of mammalia in China—*Manis javanica*. Acta Zootaxonom. Sin. 30 (2), 440–443. [In Chinese].

Yang, L., Su, C., Zhang, F.-H., Wu, S.-B., Ma, G.-Z., 2010. Age Structure and Parasites of Malayan Pangolin (*Manis javanica*). J. Econ. Anim. 14 (1), 22–25. [In Chinese].

Zhang, F., Wu, S., Yang, L., Zhang, L., Sun, R., Li, S.S., 2015. Reproductive parameters of the Sunda pangolin, *Manis javanica*. Folia Zool. 64 (2), 129–135.

ZSL, 2017. Sunda Pangolin Monitoring Protocol - Thailand V.1.0. Zoological Society of London, UK.

CHAPTER

7

Philippine pangolin *Manis culionensis* (de Elera, 1915)

Sabine Schoppe[1,2], *Lydia K.D. Katsis*[2], *Dexter Alvarado*[1] *and Levita Acosta-Lagrada*[2,3]

[1]Katala Foundation Inc., El Rancho, Puerto Princesa City, Philippines [2]IUCN SSC Pangolin Specialist Group, ℅ Zoological Society of London, Regent's Park, London, United Kingdom [3]Palawan Council for Sustainable Development Staff (PCSDS), Puerto Princesa City, Philippines

OUTLINE

Taxonomy

Previously included in the genus *Pholidotus* (de Elera, 1915) and subgenus *Paramanis* (Schlitter, 2005), the species is here included in the genus *Manis* based on morphological and genetic evidence (Gaubert et al., 2018; Chapter 2). Despite early classification as a distinct species (de Elera, 1915; Lawrence, 1939; Sanborn, 1952), much of the subsequent literature did not distinguish this species from the Sunda pangolin (*Manis javanica*), or considered it to be a subspecies (Corbet and Hill, 1992; Heaney et al., 1998; Pocock, 1924; Taylor, 1934). Feiler (1998) proposed a series of morphological characteristics distinguishing the two species, but did not consider many of the observations made by Lawrence (1939).

Pangolins
DOI: https://doi.org/10.1016/B978-0-12-815507-3.00007-1

Gaubert and Antunes (2005) classified this species as distinct from the Sunda pangolin based on six discrete morphological characters. They include (1) the number of scale rows across the middle of the back (following a line perpendicular to the anteroposterior axis): 19–21 in the Philippine, and 15–18 in the Sunda pangolin (though see *Description*); (2) uniformly small scales in the nuchal, scapular, and postscapular regions in the Philippine pangolin, which are larger in the Sunda species; (3) the ratio of the nasal bone to total skull length, which is smaller in the Philippine (<1/3) and larger (>1/3) in the Sunda pangolin; (4) the posterior region of the palatine bone which is weaker in the Philippine pangolin, i.e., is not ventrally inflated and has short lateral walls: it is strong, ventrally inflated and has large lateral walls in the Sunda pangolin; (5) the posterior extension of the zygomatic process, which is short and not posterior to the sphenopalatine foramen in the Philippine, but long and posterior to the sphenopalatine foramen in the Sunda pangolin; and (6) the ratio of head and body to tail length, which is smaller in the Philippine (1.11 ± 0.03, mean $\pm$ SD, $n = 5$) compared to the Sunda pangolin (1.25 ± 0.13, $n = 20$; though see *Description*). Feiler (1998) and Lawrence (1939) suggested 15 additional morphological features to distinguish the two species, but Gaubert and Antunes (2005) did not support their utility. Delimitation of the Philippine from the Sunda pangolin was supported by genetic analyses (see Gaubert et al., 2018). The species is monotypic.

Records from archeological sites on Palawan Island date back 5000–7000 years (Gaubert, 2011). The split between the Philippine and Sunda pangolin is speculated to have occurred when Early Pleistocene land bridges between Borneo and Palawan were submerged by rising sea levels about 500,000–800,000 years ago (Gaubert and Antunes, 2005).

Etymology: *Manis* (see Chapter 4); the epithet *culionensis* references the island of Culion in the Philippines.

Description

The Philippine pangolin is a medium-sized mammal, typically weighing 4–7 kg, and reaches a total length of 100–130 cm (Table 7.1). The species appears to be sexually dimorphic, with males larger and heavier than females (Schoppe et al., in prep. a; Table 7.1). Morphologically, the species is very similar to the Sunda pangolin. The body is covered in keratinous, overlapping scales, which grow from the skin and are rounded. Exceptions include the ventral surface, inner sides of the limbs, and parts of the head and face which are pinkish in color and covered in dense white hair (Fig. 7.1). A small number of thick hairs or bristles grow from the base of the scales. The tail is approximately 90% of head-body length (Gaubert and Antunes, 2005) and is covered in scales both dorsally and ventrally. An exception is the tail tip on the ventral side, which lacks a median scale and is replaced by a cutaneous pad. There are 19–21 transversal scale rows on the body (Gaubert and Antunes, 2005), and 28–32 rows of scales on the tail margins (Table 7.1). Total scales number approximately 850–1000 (Table 7.1) and vary in size and shape. Scales on the dorsum are broad and rhomboid-shaped, and twice as wide as the first row of postscapular scales (Gaubert, 2011). On the limbs, scales are graduated and the hindlimbs have a central keel terminating in a sharp point: the scales on the tail margins are also sharply pointed. Like in other Asian pangolins, the medio-dorsal scale row continues to the tail tip (Jentink, 1882). Scale color varies from creamy-orange, dirty yellowish-white to dark brown (Fig. 7.1), and may be uniform or individuals may possess scales of variable color (Gaubert and Antunes, 2005), which is

TABLE 7.1 Philippine pangolin morphometrics.

Measurement			Country	Source(s)
Weight	Weight (♂)	4.9 (2.7–7.3) kg, n = 9	Philippines	S. Schoppe, unpubl. data
	Weight (♀)	3.2 (3–3.5) kg, n = 4	Philippines	S. Schoppe, unpubl. data
	Weight unsexed	3.6 (2.2–5.9) kg, n = 21	Philippines	A. Ponzo and S. Schoppe, unpubl. data
Body	Total length (♂)	1067 (840–1330) mm, n = 8	Philippines	S. Schoppe, unpubl. data
	Total length (♀)	961 (828–1030) mm, n = 4	Philippines	S. Schoppe, unpubl. data
	Head-body length (♂)	582 (450–740) mm, n = 8	Philippines	S. Schoppe, unpubl. data
	Head-body length (♀)	533 (470–563) mm, n = 4	Philippines	S. Schoppe, unpubl. data
	Tail length (♂)	486 (39–59) mm, n = 8	Philippines	S. Schoppe, unpubl. data
	Tail length (♀)	428 (358–470), n = 4	Philippines	S. Schoppe, unpubl. data
Vertebrae	Total number of vertebrae	No data		
	Cervical	No data		
	Thoracic	No data		
	Lumbar	No data		
	Sacral	No data		
	Caudal	29–30		Gaubert, 2011
Skull	Length (unsexed)	60–95 mm	Philippines	Gaubert, 2011
	Breadth across zygomatic processes	No data		
Scales	Total number of scales	940 (854–999), n = 3	Philippines	Ullmann et al., 2019
	No. of scale rows (transversal, body)	19–21, n = 9	Philippines	Gaubert and Antunes, 2005
	No. of scale rows (longitudinal, body)	No data		
	No. of scales on outer margins of tail	28–32, n = 4	Philippines	L. Katsis and D. Challender unpubl. data; T. Ullmann, unpubl. data
	No. of scales on median row of tail	28–32, n = 4	Philippines	L. Katsis and D. Challender unpubl. data; T. Ullmann, unpubl. data
	Scales (wet) as proportion of body weight	No data		
	Scales (dry) as proportion of body weight	No data		

FIGURE 7.1 Adult female Philippine pangolin: this species is semi-arboreal. *Photo credit: Dexter Alvarado.*

sometimes limited to scale margins and/or the tip (e.g., darker at the tip; L. Katsis, pers. obs.). Observations indicate that juveniles may possess several white-translucent scales on the distal part of the tail, which may darken with age (D. Alvarado, pers. obs.).

The head is conical in shape and largely naked, with the exception of small scales on the forehead, which extend anteriorly and terminate before the rhinarium. The face is pinkish-gray and the rhinarium is a darker pink-brown. Ear pinnae are present, but like in the Sunda pangolin comprise a subvertical ridge of integument posterior to the auditory orifice, and are less pronounced than in the Chinese pangolin (*M. pentadactyla*). The eyes are small with a dark iris and surrounded by thick swollen eyelids. As with all pangolins, this species lacks teeth and has a long tongue that retracts into a pouch in the throat. The oral cavity is small.

The species is pentadactylous with five digits on the fore- and hindfeet. Each digit terminates with a claw, the largest (and strongest) of which is the middle. The second and fourth claws are smaller but of approximately equal size. The first and fifth digits (and claws) are vestigial and essentially non-functional. Digit and claw length and disposition on the hindfeet approximates the forefeet. It is presumed that the species possesses a pair of anal glands in the circumanal area. The genetalia are similar to the Sunda pangolin; females have two pectoral mammae.

The Philippine pangolin is easily confused with the Sunda pangolin and the two species are hard to differentiate. Gaubert and Antunes (2005) posited that the species can be distinguished by a number of characteristics (see *Taxonomy*), including the number of transversal scale rows: 19–21 in the Philippine and 15–18 in the Sunda pangolin. However, other sources (Frechkop, 1931; Mohr, 1961; Wu et al., 2004; L. Katsis and D. Challender, unpubl. data) indicate that the Sunda pangolin possesses 15–19 transversal scale rows i.e., there is overlap between the species. Gaubert and Antunes (2005) also assert that the ratio of head-body to tail length is a diagnostic characteristic, but the ratio of individuals measured for this chapter (1.21 ± 0.06; mean $\pm$ SD, $n = 12$; see Table 7.1), more closely approximates that of the Sunda pangolin (see Gaubert and Antunes, 2005). Number of transversal scale rows (assuming they do not number 19), combined with the size of scales in the nuchal, scapular and post-scapular regions (see Gaubert and Antunes, 2005) may be used to differentiate the species. The number of scales on the margins of the tail, if they number 31–32, is also a diagnostic characteristic of the Philippine pangolin: the Sunda pangolin possesses up to 30 scale rows on the tail margins. These parameters are based on a small sample size ($n = 10$) for the Philippine pangolin, and further research could reveal greater variation in scale number, and potentially greater overlap with the Sunda pangolin. Dependent on the purpose of differentiation between the species, DNA-based methods may be appropriate (see Luczon et al., 2016).

The Philippine pangolin may also be confused with the Chinese pangolin. Distinguishing features of the former include smaller and less prominent ear pinnae, smaller and shorter claws on the forefeet, which are noticeably larger in the Chinese species, and the number of transversal scale rows and scales on the tail margins, which number 14–18 and 14–20 on the Chinese species respectively (see Chapter 4), and are more numerous in the Philippine pangolin (Table 7.1).

Distribution

Endemic to the Palawan faunal region in the Philippines, including Palawan Island, the Calamian Islands, and smaller surrounding islands (Fig. 7.2; Bourns and Worcester, 1894; de Elera, 1915; Everett, 1889; Heaney et al., 1998; Lawrence, 1939). Records from mainland Palawan include fossils found in the north of the island from the Terminal Pleistocene to Late Holocene (Lewis et al., 2008; Piper et al., 2011). The species occurs on Dumaran Island (Schoppe and Alvarado, 2016) and Batas Island in Taytay (Acosta and Schoppe, 2018; Schoppe et al., 2017) and Indigenous peoples indicate that the species occurs on islands off El Nido, including Lagen Island, Tagnipa, and Nagbilisong Islands in San Vicente (Acosta and Schoppe, 2018). The species has apparently been introduced to Apulit Island in El Nido, and possibly Lagen Island (S. Schoppe, unpubl. data).

Within the Calamian Islands, north of mainland Palawan, the species occurs on Culion (de Elera, 1915; Heaney et al., 1998; Hollister, 1913), Busuanga (Hoogstraal, 1951), and Calauit Islands (Alviola, 1998). Interviews with local communities conducted between 2010 and 2015 confirmed the species' presence on Busuanga, Culion, Calauit and Maglalambay (Paguntalan et al., 2010, 2012, 2015). Similarly, in interviews conducted in 2006 local people suggested that the species occurs on 12 islands dispersed across Busuanga, Calauit, and Culion, and at least nine smaller islands from the Calamian island group (Rico and Oliver, 2006). The species is apparently absent from Coron Island (Acosta and Schoppe, 2018; Rico and Oliver, 2006; Paguntalan et al., 2010, 2012, 2015) but is distributed in the municipality of Coron on Busuanga Island.

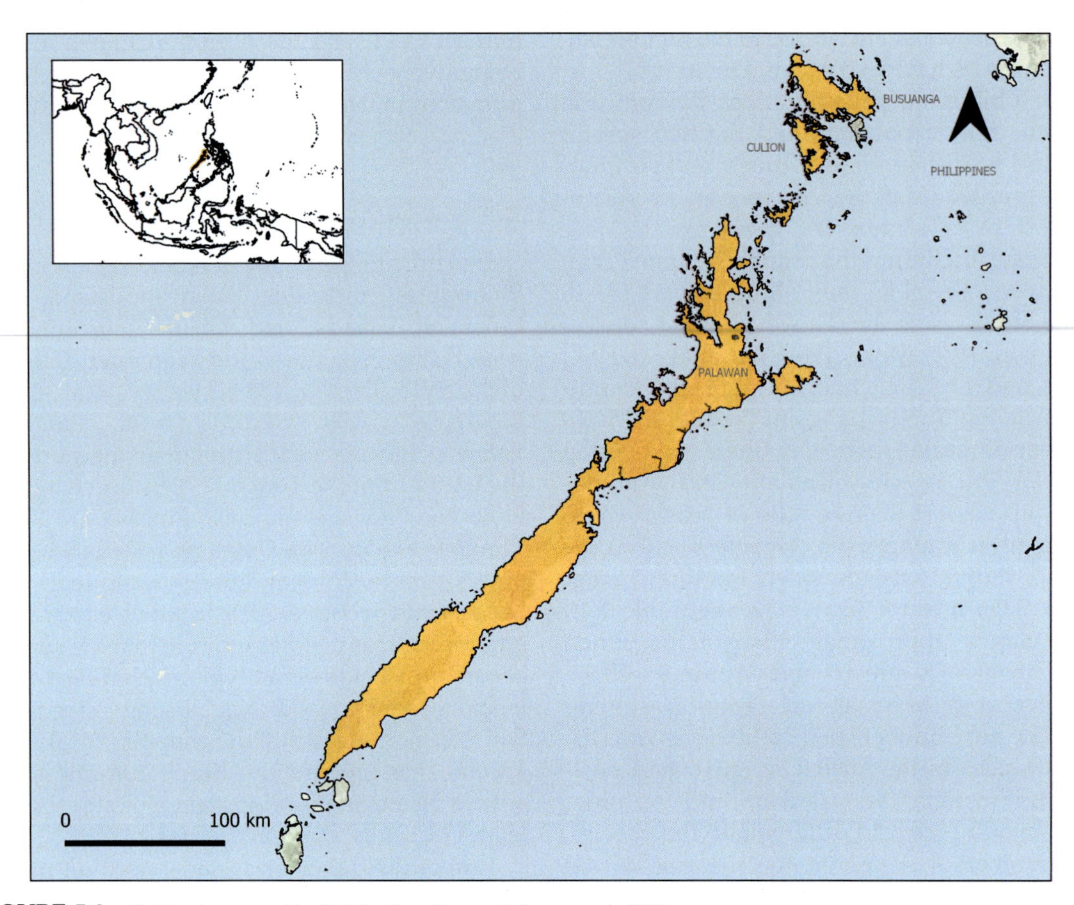

FIGURE 7.2 Philippine pangolin distribution. *Source: Schoppe et al., 2019.*

There is uncertainty over distribution on Balabac Island, south of Palawan. Steere (1888) reported that the species was absent and there are no records in the literature. Indigenous peoples from southern Palawan also refuted the species' occurrence on the island (Acosta and Schoppe, 2018). However, hunters in the Palawan tribe and local people from Balabac report that the species is present but in very low numbers (Schoppe and Cruz, 2008).

Habitat

The Philippine pangolin primarily inhabits primary and secondary lowland forests (Heaney et al., 1998; Hoogstraal, 1951), but uses a diverse range of habitats (Fig. 7.3). It has been recorded in lowland grassland-forest mosaics and logged-over lowland forest, in which most of the merchantable timber had been extracted (Esselstyn et al., 2004), agricultural ecosystems (Acosta-Lagrada, 2012; Schoppe and Cruz, 2009), coastal areas near beach forest and mangroves, and in riverine forest (Marler, 2016). There are records from up to 2015 m asl (Acosta-Lagrada, 2012). The species is thought to have an affinity to strangler figs (*Ficus* spp.) as the fruits of these trees attract ants, and the roots provide structures that pangolins can use for shelter

FIGURE 7.3 The Philippine pangolin inhabits primary and secondary lowland forest but uses a diverse range of habitats including grassland-forest mosaics and beach forest. *Photo credit: Dexter Alvarado.*

(Schoppe and Cruz, 2009). The species has been found burrowing under other tree species, including bishop wood (*Bischofia javanica*) and *Dysoxylum* spp. (Acosta-Lagrada, 2012). The diverse range of habitats used suggests that the species is adaptable to habitat degradation and modification providing there is an adequate prey base (Acosta-Lagrada, 2012): this is similar to the Sunda pangolin, which persists in highly degraded forest (see Chapters 6 and 26). Acosta-Lagrada (2012) suggested that the species may occur at higher densities in primary forest but this requires further research and verification (see Chapter 34), as does the ability of the species to persist long-term in isolated blocks of heavily modified habitats (e.g., monoculture plantations).

Ecology

The species is semi-arboreal but there is little knowledge of its ecology, and most of what is known is based on research conducted between 2010 and 2018. The species likely lives in relatively stable home ranges. Minimum Convex Polygon (MCP) home range estimates for three males were 59, 96, and 120 ha (mean = 61.6 ha) with core areas (where individuals spent ≥50% of the time) of 29, 45, and 68 ha respectively (mean = 47.3 ha; Schoppe and Alvarado, 2016; Schoppe et al., 2017; in prep. b). The smallest home range estimate was for a young male that weighed 1.5 kg (and was likely recently independent – see Chapter 6) at the onset of monitoring. Home ranges did not overlap, suggesting males are territorial (Schoppe et al., in prep. b). Equivalent estimates for two females based on 257 days of radio-tracking were 47 and 75 ha, with core areas of 12 and 18 ha, respectively. During the wet season (June–November), home ranges of two females and a young male did not overlap, but did partially during the dry season (December–May), which may be related to access to permanent water sources (Schoppe and Alvarado, 2016; Schoppe et al., in prep. b). Males traveled a mean distance of 4.2 ± 0.6 km (range 3–5.3 km) within a 24 h period, and females, 3.1 ± 0.6 km (range 0–4 km; Schoppe et al., 2017; in prep. b). Research using radio-telemetry on six wild-caught individuals has provided initial insights into homing behavior. Three males (including one young male; weight = 1.5 kg) and three females were released 1.1–3.7 km away from sites of capture. After release, the three females and the young male established new home ranges near the release site, whereas the two adult males returned to their respective home ranges

(Schoppe and Alvarado, 2016; Schoppe et al., 2017; in prep. b).

Shelter takes the form of burrows on the forest floor, tree hollows (or in the fork of tree branches), between tree buttresses or roots, and dens under or between large rocks. Acosta-Lagrada (2012) suggested that dens are located in areas away from human disturbance, on slopes of 36° to >50° and 100–200 m from water sources, which is similar to the Chinese pangolin (Wu et al., 2003; see Chapter 4). Observations suggest that the species prefers arboreal resting sites during the wet season and terrestrial sites during the dry season, which is potentially related to avoiding dens being flooded in the former (Schoppe et al., in prep. b). There is little knowledge of shelter use (e.g., site fidelity) but a male pangolin was observed to return to the same den for 2–3 consecutive nights before moving to another site, eventually returning to the previous den after about a week (S. Schoppe, unpubl. data).

The Philippine pangolin is known to forage both terrestrially and in trees but there is little knowledge of prey species or foraging behavior. Acosta-Lagrada (2012) recorded consumption of two ant species, *Odontomachus infandus* and *Diacamma* sp., and one termite species, *Nasutitermes* sp. Schoppe and Alvarado (2015a) observed the species predating frequently on the weaver ant (*Oecophylla smaragdina*), a species preferred by the Sunda pangolin (Lim, 2007; see Chapter 6). Lim (2007) noted that the Sunda pangolin predated on 11 ant genera, which do occur in the Philippines (General and Alpert, 2012), and it is likely that the diet is similar to the Sunda pangolin, though only two of these genera are known to occur in Palawan, *Polyrhachis* and *Oecophylla*.

Predators likely include pythons (e.g., reticulated python [*Malayopython reticulatus*]; Lim and Ng, 2008). Ecto-parasites include the tick *Amblyomma javanense* (Corpuz-Raros, 1993; Jaffar et al., 2018), and preliminary research indicates that 86% of Philippine pangolins host this species (n = 14; S. Schoppe, unpubl. data).

Behavior

Solitary and predominantly nocturnal, the species rests during the day and is active at night (Schoppe and Alvarado, 2015a, 2016; Schoppe et al., in prep. b). Based on research using telemetry, peak activity for six study animals over an 11-day period was 2300–0400 (Schoppe et al., in prep. b). A young, independent male monitored for 42 days had an identical peak activity period, and was active for 12 hours a day (mean), and traveled a mean distance of 3.6 ± 1.1 km (range 0–5 km) within a 24 hour period (Schoppe and Alvarado, 2015a, 2016; Schoppe et al., in prep. b). Preliminary research suggests that activity patterns may be influenced by the lunar phase, with individuals being active for longer periods during full moon (Schoppe and Alvarado, 2015a). Schoppe and Alvarado (2015a) note that a young male emerged from its den as early as 1200 noon during the new moon and returned by midnight, but during full moon (and between first and last quarter) emerged at dusk and returned in the early hours of the morning.

In the dry season, individuals will immediately search for water after emerging from a burrow or other resting place, and drink prior to foraging; during the wet season, individuals will forage first before drinking and retreating to a resting place (Schoppe et al., in prep. b). In the dry season, an individual was observed to stand in a stream and drink for 30–60 min before returning to its resting place (Schoppe et al., in prep. b).

The species is quadrupedal. When walking the head and tail are held below the level of the body, and the forelimbs face inward with the claws perpendicular to the ground, and the weight is taken on the knuckles and outer margins of the claws (i.e., folded up wrists). Semi-

FIGURE 7.4 Philippine pangolins are adept climbers and use their strong tail for support. *Photo credit: Dexter Alvarado.*

arboreal, the species is an adept climber, using its long tail as an extra limb (Fig. 7.4). Like the Sunda pangolin, the species moves its limbs in pairs when climbing tree trunks, the incredibly strong tail providing grip and bearing weight, and the species will descend trees head first (Fig. 7.4).

The Philippine pangolin predates on ants and termites, both in trees and on the ground, and locates prey using its acute olfactory senses (Acosta-Lagrada, 2012; Schoppe and Alvarado, 2015a; Schultze, 1914). Having located an ant nest or termitaria the strong forelimbs are used to break into these structures and the long tongue is used to extract ants or termites of all life stages (e.g., eggs, adults). The nests are not destroyed, enabling insect colonies to recover, but feeding sites are returned to during consecutive nights (Schoppe et al., in prep. b). In addition to searching for ant nests and termitaria, the species will search for prey among leaf litter and rotten wood on the forest floor. Schultze (1914) suggested that the Philippine pangolin may have a higher reliance on arboreal termite nests in the dry season, and terrestrial termite nests in the wet season, as terrestrial nests may be difficult to penetrate during the dry season; however, this requires verification.

The species is sensitive to human presence and typically responds by fleeing (Schoppe et al., in prep. b). Otherwise, when sensing a threat, Philippine pangolins will flee if the threat is distant, or roll up into a ball if it is in close proximity (Fig. 7.5).

FIGURE 7.5 A male Philippine pangolin assessing whether it is safe to uncurl: like all pangolins, this species curls up into a ball when threatened. *Photo credit: Sabine Schoppe.*

Vocalizations are limited to a "hissing" sound when individuals are threatened (Schultze, 1914; Schoppe et al., in prep. a).

Ontogeny and reproduction

Much of what is known is based on anecdotal evidence or presumed similarities with the Sunda pangolin (see Chapter 6). The Indigenous people of the Tagbanua, Batak and Palawan tribes report that the species has one offspring, but two fetuses have occasionally been encountered when the animals have been slaughtered (Acosta and Schoppe, 2018). It is presumed that breeding is aseasonal like the Sunda pangolin (Schoppe et al., in prep. a). However, local farmers in the species' range have reported that females with young are typically seen in August (Schoppe and Cruz, 2009). In a telemetry study, a pregnant female was caught in September and observed with the young in November (Schoppe and Alvarado, 2016). There is little knowledge of estrus cycles, gestation, young development and maternal care, weaning, age at sexual maturity or population structure, which requires research, but there are likely strong similarities with the Sunda pangolin. Adult sex ratios at six sites in Palawan ranged from 1:0.6 to 1:1, with an average of 1:0.8 in favor of males (Schoppe et al., in prep. a).

Population

There is limited knowledge of, and quantitative data on, Philippine pangolin populations. Schoppe et al. (in prep. a) estimated density at six sites representing northeastern islands, central Palawan, and southern Palawan, using systematic searches of direct signs, primarily with dogs. Across these sites, estimated mean adult density was 2.5 ± 1.4 adults/km^2 (mean $\pm$ SD). Population density was highest (3.5 and 4.0 adults/km^2) on the islands in northeastern Palawan, which is potentially due to the remoteness of the islands and their protected status (Schoppe et al., in prep. a). Three central Palawan (Puerto Princesa City) sites had a mean adult density of 1.8 ± 1.61 adult/km^2 (Schoppe et al., in prep. a) and included one site where the species was not recorded (Schoppe and Alvarado, 2015b); mean density in Rizal in southern Palawan was 2 adults/km^2.

Other evidence indicates that the species is not evenly distributed across its range, is increasingly rare, and declining (Bayron, 2014; Schoppe et al., 2017; in prep. a). Indigenous peoples report the species was common in the late 1960s when human population density was lower and forest cover much higher (Acosta and Schoppe, 2018). Pangolins were frequently seen in the 1970s–1980s, particularly in coastal and lowland areas, including in Aborlan (J. Fabello, pers. comm.). Rabor (1965) and Heaney et al. (1998) noted that the species was rare. Interview-based research indicates population declines. Multiple Indigenous peoples groups from the Tagbanua, Batak and Palaw'an Tribes, and Cuyunon communities from Palawan and the Calamian Islands estimated population declines of 85% in the south of the species' range and 95% in the north between 1980 and 2018 (Acosta and Schoppe, 2018).

Interviews with hunters and traders indicate the species is rare in southern Palawan and more common in the north (Schoppe and Cruz, 2009), while ethno-biological surveys and similar research suggests the species is increasingly rare and that sightings are uncommon (Acosta-Lagrada, 2012; Rico and Oliver, 2006). Local hunters report a decline in the number of pangolins caught per unit effort compared to the 1970s (Schoppe, 2013). One hunter in Brooke's Point, southern Palawan, reported that in 1992 it was possible to catch 3–4 pangolins a week, but in 2012, similar effort resulted in only one pangolin a month (Acosta-Lagrada, 2012). Bayron (2014) reported similar results in Roxas, Puerto Princesa City and Quezon.

Status

The Philippine pangolin is listed as Critically Endangered on The IUCN Red List of Threatened Species (Schoppe et al., 2019) and available evidence indicates declines across its range (see *Population*). However, there is a lack of quantitative data on populations. The species is listed as Critically Endangered in Palawan Province and Endangered at the national level in the Philippines. It is afforded protection under the Philippine Wildlife Act 9147, which prohibits the exploitation of threatened species, including by indigenous peoples for traditional use. The entirety of Palawan Province was declared a game refuge and bird sanctuary in 1969 and the species occurs in a number of protected areas, including the Mount Mantalingahan Protected Landscape and Puerto Princesa Underground River National Park. The species is included in CITES Appendix I.

Threats

The principle threat to the Philippine pangolin is overexploitation. This includes local use and consumption and international trafficking and use of the species and its body parts, and is compounded by the species' restricted geographic range. Habitat loss poses an additional, indirect threat.

Pangolins have been hunted on Palawan for at least 6000 years (Lewis et al., 2008) and Indigenous peoples have traditionally used, and continue to use the species for subsistence, medicinal, and ritualistic purposes. This includes the consumption of meat as a protein source, often with perceived health benefits including the treatment of gastrointestinal diseases, inflammation, asthma, and flatulence (Eder, 1987; Esselstyn et al., 2004; Estrada et al., 2015). The scales are used (in powered form) to treat asthma (Esselstyn et al., 2004) and prostate-related illnesses (Acosta-Lagrada, 2012), while the Palaw'an in Bataraza use the scales on belts to treat back pain, and multiple Indigenous communities use them to deter bad spirits (see Chapter 12; Acosta and Schoppe, 2018; Bayron, 2014; Estrada et al., 2015). In Narra, burnt scales are used to treat mothers who have just given birth, and to stop thunder (Acosta-Lagrada, 2012; Acosta and Schoppe, 2018). At Brooke's Point, Indigenous peoples will drink the blood for its believed rejuvenation properties (Acosta-Lagrada, 2012).

The species is poached for trade in its blood, meat, skin and scales within the Philippines (Cruz et al., 2007; Esselstyn et al., 2004). Bayron (2014) reported that poaching levels were particularly high in northern Palawan, and estimated that 448 pangolins were poached in three municipalities (Puerto Princesa City, Roxas and Quezon) in Palawan in 2012 alone. Poaching takes place by local people, but also by individuals from neighboring areas (barangays), including with the use of dogs (Acosta and Schoppe, 2018; Bayron, 2014). It takes place to meet consumer demand in major cities, and analysis of seizure data suggests there is substantial domestic demand for pangolin meat and scales in cities including Metro Manila (Gomez and Sy, 2018). Indications from trade research and insights from indigenous peoples are that exploitation for domestic and international trade is the main, contemporary driver of pangolin offtake, as opposed to local use (Lacerna and Widmann, 2008; Schoppe and Cruz, 2009).

The CITES trade data indicate that approximately 10,000 pangolins, in the form of skins, that were traded internationally in the 1980s, originated in or were exported from the Philippines. Despite a decline in international trade in pangolin skins post-2000 (see Chapter 16), international trade has otherwise continued, albeit illegally. Gomez and Sy (2018) report that during 2001–2017 at least 39 seizures took place in which the Philippines was implicated as a source country or place of seizure. This included an estimated 667 Philippine pangolins, primarily comprising quantities of meat and scales, and dead individuals, though actual illegal trade volumes were likely higher. Most of these seizures took place in Palawan, but seizures were also made on Luzon, Mindoro, Negros and Tablas (Gomez and Sy, 2018). Exact trafficking routes are not well documented but the species is illegally traded to East Malaysia (e.g., Kudat, Sandakan) and Peninsular Malaysia for subsequent export to China, including for use in traditional medicines (Pantel and Anak, 2010; Schoppe and Cruz, 2009). However, it is suspected that Philippine pangolin scales may be exported and then subsequently imported after processing into Traditional Chinese Medicine (TCM) for domestic consumption (E. Sy, pers. comm.). The Philippines has further been implicated in highly organized international pangolin trafficking (see Luczon et al., 2016). There remains a strong financial incentive for local people and indigenous peoples to poach and sell pangolins into illegal trade, which is apparently exacerbated by issues over their legitimate access to, and use of, non-timber forest products for livelihood purposes. This is compounded by law enforcement being a low priority for enforcement and judicial agencies in the Philippines, combined with a lack of human and technical resources, resulting in little deterrent to would-be poachers and traffickers, and the perception that illegal trade in pangolins is low risk.

Habitat loss poses an indirect threat. It results in the opening up of previously inaccessible areas through the creation of road networks for commercial concessions, which in turn makes pangolins susceptible to hunting and poaching: it also increases the likelihood of roadkill. As with the Sunda pangolin, there is a need for research into how the Philippine pangolin uses natural and modified habitats (e.g., monoculture plantations) and the ability of the species to persist in highly modified isolated habitat blocks in the long-term.

References

Acosta-Lagrada, L.S., 2012. Population density, distribution and habitat preferences of the Palawan Pangolin, *Manis culionensis* (de Elera, 1915). M.Sc. Thesis, University of the Philippines Los Baños, Laguna, Philippines.

Acosta, D., Schoppe, S., 2018. Proceedings of the Stakeholder Workshop on the Palawan Pangolin – Balintong, Bulwagang Princesa Tourist Inn, Puerto Princesa City, 17 February 2018, Katala Foundation Inc, Puerto Princesa City, Palawan, Philippines, pp. 1–10.

Alviola, P.L. III, 1998. Land vertebrates of Calauit Island, Palawan, Philippines. Asia Life Sci. 7 (2), 157–170.

Bayron, A.C., 2014. Trade dynamics of Palawan pangolin *Manis culionensis*. B.Sc. Thesis, Western Philippines University, Palawan, Philippines.

Bourns, F.S., Worcester, D.C., 1894. Preliminary Notes on the Birds and Mammals Collected by the Menage Scientific Expedition to the Philippine Islands. Minnesota Academy of Sciences, Minneapolis.

Corbet, G.B., Hill, J.E., 1992. The Mammals of the Indomalayan Region: A Systematic Review. Oxford University Press, Oxford.

Corpuz-Raros, L.A., 1993. A checklist of Philippine mites and ticks (Acari) associated with vertebrates and their nests. Asia Life Sci. 2 (2), 177–200.

Cruz, R.M., van den Beukel, D.V., Lacerna-Widmann, I., Schoppe, S., Widmann, P., 2007. Wildlife trade in Southern Palawan, Philippines. Banwa 14 (1), 12–26.

de Elera, C.D., 1915. Contribución a la Fauna Filipina. Colegio de Santo Tomás, Manila.

Eder, J.F., 1987. On the Road to Tribal Extinction: Depopulation, Deculturation, and Adaptive Well-Being Among the Batak of the Philippines. University of California Press, Berkeley.

Esselstyn, J.A., Widmann, P., Heaney, L.R., 2004. The mammals of Palawan Island, Philippines. Proc. Biol. Soc. Washington 117 (3), 271–302.

Estrada, Z.J.G., Panolino, J.G., De Mesa, T.K.A., Abordo, F. C.B., Labao, R.N., 2015. An ethnozoological study of the medicinal animals used by the Tagbanua tribe in Sitio Tabyay, Cabigaan, Aborlan, Palawan. In: Matulac, J.L.S., Cabrestante, M.P., Palon, M.P., Regoniel, P.A., Gonzales, B.J., Devanadera, N.P. (Eds.), Proceedings of the 2nd Palawan Research Symposium 2015. National Research Forum on Palawan Sustainable Development: Science, Technology and Innovation for Sustainable Development, 9–10 December, Puerto Princesa City, Palawan, Philippines, pp. 123–128.

Everett, A.H., 1889. Remarks on the zoo-geographical relationships of the island of Palawan and some adjacent islands. J. Zool. 57 (2), 220–228.

Feiler, A., 1998. Das Philippinen-Schuppentier, *Manis culionensis* Elera, 1915, eine fast vergessene Art (Mammalia: Pholidota: Manidae). Zoologische Abhandlungen - Staatliches Museum Für Tierkunde Dresden 50, 161–164.

Frechkop, S., 1931. Notes sur les mammifères. VI. Quelques observations sur la classification des pangolins (Manidae). Bull. du Musee royal d'Histoire naturelle de Belgique VII (22), 1–14.

Gaubert, P., 2011. Family Manidae. In: Wilson, D.E., Mittermeier, R.A. (Eds.), Handbook of the Mammals of the World, vol. 2. Hoofed Mammals. Lynx Edicions, Barcelona, pp. 82–103.

Gaubert, P., Antunes, A., 2005. Assessing the taxonomic status of the Palawan pangolin *Manis culionensis* (Pholidota) using discrete morphological characters. J. Mammal. 86 (6), 1068–1074.

Gaubert, P., Antunes, A., Meng, H., Miao, L., Peigne, S., Justy, F., et al., 2018. The complete phylogeny of pangolins: scaling up resources for the molecular tracing of the most trafficked mammals on earth. J. Hered. 109 (4), 347–359.

General, D.M., Alpert, G.D., 2012. A synoptic review of the ant genera (Hymenoptera, Formicidae) of the Philippines. ZooKeys 200, 1–111.

Gomez, L., Sy, E.Y., 2018. Illegal pangolin trade in the Philippines. TRAFFIC Bull. 30 (1), 37–40.

Heaney, L.R., Balete, D.S., Dollar, M.L., Alcala, A.C., Dans, A.T.L., Gonzales, P.C., et al., 1998. A synopsis of the mammalian fauna of the Philippine Islands. Fieldiana (Zool.) 88, 1–61.

Hollister, N., 1913. A review of the Philippine land mammals in the United States National Museum. Proc. United States Natl. Mus. 46, 299–341.

Hoogstraal, H., 1951. Philippine zoological expedition, 1946–1947. Narr. Itinerary. Fieldiana (Zool.) 33, 1–86.

Jaffar, R., Low, M.R., Maguire, R., Anwar, A., Cabana, F., 2018. WRS Husbandry Manual for the Sunda Pangolin (*Manis javanica*), first ed. Wildlife Reserves Singapore, Singapore.

Jentink, F.A., 1882. Note XXV. Revision of the Manidae in the Leyden Museum. Notes from the Leyden Museum IV, 193–209.

Lacerna, I.D., Widmann, P., 2008. Biodiversity utilization in a Tagbanua community, Southern Palawan, Philippines. In: Widmann, I.L., Widmann, P., Schoppe, S., van den Beukel, D.V., Espeso, M. (Eds.), Conservation Studies on Palawan Biodiversity: A Compilation of Researches Conducted in Cooperation With or Initiated by Katala Foundation Inc. Katala Foundation Inc, Puerto Princesa City, Palawan, pp. 158–170.

Lawrence, B.L., 1939. Collections from the Philippine Islands. Mammals. Bull. Mus. Comp. Zool. 86, 28–73.

Lewis, H., Paz, V., Lara, M., Barton, H., Piper, P., Ochoa, J., et al., 2008. Terminal Pleistocene to mid-Holocene occupation and an early cremation burial at Ille Cave, Palawan, Philippines. Antiquity 82 (316), 318–335.

Lim, N.T.-L, 2007. Autoecology of the Sunda Pangolin (*Manis javanica*) in Singapore. M.Sc. Thesis, National University of Singapore, Singapore.

Lim, N.T.-L., Ng, P.K.L., 2008. Predation on *Manis javanica* by *Python reticulatus* in Singapore. Hamadryad 32 (1), 62–65.

Luczon, A.U., Ong, P.S., Quilang, J.P., Fontanilla, I.K.C., 2016. Determining species identity from confiscated pangolin remains using DNA barcoding. Mitochondrial DNA Part B 1 (1), 763–766.

Marler, P.N., 2016. Camera trapping the Palawan Pangolin *Manis culionensis* (Mammalia: Pholidota: Manidae) in the wild. J. Threat. Taxa 8 (12), 9443–9448.

Mohr, E., 1961. Schuppentiere. Neue Brehm-Bucherei. A. Ziemsen Verlag, Wittenberg Lutherstadt.

Paguntalan, L.J., Gomez, R.K., Oliver, W., 2010. Threatened Species of the Calamian Islands: Developing an Integrated Regional Biodiversity Conservation Strategy in a Global Priority Area, Phase II - Biological Survey. Philippines Biodiversity Conservation Foundation. Katala Foundation Inc., and National Geographic Society Conservation Trust.

Paguntalan, L.J., Jakosalem, P.G., Oliver, W., 2012. Threatened Species of the Calamian Islands: Developing an Integrated Regional Biodiversity Conservation Strategy in a Global Priority Area, Phase

II - Biological Survey. A report submitted to National Geographic Society.

Paguntalan, L.J., Jakosalem, P.G., Oliver, W., Reintar, A.R., Doble, K.J., 2015. Threatened Species of the Calamian Islands: Developing an Integrated Regional Biodiversity Conservation Strategy in a Global Priority Area, Phase II - Follow-up Biological Survey. A report submitted to Community Centered Conservation and GIZ - Protected Areas Management Enhancement Project (GIZ–PAME).

Pantel, S., Anak, A.N., 2010. A preliminary assessment of Sunda pangolin trade in Sabah. TRAFFIC Southeast Asia, Petaling Jaya, Selangor, Malaysia.

Pocock, R.I., 1924. The external characters of the pangolins (Manidae). Proc. Zool. Soc. Lond. 94 (3), 707–723.

Piper, P.J., Ochoa, J., Robles, E.C., Lewis, H., Paz, V., 2011. Palaeozoology of Palawan Island, Philippines. Quat. Int. 233 (2), 142–158.

Rabor, D.S., 1965. Threatened species of small mammals in tropical South East Asia. The problem in the Philippines. In: Talbot, L.M., Talbot, M.H. (Eds.), Proceedings of the Conference on Conservation of Nature and Natural Resources in Tropical South East Asia, Bangkok, Thailand, 29 November – 4 December 1965. IUCN, Morges, Switzerland, pp. 272–277.

Rico, E., Oliver, W., 2006. Threatened Species of the Calamian Islands: Developing an Integrated Regional Biodiversity Conservation Strategy in a Global Priority Area, Phase I, Findings and Recommendations of an Islands-Wide Ethnobiological Survey. A report submitted to National Geographic Society.

Sanborn, C.C., 1952. Philippine zoological expedition 1946–1947. Mammals. Fieldiana (Zool.) 33, 89–158.

Schlitter, D.A., 2005. Order Pholidota. In: Wilson, D.E., Reeder, D.M. (Eds.), Mammal Species of the World: A Taxonomic and Geographic Reference, third ed. Johns Hopkins University Press, Baltimore, pp. 530–531.

Schoppe, S., 2013. Catch me if you can! How can we monitor populations of the Philippine Pangolin? Oral presentation, 1st IUCN SSC Pangolin Specialist Group Conservation Conference, 24 – 27 June 2013, Wildlife Reserves Singapore, Singapore.

Schoppe, S., Cruz, R., 2008. Armoured but endangered: Galvanizing action to mitigate the illegal trade in Asian Pangolins: The situation in Palawan, Philippines. A report submitted to TRAFFIC by Katala Foundation Incorporated, pp. 1–16.

Schoppe, S., Cruz, R., 2009. The Palawan Pangolin *Manis culionensis*. In: Pantel, S., Chin, S.-Y., (Eds.), Proceedings of the Workshop on Trade and Conservation of Pangolins Native to South and Southeast Asia, 30 June – 2 July 2008, Singapore Zoo, Singapore. TRAFFIC Southeast Asia, Petaling Jaya, Selangor, Malaysia, pp. 176–188.

Schoppe, S., Alvarado, D., 2015a. Conservation Needs of the Palawan Pangolin *Manis culionensis* – Phase II (Extension), Final Scientific and Financial Report Submitted to Wildlife Reserves Singapore. Katala Foundation Inc, Puerto Princesa City, Palawan, Philippines, pp. 1–36.

Schoppe, S., Alvarado, D., 2015b. Conservation Needs of the Palawan Pangolin *Manis culionensis* – Phase I. Final Scientific and Financial Report to Wildlife Reserves Singapore. Katala Foundation Inc, Puerto Princesa City, Palawan, Philippines, pp. 1–30.

Schoppe, S., Alvarado, D., 2016. Movements of the Palawan Pangolin *Manis culionensis* – Final Project Report Submitted to Wildlife Reserves Singapore. Katala Foundation Inc, Puerto Princesa City, Palawan, Philippines, pp. 1–16.

Schoppe, S., Alvarado, D., Luz, S., 2017. Movement patterns of the Palawan Pangolin *Manis culionensis*. Poster presentation at 26th Philippine Biodiversity Symposium, 18–21 July 2017, Ateneo de Manila University, Quezon City, Philippines.

Schoppe, S., Katsis, L., Lagrada, L., 2019. *Manis culionensis*. *The IUCN Red List of Threatened Species* 2019: e.T136497A 123586862. Available from: <http://dx.doi.org/10.2305/ IUCN.UK.2019-3.RLTS.T136497A123586862.en>.

Schoppe, S., Alvarado, D., Luz, S., in prep. a. First data on the population density of the Palawan Pangolin *Manis culionensis* from Palawan, Philippines. Katala Foundation Inc., Puerto Princesa City, Palawan, Philippines.

Schoppe, S., Alvarado, D., Luz, S. in prep. b. Home range and homing of the Palawan Pangolin *Manis culionensis*. Katala Foundation Inc., Puerto Princesa City, Palawan, Philippines.

Schultze, W., 1914. Notes on the Malay pangolin, *Manis javanica* Desmarest. Philippine J. Sci. 1, 93.

Steere, J.B., 1888. A month in Palawan. Am. Nat. 22 (254), 142–145.

Taylor, E.H., 1934. Philippine land mammals. Bur. Sci. Mongr. (Manila) 30, 1–548.

Ullmann, T., Veríssimo, D., Challender, D.W.S., 2019. Evaluating the application of scale frequency to estimate the size of pangolin scale seizures. Ecol. Conserv. 20, e00776.

Wu, S.B., Liu, N.F., Ma, G.Z., Xu, Z.R., Chen, H., 2003. Studies on habitat selection by Chinese pangolin (*Manis pentadactyla*) in winter in Dawuling Natural Reserve. Acta Ecol. Sin. 23 (6), 1079–1086. [In Chinese].

Wu, S., Liu, N., Zhang, Y., Ma, G., 2004. Physical measurement and comparison for two species of pangolins. Acta Theriol. Sin. 24 (4), 361–364. [In Chinese].

CHAPTER

8

Black-bellied pangolin *Phataginus tetradactyla* (Linnaeus, 1766)

Maja Gudehus[1], *Darren W. Pietersen*[2,3], *Michael Hoffmann*[4], *Rod Cassidy*[1], *Tamar Cassidy*[1], *Olufemi Sodeinde*[5], *Juan Lapuente*[6], *Brou Guy-Mathieu Assovi*[7] *and Matthew H. Shirley*[8]

[1]Sangha Pangolin Project, Dzanga-Sangha, Central African Republic [2]Mammal Research Institute, Department of Zoology and Entomology, University of Pretoria, Hatfield, South Africa [3]IUCN SSC Pangolin Specialist Group, ℅ Zoological Society of London, Regent's Park, London, United Kingdom [4]Conservation and Policy, Zoological Society of London, Regent's Park, London, United Kingdom [5]Department of Biological Sciences, New York City College of Technology, City University of New York, Brooklyn, NY, United States [6]Comoé Chimpanzee Conservation Project, Comoé Research Station, Côte d'Ivoire & Animal Ecology and Tropical Biology, Biozentrum, Universität Würzburg Tierökologie und Tropenbiologie (Zoologie III), Würzburg, Germany [7]Université Felix Houphouët-Boigny, Abidjan, Côte d'Ivoire [8]Tropical Conservation Institute, Florida International University, North Miami, FL, United States

OUTLINE

Pangolins
DOI: https://doi.org/10.1016/B978-0-12-815507-3.00008-3

Taxonomy

Previously included in the genus *Manis* (Meester, 1972; Schlitter, 2005) and *Uromanis* (Kingdon, 1997; McKenna and Bell, 1997; Pocock, 1924), the species is here included in *Phataginus* based on both morphological (Gaudin et al., 2009) and genetic (du Toit et al., 2017; Gaubert et al., 2018; Hassanin et al., 2015) evidence, in agreement with previous authors (Grubb et al., 1998; Kingdon and Hoffmann, 2013). Based on mitogenomic distances between the genera *Manis, Smutsia*, and *Phataginus*, and morphological traits unique to *Phataginus, P. tetradactyla* is included in the subfamily Phataginae to designate small African pangolin species (Gaubert et al., 2018; see Chapter 2). In much of the early literature, the species name *longicaudatus* is used, but this name is unavailable (Mohr, 1961). The type locality is "American Australia" (West Africa; Linnaeus, 1766). No subspecies are recognized. There are no data on chromosome number.

Synonyms: *Pholidotus longicaudatus* (Brisson, 1756; unavailable), *Pholidotus longicaudatus* (Brisson, 1762; unavailable), *Manis tetradactylus* (Linnaeus, 1766), *Manis macroura* (Erxleben, 1777), *Phataginus ceonyx* (Rafinesque, 1820), *Manis africana* (Desmarest, 1822), *Manis tetradactyla* (Gray, 1843), *Manis guineensis* (Fitzinger, 1872), *Manis longicauda* (Sundevall, 1842), *Manis longicaudata* (Sundevall, 1842, *sic., fide* Pocock, 1924), *Manis senegalensis* (Fitzinger, 1872), *Manis hessi* (Noack, 1889), *Pholidotus tetradactyla* (Sclater, 1901), *Uromanis longicaudata* (Pocock, 1924), and *Manis longicaudatus* (Rosevear, 1953).

Etymology: *Phataginus* is derived from "*phatagen*", an East Indian name for the pangolin; the species name *tetradactyla* references the four (*tetra-*) non-vestigial digits (*-dactyla*) on the feet (Gotch, 1979).

Description

The black-bellied pangolin (*Phataginus tetradactyla*), is a small, arboreal African pangolin with a body weight of 1.1–3.6 kg and a total body length of up to 120 cm (Table 8.1). The common name stems from the black skin and hairs that cover most of the ventrum, face, inner sides of the limbs and upper forelimbs and forefeet (Hatt, 1934; Jentink, 1882). The extremely long prehensile tail makes up to two-thirds of the total body length (55–75 cm; see Table 8.1; Hatt, 1934; Tahiri-Zagrët, 1970a). It has a bare, sensitive skin pad on the ventral tip, which is truncated, and the tail easily supports the weight of the animal (Jentink, 1882; Kingdon, 1971; Pocock, 1924). There are 47 caudal vertebrae (Jentink, 1882), a record among extant mammals. Head-body length ranges between 28 and 50 cm (Table 8.1).

The species is covered with large, overlapping scales that are dark brown at the base, with yellow or golden edges, and grow from the skin in a grid-like arrangement (Fig. 8.1). Each scale has slight longitudinal striations (Dorst and Dandelot, 1970; Happold, 1987), and are "harder" (i.e., thicker and more robust) than those of the white-bellied pangolin (*P. tricuspis*) and other similar sized species of pangolin (Tahiri-Zagrët, 1970a; M. Shirley and G.-B.M. Assovi, pers. obs.). The scales cover the surface of the body dorsally and laterally but are absent from parts of the face, the throat, ventrum, inner fore- and hindlimbs, and upper portions of the forelimbs (Jentink, 1882; Kingdon, 1971; Pocock, 1924). Scales on the flanks are keeled (Fig. 8.1; Hatt, 1934). No hairs project between the scales. There are 10–13 transversal, and 13 longitudinal scale rows on the body (Table 8.1), the largest of which are on the mid-dorsum and measure up to 67×50 mm (O. Sodeinde, unpubl. data). The lower sections of the forelimbs are protected by pairs of very large post-scapular scales (Fig. 8.1; Hatt, 1934). The scales on the forehead, fore- and hindlimbs, and median ventral tail row are the smallest. The elongated tail has 42–47 sharply-pointed scales along its edges (Table 8.1). There are 35–41 scales on the median dorsal scale row of the tail

TABLE 8.1 Black-bellied pangolin morphometrics.

Measurement			Country	Source(s)
Weight	Weight (♂)	2.83 (2.2–3.6) kg, n = 4	Southeast Nigeria	Kingdon and Hoffmann, 2013
		1.41 (1.12–1.65) kg, n = 12	Central African Republic	R. and T. Cassidy, unpubl. data
	Weight (♀)	2.74 (2.6–3.1) kg, n = 3	Southeast Nigeria	Kingdon and Hoffmann, 2013
		1.45 (1.3–1.7) kg, n = 4	Central African Republic	R. and T. Cassidy, unpubl. data
Body	Total length (♂)	874 (810–937) mm, n = 3	Democratic Republic of the Congo	Hatt, 1934
		985 (920–1070) mm, n = 11	Central African Republic	R. and T. Cassidy, unpubl. data
		1007 (820–1210) mm, n = 7	Côte d'Ivoire	Tahiri-Zagrët, 1970a
	Total length (♀)	851.4 (755–930) mm, n = 5	Democratic Republic of the Congo	Hatt, 1934
		1040 (970–1110) mm, n = 4	Central African Republic	R. and T. Cassidy, unpubl. data
		1015 (930–1100) mm, n = 2	Côte d'Ivoire	Tahiri-Zagrët, 1970a
	Total length (unsexed)	1115 (1060–1170) mm, n = 2	Côte d'Ivoire	Tahiri-Zagrët, 1970a
	Head-body length (♂)	314 (286–342) mm, n = 4	Southeast Nigeria	Kingdon and Hoffmann, 2013
		398 (370–460) mm, n = 10	Central African Republic	R. and T. Cassidy, unpubl. data
	Head-body length (♀)	302 (292–311) mm, n = 3	Southeast Nigeria	Kingdon and Hoffmann, 2013
		420 (370–500) mm, n = 4	Central African Republic	R. and T. Cassidy, unpubl. data
	Tail length (♂)	613.3 (560–645) mm, n = 3	Democratic Republic of the Congo	Hatt, 1934
		641 (594–707) mm, n = 4	Southeast Nigeria	Kingdon and Hoffmann, 2013
		658 (520–690) mm, n = 10	Central African Republic	R. and T. Cassidy, unpubl. data
		643 (600–710) mm, n = 7	Côte d'Ivoire	Tahiri-Zagrët, 1970a
	Tail length (♀)	560.6 (505–623) mm, n = 5	Democratic Republic of the Congo	Hatt, 1934
		633 (606–670) mm, n = 3	Southeast Nigeria	Kingdon and Hoffmann, 2013
		620 (600–670) mm, n = 4	Central African Republic	R. and T. Cassidy, unpubl. data
	Tail length (unsexed)	705 (660–750) mm, n = 2	Côte d'Ivoire	Tahiri-Zagrët, 1970a
Vertebrae	Total number of vertebrae	>70	Liberia	Frechkop, 1931; Jentink, 1882
	Cervical	7	Liberia	Jentink, 1882
	Thoracic	13	Liberia	Jentink, 1882; Mohr, 1961

(Continued)

TABLE 8.1 (Continued)

Measurement			Country	Source(s)
	Lumbar	5–6	Liberia	Jentink, 1882; Mohr, 1961
	Sacral	2–3	Liberia	Jentink, 1882; Mohr, 1961
	Caudal	46–47	Liberia	Frechkop, 1931; Jentink, 1882; Mohr, 1961
Skull	Length (♂)	70 (69.1–71) mm, n = 2	Democratic Republic of the Congo	Hatt, 1934
	Length (♀)	65.3 (62.1–68.7) mm, n = 7	Democratic Republic of the Congo	Hatt, 1934
	Breadth across zygomatic processes (♂)	24 (23.6–24.4) mm, n = 2	Democratic Republic of the Congo	Hatt, 1934
	Breadth across zygomatic processes (♀)	24.5 (22.8–26.9) mm, n = 7	Democratic Republic of the Congo	Hatt, 1934
Scales	Total number of scales	588 (542–637), n = 10	Cameroon, unknown	Ullmann et al., 2019
	No. of scale rows (transversal, body)	10–13		Frechkop, 1931; Mohr, 1961
	No. of scale rows (longitudinal, body)	13		Frechkop, 1931; Jentink, 1882
	No. of scales on outer margins of tail	42–47		Frechkop, 1931; Jentink, 1882; Mohr, 1961
	No. of scales on median row of tail	35–41		Frechkop, 1931; Jentink, 1882
	Scales (wet) as proportion of body weight	No data		
	Scales (dry) as proportion of body weight	No data		

which is replaced by two rows of 9–10 scales towards the tail tip (Fig. 8.2; Frechkop, 1931; Jentink, 1882). The total number of scales is 542–637 (Table 8.1). Scales may lack pigmentation, seemingly sporadically, but usually on the tail or flanks (Fig. 8.2; Hatt, 1934). In older animals, scales (e.g., on the ventral tail surface) may be worn to the point of having no free edge (Hatt, 1934). Observations of aged individuals in the wild with missing tail scales suggests that the species may permanently lose scales as they age (M. Shirley and B.G.-M. Assovi, pers. obs.).

FIGURE 8.1 Black-bellied pangolin in Central African Republic. Visible are the large post-scapular scales on the lower forelimbs, and the keeled scales on the left flank and hindlimb. *Photo credit: Alex Ley.*

Like other pangolins, the skull demonstrates features associated with a myrmecophagous diet (see Chapter 1). Like the white-bellied pangolin, but unlike other pangolins, the lacrimal bone is present (Emry, 1970). The head is long, slender and naked except for scales on the forehead and coarse black hair (5–10 mm) on the cheeks and throat (Gaubert, 2011; Kingdon and Hoffmann, 2013). The snout is naked and nostrils moist and slightly downturned, and similar in color to the face (Kingdon and Hoffmann, 2013). The eyes are large and beady with dark irises, and are protected by thick, swollen eyelids. The black-bellied pangolin has ear pinnae that are greatly reduced in size and comprise fleshly ridges which border the auditory orifices (Kingdon and Hoffmann, 2013). Like other pangolins, the species is edentate (i.e., lacks teeth).

The tongue is 16–18 cm in length and at the distal end is flattened but has an oval cross-section (Gaubert, 2011). The tongue extends anteriorly from the xiphisternum, a bifurcated, cartilaginous structure formed from the last pair of ribs, which is its point of attachment in the abdomen (Doran and Allbrook, 1973; Sikes, 1966). From the posterior border of the ribs the tongue extends along the ventral surface of the body to the right iliac fossa, and from there proceeds dorsally and then anteriorly, ending in a spatulate cartilaginous structure at the dorsal border of the diaphragm (see Chapter 1; Heath, 2013). The tissues in the thorax form a glossal tube in which the tongue is housed, and which proceeds through the neck to the oral cavity (Doran and Allbrook, 1973).

The fore- and hindfeet have five digits with strong, curved claws that are specialized for

FIGURE 8.2 Black-bellied pangolin climbing a tree in Central African Republic. This figure shows the extremely long tail in this species, the termination of the mediodorsal scale row near the distal end of the tail, and several scales that lack pigmentation. *Photo credit: Maja Gudehus.*

breaking into ant nests. On the forefeet the first digit is vestigial—and essentially functionless—giving the appearance of only four digits, while the fifth digit is nearly as long as the fourth and better developed than in other pangolin species (Pocock, 1924). The third digit and claw is the largest and strongest. The hindfeet are elongated, resembling the forefeet to an extent, with four long, well-developed claws of approximately equal length; the innermost claw is vestigial (Pocock, 1924). Both fore- and hindfeet have well developed skin pads, which in the hindfeet extend from the level of the first digit to the heel (Pocock, 1924).

A pair of anal glands are situated on the lateral borders of the anus and produce a strong, musky scent that may be used in scent-marking, and perhaps in defense when threatened, but this has not been observed (Pocock, 1924). Pangolin hunters in Côte d'Ivoire claim to be able to detect black-bellied pangolins via this scent (M. Shirley and B.G.-M. Assovi, unpubl. data). The female's vulva is situated just anterior to the anus; they have two pectoral nipples. In males, the testes are situated in the inguinal area and do not descend into a scrotum (see Chapter 1).

Body temperature is regulated at 30–36 °C; average resting metabolic rate has been estimated at 160.2 ml O_2/kg/h (Hildwein, 1974).

Despite significant differences in appearance, black-bellied pangolins are often confused with the sympatric white-bellied pangolin. They can most easily be distinguished by skin color and scale size and coloration. Scales are proportionally larger on the black-bellied pangolin and dark brown with yellow-golden edging as opposed to gray- or yellowish-brown on the white-bellied pangolin, and the scales may be noticeably longer than they are wide on the latter. The black-bellied pangolin has two very large post-scapular scales (Fig. 8.1), which are not enlarged and are inferior in the white-bellied pangolin. The black-bellied pangolin is distinguished from the two terrestrial African pangolins by its smaller size, extremely long tail, skin color, and scale coloration.

Distribution

The black-bellied pangolin is distributed throughout the forested regions of West and Central Africa, from Sierra Leone to the eastern limits of the Congo Basin (Fig. 8.3). Grubb et al. (1998) refuted records of the species from west of Sierra Leone, including from Senegal, The Gambia, and Guinea-Bissau (e.g., Frade, 1949; Meester, 1972; Schlitter, 2005), and suggested the most westerly confirmed record is

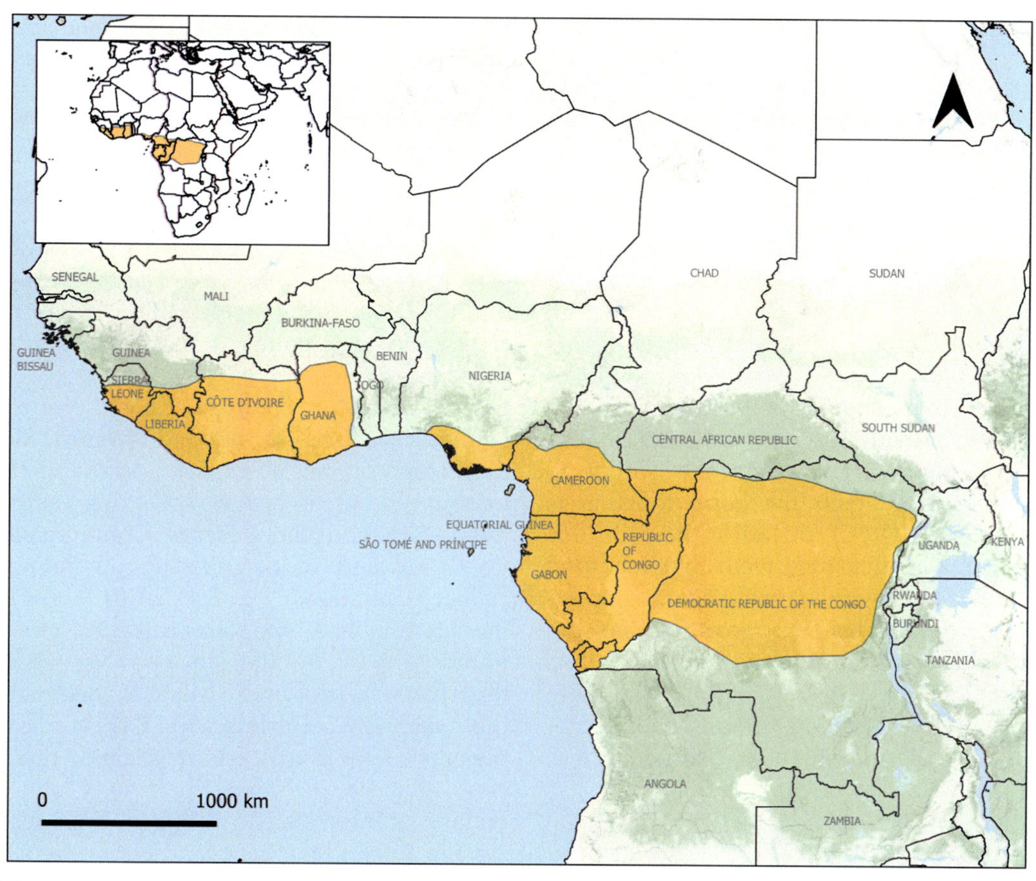

FIGURE 8.3 Black-bellied pangolin distribution. *Source: Ingram et al., 2019a.*

from the Western Area Peninsula Forest National Park in Sierra Leone. Reiner and Simões (1999) considered their presence in Guinea-Bissau possible, but there is no confirmatory evidence. There are confirmed records from a number of forest reserves in southeastern Guinea (Barnett and Prangley, 1997; Barrie and Kanté, 2006), Liberia (Allen and Coolidge, 1930), southern Côte d'Ivoire (Rahm, 1956), and southwestern Ghana, west of the Volta (Grubb et al., 1998). The species was recorded in Comoé National Park in northern Côte d'Ivoire for the first time in 2016 (J. Lapuente and K.E. Linsenmair, unpubl. data) and there are reports of the species occurring at other sites in northern Côte d'Ivoire. If accurate, it is reasonable to assume the species may also occur in northern Ghana.

There is then an apparent gap in distribution until southwestern Nigeria (Angelici et al., 2001; Happold, 1987), from where the species ranges eastward through southern Cameroon (Jeannin, 1936) and southwestern Central African Republic to eastern Democratic Republic of the Congo (DRC; Hatt, 1934; Rahm, 1966; Schouteden, 1944), the easternmost limit of their distribution. Blench and Dendo (2007) report that in Nigeria the black-bellied pangolin is more common in the southeast than in the southwest. Southwards the species ranges through Gabon and Republic of Congo (Malbrant and MacLatchy, 1949). Further investigation is warranted in the Semuliki Valley of Uganda, a well-known refuge of Congolese fauna and flora in East Africa where both the giant (*Smutsia gigantea*) and white-bellied pangolin occur.

The presence of the species in Angola is unclear, and deserves some brief discussion. There is no mention of the species in several major reference works (Beja et al., 2019; Bocage, 1889, 1890; Hill and Carter, 1941; Machado, 1969; Thomas, 1904). Monard (1935) notes that they should occur south to the Kunene but provides no material basis for this statement. Mohr (1961) noted they occur to Moçamedes, but again the basis for this is unclear. Feiler (1990) includes them in his checklist, noting that they were first recorded prior to 1990, but without reference to a specimen or a locality. There are records of the species from the Bas-Congo province of Republic of Congo (Schouteden, 1944), and hence there seems no reason that the species should not occur at least in Cabinda (Kingdon and Hoffmann, 2013; Taylor et al., 2018).

Habitat

Specific studies documenting habitat use by the black-bellied pangolin are limited. The species is the most arboreal of the African pangolins and is often found in riparian and swamp forests, typically in habitats dominated by palms (including rattans) and specialized swamp trees, such as *Uapaca* sp., *Pseudospondis* sp., and *Mitragina* sp. (Happold, 1987; Kingdon and Hoffmann, 2013; Pagès, 1970), as well as in primary forests and forest-savanna mosaic. Records in the catalogues of the American Museum of Natural History and the Smithsonian National Museum of Natural History indicate that most individuals in these collections were caught in primary forests. This is also the case in Central African Republic and northern Côte d'Ivoire, where the black-bellied pangolin has most frequently been found in primary, closed canopy forest, away from swamp and riverine forests (R. and T. Cassidy, unpubl. data; J. Lapuente and K.E. Linsenmair, pers. obs.). In the forest-savanna mosaic of Côte d'Ivoire's Comoé National Park, the species inhabits forests dominated by *Annogeissus leiocarpus*, *Dialium guineense*, *Diospyros abissinica*, and *Drypetes floribunda* which are far from rivers and lack palms (J. Lapuente and K.E. Linsenmair, pers. obs.).

In contrast, in southern Côte d'Ivoire, the species is found in flooded swamp forests dominated by palms, *Raphia hookeri* and *Hallea ledermanii*, such as in the Forets des Marais de Tanoe-Ehy and the Reserve Communautaire de Dodo, and in neighboring oil palm and rubber plantations (Fig. 8.4). In these habitats, the black-bellied pangolin can also often be found foraging in the shrubby vegetation at the riverside, not more than a meter above the water (M. Shirley and B.G.-M. Assovi, pers. obs.). In southeastern Nigeria, the species has been recorded in primary and secondary rainforests, in regenerating forests, swamp forests, and farmlands (Angelici et al., 1999, 2001; Luiselli et al., 2015). Hunters and forest workers in southwestern Nigeria report harvesting pangolins, including the black-bellied pangolin on occasion, in abandoned or infrequently harvested oil palm plantations in secondary forests (Sodeinde and Adedipe, 1994).

In all forest types, the black-bellied pangolin appears to predominantly use the crown strata more than others, which perhaps contributes to the lack of observations. Preliminary observations suggest that the species is more abundant in less disturbed habitats and those less frequented by people (M. Shirley and B.G.-M. Assovi, pers. obs.). Further research is needed to understand habitat requirements and selection, and their potential influence on the distribution of this species.

FIGURE 8.4 Adult female black-bellied pangolin in the flooded swamp forests of southeastern Côte d'Ivoire. *Photo credit: Matthew H. Shirley.*

Ecology

The black-bellied pangolin is predominantly diurnal, though does demonstrate some nocturnal activity (Booth, 1960; Carpaneto and Germi, 1989; Pagès, 1970; R. and T. Cassidy, unpubl. data). Records of two black-bellied pangolins monitored daily over a period of two years, and post-release observations of eight individuals confiscated from illegal trade in Central African Republic, support the assertion that the species is all but strictly diurnal (R. and T. Cassidy, pers. obs.).

Only two efforts have been made to estimate the home range size of the species, one using over two years of near daily follow data for two rehabilitated and released individuals (9–12 months of age) in Central African Republic (R. and T. Cassidy, unpubl. data), and a nascent study of six radio-tagged individuals tracked for two months in Côte d'Ivoire. In Central African Republic the home ranges were very small, comprising 48 and 12 ha (95% and 50% Kernel Use Density [KUD], respectively), and ranged from 36.3 to 56.2 ha (95% KUD) and 7.2–15.01 ha (50% KUD) from the first to second year respectively. In addition to little variation in annual home range, there was little apparent shift in the area of forest used, the individuals ranging over an area of 32.4 ha (95% Minimum Convex Polygon [MCP], roughly 1.2 × 0.5 linear km). In Côte d'Ivoire, preliminary individual home ranges for six individuals (comprising both sexes) averaged 9.27 ha (95% MCP), ranging from 0.13 to 25.9 ha (M. Shirley and B.G.-M. Assovi, unpubl. data). Within home ranges, individuals appear to use fixed routes and refuge holes on a continuous basis (Pagès, 1970). The species sleeps curled-up in tree hollows, tree ferns, or in tangles of lianas, and occasionally in hollowed-out insect nests.

Like all pangolins, the species is myrmecophagous. However, unlike other species, the black-bellied pangolin appears to persist on a diet of tree ants (Fig. 8.5) and does not seem to prey on termites extensively, or even at all (Kingdon and Hoffmann, 2013; R. and T. Cassidy, pers. obs.; M. Shirley and B.G.-M. Assovi, pers. obs.). *Crematogaster* spp. and *Cataulacus* spp. have been reported as preferred species (Kingdon and Hoffmann, 2013), while observations in Central African Republic indicate that the species' diet consists of at least seven different ant species, including *Cataulacus guineensis, Oecophylla longinoda* and *Polyrhachis* spp., including their eggs, larval and pupal stages (M. Gudehus, unpubl. data). Sweat bees (Halictidae) are also occasionally taken (R. and T. Cassidy, pers. obs.). The frequency of feeding bouts and proportion of time spent feeding has only been recorded for a period of two weeks, but observations suggest that the black-bellied pangolin spends most of its time resting or foraging for prey (R. and T. Cassidy, pers. obs.).

FIGURE 8.5 Black-bellied pangolin breaking into an arboreal ant nest. This species appears to predate almost exclusively on ants. *Photo credit: Michael Lorentz.*

The thermal ecology of pangolins in general is not well known, including body temperature and its impact on ecology. Preliminary insights from research in Côte d'Ivoire appear to refute the suggestion that the black-bellied pangolin spends time basking at the forest crown for thermoregulatory, or potentially, other reasons. Temperature and light data loggers attached to black-bellied pangolins in 2018 indicated that the tagged animals were exposed to an average of 110–170 lumens during daylight hours, ranging from 0– <1000, well below ambient luminosity in direct sunlight (3000–4000 lumens) (M. Shirley and B.G.-M. Assovi, unpubl. data). Similarly, the ambient temperature around the tagged animals averaged 30 °C, while comparable temperatures under the forest canopy and in direct sunlight averaged 27.5 °C and 38 °C respectively, suggesting that the species spends virtually all its time below crown level. At night, the ambient temperature around the pangolins ranged from 3 to 8 °C higher than the surrounding environment, but dropped 1–4 °C below the daytime average, suggesting an elevated microclimate in their den sites (M. Shirley and B.G.-M. Assovi, unpubl. data). These results are preliminary and further research is needed to better understand the thermal ecology of the species.

Leopard (*Panthera pardus*) and spotted hyaena (*Crocuta crocuta*) are likely the main natural predators, with black-bellied pangolin remains found in leopard scats at several sites in Gabon (Henschel et al., 2005, 2011) and in Comoé National Park in northern Côte d'Ivoire (J. Lapuente and K.E. Linsenmair, pers. obs.). Diurnal behavior and small size likely make the species vulnerable to predation by raptors and chimpanzees (*Pan troglodytes;* Kingdon and Hoffmann, 2013), African rock pythons (*Python sebae*), and ratels (*Mellivora capensis*), among other species.

The tick *Amblyomma compressum* has been found on individuals in Central African Republic and Côte d'Ivoire (Tahiri-Zagrët, 1970b, A. Kotze and E. Suleman, pers. comm.) and *Ixodes rasus* on individuals in Gabon (Sikes, 1966; see also Chapter 29).

Behavior

There is little knowledge of black-bellied pangolin behavior because the species has been subject to limited study. Like other pangolins, the species is solitary, with the only lasting social bonds occurring between a mother and present offspring. Hunters in Côte d'Ivoire occasionally find two individuals together but it is unclear whether they are mother and offspring or a courting pair (M. Shirley and B.G.-M. Assovi, pers. obs.).

The black-bellied pangolin is quadrupedal and being almost strictly arboreal, is an excellent climber. The fore- and hindlimbs are moved in pairs as they hump their way up vertical tree trunks quite rapidly, their movement resembling that of a caterpillar. With the clawed forelimbs probing for grip, they grab the tree trunk or branch with the front feet, hook the tail around the trunk or branch and bring up and anchor the hindfeet close behind. The prehensile tail provides a versatile climbing aid, clinging to branches or extending out as an apparent ballast, effectively serving as a fifth limb. In this way, the black-bellied pangolin can stretch across open spaces between branches, often relying on creepers (Fig. 8.6). The long, flat ventral surface of the tail and the sharp marginal points press against tree trunks to support the animal's weight. When descending trees too large to grip, black-bellied pangolins have been observed descending in spirals head first, aided by the grip provided by the long, keeled tail curled around the trunk (R. and T. Cassidy, pers. obs.).

The species may forage preferentially in trees and other vegetation (Fig. 8.4). This contrasts with the sympatric white-bellied pangolin, which is semi-arboreal, nocturnal and forages both in trees and on the ground (Booth, 1960; Carpaneto and Germi, 1989; Kingdon and Hoffmann, 2013). The black-bellied pangolin appears to use olfactory senses to locate arboreal ant nests, using its strong claws to break open branches and ant nests to access prey (Fig. 8.5). Columns of foraging ants moving along tree branches will also be consumed, the species using its long, salivary-coated tongue to "lick up" the ants (Kingdon and Hoffmann, 2013; R. and T. Cassidy, pers. obs.). There have been few observations of the species drinking, but initial observations indicate that black-bellied pangolins drink from little hollows of collected rain or dew water in trees rather than descending trees to frequent rivers or lakes (R. and T. Cassidy, pers. obs.).

FIGURE 8.6 Black-bellied pangolin in Central African Republic. The species uses all four limbs and its very long, prehensile tail when climbing. *Photo credit: Maja Gudehus.*

The species has been observed walking on the ground in Central African Republic, Republic of Congo, and Côte d'Ivoire, but it is unknown how frequently this behavior occurs (Wilderness Wildlife Trust, unpubl. data; R. and T. Cassidy, pers. obs.; J. Lapuente and K. E. Linsenmair, pers. obs.). In Central African Republic, there are observations of individuals crossing roads without overhead connectivity, and walking to the nearest tree after having fallen to the ground (i.e., out of a tree; R. and T. Cassidy, pers. obs.). The crossing of roads has also been observed in females carrying young (R. and T. Cassidy, pers. obs.; J. Lapuente and K.E. Linsenmair, pers. obs.). Black-bellied pangolins are known to swim to cross rivers and move from tree to tree in flooded forest habitats where overhead connectivity does not exist (M. Shirley and B.G.-M. Assovi, pers. obs.). In Côte d'Ivoire, the species has been observed crossing rivers up to 20 m wide (M. Shirley and B.G.-M. Assovi, pers. obs.).

When threatened, the species will roll into a ball with the scales acting as armor, and remain motionless, sometimes for up to several hours after sensing a threat (Kingdon and Hoffmann, 2013; R. and T. Cassidy, pers. obs.). They may actively move further into dense vegetation and liana tangles to avoid detection, resuming normal activity when the threat has passed (M. Shirley and B.G.-M. Assovi, pers. obs.). The species is reportedly very shy, and markedly so compared to other pangolin species. However, rescued animals, particularly younger individuals, have occasionally sought contact with other pangolins and human caregivers (R. and T. Cassidy, pers. obs.).

Ontogeny and reproduction

Little is known about reproduction. Being solitary, it is assumed that males and females only come together to mate. The species follows urine and glandular scent trails and it is likely that males monitor the condition of females in this way (Kingdon and Hoffmann, 2013). Pagès (1972a) reports that pre-reproductive behavior simulates aggression, with the male and female standing chest-to-chest. This is followed by the female submitting to the male and clinging to his tail (as an infant clings to the mother's tail) prior to copulation. During copulation, the male and female's tails are entwined.

Gestation period is estimated to be 140 days and the black-bellied pangolin gives birth to single offspring (Pagès, 1970, 1972b). Kingdon and Hoffmann (2013) report that breeding is aseasonal and continuous and that females conceive within 9–16 days of parturition. Neonate black-bellied pangolins weigh 100–150 g (Tahiri-Zagrët, 1970a) and measure 300–350 mm in length (Kingdon and Hoffmann, 2013). Young are born in a tree hole, in which they stay for the first few days, while the mother forages but returns periodically to nurse the young. Thereafter, the juvenile will cling to the base of the mother's tail and accompany her to another tree hollow and, and as the juvenile grows, will forage for prey with the mother. The young starts ingesting live prey at about two weeks old. Exact weaning age is unknown but juveniles leave their mother on arrival of the next offspring. Pagès (1975) reports that young black-bellied pangolins wander for 4–5 months before establishing a home range. Full adult size is attained at around 15 months (Pagès, 1970, 1972b), and the species is thought to reach sexual maturity at around two years of age. Lifespan in the wild is unknown, and no animals have been maintained long-term in captivity.

Population

There are no quantitative data on black-bellied pangolin abundance. Whilst not a formal assessment of abundance, six individuals were captured for a radio-telemetry study within a 400 ha area in southern Côte d'Ivoire in 2018–2019, suggesting a minimum density of 0.015 individuals/ha in a swamp forest habitat with a low, but constant, level of subsistence hunting pressure (M. Shirley and B.G.-M. Assovi, pers. obs.). Notably, four of these pangolins were caught within one week along a small river less than 750 m apart within a 15 ha area of forest, suggesting that they can occur at higher densities (e.g., 0.26 individual/ha). This is the least frequently recorded of all African pangolin species, possibly reflecting its occurrence in little-penetrated habitats, its selection of canopy habitats, and/or its shy and rare nature and low natural density (Kingdon and Hoffmann, 2013).

Status

Due to a dearth of research on the black-bellied pangolin, populations are not known at any scale – site, country, or globally. The species is listed as Vulnerable on The IUCN Red List of Threatened Species, with the global population considered to be declining (Ingram et al., 2019a). In 2016, the species was listed as Endangered at the national level in Uganda using the IUCN Red List Categories and Criteria (Kityo et al., 2016). Like other pangolins, this species is included in CITES Appendix I, and all range states are signatories to the Convention. The black-bellied pangolin is afforded protection through national legislation in range states, which typically prohibits exploitation. Exceptions include Gabon, Republic of Congo, and Sierra Leone, where it is legal to hunt and trade this species under certain conditions, including the holding of permits for hunting and transportation, restrictions on location (e.g., not in protected areas), and seasonality (i.e., open vs. closed seasons).

Threats

As with other pangolin species, the primary threats facing the black-bellied pangolin are anthropogenic in origin. Habitat loss and degradation have likely posed the greatest threat to this species historically (see Megevand, 2013). Deforestation throughout West Africa, in particular, is rampant with some countries having lost as much as 98% of their natural forest cover (e.g., Côte d'Ivoire and Ghana). Contemporarily, the rapidly growing human population and land conversion for agriculture, on-going forestry extraction, and increasing access to habitat in West and Central Africa are apparently facilitating increases in poaching rates (Mayaux et al., 2013).

Despite being protected by legislation in range countries, law enforcement is typically problematic and ineffective for a number of reasons including a lack of resources and capacity, both human and technical (Challender and Waterman, 2017), and the black-bellied pangolin is threatened by overexploitation. The species has been, and is, extensively hunted and poached for the bushmeat and traditional African medicine trades and openly displayed in markets across its range (see Chapter 15). Anadu et al. (1988) recorded arboreal pangolins being the eighth most preferred mammal among consumers in southwestern Nigeria. Boakye et al. (2016) reported that the black-bellied pangolin comprised 18% of 98 pangolins traded by chop-bar operators, wholesalers and farmer-hunters in Ghana between 2013 and 2014 (see also Chapter 15). In Gabon, the species continues to be hunted and traded and prices are increasing, and there is a persistent demand for arboreal pangolins, including emerging consumption by immigrant Asian populations (Mambeya et al., 2018).

There has been little recorded international trade in this species historically, which has been limited to exports of small numbers of live animals and other derivatives (Challender and Waterman, 2017). However, the CITES trade data indicate the export of 200 live black-bellied pangolins from Nigeria to China in 2015 for captive breeding purposes (Chapters 16 and 32). In 2008, intercontinental trafficking of African pangolins to Asia emerged, primarily involving scales, and between 2008 and 2019 it has ostensibly increased, and included the black-bellied pangolin (see Chapter 16: Mwale et al., 2017). Estimating the number of individuals involved is problematic, partly because of poor reporting, but scales from this species have been found in major shipments that have been seized en route to Asian markets. For example, in April 2019 12.9 tonnes of scales, involving an estimated ~35,000 African pangolins, were seized in Singapore en route from Nigeria to Vietnam (Liu, 2019; see Chapter 16 for conversion parameters). Cameroon and Nigeria have been implicated as major exporters (Ingram et al., 2019b) and pangolins in West and Central Africa are being targeted specifically for their scales for illegal export to Asia (see Chapter 16). This is a major threat to the species. Illegal trade in the black-bellied pangolin to Europe for bushmeat consumption has also been recorded (Chaber et al., 2010).

References

Allen, G.M., Coolidge, H.J. Jr., 1930. Mammals of Liberia. In: Strong, R.P. (Ed.), The African Republic of Liberia and the Belgian Congo, 2. Contributions of the Department of Tropical Medicine and the Institute for Tropical Biology and Medicine, pp. 569–622.

Anadu, P.A., Elamah, P.O., Oates, J.F., 1988. The bushmeat trade in southwestern Nigeria: a case study. Human Ecol. 16 (2), 199–208.

Angelici, F.M., Grimod, I., Politano, E., 1999. Mammals of the Eastern Niger Delta (Rivers and Bayelsa States, Nigeria): an environment affected by a gas-pipeline. Folia Zool. 48 (4), 249–264.

Angelici, F., Egbide, B., Akani, G., 2001. Some new mammal records from the rainforests of south-eastern Nigeria, Hystrix, Ital. J. Mammol. 12 (1), 37–43.

Barnett, A.A., Prangley, M.L., 1997. Mammalogy in the Republic of Guinea: an overview of research from 1946 to 1996, a preliminary check-list and a summary of research recommendations for the future. Mammal Rev. 27 (3), 115–164.

Barrie, B., Kanté, S., 2006. A rapid survey of the large mammals in Déré, Diécké and Mt. Béro classified forests in Guinée-Forestière, Southeastern Guinea. In: Wright, H.E., McCullough, J., Alonso, L.E., Diallo, M. S. (Eds.), A Rapid Biological Assessment of Three Classified Forests in Southeastern Guinea. RAP Bulletin of Biological Assessment, 40. Conservation International, Washington, D.C., pp. 189–194.

Beja, P., Pinto, P.V., Veríssimo, L., Bersacola, E., Fabiano, E., Palmeirim, J.M., et al., 2019. The Mammals of Angola. In: Huntley, B., Russo, V., Lages, F., Nuno-Ferrand, N. (Eds.), Biodiversity of Angola, Science & Conservation: A Modern Synthesis. Springer, Cham, pp. 357–443.

Blench, R., Dendo, M., 2007. Mammals of the Niger Delta, Nigeria.

Boakye, M.K., Kotzé, A., Dalton, D.L., Jansen, R., 2016. Unravelling the pangolin bushmeat commodity chain and the extent of trade in Ghana. Human Ecol. 44 (2), 257–264.

Bocage, J.V.B. du., 1889. Mammiferes d'Angola et du Congo. Jornal de Sciências, Mathemáticas, Physicas e Naturaes, Lisboa 2 (1), 8–32, 174–185.

Bocage, J.V.B. du., 1890. Mammiferes d'Angola et du Congo. Jornal de Sciências, Mathemáticas, Physicas e Naturaes, Lisboa 2 (2), 1–32.

Booth, A.H., 1960. Small Mammals of West Africa. Longmans, London.

Brisson, M.-J., 1756. Le Regne Animale Divisé En IX Classes. Chez Cl. Jean-Baptiste Bauche, Paris.

Brisson, M.-J., 1762. Regnum Animale in Classes IX. Apud Theodorum Haak, Lugduni Batavorum.

Carpaneto, G.M., Germi, F.P., 1989. The mammals in the zoological culture of the Mbuti Pygmies in north-eastern Zaire. Hystrix 1 (1), 1–83.

Chaber, A., Allebone-Webb, S., Lignereux, Y., Cunningham, A., Rowcliffe, J.M., 2010. The scale of illegal meat importation from Africa to Europe via Paris. Conserv. Lett. 3 (5), 317–323.

Challender, D., Waterman, C., 2017. Implementation of CITES Decisions 17.239 b) and 17.240 on Pangolins (*Manis* spp.), CITES SC69 Doc. 57 Annex. Available from <https://cites.org/sites/default/files/eng/com/sc/69/E-SC69-57-A.pdf>. [August 2, 2018].

Desmarest, A.G., 1822. Mammalogie ou description des Espèces de Mammifères. Second partie, contenant Les Ordres des Rongeurs, des Édentatés, des Pachydermes, des Ruminans et des Cétacés. Chez Mme Veuve Agasse, Imprimeur-Libraire, Paris.

Doran, G.A., Allbrook, D.B., 1973. The tongue and associated structures in two species of African pangolins, *Manis gigantea* and *Manis tricuspis*. J. Mammal. 54 (4), 887–899.

Dorst, J., Dandelot, P., 1970. A Field Guide to the Larger Mammals of Africa. Collins, London.

du Toit, Z., du Plessis, M., Dalton, D.L., Jansen, R., Grobler, J.P., Kotze, A., 2017. Mitochondrial genomes of African pangolins and insights into evolutionary patterns and phylogeny of the family Manidae. BMC Genom. 18, 746.

Emry, R.J., 1970. A North American Oligocene pangolin and other additions to the Pholidota. Bull. Am. Museum Nat. Hist. 142, 459–510.

Erxleben, U.C.P., 1777. Systema Regna Animalis, classis 1, Mammalia. Lipsize, Impensis Weygandianus.

Feiler, A., 1990. Distribution of mammals in Angola and notes on biogeography. In: Peters, G., Hutterer, R.

(Eds.), Vertebrates in the Tropics. Museum Alexander Koenig, Bonn, pp. 221–236.

Fitzinger, L.J., 1872. Die naturliche familie der schuppenthiere (Manes). Sitzungsberichte der Kaiserlichen Akademie der Wissenschaften. Mathematisch-Naturwissenschaftliche Classe, CI., LXV, Abth. I, 9–83.

Frade, F., 1949. Algumas novidades para a fauna da Guiné Portuguesa (aves e mamíferos), 4. Junta das Missões Geográficas e de Investigaçõeses Colóniais (Lisboa), Estudos de Zoología, Anais, pp. 165–186.

Frechkop, S., 1931. Notes sur les mammifères. VI. Quelques observations sur la classification des pangolins (Manidae). Bulletin du Musee royal d'Histoire naturelle de Belgique VII (22), 1–14.

Gaubert, P., 2011. Family Manidae. In: Wilson, D.E., Mittermeier, R.A. (Eds.), Handbook of the Mammals of the World, vol. 2. Hoofed Mammals. Lynx Edicions, Barcelona, pp. 82–103.

Gaubert, P., Antunes, A., Meng, H., Miao, L., Peigné, S., Justy, F., et al., 2018. The complete phylogeny of pangolins: scaling up resources for the molecular tracing of the most trafficked mammals on Earth. J. Hered. 109 (4), 347–359.

Gaudin, T.J., Emry, R.J., Wible, J.R., 2009. The phylogeny of living and extinct pangolins (Mammalia, Pholidota) and associated taxa: a morphology based analysis. J. Mammal. Evol. 16 (4), 235–305.

Gotch, A.F., 1979. Mammals – Their Latin Names Explained. A Guide to Animal Classification. Blandford Press, Poole.

Gray, J.E., 1843. List of the Specimens of Mammalia in the Collection of the British Museum. George Woodfall and Son, London.

Grubb, P., Jones, T.S., Davies, A.G., Edberg, E., Starin, E.D., Hill, J.E., 1998. Mammals of Ghana, Sierra Leone and the Gambia. Trendrine Press, Zennor, Cornwall.

Happold, D.C.D., 1987. The Mammals of Nigeria. Clarendon Press, Oxford.

Hassanin, A., Hugot, J.-P., van Vuuren, B.J., 2015. Comparison of mitochondrial genome sequences of pangolins (Mammalia, Pholidota). C. R. Biol. 338 (4), 260–265.

Hatt, R.T., 1934. The pangolin and aard-varks collected by the American Museum Congo expedition. Bull. Am. Museum Nat. Hist. 66, 643–671.

Heath, M.E., 2013. Order Pholidota – Pangolins. In: Kingdon, J., Hoffmann, M. (Eds.), Mammals of Africa, vol. V, Carnivores, Pangolins, Equids, Rhinoceroses. Bloomsbury Publishing, London, pp. 384–386.

Henschel, P., Abernethy, K.A., White, L.J.T., 2005. Leopard food habits in the Lopé National Park, Gabon, Central Africa. Afr. J. Ecol. 43 (1), 21–28.

Henschel, P., Hunter, L.T.B., Coad, L., Abernethy, K.A., Mühlenberg, M., 2011. Leopard prey choice in the Congo Basin rainforest suggests exploitative competition with human bushmeat hunters. J. Zool. 285 (1), 11–20.

Hildwein, G., 1974. Resting metabolic rates in pangolins (Pholidota) and squirrels of equatorial rain forests. Arch. Sci. Physiol. 28, 183–195.

Hill, J.E., Carter, T.D., 1941. The mammals of Angola, Africa. Bull. Am. Museum Nat. Hist. 78, 1–211.

Ingram, D.J., Shirley, M.H., Pietersen, D., Godwill Ichu, I., Sodeinde, O., Moumbolou, C., et al., 2019a. *Phataginus tetradactyla. The IUCN Red List of Threatened Species* 2019: e.T12766A123586126. Available from: <http://dx.doi.org/10.2305/IUCN.UK.2019-3.RLTS.T12766A123586126.en>.

Ingram, D.J., Cronin, D.T., Challender, D.W.S., Venditti, D. M., Gonder, M.K., 2019b. Characterising trafficking and trade of pangolins in the Gulf of Guinea. Glob. Ecol. Conserv. 17, e00576.

Jeannin, A., 1936. Les mammiféres sauvages du Cameroun. Encyclopedie Biol. 16, 116–130.

Jentink, F.A., 1882. Note XXV. Revision of the Manidae in the Leyden Museum. Notes from the Leyden Museum IV, 193–209.

Kingdon, J., 1971. East African mammals. An Atlas of evolution in Africa, Primates, Hyraxes, Pangolins, Protoungulates, Sirenians, vol. I. Academic Press, London.

Kingdon, J., 1997. The Kingdon Field Guide to African Mammals. Academic Press, London.

Kingdon, J., Hoffmann, M., 2013. *Phataginus tetradactyla* - Long-tailed Pangolin. In: Kingdon, J., Hoffmann, M. (Eds.), Mammals of Africa, vol. V, Carnivores, Pangolins, Equids, Rhinoceroses. Bloomsbury Publishing, London, pp. 389–391.

Kityo, R., Prinsloo, S., Ayebere, S., Plumptre, A., Rwetsiba, A., Sadic, W., et al., 2016. *Phataginus tetradactyla*. Available from: <https://www.nationalredlist.org/species-information/?speciesID=263608>. [May 28, 2019].

Linnaeus, C., 1766. Systema Naturæ Per Regna Tria Naturæ, Secundum Classes, Ordines, Genera, Species, Cum Characteribus, Differentiis, Synonymis, Locis. Tomus I. Editio Duodecima, reformata. Salvius, Stockholm.

Liu, V., 2019. World Record Haul of Pangolin Scales Worth $52 Million Seized From Container at Pasir Panjang. Available from: <https://www.straitstimes.com/singapore/environment/record-haul-of-pangolin-scales-worth-52-million-seized-from-container-atpasir?fbclid=IwAR2mQJIf80UPz_InEScc5YZ-2HcFU2n3cFO5K2vT3lsat4Yng8iqTEhJes>. [May 28, 2019].

Luiselli, L., Amori, G., Akani, G.C., Eniang, E.A., 2015. Ecological diversity, community structure and conservation of Niger Delta mammals. Biodivers. Conserv. 24 (11), 2809–2830.

Machado, A de. B., 1969. Mamiferos de Angola ainda não citados ou pouco conhecidosa. Publições culturais da Companhia de Diamantes de Angola 46, 93–232.

Malbrant, R., MacLatchy, A., 1949. Faune de l'Équateur Africain Français. Tome II: Mammifères. Paul Lechevalier, Paris.

Mambeya, M.M., Baker, F., Momboua, B.R., Pambo, A.F.K., Hega, M., Okouyi, V.J.O., et al., 2018. The emergence of a commercial trade in pangolins from Gabon. Afr. J. Ecol. 56 (3), 601–609.

Mayaux, P., Pekel, J.-F., Desclee, B., Donnay, F., Lupi, A., Achard, F., et al., 2013. State and evolution of the African rainforests between 1990 and 2010. Philos. Trans. R. Soc. B 368 (1625), 20120300.

McKenna, M.C., Bell, S.K. (Eds.), 1997. Classification of Mammals: Above the Species Level. Columbia University Press, New York.

Meester, J., 1972. Order Pholidota. In: Meester, J., Stzer, H. W. (Eds.), The Mammals of Africa: An Identification Manual, Part 4. Smithsonian Institution Press, Washington, D.C., pp. 1–3.

Megevand, C., 2013. Deforestation Trends in the Congo Basin: Reconciling Economic Growth and Forest Protection. World Bank, Washington, D.C.

Mohr, E., 1961. Schuppentiere. Neue Brehm-Bucherei. A. Ziemsen Verlag, Wittenberg Lutherstadt.

Monard, A., 1935. Contribution à la Mammalogie d'Angola et Prodrome d'une Faune d'Angola. Arquivos do Museu Bocage 6, 1–314.

Mwale, M., Dalton, D.L., Jansen, R., De Bruyn, D., Pietersen, D., Mokgokong, P.S., et al., 2017. Forensic application of DNA barcoding for identification of illegally traded African pangolin scales. Genome 60 (3), 272–284.

Noack, T., 1889. Beiträge zur Kenntniss der Saugethierfauna von Sud-und Sudwest-Afrika. In: Spengel, J.W. (Ed.), Zoologische Jahrbücher. Abtheilung für Systematik, Geograhpie und Biologie Der Thiere, fourth ed. Fischer, Jena.

Pagès, E., 1970. Sur l'écologie et les adaptations de l'oryctérope et des pangolins sympatriques du Gabon. Biol. Gabon. 6, 27–92.

Pagès, E., 1972a. Comportement agressif et sexuel chez les pangolins arboricoles (*Manis tricuspis* et *M. longicaudata*). Biol. Gabon 8, 3–62.

Pagès, E., 1972b. Comportement maternal et developpement du jeune chez un pangolin arboricole (*M. tricuspis*). Biol. Gabon. 8 (1), 63–120.

Pagès, E., 1975. Étude éco-éthologique de *Manis tricuspis* par radio-tracking. Mammalia 39, 613–641.

Pocock, R.I., 1924. The external characters of the pangolins (Manidae). Proc. Zool. Soc. Lond. 94 (3), 707–723.

Rafinesque, C.S., 1820. Sur le genre *Manis* et description d'une nouvelle espèce: *Manis ceonyx*. Annales Générales des Sciences Physiques 7, 214–215.

Rahm, U., 1956. Notes on Pangolins of the Ivory Coast. J. Mammal. 37 (4), 531–537.

Rahm, U., 1966. Les mammifères de la forêt équatoriale de l'est du Congo. Annales du Musée Royal de l'Afrique Centrale, Sciences Zoologiques 149, 39–121.

Reiner, F., Simões, P., 1999. Mamíferos selvagens de Guiné-Bissau. Centro Portugues de Estudos dos Mamiferos Marinhos, Lisbao, Portugal.

Rosevear, D.R., 1953. Checklist and Atlas of Nigerian Mammals, With a Foreword on Vegetation. The Government Printer, Lagos.

Sclater, W.A., 1901. The Fauna of South Africa. R.H. Porter, London.

Schlitter, D.A., 2005. Order Pholidota. In: Wilson, D.E., Reeder, D.M. (Eds.), Mammal Species of the World: A Taxonomic and Geographic Reference, third ed. Johns Hopkins University Press, Baltimore, pp. 530–531.

Schouteden, H., 1944. De zoogdieren van Belgisch Congo en van Ruanda-Urundi I. – Primates, Chiroptera, Insectivora, Pholidota. Annalen van het Museum van belgisch Congo. C. Dierkunde. Reeks II. Deel III. Aflevering 1, 1–168.

Sikes, S.K., 1966. The tricuspid tree pangolin (*Manis tricuspis*): Its remarkable tongue complex. Niger. Field 31, 99–110.

Sodeinde, O.A., Adedipe, S.R., 1994. Pangolins in southwest Nigeria – current status and prognosis. Oryx 28 (1), 43–50.

Sundevall, C.J., 1842. Om slägtet Sorex, med några nya arters beskrifning. Kungliga Vetenskapsakademien, Stockholm.

Tahiri-Zagrët, C., 1970a. Les Pangolins de Côte d'Ivoire II. – Les Especes et Leurs Repartitions Geographiques. Annales de l'Universite d'Abidjan, Series III, Fasicule 1, 223–244.

Tahiri-Zagrët, C., 1970b. Les Pangolins de Côte d'Ivoire III. – Observations Ethologiques. Annales de l'Universite d'Abidjan, Series E, III, Fasicule 1, 245–252.

Taylor, P.J., Neef, G., Keith, M., Weier, S., Monadjem, A., Parker, D.M., 2018. Tapping into technology and the biodiversity informatics revolution: updated terrestrial mammal list of Angola, with new records from the Okavango Basin. ZooKeys 779, 51–88.

Thomas, O., 1904. On mammals from northern Angola collected by Dr. W. J. Ansorge. Ann. Mag. Nat. Hist. 7 (13), 405–421.

Ullmann, T., Veríssimo, D., Challender, D.W.S., 2019. Evaluating the application of scale frequency to estimate the size of pangolin scale seizures. Glob. Ecol. Conserv. 20, e00776.

CHAPTER 9

White-bellied pangolin *Phataginus tricuspis* (Rafinesque, 1820)

Raymond Jansen[1,2], Olufemi Sodeinde[3], Durojaye Soewu[4], Darren W. Pietersen[5,6], Daniel Alempijevic[7] and Daniel J. Ingram[8]

[1]Department of Environmental, Water and Earth Sciences, Tshwane University of Technology, Pretoria, South Africa [2]African Pangolin Working Group, Pretoria, South Africa [3]Department of Biological Sciences, New York City College of Technology, City University of New York, Brooklyn, NY, United States [4]Department of Fisheries and Wildlife Management, College of Agriculture, Ejigbo Campus, Osun State University, Osogbo, Nigeria [5]Mammal Research Institute, Department of Zoology and Entomology, University of Pretoria, Hatfield, South Africa [6]IUCN SSC Pangolin Specialist Group, ℅ Zoological Society of London, Regent's Park, London, United Kingdom [7]Integrative Biology, Florida Atlantic University, Boca Raton, FL, United States [8]African Forest Ecology Group, Biological and Environmental Sciences, University of Stirling, Stirling, United Kingdom

OUTLINE

Pangolins
DOI: https://doi.org/10.1016/B978-0-12-815507-3.00009-5

Taxonomy

Previously included in the genus *Manis* (Meester, 1972; Schlitter, 2005), the species is here included in *Phataginus* based on both morphological (Gaudin et al., 2009) and genetic (du Toit et al., 2017; Gaubert et al., 2018) evidence. Hatt (1934) reported that the species exhibits slight morphological variation across its range, including in scale pigmentation and hair length (e.g., on the ventrum). Allen and Loveridge (1942) and Meester (1972) proposed two distinct subspecies based on morphological analyses, *P. tricuspis tricuspis* (range except Uganda) and *P. t. mabirae* (Uganda). These subspecies are not considered valid here following Kingdon and Hoffmann (2013).

The species is included in the subfamily Phataginninae to designate small African pangolins based on mitogenomic distances between the genera *Manis*, *Smutsia*, and *Phataginus*, and morphological traits unique to *Phataginus* (see Chapter 2; Gaubert et al., 2018). Gaubert et al. (2016) identified six geographic lineages within the white-bellied pangolin that were identified as Evolutionarily Significant Units (ESUs) and may warrant species or subspecies status; this requires further research. These lineages were partitioned into Central Africa, Gabon, Dahomey Gap, Ghana, western Africa, and western Central Africa (see Chapter 2). Hassanin et al. (2015) discovered a high genetic nucleotide divergence between a white-bellied pangolin from Gabon and specimens from Cameroon, Ghana and Nigeria, and the potential for a new taxon in Gabon requires further research.

The type locality is given as "Guinée", West Africa (Rafinesque, 1820). Chromosome number is not known. Synonyms: *Manis multiscutata* (Gray, 1843), *Manis tridentata* (Focillon, 1850), *P. t. mabirae* (Allen and Loveridge, 1942).

Etymology: *Phataginus* (see Chapter 8); the species name refers to the three (*tri-*) points or cusps (*-cuspis*) on the scales (Gotch, 1979).

Description

The white-bellied pangolin (*Phataginus tricuspis*) is a small, semi-arboreal African pangolin with a body weight of 1–3 kg and a total length of about 100 cm (Table 9.1). It is the lightest pangolin species, marginally lighter than the broadly sympatric black-bellied pangolin (*P. tetradactyla*). The species is not sexually dimorphic but males are reported to be slightly longer and heavier than females (Pagès, 1968). Up to 60% of the total body length comprises the long, prehensile tail (35–60 cm); head-body length ranges between 25 and 38 cm, potentially longer (Table 9.1). The tail is flat ventrally and rounded dorsally, and has a naked cutaneous pad at the tip (which is truncated) on the ventral side that replaces two median and two lateral scales (Pocock, 1924). The sensory tail pad contains many mechanoreceptor nerve endings that aid sensitivity in terms of touch and grip (Doran and Allbrook, 1973), and the tail functions as a fifth limb when the species is climbing (see *Behavior*). There are 41 caudal vertebrae (Table 9.1).

The body is covered in small, overlapping scales that grow from the epidermis in a lattice-like arrangement and are tricuspid (Jentink, 1882; Rahm, 1956). The scales cover the dorsal and lateral surfaces of the body, the fore- (except the upper portions) and hindlimbs, the tail (dorsally and ventrally), and the neck and head, but are absent from the anterior of the face, the ventral body surface and inner-side of the limbs. On the upper portion of the forelimbs (elbow to wrist) dense (sometimes long) brown hair replaces the scales (Pocock, 1924). There are 18–22

TABLE 9.1 White-bellied pangolin morphometrics.

Measurement			Country	Source(s)
Weight	Weight (♂)[a]	1.67 (1.2–2.3) kg, n = 8	Southwest Nigeria	FMNH[b]
		2.36 (1.74–2.86) kg, n = 4	Southeast Nigeria	Kingdon and Hoffmann, 2013
	Weight (♀)	1.71 (1.2–2.2) kg, n = 8	Southwest Nigeria	FMNH[a]
		2.6 (1.94–2.88) kg, n = 11	Southeast Nigeria	Kingdon and Hoffmann, 2013
Body	Total length (♂)	793.2 (617–1027) mm, n = 25	Democratic Republic of the Congo	Hatt, 1934
	Total length (♀)	768.4 (630–920) mm, n = 25	Democratic Republic of the Congo	Hatt, 1934
	Head-body length (♂)	350 (330–380) mm, n = 7	Côte d'Ivoire, Democratic Republic the of Congo, Liberia	FMNH[b]
		319 (254–375) mm, n = 17	Southeast Nigeria	Kingdon and Hoffmann, 2013
	Head-body length (♀)	333 (308–367) mm, n = 8	Côte d'Ivoire, Democratic Republic of the Congo, Ghana, Liberia	FMNH[b]
		310 (265–351) mm, n = 14	Southeast Nigeria	Kingdon and Hoffmann, 2013
	Tail length (♂)	469.6 (360–607) mm, n = 25	Democratic Republic of the Congo	Hatt, 1934
	Tail length (♀)	460.4 (350–590) mm, n = 25	Democratic Republic of the Congo	Hatt, 1934
Vertebrae	Total number of vertebrae	69		Frechkop, 1931; Jentink, 1882
	Cervical	7		Jentink, 1882; Mohr, 1961
	Thoracic	13		Jentink, 1882; Mohr, 1961
	Lumbar	6		Jentink, 1882; Mohr, 1961
	Sacral	2		Jentink, 1882; Mohr, 1961
	Caudal	41		Jentink, 1882; Mohr, 1961
Skull	Length (♂)	72.8 (63.8–80.8) mm, n = 22	Democratic Republic of the Congo	Hatt, 1934
	Length (♀)	68.7 (58.5–79.2) mm, n = 20	Democratic Republic of the Congo	Hatt, 1934

(*Continued*)

TABLE 9.1 (Continued)

	Measurement		Country	Source(s)
	Breadth across zygomatic processes (♂)	27.3 (22.7–32) mm, n = 22	Democratic Republic of the Congo	Hatt, 1934
	Breadth across zygomatic processes (♀)	25.4 (20.2–29.3) mm, n = 20	Democratic Republic of the Congo	Hatt, 1934
Scales	Total number of scales	935 (794–1141), n = 25	Cameroon, Nigeria, Sierra Leone, Uganda, unknown	Ullmann et al., 2019; D. Soewu, unpubl. data
	No. of scale rows (transversal, body)	18–22	Democratic Republic of the Congo	Frechkop, 1931; Hatt, 1934
	No. of scale rows (longitudinal, body)	19–25	Democratic Republic of the Congo	Frechkop, 1931; Hatt, 1934; Jentink, 1882
	No. of scales on outer margins of tail	35–40	Democratic Republic of the Congo	Frechkop, 1931; Hatt, 1934; Jentink, 1882
	No. of scales on median row of tail	30–33	Democratic Republic of the Congo	Frechkop, 1931; Hatt, 1934
	Scales (wet) as proportion of body weight	0.15–0.20		African Wildlife Foundation, unpubl. data
	Scales (dry) as proportion of body weight	No data		

[a]*Males are reported to be slightly heavier than females; these data are based on a small sample size.*
[b]*FMNH=Florida Museum of Natural History.*

transversal and 19–25 longitudinal scale rows on the body, and 35–40 rows of scales on the outer margins of the tail that are kite-shaped (Frechkop, 1931; Hatt, 1934; Jentink, 1882). The medio-dorsal row of scales on the tail is interrupted towards the tail tip and is replaced by two rows of 3–6 scales (Jentink, 1882). Total scales number between ~790 and 1140 (Table 9.1). The largest scales are on the dorsal surface and measure up to 47 × 26 mm; single scales weigh up to 0.66 g (O. Sodeinde, unpubl. data). The scales are uniform in color but vary substantially from brownish-gray to reddish- or yellowish-brown, and the distal edge may color faint yellow, especially in older individuals (Figs. 9.1–9.3; Rahm, 1956; R. Jansen, pers. obs.). The scales grow in length more rapidly than in width: scales from the mid-dorsal region grow from being approximately as long as wide, to four times length versus width. The scales are laterally striated (Fig. 9.2) and keeled on the flanks and limbs, and are less robust (i.e., are flimsy) compared to scales on the black-bellied pangolin and other pangolin species. The striations and cusps wear over time, sometimes completely, and the cusps may break off with age (Hatt, 1934; R. Jansen, pers. obs.). Unlike Asian pangolins, hairs do not project between the scales.

The skull demonstrates morphological features related to adaption to a myrmecophagous diet (Chapter 1; see Hatt, 1934; Heath, 2013).

FIGURE 9.1 White-bellied pangolin foraging for prey in Central African Republic. The cusps on the scales can clearly be seen on this individual. *Photo credit: Alex Ley.*

As in the black-bellied pangolin, the lacrimal bone is present – it is absent in other pangolin species (Emry, 1970). The head is conical-shaped and the muzzle is broader than that of the black-bellied pangolin (Hatt, 1934). The scales extend to the forehead but are absent from the face, which is covered in sparse hairs (Fig. 9.3). The skin around the snout and eyes is pinkish brown, with a black patch beneath the eyes; the lips are pinkish (Hatt, 1934). The irises are dark and the eyes large and bulbous with thick, swollen lids (Hatt, 1934). The rhinarium is well defined, naked and moist (Pocock, 1924). The ear pinnae are totally suppressed (Fig. 9.3), but the auditory orifice can be partially closed by a bordering small flange (Pocock, 1924). The belly is pale grayish-white and is covered with white hair up to 20 mm in length, which can be dense (Hatt, 1934).

The anatomy is highly adapted to a myrmecophagous diet. The tongue is 30 cm in length and, as in other pangolin species, is attached to the caudal end of the xiphoid process (xiphisternum) in the abdomen (Doran and Allbrook, 1973). In this species, the proximal region of the tongue is "U-shaped" and extends caudally to the right iliac fossa before turning cranially, terminating under the right side of the diaphragm (Chan, 1995; Heath, 2013). From the abdomen, the tongue passes through the thorax and neck to the oral cavity which is small (Doran and Allbrook, 1973). Large salivary glands located in the pharyngeal and cervical regions produce an alkaline mucus (pH 9–10; Fang, 1981) which is secreted into the sheath housing the tongue and onto the tongue itself (Doran and Allbrook, 1973; Heath, 2013). During feeding, the tongue is repeatedly extended and retracted and the anterior portion has a high density of fascicles, suggesting a sensory (i.e., prey location) rather than gustatory function; the tongue also lacks papillae

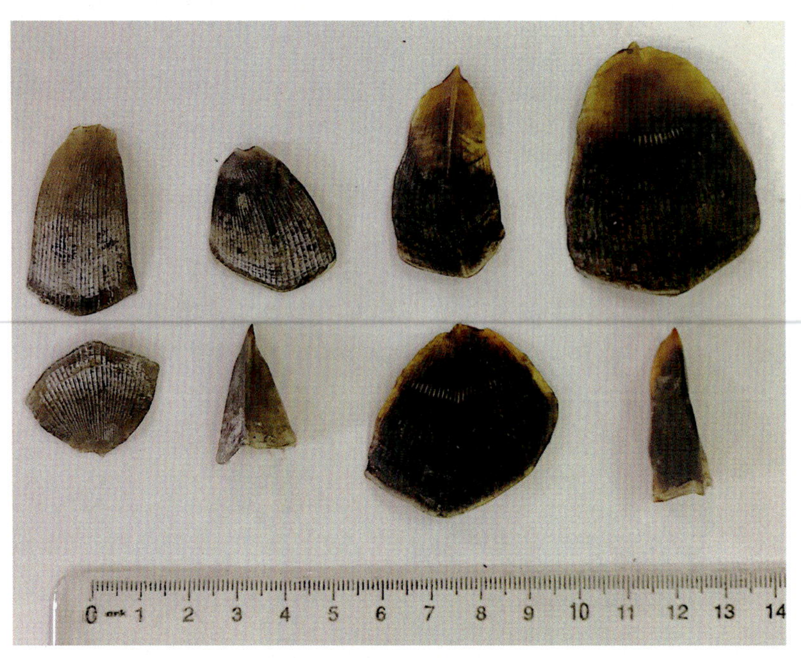

FIGURE 9.2 Scales of the white- (left) and black-bellied pangolin (right). White-bellied pangolin: top left – dorsal, post-scapular; top right – dorsal, scapular; bottom left – dorsal, tail; bottom right – marginal, tail. Black-bellied pangolin: top left – dorsal, marginal; top right – dorsal, post-scapular; bottom left – dorsal, tail; bottom right – marginal, tail. *Photo credit: Matthew H. Shirley.*

FIGURE 9.3 Aged white-bellied pangolin showing worn scales. *Photo credit: Frank Kohn.*

indicative of its transport function (Ofusori et al., 2008). The hyoid bone serves to scrape prey from the tongue and direct it down the esophagus to the stomach where it is masticated (Doran and Allbrook, 1973). The gizzard-like stomach is lined by keratinized stratified squamous epithelium and dense collagen fibers, which offer protection against ulceration by the hard chitinous parts of ants and termites (Ofusori et al., 2008).

The forelimbs are slightly shorter than the hindlimbs (Sodeinde et al., 2002); both terminate in five digits. On the forefeet, the first digit is vestigial; the second, third and fourth digits have long claws, the third being the longest and most extended (Fig. 9.3; Pocock, 1924; Rahm, 1956). The fifth digit is longer than the first and stems from the base of the fourth digit (Pocock, 1924). The claws on the forefeet are slightly longer than on the hindfeet. As in the black-bellied pangolin, the hindfeet have four long, well-developed claws of approximately equal length, and the innermost claw is vestigial (Pocock, 1924). Both fore- and hindfeet are placed palm downwards and have well developed skin pads, extending from the level of the first digit to the heel in the hindfeet (Pocock, 1924).

Large anal glands are situated to the side of the anus, which produce a white secretion and foul-smelling odor and are important to the species' ecology (see *Ecology*; Pagès, 1968; Pocock, 1924). The perineum is short and in the female the vulva is anterior to the anus; the female has one pair of pectoral mammae. Male testes descend to the inguinal area.

Body temperature is regulated at 27–34 °C, potentially higher (see Heath and Hammel, 1986), and fluctuates with activity, being higher when individuals are active; resting metabolic rate has been estimated at 202.2 ml O_2/kg/h (Hildwein, 1974; Jones, 1973).

The white-bellied pangolin may be confused with its congener the black-bellied pangolin. Distinguishing features include the absence of very large post-scapular scales in the former, in addition to skin color, scale size and coloration (Fig. 9.2), and scale number (see Table 9.1). Other differences include the broader muzzle in the white-bellied, and proportionally longer tail in the black-bellied pangolin (Chapter 8). The species is distinguishable from the two larger African pangolins by its smaller size, longer tail, and scale coloration and morphology: dorsal scales on the white-bellied pangolin are typically longer than they are wide.

Distribution

A widely distributed species in West and Central Africa (Fig. 9.4). Recorded from Guinea-Bissau (Cantanhez National Park; Bout and Ghiurghi, 2013) and Guinea in West Africa (Ziegler et al., 2002) through Sierra Leone (Boakye et al., 2016a; Grubb et al., 1998), Liberia (Allen and Coolidge, 1948; Verschuren, 1982), Côte d'Ivoire (Rahm, 1956) and Ghana (Boakye et al., 2016b; Grubb et al., 1998; Ofori et al., 2012). There are no confirmed records in Senegal or The Gambia (Grubb et al., 1998). There are fewer records eastward in Togo and Benin, but the species has been recorded in southern Togo (Amori et al., 2016). Apparently infrequently recorded in Benin, there are records from the Monts Kouffé protected forest in 1978, in bushmeat markets in central Benin in the late 1970s (Sayer and Green, 1984), and from an ecological study undertaken in Lama Forest Reserve in the south in the mid-2000s (Akpona et al., 2008). There are unverified reports from northern Benin (H. Akpona, pers. comm.).

The species occurs in southern Nigeria (Angelici et al., 1999a; Sodeinde and Adedipe, 1994), and Cameroon (Allen and Loveridge, 1942; Jeannin, 1936), Equatorial Guinea (Kümpel, 2006), including Bioko island (Albrechtsen et al., 2007), Gabon (Pagès, 1965, 1975) and Republic of Congo, though published records are dated (Hatt, 1934; Swiacká, 2018; R. Jansen, pers. obs.). There are no records from Chad, but the species occurs across southern Central African Republic and in suitable habitat in Democratic Republic of the Congo (DRC; e.g., Garamba National Park; Monroe et al., 2015; van Vliet et al., 2015). There are records from Burundi and Rwanda (Verschuren, 1987) and Uganda, including in Semuliki National Park (Kityo, 2009; Treves et al., 2010; S. Nixon, pers. comm.). In Tanzania, the species is known from only two locations: the Minziro Forest Reserve on the northwestern border with Uganda, and close to Bukoba (Foley et al., 2014). The extreme eastern distribution comprises southwestern Kenya, including in the Kakamega Forest Reserve (Roth and Cords, 2015). Northern Mozambique has previously been suggested as the species' eastern limit (Smithers and Lobão Tello, 1976), but this has been rejected (Ansell, 1982; Kingdon and Hoffmann, 2013).

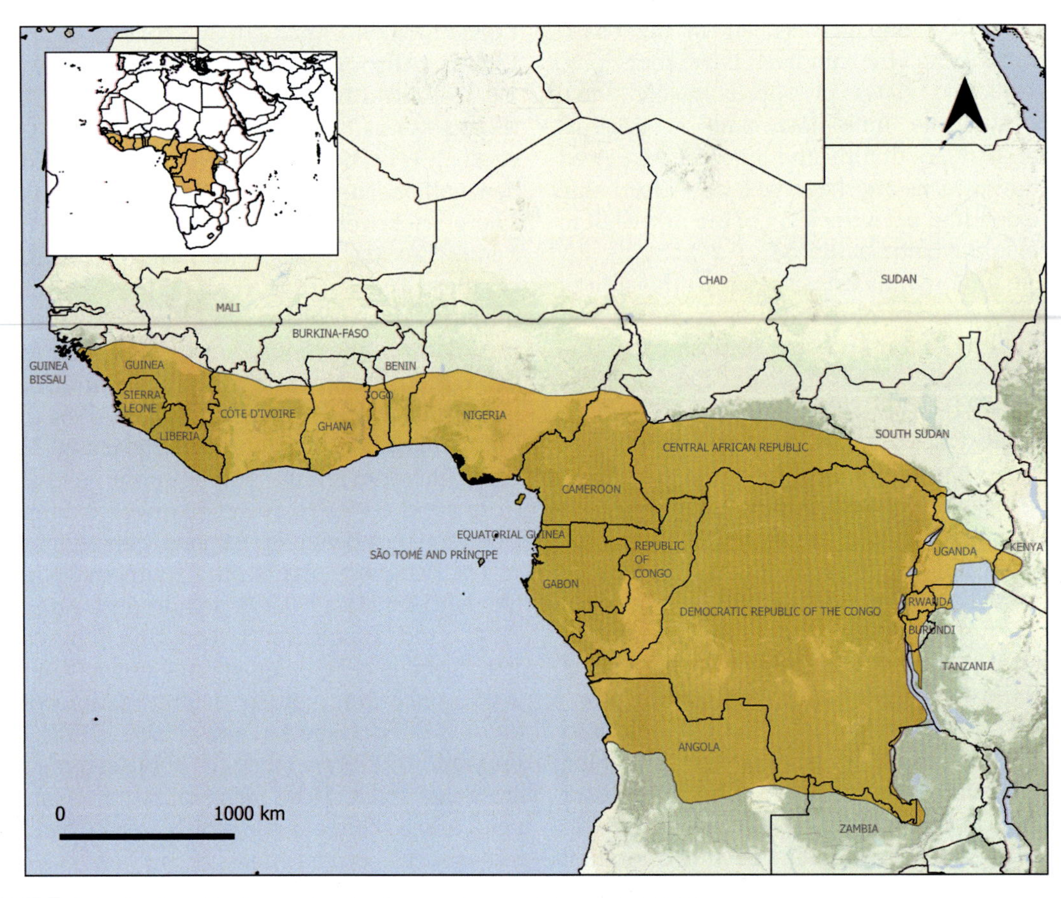

FIGURE 9.4 White-bellied pangolin distribution. *Source: Pietersen et al., 2019.*

To the south, the species occurs in northern Angola, including Cabinda, and there are recent records from Cangandala National Park (Beja et al., 2019; Hill and Carter, 1941). The southern distributional limit has previously been considered Mwinilunga District in northwestern Zambia (Cotterill, 2002) and there have been doubts about the species' occurrence in far south DRC (Schouteden, 1948). However, there are confirmed records from northwestern Zambia (near Solwezi) and central Zambia (Serenje) close to the DRC border from 2016 to 2018 indicating that the species occurs further south than previously recorded (Fig. 9.4; D. Pietersen, unpubl. data).

Habitat

The white-bellied pangolin predominantly inhabits moist tropical lowland forests and secondary forests (Angelici et al., 1999b; Happold, 1987; Kingdon, 1971), but also savanna-forest mosaics, dense woodland (including miombo woodland [*Brachystegia-Julbernardia*]), and riparian forests (Kingdon, 1997). The species also occurs in modified habitats including commercial plantations (e.g., teak [*Tectona grandis*], oil palm), in particular those that are little-used or abandoned, fallows, and farmland (e.g., areas of former lowland rainforest; Angelici et al., 1999b; Sodeinde and Adedipe, 1994). In Benin, Akpona et al. (2008) found no statistically significant difference in observations of the species between natural forest and plantations, though most individuals (70%) were observed in the former. This study suggested that the species prefers closed forest habitats, and based on the distribution of observed individuals, may be more sensitive to forest age rather than composition (Akpona et al., 2008). In contrast, a study investigating camera trap records across multiple study sites within the species' distribution (Khwaja et al., 2019), estimated a higher probability of occupancy outside of protected areas. Although using a coarse measure of protected status, this supports the notion that the species is adaptable to modified and degraded habitats. Odemuni and Ogunsina (2018) note that particular woody plants such as *Vitex doniana* may influence white-bellied pangolin distribution (e.g., in plantations), as the fruit they bear attracts large quantities of ants.

The occurrence of the species in modified, artificial and degraded landscapes suggests a level of adaptability to these habitats, providing basic needs such as an adequate prey base, suitable den sites, and tolerable levels of exploitation are met. However, use of these habitats, including the ability of the species to persist long-term, and reproduce, has not been well documented. Use of modified habitats, together with associated impacts on distribution and densities, is thus an avenue for future research.

The species is broadly sympatric with the black-bellied pangolin. There is some habitat overlap, but the latter is reported to show a preference for swamp forest (Angelici et al., 2001; Kingdon, 1997), and for arboreal ants, as opposed to termite prey (Chapter 8). Ecological differences between the two species have not been fully elucidated.

Ecology

Elizabeth Pagès pioneered ecological research on tropical African pangolins in the late 1960s and 1970s and much existing knowledge of white-bellied pangolin ecology (and behavior) stems from this research. Pagès (1975) used radio-telemetry and determined that the species occupies home ranges that vary in size and by sex, and may change temporally. Males generally occupy a larger home range (20–30 ha) than females (3–4 ha) and male home ranges, which are mutually exclusive, overlap with those of several (up to 10) females, which may intersect, suggesting a polygynous social structure. Evidence indicates that males are territorial (Pagès, 1975). Females explore only a small portion of their home range each night, in contrast to males that traverse larger sections of their range, including that of several females. Females spend less time foraging than males and cover a smaller area in doing so; Pagès (1975) estimated that females travel a mean of 400 m per night compared to 700 m in males, and males may travel up to 1.8 km a night. Activity is dependent on conditions (e.g., season and weather) and individuals may be active for up to 10.5 hours a night (Pagès, 1975). In Gabon, in the dry season (May–June), females forage, on average, for 5 hours a night, dropping to 2.45 hours in the wet season (January–April); males forage for 6.45 hours a night in the dry season, reducing to 3.45 in the rainy season (Pagès, 1975). This variation is reportedly because prey are more abundant in the wet season (Pagès, 1975). The species appears to follow regular arboreal routes through home ranges and will actively scent-mark using secretions from the anal glands and/or urine (Pagès, 1968).

In Gabon, Pagès (1975) identified two types of burrows used by white-bellied pangolins. The first, used only occasionally, is a burrow of 20–40 cm dug into the soil substrate or frequently into a termite mound, where the animal may feed before resting. The second is a resting place in a tree, such as a tree hollow (e.g., the hollow trunk of a dead tree), which is usually 10–15 m above the ground, in the fork of tree branches, or curled up amongst epiphytes. In Benin, Akpona et al. (2008) found that most refuges used were in trees and observed a preference for two tree species in particular, velvet tamarin (*Dialium guineneesis*) and kapok (*Ceiba pentandra*) in closed forest. While females will use resting places for extended periods (e.g., weeks at a time), males will use a new site nearly every night (Pagès, 1975).

The species is macrosmatic and uses olfaction to search for arboreal and terrestrial ant nests and termitaria (Pagès, 1975). The nests and termitaria are not destroyed and thereby provide a continuous source of prey. The species will also attack a column of ants traversing a tree trunk or branch and search for prey in dead branches and rotten tree trunks. Due to greater availability, it is likely that terrestrial ants and termites comprise the bulk of the diet (Pagès, 1975). Prey species include army ants (*Dorylus* and *Myrmicaria*), other ant genera including *Camponotus*, *Cataulacus*, *Oecophylla* and *Crematogaster* spp. (Pagès, 1970), and adults and nymphs of the termite genera *Nasutitermes* and *Microcerotermes* (Kingdon, 1971; Pagès, 1975). Leaf-processing mushroom termites (Macrotermitinae) are also preyed upon (Pagès, 1975).

There are numerous predators. White-bellied pangolin remains have been found in leopard (*Panthera pardus*) scat at various sites in Gabon (Henschel et al., 2005, 2011; Pagès, 1970) and other likely predators include African golden cat (*Profelis aurata*), African rock python (*Python sebae*), jackals (*Canis* spp.), ratels (*Mellivora capensis*), chimpanzees (*Pan troglodytes*), large owls, and possibly eagles (Kingdon and Hoffmann, 2013; Pagès, 1970). Ausden and Wood (1990) reported a marsh mongoose (*Atilax paludinosus*) eating the remains of a white-bellied pangolin in Sierra Leone, presumably a random scavenging event on a carcass.

White-bellied pangolins host internal and external parasites. Ticks from the genus *Amblyomma* are associated with the species (Allen and Loveridge, 1942; Ntiamoa-Baidu et al., 2005; Orhierhor et al., 2017) and Sodeinde and Soewu (2016) extracted acanthocephalans belonging to two genera, *Macracanthorhyncus* and *Oncicola*, from the gastrointestinal tract of individuals in Nigeria (see also Chapter 29).

Behavior

Primarily nocturnal, the species is semi-arboreal, spending time on the ground and in the trees (Pagès, 1975) and is less arboreal than the black-bellied pangolin (see Chapter 8). Although solitary, females are observed with young (Fig. 9.5), and Pagès (1965) often found pairs curled up together in tree hollows high off the ground. In Gabon, the species was recorded foraging mostly on the ground (Pagès, 1975), in contrast to Benin, where Akpona et al. (2008) observed the opposite – the species foraging largely in trees in a semi-deciduous forest. The same variation is apparent in DRC. In Lomami National Park, the species was detected most frequently on the ground at the edge of regenerating garden plots in forest adjacent to villages in the buffer zone, but were detected more often in the canopy in forest within the park's interior (D. Alempijevic, unpubl. data).

White-bellied pangolins spend most of the day sheltering in tree hollows, in the forked branch of a tree, or curled up amongst epiphytes. They are active nocturnally, devoting most of their active time to foraging for prey. However, they have been observed active diurnally (Jones, 1973). Pagès (1975) found that females and juveniles were active between 1900 and 2130, with males active over a wider interval, up until 0400 on occasion (Pagès, 1975). Camera trap studies in the Dja Biosphere Reserve, Cameroon found the species was active between 2000 and 0400 over 100 camera trap nights (ZSL, unpubl. data). In Lomami National Park, the species was recorded by video and was active between 1800 and 0500, with peak activity at 0300 (D.

FIGURE 9.5 White-bellied pangolin juvenile being "taxied" around on its mother's tail in Cameroon. *Photo credit: Jiri Prochazka/Shutterstock.com.*

Alempijevic, unpubl. data). Conspecifics were never recorded together, with the exception of a female carrying its offspring (D. Alempijevic, unpubl. data).

Generally quadrupedal, the species uses all four limbs for locomotion on the ground and in trees, and is an excellent climber. When walking most weight is taken on the hind legs, but all four feet are placed palm downwards (Kingdon, 1971). This contrasts with the giant and Asian pangolins where weight is taken on the margins of the forefeet. Hatt (1934) reports that the species' articulation and movement is fast for a pangolin; Pagès (1970) reports that white-bellied pangolins move at speeds of 1–1.5 km/h. When climbing, the fore- and hindlimbs move as pairs and the prehensile tail, with its sensitive distal pad, acts as a fifth limb, searching for and wrapping around branches – this movement resembles that of a caterpillar (Hatt, 1934; Pagès, 1970). There is enormous strength in the tail, which can serve as a counterweight, and can support the species' body weight enabling foraging to take place on the underside of tree branches and is generally used for support when climbing (Hatt, 1934; Pagès, 1965, 1970; D. Alempijevic, pers. obs.). When climbing large trees, the body is pressed against the tree trunk and the flat, ventral surface of the tail and the pointed scales along the tail margins take the animal's weight. Pagès (1965) reports that the species "spirals" down large tree trunks, taking the weight on the tail as it goes. Grubb et al. (1998) suggest that white-bellied pangolins may roll into a ball when in trees before dropping to the ground, potentially as a means of evading a threat.

The white-bellied pangolin has poor vision (except in close proximity) but excellent olfactory senses; on locating an ant nest or termitaria the clawed forefeet are proficient at tearing them apart, and the long tongue, which is coated in a very sticky mucous-like saliva, is used in a rapid "protrude and withdraw" motion to ingest prey (Heath, 2013; Kingdon, 1971). Pagès (1975) noted that the species consumes prey in small successive quantities invariably without digging. Kingdon (1971) reports that the species will shiver the scales in order to loosen ants or termites clambering over the body, and erect and depress the scales to dislodge prey, both actions facilitating subsequent consumption of said prey.

Scent-orientated, the species displays scent-marking behavior, using a combination of urine and secretions from the anal glands (Pagès, 1968). In Lomami National Park, DRC, scent-marking was recorded in 26% of observations in the mid-high canopy (>9 m height) but never in the understory or on the ground during a multi-strata camera trap study (D. Alempijevic, unpubl. data). This behavior entailed individuals walking along a branch, then stooping to place the rear-end of the body directly on the branch, the hind legs dangling down either side, and then scooting along for the length of 3–4 paces, presumably while releasing urine or scent from the anal glands; alternatively individuals would place their entire ventral surface on a branch and slide along; in both cases the behavior is presumably to smear the scent further (D. Alempijevic, unpubl. data). In captivity in Gabon both males and females scent-marked at the base of trees and at numerous points on the tree trunk, on branches, and at the fork of branches (Pagès, 1968).

The species is sensitive to sounds and vibration and when threatened either climbs a nearby tree or curls up into a tight ball. Although protected by the armor of scales, white-bellied pangolin scales are not as thick and robust as in other species of pangolin and are insufficient against some predators (e.g., leopard). The species' preference for dense vegetation likely offers some protection from detection by larger predators (Kingdon and Hoffmann, 2013); if detected they likely respond by releasing repulsive secretions from the anal glands and potentially defecating and urinating (Kingdon, 1971). The species is an able swimmer, adopting a "doggy-style" paddling approach and using body undulations to move through the water; the tail is not used but hits the water regularly (Pagès, 1970).

Ontogeny and reproduction

Breeding is continuous. In Gabon, adult females found were seldom not pregnant (Pagès, 1965, 1972a, 1975). Tahiri-Zagret (1968) estimated a mean estrus cycle at 9 days but this was highly variant (range: 3–29 days). Males will seek out females using scent trails, and sexual behavior is reportedly elaborate. Males and females simulate aggression, including standing chest to chest, before the female submits, and prior to mating, the pair move to a tree; during copulation the male and female's tail are intertwined (Pagès, 1972b). The species usually gives birth to one young at parturition; twins are thought to be rare. The gestation period has been estimated at 140–150 days (Pagès, 1972a). However, observations of captive white-bellied pangolins at Brookfield Zoo in the United States suggest a gestation period of around eight months (~209 days; Kersey et al., 2018). This suggests the possibility of delayed embryo implantation or embryonic diapause, where the embryo is implanted into the uterus only under favorable conditions (e.g., when prey is abundant). Considerable variation in gestation period has been observed in other species of pangolin (e.g., Chinese pangolin [*Manis pentadactyla*]) and requires further research. Post-partum estrus occurs 9–16 days after parturition (Pagès, 1972a).

At birth, neonates weigh ~100 g (Menzies, 1967) and are ~290 mm in length (Hatt, 1934), and well developed (Kingdon, 1971). The skin is pink and young individuals are hairless, except a ring around the eyes; the hair starts to become visible three weeks after birth (Rahm, 1956). As with other pangolins, the mother will sleep curled around the young, which aids in nursing, but at night will leave for short periods in order to forage for prey (Kingdon, 1971). In the first week after birth females, in estrus, leave the young and mate again and are pregnant for nearly the entire period of maternal care (Pagès, 1972a). The offspring will accompany the mother to forage as it grows and will be "taxied" around on the base of the mother's tail (Fig. 9.5). The age at which young white-bellied pangolins start predating on ants and termites is not known, but in captivity one young animal displayed interest in solid food at five weeks old (Menzies, 1967). Young are likely to become independent at 3–5 months old (upper limit 6 months), which may coincide with the next parturition. Sub-adults are errant and wander over large areas until a home range is established; they reach adult size by ~18 months old (Pagès, 1972a). Longevity in the wild is not known. In captivity, an individual at San Diego Zoo lived to an estimated 10 years of age.

Population

The most frequently encountered African pangolin species but apparently uncommon across its range. There are few quantitative data on white-bellied pangolin abundance and no population estimates exist at the site, national or global level. In the Lama Forest Reserve, Benin, Akpona et al. (2008) estimated a density of 0.84 individuals/km^2 during the dry season in natural forest and monoculture plantation. A review of Afrotropical forest mammals suggested a density of 10.9 individuals/km^2 (Fa and Purvis, 1997), but this would seem to be an overestimate based on a small sample size and inferred density. Based on observations in secondary growth forest in Uganda, Kingdon and Hoffmann (2013) suggest the species occurs at relatively high densities in suitable habitat. In contrast, Laurance et al. (2006) found that pangolins increased in abundance outside of protected oil concessions in Gabon, possibly in response to greater forest disturbance within concessions. In a multistrata camera trap study in Lomami National Park, DRC, 74 records of the species were accumulated over 14 months (7499 camera trap nights) between 2016 and 2018, with the highest detection frequency on the ground in the buffer zone; in the forest interior the detection frequency was highest in the canopy (15–30 m; D. Alempijevic, unpubl. data).

Most other accounts suggest that the species is declining. In Ghana, hunters in villages in the Ashanti region of the Upper Guinea Forest Ecosystem reported in 2011 that white-bellied pangolins were rare (Alexander et al., 2015), though they were considered common by more than 70% of hunters (n = 35) in the Akposa Traditional Area in the Volta Region (Emieaboe et al., 2014). In southern Benin, hunters considered the species to be rare in 2007–2008 (Djagoun and Gaubert, 2009). Soewu and Adekanola (2011) reported that the majority of traditional medicine practitioners of the Awori people in Ogun State, Nigeria believe that populations are declining, and inferred a decline in the size of individuals caught. In southwestern Nigeria, hunters have reported that the species is increasingly rare (Sodeinde and Adedipe, 1994). In Uganda, the white-bellied pangolin is thought to be declining rapidly (Kityo et al., 2016), and declines have been reported for Ghana and Guinea (Bräutigam et al., 1994).

Status

The white-bellied pangolin is categorized as Endangered on The IUCN Red List of Threatened Species (Pietersen et al., 2019). The species was assessed as Endangered in Uganda in 2016 using the IUCN Red List Categories and Criteria (Kityo et al., 2016). No other national or regional assessments have been undertaken. There is a recognized lack of data on populations at all levels (site, national and global) and addressing this is a research priority (see Chapter 34). The species is afforded protection through national legislation in most range states: in some range countries the species is fully protected which prohibits exploitation. In Gabon, Republic of Congo, and Sierra Leone, the species may be hunted and traded legally subject to specific conditions, including that a permit is acquired from the relevant authorities. The species is not afforded legislative protection in Burundi, Kenya or Liberia (Challender and Waterman, 2017). Like other pangolins, this species is included in CITES Appendix I.

Threats

The main threats to the white-bellied pangolin are anthropogenic and comprise habitat destruction and alteration, and overexploitation for local and international use, including international trafficking to Asia.

The destruction, transformation and degradation of natural tropical forest across West and Central Africa has likely been the biggest threat to the white-bellied pangolin historically, and remains present (Megevand, 2013; Ofori et al., 2012). In parts of the species' range (e.g., Côte d'Ivoire and Ghana) loss of natural forest cover has been extremely high (Achard et al., 2014; Megevand, 2013). This is a consequence of logging activities, the establishment of agricultural crops (including shifting agriculture) and monoculture plantations (e.g., oil palm and cocoa [*Theobroma cacao*]), and a rapidly growing human population and urban development, which increases proximity to natural areas, and access to pangolin habitat (Mayaux et al., 2013; Sodeinde and Adedipe, 1994). Although white-bellied pangolins appear able to adapt to some degree of habitat modification, this has not been fully elucidated and further research is needed, in particular comparative ecological studies between natural and modified habitats, and the ability of the species to persist in such habitats long-term (Akpona et al., 2008; Sodeinde and Adedipe, 1994).

Unsustainable and intensive exploitation for local use and international trafficking is a major threat. The species has been used throughout human history as bushmeat and is sold in high numbers at bushmeat markets and roadsides, and traded for medicinal use (see Chapter 15). Fa et al. (2006) reported white-bellied pangolins to be the fourth most abundant species in bushmeat markets in Cameroon between 2002 and 2003. Boakye et al. (2016b) recorded 341 pangolins illegally traded between September 2013 and January 2014 in five regions of Ghana, 82% of which were white-bellied pangolins. Ingram et al. (2018) estimated that in Central Africa alone between 0.42 and 2.71 million pangolins (most likely 0.42 million) were harvested annually between 1975 and 2014, primarily involving the white-bellied pangolin, and that exploitation appears to be increasing. As a proportion of overall hunting catch, pangolins increased significantly from 0.04% in 1972 to 1.83% in 2014, and annual pangolin catch increased by ~150% pre- (1975–1999) and post-2000 (2000–2014; Ingram et al., 2018). This corroborates earlier research, which estimated annual offtake in the region to involve ~425,000 pangolins annually, the majority of which likely comprised white-bellied pangolins (Fa and Peres, 2001). This trade drives exploitation of the species. Fa et al. (2006) reported that white-bellied pangolins were the fourth most harvested species across 47 sites in Cameroon between 2002 and 2003. Kümpel (2006) found similar results in Equatorial Guinea. The retail price of smaller African pangolins in urban markets in parts of West and Central is reportedly increasing, and more than doubled between 1993 and 2014 (Mambeya et al., 2018), and may reflect increasing demand or increasing rarity (Ingram et al., 2018).

Various body parts of the white-bellied pangolin are used in traditional medicine, which drives harvest of the species (see Chapter 15). In Benin, Djagoun et al. (2012) found that 26.4% of medicinal traders surveyed sold white-bellied pangolin parts. Sodeinde and Adedipe (1994) reported that traditional medicines practitioners in southwestern Nigeria, when using pangolins, used this species exclusively (see Chapter 15).

There has been limited international trade in white-bellied pangolins reported to CITES historically, especially compared to the Asian species. However, trade dynamics changed in the 2010s and international trafficking is a major, contemporary threat. Prior to 2000, international trade largely comprised small numbers of live animals (Challender and Waterman, 2017). The 2010s saw the advent of international trade in commercial quantities of white-bellied pangolin scales (involving ~7000 animals between 2013 and 2016) that were mainly exported from DRC and Republic of Congo to China, and animals reportedly exported for commercial captive breeding purposes (Chapter 16). This decade also saw the emergence of intercontinental trafficking of African pangolin scales to Asia, in particular from the Gulf of Guinea (Ingram et al., 2019). Estimates suggest that between 2000 and July 2019, international trafficking involved scales from more than 500,000 African pangolins, mainly between 2015 and 2019, the majority of which were likely from white-bellied pangolins (see Chapter 16).

References

Achard, F., Beuchle, R., Mayaux, P., Stibig, H.-J., Bodart, C., Brink, A., et al., 2014. Determination of tropical deforestation rates and related carbon losses from 1990 to 2010. Glob. Change Biol. 20 (8), 2540–2554.

Akpona, H.A., Djagoun, C.A.M.S., Sinsin, B., 2008. Ecology and ethnozoology of the three-cusped pangolin *Manis tricuspis* (Mammalia, Pholidota) in the Lama forest reserve, Benin. Mammalia 72 (3), 198–202.

Albrechtsen, L., Macdonald, D.W., Johnson, P.J., Castelo, R., Fa, J.E., 2007. Faunal loss from bushmeat hunting: empirical evidence and policy implications in Bioko Island. Environ. Sci. Policy 10 (7–8), 654–667.

Alexander, J.S., McNamara, J., Rowcliffe, J.M., Oppong, J., Milner-Gulland, E.J., 2015. The role of bushmeat in a West African agricultural landscape. Oryx 49 (4), 643–651.

Allen, G.L., Loveridge, A., 1942. Scientific results of a fourth expedition to forested areas in East and Central Africa. I: Mammals. Bull. Museum Comp. Zool. Harv. Coll. 89 (4), 147–213.

Allen, G.M., Coolidge Jr., H.J., 1948. Mammals of Liberia. Harvard University Press, Massachusetts.

Amori, G., Segniagbeto, G.H., Decher, J., Assou, D., Gippoliti, S., Luiselli, L., 2016. Non-marine mammals of Togo (West Africa): an annotated checklist. Zoosystema 38 (2), 201–244.

Angelici, E.M., Luiselli, L., Politano, E., Akani, G.C., 1999a. Bushmen and mammal-fauna: a survey of the mammals traded in bush-meat markets of local people in the rainforests of south-eastern Nigeria. Anthropozoologica 30, 51–58.

Angelici, F.M., Grimod, I., Politano, E., 1999b. Mammals of the Eastern Niger Delta (Rivers and Bayelsa States, Nigeria): an environment affected by a gas-pipeline. Folia Zool. 48 (4), 249–264.

Angelici, F.M., Egbide, B., Akani, G., 2001. Some new mammal records from the rainforests of south-eastern Nigeria. Hystrix, Ital. J. Mammol. 12 (1), 37–43.

Ansell, W.F.H., 1982. The Mammals of Zambia, Issue 1. National Parks and Wildlife Service, Zambia.

Ausden, M., Wood, P., 1990. *The Wildlife of the Western Area Forest Reserve, Sierra Leone*. Forestry Division of the Government of Sierra Leone, Conservation Society of Sierra Leone, International Council for Bird Preservation, Royal Society for the Protection of Birds, Freetown, Sierra Leone.

Beja, P., Pinto, P.V., Veríssimo, L., Bersacola, E., Fabiano, E., Palmeirim, J.M., et al., 2019. The Mammals of Angola.

In: Huntley, B.J., Russo, V., Lages, F., Nuno-Ferrand, N. (Eds.), Biodiversity of Angola, Science & Conservation: A Modern Synthesis. Springer, Cham, pp. 357–443.

Boakye, M.K., Pietersen, D.W., Kotzé, A., Dalton, D.L., Jansen, R., 2016a. Ethnomedical use of African pangolins by traditional medical practitioners in Sierra Leone. J. Ethnobiol. Ethnomed. 10, 76.

Boakye, M.K., Kotzé, A., Dalton, D.L., Jansen, R., 2016b. Unravelling the pangolin bushmeat commodity chain and the extent of trade in Ghana. Hum. Ecol. 44 (2), 257–264.

Bout, N., Ghiurghi, A., 2013. Guide des Mammiferes Du Parc National De Cantanhez, Guinée-Bissau. Acção para o Desenvolvimento, Guinêe-Bissau, Associazione Interpreti Naturalistici ONLUS, Italie.

Bräutigam, A., Howes, J., Humphreys, T., Hutton, J., 1994. Recent information on the status and utilization of African pangolins. TRAFFIC Bull. 15 (1), 15–22.

Chan, L.-K., 1995. Extrinsic lingual musculature of two pangolins (Pholidota: Manidae). J. Mammal. 76 (2), 472–480.

Challender, D., Waterman, C., 2017. Implementation of CITES Decisions 17.239 b) and 17.240 on Pangolins (*Manis* spp.), CITES SC69 Doc. 57 Annex. Available from <https://cites.org/sites/default/files/eng/com/sc/69/E-SC69-57-A.pdf>. [December 18, 2018].

Cotterill, F.P.D., 2002. Biodiversity conservation in the Ikelenge Pedicle, Mwinilunga District, Northwest Zambia. Occasional publication in Biodiversity No. 10, Biodiversity Foundation for Africa, Bulawayo, Zimbabwe.

Doran, G.A., Allbrook, D.B., 1973. The tongue and associated structures in two species of African pangolins, *Manis gigantea* and *Manis tricuspis*. J. Mammal. 54 (4), 887–899.

Djagoun, C.A.M.S., Gaubert, P., 2009. Small carnivorans from southern Benin: a preliminary assessment of diversity and hunting pressure. Small Carnivore Conserv. 40, 1–10.

Djagoun, C.A., Akpona, H.A., Mensah, G.A., Nuttman, C., Sinsin, B., 2012. Wild mammals trade for zootherapeutic and mythic purposes in Benin (West Africa): capitalizing species involved, provision sources, and implications for conservation. In: Nóbregra Alves, R.R., Rosa, I.L. (Eds.), Animals in Traditional Folk Medicine. Springer, Berlin, Heidelberg, pp. 367–381.

du Toit, Z., du Plessis, M., Jansen, R., Grobler, J.P., Kotzé, A., 2017. Mitochondrial genomes of African pangolins and insights into evolutionary patterns and phylogeny of the family Manidae. BMC Genom. 18, 746.

Emieaboe, P.A., Ahorsu, K.E., Gbogbo, F., 2014. Myths, taboos and biodiversity conservation: the case of hunters in a rural community in Ghana. Ecol., Environ. Conserv. 20 (3), 879–886.

Emry, R.J., 1970. A North American Oligocene pangolin and other additions to the Pholidota. Bull. Am. Museum Nat. Hist. 142, 459–510.

Fa, J.E., Purvis, A., 1997. Body size, diet and population density in Afrotropical Forest Mammals: a comparison with neotropical species. J. Anim. Ecol. 66 (1), 98–112.

Fa, J.E., Peres, C.A., 2001. Game vertebrate extraction in African and Neotropical forests: an intercontinental comparison. In: Reynolds, J.D., Mace, G.M., Radford, K. H., Robinson, J.G. (Eds.), Conservation of Exploited Species. Cambridge University Press, Cambridge, pp. 203–241.

Fa, J.E., Seymore, S., Dupain, J., Amin, R., Albrechtsen, L., Macdonald, D., 2006. Getting to grips with the magnitude of exploitation: bushmeat in the Cross-Sanaga rivers region, Nigeria and Cameroon. Biol. Conserv. 129 (4), 497–510.

Fang, L.-X., 1981. Investigation on pangolins by following their trace and observing their cave. Nat., Beijing Nat. Hist. Museum 3, 64–66. [In Chinese].

Focillon, A.D., 1850. Du genre Pangolin (*Manis* Linn.) et de deux nouvelles espèces de ce genre. Revue et Mag. de Zool. 2 Sér., Tome VII, 465–474, 513–534.

Foley, C., Foley, L., Lobora, A., De Luca, D., Msuha, M., Davenport, T.R.B., et al., 2014. A Field Guide to the Larger Mammals of Tanzania. Princeton University Press, Princeton.

Frechkop, S., 1931. Notes sur les mammifères. VI. Quelques observations sur la classification des pangolins (Manidae). Bulletin du Musee royal d'Histoire naturelle de Belgique VII (22), 1–14.

Gaubert, P., Njiokou, F., Ngua, G., Afiademanyo, K., Dufour, S., Malekani, J., et al., 2016. Phylogeography of the heavily poached African common pangolin (Pholidota, *Manis tricuspis*) reveals six cryptic lineages as traceable signatures of Pleistocene diversification. Mol. Ecol. 25 (23), 5975–5993.

Gaubert, P., Antunes, A., Meng, H., Miao, L., Peigné, S., Justy, F., et al., 2018. The complete phylogeny of pangolins: scaling up resources for the molecular tracing of the most trafficked mammals on Earth. J. Hered. 109 (4), 347–359.

Gaudin, T.J., Emry, R.J., Wible, J.R., 2009. The phylogeny of living and extinct pangolins (Mammalia, Pholidota) and associated taxa: a morphology based analysis. J. Mammal. Evol. 16 (4), 235–305.

Gotch, A.F., 1979. Mammals – Their Latin Names Explained. A Guide to Animal Classification. Blandford Press, Poole.

Gray, J.E., 1843. Proceedings of Zoological Society of London 11 (1), 20–22.

Grubb, P., Jones, T.S., Davies, A.G., Edberg, E., Starin, E.D., Hill, J.E., 1998. Mammals of Ghana, Sierra Leone and the Gambia. Trendrine Press, Zennor, Cornwall.

Happold, D.C.D., 1987. The Mammals of Nigeria. Clarendon Press, Oxford.

Hassanin, A., Hugot, J.-P., van Vuuren, B.J., 2015. Comparison of mitochondrial genome sequences of

pangolins (Mammalia, Pholidota). C. R. Biol. 338 (4), 260–265.
Hatt, R.T., 1934. The pangolin and aard-varks collected by the American Museum Congo expedition. Bull. Am. Museum Nat. Hist. 66, 643–671.
Heath, M.E., 2013. Order Pholidota – Pangolins. In: Kingdon, J., Hoffmann, M. (Eds.), Mammals of Africa, vol. V, Carnivores, Pangolins, Equids, Rhinoceroses. Bloomsbury Publishing, London, pp. 384–386.
Heath, M.E., Hammel, H.T., 1986. Body temperature and rate of O_2 consumption in Chinese pangolins. Am. J. Physiol.-Regul., Integr. Comp. Physiol 250 (3), R377–R382.
Henschel, P., Abernethy, K.A., White, L.J.T., 2005. Leopard food habits in the Lopè National Park, Gabon, Central Africa. Afr. J. Ecol. 43 (1), 21–28.
Henschel, P., Hunter, L.T.B., Coad, L., Abernethy, K.A., Mühlenberg, M., 2011. Leopard prey choice in the Congo Basin rainforest suggests exploitative competition with human bushmeat hunters. J. Zool. 285 (1), 11–20.
Hildwein, G., 1974. Resting metabolic rates in pangolins (Pholidota) and squirrels of equatorial rain forests. Arch. Sci. Physiol. 28, 183–195.
Hill, J.E., Carter, T.D., 1941. The mammals of Angola, Africa. Bull. Am. Museum Nat. Hist. 78, 1–211.
Ingram, D.J., Coad, L., Abernethy, K.A., Maisels, F., Stokes, E.J., Bobo, K.S., et al., 2018. Assessing Africa-wide pangolin exploitation by scaling local data. Conserv. Lett. 11 (2), e12389.
Ingram, D.J., Cronin, D.T., Challender, D.W.S., Venditti, D. M., Gonder, M.K., 2019. Characterising trafficking and trade of pangolins in the Gulf of Guinea. Glob. Ecol. Conserv. 17, e00576.
Jeannin, A., 1936. Les mammiféres sauvages du Cameroun. Encyclopedie Biol. 16, 116–130.
Jentink, F.A., 1882. Note XXV. Revision of the Manidae in the Leyden Museum. Notes from the Leyden Museum IV, 193–209.
Jones, C., 1973. Body temperatures of *Manis gigantea* and *Manis tricuspis*. J. Mammal. 54 (1), 263–266.
Kersey, D., Guilfoyle, C., Aitken-Palmer, C., 2018. Reproductive hormone monitoring of the tree pangolin (*Phataginus tricuspis*). Chicago International Symposium on Pangolin Care and Conservation, Brookfield Zoo, Chicago, IL, 23–25 August 2018.
Kingdon, J., 1971. East African mammals. An Atlas of evolution in Africa, Primates, Hyraxes, Pangolins, Protoungulates, Sirenians, vol. I. Academic Press, London.
Kingdon, J., 1997. The Kingdon Field Guide to African Mammals. Academic Press, London.
Kingdon, J., Hoffmann, M., 2013. *Phataginus tricuspis* Tree Pangolin. In: Kingdon, J., Hoffmann, M. (Eds.), Mammals of Africa, vol. V, Carnivores, Pangolins, Equids and Rhinoceroses. Bloomsbury Publishing, London, pp. 391–395.
Khwaja, H., Buchan, C., Wearn, O.R., Bahaa-el-din, L., Bantlin, D., Bernard, H., et al., 2019. Pangolins in global camera trap data: implications for ecological monitoring, Glob. Ecol. Conserv. 20, e00769.
Kityo, R., 2009. Mammals: information on the mammal diversity for the Opeta-Bisina and Mburo-Nakivale wetland systems, eastern and western Uganda. In: Odul, M. O., Byaruhanga, A. (Eds.), Ecological Baseline Surveys of Lake Bisina-Opeta Wetland System, Lake Mburo-Nakivale Wetland System. Nature Uganda, East African Natural History Society, Kampala, pp. 85–98.
Kityo, R., Prinsloo, S., Ayebare, S., Plumptree, A., Rwetsiba, A., Sadic, W., et al., 2016. Nationally Threatened Species of Uganda. Available at: <http://www.nationalredlist.org/files/2016/03/National-Redlist-for-Uganda.pdf>. [June 21, 2018].
Kümpel, N.F., 2006. Incentives for Sustainable Hunting of Bushmeat in Rio Muni, Equatorial Guinea. Ph.D. Thesis, Imperial College London, London, UK.
Laurance, W.F., Croes, B.M., Tchignoumba, L., Lahm, S.A., Alonso, A., Lee, M.E., et al., 2006. Impacts of roads and hunting on central African rainforest mammals. Conserv. Biol. 20 (4), 1251–1261.
Mambeya, M.M., Baker, F., Momboua, B.R., Pambo, A.F.K., Hega, M., Okouyi, V.J.O., et al., 2018. The emergence of a commercial trade in pangolins from Gabon. Afr. J. Ecol. 56 (3), 601–609.
Mayaux, P., Pekel, J.-F., Desclee, B., Donnay, F., Lupi, A., Achard, F., et al., 2013. State and evolution of the African rainforests between 1990 and 2010. Philos. Trans. R. Soc. B 368 (1625), 20120300.
Megevand, C., 2013. Deforestation Trends in the Congo Basin: Reconciling Economic Growth and Forest Protection. World Bank, Washington, D.C.
Menzies, J.I., 1967. A preliminary note on the birth and development of a small-scaled tree pangolin *Manis tricuspis*. Int. Zoo Yearbook 7 (1), 114.
Meester, J., 1972. Order Pholidota. In: Meester, J., Stzer, H. W. (Eds.), The Mammals of Africa: An Identification Manual, Part 4. Smithsonian Institution Press, Washington, D.C., pp. 1–3.
Mohr, E., 1961. Schuppentiere. Neue Brehm-Bucherei. A. Ziemsen Verlag, Wittenberg Lutherstadt.
Monroe, B.P., Doty, J.B., Moses, C., Ibata, S., Reynolds, M., Carroll, D., 2015. Collection and utilization of animal carcasses associated with zoonotic disease in Tshuapa District, the Democratic Republic of the Congo, 2012. J. Wildlife Dis. 51 (3), 734–738.
Ntiamoa-Baidu, Y., Carr-Saunders, C., Matthews, B.E., Preston, P.M., Walker, A.R., 2005. Ticks associated with

wild mammals in Ghana. Bull. Entomol. Res. 95 (3), 205–219.

Odemuni, O.S., Ogunsina, A.M., 2018. Pangolin habitat characterization and preference in Old Oyo National Park, Southwest Nigeria. J. Res. For., Wildlife Environ. 10 (2), 56–64.

Ofori, B.Y., Attuquayefio, D.K., Owusu, E.H., 2012. Ecological status of large mammals of a moist semi-deciduous forest of Ghana: implications for wildlife conservation. J. Biodivers. Environ. Sci. 2 (2), 28–37.

Ofusori, D.A., Caxton-Martins, E.A., Keji, S.T., Oluwayinka, P.O., Abayomi, T.A., Ajayi, S.A., 2008. Microarchitectural adaptations in the stomach of African Tree Pangolin (*Manis tricuspis*). Int. J. Morphol. 26 (3), 701–705.

Orhierhor, M., Okaka, C.E., Okonkwo, V.O., 2017. A survey of the parasites of the African white-bellied pangolin, *Phataginus tricuspis* in Benin City, Edo State, Nigeria. Nigerian. J. Parasitol. 38 (2), 266.

Pagès, E., 1965. Notes sur les pangolins du Gabon. Biol. Gabon. 1, 209–237.

Pagès, E., 1968. Les glandes odorantes des pangolins arboricoles (*M. tricuspis* et *M. longicaudata*): morphologie, développement et rôles. Biol. Gabon. 4, 353–400.

Pagès, E., 1970. Sur l'ecologie et les adaptations de l'orycteope et des pangolins sympatriques du Gabon. Biol. Gabon. 6, 27–92.

Pagès, E., 1972a. Comportement maternal et développement due jeune chez un pangolin arboricole (*M. tricuspis*). Biol. Gabon. 8, 63–120.

Pagès, E., 1972b. Comportement aggressif et sexuel chez les pangolins arboricoles (*Manis tricuspis* et *M. longicaudata*). Biol. Gabon. 8, 3–62.

Pagès, E., 1975. Étude éco-éthologique de *Manis tricuspis* par radio-tracking. Mammalia 39, 613–641.

Pietersen, D., Moumbolou, C., Ingram, D.J., Soewu, D., Jansen, R., Sodeinde, O., et al., 2019. *Phataginus tricuspis. The IUCN Red List of Threatened Species* 2019: e.T12767A123586469. Available from: <http://dx.doi.org/10.2305/IUCN.UK.2019-3.RLTS.T12767A123586469.en>.

Pocock, R.I., 1924. The external characters of the pangolins (Manidae). Proc. Zool. Soc. Lond. 94 (3), 707–723.

Rahm, U., 1956. Notes on Pangolins of the Ivory Coast. J. Mammal. 37 (4), 531–537.

Rafinesque, C.S., 1820. Sur le genre *Manis* et description d'une nouvelle espèce: *Manis ceonyx*. Annales Générales des Sciences Physiques 7, 214–215.

Roth, A.M., Cords, M., 2015. Some nocturnal and crepuscular mammals of Kakamega Forest: photographic evidence. J. East Afr. Nat. Hist. 104 (1–2), 213–225.

Sayer, J.A., Green, A.A., 1984. The distribution and status of large mammals in Benin. Mammal Rev. 14 (1), 37–50.

Schlitter, D.A., 2005. Order Pholidota. In: Wilson, D.E., Reeder, D.M. (Eds.), Mammal Species of the World: A Taxonomic and Geographic Reference, third ed. Johns Hopkins University Press, Baltimore, pp. 530–531.

Schouteden, H., 1948. Fauna de Congo Belge et du Ruanda-Urundi. I Mammiféres. Annales du Musée du Congo Belge, Zoologie 8 (1), 1–331.

Smithers, R.H.N., Lobão Tello, J.L.P., 1976. Check List and Atlas of the Mammals of Mozambique, Trustees of the National Museums and Monuments of Rhodesia, Salisbury, Zimbabwe.

Sodeinde, O.A., Adedipe, S.R., 1994. Pangolins in southwest Nigeria – current status and prognosis. Oryx 28 (1), 43–50.

Sodeinde, O.A., Adefuke, A.A., Balogun, O.F., 2002. Morphometric analysis of *Manis tricuspis* (Pholidota-Mammalia) from southwestern Nigeria. Glob. J. Pure Appl. Sci. 8, 7–13.

Sodeinde, O.A., Soewu, D.A., 2016. Pangolins in Nigeria: Their biology and ecology and challenges to their conservation. Poster Presented at the 3rd African Congress for Conservation Biology (ACCB, 2016), September 2016, El-Jadida, Morocco.

Soewu, D.A., Adekanola, T.A., 2011. Traditional medical knowledge and perceptions of pangolins (*Manis* sps [sic]) among the Awori People, Southwestern Nigeria. J. Ethnobiol. Ethnomed. 7, 25.

Swiacká, M., 2018. Market survey and population characteristics of three species of pangolins (Pholidota) in the Republic of Congo. M.Sc. Thesis, Czech University of Life Sciences Prague, Prague, Czech Republic.

Tahiri-Zagret, C., 1968. Étude du Cycle Œstrien du pangolin *Manis tricuspis* (Rafinesque) Pholidotes. Annales de l'Universitié d'Abidjan 4, 129–141.

Treves, A., Wima, P., Plumptre, A.J., Isoke, S., 2010. Camera-trapping forest-woodland wildlife of western Uganda reveals how gregariousness biases estimates of relative abundance and distribution. Biol. Conserv. 143 (2), 521–528.

Ullmann, T., Veríssimo, D., Challender, D.W.S., 2019. Evaluating the application of scale frequency to estimate the size of pangolin scale seizures. Glob. Ecol. Conserv. 20, e00776.

van Vliet, N., Nebesse, C., Nasi, R., 2015. Bushmeat consumption among rural and urban children from Province Orientale, Democratic Republic of Congo. Oryx 49 (1), 165–174.

Verschuren, J., 1982. Hope for Liberia. Oryx 16 (5), 421–427.

Verschuren, J., 1987. Liste commentée des Mammifères des Parcs Nationaux du Zaïre, du Rwanda et du Burundi. Bulletin de L'institut Royal de Sciences Naturelles de Belgique, Biologie 57, 17–39.

Ziegler, S., Nikolaus, G., Hutterer, R., 2002. High mammalian diversity in the newly established National Park of upper Niger, Republic of Guinea. Oryx 36 (1), 73–80.

CHAPTER 10

Giant pangolin *Smutsia gigantea* (Illiger, 1815)

Michael Hoffmann[1], Stuart Nixon[2], Daniel Alempijevic[3], Sam Ayebare[4], Tom Bruce[1], Tim R.B. Davenport[5], John Hart[6], Terese Hart[6], Martin Hega[7], Fiona Maisels[4,8], David Mills[9,10] and Constant Ndjassi[11]

[1]Conservation and Policy, Zoological Society of London, Regent's Park, London, United Kingdom [2]Field Programmes, North of England Zoological Society, Chester Zoo, Chester, United Kingdom [3]Integrative Biology, Florida Atlantic University, Boca Raton, FL, United States [4]Wildlife Conservation Society, Bronx, NY, United States [5]Tanzania Program, Wildlife Conservation Society (WCS), Zanzibar, Tanzania [6]Lukuru Foundation, Kinshasa, Democratic Republic of the Congo [7]Gabon Program, Wildlife Conservation Society (WCS), Libreville, Gabon [8]Biological and Environmental Sciences, University of Stirling, Stirling, United Kingdom [9]School of Life Sciences, University of Kwazulu-Natal, Durban, South Africa [10]Panthera, New York, NY, United States [11]Liberia Programme, Fauna and Flora International, Monrovia, Liberia

OUTLINE

Pangolins
DOI: https://doi.org/10.1016/B978-0-12-815507-3.00010-1

Taxonomy

Previously included in the genus *Manis* (Meester, 1972; Schlitter, 2005) and *Phataginus* (e.g., Grubb et al., 1998), the species is here included in *Smutsia* based on both morphological (Gaudin et al., 2009) and genetic (Gaubert et al., 2018; Hassanin et al., 2015) evidence, following a synthetic classification of extant pangolins (Chapter 2), and in agreement with previous authors (Allen, 1939; Kingdon et al., 2013; Pocock, 1924). Synonyms: *africanus* (Gray, 1865), *wagneri* (Fitzinger, 1872). The type locality is given as "River Niger." There are no data on chromosome number.

Etymology: *Smutsia* is derived from Johannes Smuts, an early 19th century South African naturalist; the species name *gigantea* refers to the Latin *gigas*, meaning giant (Gotch, 1979).

Description

The giant pangolin (*Smutsia gigantea*) is the largest of all living pangolin species with a body weight in excess of 30 kg (possibly up to 40 kg in older individuals), and a total body length between 140 and 180 cm (Table 10.1). The body is covered in keratinous, overlapping scales (usually more than 17 transversal rows) that grow from the skin in an obliquely angled, linear grid, with the largest scales along the middle of the back, flanks, shoulders and thighs (Fig. 10.1). In common with other African pangolins, no hairs project between the scales. The scales on the limbs are graduated. The margins of the tail are bordered by 15–19 rows of sharply pointed scales that interdigitate with the large, thick scales (12–15 rows) of the convex dorsal surface and the smaller, thinner scales of the slightly concave lower surface (see Table 10.1). Total scale number ranges from 446 to 664 (Table 10.1). The tip of the tail is slightly pointed and protected by thick, closely interlocking scales. The pale, pinkish-gray head is naked except for closely fitting scales covering the forehead and upper muzzle and fringes of short hair around the eyes and auditory meatus (Kingdon, 1971). The belly is naked, and light pink to pale gray in coloration. When fully curled up, the large, broad muscular tail provides adequate protective covering for the unprotected underside and can stretch between infolded legs over the tucked-in head to the shoulders.

The muzzle is distinctly elongated and there are vestigial ear cartilages beneath the skin but no external pinnae. External auditory canals can be closed by lateroventral folds of skin when the animal curls up. The small black eyes are surrounded by well-muscled, swollen eyelids. The skull is composed of exceptionally dense, thickened bone (likely protective in function) and, in common with other pangolins, several cranial bones are absent, notably the zygomatic arch, a reduction that is associated with effective elimination of the chewing muscles (temporalis and masseter) (Kingdon, 1971; see Chapter 1). The jaws lack any teeth. The tongue is long and strap-like, changing in cross-sectional shape through its length (70 cm fully extended) and can extend out of the mouth for more than 30 cm (Doran and Allbrook, 1973). As with all pangolins, the tongue retracts into a pouch in the throat. Rapid extension and retraction of the tongue is further assisted by the tongue's base extending back into the chest cavity where muscular extensions of the tongue base attach to a spatulate, cartilaginous extension of the sternum (Chapter 1). This cartilage and its muscular attachment are free to slide and swing about along the walls of the abdominal cavity with every extension and retraction of tongue and muscle. Food passes down directly to the pyloric region where a structure resembling the gizzard of a bird grinds up ants and termites with small stones and sand; this structure is

TABLE 10.1 Giant pangolin morphometrics.

Measurement			Country	Source(s)
Weight	Weight (♂)	32.1 kg, n = 1	Uganda	S. Nixon and N. Matthews, unpubl. data
	Weight (unsexed)	33 kg, n = 1	Uganda	Kingdon, 1971
		28.8 kg, n = 119	Gabon	Mambeya et al., 2018
Body	Total length (♂)	1438 (1370–1530) mm, n = 5	Democratic Republic of the Congo	Hatt, 1934
		1798 mm, n = 1	Uganda	S. Nixon and N. Matthews, unpubl. data
	Total length (♀)	1298 (1185–1365) mm, n = 7	Democratic Republic of the Congo	Hatt, 1934
	Total length (unsexed)	1255 mm, n = 1	Uganda	Uganda National Museum (measured by S. Nixon, unpubl. data)
		1710 mm, n = 1	Liberia	Allen and Coolidge, 1930
	Head-body length (♂)	1088 mm, n = 1	Uganda	S. Nixon and N. Matthews, unpubl. data
	Head-body length (♀)	660 mm, n = 1	Democratic Republic of the Congo	Rahm, 1966
	Tail length (♂)	674 (650–700) mm, n = 5	Democratic Republic of the Congo	Hatt, 1934
		710 mm, n = 1	Uganda	S. Nixon and N. Matthews, unpubl. data
	Tail length (♀)	596 (545–675) mm, n = 7	Democratic Republic of the Congo	Hatt, 1934
		670 mm, n = 1	Democratic Republic of the Congo	Rahm, 1966
Vertebrae	Total number of vertebrae	55–57		Frechkop, 1931; Jentink, 1882
	Cervical	7		Jentink, 1882
	Thoracic	14		Jentink, 1882; Mohr, 1961
	Lumbar	5		Jentink, 1882; Mohr, 1961
	Sacral	3–4		Jentink, 1882; Mohr, 1961
	Caudal	26–27		Jentink, 1882; Mohr, 1961
Skull	Length (♂)	152 (148–162) mm, n = 4	Democratic Republic of the Congo	Hatt, 1934
	Length (♀)	142 (134–148) mm, n = 6	Democratic Republic of the Congo	Hatt, 1934

(*Continued*)

TABLE 10.1 (Continued)

	Measurement		Country	Source(s)
	Breadth across zygomatic processes (♂)	49.5 (49–50) mm, n = 4	Democratic Republic of the Congo	Hatt, 1934
	Breadth across zygomatic processes (♀)	47 (45–49) mm, n = 6	Democratic Republic of the Congo	Hatt, 1934
Scales	Total number of scales	567 (446–664), n = 10	Cameroon, Uganda, unknown	Ullmann et al., 2019; Uganda National Museum (measured by S. Nixon, 2018); S. Nixon and N. Matthews, unpubl. data
	No. of scale rows (transversal, body)	17		Frechkop, 1931
	No. of scale rows (longitudinal, body)	13–17		Frechkop, 1931; Jentink, 1882
	No. of scales on outer margins of tail	15–19		Frechkop, 1931; Jentink, 1882
	No. of scales on median row of tail	11–15		Frechkop, 1931; Jentink, 1882
	Scales (wet) as proportion of body weight	No data		
	Scales (dry) as proportion of body weight	No data		

FIGURE 10.1 A giant pangolin in Uganda – the range in scale size is visible. Photo taken on a camera trap; these are helping to provide an understanding of the species' behavior and ecology. *Photo credit: Naomi Matthews.*

particularly large and well developed in this species (Doran and Allbrook, 1973; Hatt, 1934; Mohr, 1961). There is no cecum.

Both fore- and hindfeet have five digits. On the forefeet, the claws on the outer two digits are vestigial, while the middle claw is the longest and most robust. Claws can be extended forwards and splay when digging but fold backwards and inwards when the animal is walking. On the hindfeet, claws are short and blunt and the feet appear stubby. Large anal glands, especially in the male, produce a white waxy secretion with a powerful odor (Hatt, 1934; Kingdon, 1971). Females have two pectoral nipples.

Body temperature is regulated at 32–34.5 °C, slightly below the usual range of mammals (Jones, 1973).

The species with which the giant pangolin may readily be confused is its congener Temminck's pangolin (*Smutsia temminckii*). However, the latter is considerably smaller, with a shorter, more rounded tail; fewer rows of body scales (12 transversal and 11–13 longitudinal rows) and on outer (11–13) and middle (4–7) of tail (Frechkop, 1931; Jentink, 1882). Further, because Temminck's pangolins are mostly bipedal, and do little digging, their forelimbs are considerably reduced compared with giant pangolins, which have a preponderance of bone and muscle at the front end (Kingdon, 1971). This difference is also noticeable in the structure of the pelvis, which is robust, elongated and more horizontal in the former, whereas the pelvis is more lightly built, shorter and more vertical in the latter (Kingdon, 1971).

Distribution

Widely distributed species in equatorial Africa (Fig. 10.2). Recorded from Senegal (Dupuy, 1968), Guinea-Bissau (Frade, 1949; Reiner and Simões, 1999), Guinea (Barnett and Prangley, 1997), Sierra Leone (Grubb et al., 1998), Liberia (Allen and Coolidge, 1930; Coe, 1975), Côte d'Ivoire (Rahm, 1956) and Ghana, as far north as Mole National Park (Grubb et al., 1998). The species then appears to be somewhat discontinuously distributed between southeastern Ghana and eastern Nigeria. There is no recent information from Togo (Amori et al., 2016), although Grubb et al. (1998) map older records from Ghana near Fazao-Malfakassa National Park on the border with Togo. In Benin, Sayer and Green (1984) recorded the species from Batia on the border of Pendjari National Park in the north in the 1970s (Verschuren 1988 documented them as present in Pendjari) and referred to sightings in neighboring Burkina Faso and Niger (Poche, 1973 observed them in Niger's W National Park). Akpona and Daouda (2011) noted that while formerly reported from most areas in northern Benin, it may only persist in Pendjari and W National Parks. However, they noted further that the species had not been seen in recent surveys, and this would appear to be unchanged based on ongoing camera trapping work across W-Arly-Pendjari (Harris et al., 2019; ZSL unpubl. data). Although always hypothesized to occur in Nigeria, material evidence has been lacking (Happold, 1987; Rosevear, 1953). However, the species was camera trapped in Gashaka Gumti National Park in the east of the country in 2016 (S. Nixon, unpubl. data); ongoing surveys in Cross River National Park have failed to record the species (A. Dunn, unpubl. data).

From eastern Nigeria, the species is then more continuously distributed through southern Cameroon and mainland Equatorial Guinea to Uganda (Hatt, 1934; Jeannin, 1936; Kingdon, 1971; Kingdon et al., 2013; Malbrant and MacLatchy, 1949; Rahm, 1966; Schouteden, 1944, 1948). Giant pangolins may occur in sympatry with Temminck's pangolin in Murchison Falls National Park, Uganda, where both species have been recorded.

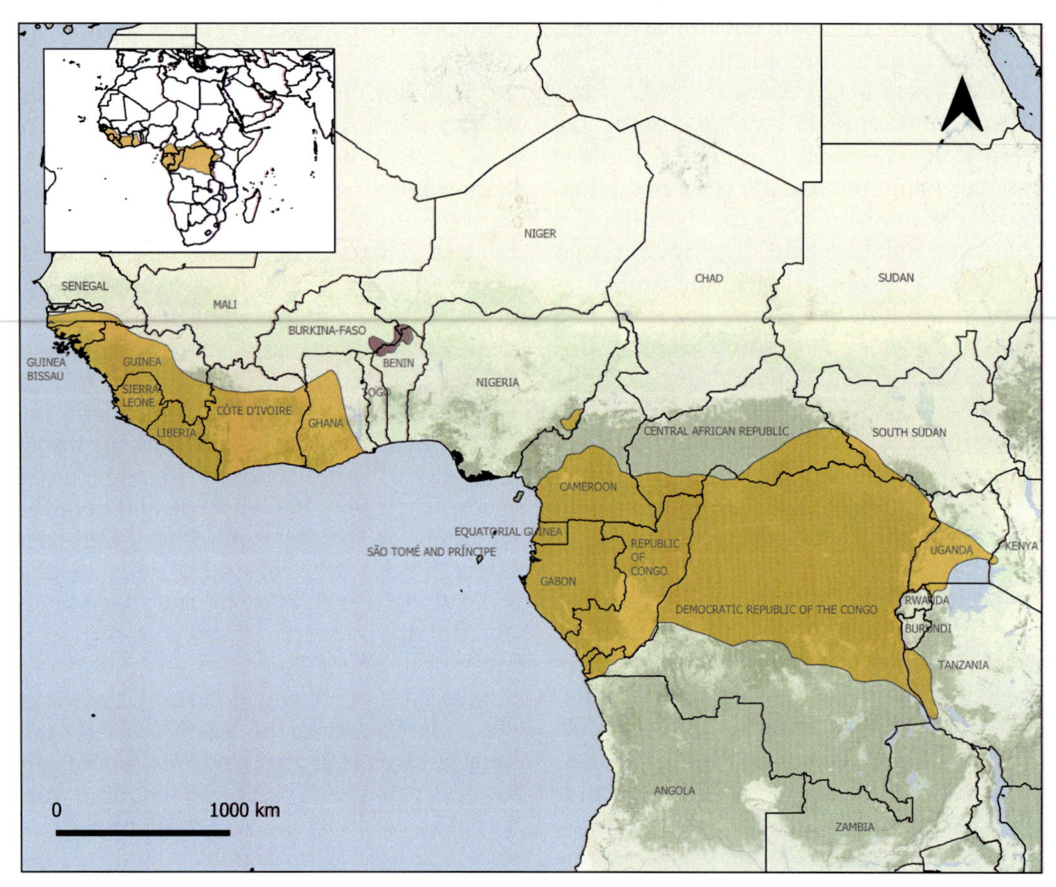

FIGURE 10.2 Giant pangolin distribution. Purple shading indicates areas where the species is possibly extinct. *Source: Nixon et al., 2019.*

However, it is possible that the Victoria Nile, which bisects the northern and southern sectors of the park may represent the distributional limits of each species in this part of Uganda, although further work is needed to investigate the distribution of the species in this region. The species has been observed in far west Kenya, close to the Uganda border (Kingdon, 1971), but recent records are lacking. West of the Albertine Rift, in western Tanzania, the species is confirmed from the Mahale Mountains on the edge of Lake Tanganyika and Minziro Forest Nature Reserve on the border with Uganda, and has also been camera trapped in Gombe National Park (A. Collins, unpubl. data). There is also a camera trap record from the Issa Valley (Greater Mahale Ecosystem Research & Conservation/MPI-EVA, unpubl. data). It seems likely that they should occur in Tembwa, Ntakatta, and other forested sites along the lakeshore to the south. The species was believed extinct in Rwanda until camera trap information confirmed its presence in Akagera National Park in the northeast in 2016 (D. Bantlin, unpubl. data). There are no records from Burundi (Kingdon et al., 2013).

The northern limits of the distribution are not well understood. Fischer et al. (2002) recorded a giant pangolin at the edge of gallery forest in savanna woodland in the southern part of the Comoé National Park in Côte d'Ivoire, and it has also been recorded at about 7°N in Central African Republic, contiguous with the South Sudan border, in the Zemongo Faunal Reserve (Roulet et al., 2007). Both country records are further north than is shown by previous authors (e.g., Gaubert, 2011; Kingdon, 1997; Kingdon et al., 2013). In 2015, the presence of giant pangolin was confirmed in southwestern South Sudan, near the border with the Democratic Republic of the Congo (DRC) (D. Reeder, unpubl. data). At the southern limits, giant pangolins occur widely across the entire forested region of Congo's central cuvette, south to the right banks of the Kasai River and Sankuru Rivers (J. Thompson, unpubl. data). The species occurs widely in the forests between the Lualaba and Congo Rivers south to almost 3°S (Lukuru Foundation, unpubl. data).

As with the black-bellied pangolin (*Phataginus tetradactyla*), the presence of this species in Angola is unclear. There is no mention of them in Bocage (1889, 1890), Monard (1935), Hill and Carter (1941) or Machado (1969); Feiler (1990) includes them, but without any indication to source. However, there are several passing references to their presence in the forests of Cabinda, particularly Maiombe, so it seems reasonable to include them in the country's fauna (Kingdon et al., 2013; Beja et al., 2019 and see references therein).

Note that previous reference to this species occurring on the island of Bioko (e.g., Gaubert, 2011; Kingdon et al., 2013) stem from records of carcasses imported from the mainland (Hoffmann et al., 2015; and see Ingram et al., 2019).

Habitat

The giant pangolin inhabits primary and secondary forest formations, forest-savanna mosaics, seasonally inundated swamp forests, gallery forest, wooded savanna and wet grasslands. It occurs across a wide altitudinal gradient from slightly above sea level in the lowland coastal forests of West Africa to at least 2220 m (Mount Mutenda in the Tayna Nature Reserve in the Albertine Rift of eastern DRC; S. Nixon unpubl. data). In Uganda, Kingdon (1971) noted that they were widely distributed in the forest-savanna-cultivation mosaic typical of the western and southern parts. At the time, animals were known to range from the exposed, ironstone plateaus that cap most low hills down through cultivated and forested hillsides to the swamp forests and papyrus (*Cyperus papyrus*) beds that filled valley bottoms. Recent confirmation of the presence of giant pangolins in the 70 km^2 Ziwa Rhino Sanctuary in central Uganda, indicates that the species can persist in isolated habitats within this mosaic habitat type, when threats are reduced, although it seems unlikely that they would occur widely in cultivated mosaics. None were observed in farmland mosaics adjacent to Kibale National Park, Uganda, during three months of camera trap surveys (D. Mills, unpubl. data). In Mahale National Park, Tanzania, pangolins were found in solid stands of bamboo (*Oxytenanthera abyssinica*) surrounded by miombo woodland (Foley et al., 2014). Giant pangolin may be dependent on permanent water sources (Kingdon et al., 2013). When local guides in the Dja Faunal Reserve, Cameroon, were asked to identify likely pangolin burrows, six of the eight putative burrows were within 100 m of swamp habitat (Bruce et al., 2018). Images captured of an adult pangolin also showed it had water on its legs, suggesting that swamp habitat is visited by the species.

Ecology

Much of what is known about giant pangolin ecology and behavior is based on work undertaken some 50 years ago in Gabon by Elisabeth Pagès. Pagès (1970) observed that giant pangolins live in relatively stable and well-defined home ranges, frequenting a series of well-visited feeding sites and rest sites. However, she noted that daily travel distances can number several kilometers in search of food. Low trapping rates of giant pangolins in discrete areas from recent camera trap studies in Nigeria and western Uganda suggest that home ranges may be relatively large in these sites (S. Nixon, unpubl. data). On the contrary, Kingdon (1971) suggested that home ranges are likely to be quite small based on observations of a male in Uganda that traversed a very limited area and regularly used a semi-exposed resting-place in a termite mound for about two years.

Shelter can take the form of any suitable and secure hiding spaces, between tree buttresses, under fallen trees, dense thickets, caves and burrows at the base of trees and roots, or partially opened termite mounds and old aardvark (*Orycteropus afer*) burrows (Fig. 10.3). Pangolins in central Gabon also widened and exposed abandoned subsoil termite galleries, creating burrows with an elongated but simple network of chambers although rarely to a depth greater than 1 m (Pagès, 1970); this is in contrast to Kingdon (1971) who noted that giant pangolins could excavate burrows up to 40 m long and 5 m deep. Research in Ziwa Rhino Sanctuary has revealed that individual pangolins use a network of multi-species burrows located within their home range, although burrow use appears irregular and infrequent (S. Nixon and N. Matthews, unpubl. data). Burrows often have multiple entrances and larger chambers for sleeping or to enable the pangolin to turn around or exit the complex facing forwards, possibly to protect against predation, but also because the angling of the scales may hinder backwards locomotion in confined tunnels (Pagès, 1970). Burrow entrances are sometimes sealed from within (Kingdon, 1971). Pagès (1970) documented that Gabonese hunters reported finding single and multiple pangolins (likely mother and offspring) together in burrows. Several observers have recorded African brush-tailed porcupine (*Atherurus africanus*) and pouched rats (*Cricetomys* spp.) entering active burrows on the same evening as giant pangolins, suggesting some degree of cohabitation (Bruce et al., 2018; S. Nixon and N. Matthews, unpubl. data). In Kibale National Park, D. Mills (unpubl. data) recorded a marsh mongoose (*Atilax paludinosus*) entering a cave at the base of a tree, and pangolins emerging a

FIGURE 10.3 Two giant pangolins in a burrow in Gabon. The species will shelter in burrows as well as between tree buttresses, under fallen trees, in caves and in dense thickets. *Photo credit: WCS Gabon.*

couple of hours later, suggesting they must have been in the cave at the same time.

The giant pangolin feeds predominantly on termites, notably *Macrotermes, Pseudocanthotermes, Odontotermes, Cubitermes, Apicotermes, Protermes*, and ants, such as *Palthothyreus* and *Dorylus* (Pagès, 1970; Vincent, 1964). Bequaert (1922) recorded the species feeding on the following ant species: *Camponotus manidis* and *C. foraminosus, Crematogaster impressa, Dorylus* sp., *Tetramorium aculeata, T. opacum, Myrmycaria eumenoides, Pheidole punctulata, Plagiolepis tenella, Polyrhachis concava*, and *Oecophylla longinoda*. Termites and ants are harvested by very rapid extrusions and retractions of the tongue, which is coated with an exceptionally viscous saliva (the pangolins' dependence on frequent drinking may well be influenced by their copious production of specialized saliva). Other insects may also be taken; L. S. B. Leakey (in Kingdon, 1971) reported watching a giant pangolin, semi-submerged in lakeside water, whipping its tongue over the surface of the water to corral and ingest water beetles (Dytiscidae). Giant pangolin dung observed in Gabon also contained the chitinous fragments of beetles (Pagès, 1970).

Major non-human predators include lion (*Panthera leo*) and probably leopard (*P. pardus*) and large pythons (*Python* spp.; Pagès, 1970). Recorded parasites include nematodes (*Ancylostoma* sp.) and ticks (*Amblyomma compressum*; Mohapatra et al., 2016; Uilenberg et al., 2013; see Chapter 29).

Behavior

Normally solitary, although females are sometimes observed accompanied by their young, while Kingdon (1971) reports a pair together with a single young animal. In the Dja Faunal Reserve, camera trap surveys have twice captured two adult animals walking one after another suggesting close associations (Bruce et al., 2018). In Ziwa Rhino Sanctuary, camera traps have regularly recorded several individuals using the same areas and burrow sites during the same time period indicating that home ranges may overlap considerably and that burrows aren't specific to individual pangolins (S. Nixon and N. Matthews, unpubl. data). Social relations are primarily mediated by scent, and the enlarged anal glands in males and associated marking behavior suggest they may be territorial. In the Dja Faunal Reserve, a male was observed apparently wiping his anal glands close to the entrance of a burrow (Bruce et al., 2018).

Kingdon (1971) found captive animals to be most active between midnight and 0500, while hunters in Liberia report activity patterns from 2000 to early morning (Fig. 10.4; Kingdon et al., 2013). Within the Dja Faunal Reserve, across two large-scale camera trap surveys, 32/33 encounters were recorded at night, with one detection at 0700 (Bruce et al., 2017; ZSL, unpubl. data). Similarly, in Kibale National Park, based on 46 observations, most animals had left their burrows by around 2000 and the latest they returned was 0530 (D. Mills, unpubl. data). On the other hand, camera trapping in the Batéké Plateau National Park, Gabon, showed clear diurnal activity patterns for giant pangolin (Hedwig et al., 2018), a pangolin was detected in full daylight at around 0830 in Semuliki National Park, Uganda (S. Nixon et al., unpubl. data) and diurnal behavior is also reported elsewhere (see Bruce et al., 2017). This may reflect hunting pressure, as other species (such as elephants [*Loxodonta africana*] and chimpanzees [*Pan troglodytes*]) become more nocturnal in areas where they are threatened (Krief et al., 2014; Maisels et al., 2015; Wrege et al., 2010). Giant pangolins may remain inactive for long periods sometimes for up to weeks at a time (Bequaert, 1922; Kingdon, 1971).

In contrast to the bipedal Temminck's pangolin, giant pangolins are quadrupedal. They

FIGURE 10.4 Principally nocturnal, giant pangolins are active at night. *Photo credit: David Mills.*

are capable diggers with heavy, powerful, clawed forelimbs that enable them to expose subterranean termite colonies. Pagès (1975) reports that giant pangolin studied in the central forests of Gabon rarely destroyed aboveground termite nests, primarily feeding on subsoil colonies to a depth of 50 cm. Once termites are exposed, they are swiftly "licked" up by the tongue from the surface of the disturbed soil. The tail is sufficiently long and robust to serve as a prop while all four limbs are engaged in digging (Kingdon, 1971). When alarmed the tail can also be used to violently club the attacker against the larger scales of the back (Pagès, 1970) and can also be used as a club when harassed, making rapid side movements that splay and "scissor" the long, very sharp scales near the tail's base. These movements and the blades and points on the lateral scales of the tail can catch or cut any animal attempting to probe into a curled-up pangolin. Booth (1960) reported a hunter's encounter with a giant pangolin with the head of a dead leopard held fast in the curl of its tail.

Although giant pangolins hiss and grunt loudly when disturbed or harassed (Kingdon, 1971; T.R.B. Davenport, unpubl. data), they appear to be largely mute.

There is some possible evidence for a polyspecific association between bay duiker (*Cephalophus dorsalis*) and the giant pangolin. On two separate occasions, one month apart, a bay duiker in the Dja Faunal Reserve was camera trapped closely following a giant pangolin (ZSL, unpubl. data). The reasons for and the strength of this association remain unclear, but remains of ants and termites have been found in bay duiker stomachs (Kingdon and Feer, 2013) and it is possible that bay duiker may feed on remains from giant pangolin diggings or scat.

Ontogeny and reproduction

Breeding is aseasonal. Two birth records of single young from Uganda are from September and October (Kingdon, 1971). Hatt (1934) reported two fetuses in specimens from

FIGURE 10.5 Giant pangolins are solitary, although females are sometimes observed accompanied by their young. *Photo credit: Naomi Matthews.*

DRC in November and December, one measuring 240 mm and the other 290 mm. A mother and nursing young were recorded in August (D. Mills, unpubl. data) and a mother and young in September in Uganda (Fig. 10.5; S. Nixon and N. Matthews, unpubl. data). Total length and weight of newborns is up to 450 mm and just over 500 g, respectively, and young are born with a full complement of soft scales and eyes open (Kingdon, 1971). A one-day-old pangolin placed near its mother immediately worked its way onto the base of her tail; the tail of a newborn is strongly prehensile and the animal has a very strong clinging reflex (Kingdon, 1971) able to hang suspended from its mother while nursing (D. Mills, unpubl. data). Young do not usually become independent until the next infant is born. Longevity in the wild or captivity is not known.

Population

The quantity and availability of termites and ants, proximity to water, levels of predation from humans and large carnivores, available habitat, and frequency of fires are likely influences on abundance and occupancy (see Khwaja et al., 2019). Available data suggest the species is nowhere common and generally rare, but the advent of camera traps is helping shed some light on the abundance of this species (Khwaja et al., 2019). In Mahale National Park, seven individuals were detected in 653 camera trap nights suggesting they may be relatively common; however, in Minziro Forest Nature Reserve, where human pressure is higher, only two individuals were captured in 1500 camera trap nights (Foley et al., 2014). In Uganda, between October 2014 and June 2016, the species was detected six times in 750 camera trap nights in Budongo Forest Reserve, once in 1600 camera trap nights in Bugungu Wildlife Reserve, once in 264 camera trap nights in Karuma Wildlife Reserve and once in 1485 camera trap days in Murchison Falls National Park (S. Ayebare, unpubl. data). In Kibale National Park, a recent and intensive dry-season camera trap survey focused on giant pangolin in one study site detected the species only once in more than 6000 camera trap nights (Chester Zoo, unpubl. data), suggesting either a reduction in population size since studies between 2010 and 2013 (D. Mills, unpubl. data) or seasonal differences in habitat use and ranging patterns. In the Lomami National Park, DRC, giant pangolins were recorded between

2012 and 2018 from five of eight camera trap surveys (20 cameras per site, placed near ground level on a 1 km grid for 50 days; Lukuru Foundation, unpubl. data). Encounter rates were low, with individuals recorded at only 10 of 220 camera locations (0.1 encounters per 100 camera trap days). This contrasts with the aardvark, which was recorded from six of eight surveys, at 34 camera locations, for a total of 77 encounters over the same grids. Six of 10 records of giant pangolins occurred on camera trap grids where aardvarks were not recorded, suggesting that these two insectivorous species may have differing ecological requirements in this habitat. In Ziwa Rhino Sanctuary, a grassland site, both giant pangolins and aardvarks are recorded regularly by camera traps in the survey area at the same time and have been observed using the same burrow networks on different nights. In the Dja Faunal Reserve, separate camera trapping surveys recorded 23 individual records in 3371 trap nights (ZSL, unpubl. data) compared with 10 in 3725 trap nights (Bruce et al., 2017) in two different areas of the reserve. Although not a substantial difference, the area with fewer events is presumed to be undergoing more intensive human activity, due to villages being relatively close by on the other side of the Dja River (T. Bruce et al., unpubl. data). In the Kwano region of Gashaka Gumti National Park, the species was captured only twice in 3350 trapping nights. Although poaching levels were low in the Kwano region during the survey period, seasonal burning is widespread and intensive in the mosaic habitat. The species was only captured at the height of the wet season (S. Nixon, unpubl. data). In general, the trapping effort required to monitor populations at any given study site using existing methods appears prohibitively high; Khwaja et al. (2019) estimated that a minimum of 100 locations would need to be monitored for at least 6 months to obtain a reasonably precise occupancy estimate for giant pangolin in undisturbed habitat.

Status

The giant pangolin is listed as Endangered on The IUCN Red List of Threatened Species (Nixon et al., 2019) and the species is declining across its range. However, as noted, there is a paucity of quantitative data on populations. It was included in Appendix I of CITES in 2016. The giant pangolin is also afforded protection under national legislation in most of its range states, which typically prohibits exploitation. Traditional customs provide the species some level of protection in portions of its range, as in Dekese Territory in central DRC where it is considered a protected ancestral animal and seldom hunted. If one is killed, the traditional leaders require special concessions to honor the ancestors. A dead pangolin was displayed on a ceremonial altar or "etuka" in a village near Dekese followed by its consumption at a ceremonial communal meal (J. Thompson, unpubl. data). Likewise, some Baka communities around Lobéké did not hunt or eat them in the 1990s, regarding them as strong taboo (T.R.B. Davenport, unpubl. data) although this may have changed as the value of their scales to the Asian market has become known, and as large numbers of people from elsewhere in Cameroon (and other Central African countries) have moved into the area to work in the many logging concessions.

Threats

There are two main threats to the giant pangolin, both of which are anthropogenic: overexploitation for local use and illegal trade to Asia, and the loss and decline in extent and quality of suitable habitat.

Unsustainable hunting and poaching for bushmeat and traditional medicine is a major threat to the species and is compounded by a rapidly growing illegal international trade from Africa to Asia to supply the demand for pangolin scales. This has seen their prevalence in bushmeat markets increase rapidly. Colyn et al. (1987) found that this species comprised one-tenth of the total number of pangolins (~100) on sale as bushmeat in rural areas around Kisangani, DRC. The number of giant pangolins observed on sale for bushmeat at Kisangani market increased seven-fold between 2002 and 2008–09 (van Vliet et al., 2012). Similarly, in 2000–06, although giant pangolins were recorded in trade in Gabon, the species did not appear in the eighteen most commonly traded species from a comprehensive survey across Gabon, but by 2014 they were the seventh most traded species, with prices increasing 211% in Libreville, despite being fully protected (Mambeya et al., 2018). Ingram et al. (2019) documented an increase in the number and price of giant pangolins observed in Malabo market on Bioko, Equatorial Guinea, between 1997 and 2017. As noted under *Distribution*, giant pangolins do not occur on Bioko and did not appear in the Malabo market until 2004 after imports began from mainland Equatorial Guinea; from 2007, imports have mainly come from Cameroon. In Salonga National Park, DRC, pangolin scales are now commonly found in hunters' camps, whereas as little as five years ago they were seldom recorded (J. Eriksson, unpubl. data).

Large volumes of scales destined for the international trade are seized throughout the range. Some 2270 kg of scales from giant pangolin were seized in seven seizures in Uganda in 2014 and 2015. A total of 1670 kg of scales were also seized in Kenya in eight seizures between 2013 and 2016 and which had alleged end-destinations of China, Thailand and Vietnam (Challender and Waterman, 2017). It is likely that many, if not most, of the scales in these reported seizures were from DRC. A large seizure (>2000 kg) of giant pangolin scales was made by authorities in Kisangani in 2017 (T. Hart, unpubl. data). Cameroon seized 1000 kg of scales in 2014, which had an alleged destination of China. Finally, Thailand seized 1066 kg of scales in 2017 that allegedly originated from DRC and had transited through Kenya to Thailand with an alleged destination of Lao PDR (Challender and Waterman, 2017; see also Chapter 16). Between 1975 and 2016, there was some legal trade reported to CITES, mainly scales to China and smaller numbers of individuals for captive breeding (see Chapter 16).

Giant pangolins are hunted and poached (Fig. 10.6) by a variety of methods, both targeted and opportunistic. In central Gabon, experienced hunters could detect and track the species to regularly used sleeping sites following footprints and feeding signs (Pagès, 1970). In the Lomami region of central DRC, T. Hart (unpubl. data) reports hunters using trained dogs to locate active and occupied rest sites where pangolins are then excavated and killed with a machete blow. Similarly, there are reports of hunting giant pangolins in burrows with dogs in Mbam et Djerem National Park (I. Goodwill, unpubl. data) and surveys in 2018 in coastal Gabon have revealed evidence of targeted hunting with occupied burrows being excavated during the day to remove pangolins (Chester Zoo, unpubl. data). In the Lubutu region of DRC, where poaching pressure is high, giant pangolins are caught opportunistically in snares set primarily for small ungulates such as duikers (S. Nixon, unpubl. data), and the same problem appears to be reported from other protected areas, such as Kibale National Park (R. Wrangham, unpubl. data). In Maiko National Park, as well as hunting pangolins with dogs, there is direct evidence of targeted hunting with hunters setting specialized snares in the mouths of occupied burrows (Chester Zoo, unpubl. data).

The second main threat is habitat related. The species occurs in primary and secondary forest, but its extent of occupancy and the quality of available habitat is understood to have declined in recent decades as a result of forest loss, degradation and fragmentation. Mayaux et al. (2013) estimated net forest loss between 2000 and 2010 of 0.1% in Central Africa and 0.3% in West

FIGURE 10.6 A poached giant pangolin in Gabon. Unsustainable exploitation for bushmeat and traditional medicine is a major threat, which is compounded by illicit international trade to Asia. *Photo credit: Martin Hega.*

Africa. Hansen et al. (2013) reported that tropical forest loss increased between 2000 and 2012. The drivers of forest loss include the expansion of agriculture and croplands, including shifting cultivation; commercial logging, including the clear cutting of forest sites; expansion of urban infrastructure and increasing human population densities, culminating in an "agriculture-population" combination driving deforestation (Mayaux et al., 2013).

References

Akpona, H.A., Daouda, I.-H., 2011. Mammifères myrmécophages. In: Neuenschwander, P., Sinsin, B., Goergen, G. (Eds.), Protection de la Nature en Afrique de l'Ouest: Une Liste Rouge pour le Bénin. Nature Conservation in West Africa: Red List for Benin. International Institute of Tropical Agriculture, Ibadan, Nigeria, pp. 298–303.

Allen, G.M., 1939. A checklist of African mammals. Bull. Museum Comp. Zool. Harv. Coll. 83, 1–763.

Allen, G.M., Coolidge Jr., H.J., 1930. Mammals of Liberia. In: Strong, R.P. (Ed.), The African Republic of Liberia and the Belgian Congo, 2. Contributions of the Department of Tropical Medicine and the Institute for Tropical Biology and Medicine, pp. 569–622.

Amori, G., Segniagbeto, G.H., Decher, J., Assou, D., Gippoliti, S., Luiselli, L., 2016. Non-marine mammals of Togo (West Africa): an annotated checklist. Zoosystema 38 (2), 201–244.

Barnett, A.A., Prangley, M.L., 1997. Mammalogy in the Republic of Guinea: an overview of research from 1946 to 1996, a preliminary check-list and a summary of research recommendations for the future. Mammal Rev. 27 (3), 115–164.

Beja, P., Pinto, P.V., Veríssimo, L., Bersacola, E., Fabiano, E., Palmeirim, J.M., et al., 2019. The Mammals of Angola. In: Huntley, B.J., Russo, V., Lages, F., Nuno-Ferrand, N. (Eds.), Biodiversity of Angola, Science & Conservation: A Modern Synthesis. Springer, Cham, pp. 357–443.

Bequaert, J., 1922. The predaceous enemies of ants. Bull. Am. Museum Nat. Hist. 45, 271–331.

Bocage, J.V.B. du., 1889. Mammiferes d'Angola et du Congo. Jornal de Sciências, Mathemáticas, Physicas e Naturaes, Lisboa 2 (1), 8–32, 174–185.

Bocage, J.V.B. du., 1890. Mammiferes d'Angola et du Congo. Jornal de Sciências, Mathemáticas, Physicas e Naturaes, Lisboa 2 (2), 1–32.

Booth, A.H., 1960. Small Mammals of West Africa. Longmans, London.

Bruce, T., Wacher, T., Ndinga, H., Bidjoka, V., Meyong, F., Ngo Bata, M., et al., 2017. Camera-Trap Survey for Larger Terrestrial Wildlife in the Dja Biosphere Reserve, Cameroon. Yaoundé, Cameroon: Zoological

Society of London (ZSL) and Cameroon Ministry of Forests and Wildlife (MINFOF).

Bruce, T., Kamta, R., Mbobda, R.B.T., Kanto, S.T., Djibrilla, D., Moses, I., et al., 2018. Locating giant ground pangolins (*Smutsia gigantea*) using camera traps on burrows in the Dja Biosphere Reserve, Cameroon. Trop. Conserv. Sci. 11, 1–5.

Challender, D., Waterman, C., 2017. Implementation of CITES Decisions 17.239 b) and 17.240 on Pangolins (*Manis* spp.), CITES SC69 Doc. 57 Annex. Available from <https://cites.org/sites/default/files/eng/com/sc/69/E-SC69-57-A.pdf>. [2 August 2018].

Coe, M., 1975. Mammalian ecological studies of Mount Nimba. Liberia Mammal. 39 (4), 523–587.

Colyn, M., Dudu, A., Mankoto, M.M., 1987. Données sur l'exploitation du 'petit et moyen gibier' des forêts ombrophiles du Zaïre. In: Proceedings of International Symposium on Wildlife Management in Sub-Saharan Africa. International Foundation for the Conservation of Game, pp. 110–145.

Doran, G.A., Allbrook, D.B., 1973. The tongue and associated structures in two species of African pangolins, *Manis gigantea* and *Manis tricuspis*. J. Mammal. 54 (4), 887–899.

Dupuy, A.R., 1968. Sur la première capture au Sénégal d'un grand Pangolin *Smutsia gigantea*. Mammalia 32 (1), 131–132.

Feiler, A., 1990. Distribution of mammals in Angola and notes on biogeography. In: Peters, G., Hutterer, R. (Eds.), Vertebrates in the Tropics. Museum Alexander Koenig, Bonn, pp. 221–236.

Fischer, F., Gross, M., Linsenmair, K., 2002. Updated list of the larger mammals of the Comoé National Park, Ivory Coast. Mammalia 66 (1), 83–92.

Fitzinger, L.J., 1872. Die naturliche familie der schuppenthiere (Manes). Sitzungsberichte der Kaiserlichen Akademie der Wissenschaften. Mathematisch-Naturwissenschaftliche Classe, CI., LXV, Abth. I, 9–83.

Foley, C., Foley, L., Lobora, A., De Luca, D., Msuha, M., Davenport, T.R.B., et al., 2014. A Field Guide to the Larger Mammals of Tanzania. Princeton University Press, Princeton.

Frade, F., 1949. Algumas novidades para a fauna da Guiné Portuguesa (aves e mamíferos). Anais: Junta das Missões Geográficas e de Investigaçõeses Colóniais (Lisboa), Estudos de Zoologia 4, 165–186.

Frechkop, S., 1931. Notes sur les mammifères. VI. Quelques observations sur la classification des pangolins (Manidae). Bulletin du Musee royal d'Histoire naturelle de Belgique VII (22), 1–14.

Gaubert, P., 2011. Family Manidae. In: Wilson, D.E., Mittermeier, R.A. (Eds.), Handbook of the Mammals of the World, vol. 2. Hoofed Mammals. Lynx Edicions, Barcelona, pp. 82–103.

Gaubert, P., Antunes, A., Meng, H., Miao, L., Peigné, S., Justy, F., et al., 2018. The complete phylogeny of pangolins: scaling up resources for the molecular tracing of the most trafficked mammals on Earth. J. Hered. 109 (4), 347–359.

Gaudin, T.J., Emry, R.J., Wible, J.R., 2009. The phylogeny of living and extinct pangolins (Mammalia, Pholidota) and associated taxa: a morphology based analysis. J. Mammal. Evol. 16 (4), 235–305.

Gotch, A.F., 1979. Mammals – Their Latin Names Explained. A Guide to Animal Classification. Blandford Press, Poole.

Gray, J.E., 1865. 4. Revision of the genera and species of entomophagous Edentata, founded on the examination of the specimens in the British Museum. Proc. Zool. Soc. Lond. 33 (1), 359–386.

Grubb, P., Jones, T.S., Davies, A.G., Edberg, E., Starin, E.D., Hill, J.E., 1998. Mammals of Ghana, Sierra Leone and the Gambia. Trendrine Press, Zennor, Cornwall.

Hansen, M.C., Potapov, P.V., Moore, R., Hancher, M., Turubanova, S.A., Tyukavina, A., et al., 2013. High-resolution global maps of 21st-century forest cover change. Science 342 (6160), 850–853.

Happold, D.C.D., 1987. The Mammals of Nigeria. Clarendon Press, Oxford.

Harris, N.C., Mills, K.L., Harissou, Y., Hema, E.M., Gnoumou, I.T., VanZoeren, I.J., et al., 2019. First camera survey in Burkina Faso and Niger reveals human pressures on mammal communities within the largest protected area complex in West Africa. Conserv. Lett. 2019, e12667.

Hassanin, A., Hugot, J.-P., van Vuuren, B.J., 2015. Comparison of mitochondrial genome sequences of pangolins (Mammalia, Pholidota). C. R. Biol. 338 (4), 260–265.

Hatt, R.T., 1934. The pangolin and aard-varks collected by the American Museum Congo expedition. Bull. Am. Museum Nat. Hist. 66, 643–671.

Hedwig, D., Kienast, I., Bonnet, M., Curran, B.K., Courage, A., Boesch, C., et al., 2018. A camera trap assessment of the forest mammal community within the transitional savannah-forest mosaic of the Batéké Plateau National Park, Gabon. Afr. J. Ecol. 56 (4), 777–790.

Hill, J.E., Carter, T.D., 1941. The mammals of Angola, Africa. Bull. Am. Museum Nat. Hist. 78, 1–211.

Hoffmann, M., Cronin, D.T., Hearn, G., Butynski, T.M., Do Linh San, E., 2015. A review of evidence for the presence of Two-spotted Palm Civet *Nandinia binotata* and four other small carnivores on Bioko, Equatorial Guinea. Small Carnivore Conserv. 52 & 53, 13–23.

Ingram, D.J., Cronin, D.T., Challender, D.W.S., Venditti, D. M., Gonder, M.K., 2019. Characterising trafficking and trade of pangolins in the Gulf of Guinea. Glob. Ecol. Conserv. 17, e00576.

Jeannin, A., 1936. Les mammiféres sauvages du Cameroun. Encyclopedie Biol. 16, 116–130.

Jentink, F.A., 1882. Note XXV. Revision of the Manidae in the Leyden Museum. Notes from the Leyden Museum IV, 193–209.

Jones, C., 1973. Body temperatures of *Manis gigantea* and *Manis tricuspis*. J. Mammal. 54 (1), 263–266.

Khwaja, H., Buchan, C., Wearn, O.R., Bahaa-el-din, L., Bantlin, D., Bernard, H., et al., 2019. Pangolins in global camera trap data: implications for ecological monitoring. Glob. Ecol. Conserv. 20, e00769.

Kingdon, J., 1971. East African mammals. An Atlas of evolution in Africa, Primates, Hyraxes, Pangolins, Protoungulates, Sirenians, vol. I. Academic Press, London.

Kingdon, J., 1997. The Kingdon Field Guide to African Mammals. Academic Press, London.

Kingdon, J., Feer, F., 2013. Bay duiker *Cephalophus dorsalis*. In: Kingdon, J., Hoffmann, M. (Eds.), Mammals of Africa, vol. VI, Hippopotamuses, Pigs, Deer, Giraffe and Bovids. Bloomsbury Publishing, London, pp. 294–298.

Kingdon, J., Hoffmann, M., Hoyt, R., 2013. *Smutsia gigantea* Giant Ground Pangolin. In: Kingdon, J., Hoffmann, M. (Eds.), Mammals of Africa. vol. V, Carnivores, Pangolins, Equids, Rhinoceroses. Bloomsbury Publishing, London, pp. 396–399.

Krief, S., Cibot, M., Bortolamiol, S., Seguya, A., Krief, J.-M., Masi, S., 2014. Wild Chimpanzees on the Edge: Nocturnal Activities in Croplands. PLoS One 9 (10), e109925.

Machado A de B., 1969. Mamiferos de Angola ainda não citados ou pouco conhecidosa. Publições culturais da Companhia de Diamantes de Angola 46, 93–232.

Maisels, F., Fishlock, V., Greenway, K., Wittemyer, G., Breuer, T., 2015. Detecting threats and measuring change at bais: a monitoring framework. In: Fishlock, V., Breuer, T. (Eds.), Studying Forest Elephants. Neuer Sportverlag, Stuttgart, Germany, pp. 144–155.

Malbrant, R., MacLatchy, A., 1949. Faune de l'Équateur Africain Français. Tome II: Mammifères. Paul Lechevalier, Paris.

Mambeya, M.M., Baker, F., Momboua, B.R., Pambo, A.F.K., Hega, M., Okouyi, V.J.O., et al., 2018. The emergence of a commercial trade in pangolins from Gabon. Afr. J. Ecol. 56 (3), 601–609.

Mayaux, P., Pekel, J.-F., Desclee, B., Donnay, F., Lupi, A., Achard, F., et al., 2013. State and evolution of the African rainforests between 1990 and 2010. Philos. Trans. R. Soc. B 368 (1625), 20120300.

Meester, J., 1972. Order Pholidota. In: Meester, J., Stzer, H. W. (Eds.), The Mammals of Africa: An Identification Manual, Part 4. Smithsonian Institution Press, Washington, D.C., pp. 1–3.

Mohapatra, R.K., Panda, S., Nair, M.V., Acharjyo, L.N., 2016. Check list of parasites and bacteria recorded from pangolins (*Manis* sp.). J. Parasit. Dis. 40 (4), 1109–1115.

Mohr, E., 1961. Schuppentiere. Neue Brehm-Bucherei. A. Ziemsen Verlag, Wittenberg Lutherstadt.

Monard, A., 1935. Contribution à la Mammalogie d'Angola et Prodrome d'une Faune d'Angola. Arquivos do Museu Bocage 6, 1–314.

Nixon, S., Pietersen, D., Challender, D., Hoffmann, M., Godwill Ichu, I., Bruce, T., et al., 2019. *Smutsia gigantea. The IUCN Red List of Threatened Species* 2019: e.T12762A 123584478. Available from: <http://dx.doi.org/10.2305/IUCN.UK.2019-3.RLTS.T12762A123584478.en>.

Pagès, E., 1970. Sur l'écologie et les adaptations de l'oryctérope et des pangolins sympatriques du Gabon. Biol. Gabon. 6, 27–92.

Pagès, E., 1975. Étude éco-éthologique de *Manis tricuspis* par radio-tracking. Mammalia 39, 613–641.

Poche, R., 1973. Niger's threatened Park W. Oryx 12, 216–222.

Pocock, R.I., 1924. The external characters of the pangolins (Manidae). Proc. Zool. Soc. Lond. 94 (3), 707–723.

Rahm, U., 1956. Notes on Pangolins of the Ivory Coast. J. Mammal. 37 (4), 531–537.

Rahm, U., 1966. Les mammifères de la forêt équatoriale de l'est du Congo. Annales du Musée Royal de l'Afrique Centrale, Sciences Zoologiques 149, 39–121.

Reiner, F., Simões, P., 1999. Mamíferos selvagens de Guiné-Bissau. Centro Portugues de Estudos dos Mamiferos Marinhos, Lisbao, Portugal.

Roulet, P.A., Pelissier, C., Patek, G., Beina, D., Ndallot, J., 2007. Projet Zemongo- Un aperçu du contexte écologique et de la pression anthropique sur les resources naturelles de la Réserve de Faune de Zemongo, Préfecture du Haut - Mbomou, République Centrafricaine. Rapport final de la mission du 15 janvier au 19 mars 2006. MEFCP, Bangui, CAR, pp. 1–78.

Rosevear, D.R., 1953. Checklist and Atlas of Nigerian Mammals, With a Foreword on Vegetation. The Government Printer, Lagos.

Sayer, J.A., Green, A.A., 1984. The distribution and status of large mammals in Benin. Mammal Rev. 14 (1), 37–50.

Schlitter, D.A., 2005. Order Pholidota. In: Wilson, D.E., Reeder, D.M. (Eds.), Mammal Species of the World: A Taxonomic and Geographic Reference, third ed. Johns Hopkins University Press, Baltimore, pp. 530–531.

Schouteden, H., 1944. De zoogdieren van Belgisch Congo en van Ruanda-Urundi I. – Primates, Chiroptera, Insectivora, Pholidota. Annalen van het Museum van belgisch Congo. C. Dierkunde. Reeks II. Deel III. Aflevering 1, 1–168.

Schouteden, H., 1948. Faune de Congo Belge et du Ruanda-Urundi. I. Mammifères. Annales du Musée du Congo Belge, Zoologie 8 (1), 1–331.

Uilenberg, G., Estrada-Pena, A., Thal, J., 2013. Ticks of the Central African Republic. Exp. Appl. Acarol. 60, 1–40.

Ullmann, T., Veríssimo, D., Challender, D.W.S., 2019. Evaluating the application of scale frequency to estimate the size of pangolin scale seizures. Glob. Ecol. Conserv. 20, e00776.

van Vliet, N., Nebesse, C., Gambalemoke, S., Akaibe, D., Nasi, R., 2012. The bushmeat market in Kisangani, Democratic Republic of Congo: implications for conservation and food security. Oryx 46 (2), 196–203.

Verschuren, J., 1988. Notes d'Ecologie, principalement des mammiferes, du Pare National de la Pendjari, Benin. Bulletin de l'Institut Royal Des Sciences Naturelles de Belgique 58, 185–206.

Vincent, F., 1964. Quelques observations sur les pangolins (Pholidota). Mammalia 28, 659–665.

Wrege, P.H., Rowland, E.D., Thompson, B.G., Batruch, N., 2010. Use of acoustic tools to reveal otherwise cryptic responses of forest elephants to oil exploration. Conserv. Biol. 24 (6), 1578–1585.

CHAPTER

11

Temminck's pangolin *Smutsia temminckii* (Smuts, 1832)

Darren W. Pietersen[1,2], *Raymond Jansen*[3,4], *Jonathan Swart*[5], *Wendy Panaino*[6], *Antoinette Kotze*[7,8], *Paul Rankin*[9] *and Bruno Nebe*[10]

[1]Mammal Research Institute, Department of Zoology and Entomology, University of Pretoria, Hatfield, South Africa [2]IUCN SSC Pangolin Specialist Group, ℅ Zoological Society of London, Regent's Park, London, United Kingdom [3]Department of Environmental, Water and Earth Sciences, Tshwane University of Technology, Pretoria, South Africa [4]African Pangolin Working Group, Pretoria, South Africa [5]Welgevonden Game Reserve, Vaalwater, South Africa [6]Brain Function Research Group, School of Physiology and Centre for African Ecology, School of Animal, Plant and Environmental Sciences, University of the Witwatersrand, Johannesburg, South Africa [7]National Zoological Garden, South African National Biodiversity Institute, Pretoria, South Africa [8]Genetics Department, University of the Free Sate, Bloemfontein, South Africa [9]Deceased [10]Pangolin Research Mundulea, Swakopmund, Namibia

OUTLINE

Pangolins
DOI: https://doi.org/10.1016/B978-0-12-815507-3.00011-3

Taxonomy

Previously included in the genera *Manis* (Meester, 1972; Schlitter, 2005) and *Phataginus* (e.g., Grubb et al., 1998) this species is here included in the genus *Smutsia* based on morphological (Gaudin et al., 2009) and genetic evidence (du Toit et al., 2017; Gaubert et al., 2018), and following a synthetic classification of extant pangolins (see Chapter 2). Synonyms: *Phatages hedenborgii* (Fitzinger, 1872). The type locality is Kuruman, Northern Cape, South Africa. No subspecies are recognized. Chromosome number is not known.

Etymology: *Smutsia* (see Chapter 10); *temminckii* refers to Prof C.J. Temminck (1778–1858), a Dutch zoologist (Gotch, 1979).

Description

Temminck's pangolin (*Smutsia temminckii*) is a medium-sized, stout species with a body weight of around 9–10 kg and a total body length of up to 140 cm, including a thickset tail that is slightly shorter than total head-body length (Table 11.1). There appears to be some variation in size across the species' range. In Sudan, Sweeney (1974) recorded a male weighing 21 kg. In South Africa, individuals in the Kalahari are typically 25–30% smaller than in mesic eastern parts of the country (Table 11.1; Pietersen et al., 2014a). Adult males tend to be larger and heavier than females but there is no apparent sexual dimorphism (Heath and Coulson, 1997a; Pietersen, 2013; D.W. Pietersen, unpubl. data). There is a linear correlation between total length and mass (Jacobsen et al., 1991).

The body is covered in large, overlapping scales comprised of keratin, which cover the dorsal and lateral surfaces, the limbs, both the dorsal and ventral surfaces of the tail, and the forehead (Fig. 11.1). Scales are absent from the head (excluding the forehead), the ventral surface of the body, and the inside of the limbs, which are covered in soft white skin with sparse, short hairs (~5 mm). As with other African pangolins, no hairs project between the scales. There are 12 transverse, and 11–13 longitudinal scale rows on the body, and 11–13 rows of scales on the tail margins (Table 11.1; Frechkop, 1931; Jentink, 1882). The scales on the tail margins are pointed and cover the edge of the tail both dorsally and ventrally. The tail has five scale rows proximally, reducing to four towards the distal end, resulting in the last nine scale rows comprising four scales each (Jentink, 1882; D.W. Pietersen, unpubl. data). Total scales number between ~340 and 420 (Table 11.1). Scales on the forehead and upper (e.g., elbow to wrist) portions of the limbs are the smallest; those on the dorsum and tail are the largest (Fig. 11.1). The scales are directed to the posterior with the exception of those on the hindlimbs, which point down (Swart, 2013). In adults, the distal edge of the scales is rounded due to continuous abrasion against the underlying scales, but very sharp. In juveniles and sub-adults, the scales end in a pronounced medial point or cusp, which wears and may break off with age. The scales vary in color geographically, ranging from slate-gray to dark brown and yellow-brown, and may terminate with ivory colored edges (Fig. 11.1). In older animals, scales are as thick at the distal edge as they are proximally (D.W. Pietersen, unpubl. data). The skin and scales weigh 33–35% of the total body mass (Kingdon, 1971; Pietersen, 2013), and approximately 25% of body mass if dried and the interstitial tissue removed (D.W. Pietersen and Tikki Hywood Foundation, unpubl. data).

The skull is pear-shaped, being widest just behind the ear opening and tapering gradually to the narrow muzzle (Fig. 11.2); the taper is more abrupt from the ear openings to the occiput. There is no visible neck (Swart, 2013). The facial skin is dark gray, and the eyes are small, dark and bulbous and have a nictitating membrane in addition to eyelids, which offers protection from swarming prey (Fig. 11.2);

TABLE 11.1 Temminck's pangolin morphometrics.

Measurement			Country	Source(s)
Weight	Weight (♂)	9.3 (2.5–16.1) kg, n = 29	South Africa, Zimbabwe	Coulson, 1989; Heath and Coulson, 1997a,b; Jacobsen et al., 1991; Swart et al., 1999
		6.0 (2.6–10.6) kg, n = 50	Kalahari, Northern Cape Province, South Africa	D.W. Pietersen, unpubl. data
		21 kg, n = 1	Sudan	Sweeney, 1974
	Weight (♀)	9.0 (4.6–15.8) kg, n = 28	South Africa, Zimbabwe	Coulson, 1989; Heath and Coulson, 1997a,b; Jacobsen et al., 1991; Swart et al., 1999
		5.6 (2.5–10.2) kg, n = 28	Kalahari, Northern Cape Province, South Africa	D.W. Pietersen, unpubl. data
Body	Total length (♂)	836 (634–1049) mm, n = 18	Zimbabwe	Coulson, 1989
		997 (690–1240) mm, n = 50	Kalahari, Northern Cape Province, South Africa	D.W. Pietersen, unpubl. data
	Total length (♀)	827 (720–925) mm, n = 15	Zimbabwe	Coulson, 1989
		984 (640–1250) mm, n = 28	Kalahari, Northern Cape Province, South Africa	D.W. Pietersen, unpubl. data
	Total length (unsexed)	1014 (587–1403) mm, n = 10	Zimbabwe	Heath and Coulson, 1998
	Head-body length (♂)	431 (297–565) mm, n = 18	Zimbabwe	Coulson, 1989
	Head-body length (♀)	458 (350–677) mm, n = 15	Zimbabwe	Coulson, 1989
	Tail length (♂)	405 (290–585) mm, n = 18	Zimbabwe	Coulson, 1989
	Tail length (♀)	370 (223–440) mm, n = 15	Zimbabwe	Coulson, 1989
Vertebrae	Total number of vertebrae	48		Jentink, 1882
	Cervical	7		Jentink, 1882
	Thoracic	11–12		Jentink, 1882; Mohr, 1961
	Lumbar	5–6		Jentink, 1882; Mohr, 1961
	Sacral	3		Jentink, 1882; Mohr, 1961
	Caudal	21–24		Jentink, 1882; Mohr, 1961

(Continued)

TABLE 11.1 (Continued)

Measurement			Country	Source(s)
Skull	Length (♂)	84.9 (72–98.3) mm, n = 11		Coulson, 1989
	Length (♀)	81.2 (71.8–89) mm, n = 6		Coulson, 1989
	Breadth across zygomatic processes	No data		
Scales	Total number of scales	382 (343–422), n = 6	South Africa, unknown	Ullmann et al., 2019
	No. of scale rows (transversal, body)	12		Frechkop, 1931
	No. of scale rows (longitudinal, body)	11–13		Frechkop, 1931; Jentink, 1882
	No. of scales on outer margins of tail	11–13		Frechkop, 1931; Jentink, 1882
	No. of scales on median row of tail	4–7		Frechkop, 1931; Jentink, 1882
	Scales (wet) as proportion of body weight	34.3 ± 3.4%, n = 18		Pietersen, 2013
	Scales (dry) as proportion of body weight	25%		D.W. Pietersen and Tikki Hywood Foundation, unpubl. data

FIGURE 11.1 Temminck's pangolin in the Kalahari, South Africa. The ivory colored scale tips are noticeable in this animal. *Photo credit: D.W. Pietersen.*

FIGURE 11.2 Close-up of the face of a Temminck's pangolin, showing the small scales on the forehead. The claws on the left hind foot are also visible. *Photo credit: D.W. Pietersen.*

eyesight is poor. The nose is moist. External ear openings are present, and large, although the ear pinnae are vestigial; the ear openings have soft, fluffy hair inside. Like other pangolins, teeth are absent, but two small pseudoteeth are present on the mandible anteriorly. The jaw is delicate and the mouth opening very small. The tongue is attached to the caudal end of the cartilaginous xiphisternum (formed from the last pair of cartilaginous ribs) in the abdomen (see Chapter 1; Doran and Allbrook, 1973; Heath, 1992). The xiphisternum extends posteriorly to the iliac fossa before turning dorsally and extending anteriorly along the dorsal wall of the abdomen before terminating in a spatulate-like sac against the diaphragm. The xiphisternum usually extends to the right iliac fossa, rarely to the left. In the throat region, the tongue is housed in a glossal tube, which together extend through the neck to the oral cavity (Doran and Allbrook, 1973). The tongue is 40–60 cm in length, and can be longer than the head and body length combined (Kingdon, 1971), extending out of the mouth by 20–40 cm; the glossal tube likely lubricates the tongue with a mucus membrane which is used to "capture" prey (Swart, 2013). As the tongue is withdrawn, the hyoid functions to scrape prey from the tongue directing it down the esophagus to the pyloric region of the stomach where it is masticated (Doran and Allbrook, 1973; Weber, 1892). The submandibular salivary glands are greatly enlarged and displaced caudally to reside in the pharyngeal and cervical regions (see Heath, 1992).

Both fore- and hindlimbs have five digits. The forelimbs are highly muscular and terminate with three long, sharp claws (~60 mm) in the center and much smaller claws (~30 mm) on the first and fifth digits (Swart, 2013). The hindlimbs are columnar with a cushioned pad that is widest at the front and tapers slightly towards the rear, resulting in footprints that resemble those of a miniature elephant (Swart, 2013). Compared to most other pangolin species, the claws on the hindfeet are much reduced (Fig. 11.2). Unlike other pangolins, Temminck's pangolin takes most of the weight on its hindlimbs when in locomotion and its pelvis is more vertical than in other species (Kingdon, 1971). The tail is broad, flat and muscular, and heavy (Swart, 2013), with numerous tendons attached to the caudal vertebrae affording it enormous strength. Convex above and slightly concave below, the tail forms a tight fit against the body, covering the head, shoulders and limbs when the pangolin rolls up (Swart, 2013).

Temminck's pangolins may occasionally build up subcutaneous fat deposits. They are first deposited in the scapular regions, before extending down the dorsum and onto the dorsal surface of the tail. These subcutaneous fat reserves may play a role in insulation, as scales afford very little thermal insulation (Heath and Hammel, 1986; McNab, 1984; Weber et al., 1986). Intra-abdominal fat deposits are rare and are used first in times of nutritional stress (Pietersen, 2013).

Anal glands are present laterally on either side of the anus. These produce a pungent musk-scented liquid that may play a role in territorial marking (Pocock, 1924; D.W. Pietersen, pers. obs.). Females have two pectoral nipples.

The female's vulva is situated just anterior to the anus; the male's testes are situated in the inguinal area and do not descend into a scrotum (see Chapter 1).

The body temperature is lower than other eutherian mammals. A free-ranging adult female remotely monitored over 34 days in the Kalahari region of South Africa during winter had a body temperature averaging 32–35 °C (min–max, 29.5–35.4 °C; Pietersen, 2013), with similar values recorded elsewhere in the Kalahari (W. Panaino, unpubl. data). The body temperature showed a predictive cyclical pattern, rapidly increasing and peaking just prior to the onset of activity before dropping slightly (0.8–1.2 °C) to an active temperature of 33–34 °C, but was characterized by minor peaks and troughs which are thought to be related to activity (Pietersen, 2013). The body temperature steadily decreased with inactivity in a burrow, reaching its lowest point just prior to it increasing with the onset of activity (Pietersen, 2013). This cyclical pattern was not observed in another Kalahari population during the summer months (W. Panaino, unpubl. data). Resting metabolic rate is about half that of other eutherian mammals of similar size; the rate among four free-ranging individuals in eastern South Africa averaged 140.4 ml O_2/kg/h (Swart, 2013).

Distribution

Widely distributed species primarily in Southern and East Africa (Fig. 11.3). Distribution is patchy and is determined by the presence and abundance of suitable prey species and burrows or denning sites (Pietersen et al., 2016a), but also reflects incompatible habitat changes such as crop agriculture (Coulson, 1989; Pietersen et al., 2019); overexploitation has caused local extinctions or greatly reduced populations in some areas.

The northern boundaries of the species' distribution are not well known. Recorded from Ennedi in northeastern Chad (Malbrant, 1952) and there is a record from Ouanda Djallé in northeastern Central African Republic; the species reportedly occurs widely in this region (Malbrant, 1952). Recorded from Kadugli in the Nuba Mountains of southern Sudan (Sweeney, 1956, 1974) and the species has been collected in the Sennar region, close to the Ethiopian border (Yalden et al., 1996). It is likely that Temminck's pangolin occurs on the western border regions of Ethiopia (Yalden et al., 1996) and Schloeder and Jacobs (1996) confirm the presence of the species in the Omo River basin in the southwest of the country.

Widely distributed in East Africa, including Kenya, except the east and northeast of the country, Tanzania (Foley et al., 2014; Swynnerton and Hayman, 1950), Uganda (Bere, 1962), Burundi, and Rwanda (Dorst and Dandelot, 1972; Kingdon, 1971). Both Temminck's and giant (*S. gigantea*) pangolins occur in Murchison Falls National Park in Uganda, potentially separated by the Victoria Nile (the former on the north and the latter on the south bank), but it is possible that they are sympatric (S. Nixon, pers. comm.) and this requires investigation.

In Malawi, Temminck's pangolin has primarily been recorded in the south but it is believed to occur throughout the country (Ansell and Dowsett, 1988; Smithers, 1966; Sweeney, 1959). Although absent from large tracts of central and northern Zambia, there are records in western, southern, central and eastern regions of the country (Ansell, 1960, 1978; Smithers, 1966). The species does not occur in the forested regions of extreme northwestern Zambia.

The western limits of the distribution are reached in Namibia and central and southern Angola. The species is widespread in Namibia, with the exception of the arid coastal regions (Shortridge, 1934; Stuart, 1980). In Angola, Temminck's pangolin occurs in central and southern regions, and there are records from Benguela, Bié, Caconda, Cuanza-Sul, Chitaeu,

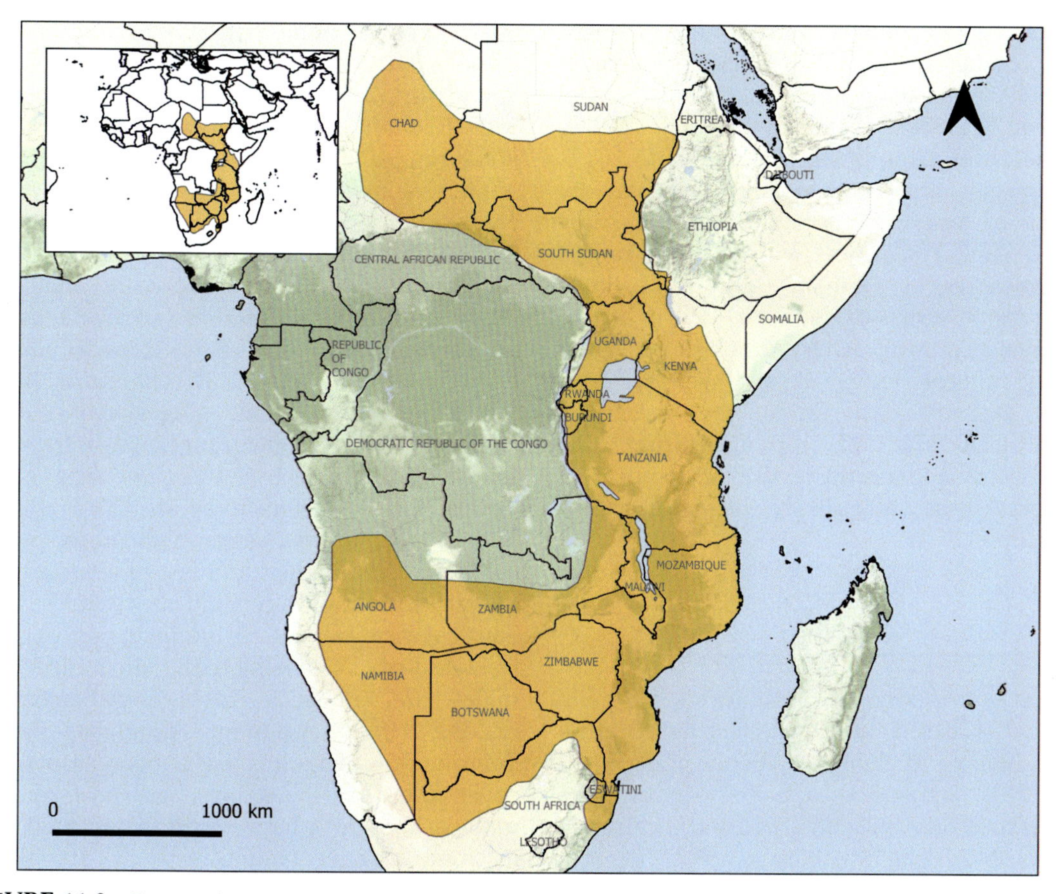

FIGURE 11.3 Temminck's pangolin distribution. *Source: Pietersen et al., 2019.*

Cuando-Cubango, Huíla, Mombolo and areas adjacent to Namibe (Beja et al., 2019; Hill and Carter, 1941; Meester, 1972; Monard, 1935).

To the south, the species occurs throughout Mozambique (Smithers and Tello, 1976; Spassov, 1990), Zimbabwe (Coulson, 1989) and Botswana (Smithers, 1971). There are no contemporary records for eSwatini. In South Africa, the species occurs in western, northern and eastern provinces (Jacobsen et al., 1991; Kyle, 2000; Pietersen et al., 2016a; Rautenbach, 1982; Swart, 1996). Overexploitation has caused local extinctions or greatly reduced populations in some areas, including KwaZulu-Natal Province, South Africa and eSwatini (Friedmann and Daly, 2004; Kyle, 2000; Monadjem, 1998; Ngwenya, 2001; Pietersen et al., 2014a).

Habitat

Occurs in arid and mesic savannas, floodplains (Heath and Coulson, 1997a), woodlands (including mopane [*Colophospermum mopane*], miombo [*Brachystegia-Julbernardia*], mixed marula [*Sclerocarya birrea*] – bush willow [*Combretum* spp.] and broad-leaved woodland; Heath and Coulson, 1997a; Smithers, 1966), thorn (*Vachellia* spp. and *Senegalia* spp.)

thickets, gallery forest and duneveld grassland, where annual rainfall averages 250–1400 mm (Coulson, 1989; Heath and Coulson, 1997a; Pietersen et al., 2016a; Skinner and Chimimba, 2005; Swart, 2013). The species does not inhabit closed-canopy forests or true deserts and does not show strong habitat selection within home ranges (Pietersen et al., 2014a). Altitudinal limits range from near sea level to 1700 m above sea level (Coulson, 1989). Although occurring widely in protected areas and well-managed game and livestock farms, the species is absent from areas of crop agriculture, presumably due to pesticide use or removal of the prey base, and likely increased direct persecution.

Ecology

Most knowledge of Temminck's pangolin ecology is based on research conducted in South Africa and Zimbabwe between the 1980s and 2010s. The species appears to hold home ranges that vary with both age and locality. In the South African Kalahari region, Minimum Convex Polygon (MCP) home range estimates for adults were $10.0 \pm 8.9\ km^2$, and $7.1 \pm 1.1\ km^2$ for sub-adults (Pietersen et al., 2014a). Males and females had closely matching home range sizes, with each male's home range overlapping that of a single female, suggesting a monogamous mating system (Pietersen et al., 2014a). In contrast, in Zimbabwe home ranges varied between 0.2 and $23.4\ km^2$ with males having a larger home range which overlapped with those of several females, albeit temporarily, and is presumably related to breeding behavior (Heath and Coulson, 1997a). In northeastern South Africa, home ranges varied between 1.3 and $7.9\ km^2$ (van Aarde et al., 1990). In eastern South Africa (Kruger National Park region), males had home ranges of $9.28–22.98\ km^2$, which overlapped the home ranges of up to five adult females, suggesting a polygynous mating system (Swart, 2013). Females had home ranges of $0.65–6.66\ km^2$, which overlapped with the home ranges of up to three adult males (Swart, 2013). One male in eastern South Africa (which had the largest home range in the study, $22.98\ km^2$) was located on 234 separate tracking days over a period of two years. The home range overlapped with those of five females, on average by 39% (9–100%) of each female's home range (Swart, 2013). The same study found that female home ranges may overlap by up to 34%. Home ranges of both sexes reportedly increase by ~4% in the wet season (Swart, 2013). Pietersen et al. (2014a) questioned whether the larger male home range estimates and higher number of overlapping female home ranges in Zimbabwe and eastern South Africa are due to the inclusion of transient males. Dispersing pangolins, especially males, are known to cover large distances, making this argument plausible. Male Temminck's pangolins have been known to cover 32–81 km in 20 days, and the farthest a male is known to have dispersed was 300 km in four months (van Aarde et al., 1990). Females appear to have shorter dispersal distances (<100 km, Pietersen et al., 2014a). Further research into mating systems, which may vary regionally, and dispersal in this species is required.

Males do not appear to be inherently territorial (though they do scent-mark, see *Behavior*) but are intolerant of other mature males in their home range and will aggressively attack interlopers. This may take the form of wrestling with the tail and scratching at the opponent with the forelimbs and claws. These battles sometimes last for hours until one pangolin relents and leaves the area (Swart, 2013; W. Panaino, pers. obs.).

Home ranges are used on a continuous basis, with individuals typically spending consecutive days in a particular burrow before moving to another, seemingly in an ad hoc manner (Heath and Coulson, 1997a; Pietersen et al., 2014a; Swart, 2013). Swart (2013) reported that in the Kruger National Park region males use burrows for up to 16 consecutive days (mean = 2.3 days, n = 4), and females up to 75 days (mean = 5 days, n = 8), returning to only 18% and 23% of dens used previously respectively. The species tends not to dig burrows, instead using the abandoned burrows of other species including aardvark (*Orycteropus afer*), springhares (*Pedetes capensis*) Cape porcupines (*Hystrix africaeaustralis*) or warthog (*Phacochoerus* spp.), among others, modifying them to some degree (Swart, 2013). Sweeney (1974) described a burrow that was 3–5 m long, sloped steeply downwards and terminated about 1 m below the soil surface. Burrows used in the Kalahari region ranged from 1.2 to 12 m in length and terminated 0.5–5 m below the soil surface (D.W. Pietersen, unpubl. data). The species also rests in rock crevices, caves, termitaria, or in thickets or piles of driftwood (Heath, 1992; Heath and Coulson, 1997a; Jacobsen et al., 1991; Pietersen et al., 2014a). Rarely, dispersing individuals may dig a burrow in soft sand, but these are typically not very deep (<0.5 m in depth) and do not afford much protection (D.W. Pietersen and W. Panaino, pers. obs.). In the Kalahari, this species has been observed sleeping in hollows in small epigeal termitaria following the excavation and consumption of its inhabitants (D.W. Pietersen and W. Panaino, pers. obs.); similar behavior has been observed in rehabilitated animals in the South African Highveld (R. Jansen, pers. obs.). Old burrows are abandoned as they decay. Temminck's pangolins are known to return to established home ranges if translocated, a behavior apparently most well developed in adults, though the distance over which this occurs is presumably limited (Heath and Coulson, 1997b; P. Rankin, unpubl. data).

The species is entirely myrmecophagous and the distance covered in search of prey depends on both habitat and prey availability (Swart, 2013). In eastern South Africa, Swart (2013) recorded males covering a distance of 202–3791 m per night and females, 40–2176 m per night. Temminck's pangolin is highly prey selective including geographically, with different prey species having been predated on in different parts of South Africa and Zimbabwe (Coulson, 1989; Jacobsen et al., 1991; Richer et al., 1997; Swart et al., 1999). In total, 30 ant and 10 termite species are reported to be preyed upon, with many other species, including common taxa, ignored (Coulson, 1989; Jacobsen et al., 1991; Pietersen et al., 2016b; Richer et al., 1997; Swart et al., 1999; W. Panaino, unpubl. data). Swart et al. (1999) report that Temminck's pangolin prefers larger prey (>5 mm in length) and at Sabi Sands Wildtuin in South Africa six species >5 mm in length comprised 97% of the diet: *Anoplolepis custodiens*, *Myrmicaria natalensis*, *Camponotus cinctellus*, *Polyrhachis schistacaea*, *Hodotermes mossambicus* and *Camponotus* sp.–*maculatus*-group. One species, *A. custodiens*, comprised 77% of the overall diet and was the most important prey species throughout the year (Swart et al., 1999). Swart et al. (1999) and Swart (2013) assert that *A. custodiens* is the key species in the diet of Temminck's pangolin in Southern Africa, and is likely important in determining the distribution of *S. temminckii* in the region. This selectivity may be because the nest galleries of *A. custodiens* are close to the soil surface, meaning they are easily accessible to pangolins (Swart et al., 1999). Jacobsen et al. (1991) reported the species predating on ants (*Acantholepsis capensis*, *A. custodiens*, *Camponotus* spp., *Crematogaster amita*, *Monomorium albopilosum*, *Myrmicaria natalensis*, *Pheidole megacephala*, *P. schistacea*, *Tapenonia luteum*, *Technomyrmex albipes*) and termites (*Odontotermes badius*, *Trinervitermes rhodesiensis*) in eastern South Africa.

Predators include African lion (*Panthera leo*), leopard (*P. pardus*), spotted hyaenas (*Crocuta crocuta*) and ratels (*Mellivora capensis*). There is a record of a Nile crocodile (*Crocodylus niloticus*) predating on a Temminck's pangolin (Coulson, 1989) and an African elephant (*Loxodonta africana*) crushing a pangolin in an apparent musth rage (R. Jansen, unpubl. data). African lions and leopards are observed "playing" with pangolins that have rolled into a defensive ball, but the scales and round shape make it challenging for predators to grasp them and these interactions rarely result in the death of the pangolin. Spotted hyaenas and lions typically prey on pangolins opportunistically. In some areas ratels appear to specialize on pangolins, or at least know how to effectively tackle and consume them (Swart, 2013; B. Nebe and W. Panaino, pers. obs.). Young pangolins are especially vulnerable to predation because of their smaller size and their scales are weaker than those of adults.

Temminck's pangolins frequently host pangolin mites *Manitherionyssus heterotarsus*, and occasionally soft (*Ornithodoros moubata* and *O. compactus*), hard (*Rhipicephalus theileri*) and other tick species (Jacobsen et al., 1991; D.W. Pietersen, W. Panaino and T. Radebe, unpubl. data; see Chapter 29).

Behavior

Solitary, with males and females coming together briefly to mate, and the only lasting social bonds exist between females and their dependent offspring (Swart, 2013; D.W. Pietersen, unpubl. data). Adapted to a diet of rather poor nutritional value, the species spends most of the time resting in a burrow or similar place of shelter and is typically active for only a small proportion of the day or night (Swart, 2013). Temminck's pangolin is predominantly nocturnal, but is also crepuscular and diurnal (Jacobsen et al., 1991; Richer et al., 1997; Swart, 2013), and activity may vary according to season, prey availability, and to avoid predation (Pietersen et al., 2014a). In warm mesic environments (e.g., eastern South Africa) and during summer, the species is nocturnal, conserving energy and moisture by limiting exposure to high temperatures (Jacobsen et al., 1991; Pietersen et al., 2014a, 2016a; Swart, 2013; Wilson, 1994). In the Kalahari region, activity, including that of juveniles, may be entirely diurnal or may start during daylight hours and extend into the night during winter, presumably to avoid the extremely cold nighttime temperatures, and conserve energy (Pietersen et al., 2014a; W. Panaino unpubl. data). In these circumstances, emergence times from burrows is closely associated with minimum ambient temperature (Pietersen et al., 2014a). This activity is also thought to be a response to nutritional stress as ants are less active and congregate deeper in their nest chambers in winter and at night, being less accessible and necessitating pangolins to expend more energy to prey on them (Pietersen et al., 2014a, 2016b). Research in South Africa and Zimbabwe suggests that juveniles and sub-adults in particular, are diurnal or crepuscular foragers, and behave this way to avoid nocturnal predators (e.g., African lion), being more vulnerable to predation than adults (Richer et al., 1997; Swart, 1996, 2013).

In eastern South Africa (Sabi Sand Wildtuin), activity periods averaged 3.9 ± 1.9 hours per night and ranged from 0.6 to 8.3 hours, with foraging behavior commencing at 2025 (mean; range = 1434–0215) for adults and 1802 (mean; range = 1430–2015) for sub-adults (Swart, 1996). In the Kalahari, activity periods lasted for an average of 5.7 ± 2.0 hours (mean $\pm$ SD), ranging from 1 to 12 hours, and did not show seasonal variation (Pietersen et al., 2014a). The most frequent emergence times among three adults were between 1600 and 1800, but these individuals often emerged as early as 1100 (Pietersen et al., 2014a). Based on research in Sabi Sand Wildtuin,

FIGURE 11.4 Temminck's pangolin is the only bipedal species of pangolin and uses its tail as a counter-balance. *Photo credit: Paula French/Shutterstock.com.*

Swart (1996) reported that active time is correlated with feeding intensity (feeding time per hour), i.e., when feeding intensity is higher, the species is active for shorter periods. However, variation in the duration of activity in different environments may be related to prey availability and requires further investigation.

Unlike other pangolins, Temminck's pangolin is bipedal, walking and bearing the weight on its hind legs, with the forelimbs tucked up towards the chest, and the tail held off the ground and used as a counter-balance (Fig. 11.4). They are capable of climbing and can clamber over fallen logs and other debris with ease, and the species is an accomplished swimmer (Kingdon, 1971; P. Rankin, pers. obs.). When climbing steep embankments or rocky ridges, the forelimbs are used to scale obstacles and the tail may be used to anchor the animal or help push it forward. The species often stands erect on its hindlimbs in a near-vertical position, balancing on the broad tail, and sniffing the air to survey the surroundings.

Most active time is spent foraging, mainly in dense undergrowth, and a smaller proportion of time spent actually feeding, with estimates ranging from 7% to 20%, varying between habitat type (Richer et al., 1997; Swart et al., 1999). Temminck's pangolin is macrosmatic, using acute olfactory senses to locate prey, even beneath the soil surface (Swart, 2013). When foraging, the species follows a haphazard path, with the nose held close to the ground while continuously sniffing to locate prey; sniffing intensity increases closer to the prey source (Swart, 2013). The muscular forelimbs and strong front claws are used to tear open terrestrial ant nests, scratch away cartilaginous material from termite mounds, and tear away bark on dead trees when foraging, enabling access to prey. At epigeal ant nests, the species typically digs only shallow holes (5–10 cm deep), removing the soil in the immediate vicinity before inserting the long, salivary tongue (see Fig. 11.5) to capture ants located deep within the nest chambers (Pietersen et al., 2016b; Swart et al., 1999; Fig. 11.5). Feeding bouts in the South African lowveld are typically short, averaging 40 s and seldom exceed one minute (Swart et al., 1999), and are probably terminated in response to the chemical and physical defenses (e.g., swarming behavior) employed by the prey. In the

FIGURE 11.5 Temminck's pangolin uses its long tongue to pick up prey. *Photo credit: Francois Meyer.*

Kalahari, feeding bouts are longer, sometimes lasting for several minutes (D.W. Pietersen and B. Nebe, unpubl. data). Short feeding bouts likely ensure that an ant or termite colony is never annihilated, affording the species the opportunity to periodically revisit the same nest. While foraging, nictitating membranes cover the eyes, and the nostrils and ears are closed (Swart, 2013). Feeding sites vary between epigeal ant nests and those at the base of trees and shrubs (Richer et al., 1997). The species does not usually feed at epigeal termite nests, largely owing to the hard exterior making the nests impenetrable, but will feed at termitaria after heavy rains have softened the exterior, or after other species (e.g., aardvark) have partially excavated a nest. Largely water-independent, the species obtains moisture from prey, but will opportunistically drink freestanding water (W. Panaino, pers. obs.), especially in captivity.

Both males and females scent-mark by regularly depositing small volumes of urine as they walk (D.W. Pietersen and W. Panaino, pers. obs.) and may defecate and drag the tail through the feces in order to spread it further. The species also defecates at burrow entrances and in burrows (D.W. Pietersen, F. Meyer and R. Jansen, pers. obs.), which may serve an additional territorial function. Males are known to raise a hind leg enabling them to squirt a small volume of urine onto objects including trees and rocks (Swart, 2013). The anal glands are also thought to play a role in scent-marking. In addition to sleeping burrows, abandoned burrows of other species are regularly visited on nocturnal excursions, with the pangolin entering the burrow and scent-marking before reemerging (Swart, 2013; D.W. Pietersen and W. Panaino, pers. obs.). Feces are usually buried, which may act as a mechanism to avoid detection by predators but may also serve as a territorial marker; buried feces have been dug up by other pangolins (and aardvark) months or even years after initial deposition (D.W. Pietersen, pers. obs.). Frequently used dens are cleaned periodically with accumulated feces moved to the burrow entrance.

FIGURE 11.6 Temminck's pangolin rolling in herbivore dung. This may have an anti-parasitic role. *Photo credit: D.W. Pietersen.*

Temminck's pangolin is fond of wallowing in mud and will also roll in herbivore dung and urine (Swart, 2013; D.W. Pietersen and W. Panaino, pers. obs.). On occasion, they will lie on their side next to the fresh dung and grasp it between their fore- and hindfeet, before rolling onto their back and crumbling the boll with their feet. Curling around the crumbled boll, excrement covering the ventral surface, they will simultaneously writhe in the dung pile coating the dorsal surface (Swart, 2013; D. Pietersen and W. Panaino, pers. obs.). They also create wallows by loosening coarse sand and urinating in it (Swart, 2013; D.W. Pietersen pers. obs.; Fig. 11.6). This behavior is believed to have an anti-parasitic role, and may serve to disguise the pangolin's scent from its prey, potentially delaying defensive actions from prey while foraging.

When threatened, the species will freeze and when motionless is often well camouflaged. Otherwise, Temminck's pangolin rolls up into a tight ball with the vulnerable head, legs and unscaled ventral surface protected, presenting predators with an almost impenetrable armor of scales. When curled up, the forelimbs tightly grasp the hindlimbs, and the tail is curled around the body, and cannot be prized open by predators, or human beings. The broad, muscular tail may be moved slowly across the dorsal surface in a scything motion and when a foreign object is located e.g., an errant hand or predator's paw, it is lashed at with the tail. The scale edges are sharp and are an effective defense.

Individuals communicate by vibrating their bodies when curled up, which is audible to the human ear, and this behavior is likely exhibited during threat displays between rivals (Swart, 2013; D.W. Pietersen, W. Panaino and R. Jansen, pers. obs.). Males may hiss when approaching females.

Ontogeny and reproduction

Females give birth to a single young once a year (van Ee, 1966; W. Panaino, unpubl. data), possibly only every second year. Twins have been recorded, but rarely (e.g., Jacobsen et al., 1991). Sweeney (1974) reported that a female carrying a well-developed fetus was still lactating and had a juvenile in tow suggesting that females may become pregnant while nursing current young, as in some other species of pangolin. In contrast, in the Kalahari region of South Africa, a female observed for over 5 years only gave birth twice suggesting that breeding is biennial (D.W. Pietersen, unpubl. data).

Breeding is aseasonal (Ansell, 1960; Coulson, 1989; Jacobsen et al., 1991; Smithers, 1971; Swart, 2013) although a seasonal peak has been observed in some populations (D. Pietersen and W. Panaino, unpubl. data). Males locate receptive females while foraging, presumably through a scent trail left by the female (Swart, 2013; D.W. Pietersen, pers. obs.). The male approaches the female cautiously, sniffing continuously (and may hiss) while circling and clambering over her (Swart, 2013). The male mounts the female from the side, forcing his tail below hers to ensure alignment of the genitals, and will curl around her tail to prevent being dislodged (Swart, 2013; van Ee, 1978; D.W. Pietersen, unpubl. data). From limited field observations, the female carries the male to a burrow in this position, where they remain continuously for 24–48 hours (Swart, 2013; D.W. Pietersen, unpubl. data). Gestation period is 105–140 days, after which the female gives birth to one young, typically in a burrow (Swart, 2013; van Ee, 1966; D.W. Pietersen, unpubl. data). The young's eyes are open at parturition, and the scales are soft and pink (Swart, 2013). Neonates measure about 150 mm and weigh 340 g when born (van Ee, 1966).

Post-partum care is not well understood and reports vary. The female initially leaves the young unattended for short periods to forage, returning at intervals to nurse (Swart, 2013). Within a week of parturition, the female will move the juvenile from the birthing burrow to another burrow (W. Panaino, unpubl. data), the juvenile grasping onto the scales at the base of the mother's tail (van Ee, 1978; D. W. Pietersen and W. Panaino, pers. obs.). Kingdon (1971) reports that in East Africa young start riding on the mother's back after a month. As they get older, juveniles make exploratory ventures outside burrows (Swart, 2013; D.W. Pietersen and W. Panaino, pers. obs.). In the Kalahari, juveniles start accompanying the mother during her feeding forays by riding on the base of her tail (D.W. Pietersen, unpubl. data). In the South African lowveld and elsewhere in the Kalahari, females never foraged while moving young between burrows and the young never accompanied the mother on feeding forays (J. Swart and W. Panaino, pers. obs.). As the young nears independence, it alternates between riding on the mother's tail and feeding on its own near the mother, until it becomes independent at about 3 months old (D.W. Pietersen, pers. obs.). Other observers report that independence is reached at 4.5–6 months in the Kalahari (W. Panaino, unpubl. data) and 6–7 months in the South African lowveld, but with records of offspring as old as 10–12 months having been observed riding on their mothers' backs and sharing burrows with their mother (Smithers, 1983; Swart, 2013; W. Panaino, unpubl. data). In one instance, a young Temminck's pangolin in the Kalahari was found to be sharing a burrow with its father for a month following independence from its mother (D.W. Pietersen, unpubl. data). In the lowveld, young of this species learn to forage from an early age by trial and error without guidance and were accomplished foragers by the time of independence (J. Swart, pers. obs.).

If threatened, the mother will curl around the juvenile, completely shielding it. Large juveniles are only partially encircled, curling around the female at a right angle with the forequarters, including the head, enclosed by the female (Skinner and Chimimba, 2005).

Young pangolins establish a home range within the maternal home range, typically focusing their activity on a small portion of it, likely the area they are most familiar with (Heath and Coulson, 1997a; Pietersen et al., 2014a; Swart, 2013). Newly independent Temminck's pangolins remain in their mother's home range for about 12 months, before dispersing further afield (Pietersen et al., 2014a; Swart, 2013).

Females are thought to reach sexual maturity as they approach 2 years old, but as they are generally in the process of dispersing at this age, it is likely that reproduction does not start until 3 or 4 years of age and a home range has been established. Males also likely reach sexual maturity as they approach 2 years old, but appear to rove for several years, and may only establish a home range (and likely start breeding) at 6 or 7 years of age (Pietersen, 2013). Further research is required to determine whether these males contribute to the genepool before they establish home ranges.

Longevity in the wild is not known. One individual lived in captivity for about three years (Hoyt, 1987; Van Ee, 1966). The seemingly late onset of reproduction and slow reproductive rate suggest that the species may live for 20–30 years (Swart, 2013; D.W. Pietersen, unpubl. data; Tikki Hywood Foundation, unpubl. data). An Indian pangolin (*Manis crassicaudata*), which is of similar size, biology, and ecology, lived for more than 19 years in captivity (Hoyt, 1987).

Population

Population estimates are lacking across most of the species' range, owing to the secretive and predominantly nocturnal habits, and a lack of research. In eastern and western South Africa, and Zimbabwe, the species occurs at estimated densities of 0.12–0.16 reproductively active individuals/km^2, and absolute densities of 0.23–0.31 individuals/km^2 (Pietersen et al., 2014a; Swart, 2013). Based on area of occupancy and estimated densities, South Africa's population was estimated to be 16,000–24,000 mature individuals in 2016 (Pietersen et al., 2016a). No population estimates are available for any other range countries, regions, or globally. Populations are suspected to be declining globally, driven primarily by threats including accidental electrocution on electrified fences, overexploitation for illicit local and international trade, and habitat loss (see *Threats*).

Status

Temminck's pangolin is listed as Vulnerable on The IUCN Red List of Threatened Species (Pietersen et al., 2019) and Vulnerable on the South African Red List of Mammals (Pietersen et al., 2016a). No other national or regional assessments have been undertaken. The species is afforded protection under national wildlife legislation in most range states, which typically prohibits exploitation, and in 2016 was included in CITES Appendix I.

Threats

All threats are anthropogenic. Accidental electrocution on electric fences and overexploitation for local use and international trafficking are the main threats. Electrocution on electric fences is most prevalent in South Africa and Namibia where electric fences are common on both game and livestock farms (Beck, 2008; Pietersen et al., 2014b; van Aarde et al., 1990; B. Nebe, pers. obs.). However, there are reports of individuals being electrocuted in Uganda and this threat is likely to occur wherever pangolins occupy areas with electric fences. Temminck's pangolins walk on their hind legs with the front limbs and tail held off the ground. When the soft, unprotected belly contacts an electric fence, the pangolin receives an electric shock and curls into a protective ball, often inadvertently curling around the electrified wire. This results in a continuous electrical pulse that causes terminal damage; some individuals succumb to exposure from remaining trapped on the wire. Pietersen et al. (2016a) estimate that in South Africa these fences electrocute between 377 and 1028 Temminck's pangolins annually (though the actual number is likely higher). This accounts for 2–13% of South Africa's population (Pietersen et al., 2016a).

Temminck's pangolin is used extensively in traditional African medicine and plays an important role in cultural rituals and as talismans (see Chapter 15; Baiyewu et al., 2018; Bräutigam et al., 1994). In South Africa, the species is highly sought-after in commercial herbal medicine markets (Cunningham and Zondi, 1991). While such use may have been sustainable historically, evidence suggests that this is no longer the case, and the species has been extirpated from parts of its range because of overexploitation (e.g., KwaZulu-Natal Province, South Africa; see Pietersen et al., 2016a). In East Africa, the species is locally referred to as "*Bwana mganga*", meaning "Mister Doctor", alluding to the fact that most of the species' derivatives are used in cultural rituals and traditional medicine (see Chapters 12 and 14; Wright, 1954). It is possible that poaching rates increase during periods of drought as pangolins may be under nutritional stress and forage for longer periods as well as diurnally, and less vegetation cover makes them more visible, and therefore easier to capture.

There has been little international trade in the species reported to CITES historically (see Chapter 16). However, there has been an apparent increase in illegal, international trade in Temminck's pangolin between around 2008 and 2019, involving both live animals and scales (Challender and Hywood, 2012; Shepherd et al., 2017). Available data suggest that this has involved at least 114 individuals (Chapter 16), but the actual number is likely higher. Of concern is the seemingly insatiable demand for pangolin scales in East and Southeast Asia, with numerous seizures of shipments in the last ten years implicating range states in Southern and East Africa (Heinrich et al., 2017), and illicit trade in pangolins moving in to Southern Africa (Pietersen et al., 2014a; Shepherd et al., 2017).

Habitat loss because of land transformation for subsistence and commercial crop agriculture, including shifting agriculture, smallholder farming and agro-industry poses a threat, though a minor one in isolation, but crucially, increased human presence in these areas likely results in higher levels of poaching (Pietersen et al., 2016a).

Other threats include the death of individuals from traffic collisions on roads, hunting or poaching for consumption as a source of protein (see Lindsey et al., 2011), and capture for presentation as a gift to persons in power, including chiefs and state presidents. Capture in gin traps set for small carnivores has also been reported from western South Africa but may be more widespread (Pietersen et al., 2014b; van Aarde et al., 1990). Artisanal mines and large, open water canals (B. Nebe, unpubl. data) pose a local threat where they occur.

References

Ansell, W.F.H., 1960. Mammals of Northern Rhodesia. The Government Printer, Lusaka.

Ansell, W.F.H., 1978. The Mammals of Zambia. National Parks and Wildlife Services, Chilanga, Zambia.

Ansell, W.F.H., Dowsett, R.J., 1988. Mammals of Malawi: An Annotated Check List and Atlas. Trendrine Press, Zennor, Cornwall.

Baiyewu, A.O., Boakye, M.K., Kotzé, A., Dalton, D.L., Jansen, R., 2018. Ethnozoological survey of the traditional uses of Temminck's Ground Pangolin (*Smutsia temminckii*) in South Africa. Soc. Anim. 26, 1–20.

Beck, A., 2008. Electric Fence Induced Mortality in South Africa. M.Sc. Thesis, University of the Witwatersrand, Johannesburg, South Africa.

Beja, P., Pinto, P.V., Veríssimo, L., Bersacola, E., Fabiano, E., Palmeirim, J.M., et al., 2019. The Mammals of Angola. In: Huntley, B.J., Russo, V., Lages, F., Nuno-Ferrand, N. (Eds.), Biodiversity of Angola, Science & Conservation: A Modern Synthesis. Springer, Cham, pp. 357–443.

Bere, R.M., 1962. The wild mammals of Uganda and neighbouring regions of East Africa. In Association With the East African Literature Bureau. Longmans, London.

Bräutigam, A., Howes, J., Humphreys, T., Hutton, J., 1994. Recent information on the status and utilisation of African pangolins. TRAFFIC Bull. 15 (1), 15–22.

Challender, D.W.S., Hywood, L., 2012. African pangolins under increased pressure from poaching and intercontinental trade. TRAFFIC Bull. 24 (2), 53–55.

Coulson, M.H., 1989. The pangolin (*Manis temminckii* Smuts, 1832) in Zimbabwe. Afr. J. Ecol. 27 (2), 149–155.

Cunningham, A.B., Zondi, A.S., 1991. Use of Animal Parts for the Commercial Trade in Traditional Medicines. Institute of Natural Resources, University of Natal, South Africa, Working paper 76.

Doran, G.A., Allbrook, D.B., 1973. The tongue and associated structures in two species of African pangolins, *Manis gigantea* and *Manis tricuspis*. J. Mammal. 54 (4), 887–899.

Dorst, J., Dandelot, P., 1972. A Field Guide to the Larger Mammals of Africa. Collins, London.

du Toit, Z., du Plessis, M., Dalton, D.L., Jansen, R., Grobler, J.P., Kotzé, A., 2017. Mitochondrial genomes of African pangolins and insights into evolutionary patterns and phylogeny of the family Manidae. BMC Genom. 18, 746.

Fitzinger, L.J., 1872. Die naturliche familie der schuppenthiere (Manes). Sitzungsberichte der Kaiserlichen Akademie der Wissenschaften. Mathematisch-Naturwissenschaftliche Classe, CI., LXV, Abth. I, 9–83.

Frechkop, S., 1931. Notes sur les mammifères. VI. Quelques observations sur la classification des pangolins (Manidae). Bulletin du Musee royal d'Histoire naturelle de Belgique VII (22), 1–14.

Friedmann, Y., Daly, B. (Eds.), 2004. Red Data Book of the Mammals of South Africa: A Conservation Assessment. Conservation Breeding Specialist Group Southern Africa, IUCN Species Survival Commission, Endangered Wildlife Trust, South Africa.

Foley, C., Foley, L., Lobora, A., De Luca, D., Msuha, M., Davenport, T.R.B., et al., 2014. A Field Guide to the Larger Mammals of Tanzania. Princeton University Press, Princeton.

Gaubert, P., Antunes, A., Meng, H., Miao, L., Peigné, S., Justy, F., et al., 2018. The complete phylogeny of pangolins: scaling up resources for the molecular tracing of the most trafficked mammals on Earth. J. Hered. 109 (4), 347–359.

Gaudin, T.J., Emry, R.J., Wible, J.R., 2009. The phylogeny of living and extinct pangolins (Mammalia, Pholidota) and associated taxa: a morphology based analysis. J. Mammal. Evol. 16 (4), 235–305.

Gotch, A.F., 1979. Mammals – Their Latin Names Explained. A Guide to Animal Classification. Blandford Press, Poole.

Grubb, P., Jones, T.S., Davies, A.G., Edberg, E., Starin, E.D., Hill, J.E., 1998. Mammals of Ghana, Sierra Leone and the Gambia. Trendrine Press, Zennor, Cornwall.

Heath, M.E., 1992. *Manis temminckii*. Mammal. Sp. 415, 1–5.

Heath, M.E., Hammel, H.T., 1986. Body temperature and rate of O_2 consumption in Chinese pangolins. Am. J. Physiol.-Regul., Integr. Comp. Physiol. 250 (3), R377–R382.

Heath, M.E., Coulson, I.M., 1997a. Home range size and distribution in a wild population of Cape pangolins, *Manis temminckii*, in north-west Zimbabwe. Afr. J. Ecol. 35 (2), 94–109.

Heath, M.E., Coulson, I.M., 1997b. Preliminary studies on relocation of Cape pangolins *Manis temminckii*. South Afr. J. Wildlife Res. 27 (2), 51–56.

Heath, M.E., Coulson, I.M., 1998. Measurements of length and mass in a wild population of Cape pangolins (*Manis temminckii*) in north-west Zimbabwe. Afr. J. Ecol. 36 (3), 267–270.

Heinrich, S., Wittman, T.A., Ross, J.V., Shepherd, C.R., Challender, D.W.S., Cassey, P., 2017. The Global Trafficking of Pangolins: A Comprehensive Summary of Seizures and Trafficking Routes From 2010–2015. TRAFFIC, Southeast Asia Regional Office, Petaling Jaya, Selangor, Malaysia.

Hill, J.E., Carter, T.D., 1941. The mammals of Angola, Africa. Bull. Am. Museum Nat. Hist. 78, 1–211.

Hoyt, R., 1987. Pangolins: Past, Present and Future. AAZPA National Conference Proceedings, pp. 107–134.

Jacobsen, N.H.G., Newbery, R.E., De Wet, M.J., Viljoen, P. C., Pietersen, E., 1991. A contribution of the ecology of the Steppe Pangolin *Manis temminckii* in the Transvaal. Zeitschrift für Säugetierkunde 56 (2), 94–100.

Jentink, F.A., 1882. Note XXV. Revision of the Manidae in the Leyden Museum. Notes from the Leyden Museum IV, 193–209.

Kingdon, J., 1971. East African mammals. An Atlas of evolution in Africa, Primates, Hyraxes, Pangolins, Protoungulates, Sirenians, vol. I. Academic Press, London.

Kyle, R., 2000. Some notes on the occurrence and conservation status of *Manis temminckii*, the pangolin, in Maputaland, Kwazulu/Natal. Koedoe 43, 97–98.

Lindsey, P.A., Romanach, S.S., Tambling, C.J., Charter, K., Groom, R., 2011. Ecological and financial impacts of illegal bushmeat trade in Zimbabwe. Oryx 45 (1), 96–111.

Malbrant, R., 1952. Fauna du centre Africain Français (Maniferes et Oiseaux), second ed. Paul Lechevalier, Paris.

McNab, B.K., 1984. Physiological convergence amongst ant-eating and termite-eating mammals. J. Zool. 203 (4), 485–510.

Meester, J., 1972. Order Pholidota. In: Meester, J., Stzer, H. W. (Eds.), The Mammals of Africa: An Identification Manual, Part 4. Smithsonian Institution Press, Washington, D.C., pp. 1–3.

Mohr, E., 1961. Schuppentiere. Neue Brehm-Bucherai. A. Ziemsen Verlag, Wittenberg Lutherstadt.

Monadjem, A., 1998. The Mammals of Swaziland. Conservation Trust of Swaziland and Big Games Parks.

Monard, A., 1935. Contribution à la Mammologie d'Angola et Prodrome d'une Faune d'Angola. Arquivos do Museu Bocage 6, 1–314.

Ngwenya, M.P., 2001. Implications of the Medicinal Animal Trade for Nature Conservation in Kwazulu-Natal. Ezemvelo KZN Wildlife Report No. Na/124/04.

Pietersen, D.W., 2013. Behavioural Ecology and Conservation Biology of Ground Pangolins (*Smutsia temminckii*) in the Kalahari Desert. M.Sc. Thesis, University of Pretoria, Pretoria, South Africa.

Pietersen, D.W., McKechnie, A.E., Jansen, R., 2014a. Home range, habitat selection and activity patterns of an arid-zone population of Temminck's ground pangolins, *Smutsia temminckii*. Afr. Zool. 49 (2), 265–276.

Pietersen, D.W., McKechnie, A.E., Jansen, R., 2014b. A review of the anthropogenic threats faced by Temminck's ground pangolin, *Smutsia temminckii*, in southern Africa. South Afr. J. Wildlife Res. 44 (2), 167–178.

Pietersen, D., Jansen, R., Swart, J., Kotze, A., 2016a. A conservation assessment of *Smutsia temminckii*. In: Child, M.F., Roxburgh, L., Do Linh San, E., Raimondo, D., Davies-Mostert, H.T. (Eds.), The Red List of Mammals of South Africa, Swaziland and Lesotho. South African National Biodiversity Institute and Endangered Wildlife Trust, South Africa.

Pietersen, D.W., Symes, C.T., Woodborne, S., McKechnie, A.E., Jansen, R., 2016b. Diet and prey selectivity of the specialist myrmecophage, Temminck's Ground Pangolin. J. Zool. 298 (3), 198–208.

Pietersen, D., Jansen, R., Connelly, E., 2019. *Smutsia temminckii. The IUCN Red List of Threatened Species* 2019: e.T12765A123585768. Available from: <http://dx.doi.org/10.2305/IUCN.UK.2019-3.RLTS.T12765A123585768.en>.

Pocock, R.I., 1924. The external characters of the pangolins (Manidae). Proc. Zool. Soc. Lond. 94 (3), 707–723.

Rautenbach, I.L., 1982. Mammals of the Transvaal. Ecoplan Monograph 1. Pretoria, South Africa.

Richer, R., Coulson, I., Heath, M., 1997. Foraging behaviour and ecology of the Cape pangolin (*Manis temminckii*) in north-western Zimbabwe. Afr. J. Ecol. 35 (4), 361–369.

Schlitter, D.A., 2005. Order Pholidota. In: Wilson, D.E., Reeder, D.M. (Eds.), Mammal Species of the World: A Taxonomic and Geographic Reference, third ed. Johns Hopkins University Press, Baltimore, pp. 530–531.

Schloeder, C.A., Jacobs, M.J., 1996. A report on the occurrence of three new mammal species in Ethiopia. Afr. J. Ecol. 34 (4), 401–403.

Shepherd, C.R., Connelly, E., Hywood, L., Cassey, P., 2017. Taking a stand against illegal wildlife trade: the Zimbabwean approach to pangolin conservation. Oryx 51 (2), 280–285.

Shortridge, G.C., 1934. The Mammals of South West Africa. William Heinemann Publishers, London.

Skinner, J.D., Chimimba, C.T., 2005. The Mammals of the Southern African Subregion. Cambridge University Press, Cambridge.

Smithers, R.H.N., 1966. The Mammals of Rhodesia, Zambia and Malawi. Collins, London.

Smithers, R.H.N., 1971. The Mammals of Botswana, 4. Museum Memoirs of the National Museums and Monuments, Rhodesia, pp. 1–340.

Smithers, R.H.N., 1983. The Mammals of the Southern African Subregion. University of Pretoria, Pretoria.

Smithers, R.H.N., Lobão Tello, J.L.P., 1976. Check List and Atlas of the Mammals of Mozambique, Trustees of the National Museums and Monuments of Rhodesia, Salisbury, Zimbabwe.

Smuts, J., 1832. Enumerationem Mammalium Capensium, Dessertatio Zoologica Inauguralis. J.C. Cyfveer, Leidae.

Spassov, N., 1990. On the presence and specific position of pangolins (Gen. Manis L.: Pholidota) in north Mozambique. Hist. Nat. Bulg. 2, 61–64.

Stuart, C.T., 1980. The distribution and status of *Manis temminckii* Pholidota Manidae. Säugetierkundliche Mitteilungen 28, 123–129.

Swart, J., 1996. Foraging Behaviour of the Cape Pangolin *Manis temminckii* in the Sabi Sand Wildtuin. M. Sc. Thesis, University of Pretoria, Pretoria, South Africa.

Swart, J., 2013. *Smutsia temminckii* Ground Pangolin. In: Kingdon, J., Hoffmann, M. (Eds.), Mammals of Africa, vol. V, Carnivores, Pangolins, Equids, Rhinoceroses. Bloomsbury Publishing, London, pp. 400–405.

Swart, J.M., Richardson, P.R.K., Ferguson, J.W.H., 1999. Ecological factors affecting the feeding behaviour of pangolins (*Manis temminckii*). J. Zool. 247 (3), 281–292.

Sweeney, R.C.H., 1956. Some notes on the feeding habits of the ground pangolin, *Smutsia temminckii* (Smuts). Ann. Mag. Nat. Hist. 9 (108), 893–896.

Sweeney, R., 1959. A Preliminary Annotated Check-List of the Mammals of Nyasaland. The Nyasaland Society, Blantyre.

Sweeney, R.C.H., 1974. Naturalist in the Sudan. Taplinger Publishing Co, New York.

Swynnerton, G.H., Hayman, R.W., 1950. A check-list of the land mammals of the Tanganyika Territory and the Zanzibar Protectorate. J. East Afr. Nat. Hist. Soc. 20 (6&7), 274–392.

Ullmann, T., Veríssimo, D., Challender, D.W.S., 2019. Evaluating the application of scale frequency to estimate the size of pangolin scale seizures. Glob. Ecol. Conserv. 20, e00776.

van Aarde, R.J., Richardson, P.R.K., Pietersen, E., 1990. Report on the Behavioural Ecology of the Cape Pangolin (*Manis temminckii*). Mammal Research Institute, University of Pretoria, Internal Report.

van Ee, C.A., 1966. A note on breeding the Cape pangolin *Manis temminckii* at Bloemfontein Zoo. Int. Zoo Yearbook 6 (1), 163–164.

van Ee, C.A., 1978. Pangolins can't be bred in captivity. Afr. Wildlife 32, 24–25.

Weber, M., 1892. Beitrage zur Anatomie und Entwickelungsge-schichte der Genus *Manis*. Mit tafel I-IX. Zoologische Ergenbnisse. Einer Reise in Niederländisch OstIndien, Band II, Leiden, pp. 1–116.

Weber, R.E., Heath, M.E., White, F.N., 1986. Oxygen binding functions of blood and hemoglobin from the Chinese Pangolin, *Manis pentadactyla*: Possible implications of burrowing and low body temperature. Respir. Physiol. 64 (1), 103–112.

Wilson, A.E., 1994. Husbandry of pangolins *Manis* spp. Int. Zoo Yearbook 33 (1), 248–251.

Wright, A.C.A., 1954. The magical importance of pangolins among the Basukuma. Tanzan. Notes. Rec. 36, 71–72.

Yalden, D.W., Largen, M.J., Kock, D., Hillman, J.C., 1996. Catalogue of the mammals of Ethiopia and Eritrea. 7. Revised checklist, zoogeography and conservation. Trop. Zool. 9 (1), 73–164.

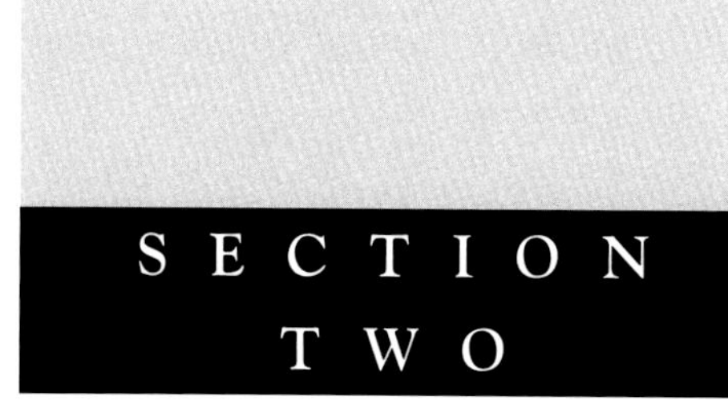

SECTION TWO

Cultural Significance, Use and Trade

Overview

Section Two examines the valuation, use and trade of pangolins in Africa and Asia, as well as in Europe, and globally concerning international trade and trafficking. Chapter 12 explores the significance of pangolins in symbolism, mythology and ritual in Africa and Asia, and describes in detail a wide range of beliefs and practices on each continent, involving consumptive and non-consumptive use of the species, and associated anthropological debate. Chapter 13 examines the construction of pangolins as exotic animals in natural history in early modern Europe (16th–18th centuries), and presents accounts of the first encounters with pangolins by Europeans, which challenged traditional ways of seeing the world. Chapters 14 and 15 discuss the myriad ways pangolins and their body parts and derivatives have been, and are, used in Asia and Africa respectively. This includes the consumption of pangolin meat, and use of the scales, but also other body parts in various medicinal and cultural applications. These chapters also discuss the impact of such exploitation on populations. The final chapter in this section (Chapter 16) examines international trade and trafficking in pangolins and their parts from 1900 to 2019. It draws on historical accounts, data from CITES, the Convention on International Trade in Endangered Species of Wild Fauna and Flora, and illegal trade data from the last 20 years, to characterize legal and illegal international trade in this period. The chapter provides an updated account of pangolin trafficking dynamics, including the species involved, trafficking routes, and available knowledge of the impact exploitation for international trade and trafficking has had, and is having, on populations. It concludes that multi-faceted interventions implemented at multiple levels, from the local to global, are needed to mitigate the threat to pangolins from overexploitation. These are discussed in more detail in Section Three.

CHAPTER

12

Symbolism, myth and ritual in Africa and Asia

Martin T. Walsh

Wolfson College, University of Cambridge, Cambridge, United Kingdom

OUTLINE

Introduction

This chapter reviews what is known about religious and related beliefs and practices pertaining to pangolins in Africa and Asia, including their symbolic uses and the ways in which they feature in myth and ritual. It is admittedly an incomplete review. There are clearly many knowledge gaps – beliefs and practice waiting to be described, ethnographic details filled in, and symbolic uses convincingly interpreted and understood. One thing is certain: people often find the appearance and behavior of scaly anteaters interesting; intriguing even. In the formulation made famous by followers of the anthropologist Claude Lévi-Strauss (1963), they are "good to think with", indeed perhaps exceptionally so. As a result, pangolins have attracted more attention in human societies than many other animals of comparable size, abundance and visibility. They have also attracted the interest of anthropologists and other social scientists, and though this stems from one particular account, it has not prompted as much in-depth empirical research as might otherwise have been expected.

Pangolins
DOI: https://doi.org/10.1016/B978-0-12-815507-3.00012-5

This review could have been organized in different ways. The following sections proceed geographically, continent by continent and case by case, drawing out thematic linkages where possible. The "ethnographic present" of different sources varies considerably, as dates of publication and other information indicate. The pangolin species involved are identified, where this is known. This illustrates one of the gaps in knowledge alluded to above: much more is known about the practices and beliefs attaching to some species than others. This and other points will be revisited in the concluding section of the chapter, which will summarize what is and isn't known about the cultural significance of pangolins and the reasons behind it.

Africa

The Lele pangolin cult

The case that has generated most of the scholarly interest in the ritual use and symbolic significance of pangolins is Mary Douglas's description and analysis of a Lele fertility cult that focuses on the white-bellied pangolin (*Phataginus tricuspis*) (1957; 1963). The matrilineal Lele are speakers of a Bantu language; their traditional territory lies just south of the Central African rainforest, south and west of the Kasai river in what is now the Democratic Republic of the Congo (DRC). Douglas conducted anthropological fieldwork there in 1949–50 and 1953, and her description of the pangolin and other cults refers mostly to that period.

All these cults were organized primarily on a village basis, but were uniform across a wide area. They were mostly directed towards fertility, good hunting, and the suppression of sorcery. The pangolin cult was the only one that had an animal as its object; it was also the most exclusive cult. In the village of Yenga-Yenga ("South Homba") in which Douglas conducted most of her research, only four out of forty adult men were initiates (Douglas, 1957; Fardon, 1999). "The candidate had to have begotten a male and a female child, by the same wife; he himself had to be a member of one of the founding clans of the village, and his father also; the wife through whom he qualified had herself to be a member of a founding clan. The result of such stipulations was that very few men were ever eligible for initiation, even when the rules were relaxed to include the founding clans of brother villages" (Douglas, 1963). Cult membership therefore emphasized the prestige of the founding clans, and of mature men within them. These "Pangolin Men", *Bina Luwawa*, could only hold an initiation when they killed a pangolin and feasted upon it as part of the ceremony. "They were then invested with power over hunting and over fertility of women, and over a wide related range of human ills and misfortunes. Their special responsibility was the ritual for the removal of the village to a new site, when a Pangolin Man, with his wife, had to be the first to sleep on the new ground" (Douglas, 1963). Some eligible men, however, were afraid of the consequences of this ritual role, and chose not to be initiated into the cult.

Douglas describes a series of hunting rites that took place in Yenga-Yenga in the dry season of 1953, at a time when accusations of sorcery were rife and tensions running high in the village. Following the killing of two pangolins, sexual intercourse in the village was banned until they had been eaten and a successful hunt held. Preparations began to be made for a feast, but were delayed by various quarrels and ill omens. The pangolin feast and initiation rite were finally held after a third pangolin had been killed. People spoke of it as though it had voluntarily offered itself for sacrifice and had honored the village by choosing it (Douglas, 1957, 1963). Here is Douglas's brief account of the ritual: "5 *September*. The

Pangolin feast and initiation rite were eventually held. I was unfortunately unable to see the rites. I was told that emphasis was laid on the chiefship of the pangolin. We call him *kum* [master or chief] they said, because he makes women conceive. They expressed shame and embarrassment at having eaten a *kum*. No one is allowed to see the pangolins being roasted over the fire. The tongues, necks, ribs, and stomachs were not eaten, but buried under a palm-tree whose wine thenceforth becomes the sole prerogative of the Begetters [men and women who have had both male and female children]. Apparently, the new initiate was made to eat some of the flesh of the first two pangolins which were in process of decay; the more rotten parts, together with the scales and bones, were given to the dogs. The senior initiates ate the flesh of the more recently killed animal. All were confident that the hunt on the following day would be successful" (1957). It was not, and some days passed before a successful hunt was held. Only then was the prohibition on sexual intercourse lifted.

Pangolins as paradigms of symbolic significance

This incomplete account would be unremarkable if it had not been for Douglas's attempt to explain the role of pangolins and other animals in Lele religious symbolism. Women are forbidden to touch the white-bellied pangolin, and like similar avoidances, Douglas ascribes this in part to the perceived anomalous nature of these unusual animals. "The pangolin is described by the Lele in terms in which there is no mistaking its anomalous character. They say: 'In our forest there is an animal with the body and tail of a fish, covered in scales. It has four little legs and it climbs in the trees'" (1957). Its resemblance to an aquatic creature connects it to the spirits that are thought to dwell in the deepest and dampest parts of the forest and are believed to control human fertility.

It is also said to be anomalous in other ways. "Unlike other animals, it does not shun men but offers itself patiently to the hunter. If you see a pangolin in the forest, you come up quietly behind it and smack it sharply on the back. It falls off the branch and, instead of scuttling away as other animals would do, it curls into a tightly armored ball. You wait quietly until it eventually uncurls and pokes its head out, then you strike it dead. Furthermore, the pangolin reproduces itself after the human rather than the fish or lizard pattern, as one might expect from its appearance. Lele say that, like humans, it gives birth to one child at a time. This in itself is sufficiently unusual to mark the pangolin out from the rest of the animal creation and cause it to be treated as a special kind of link between humans and animals" (Douglas, 1957).

Douglas adds, "In this respect the pangolin would seem to stand towards humans as parents of twins stand towards animals. Parents of twins and triplets are, of course, regarded as anomalous humans who produce their young in the manner of animals" (1957). Parents of twins were also treated as ritually significant by the Lele: they were subject to special rites, and became twin diviners with power over hunting and fertility, in particular other multiple births in the village. Indeed, they were also involved in the events of August and September 1953 that Douglas reported in her subsequent paper (1957) and monograph (1963).

Douglas developed these ideas about ethnotaxonomy and anomalous categories into a general theory of symbolic significance, popularized in her book *Purity and Danger* (1966), and elaborated further in *Natural Symbols* (1970). The first of these is one of the most frequently cited anthropological texts, and has had an influence far beyond that discipline. It brought pangolins to the attention of many researchers who had otherwise never heard of them, and helped to spawn a large literature on animal classification and symbolism that also drew on Lévi-Strauss's structural approach to the study of totemism and "primitive thought" (1963; 1966). Since then these "anomalous" anteaters have occupied a special place in anthropological discussions of animal symbolism (e.g., Bulmer, 1967; Ellen, 1972; Lewis, 1991; Morris, 1998; Richards, 1993; Sperber, 1996; Tambiah, 1969; Wijeyewardene, 1968; Willis, 1974).

Critical commentary on Douglas's interpretation of her Lele material has focused on the classificatory status and cultural significance of different species of pangolin among the Lele and their neighbors (Douglas, 1990, 1993, 1999; Ellen, 1994; Fardon, 1993; de Heusch, 1985, 1993; Lewis, 1991, 1993a, 1993b, 2003). In some respects, Douglas anticipated these criticisms herself. In her original paper (1957), she confessed that she did not know how the Lele differentiated between the white-bellied pangolin, which they called *luwawa*, and the giant pangolin (*Smutsia gigantea*), which was known as *yolabondu* and was not the object of a cult, though it was avoided by pregnant women. She speculated that the reasons for their different treatment might be historical, given the existence of other pangolin cults in the Congo.

A more serious difficulty for her analysis was signposted in *Purity and Danger* (1966), where she noted that "It may well seem that I have made too much of the Lele pangolin cult", admitting that such cults "can have many different levels and kinds of meaning", and that Lele themselves had not volunteered the interpretation that she arrived at. When Douglas revisited the Lele in 1988, she discovered that Christianity had spread and the pangolin cult had been outlawed (Douglas, 1990). While this circumstance allowed her to collect previously secret information, it also means that the cult can only be studied historically, unless it has persisted in some communities. The resulting lack of ethnographic detail has made it difficult to reanalyze the Lele case, except in general theoretical terms and with reference to other studies of symbolically significant animals.

Central and West African transformations

As Douglas was aware, pangolins are accorded symbolic and ritual significance elsewhere in Central Africa (1957, 1990; see Chapter 15). Although Lele practice may have been unique, similar preoccupations occur among different Bantu-speaking ethnic groups in and around the rainforest. Like the Lele, the Bembe of Kivu in eastern DRC focus their attention on the white-bellied pangolin, the species which is most common in their territory. In Bembe cosmology it is viewed as a culture hero, one of a small number of animals that mediate between the world of the dead and the living. It is associated with death because it is nocturnal, lives in underground burrows, and feeds on termites, which are themselves associated with corpses and perhaps the souls of the dead. At the same time, Bembe note that these small pangolins can effectively resist even the most powerful natural predator by coiling into a spiral. They behave like monkeys by standing upright and resemble birds and bats when they climb and hang on trees, striving for the light and (by implication) culture that is the antithesis of darkness and death. It is said that women learned how to carry and nurture their children by observing pangolins, and that people learned how to roof their houses by copying their overlapping scales, thus providing them with shelter from the rain (Gossiaux, 2000).

Among the Lega, northern neighbors of the Bembe, it is the giant pangolin, described as the "elder brother" of the white-bellied pangolin that is the subject of special rites. Again, it is treated as a culture hero and said to have taught people how to build houses – its scales evoking the leaf tiles that are used to roof houses. It is therefore forbidden to kill giant pangolins. When one is found dead in the forest it becomes the property of the powerful Bwami association, which functions to maintain social cohesion, and must then perform elaborate ceremonies to ritually cleanse the community. The meat of the pangolin is shared out and its scales added to the association's ritual paraphernalia, while some are thrown onto the roof of the house. Following the sacrifice and similar distribution of the meat of a billy goat, all the participants bathe in order to purify themselves and complete the ritual cleansing (de Heusch, 1985, summarizing Biebuyck, 1953, 1973).

Among the Hamba of north Kasai (DRC) the killing of giant pangolins is also generally prohibited. In this case, the meat is usually reserved for members of a relatively new political institution, the "Masters of the Forest" (*nkum'okunda*), who claim much of the power once held by lineage chiefs. This is a closed, male brotherhood that practices secret rites in an enclosure forbidden to women and non-initiates. If a pangolin is accidentally trapped, a heavy fine must be paid to the group. Aardvarks are subject to a similar prohibition, because, it is said, they live hidden in deep burrows and, like pangolins, only bear single offspring at a time (de Heusch, 1985). The transfer of ideas like this from one species to another with similar characteristics is relatively common.

The Tabwa, for example, who live at the southwestern end of Lake Tanganyika, pay much more attention to aardvarks than they do pangolins (*nkaka*), though they use pangolin scales in medicine, and hunters burn them at the four corners of their camps to keep lions away (Douglas, 1990; Roberts, 1986). According to Roberts, these are usually the scales of Temminck's pangolin (*Smutsia temminckii*). Tabwa aver that "the king of beasts is not the lion" but the pangolin, and because of their apparent strength and other qualities, their scales often feature in the material culture of this region (Roberts, 2009). *Nkaka* is the name of a pattern that is inscribed on different cultural objects as well as women's bodies, as it is among the Luba of southeastern DRC, where the motif functions "as a cognitive cuing structure that triggers a barrage of associations, contexts, and meanings." It "appears on objects whose purpose is to contain power", including pots that can contain ancestral spirits and "virtually all Luba royal emblems in which the ruler's possessing spirit resides." *Nkaka* also names a beaded headdress worn by certain classes of diviner when they enter a state of spirit possession, "to catch and hold the spirit within" (Roberts, 2013).

In some West African societies, pangolins are no less important, though they may be put to different intellectual uses. The anthropologist Ariane Deluz has discussed an intriguing set of ideas about the giant pangolin among the Guro, speakers of a Mande language in Côte d'Ivoire. The pangolin (*zè* or *zèginè*) features in a song, recorded in the mid-1970s, that outlines the mythical charter of a clan. In this tale, a man called Yuro climbs up a chain into the sky and returns to earth with a pangolin that has a much-desired war fetish in its belly. However, his daughter Na subsequently becomes pregnant, and blames this on the pangolin, which leaves its "shell" behind and returns to the sky when Yuro threatens to kill it. Na's child becomes the founder of the clan, and is referred to as "descendant of the pangolin", though the implication is that he was the product of an incestuous union. This tale is said to be echoed in Guro fertility statues, while its social and psychoanalytic implications are drawn out by Deluz and a colleague (Deluz, 1994).

Ideas about fertility are evidently a part of this tale, which can also be related to the

Sangu case discussed in the next section. A systematic compilation, comparison and analysis of examples such as these has never been undertaken, and would be a difficult task, given the incompleteness of much of the ethnography. Pangolins are accorded cultural significance throughout their African range, especially among peoples speaking Niger-Congo languages and in the Bantu diaspora. In some cases, this may be because of borrowing, but it is also possible that some of the ideas and attitudes relating to pangolins were carried by early migrants from the rainforest, along with the vocabulary describing them (Vansina, 1990; Walsh, 1995/96). As discussed, beliefs and practices can also jump from one species to another, and a full study would also have to take this into account.

Pangolin prediction and ritual sacrifice

As might be expected, similar themes recur among the speakers of Bantu languages in East and Southern Africa, with new elements added. The association of pangolins with political leadership and fertility that is evident among the Lele also has parallels far to the east, among different peoples in the Southern Highlands of Tanzania. The Sangu of south-western Tanzania traditionally treat the sighting of a Temminck's pangolin – the only species found in Usangu – as an event of great importance. Pangolins are believed to fall from the sky, being sent to the earth by the ancestors. When a man or woman encounters a pangolin in the bush, it is said that the animal will latch onto them, and follow them back into the village. When this happens the Sangu chief and his ritual specialist must be informed, setting in motion a ritual process that parallels the rituals of twin birth. The pangolin and the person it followed are secluded for a period, during which they are joined by others singing and dancing the songs of twin birth. At times they dance naked, and the pangolin is said to join in, standing and swaying on its hind legs. It is claimed that pangolins sometimes shed tears while they are dancing, and this is interpreted as an omen of good rains in the coming year. If the pangolin's eyes remain dry, then it is taken as a sign that drought will ensue (Bilodeau, 1979; Walsh, 1995/96).

Details of the confinement and its duration vary considerably, as do descriptions of what happens next. On one account the period of seclusion lasts for a whole day, following which the pangolin is dressed in a black cloth and turban brought by the ritual specialist. They are then taken down to the nearest river together with a group of elders who take a sheep with them. They first sit by the river and sway from side to side, after which the sheep is killed and the meat is roasted over an open fire and shared out. A hole is then dug for the pangolin and it is made to sit in this on top of the fleece of the newly slaughtered sheep. The hole is then covered up with earth and the pangolin buried alive, concluding the ritual. On other accounts, the seclusion lasts much longer – up to three weeks – and the pangolin is killed before being buried in the black cloth purchased by the Sangu chief (Walsh, 1995/96). These different accounts may reflect the rarity of the ritual and variations in practice, as well as lapses in memory. Like the Lele pangolin feast, none of these ceremonies has been observed directly by a researcher.

A Sangu folktale recorded in 1975 reinforces the association between pangolins, chiefs and fecundity. In this tale a young woman, who refuses to follow her sisters by marrying a wild animal, is pursued instead by a magical tree which becomes her husband. At night, the tree metamorphoses into a chief by shedding its scaly bark in the form of a pangolin. At dawn, it becomes a tree again after calling for the pangolin to return and cover it. This strange relationship continues until one night the woman, who by this time has a child by her magical husband, kills the pangolin by burning it on a fire. The next morning the tree is unable to resume its usual daytime form, and only the chief remains. People rejoice at this and the woman's sisters divorce their animal husbands and join her as wives of the chief (Bilodeau, 1979). This concludes the tale, which though not specifically related to the Sangu pangolin ritual, implies that the sacrifice of pangolins is necessary to restore the normal balance of the world and the status of chiefs as guarantors of its fertility – the alternative being an upside-down world in which animals rule instead. It is important to note, however, that this is the researcher's interpretation, rather than one explicitly formulated by Sangu themselves (Walsh, 1995/96).

It is also important to add that not all Sangu know about the pangolin ritual, especially younger people who have no experience of it. In the early 1980s, one young member of the Sangu royal family claimed that pangolins are killed and buried in cattle enclosures as a protective charm to prevent the cattle from being "startled", a practice perhaps also known to Sukuma immigrants from the north of Tanzania, where pangolin scales and the burned ashes of their bones and bodies are highly valued as charms against the attacks of snakes and wild animals (Wright, 1954). Immigrant Nyakyusa farmers from the south professed ignorance of pangolins and their possible ritual or magical uses, while a group of Ndali, also from the Nyasa-Tanganyika corridor, incensed their Sangu hosts by killing and eating a pangolin (Walsh, 1995/96).

Similar variations in practice have been recorded in the Great Ruaha valley, downstream and east of Usangu, where Hehe, Gogo and other long-term residents also treat encounters with Temminck's pangolins as auspicious events. When this happens far from home, the recommended course of action is to spit on a leaf and stick this on the scales of the curled-up animal, asking for a blessing and successful journey. When a pangolin is found nearer to home, the preferred practice is to take it back to the village and then inform senior members of the community, especially local ritual or political leaders who can organize a communal divination ceremony. The pangolin is dressed in a black cloth and surrounded by different products, such as foodstuffs, water, weapons, and medicine. To the accompaniment of singing and dancing, the pangolin is asked to foretell the future by choosing between them, whereupon its actions are interpreted by the ritual specialist or chief leading the ceremony. If, for example, the pangolin moves towards a pile of grain, then this is thought to presage a good harvest of the crop in the coming year. Other observed behavior is also fed into the prediction: if the pangolin "cries", then this may be interpreted as a bad omen. In practice, the pangolin's "choices" are often reinterpreted in the light of future events, retrospectively fitted in hindsight to match significant occurrences at different scales, including global or community-wide disasters, good or bad harvests, and the outcome of public or domestic disputes (Walsh, 2007).

The practice of pangolin divination varies from one village to another, but the basic pattern is the same. In some instances, however, it is not carried out as it should be. Recorded cases include the seizure of the pangolin by a government Game Assistant, a botched attempt by an unqualified village official to hold the ritual himself, the escape of the pangolin, another ritual that failed because the pangolin had been beaten beforehand, and many cases in which immigrants to the area killed a pangolin to eat it and/or sell its scales.

In the Great Ruaha valley, pangolins traditionally provide the ingredients for different kinds of medicine and magic. Pangolin scales have the greatest number of uses, both curative and protective, being used in preparations to treat whitlow, stiff necks, back pain, pneumonia, children's convulsions and rashes, and to protect hunters and their camps against wild animals, park rangers and other sources of misfortune (see Chapter 15). Other parts of the pangolin, including the heart and larynx, are also used for protective purposes, as well as to magically stupefy animals that are being hunted. But Hehe and Gogo speakers generally agree that pangolins should not be deliberately killed in order to obtain scales and other body parts for these purposes. Rather they should be taken from animals that have died naturally or sacrificed ritually, just as the consumption of pangolin meat should also be limited to ritual contexts (Walsh, 2007).

The ritual performances and practices which accompany and then follow pangolin divination also vary considerably. In Gogo-speaking communities they are often linked to rainmaking rites using special rain-stones. In some cases, they are intertwined with girls' puberty rites and involve recent initiates. Elsewhere, the rites of twin birth and twin parenthood are a significant component. The fate of the pangolin itself is also subject to differences in local custom. Whereas the animal is often returned to the bush after the ritual has ended, in some places it is taken to a dry water course, beaten to death, and then roasted and eaten by the ritual specialist and perhaps other participants. If these rites have anything in common, it is their connection to ideas of fertility and the well-being of the land and its people – hence descriptions of the pangolin as a "chief" (*mtwa*) and "the owner of the rain" (*munyamdonya*; Walsh, 2007).

Scattered reports indicate that pangolin divination is practiced in places as far apart as the shores of Lake Tanganyika (Walsh, 1997) and the Indian Ocean coast (Keregero, 1998), while the sacrifice of pangolins has also been described among different peoples, including the Nyaturu of central Tanzania (Jellicoe, 1978). The association of pangolins with chiefship, fertility and wealth is known from much further afield, including among some Bantu-speaking groups in Southern Africa (for examples see Chapter 15). Among the Lovedu (Lobedu) of Limpopo Province in South Africa, Temminck's pangolins traditionally belong to the queen, and must be captured alive. Their fat is used in making rain medicine, which she uses in her role as a rainmaker, sacrificing a black sheep from time to time to add potency to the charm (de Heusch, 1985, citing Krige and Krige, 1943). The precolonial Shona of Zimbabwe were prohibited from killing this species; as well as incurring material penalties, people who broke this taboo were also believed to incur the wrath of the ancestors. Anyone who found a pangolin was obliged to hand it over to the local chief, in the belief that this would ensure prosperity for all. Despite the introduction of strict anti-poaching laws, this practice has continued (see Chapter 15) and even been extended, with many people after independence trying to curry favor by presenting pangolins to the country's former president, Robert Mugabe (Duri, 2017).

Pangolin products as medicine and magic

The principle behind many of these ideas and practices seems to be analogy with the observed or supposed characteristics and behavior of pangolins rather than any perception of them as anomalous creatures. This is especially the case with widespread belief in the potency of pangolin scales and other body parts, found throughout the range of all the African species, including the black-bellied pangolin (*P. tetradactyla*). Pangolin body parts are put to a dazzling array of medicinal and magical uses (see Chapter 15). In a survey of traditional Awori medical practitioners in southwestern Nigeria, pangolin products were reportedly used to treat 47 different conditions, 15 of them using scales (Soewu and

Adekanola, 2011; see also Soewu and Ayodele, 2009). A survey in Sierra Leone identified 22 pangolin body parts, most often scales, being prescribed for the treatment of 59 diseases and ailments (Boakye et al., 2014). Similarly heterogeneous, but shorter lists can be found in the literature (e.g., Akpona et al., 2008, on Lama forest reserve in Benin; Setlalekgomo, 2014, on Kweneng district in Botswana).

Some examples of the protective and other medicinal uses of pangolin scales have already been mentioned (see also Chapter 15). In the past, scales were often obtained as a by-product of ritual sacrifice. The best-known illustration of this is Wright's (1954) description of a pangolin impaled by a Sukuma chief and left to decay on a pole opposite the entrance to his enclosure: "As the scales [...] rot off, they are given away by the Chief to visitors to his kraal in exchange for presents made; the market value of each scale being about a shilling. [...] These scales if worn round the neck or leg are considered as a sovereign defence against the bite of snakes and other wild beasts. They are also tied by old men to their limbs during rheumatic attacks, and around the body of young children who suffer from a swollen spleen. [Scales are] also an important 'activator' [...] in the fertility medicine, which is mixed with the seed for planting at the beginning of the year. The decayed body and bones of the pangolin are finally burnt and the ashes carefully preserved in a pot in the Chief's house. Small quantities of this ash may be released from time to time to peasants suffering from the attacks of lions. A little of this medicine burnt in the kraal fire is believed to be a most powerful deterrent to lions" (Wright, 1954). According to another account, some Sukuma believed that pangolin scales conferred invisibility, drawing an analogy with the animal's apparent rarity and lack of visibility (Cory, 1949). As this indicates, there was a market in pangolin scales during the colonial period, and probably before. As evidence from across Africa suggests, the direct killing of pangolins for their scales now poses a considerable threat to their survival, as documented elsewhere in this volume (see Chapters 15 and 16).

Asia

Asian alterities

Pangolins also occupy a special place in the beliefs and rituals of peoples living in the range of the four Asian species of pangolin, though again the evidence is very uneven. Most reports focus on the use of pangolin products, and especially scales, in traditional medicine and magic. As in Africa, scales and other parts of pangolins are put to a wide variety of uses, some examples of which are given in Chapter 14. If this review is any guide, the range of ideas about pangolins and the magical efficacy of their scales and other body parts is even more diverse in South and East Asia, reflecting perhaps the greater linguistic and cultural diversity of the peoples across the continent who are familiar with pangolins. Some of the same concerns are evident, for example around protection and procreation, but there are also many notions and practices that differ. Underlying this variation, however, the same logic of analogy is often at work, whereby pangolins and their products are invested with meaning and power on the basis of local understandings of their behavior and functions – hence the common link between the protective role of scales and their use in healing and warding off evil.

While the focus of many contemporary studies is on the trade and trafficking in pangolins and their scales, relatively little detail is available on the role played by pangolins in local ethnobiologies, cosmologies and ritual practice. One exception is Alex Aisher's (2016) study of the beliefs and practices that support pangolin conservation among the Nyishi people of Arunachal Pradesh in northeastern India. Aisher argues that whereas the alterity and taxonomic anomalousness of pangolins makes them a good fit for Traditional Chinese Medicine (TCM) and so helps drive the international trade, among the Nyishi it has quite different effects. In Nyishi cosmology, *sechik*, the Chinese pangolin (*Manis pentadactyla*), is spoken of as the child of Buru, the water spirit: fear of the vengeance of this and other "master-spirits" discourages locals from hunting pangolins, especially when they and other spiritually significant animals are being hunted by outsiders without regard to local custom (Aisher, 2016).

Local practice may help protect pangolins, but it does not always help protect them against over-exploitation by external hunters and traders. Otherwise, local practice may itself be part of the problem. The offerings that indigenous people in Palawan have to make before hunting (and eating) the Philippine pangolin (*M. culionensis*) have nonetheless provided them with diminishing returns, at least as reported by hunters in Barangay Malis (Acosta-Lagrada, 2012). Pangolins are reported to have a variety of medicinal and magical uses in Palawan (see Chapters 7, 14), including the wearing of their scales to protect against the shape-shifting *aswang*, a blood-sucking, fetus- and viscera-consuming Philippine version of the Old Hag (Estrada et al., 2015; Nadeau, 2011). Ironically, they have also been mistaken for this vampiric ghoul, as in the case of a pangolin found perambulating on a rooftop in Pasay one night in March 2015 (Frialde, 2015).

Pangolin prohibitions in Peninsular Malaysia

Some of the best information comes from Southeast Asia, and one of the most interesting cases can be found in a study of the religion of the Temiar, an Orang Asli group in the uplands of northern Peninsular Malaysia (Benjamin, 2014). Their indigenous religion centers on public spirit-mediumship ceremonies, performed at night with choral singing, dance and trance. They have an elaborate set of cosmological ideas and dietary rules, recalling in some respects Lele practice as described by Mary Douglas, who complained that other ethnographers had not followed her by investigating food prohibitions in the same way (Douglas, 1990). Pangolins play an important role in the Temiar system, and are subject to more prohibitions than other tabooed animals, in relation to which the pangolin is described as the senior headman (Benjamin, 2014).

The species in question is the Sunda pangolin (*M. javanica*). Like some other animals, pangolins are forbidden to young children, their mothers and midwives. On top of this, "If a pangolin is killed, it must be brought into the village circumspectly so that the carcass is not carried nearer to any other house than to the house where it will finally be cooked. Once cooked, a portion of the meat must be distributed to anyone who is accustomed to eat it (essentially those for whom it is not [tabooed]) and who saw the carcass being brought home by the hunters. Eating the flesh is fraught with danger. In the words of an informant, 'we must not drop the bones down to the ground, but eat carefully by incinerating them in the fire; and then we must burn up the bamboo sections we ate from'. The sanction for contravening any of these rules is believed to be rapid death of the miscreant. [This degree of taboo] also attaches to the pangolin while it is still alive: it is one of the species classed as [...] 'non-tameable' [...], and to attempt to tame it as a village pet is believed to cause death [...]. In myth too, the pangolin (*wejwooj*) seems to bear a special relationship to the human domain. In the creation story [...]. Wejwooj is responsible for planting the *təlayaak* tree, which ever since has been used as the best source of fertile wood ash in which to plant tobacco" (Benjamin, 2014).

This is only one part of a complex system, and the ethnographer, Geoffrey Benjamin, honestly admits that "the whole question of the interrelationship between everyday ritual observances towards natural species and their appearance in myth is a topic on which I do not have sufficient data from which to generalize" (Benjamin, 2014). Again, this recalls the ethnography of the Lele, as well as the somewhat obscure relation between Sangu practice and their myth-like folktale featuring a pangolin. There is an apparent link between pangolins and fertility, and a recognition of the part played by pangolins in establishing human culture, but Benjamin makes no attempt to explain their role. At the same time, there are few accounts of the role of pangolins in ritual practice and myth that are as detailed as this.

Other Orang Asli groups in Peninsular Malaysia are known to have ideas and practices that can be related to these. The Batek have a "taboo on women eating pangolins during pregnancy, the prohibition this time connected to the animal's defensive tendency of rolling up into a tight ball when threatened, which [...] is seen as raising the analogous possibility of constriction impeding an easy birth" (Tacey, 2013). Among the Mah Meri (Ma' Betisék) of Carey Island, off the west coast of Selangor, "The *kondok* or pangolin, (*Manis javanica*, Linn) is said to have been formed from the after-birth of humans. The Ma' Betisék normally wrap the after-birth in a mat and bury it in the ground. Hence the scales of the pangolin resemble the texture and weaving patterns of the pandanus mat" (Karim, 1981). This is yet another example of analogy at work.

In Perak, pangolins are said to be able to kill elephants by biting their feet, or by coiling themselves around their trunks and suffocating them. According to Skeat, this story was also told in Selangor with additional details: "Thus it is said that the 'Jawi-jawi' tree (a kind of banyan) is always avoided by elephants because it was once licked by the armadillo [sic]. The latter, after licking it, went his way, and 'the elephant coming up was greatly taken aback by the offensive odor, and swore that he would never go near the tree again. He kept his oath, and his example has been followed by his descendants, so that to this day the 'Jawi-jawi' is the one tree in the forest which the elephant is afraid to approach'" (Skeat, 1900, citing an earlier source). Knowledge of these tales among some Temiar and Lanoh is said to have led them to believe that pangolins have supernatural powers. "As a result, there are many in the Orang Asli community that wear pangolin scales to ward off evil spirits." On one account, "the scales of the pangolin are burned and kept indoors to ward off elephants which come into the reserve, because burnt pangolin scales have an extremely strong smell, as well as because of the fable that stated the elephant was killed by a pangolin" (Yahaya, 2014).

Pangolin imagery in Indonesia

While there is insufficient information, linguistic or otherwise, to enable the tracing of the history of ideas about pangolins in Peninsular Malaysia, a very different situation exists in central Java, where images of the Sunda pangolin have been carved into the 9th century Hindu Shaivite temple complex, Candi Loro Jonggrang, in Prambanan village. A large pangolin was interred in a central location below a statue of Nandi the bull, Siwa's devotee and celestial vehicle, while a second one was carved on the balustrade of the main temple devoted to Siwa, "curled before Rama in one of the bas-relief panels of the Ramayana narrative" that encircle the building (Totton, 2011). Pangolins also feature in the *Ramayana kakawin*, the Old Javanese version of the Sanskrit epic, written circa 870 CE; in it they are extolled for driving away venomous snakes as well as being served up as part of the celebratory menu when Rama is eventually crowned. Mary-Louise Totton argues that the pangolin's representation at Loro Jonggrang goes beyond this, and alludes to the "docile nature of the pangolin and its tendency to roll up into a protective ball", which "when threatened allows it to wait out danger." "The pangolin curls before Rama, acting as a submissive devotee who offers a warning and a strategy of surviving: wait out the danger ahead" (Totton, 2011). She gives a similar interpretation to an episode featuring a pangolin in a later version of the epic poem, and speculates on a series of connections between pangolins and ideas about agricultural fertility – drawing parallels with the Lele and other African beliefs and practices.

Totton also draws a parallel with more recent events. The story is best told in her words: "A Malay and western Indonesian folktale reveals another attribute of the pangolin. A pangolin lies on the ground and raises its scales. Thinking it dead, ants swarm over it. Soon the larger animal snaps down its scales and ambles over to a nearby pond to drown the ants and eat them [...]. This folklore sets up the pangolin as a clever trickster who uses passive treachery to achieve its goals. [...] During the heated Indonesian pre-election period of 1999, Megawati Sukarnoputri was accused by many of being too passive. A newspaper article that came to her defence was headlined: 'Megawati, Api Api Trengiling Mati' (Megawati, in the spirit of a dead pangolin) [...]. In the article, a Javanese psychic sums up Megawati's political gestures as 'layered in meaning' and how Mega was channeling her father, as the pangolin also recalls past wisdom. He reminds the readers of the ways of the pangolin. At first glance Mega appears weak and submissive, but he says, like the pretending pangolin, this is only a clever strategy to outlast the opposition. Eerily, once again the pangolin lends its essence to a bull, as we look at the paper's follow-up story. The pangolin strategy is reiterated under the potent icon of Megawati's party, the black bull. A Mega supporter swathed in this symbolism dominates the page, a visual pairing of the brave bull icon of her political party (the people) supported by the spirit of a pangolin (the wise leader) [...]. The pangolin rhetoric is effective because the public has a base for understanding this reference" (Totton, 2011). Recalling the Zimbabwean case, this reinforces the observation that as well as reflecting past practice, pangolin symbolism can also be given contemporary meaning.

Conclusion

It is important to emphasize that pangolin symbolism can have both traditional roots and current political resonance. The significance and use of pangolins in any one place are liable to be contested by different groups and can change over time: the impact of globalization and the recent expansion of the trade and trafficking in pangolin products is the most obvious example of this (see Chapter 16), but not the only one. The available evidence suggests that many of the beliefs and practices relating to pangolins in sub-Saharan Africa are historically connected, especially among the speakers of Bantu and other Niger-Congo languages whose territories cover much of the range of the four African pangolin species. The notions of chiefship and fecundity that often attach to pangolins have also been subject to many transformations, and in some cases, have jumped from pangolins to other burrowing animals, including aardvarks and porcupines. In Asia, it is harder to discern such connections, given that the four Asian species are known to peoples of divergent origins speaking languages belonging to very different families. Nonetheless, some regional historical relationships can be discerned, and it is quite possible further research will reveal patterns of borrowing that have followed past as well as present routes of the trade in pangolins and their products.

At the same time, a number of parallels can be perceived between some of the beliefs and practices pertaining to pangolins in Asia and those that are prevalent in parts of Africa. These are no doubt cases of convergence, deriving from similarities in the observed features and behaviors of pangolins and the limited range of human concerns to which they can be compared and applied. The basis for many of these applications is analogy (Descola, 2013; Kohn, 2015; and cf. Aisher, 2016), and the attribution of protective properties to pangolin scales when detached and used as charms is just one example of this. Pangolins, it seems, are not only "good to think with", but good to think with analogically. Their unusual features – unusual, that is, to human eyes – make them excellent subjects for human thought and action, for symbolizing, myth-making, and ritual practice, including different varieties of medicine and magic. At the heart of this is their alterity or otherness to people (Aisher, 2016; Walsh, 2007), whether or not this is expressed as taxonomic anomalousness (Douglas, 1957, 1966). Changing, adapting, or working with these ways of thinking poses a challenge to conservation, and to conservation education in particular. It also poses a challenge to anthropologists and other social scientists, for whom the real "paradox of the pangolin" (Willis, 1974) is not its ethnobiological and cosmological classification, but the lack of good comparative research on exactly what people think about and do with pangolins, and how and why this is changing, both with and without the interventions of conservationists.

References

Acosta-Lagrada, L.S., 2012. Population density, distribution and habitat preferences of the Palawan Pangolin, *Manis culionensis* (de Elera, 1915). M.Sc. Thesis, University of the Philippines Los Baños, Laguna, Philippines.

Aisher, A., 2016. Scarcity, alterity and value: decline of the pangolin, the world's most trafficked mammal. Conserv. Soc. 14 (4), 317–329.

Akpona, H.A., Djagoun, C.A.M.S., Sinsin, B., 2008. Ecology and ethnozoology of the three-cusped pangolin *Manis tricuspis* (Mammalia, Pholidota) in the Lama forest reserve, Benin. Mammalia 72 (3), 198–202.

Benjamin, G., 2014. Temiar Religion, 1964-2012: Enchantment, Disenchantment and Re-enchantment in Malaysia's Uplands. NUS Press, Singapore.

Biebuyck, D., 1953. Répartitions et droits du Pangolin chez les Balega. Zaïre 7 (8), 899–934.

Biebuyck, D., 1973. Lega Culture: Art, Initiation and Moral Philosophy among a Central African People. University of California Press, Berkeley and Los Angeles.

Bilodeau, J., 1979. Sept contes Sangu dans leur context culturel et linguistique. Ph.D. Thèse, Université de la Sorbonne Nouvelle, Paris, France.

Boakye, M.K., Pietersen, D.W., Kotzé, A., Dalton, D.L., Jansen, R., 2014. Ethnomedicinal use of African pangolins by traditional medical practitioners in Sierra Leone. J. Ethnobiol. Ethnomed. 10 (76).

Bulmer, R., 1967. Why is the cassowary not a bird? A problem of zoological taxonomy among the Karam of the New Guinea Highlands. Man (New Series) 2 (1), 5–25.

Cory, H., 1949. The ingredients of magic medicines. Africa 19 (1), 13–32.

Deluz, A., 1994. Incestuous fantasy and kinship among the Guro. In: Heald, S., Deluz, A. (Eds.), Anthropology and Psychoanalysis: An Encounter through Culture. Routledge, London and New York, pp. 40–53.

Descola, P., 2013. Beyond Nature and Culture. University of Chicago Press, Chicago (Original work published 2005).

de Heusch, L., 1985. Sacrifice in. Africa: A Structuralist Approach. Manchester University Press, Manchester.

de Heusch, L., 1993. Hunting the pangolin. Man 28 (1), 159–161.

Douglas, M., 1957. Animals in Lele religious symbolism. Africa 27 (1), 46–58.

Douglas, M., 1963. The Lele of the Kasai. Oxford University Press for the International African Institute, London.

Douglas, M., 1966. Purity and Danger: An Analysis of Concepts of Pollution and Taboo. Routledge & Kegan Paul, London.

Douglas, M., 1970. Natural Symbols: Explorations in Cosmology. Penguin Books, Harmondsworth, Middlesex.

Douglas, M., 1990. The pangolin revisited: a new approach to animal symbolism. In: Willis, R.G. (Ed.), Signifying Animals: Human Meaning in the Natural World. Unwin Hyman, London, pp. 25–36.

Douglas, M., 1993. Hunting the pangolin. Man 28 (1), 161–165.

Douglas, M., 1999. Implicit Meanings: Selected Essays in Anthropology, second ed. Routledge, London and New York.

Duri, F.T.P., 2017. Development discourse and the legacies of pre-colonial Shona environmental jurisprudence: pangolins and political opportunism in independent Zimbabwe. In: Mawere, M. (Ed.), Underdevelopment, Development, and the Future of Africa. Langaa Research & Publishing Common Initiative Group, Bamenda, Cameroon, pp. 435–460.

Ellen, R., 1972. The marsupial in Nuaulu ritual behaviour. Man (New Series) 7 (2), 223–238.

Ellen, R.F., 1994. Hunting the pangolin. Man 29 (1), 181–182.

Estrada, Z.J.G., Panolino, J.G., De Mesa, T.K.A., Abordo, F. C.B., Labao, R.N., 2015. An ethnozoological study of the medicinal animals used by the Tagbanua tribe in Sitio Tabyay, Cabigaan, Aborlan, Palawan. In: Matulac, J.L.S., Cabrestante, M.P., Palon, M.P., Regoniel, P.A., Gonzales, B.J., Devanadera, N.P. (Eds.), Proceedings of the 2nd Palawan Research Symposium 2015. National Research Forum on Palawan Sustainable Development: Science, Technology and Innovation for Sustainable Development, 9–10 December. Puerto Princesa City, Palawan, Philippines, pp. 123–128.

Fardon, R., 1993. Spiders, pangolins and zoo visitors. Man 28 (2), 361–363.

Fardon, R., 1999. Mary Douglas: An Intellectual Biography. Routledge, London and New York.

Frialde, M., 2015. Pangolin rescued in Pasay. The Philippine Star, Manila. Available from: <https://www.philstar.com/metro/2015/03/08/1431127/pangolin-rescued-pasay>. [October 14, 2018].

Gossiaux, P.P., 2000. Le Bwame du Léopard des Babembe (Kivu-Congo): Rituel Initiatique et Rituel Funéraire (1ère partie). Available from <http://www.anthroposys.be/bwame1.htm>. [September 28, 2018].

Jellicoe, M., 1978. The Long Path: A Case Study of Social Change in Wahi, Singida District, Tanzania. East African Publishing House, Nairobi.

Karim, W.-J.B., 1981. Ma' Betisék Concepts of Living Things. The Athlone Press, London.

Keregero, K., 1998. Pangolin brings hope to Coast Region. The Guardian, Dar es Salaam, 1 September 3.

Kohn, E., 2015. Anthropology of ontologies. Annu. Rev. Anthropol. 44, 311–327.

Krige, J.D., Krige, E.J., 1943. The Realm of the Rain-Queen: A Study of the Pattern of Lovedu Society. Oxford University Press, London.

Lévi-Strauss, C., 1963. Totemism (R. Needham, Trans.). Beacon Press, Boston (Original work published 1962).

Lévi-Strauss, C., 1966. The Savage Mind (La Pensée Sauvage). Weidenfeld and Nicolson, London (Original work published 1962).

Lewis, I.M., 1991. The spider and the pangolin. Man 26 (3), 513–525.

Lewis, I.M., 1993a. Hunting the pangolin. Man 28 (1), 165–166.

Lewis, I.M., 1993b. Spiders, pangolins and zoo visitors. Man 28 (2), 363.

Lewis, I.M., 2003. Social and Cultural Anthropology in Perspective, third ed. Transaction Publishers, New Brunswick and London.

Morris, B., 1998. The Power of Animals: An Ethnography. Berg, Oxford.

Nadeau, K., 2011. *Aswang* and other kinds of witches: a comparative analysis. Philip. Quart. Cult. Soc. 39 (3/4), 250–266.

Richards, P., 1993. Natural symbols and natural history: chimpanzees, elephants and experiments in Mende thought. In: Milton, K. (Ed.), Environmentalism: The View From Anthropology. Routledge, London and New York, pp. 144–159.

Roberts, A.F., 1986. Social and historical contexts of Tabwa art. In: Roberts, A.F., Maurer, E.M. (Eds.), Tabwa. The Rising of a New Moon: A Century of Tabwa Art. The University of Michigan Museum of Art, Ann Arbor, pp. 1–48.

Roberts, A.F., 2009. Bugabo: Arts, Ambiguity, and Transformation in Southeastern Congo. Available from <http://www.anthroposys.be/robertspdf.pdf>. [September 28, 2018].

Roberts, M.N., 2013. The king is a woman: shaping power in Luba Royal Arts. Afr. Arts 46 (3), 68–81.

Setlalekgomo, M.R., 2014. Ethnozoological survey of the indigenous knowledge on the use of pangolins (*Manis* sps [sic]) in traditional medicine in Lentsweletau Extended Area in Botswana. J. Anim. Sci. Adv. 4 (6), 883–890.

Skeat, W.W., 1900. Malay Magic Being an Introduction to the Folklore and Popular Religion of the Malaya Peninsula. Macmillan and Co, London.

Soewu, D.A., Ayodele, I.A., 2009. Utilisation of pangolin (*Manis* sps [sic]) in traditional Yorubic medicine in Ijebu province, Ogun State, Nigeria. J. Ethnobiol. Ethnomed. 5, 39.

Soewu, D.A., Adekanola, T.A., 2011. Traditional-medical knowledge and perception of pangolins (*Manis* sps [sic]) among the Awori People, southwestern Nigeria. J. Ethnobiol. Ethnomed. 7, 25.

Sperber, D., 1996. Why are perfect animals, hybrids, and monsters food for symbolic thought? Method Theory Study Religion 8 (2), 143–169.

Tacey, I., 2013. Tropes of fear: the impact of globalization on Batek religious landscapes. Religions 4, 240–266.

Tambiah, S.J., 1969. Animals are good to think and good to prohibit. Ethnology 8 (4), 424–459.

Totton, M.-L., 2011. The pangolin: a multivalent memento in Indonesian art. Indones. Malay World 39 (113), 7–28.

Vansina, J., 1990. Paths in the Rainforests: Toward a History of Political Tradition in Equatorial Africa. James Currey, London.

Walsh, M.T., 1995/96. The ritual sacrifice of pangolins among the Sangu of south-west Tanzania. Bull. Int. Committ. Urgent Anthropol. Ethnol. Res. 37/38, 155–170.

Walsh, M.T., 1997. Mammals in Mtanga: Notes on Ha and Bembe Ethnomammalogy in a Village Bordering Gombe Stream National Park, Western Tanzania. Lake Tanganyika Biodiversity Project, Kigoma.

Walsh, M.T., 2007. Pangolins and politics in the Great Ruaha valley, Tanzania: symbol, ritual and difference / Pangolin et politique dans la vallée du Great Ruaha, Tanzanie: symbole, rituel et différence. In: Dounias, E., Motte-Florac, E., Dunham, M. (Eds.), Le symbolisme des animaux: L'animal, clef de voûte de la relation entre l'homme et la nature? / Animal Symbolism: Animals, Keystone of the Relationship Between Man and Nature? Éditions de l'IRD. Paris, pp. 1003–1044.

Wijeyewardene, G., 1968. Address, abuse and animal categories in northern Thailand. Man (New Series) 3 (1), 76–93.

Willis, R., 1974. Man and Beast. Hart-Davis, MacGibbon, London.

Wright, A.C.A., 1954. The magical importance of pangolins among the Basukuma. Tanganyika Notes Records 36, 71–72.

Yahaya, F.H., 2014. The usage of animals in the lives of the Lanoh and Temiar tribes of Lenggong, Perak. Paper presented at ICoLASS 2014 – USM-POTO International Conference on Liberal Arts and Social Sciences, Hanoi and Ha Long Bay, Vietnam, 25–29 April.

CHAPTER

13

Early biogeographies and symbolic use of pangolins in Europe in the 16th–18th centuries

Natalie Lawrence

Department of History and Philosophy of Science, University of Cambridge, Cambridge, United Kingdom

OUTLINE

Introduction

The history of pangolins in Europe is one composed of texts and skins, indistinct descriptions and unmistakeable imagery. Pangolin-like creatures started to appear in the travel accounts of European writers in the late 16th century. Descriptions of animals that can only have been pangolins were written by travelers in locations such as Ceylon, Taiwan, Siam and Guinea. These animals were called *lin* by the Siamese, *pangoelling* in China, Sumatra, Java and Malacca, *allegoe* in Malabar and *quogelo* in Guinea. At the same time, naturalists and collectors in Europe published descriptions of skins of "scaly lizards" in curiosity collections, usually with little reference to these travel accounts. The origins of these objects were indistinct, often because of confusion along shipping networks and lack of care about recording specimen

Pangolins
DOI: https://doi.org/10.1016/B978-0-12-815507-3.00013-7

provenance, but they were seen as coming from a generalized "Indies." They did not sit easily in the traditional European ways of classifying the natural world, appearing to exist between groups, so these skins became tokens of monstrous and boundary-crossing scaly creatures. The "scaly lizards" became nebulous beasts, depicted as both devils and innocents, embodying the colonial strife in the exotic regions from which they came.

This chapter discusses the early encounters with pangolins in the field by European travelers, the transport of pangolins skins and their integration into European curiosity collections. It explores the nebulous images of pangolins and their conflation with the West Indian armadillos, and how these "scaly beasts" were used as biogeographical symbols and liminal creatures on the Great Chain of Being.

Pangolin encounters: the *pangoelling*, *allegoe*, *quogelo* and *tamach*

Possibly the earliest travel account of a pangolin was written by the prolific Dutch explorer, Jan Huygen van Linschoten, in his account of his voyages in the East, *Itinerario, Voyage ofte Schipvaert ... naer Oost ofte Portugaels Indien* (1597). He described "a fish" taken from the "River of Goa", as big as "a middle sized Dogge", which ran around the hall "snorting like a hogge." It was covered in "scales a thumbs breadth, harder than Iron or Steel." When "hewed upon", it rolled into a ball and could not be prised open by force or "anie instrument." It was only when "let alone" that "hee opened himselfe and ranne away." Van Linschoten's influential work was the first extensive European publication describing Asian trade and nature as well as valuable shipping routes, but his pangolin description was not used by any subsequent authors: it is not the discovery story of the pangolin (Van Linschoten, 1885).

Another creature was described by the Dutch naturalist-physician Jacobus Bontius, stationed in Dutch Batavia, Java, in the 1630s. In his manuscripts, now at the Sherard Collection (Plant Sciences Library, Oxford University), he described an animal called "tamach by some, Larii by others", or *testudo squamata* ("scaled turtle"). The scaly sketch accompanying the description suggests this was almost certainly a pangolin. Bontius described this hole-digging "somnolent animal" with a "cold nature", as covered in carp-like scales and unlike any other land-turtle Bontius had seen. He had been given one that he "kept living some time in the water." The Javanese apparently called "this monster *taunah*, which is the same as: digger in the earth, since it digs holes in the ground along the riverbanks, and hides itself", and was therefore "an amphibian." He also described how the Chinese physicians valued "the scales very much in case of bile-disturbances, dysenterics and cholera", powdering the dried scales for use with wine or rice-water: Bontius could himself testify to their efficacy (Bontius, 1630; Bontius, 1931). Bontius' sketch and account of the *tamach* was published in *Historiae naturalis*, part of a seminal volume on "Indian nature", Willem Piso's *Indiae utriusque re naturali et medica* (Bontius, 1658).

Nearly a century later, in Siam in the 1680s, the French Jesuit missionary Guy Tachard described a reptilian, scaly *herisson* (hedgehog) that he encountered, apparently called *bicho verghonso* (shameful insect) by the Portuguese. He wrote that it "lives in the woods where it hides in holes", and he marveled at its refusal of all foodstuffs, including fruit, meat and rice, wondering how it survived at all. It had a snake-like but harmless tongue, and he found it to be cold-blooded when he dissected a freshly-killed specimen. Yet those he had seen alive rolled up when fearful and his dissection of the creature showed it had young in its uterus. He remarked that, when a mother was killed, the juvenile continued to cling to the base of the mother's tail as it did when she was alive. The engraving published with Tachard's travelogue in 1689 showed a living pangolin with a juvenile riding at the base of its mother's tail (Tachard, 1689; Fig. 13.1).

In the 1720s, two colonial descriptions of the pangolin were published. The Dutch minister in Ambon, François Valentijn, described how in "Java, Sumatra, and Malakka" there is an animal they called a *panggoeling,* or *mierenvanger* (ant-catcher). It was about the size of a dog, with a lizard-like tongue and scales. Its long claws enabled it to dig deep holes in the earth, or even stone floors, earning it the name "*Duyvel.*" These scales were "much sought after by the Chinese" and used to make armor and weapons, "being light and very hard, so not easy to pierce" (Valentijn, 1724). It was not only educated officials and naturalists who were interested in pangolins in the Dutch colonies. There is an anonymous color picture in the Dutch East India Company (*Vereenigde Oost-Indische Compagnie*, hereafter VOC, chartered 1602) archives in The Hague, probably by a merchant or soldier and contemporary to Valentijn's description. It depicts a very lifelike pangolin in rolled-up and unfurled positions, and sits amongst other heterogeneous natural history images (Nationaal Archief, The Hague: inventory 1.11.01.01, nr. 150B).

FIGURE 13.1 "L'Herisson" from Guy Tachard's *second voyage* (1689, bk. 6 p. 250). He depicted this animal in a "typical" oriental landscape with its young riding on the base of its tail.

The second account from the 1720s was written in Guinea by the French cartographer and navigator Reynaud des Marchais. He described "an animal with four feet that the Negroes call *Quogelo*" in the woods. It was "covered in scales arranged a bit like the leaves on an artichoke, but a bit more pointy." It defended itself by gathering itself into a ball, presenting nothing but its scales, which were "iron-like [*ferrées*]", "quite thick and sufficiently strong to defend against the claws and teeth of the animals that attack it." It ate delicately, extending its "extremely long tongue, covered in an unctuous and sticky liquid" to catch ants, just like the anteater. He attested that this animal was "not the least bit malign, it attacks nobody", though "the Negroes assault it with baton blows, skin it, sell its white skin and eat its flesh." This was "white and delicate", despite its diet of "musky" ants. Des Marchais suggested that it "would be a pleasure to have one of these animals privately in places where ants are inconvenient", though, as far as it is possible to tell, he never fulfilled this ambition (Labat, 1730).

Scaly lizards in cabinets of curiosity

As Europeans explored the previously "unknown" regions of the world from the 16th century, a complex global trading network developed, which brought back a wealth of material to Europe. What arrived in European trading ports from the distant corners of the globe was a vast array of *exotica* in different states of preservation and completeness. Certain objects became increasingly prevalent through the 17th century, possibly reflecting a systematization of ship cargoes: rather than a haphazard collection of potentially valuable "stuff" brought by sailors from the Indies, the flourishing market in curiosities in early modern Europe shaped the acquisition practices of commercial ventures. Objects that traveled well, were relatively compact in size and highly saleable in Europe were prioritized in ship cargos (Parsons and Murphy, 2012).

Unlike the bodies of most large animals, dried pangolin skins would have been relatively easy to transport long distances. It seems that pangolins were most frequently brought to Europe as skins, rather than as entire animals, live or dead, or processed scales, as they are often traded in African and Asian markets today (Chapters 14–16). They were already well-established goods in Asian trade networks, so were transported from East to West as part of the commercial operations of the VOC and other European trading ventures. The Dutch East Indies was becoming a touchstone for the "exotic" in Europe. The fact that pangolin skins were "from the Indies" and had a very striking appearance was sufficient to make these skins into highly saleable objects in the rapidly-expanding markets for exotic natural objects in early modern Europe (Lawrence, 2015a,b).

The skins of pangolins appear to have entered European markets as objects of luxury *naturalia* destined primarily for cabinets of curiosities. Along with saw-fish (Pristidae) rostrums, armadillo (Cingulata) carapaces, dried puffer fish (Tetraodontidae) and narwhal (*Monodon monoceros*) teeth ("unicorn horns"), pangolin skins seem to have been part of the requisite set of items to be acquired by any serious collector. Pangolin skins in fact became relatively common collection items - perhaps unsurprising, given the significant trade between the East and Europe in this period. These objects were desirable because they were bizarre, surprising forms that were unfamiliar yet highly recognizable. Very importantly, they were also highly *exotic* (Lawrence, 2015a,b).

Numerous ambiguous "scaly lizard" specimens circulated around scholarly circles through the 17th century. For example, in 1698 the secretary of the Académie Royale des

Sciences in Paris, Jean-Baptiste Du Hamel, reported that he had sent specimens to the Professor of Anatomy at the Jardin du Roi, Joseph-Guichard Duverney and to an author, Charles Perrault, which they had both examined with great interest (Du Hamel, 1701). Specimens were described in the collection catalogues of the Bolognese naturalist, Ulysse Aldrovandi; the *Wunderkammer* at Castle Gottdorf; Manfredo Settala's museum in Milan; the Royal Society collection in London and Basil Besler's collection in Nuremberg. One of several original drawings of pangolins in European collections is a detailed depiction of a rather disgruntled-looking animal with a lizard-like face in an album in Emperor Rudolf II's collection in Prague (Haupt et al., 1990).

A curiously parallel case to the pangolin is that of armadillos, which were relatively familiar in Europe by the time pangolins became collection staples. Armadillos were brought from the Americas in the early 16th century and became perhaps the most common exotic specimens in collections during this period (Brienen, 2007). The eminent Swiss naturalist, Conrad Gessner described how the armadillo was "easily transported from distant regions, because nature has armed it with a hard skin", so the "flesh inside can be easily taken out without any harm to the original shape" (Brienen, 2007; Gessner, 1554). Descriptions of armadillo carapaces in Middle Eastern markets by other contemporary authors suggest that armadillos were even transported further East by the mid-16th century, probably circulated on slave-trading routes between Guinea, the Americas and Europe. In both Europe and the Middle East, armadillo shell was used as a medicinal substance. The Spanish physician, Nicholas Monardes, reported that armadillo tail, ground and made into pellets, cured ear ailments such as tinnitus when placed into the ear canal. It is possible that pangolin scales may also have been used like this in pharmacopoeias (Monardes in Clusius, 1605).

Once stashed in cabinets, pangolin and armadillo skins were also more durable than many other specimens. At least one early modern pangolin specimen, owned by the wealthy Amsterdam apothecary Albertus Seba, still exists in the Peter the Great Museum of Anthropology and Ethnography in St. Petersburg. They were also occasionally transported as wet specimens - Seba reported receiving one from Ceylon "preserved in Arak" (Seba, 1734). Pangolins were sometimes brought alive to Europe, though there are few records of creatures in menageries. One possible example exists in an advertisement bill for the White Elephant Menagerie in Amsterdam in the 1700s, inviting visitors to see a stuffed *nigomsen duyvel*. It was said that the specimen had been killed *en route* to Amsterdam because of its troublesome escape attempts, or perhaps the creature's myrmecophagous diet made it impossible to keep alive on a long voyage. A fabulous sketch of this *duyvel*, an unmistakable pangolin, with vast front claws and wearing a gamesome expression was made by a visitor to the menagerie, Jan Velten (Pieters, 1998).

Scaly lizards in books of natural history

Along with the influx of novel material into Europe through the early modern period, there was an explosion of natural history publishing, especially that focusing on the exotic. The assortment of accounts of pangolins by travelers and officials in various colonial locations described earlier never really made it into the authoritative natural history works of the 17th and 18th centuries. Rather, most scholarly writers seem to have only discussed the skins that they came across in Europe. What is now understood as a pangolin was first included in a European natural history work in 1605, in the *Exoticorum libri decem* (Ten books of exotics) by the Professor of Botany at Leiden University, Carolus Clusius. He described the

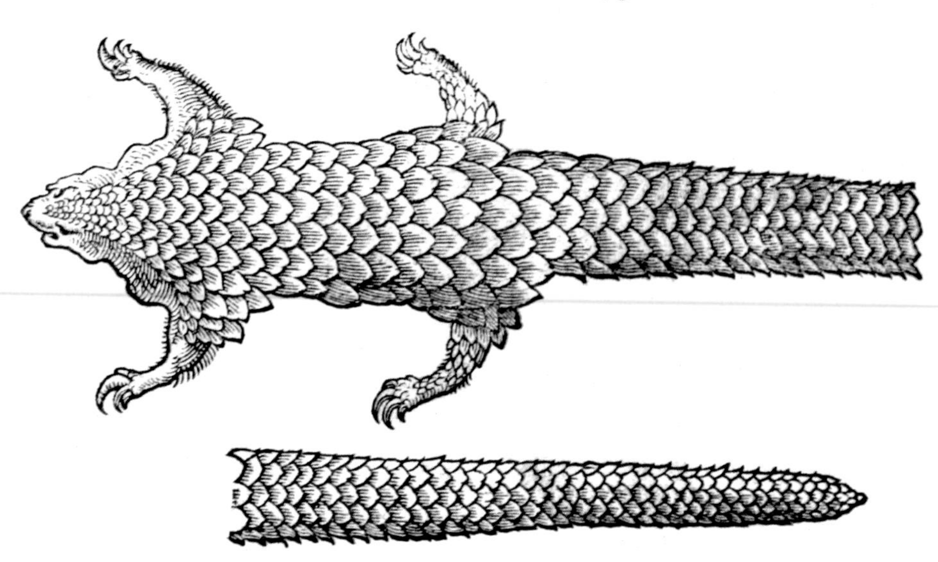

FIGURE 13.2 "Lacertus Peregrinus squamosus" or "foreign scaly lizard" from Carolus Clusius's *Exoticorum* (1605, bk. V appendix, p. 374). This is clearly a careful depiction of a skinned specimen but little is known about its identity or provenance.

skin of a *Lacertus peregrinus squamosus* (foreign scaly lizard) with a foot-long body and extraordinary two and a half foot-long tail, shown to him by a friend of his, the eminent collector, Christian Porrett (Fig. 13.2).

The erudite Clusius was certain that nothing similar had been described in a natural history before, so took great pains to describe, in exquisite detail, how it had been covered on its upper side with pointed scales of variegated colors, and its forefeet had long hooked claws. Porrett had no idea "from whence the skin of the lizard was originally brought", so "because of its rarity", he kept it "among his miscellaneous exotica." Another correspondent had sent two scales from a similar skin and Clusius had been told of another skin owned by a trader in "foreign goods" in Amsterdam. Beyond these detailed descriptions of pangolin skins, Clusius had no other material on their provenance or about this strange creature (Clusius, 1605).

The second major 17th century text dealing with this scaly lizard was Jacobus Bontius's *Historiae naturalis*. Two pangolins were described in this volume: the *tamach* described earlier, as well as *"De Lacerto Indico squamoso"* ("Scaly Indian Lizard"; Fig. 13.3). The latter was described as a "disemboweled" specimen from *Insulae Tajoán* ("Island of Taiwan") that lacked a vernacular name. This lizard-creature frequented woods and had rigid scales that it raised when aggravated. The Dutch called it the "devil of Taiwan", whose skin had a "horrible form" (Bontius, 1658). It tore apart ant nests aggressively with its clawed feet to feed, like the Brazilian *tamandoá* (anteater), and was a dish prized by the natives, like those other "great lizards of Brazil", the *leguánae* (iguana) and *tatu* (armadillo). The *Lacerto Indico* is not present in Bontius's original manuscripts in the Sherard Collection. It was included in an appendix of the *Historiae*, so this creature is most probably a later addition by Willem Piso, from an unknown source, during his production of *Indiae utriusque re naturali et medica* (Bontius, 1630; Bontius, 1658; Cook, 2007).

FIGURE 13.3 "Lacertus squamosus" from Jacobus Bontius's *Historiae Naturalis* (1658 bk. V appendix, p. 60). This image seems to have been drawn from a skinned specimen, perhaps even second-hand, judging by the posture of the animal and scaly face.

The double entry of the *tamach* and the "devil of Taiwan" in the *Historiae naturalis* demonstrates the ambiguity of the scaly lizard's identity. The text and images from Clusius and Piso's publications were used in many subsequent publications, including the catalogues for collections in which "scaly lizard" skins were described. For example, the catalogue of Basil Besler's collection and Adam Olearius's catalogue of the Castle Gottdorf cabinet both used Clusius's image, whilst some catalogues depicted new examples (Lochner and Lochner, 1716; Olearius, 1666). These ambiguous, scaly quadrupeds were scattered around European natural history texts, travelogues, catalogues and taxonomies, but little work was done to amalgamate these into a cohesive understanding.

Classifying the scaly mammals

The new objects that arrived in Europe in the 16th and 17th centuries sometimes posed a challenge to traditional European ways of seeing the world. They also had to be integrated into European classification systems, which could prove difficult. From the Middle Ages, animals had been divided using the classical Aristotelian categories: viviparous quadrupeds, birds, whales, oviparous quadrupeds and fishes. Creatures such as pangolins that transgressed the boundaries between these groups - possessing both scales and hair, with apparently semi-aquatic natures, myrmecophagia, and viviparity - required a shift in the ways the natural world was ordered. There were two aspects of classification of the "scaly lizard" that caused confusion for the scholars of early modern Europe. First, was there any difference between the scaly lizards of the East and the shelled armadillos of the West? Second, where did these creatures fit in the grander scheme of nature?

The first question - were pangolins the same type of animal as the more familiar armadillo, or *tatu*? As has been discussed, specimen provenance was often indistinct or missing entirely. One complicating factor was that the West Coast of Africa was an important transshipping center for the trade routes from both the East and West Indies, especially for goods from the Atlantic circulation. This could cause significant confusion about origins, as it certainly did with many other kinds of natural object. Often, merchants and sailors simply neglected to collect this information: strange scaly beasts were easy to confuse, especially through incomplete textual references (Smith, 2007).

In the many taxonomic works of the late 17th and early 18th centuries, and in some collection catalogues, armadillos and pangolins became explicitly conflated. As the number of armadillo types increased in later works of classification and collection catalogues, so did references to them occurring in the East Indies and Africa. Similarly, the ascribed origins of pangolins were expanded from the Orient to encompass the Occidental Indies. In some publications, the two animals were described indiscriminately. For example, in Robert Hubert's 1664 collection catalogue, he mentioned armadillos from the East and West Indies, and a "B[...]gelugey", "a creature of some parts of Africa, a kind of Lizard, that hath great scales like a fish", possibly a pangolin (Hubert, 1664).

Similarly, in Seba's luxurious catalogue (1734), his specimens of the *tatu d'Afrique* and *tatu Orientalis* were described. The two pangolin descriptions in Seba's catalogue, not only treated two animals that looked very different as one, but also said they were called "*tatoe* by the Brazilians" and "*armadillo*" by the Spanish. Jacobus Theodore Klein's taxonomy (1751) divided these "armoured and hairy" things into three: the *tatu mustelinus* (weasel-tatu) or *vivera cataphracta* (armoured ferret) or *diabolus Tajovanicus* from Taiwan; the *porcellus cataphractus* (armoured pig) or common *tatu* from the Americas and Orient; and the *tatu caninus* or *cynocephalus* from Africa. John Hill's *History of Animals* (1752) described seven types of *dasypus* (armadillo), from Africa, the East Indies, and South America. He also included the "scaly lizard" from both the East Indies and South America. Other taxonomists, such as Mathurin-Jacques Brisson in *Regnum Animale* (1762), described pangolins from Brazil, Formosa, Java and other specific locations, or else omitted the pangolin as an individual creature and described types of armadillo also called a "devil of Taiwan" or scaly armadillo.

The division or conflation of the eastern pangolins and western armadillos could be more than just a point of natural history, it often concerned political biogeographies, where creatures were intentionally grouped together to fulfill a particular agenda. In *De Indiae utriusque* (1658), Piso altered an original description of the armadillo written in 1648 in Dutch Brazil. Piso's editorial additions asserted that this "animal of extraordinary form is distributed not only in the Occidental, but also in the Oriental regions" (Piso, 1658; Piso and Marcgrave, 1648). Piso's publication was a generalized biogeography that created a metaphorical connection between East and West Indies: mapping the reach of the global Dutch trading enterprise by amassing the exotic riches from Dutch colonies into one volume. Many other Dutch publications at this time produced similar volumes that presented a generalized exotic, a "hodgepodge of all things global" (Schmidt, 2011; Schmidt, 2015).

The conflations present in many other taxonomic works were criticized by the eminent French naturalist, Georges-Louis Leclerc, Comte de Buffon (1707–88) in his *Histoire Naturelle* (1749–88, 36 vol.). Buffon distinguished between the pangolin or *manis*, and the armadillo as Old and New World creatures. He determined two types of *manis:* the short-tailed *manis* or *pangolin* and the smaller long-tailed *manis* or *phatagin* that originated in Africa and the East Indies. Buffon argued that other taxonomists relied too heavily on the descriptions by collectors such as Seba. They were often misinformed about specimen origins, using the often-unreliable purchase history of items, or accounts from travelers who ignorantly confused armadillos and pangolins. He posited that the name "scaly lizard" was overly ambiguous. He criticized Piso in particular for asserting "without any authority, that the armadillos were found in the East Indies." Buffon's project was essentially opposed to that of Piso's: to distinguish and differentiate the biogeographies of the Old and New World (Buffon, 1797; Lawrence, 2015b).

Apart from the question of whether the pangolin and armadillo were interchangeable, where exactly should such scaly beasts be placed in the wider scheme of nature? The hierarchical Great Chain of Being was the traditional structure that ordered all of Creation, starting with inanimate matter at the bottom and ascending through the living beings to angels and God at the very top. Fish, reptiles and invertebrates were somewhere near the bottom, whilst mammals and birds were higher in the Chain, all joined by infinite linking forms (Lovejoy, 1964).

Were pangolins reptiles or mammals? It seemed they were somewhere in between, so naturalists depicted them as liminal beasts on the border of these two categories. In amongst descriptions of various lizards in his *Historiae animalium* (1554), Conrad Gessner described "animals similar to serpents" in "Oriental India" with four feet and very long tails. These were similar to the *Hyuana* and *Bardato* (armadillo) of Occidental India, themselves "four footed serpents." In Aldrovandi's 1645 *De quadrupedibus digitatis*, an original woodcut of his own scaly *Lacertus Indicus* (Indian lizard) specimen, a pangolin skin, was inserted amongst the exotic lizards, described as a "congener of the *Iguana*." Similarly, Seba's catalogue pictured one pangolin extruding its long "reptilian" tongue and another in the company of several snakes (Seba, 1734).

Pangolins were also included amongst reptiles in many works of taxonomy. For example, in the *Synopsis animalium quadrupedum* (1693) by the English naturalist John Ray, the *lacertus peregrinus* ("foreign lizard") was classified amongst the egg-laying reptiles or "oviparous quadrupeds", "on account of its form." Others made a special role for them as a liminal group. Buffon had distinguished pangolins from the "lizards" because of their lack of scales on "throat, breast or belly" and the presence of hair or smooth skin on their underparts. Yet Buffon also described how "essential differences" distinguished the pangolin and armadillo "from all other quadrupeds" to such a degree that they were "an intermediate class betwixt the quadrupeds and reptiles" (Buffon, 1797). Even in the late 18th century, the naturalist Thomas Pennant described how "these animals approach so nearly the genus of Lizards, as to be the links in the chain of beings which connect the proper quadrupeds with the reptile clans" (Pennant, 1771). The pangolin remained a hybrid creature, linking disparate forms in the Great Chain.

This confusion is evident in the tortuous reclassification of the pangolin and armadillo in the Swedish naturalist, Carl Linnaeus' *Systema Naturae* (first ed. 1735). In the 1744 fourth edition, Linnaeus used morphological characters to place the *Lacertus squamosus* with the tamandua as the *Myrmecophagia* in Order one, the *Anthropomorpha*, which also included sloths, simians and humans. Seba's *tatu Africanus* and *Orientalis* joined the hedgehog in the *Erinaceus* group in Order two, the sharp-toothed *Ferae*. By the 10th edition (1758) the *manis* was in its own group, on account of its scales and lack of ears, in Order two, *Bruta*. Six types of armadillo - most ascribed a West Indian origin - were grouped as bony-shelled *Dasypus*, in Order four, *Bestiae*. By the 13th edition, the armadillos had been moved into Order two, *Bruta*, with the *manis* (Linnæus, 1744; Linnæus, 1758; Linnæus and Gmelin, 1788–9; Linnaeus [1758] is the taxonomic authority for the genus *Manis* - see Chapter 2).

Colonial monsters

In the early modern period, animals and plants still carried the explicit emblematic significances and moral meanings transmitted in oral culture and recorded in places such as medieval bestiaries or later books of emblems. New beasts that arrived often acquired symbolic meanings of various sorts, frequently linked to the places from which they had come. The pangolin or scaly lizard, brought from the exotic Indies, and encountered in many locations where European powers had strained or troublesome relationships with resident peoples and governments, became a beast that embodied ambivalent colonial relationships in its form and behavior. This meaning is clear from a close analysis of the textual descriptions of these animals.

In 17th and 18th century accounts, pangolins were both devils and innocents. On the one hand, the pangolin could be seen as an impossibly strong, aggressively spiked devil that undermined colonial infrastructures. Some naturalists characterized the animal as a "'devil" on account of its physical form and the "horrible scales" it raised "when aggravated", or because it dug up rice paddies and the foundations of European houses. The animal was like a ballistic projectile when curled up and its iron-like armor could not be pierced by European weapons (Bontius, 1658; Du Hamel, 1701; Valentijn, 1724). Impenetrability was a characteristic often attributed to armadillo armor too: it was an "armed beast" covered in plates "so hard" that "no Arrow can pierce them", the skins of which could be used for making "warlike gauntlets" (Jonston, 1678; Van Linschoten, 1885).

On the other hand, pangolins could be harmless creatures in dire need of protective armor. Des Marchais described the pangolin avoiding danger and eating nothing but troublesome ants. Indeed, he suggested that pangolins could be valuable for the control of insect infestations (Labat, 1730). Buffon described a tame and innocent creature in need of armor, being "fearful" - it used its scales to wound only animals seeking to attack it (Buffon, 1797). In this way, the pangolin's formidable armor became part of a Divine balance, where the vulnerable were given means of protection. The armadillo's shell was similarly an ambivalent characteristic, "usefull both in warre and peace" (Jonston, 1678). It was described by Samuel Purchas as a vulnerable creature, with affinities to the turtle or hedgehog, "creeping through the bryars and bushes" and "not very well able to runne", which required scaly protection for its toothsome white flesh (Purchas, 1625).

Other scholars have explored the production of politicized natural histories in which plants and animals take on the ambiguous natures of humans in colonial regions. The ethnohistorians Michael Dove and Carol Carpenter, for example, analyze mythopraxis in 17th century colonial natural histories of the *upas* tree of the East Indies. In the work of the VOC botanist Georg Eberhard Rumphius (1627–1702), the tree was the deadly "poison tree", emblematic of the colonial struggles in the region with both indigenous people and nature. Changes in colonial interactions and perceptions led to concomitant shifts in the portrayal of the *upas* (Dove and Carpenter, 2005). These were political biogeographies: exotically rich but potentially dangerous places which produced beasts that were both edible innocents and warlike; plants that could be medicinal or lethal.

These moral attributions were not insignificant: they embodied colonial anxieties and prejudices. Flora and fauna came to stand for the predicaments of colonizers, and details of their features took on the characters of regional political dilemmas. Recent scholarship illustrating the situation of the Dutch in Asia during this period makes it possible to consider the moral attributes of the pangolin alongside the struggles experienced by traders and colonists in different geographical locations. The Dutch base in Taiwan in the 1620s and 1640s, in particular, was a crucial but precarious trading position for which the Portuguese and Chinese powers also contended (Dove and Carpenter, 2005; Parthesius, 2010; Van Dyke, 1997; Weststeijn, 2014).

The *dyvel* that Valentijn described, undermining the Dutch in their attempts to claim Taiwan by digging up fields and the foundations of their buildings, was another obstacle against which they had to battle violently. The sharp-clawed East Indian *pangoelling* was a threatening beast that undermined floors and houses in the Dutch colonies with its incredibly rapid digging. It had shield-like scales on its hide that provided the raw materials from which opposing forces such as the Chinese and Javanese could create armor. Valentijn described how the Chinese used pangolin scales to make "weapon and armor skirts" [*Pantsiers en Wapenrokken van te maken*] (Valentijn, 1724).

Like Rumphius's deadly *upas* or "poison tree" that tipped the lethal darts of Ambonese natives, this timid creature's defenses supplied the armory of the forces opposing Dutch operations. For this reason, it was called the *"Ceylonsche Duyvel"* and in Taiwan, the *"Taywansche Duyvel"*, reflecting the perceived threats to Dutch attempts to obtain and maintain this important Indian Ocean trading base from the 1630s. The pangolin's ambiguity, perhaps, expressed colonial aggression and righteous European mastery experienced by the Dutch, as well as a dichotomy between the ferocity of ambitious colonizers and the fear of the potentially potent colonized.

Conclusion

Pangolins were common trade items throughout Africa and Asia in the 16th and 17th centuries, but their skins remained ambiguous and surprising when they passed through the global trade networks to the collectors and naturalists of early modern Europe. The living animals bemused European travelers no less. Though seemingly at odds with traditional taxonomic categories, the "scaly devil" was integrated into the European knowledge systems through a process of emblematisation and shifts in extant classification structures. Like many other exotic things, the characterization of pangolins in European natural history embodied both old and new ways of seeing the world in Europe. Pangolins still seem to not quite fit into categories in which the world is organized: they continue to be objects of wonder, surprise and fascination.

Acknowledgments

Thanks to Janice Thomas for English translations from the French and the Cambridge University Latin Therapy Group for help with translations from the Latin.

References

Aldrovandi, U., 1645. De quadrupedibus digitatis viviparis libri tres et de quadrupedibus digitatis oviparis libri duo. Bologna.

Bontius, J., 1630. 'Jacobi Bontii medici arcis ac civitatis Bataviae Novae in Indiis ordinarii Exoticorum Indicorum Centuria prima, 1630'. MS Sherard, 186, 28. Plant Sciences Library, Oxford, Sherard Collection.

Bontius, J., 1658. Historiae Naturalis et Medicae Indiae Orientalis Libri sex. In: Piso, W. (Ed.), De Indiae Utriusque Re Naturali Et Medica, Libri Quatuordecim. Elzevir, Amsterdam.

Bontius, J., 1931. Tropische Geneeskunde/on tropical medicine. In: Andel, M. (Ed.), Opuscula Selecta Neerlandicorum De Arte Medica. Sumptibus Societatis, Amsterdam, no. 10.

Brienen, R.P., 2007. From Brazil to Europe: the zoological drawings of Albert Eckhout and Georg Marcgraf. In: Enenkel, K.A.E., Smith, P.J. (Eds.), Early Modern Zoology: The Construction of Animals in Science, Literature and the Visual Arts. Brill, Leiden and Boston, pp. 273–315.

Brisson, M.-J., 1762. Regnum Animale in Classes IX. Apud Theodorum Haak, Lugduni Batavorum.

Buffon, G.L.L., comte de, 1797. Barr's Buffon. Buffon's Natural History. H.D. Symonds, London.

Clusius, C., 1605. Exoticorum libri decem: quibus animalium, plantarum, aromaticum, aliorumque, peregrinorum fructum historiae describuntur. Plantin Press, Leiden.

Cook, H.J., 2007. Matters of Exchange: Commerce, Medicine and Science in the Dutch Golden Age. Yale University Press, New Haven, Connecticut and London.

Dove, M., Carpenter, C., 2005. The "Poison Tree" and the changing vision of the indo-malay realm. In: Wadley, R. L. (Ed.), Histories of the Borneo Environment: Economic, Political and Social Dimensions of Change and Continuity. KITLV Press, Netherlands, pp. 183–212.

Du Hamel, J., 1701. Regiae Scientiarum Academiae Historia, second ed. Paris.

Gessner, C., 1554. Historiae animalium, vol. 2, Appendix historiae quadrupedum viviparorum et oviparorum. C. Froschoverus, Zurich.

Haupt, H., Vignau-Wilberg, T., Irblich, E., Staudinger, M., 1990. Le bestiaire de Rodolphe II. Cod. Min. 129 et 130 de la Bibliothèque nationale d'Autriche. Paris, fol. 69r.

Hill, J., 1752. An History of Animals. Thomas Osborne, London.

Hubert, R., 1664. A Catalogue of Many Natural Rarities... Collected by Robert Hubert... and Dayly to be Seen at the Place Called the Musick House at the Miter.... Tho. Ratcliffe, London.

Jonston, J. A Description of the Nature of Four-footed Beasts... Translated into English by J.P. Amsterdam, 1678.

Klein, J.T., 1751. Quadrupedum dispositio brevisque historia naturalis. Schmidt, Leipzig.

Labat, J.B., 1730. Voyage du chevalier Des Marchais en Guinée, isles voisines, et à Cayenne, fait en 1725, 1726 and 1727: Contenant une description très exacte and très étendue de ces païs..., vol. 1. G. Saugrain l'ainé, Paris.

Lawrence, N., 2015a. Assembling the dodo in early modern natural history. Br. J. Hist. Sci. 48 (3), 387–408.

Lawrence, N., 2015b. Exotic origins: the emblematic biogeographies of early modern scaly mammals. Itinerario 39 (1), 17–43.

Linnæus, C., 1744. Systema Naturæ in quo proponuntur Naturæ Regna tria secundum Classes, Ordines, Genera & Species. Editio quarta ab Auctore emendata & aucta. Accesserunt nomina Gallica. David, Paris.

Linnæus, C., 1758. Systema Naturæ Per Regna Tria Naturæ, Secundum Classes, Ordines, Genera, Species, Cum Characteribus, Differentiis, Synonymis, Locis. Tomus I. Editio decima, reformata. Salvius, Stockholm.

Linnæus, C., Gmelin, F. (Ed.), 1788-9. The Animal Kingdom, or Zoological System, of the Celebrated Sir Charles Linnæus Mammalia... being a translation of that part of the Systema Naturæ... by Professor Gmelin of Goettingen..., thirteenth ed. G.E. Beer, Leipzig.

Lochner, M.F., Lochner, J.H., 1716. Rariora musei besleriani quae olim Basilius et Michael Rupertus Besleri.... Nuremberg.

Lovejoy, A.O., 1964. The Great Chain of Being. Harvard University Press, Cambridge.

Olearius, A., 1666. Gottorffische Kunst-Cammer.... Johan Holwein, Schlesswig.

Parsons, C., Murphy, K., 2012. Ecosystems under sail, specimen transport in the eighteenth-century French and British Atlantics. Early Am. Stud. 10 (3), 503–539.

Parthesius, R., 2010. Dutch Ships in Tropical Waters: The Development of the Dutch East India Company (VOC) Shipping Network in Asia, 1595–1660. Amsterdam University Press, Amsterdam.

Pennant, T., 1771. Synopsis of Quadrupeds. J. Monk, Chester.

Pieters, F.J.M., 1998. Wonderen der Nature in de Menagerie van Blauw Jan te Amsterdam, zoals gezien door Jan Velten rond 1700. Rare and Historical Books, ETI Digital.

Piso, W., 1658. De Indiae utriusque re naturali et medica, libri quatuordecim. Elzevir, Amsterdam.

Piso, W., Marcgrave, G., 1648. Historia Naturalis Brasiliae. Amsterdam.

Purchas, S., 1625. Hakluyts posthumus, or, Purchas his Pilgrimes. Contayning a history of the world, in sea voyages, & lande-trauells.... W.Stansby, London, part 4, book 7.

Ray, J., 1693. Synopsis methodica animalium quadrupedum et serpentini generis.... London.

Schmidt, B., 2011. Collecting global icons: the case of the exotic parasol. In: Bleichmar, D., Mancall, P. (Eds.), Collecting Across Cultures: Material Exchanges In The Early Modern Atlantic World. University of Pennsylvania Press, Philadelphia, pp. 31–57.

Schmidt, B., 2015. Inventing Exoticism: Geography, Globalism, and Europe's Early Modern World. University of Pennsylvania Press, Philadelphia.

Seba, A., 1734. Locupletissimi rerum naturalium thesauri... et depingendum curavit Albertus Seba. Amsterdam.

Smith, P.J., 2007. On Toucans and Hornbills: readings in early modern ornithology from Belon to Buffon. In: Enenkel, K.A.E., Smith, P.J. (Eds.), Early Modern Zoology: The Construction of Animals in Science, Literature and the Visual Arts. Brill, Leiden and Boston, pp. 75–120.

Tachard, G., 1689. Second voyage du père Tachard et des Jésuites envoyez par le roy au royaume de Siam.... P. Mortier, Amsterdam.

Valentijn, F., 1724. Oud en Nieuw Oost-Indiën. Van Braam, Dordrecht.

Van Dyke, P.A., 1997. How and why the Dutch east India company became competitive in intra-asian trade in east Asia in the 1630s. Itinerario 21 (03), 41–56.

Van Linschoten, J.H., 1885. The Voyage of John Huyghen van Linschoten to the East Indies: From the Old English Translation of 1598.... Hakluyt Society, London.

Weststeijn, A., 2014. The VOC as a company-state: debating seventeenth-century Dutch colonial expansion. Itinerario 38 (1), 13–34.

14

Meat and medicine: historic and contemporary use in Asia

Shuang Xing[1], *Timothy C. Bonebrake*[1], *Wenda Cheng*[1], *Mingxia Zhang*[2], *Gary Ades*[3], *Debbie Shaw*[4] *and Youlong Zhou*[5]

[1]School of Biological Sciences, The University of Hong Kong, Hong Kong SAR, P.R. China [2]Xishuangbanna Tropical Botanical Garden, Chinese Academy of Sciences, Mengla, P.R. China [3]Fauna Conservation Department, Kadoorie Farm & Botanic Garden, Hong Kong SAR, P.R. China [4]IUCN SSC Pangolin Specialist Group, ℅ Zoological Society of London, Regents Park, London, United Kingdom [5]Henan University of Chinese Medicine, Zhengzhou, P.R. China

OUTLINE

Introduction

Wildlife has been used for millennia in Asia (Corlett, 2007; Donovan, 2004); Asian pangolins and their derivatives have been used consumptively in virtually every range country. This primarily entails the consumption of pangolin meat, and the use of scales in traditional medicines, including Traditional Chinese Medicine (TCM), as well as ritualistic,

Pangolins
DOI: https://doi.org/10.1016/B978-0-12-815507-3.00014-9

ornamental, and other uses. This chapter reviews the variety of uses of pangolins and their parts in Asia by sub-region (South, Southeast, and East Asia). Given the diversity of uses, this review is not exhaustive but draws on published and gray literature. Historically, pangolins have been caught and sold as a means of income generation, especially in Southeast Asia but these dynamics (e.g., the contribution selling pangolins makes to household income) are not discussed in detail. Pangolins in Asia are afforded legislative protection in every range country, which prohibits exploitation, though there are some exceptions for customary hunting by indigenous peoples (e.g., India; see D'Cruze et al., 2018). Distinguishing between historic (i.e., is no longer practiced) and contemporary use is challenging but distinctions have been made throughout the chapter where appropriate. The chapter concludes with an evaluation of the impact of exploitation for domestic use on pangolin populations.

South Asia

South Asia, which for the purposes of this chapter includes Bangladesh, Bhutan, India, Nepal, Pakistan and Sri Lanka, is inhabited by the Indian (*Manis crassicaudata*) and Chinese pangolin (*M. pentadactyla*).

Nutritional use

Pangolins in South Asia have been, and continue to be, exploited for their meat. In India there is a long history of eating pangolins for sustenance; typically local communities, including tribal communities, mainly in rural areas, hunt the animals (e.g., with dogs) and either catch them directly or locate their burrows and smoke them out. This is known to occur in the Western Ghats where the Indian pangolin is hunted (Kanagavel et al., 2016; A. Kanagavel, pers. comm.). The Kadars of the Anamali Hills in Kerala, southern India are also reported to catch and consume pangolins, which are apparently a favored food, though reports in the 1990s suggest such use was infrequent (see Anon, 1992). Pangolins have also traditionally been hunted in India as part of the Shikar Utsav hunting festival, held to celebrate Buddha Purnima. The Chinese pangolin has been, and is, hunted for its meat in northeastern India (e.g., Nagaland and Assam; Anon, 1999; D'Cruze et al., 2018). In Sri Lanka, ethno-archeological research indicates that indigenous "vedda" communities used to smoke out Indian pangolins from their burrows to catch them for food (Chandraratne, 2016; Perera et al., 2017). Khan (1984) reported that pangolins were killed in Bangladesh because of their unusual appearance and the meat (and scales) of the animals would be used, presumably consumed. Trageser et al. (2017) report that Mro hunters in the Chittagong Hill Tracts in Bangladesh hunt the Chinese pangolin, but do so specifically for commercial export of the scales. There is evidence that consumption has been replaced with selling pangolins or their parts into illegal trade elsewhere in South Asia, including northeastern India (D'Cruze et al., 2018), Nepal (Katuwal et al., 2015), and Pakistan (Mahmood et al., 2012; see also Chapter 5).

Medicinal use

The scales of both the Indian and Chinese pangolin are used for numerous medicinal applications in India, Nepal, and Pakistan (Mitra, 1998; Mohapatra et al., 2015; Perera et al., 2017; Roberts, 1977; see Table 14.1). In parts of Nepal, local people believe that wearing products made from pangolin scales help women who have reproductive problems, and the uterus of pangolins is known locally as "garvello" and is believed to prevent miscarriages (Kaspal, 2009). Finger rings fashioned from scales are worn in India and can be

TABLE 14.1 Selected medicinal and other uses of pangolin body parts in South Asia.

Body part	Use	Species	Country	Source
Medicinal				
Scales	Worn as finger rings to treat reproductive problems, prevention of miscarriages	Chinese pangolin	Nepal	Kaspal, 2009
	Worn as finger rings to treat hemorrhoids	Indian pangolin	India	Chinlampianga et al., 2013
	Tied to the lumbar region to relieve back pain	Indian pangolin	India	Bagde and Jain, 2013[a]
	To cure pneumonia		Nepal	Katuwal et al., 2015
Skin and scales	Worn around the neck to help prevent pneumonia	Chinese pangolin	India	Lalmuanpuii et al., 2013
Meat	To relieve stiff muscles	Indian pangolin	India	Chinlampianga et al., 2013
	To treat gastro-intestinal problems, as a pain killer during pregnancy, to relieve back pain		Nepal	Katuwal et al., 2015
Other				
Scales	Placed next to babies baskets to protect them from disease and bad spirits	Chinese pangolin	Nepal	Kaspal, 2009
	To avoid danger		Nepal	Katuwal et al., 2015

[a]*Bagde, N., Jain, S., 2013. An ethnozoological studies [sic] and medicinal values of vertebrate origin in the adjoining areas of Pench National Park of Chhindwara District of Madhya Pradesh, India. Int. J. Life Sci. 1 (4), 278–283.*

found in local markets, and it is believed by some local tribes that they can treat hemorrhoids (Chinlampianga et al., 2013; Mohapatra et al., 2015; Fig. 14.1). Some ethnic groups believe that wearing the skin and scales from a Chinese pangolin can help prevent pneumonia (Lalmuanpuii et al., 2013; Mohapatra et al., 2015). In Nepal, scales have also been placed next to baby baskets to protect babies and young children from disease and bad spirits, and in parts of eastern Nepal the scales are treated as a symbol of good luck (Kaspal, 2009; Katuwal et al., 2015).

Other uses

Pangolins and their parts have a range of other uses in South Asia, including for cultural and religious reasons. In the "Nyishi" tribe in the eastern Himalayas, pangolins are believed to be the children of "Dojung-Buru", a master-spirit, meaning that they are treated as sacred animals. This may discourage hunting and consumption (see Aisher, 2016) since it is believed that hunting and killing a pangolin may lead to misfortune and tragedy for those disrespecting the master-spirits (see also Chapter 12; Aisher, 2016).

Southeast Asia

The Sunda (*M. javanica*) and Chinese pangolins are native to Southeast Asia, which here encompasses Brunei Darussalam, Cambodia, Indonesia, Lao People's Democratic Republic

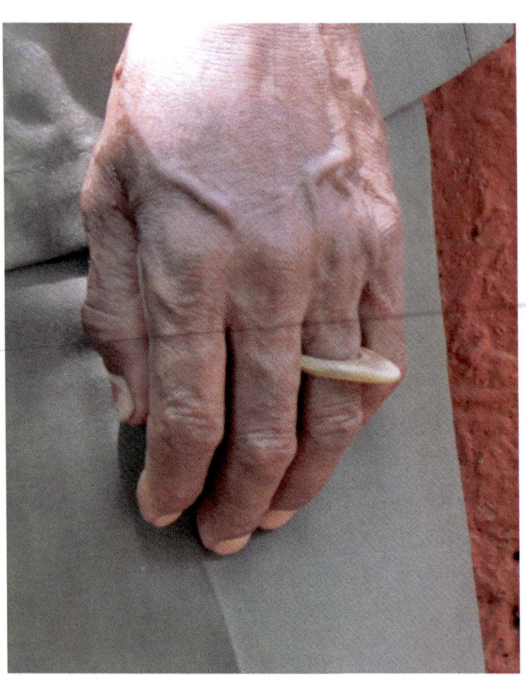

FIGURE 14.1 Pangolin scales are used to make finger rings in South Asia which are believed to treat medical conditions. *Photo credit: Rajesh Kumar Mohapatra.*

(Lao PDR), Malaysia, Myanmar, Singapore, Thailand and Vietnam. This section also includes the Philippines, which is inhabited by the endemic Philippine pangolin (*M. culionensis*).

Nutritional use

Across Southeast Asia, pangolins have historically been extensively exploited for their meat (Corlett, 2007). The Sunda pangolin has been reported as a favorite food of the Dayak in Sabah (Malaysian Borneo) and the "Orang Asli" in Peninsular Malaysia (Anon, 1992; Harrisson and Loh, 1965). Puri (2005) reported that the species may be eaten by the Penan Benalui in Sarawak (Malaysian Borneo), if encountered opportunistically. In Indonesia, the species has historically been consumed as a source of protein, and local people in Sumatra and Java believe that the flesh can cure skin diseases (see Anon, 1999). The Philippine pangolin has traditionally been hunted for subsistence purposes by, among others, the Pala'wan tribe (Acosta-Lagrada, 2012), though the species is increasingly rare (see Chapter 7). In Vietnam, both Chinese and Sunda pangolins have historically been exploited for their meat as well as scales, which have been consumed locally and traded to towns and cities (Sterling et al., 2006). As in parts of South Asia, the high price pangolins fetch along trade routes from the forest gate to end consumer means that local use is being substituted with selling the animals into illegal trade, and this appears to be widespread (MacMillan and Nguyen, 2014; Nuwer and Bell, 2014; see Chapter 6), though some local use continues to take place (e.g., close to protected areas). Consumption in urban centers takes place in high-end restaurants where pangolins are sold for high prices. With the exception of bear paw, pangolin meat was the most expensive wild meat in restaurants frequented in Ho Chi Minh City (HCMC) in 2013 (Challender et al., 2015), and in 2018 was offered for sale for USD 700/kg in HCMC (D. Challender, unpubl. data). Research conducted in 2016 suggests that >50% of pangolin meat consumers in Vietnam have eaten it because they like the taste (Save Vietnam's Wildlife, unpubl. data). Pangolins are usually ordered whole, priced by weight (Dang et al., 2009), and the animals may be killed in front of consumers in the restaurant; the blood may be mixed with alcohol and drunk before the meat is served (Dang et al., 2009). Pangolin meat is typically served boiled, grilled or stir-fried (Fig. 14.2). Some reports suggest that consumers of pangolin meat in Vietnam consider it to have medicinal properties (Dang et al., 2009; Vo, 1998), but the motivations for consumption are not well understood (see Chapter 22).

In Brunei Darussalam, the majority of the human population do not hunt or consume wildlife, including pangolins, because of their Muslim faith (wildlife is perceived as unhealthy), while many ethnic minority groups consider wild animals to be sacred and hunting and eating them is taboo (Nyawa, 2009).

Medicinal use

Chinese, Sunda and Philippine pangolin scales are used for a range of medicinal applications in Southeast Asia. In northern Myanmar, they have been used to treat stomach aches and skin rashes, and single scales are used to make necklaces to afford the wearer protection from disease (P. Aung and A.M. Phyoe, pers. comm.). Similar uses have been reported in Indonesia and Malaysia (see Anon, 1992, 1999; Pantel and Anak, 2010). An ethno-biological survey of indigenous peoples including chieftains at Barangay Malis in southern Palawan, Philippines revealed that local people believed that elders can avoid prostate-related illnesses by wearing belts made with pangolin scales: they also believed that drinking pangolin blood can bring good health (Acosta-Lagrada, 2012).

FIGURE 14.2 Pangolin meat on a menu in Vietnam. *Photo credit: Daniel W.S. Challender.*

In Vietnam, pangolin scales are used as an ingredient in Traditional Vietnamese Medicine (TVM). The official Vietnamese pharmacopeia includes scales ("xuyên sơn giáp" in Vietnamese) as an ingredient in medicines to stimulate energy, improve blood circulation, for treating ulcers, and promoting milk secretion in lactating women. Depending on the specific use, scales are grilled and ground into a powder, which may be taken with wine; burnt, powdered and applied directly to the skin; or mixed with other ingredients in water and ingested. Dang et al. (2009) report that scales are also used to treat other ailments including malaria and rheumatism. The use of pangolin derivatives in Vietnam is prohibited (see Challender and Waterman, 2017) but the scales may still be found for sale illegally in traditional medicine shops in the country (Challender et al., 2015).

Other uses

Scales and other body parts are used for a range of additional uses in Southeast Asia. This includes the tongue which is used as protection from harmful magic in parts of Indonesia (Anon, 1999) and the scales which have been used as protection from witchcraft in Malaysia (Sabah and Sarawak; Hoi-Sen, 1977). In Cambodia, local people put pangolin scales on children's necks and wrists to protect them from bad spirits (Namyi and Olsson, 2009). The Pala'wan in the Philippines have used scales to gain protection from witchcraft, and believe that burning pangolin scales can stop thunder, which may bring disease (Acosta-Lagrada, 2012). In Brunei Darussalam, ethnic tribes have historically used Sunda pangolin scales for making armor and for decoration (Nyawa, 2009).

East Asia

This section discusses use in East Asia including in China, Hong Kong Special Administrative Region (hereafter "Hong Kong"), Taiwan, and South Korea. Only the Chinese pangolin is native to East Asia (see Chapter 4).

Nutritional use

The Chinese pangolin has been used in parts of East Asia for more than a thousand years, including for the consumption of its meat, especially in China. In *Ji Lei Bian* ("Chicken Rib Chronicles"; 1139 CE), the custom of "Xin Zhou" (in present day Jiangxi Province) was to cook pangolin meat in lees (dead yeast and debris) from fermented rice wine, and the meat was sold as street food during the winter. In the Ming dynasty (1368–1644) a popular recipe, *Zhu Yu Shan Fang Za Bu* ("A recipe of Zhu Yu Mountain Village"; 1504 CE), involved marinating a pangolin in salt for two days before boiling it in water. In the Qing dynasty, Miao people in southwestern China commonly dried and smoked pangolin meat for later use (Wang, 2000). Influenced by TCM, some populations in southern China have treated pangolins as a "hot" food historically, preferring to eat it during the winter to remove the "chill" from the body (see Anon, 1992). Allen (1938) reports seeing live pangolins in markets in Canton (now Guangzhou) in winter in 1922. Similarly, the meat has been consumed with perceived "tonic" qualities, in particular in southern China (e.g., Guangdong, Guangxi, Fujian and Hainan Provinces; Wu et al., 2002), and some believe that the meat is good for the kidneys and can help remove heat and toxins from the human body (Zhang, 2009). Coggins (2003) reports that in Fujian Province the meat is sold with pieces of pangolin tongue, which is believed to have the strongest medical effect. Chinese pangolins were also favored by Chinese and aboriginal peoples in Taiwan during the mid-late 20th century and sold at restaurants as "wild game meat", but there appears to be little contemporary use for this purpose (Anon, 1992; see Chapter 36).

There is little in-depth knowledge of the motivations of contemporary pangolin meat consumers in China (see Chapter 22). Zhang (2009) reports that the meat is eaten as a delicacy and "tonic" food, while Zhang and Yin (2014) note that consumption is driven by the curiosity of trying "wild flavors" and perceived medicinal benefits. Pangolin meat may be cooked in various ways, including in a stew or soup, usually with Chinese wine and medicinal herbs such as *Ligusticum striatum, Tetrapanax papyriferus, Stemmacantha* spp., and *Akebia* spp., and with other meat including chicken or pork (Jiao, 2013; Yao, 2005). In southern China, smoked pangolin products are occasionally found in local markets in remote rural areas (M. Zhang, pers. obs.). Based on trafficking dynamics, and following declines in China's pangolin populations (see subsequent discussion), modern day consumption includes the Sunda pangolin as well (see Chapter 16).

Meat consumption also takes place in China's border regions, in particular with Lao PDR and Myanmar (Gomez et al., 2016; Zhang et al., 2017). Nijman et al. (2016) report that pangolin meat and wine are frequently found in Mong La in Myanmar, which borders Yunnan Province, and live pangolins are displayed in front of restaurants and casinos. Most of the venders and consumers are Chinese nationals living and working in Mong La (Nijman et al., 2016).

Medicinal use

The first traceable record of pangolins being used for medicinal purposes in China is the ancient Chinese herb book *Ben Cao Jing Ji Zhu* ("Variorum of Shennong's Classic of Materia Medica") in the Liang dynasty (500 CE). According to this source, pangolin scales were believed to cure ant bites, a consequence of the function of the scales protecting pangolins from such bites. This source also introduced the burning of scales as a cure for people crying hysterically during the night. During the Tang dynasty (682 CE), *Qian Jin Yao Fang* ("Supplement to the Formulas of a Thousand Gold Worth") first introduced pangolin scale powder soup with liquor for curing malaria, scales burned to ash with pig fat for curing infections from ant bites, and scales taken with herbs and minerals to expel evil spirits. The book, *Wai Tai Mi Yao* ("Arcane Essentials from the Imperial Library"; 752 CE) appears to be the first text to record the use of scales as an ingredient in TCM for stimulating milk secretion in lactating women, a use that continues to take place. In the Song dynasty (978 CE), *Tai Ping Sheng Hui Fang* ("Formulas from Benevolent Sages Compiled During the Taiping Era") added applications for unblocking blood clots in the human body and promoting blood circulation to the list of benefits pangolin derivatives purportedly provide. In 1131 CE *Chan Yu Bao Qing Ji* ("A Collection of Formulas for Birth and Raise") first applied processing pangolin scales with vinegar before using them as a stimulant for milk secretion (Hu et al., 2012). The use of scales in TCM to unblock blood clots, promote blood circulation and to help lactating women secrete milk remain the major medicinal applications of pangolin scales in China. However, pangolin products also have a wide variety of applications associated with treating gynecological diseases and have been widely used to treat infertility in women. Pills including powdered pangolin scales are considered to be useful in treating blockages of the fallopian tubes to cure infertility in TCM (Zhou and Li, 2010). Scales have also been used to treat ovarian cancer (see Au, 1981), and Yu and Hong (2016) report that they are a common ingredient in TCM to treat symptoms associated with breast cancer and lymphoma (Yu and Hong, 2016). They reportedly have application for the treatment of a broader range of medical conditions (Table 14.2). Scales have also been applied in veterinary practices in China to improve lactation in cattle after parturition and to treat mastitis (Bayin et al., 2009; Hao et al., 2017).

Scales are processed for use in TCM. First, they are sand-roasted, i.e., heated by sand at 230–250°C until they curl up and turn a beige color (Fig. 14.3). Sand with particle diameters between 0.1 and 0.2 cm are preferred (Zhou et al., 2014). The scales may then be soaked in vinegar, washed to remove excess vinegar, and sun-dried. The vinegar-soaked scales are called "Cu Shan Jia." A more recent method is to place scales in a microwave oven until they expand and curl up (Zhou et al., 2014: Fig. 14.3). These scales are referred to as "Pao Shan Jia", and are typically thought of as inferior by TCM practitioners, although the latter process is simpler and less labor intensive, and hence is preferred for industrial production (Cao and Tang, 2002; Zhou et al., 2014). Powdered scales can be further compressed into pills, "Jia Zhu" in Chinese. Scales, in both unprocessed and processed forms, may also be referred to as "Chuan Shan Jia" (穿山甲), the Chinese name for pangolins. When scales are used in TCM, they are typically combined with other herbal ingredients including *Ligusticum striatum*, *Tetrapanax papyriferus*, *Stemmacantha* spp., and *Akebia* spp. to achieve the desired, purported medicinal effect (Jiao, 2013).

The official pharmacopoeia of the People's Republic of China includes Chinese pangolin scales as an ingredient in TCM, and there is a legal market for scales in the country. In 2007 the Chinese government promulgated a notification governing the use of scales. The manufacture of medicines containing scales, and the retail of medicines and scales that have come from government stockpiles is restricted through a certification system designed to ensure that wild Chinese pangolins and their parts do not enter trade. Between 2009 and 2015, the Chinese government released an average of 26.6 tonnes of scales annually from

TABLE 14.2 Other conditions that pangolin scales reportedly have application for in TCM.

Use
Adhesive intestinal obstruction[a]
Asthma[b]
Childhood anorexia[c]
Chronic pharyngitis[d]
Coronary artery disease[e]
Frozen shoulder[f]
Goiter[g]
Growing pains[h]
Hemorrhoids[i]
Hyperlipidemia[e]
Leukopenia[j]
Malaria[k]
Osteomyelitis[l]
Pain relief[m]
Parkinson's disease[n]
Pinworm infection[o]

[a]Zhao, L., 2004. Pangolin scales cure adhesive intestinal obstruction. Shandong J. Traditional Chin. Med. 23 (12), 758–759. [In Chinese]. [b]Wang, J.P., Hu, M.L., 2012. Forty-two cases of "Er long ma xing tang" soup with modern medicine treat asthma. Traditional Chin. Med. Res. 25 (11), 28–29. [In Chinese]. [c]Sun, S., 2002. Pangolin scales cure children anorexia. J. Traditional Chin. Med. 43 (2), 95–95. [In Chinese]. [d]Chen, S., 2002. Pangolin scales cure chronic pharyngitis. J. Traditional Chin. Med. 43(2), 92–92. [In Chinese]. [e]Fan, X., 2002. Pangolin scales cure coronary artery disease and Hyperlipidemia. J. Traditional Chin. Med. 43 (4), 252. [In Chinese]. [f]Ren, L., Zhou, H., 2005. Pangolin scales cure frozen shoulder. Nei Mongol J. Traditional Chin. Med. 3, 7. [In Chinese]. [g]Li, C., 2002. Pangolin scales cure goitre. J. Traditional Chin. Med. 43 (4), 253. [In Chinese]. [h]Ma, J., 2002a. Pangolin scales cure growth pain. J. Traditional Chin. Med. 43 (2), 95. [In Chinese]. [i]Ma, J., 2002b. Pangolin scales cure hemorrhoid. J. Traditional Chin. Med. 43 (4), 254. [In Chinese]. [j]Zheng, M., 2002. Pangolin scales cure Leukopenia. J. Traditional Chin. Med. 43 (4), 252. [In Chinese]. [k]Chen, D.Q., 2002b. Pangolin scales treat malaria. J. Traditional Chin. Med. 43 (2), 92. [In Chinese]. [l]Wang, X., 2002. Pangolin scales cure osteomyelitis. J. Traditional Chin. Med. 43 (2), 95. [In Chinese]. [m]Wu, S., Nong, C., He, X., Chen, Y., Lu, Q., Wei, J., 2012. Study of water abstraction of pangolin scales on pain relieving. Guangxi Med. 34 (1), 7–9. [In Chinese]. [n]Jin, D., 2002. Pangolin scales cure Parkinson's disease. J. Traditional Chin. Med. 43 (4), 252. [In Chinese]. [o]Yin, Q., 2002. Pangolin scales cure enterobiasis. J. Traditional Chin. Med. 43 (4), 253. [In Chinese].

FIGURE 14.3 Unprocessed Sunda pangolin (*M. javanica*) scales (right) and scales processed for use in TCM/TVM (left) for sale in Vietnam. *Photo credit: Daniel W.S. Challender.*

stockpiles for the manufacture and retail of TCM (China Biodiversity Conservation and Green Development Foundation, 2016). Certification includes the use of stickers on packaging (see Fig. 14.4), and certified medicines containing scales are restricted to use and sale in 716 designated hospitals in China (China Biodiversity Conservation and Green Development Foundation, 2016). However, there appears to be substantial demand for pangolin scales and medicines containing them in China, and uncertified scales are widely available illegally in pharmacies, traditional medicines shops and unlicensed hospitals (Xu et al., 2016; Wu and Ma, 2007).

Several potential substitutes have been proposed for pangolin scales in TCM. This includes pig hooves (Liu et al., 2002), horns from Bovidae and Cervidae species (Luo et al., 2011), and dried seeds of the cowherb plant (*Vaccaria segetalis*; Wang, 2009). Other herbal ingredients with similar purported medical effects include *Paris polyphila, Ligusticum striatum, Salvia miltiorrhiza,* and *Spatholobus suberectus* stems. Wang (2009) reported that research in Taiwan in the 1990s revealed that despite agreement with a ban on the use of scales in TCM by doctors, the deep-rooted nature of the practice meant they had reservations about the effectiveness and use of substitutes.

Pangolin scales have also been used in Traditional Korean Medicine in South Korea, and large quantities of scales were imported to the country during the 1980s and 1990s (see Anon, 1992, 1999). However, there is little documented knowledge of contemporary use.

Other uses

Other uses of pangolins include for personal protection. In Hainan Province, China, amulets composed of a piece of pangolin scale are worn by children to protect them from evil spirits (Nash et al., 2016). In Hong Kong, people wear charms made of four pangolin

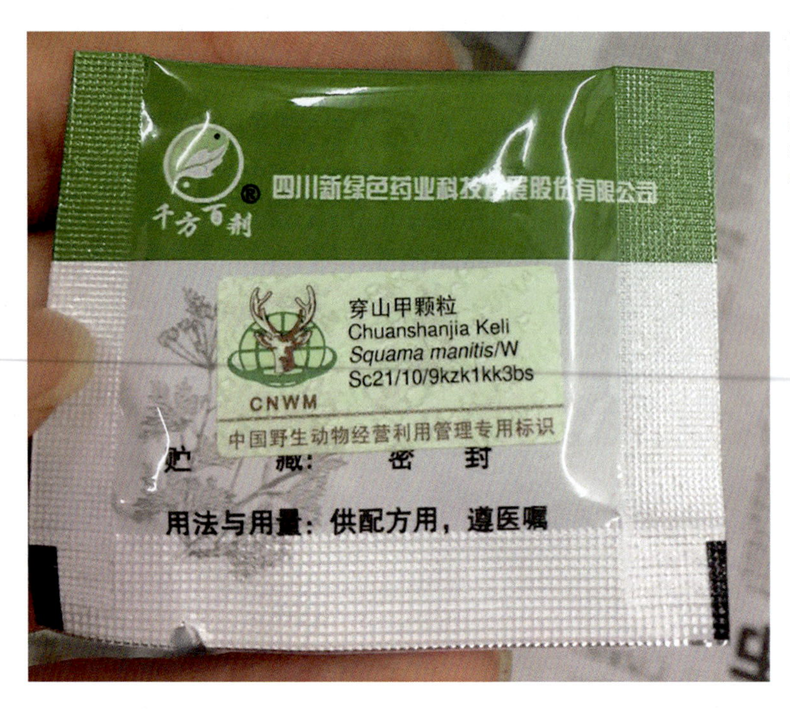

FIGURE 14.4 Certified traditional medicine containing pangolin scales on sale in China. The sticker on the packaging indicates the medicine contains scales from a certified source. *Photo credit: Shu Chen.*

scales to scare off ghosts (see Anon, 1992). Historically, pangolins were hunted and killed in Hong Kong because it was believed that they could dig up human corpses (Ye, 1985). This likely stems from their nocturnal and fossorial habits (see Chapter 4).

Pangolin use is also influenced by cultural trends. Pangolin claws featured in a popular Chinese fiction book about tomb thieves, "Daomu Biji", which was made into a 2016 film, "Time Raiders." This has apparently led to demand for pangolin claws for use on necklaces, which can be found for sale in southern China and Myanmar (M. Zhang, pers. obs.), though the extent of this demand is unknown.

Impact of local and national use on pangolin populations

There is generally a lack of quantitative data on Asian pangolin populations (Chapters 4–7), and extricating the impact of domestic, from international, use is challenging. However, available evidence indicates that populations have declined, and/or, are declining in many places, in part due to national-level (i.e., domestic) use. There is little knowledge of pangolin populations in South Asia in particular, but the Indian pangolin, which is most widely distributed in the region, is uncommon, rarely observed and occurs at low densities. This suggests that where exploitation does occur, it has a significant impact on populations (see Chapter 5). This is compounded by targeted exploitation for international trafficking (Mahmood et al., 2012). In Southeast Asia, steep population declines were inferred for the Chinese and Sunda pangolins in the 1980s and 1990s as a result of overhunting (Anon, 1992, 1999), inferences that appear correct (see Chapters 4 and 6), in part due to offtake for domestic use. The Sunda pangolin, in particular, is known to have been extirpated from lowland areas (e.g., in Myanmar and Thailand) as a result of consumption associated with human development, and targeted exploitation, including in Cambodia, Lao PDR and Vietnam (Duckworth et al., 1999; Nooren and Claridge, 2001; see Chapter 6). More data is available for East Asia. In China, Chinese pangolin populations declined steeply during the 1960s–1980s because of overexploitation for the consumption of meat and scales, which involved up to 160,000 individuals annually (Zhang, 2009; see Zhang et al., 2010). Although the species remains present, it is suspected to occur at very low densities (see Chapter 4). In Taiwan, exploitation of Chinese pangolins for the domestic leather industry in the mid-20th century involved a minimum estimate of 60,000 individuals annually, which caused populations to decline steeply (Anon, 1992). Positively, populations in Taiwan appear to be recovering (see Chapters 4 and 36). The threats facing Asian pangolins, both from local and national level use, and targeted exploitation for international trafficking, means there is an urgent need to determine the impact of this offtake on populations to inform conservation management.

Conclusion

Pangolins have a long history of use in Asia, having been used consumptively in virtually every range country. This has principally comprised the consumption of meat, with perceived medicinal benefits in East Asia, and use of the scales in traditional medicines, including TCM, among other applications. The evidence base differs between range countries, but indicates that domestic use has contributed to steep population declines in many places, especially Southeast Asia and China. Mitigating the threat of overexploitation requires a range of interventions (see Chapters 17–38), and critically, an understanding of the complex cultural contexts in which pangolins are used across Asia.

References

Acosta-Lagrada, L.S., 2012. Population density, distribution and habitat preferences of the Palawan Pangolin, *Manis culionensis* (de Elera, 1915). M.Sc. Thesis, University of the Philippines Los Baños, Laguna, Philippines.

Aisher, A., 2016. Scarcity, alterity and value: decline of the pangolin, the world's most trafficked mammal. Conserv. Soc. 14 (4), 317–329.

Allen, G.M., 1938. The Mammals of China and Mongolia. Natural History of Central Asia, vol. XI. Part I. The American Museum of Natural History, New York.

Anon, 1992. Review of Significant Trade in Animal Species included in CITES Appendix II, Detailed Review of 24 Priority Species, Indian, Malayan and Chinese Pangolin. CITES, Geneva, Switzerland.

Anon, 1999. Review of Significant Trade in Animal Species Included in CITES Appendix II, Detailed Review of 37 Species. *Manis javanica*. World Conservation Monitoring Centre, IUCN Species Survival Commission and TRAFFIC, Cambridge, UK.

Au, B., 1981. Reports of eight cases of using pangolin scale to treat ovarian cancer. Jiangxi J. Traditional Chin. Med. 3, 35. [In Chinese].

Bayin, J., Matsumoto, M., Islam, M.S., Yabuki, A., Kanouchi, H., Oka, T., et al., 2009. Promoting effects of Chinese pangolin and wild pink medicines on the mammary gland development in immature mice. J. Vet. Med. Sci. 71 (10), 1325–1330.

Cao, Z., Tang, J., 2002. Comparison between different methods in preparing pangolin scales. J. Henan Coll. Traditional Chin. Med. 17 (6), 24. [In Chinese].

Challender, D.W.S., Harrop, S.R., MacMillan, D.C., 2015. Understanding markets to conserve trade-threatened species in CITES. Biol. Conserv. 187, 249–259.

Challender, D., Waterman, C., 2017. Implementation of CITES Decisions 17.239 b) and 17.240 on Pangolins (*Manis* spp.), CITES SC69 Doc. 57 Annex. Available from <https://cites.org/sites/default/files/eng/com/sc/69/E-SC69-57-A.pdf>. [August 2, 2018].

Chandraratne, R., 2016. Some ethno-archaeological observations on the subsistence strategies of the veddas in Sri Lanka. Soc. Aff. 1 (4), 33–44.

China Biodiversity Conservation and Green Development Foundation, 2016. An Overview of Pangolin Data: When Will the Over-Exploitation of the Pangolin End? Available from: <http://www.cbcgdf.org/English/NewsShow/5011/6145.html>. [March 19, 2019].

Chinlampianga, M., Singh, R.K., Sukla, A.C., 2013. Ethnozoological diversity of Northeast India: Empirical learning with traditional knowledge holders of Mizoram and Arunachal Pradesh. Indian J. Traditional Knowledge 12 (1), 18–30.

Coggins, C., 2003. The Tiger and the Pangolin: Nature, Culture, and Conservation in China. University of Hawaii Press, Honolulu.

Corlett, R.T., 2007. The Impact of Hunting on the Mammalian Fauna of Tropical Asian Forests. Biotropica 39 (3), 292–303.

D'Cruze, N., Singh, B., Mookerjee, A., Harrington, L.A., Macdonald, D.W., 2018. A socio-economic survey of pangolin hunting in Assam, Northeast India. Nat. Conserv. 30, 83–105.

Dang, N.X., Tuong, N.X., Phong, P.H., Nghia, N.X., 2009. The Pangolin Trade in Viet Nam. TRAFFIC Southeast Asia, Unpunished report.

Donovan, D.G., 2004. Cultural underpinnings of the wildlife trade in Southeast Asia. In: Knight, J. (Ed.), Wildlife in Asia. Cultural Perspectives. Routledge Curzon, London, pp. 88–111.

Duckworth, J.W., Salter, R.E., Khounboline, K., 1999. Wildlife in Lao PDR: 1999 Status Report. IUCN, Wildlife Conservation Society, Centre for Protected Areas and Watershed Management, Vientiane, Lao PDR.

Gomez, L., Leupen, B.T.C., Heinrich, S., 2016. Observations of the Illegal Pangolin Trade in Lao PDR. TRAFFIC, Southeast Asia Regional Office, Petaling Jaya, Selangor, Malaysia.

Hao, J., Li, J., Yin, B., Sha, W., Liu, F., Li, D., et al., 2017. Effects of Chinese herbal compound formula on LDH, NAG activity in serum of dairy cows with recessive mastitis. Heilongjiang Anim. Sci. Vet. Med. 24, 52. [In Chinese].

Harrisson, T., Loh, C.Y., 1965. To scale a pangolin. Sarawak Museum J. 12, 415–418.

Hoi-Sen, Y., 1977. Scaly anteater. Nat. Malays. 2 (4), 26–31.

Hu, X., Wen, C., Xie, Z., 2012. History and application of pangolin scales. Chin. Arch. Traditional Chin. Med. 30 (3), 590–591. [In Chinese].

Jiao, H., 2013. Combination of TCM and modern medicine in stimulating milk secretion. Asian-Pac. Traditional Med. 9 (12), 134. [In Chinese].

Kanagavel, A., Parvathy, S., Nameer, P.O., Raghavan, R., 2016. Conservation implications of wildlife utilization by indigenous communities in the southern Western Ghats of India. J. Asia-Pac. Biodivers. 9 (3), 271–279.

Kaspal, P., 2009. Saving the Pangolins: Ethnozoology and Pangolin Conservation Awareness in Human Dominated Landscape. A Preliminary Report to the Rufford Small Grants Foundation. Available from: <https://www.rufford.org/rsg/projects/prativa_kaspal>. [January 17, 2019].

Katuwal, H.B., Neupane, K.R., Adhikari, D., Sharma, M., Thapa, S., 2015. Pangolins in eastern Nepal: trade and ethno-medicinal importance. J. Threat. Taxa 7 (9), 7563–7567.

Khan, M.A.R., 1984. Endangered mammals of Bangladesh. Oryx 18 (3), 152–156.

Lalmuanpuii, J., Rosangkima, G., Lamin, H., 2013. Ethnomedicinal practices among the Mizo ethnic group in Lunglei district, Mizoram. Sci. Vision 13 (1), 24–34.

Liu, Y., Zhai, J., Wang, H., Yang, Z., 2002. Research progress of using pig nails as substitutes of pangolin scales. Hebei Traditional Chin. Med. 24 (8), 624–625. [In Chinese].

Luo, J., Yan, D., Zhang, D., Feng, X., Yan, Y., Dong, X., et al., 2011. Substitutes for endangered medicinal animal horns and shells exposed by antithrombotic and anticoagulation effects. J. Ethnopharmacol. 136 (1), 210–216.

MacMillan, D.C., Nguyen, Q.A., 2014. Factors influencing the illegal harvest of wildlife by trapping and snaring among the Katu ethnic group in Vietnam. Oryx 48 (2), 304–312.

Mahmood, T., Hussain, R., Irshad, N., Akrim, F., Nadeem, M.S., 2012. Illegal mass killing of Indian pangolin (*Manis crassicaudata*) in Potohar region, Pakistan. Pak. J. Zool. 44 (5), 1457–1461.

Mitra, S., 1998. On the scales of the scaly anteater *Manis crassicaudata*. J. Bombay Nat. Hist. Soc. 95 (3), 495–498.

Mohapatra, R.K., Panda, S., Acharjyo, L.N., Nair, M.V., Challender, D.W.S., 2015. A note on the illegal trade and use of pangolin body parts in India. TRAFFIC Bull. 27 (1), 33–40.

Nash, H.C., Wong, M.H.G., Turvey, S.T., 2016. Using local ecological knowledge to determine status and threats of the Critically Endangered Chinese pangolin (*Manis pentadactyla*) in Hainan, China. Biol. Conserv. 196, 189–195.

Namyi, H., Olsson, A., 2009. Pangolin Research in Cambodia. In: Pantel, S., Chin, S.-Y. (Eds.), Proceedings of the Workshop on Trade and Conservation of Pangolins Native To South and Southeast Asia, 30 June – 2 July 2008, Singapore Zoo, Singapore. TRAFFIC Southeast Asia, Petaling Jaya, Selangor, Malaysia, pp. 172–175.

Nijman, V., Zhang, M., Shepherd, C.R., 2016. Pangolin trade in the Mong La wildlife market and the role of Myanmar in the smuggling of pangolins into China. Glob. Ecol. Conserv. 5, 118–126.

Nooren, H., Claridge, G., 2001. Wildlife Trade in Laos: the End of the Game. Netherlands Committee for IUCN, Amsterdam.

Nuwer, R., Bell, D., 2014. Identifying and quantifying the threats to biodiversity in the U Minh peat swamp forests of the Mekong Delta, Vietnam. Oryx 48 (1), 88–94.

Nyawa, S., 2009. Pangolin in Brunei Darussalam. In: Pantel, S., Chin, S.-Y. (Eds.), Proceedings of the Workshop on Trade and Conservation of Pangolins Native to South and Southeast Asia, 30 June – 2 July 2008, Singapore Zoo, Singapore. TRAFFIC Southeast Asia, Petaling Jaya, Selangor, Malaysia, pp. 25–28.

Pantel, S., Anak, N.A., 2010. A preliminary assessment of Sunda pangolin trade in Sabah. TRAFFIC Southeast Asia, Petaling Jaya, Selangor, Malaysia.

Perera, P.K.P., Karawita, K.V.D.H.R., Pabasara, M.G.T., 2017. Pangolins (*Manis crassicaudata*) in Sri Lanka: a review of current knowledge, threats and research priorities. J. Trop. For. Environ. 7 (1), 1–14.

Puri, P.K., 2005. Deadly dances in the Bornean rainforest: hunting knowledge of the Penan Benalui. Royal Netherlands Institute of Southeast Asian and Caribbean Studies Monograph Series. KITLV Press, Leiden.

Roberts, T.J., 1977. The Mammals of Pakistan. Ernest Benn Ltd, London.

Sterling, E.J., Hurley, M.M., Minh, L.D., 2006. Vietnam: A Natural History. Yale University Press, New Haven, Connecticut and London.

Trageser, S.J., Ghose, A., Faisal, M., Mro, P., Mro, P., Rahman, S.C., 2017. Pangolin distribution and conservation status in Bangladesh. Plos ONE 12 (4), e0175450.

Vo, V.C., 1998. Dictionary of Vietnamese Medicinal Fauna and Minerals. Health Publishing House, Hanoi. [In Vietnamese].

Wang, S., 2000. Ancient delicacy: pangolin for people from the south. Sichuan Cooking 5, 16. [In Chinese].

Wang, G.B., 2009. Conservation of pangolins in Taiwan. In: Pantel, S., Chin, S.-Y. (Eds.), Proceedings of the Workshop on Trade and Conservation of Pangolins Native To South and Southeast Asia, 30 June – 2 July 2008, Singapore Zoo, Singapore. TRAFFIC Southeast Asia, Petaling Jaya, Selangor, Malaysia, pp. 80–83.

Wu, S., Ma, G., Tang, M., Chen, H., Liu, N., 2002. The status of pangolin resources in China and conservation strategies. J. Nat. Resour. 17 (2), 174–180. [In Chinese].

Wu, S.B., Ma, G.Z., 2007. The status and conservation of pangolins in China. TRAFFIC East Asia Newsl. 4, 1–5. [In Chinese].

Xu, L., Guan, J., Lau, W., Xiao, Y., 2016. An Overview of Pangolin Trade in China. TRAFFIC Briefing Report. TRAFFIC, Cambridge, UK, pp. 1–10.

Yao, L., 2005. Best gift for babies. Food Sci. 2, 33. [In Chinese].

Ye, L.F., 1985. Local Animal and Plants in Hong Kong. Joint Publishing, Hong Kong.

Yu, R., Hong, H., 2016. Cancer Management With Chinese Medicine - Prevention and Complementary Treatments. World Scientific Publishing, New Jersey.

Zhang, Y., 2009. Conservation and trade control of pangolins in China. In: Pantel, S., Chin, S.-Y. (Eds.), Proceedings of the Workshop on Trade and Conservation of Pangolins Native To South and Southeast Asia, 30 June – 2 July 2008, Singapore Zoo, Singapore, TRAFFIC Southeast Asia, Petaling Jaya, Selangor, Malaysia, pp. 66–74.

Zhang, L., Li, Q., Sun, G., Luo, S., 2010. Population status and conservation of pangolins in China. Bull. Biol. 45 (9), 1–4. [In Chinese].

Zhang, L., Yin, F., 2014. Wildlife consumption and conservation awareness in China: a long way to go. Biodivers. Conserv. 23 (9), 2371–2381.

Zhang, M., Gouveia, A., Qin, T., Quan, R., Nijman, V., 2017. Illegal pangolin trade in northernmost Myanmar and its links to India and China. Glob. Ecol. Conserv. 10, 23–31.

Zhou, R., Li, H., 2010. Investigating the source of pangolin as a treat for infertility. Bull. Yunnan Chin. Traditional Med. 1, 61–63. [In Chinese].

Zhou, Z., Wang, J., Ma, X., 2014. The research progress of pangolin. Pharmacy Clin. Chin. Mater. Med. 5 (1), 54–56. [In Chinese].

CHAPTER

15

Bushmeat and beyond: historic and contemporary use in Africa

Durojaye Soewu[1,*], *Daniel J. Ingram*[2,*], *Raymond Jansen*[3,4], *Olufemi Sodeinde*[5] *and Darren W. Pietersen*[6,7]

[1]Department of Fisheries and Wildlife Management, College of Agriculture, Ejigbo Campus, Osun State University, Osogbo, Nigeria [2]African Forest Ecology Group, Biological and Environmental Sciences, University of Stirling, Stirling, United Kingdom [3]Department of Environmental, Water and Earth Sciences, Tshwane University of Technology, Pretoria, South Africa [4]African Pangolin Working Group, Pretoria, South Africa [5]Department of Biological Sciences, New York City College of Technology, City University of New York, Brooklyn, NY, United States [6]Mammal Research Institute, Department of Zoology and Entomology, University of Pretoria, Hatfield, South Africa [7]IUCN SSC Pangolin Specialist Group, ℅ Zoological Society of London, Regent's Park, London, United Kingdom

OUTLINE

* These authors contributed equally to this chapter.

Pangolins
DOI: https://doi.org/10.1016/B978-0-12-815507-3.00015-0

Introduction

Humanity has used wildlife since time immemorial for food, clothing, income and medicinal purposes (MacKinney, 1946; Milner-Gulland et al., 2003). Historically, African pangolins have been exploited for bushmeat, both for sustenance (i.e., as a protein source) and a source of income, and for the purported medicinal properties of their scales and other body parts (Boakye et al., 2016; Fa et al., 2006). This has involved all four species, the white-bellied (*Phataginus tricuspis*), black-bellied (*P. tetradactyla*), Temminck's (*Smutsia temminckii*) and giant pangolin (*S. gigantea*). This chapter discusses the many historical and contemporary uses of pangolins in Africa by geographic sub-region (West Africa, Central Africa, East Africa and Southern Africa). These uses broadly fall into three main categories (1) consumptive use as a source of food (i.e., bushmeat), (2) consumptive use of other body parts for medicinal or ethno-pharmacological applications, and (3) other uses (e.g., in spiritual remedies or as omens). The chapter does not provide in-depth detail on local trade dynamics, but uses are discussed in the context that in West and Central Africa in particular, pangolins are hunted or poached, and either consumed at a local (household or village) level, traded or exchanged for other commodities, or smoked and transported to towns and cities and sold in bushmeat markets (Cowlishaw et al., 2005; Ingram et al., 2019; Mambeya et al., 2018). Despite pangolins being protected by legislation in most range states (Challender and Waterman, 2017), though to varying degrees, law enforcement officers are often ill-trained and ill-equipped to detect pangolins, may be unaware that it is illegal to transport such products across borders, and/or may simply turn a blind eye to this illegal trade. The chapter concludes by evaluating the impact of local and national exploitation on pangolin populations.

West Africa

West Africa is inhabited by the three tropical African pangolins: white- and black-bellied, and giant pangolin. For the purposes of this chapter the region comprises Benin, Burkina Faso, Côte d'Ivoire, The Gambia, Ghana, Guinea, Guinea Bissau, Liberia, Nigeria, Senegal, Sierra Leone, and Togo.

Nutritional use

Zeba (1998) observed that in most of the local languages in West Africa, the translation of "wildlife" etymologically is "bushmeat." Pangolins have been consumed as a source of bushmeat across the region historically (Ajayi, 1978; Ordaz-Nemeth et al., 2017; Petrozzi et al., 2016), and remain in substantial demand (Boakye et al., 2015; Gonedelé Bi et al., 2017; Greengrass, 2016; Fig. 15.1). All three native pangolin species are consumed, but the white-bellied pangolin is by far the most common species of pangolin found in bushmeat markets in the region, and is subject to widespread and often intensive exploitation for this purpose (Boakye et al., 2016; Bräutigam et al., 1994; Soewu and Ayodele, 2009). Boakye et al. (2016) found that in Ghana, white-bellied pangolins constituted 82% of 341 pangolins traded along bushmeat commodity chains compared with 18% comprising black-bellied pangolins. Anadu et al. (1988) found that white- and black-bellied pangolins, grouped together as tree pangolins, ranked eighth among the mammals that consumers in southwestern Nigeria most preferred. In contrast, Hoyt (2004) noted that in Liberia, white-bellied pangolins ranked twelfth behind the giant (third) and black-bellied pangolin (eighth) in a taste preference exercise of wild mammals consumed by urban communities.

Given the substantial demand for pangolins as bushmeat, they are regularly found in

FIGURE 15.1 A descaled black-bellied pangolin (*P. tetradactyla*) being prepared for consumption. *Photo credit: African Pangolin Working Group.*

bushmeat markets, and are hunted for such markets. Anadu et al. (1988) found that white- and black-bellied pangolins constituted 0.7% of all animals traded at a market and along roadsides in southwestern Nigeria. Similarly, white-bellied pangolins comprised 0.32% of total mammal biomass traded in Swali market in Bayelsa State (Akani et al., 2015). Fa et al. (2006) estimated that approximately 28,000 kg/year of biomass representing white-bellied pangolins (approximately 10,000 individuals) was extracted for bushmeat by hunters in the Cross-Sanaga Rivers region of southeastern Nigeria. In Liberia, Bene et al. (2013) reported that white- and black-bellied pangolins constituted 1.76% and 1.35%, respectively, of hunters' total catch in Nimba County. Greengrass (2016) estimated that white-bellied pangolins constituted 2.6% and 0.78% of the total catch in two hunting camps (Neechebu and Chanedae, respectively) in the vicinity of Sapo National Park, Liberia.

Medicinal use

Pangolins are used extensively in traditional medicines in West Africa (Djagoun et al., 2012; Soewu and Adekanola, 2011). Most West Africans in rural areas depend on traditional medicine for their healthcare needs (Boakye et al., 2014; see also World Health Organization, 2013) and market stalls selling plant and animal parts for this purpose are common in both rural and urban areas (Ntiamoa-Baidu, 1987; Fig. 15.2). In Sierra Leone, 22 pangolin body parts are used to treat

FIGURE 15.2 A white-bellied pangolin (*P. tricuspis*) skin on sale at a traditional medicine market. *Photo credit: African Pangolin Working Group.*

medical conditions, and there is a similar pattern in Benin, Ghana and Nigeria where white-bellied pangolin body parts including the head, heart, blood, eyes, intestines, tongue, and scales are prescribed by traditional healers to treat a multitude of ailments (see Table 15.1). There are differences in the use of body parts between countries, but scales are prescribed for the greatest variety of medical conditions, ranging from rheumatism to leprosy (Table 15.1).

Other uses

Other uses include as spiritual remedies and as omens, which largely afford people protection or confer good fortune. A range of pangolin body parts are used (see Table 15.2). In Benin, scales of the white-bellied pangolin are used to prevent accidents and provide protection from gun shots or knife wounds. Boakye et al. (2014) found scales of all three tropical African pangolins to be important in protecting people from witchcraft, in the preparation of charms, and for warding off evil spirits among the Temne and Limba ethnic groups in Sierra Leone and the Ashanti in Ghana. In Nigeria, scales have greatest application among the Ijebus and Aworis in the southwest of the country and are used to treat mental illness, kleptomania, for the conferment of good luck and warding off witches, as in Ghana and Sierra Leone (Table 15.2). The scales may be used in isolation or with other

TABLE 15.1 Body parts of the white-bellied pangolin (*P. tricuspis*) prescribed for ailments and other conditions by traditional healers in Benin, Ghana, Nigeria, and Sierra Leone.

Body part	Ailment/condition treated	Country
Bile	Menstrual pain, scrotal mass	Ghana[a]
Bone	Skin scars, wound healing, rheumatism, joint pains and stiffness, convulsions, headache, stroke, waist pain, asthma, bed-wetting, fever, broken leg, skin rash, breast cancer	Benin[b], Ghana[a], Nigeria[c], Sierra Leone[d]
Blood	Wound healing, elephantiasis, rheumatism, stomach disease, heart disease	Sierra Leone[d]
Brain	Heart disease, stomach disease, mental illness	Sierra Leone[d]
Claws	Asthma, stretch marks, heartburn, infertility	Ghana[a], Sierra Leone[d]
Eyes	Conjunctivitis, impotence, mental illness	Ghana[a], Nigeria[c], Sierra Leone[d]
Female sex organ	Ejection of placenta	Nigeria[c]
Foot	Heel fissure, back pain, elephantiasis, athlete's foot, broken bones	Benin[b], Sierra Leone[d]
Forefoot	Impotence, elephantiasis	Sierra Leone[d]
Head	Infertility, headache, skin disease, toothache, heart disease, paralysis, stroke, asthma, hernia, fever, body aches, gonorrhea, claw hand, mental illness	Benin[b], Ghana[a], Nigeria[c], Sierra Leone[d]
Heart	Prevention of miscarriage, stomach disease, heart disease	Benin[b], Ghana[a], Sierra Leone[d]
Internal organs	Food poisoning	Nigeria[c]
Intestines	Stomach disease, headache	Benin[b], Sierra Leone[d]
Liver	Asthma	Sierra Leone[d]
Male sex organ	Hernia, headache, elephantiasis, athlete's foot, infertility, impotence	Sierra Leone[d]
Meat	To aid normal development in premature babies, stomach disease, rheumatism, epilepsy, hypertension, body pain, infertility, menstrual pains, coughing, prevention of miscarriage, convulsion, anemia, common childhood diseases	Ghana[a], Sierra Leone[d]
Oil	Skin rash, stretch marks, heel fissure, skin disease, knee pain, skin scars, heart disease, claw hand, body aches, elephantiasis	Sierra Leone[d]
Scales	Muscular pain, back pain, headache, excessive menstrual bleeding, menstrual cramps, bed-wetting, stroke, chicken pox, epilepsy, heart disease, wound healing, dry skin, skin rashes, sores, cracked heels, convulsions, arthritis, ear infection, stomach disorders, leprosy, to aid normal development in premature babies, elephantiasis, impotence, infertility, broken bones, waist pain, skin scars, stomach disease, inflammation of the navel, nail infections, arthritis, rheumatism, epilepsy, blood cleansing, stomach ulcer, stroke, venereal diseases, to ensure safe childbirth, mental illness, as an aphrodisiac/for male potency	Ghana[a], Nigeria[c], Sierra Leone[d]

(*Continued*)

TABLE 15.1 (Continued)

Body part	Ailment/condition treated	Country
Sex organ (male and female)	Infertility	Sierra Leone[d]
Skin	Dermatosis	Benin[b]
Tail	Impotence, acute hemorrhagic conjunctivitis, paralysis, claw hand, convulsions, fainting, stomach disease, elephantiasis, waist pain, heel fissure, protection against snake bites and scorpion stings	Sierra Leone[d]
Toes	Acute hemorrhagic conjunctivitis, epilepsy	Ghana[a], Sierra Leone[d]
Tongue	Asthma	Benin[b]
Thorax	Goiter	Ghana[a], Nigeria[c]
Vertebral bones	Stroke	Nigeria[c]
Waist	Prevention of miscarriage	Benin[b]
Whole animal	Prolonged or continuous menstrual bleeding, elephantiasis, leprosy	Ghana[a], Nigeria[c], Sierra Leone[d]

[a]*Boakye, M.K., Pietersen, D.W., Kotzé, A., Dalton, D.L, Jansen, R., 2016. Unravelling the pangolin bushmeat commodity chain and the extent of trade in Ghana. Hum. Ecol. 44 (2), 257–264.*
[b]*Akpona, H.A., Djagoun, C.A.M.S., Sinsin, B., 2008. Ecology of the three-cusped pangolin* Manis tricuspis *(Mammalia, Pholidota) in the Lama forest reserve, Benin. Mammal. 72 (3), 198–202.*
[c]*Soewu, D.A., Adekanola, T.A., 2011. Traditional-medical knowledge and perception of pangolins (Manis sps [sic]) among Awori People, Southwestern Nigeria. J. Ethnobiol. Ethnomed. 7, 25.*
[d]*Boakye, M.K., Pietersen, D.W., Kotzé, A., Dalton, D.L., Jansen, R., 2014. Ethnomedicinal use of African pangolins by traditional medical practitioners in Sierra Leone. J. Ethnobiol. Ethnomed. 10, 76.*

TABLE 15.2 Body parts of the white-bellied pangolin (*P. tricuspis*) prescribed for spiritual remedies by traditional healers in Benin, Ghana, Nigeria and Sierra Leone.

Body part	Prescribed for	Country
Bone	Spiritual protection, protection from witchcraft	Ghana[a], Nigeria[b], Sierra Leone[c]
Blood	Protection against witchcraft	Sierra Leone[c]
Claws	Protection from witchcraft	Ghana[a], Sierra Leone[c]
Eyes	Kleptomania, spiritual protection	Ghana[a], Nigeria[b], Sierra Leone[c]
Flesh	To confer abilities for divination, good luck, protection, safety	Nigeria[b]
Head	Financial rituals, spiritual protection, good luck, safety, kleptomania, induction in to wizard groups	Benin[d], Ghana[a], Nigeria[b], Sierra Leone[c]
Head and tail tip	Entrepreneurial prowess	Nigeria[b]
Internal organs	To treat sexual poison "magun"	Nigeria[b]

(Continued)

TABLE 15.2 (Continued)

Body part	Prescribed for	Country
Intestines	Good luck	Sierra Leone[c]
Leg	Spiritual protection, financial rituals	Ghana[a]
Limbs	Good fortune	Nigeria[b]
Limbs and internal organs	Financial rituals	Nigeria[b]
Meat	To increase intelligence, spiritual protection, financial rituals, charms for chiefs	Ghana[a], Sierra Leone[c]
Scales	To be cutlass proof, spiritual protection, protection from witchcraft, good luck, high productivity on the farm/financial rituals, kleptomania	Benin[d], Ghana[a], Nigeria[b], Sierra Leone[c]
Tail	Kleptomania, high productivity on farm, spiritual protection	Sierra Leone[c]
Toes	Spiritual protection	Ghana[a], Sierra Leone[c]
Thorax	Prevention of rain	Nigeria[b], Sierra Leone[c]
Whole animal	Financial rituals, conferring invisibility, good fortune, prosperity	Ghana[a], Nigeria[b], Sierra Leone[c]

[a]*Boakye, M.K., Pietersen, D.W., Kotzé, A., Dalton, D.L, Jansen, R., 2016. Unravelling the pangolin bushmeat commodity chain and the extent of trade in Ghana. Hum. Ecol. 44 (2), 257–264.*

[b]*Sodeinde, O.A., Adedipe, S.R., 1994. Pangolins in south-west Nigeria – current status and prognosis. Oryx 28 (1), 43–50; Sodeinde, O.A., Soewu, D. A., 1999. Pilot Study of the traditional medicine trade in Nigeria. TRAFFIC Bull. 18, 35–40; Soewu, D.A., Adekanola, T.A., 2011. Traditional-medical knowledge and perception of pangolins (Manis sps [sic]) among Awori People, Southwestern Nigeria. J. Ethnobiol. Ethnomed. 7, 25.*

[c]*Boakye, M.K., Pietersen, D.W., Kotzé, A., Dalton, D.L., Jansen, R., 2014. Ethnomedicinal use of African pangolins by traditional medical practitioners in Sierra Leone. J. Ethnobiol. Ethnomed. 10, 76.*

[d]*Akpona, H.A., Djagoun, C.A.M.S., Sinsin, B., 2008. Ecology of the three-cusped pangolin* Manis tricuspis *(Mammalia, Pholidota) in the Lama forest reserve, Benin. Mammal. 72 (3), 198–202.*

ingredients. In Nigeria, the seeds of the plant *Aframomum melegueta* are ground and mixed with powdered, roasted pangolin scales, and the powder is added to maize porridge (*akamu*) for consumption (Soewu and Adekanola, 2011).

Central Africa

Central Africa hosts the three tropical African pangolin species. For the purpose of this chapter the region comprises Angola, Cameroon, Central African Republic (CAR), Chad, Democratic Republic of the Congo (DRC), Equatorial Guinea, Gabon and Republic of Congo.

Nutritional use

Pangolins are a preferred bushmeat species in Central Africa. They are frequently recorded among the top species in taste preference studies (Kümpel, 2006). In a study in Cameroon, respondents scored the white-bellied pangolin highly on taste, ranking it third out of ten species they hunted and trapped (Wright and Priston, 2010). Offtake of pangolins relative to other species also appears to be increasing. Ingram et al. (2018) estimated that at least 0.4

million tropical African pangolins are harvested annually in Central Africa, including for bushmeat, and that the proportion of pangolins hunted of all animals increased between the 1970s and 2010s.

In Cameroon, DRC, Equatorial Guinea, Gabon and Republic of Congo, pangolins are openly sold at roadside restaurants, in bushmeat markets, and at restaurants in capital cities (e.g., Libreville, Malabo, Yaoundé; Albrechtsen et al., 2007; Cronin et al., 2015; Dethier, 1995; Dupain et al., 2012; Mambeya et al., 2018; Mbete, 2012). In Kisangani, DRC, the number of giant pangolins in the market increased seven-fold between 2002 and 2009, although the number of other medium-sized (10–50 kg) animals in the market decreased significantly; small (<10 kg) animals increased over the same period (van Vliet et al., 2012). In Cameroon, Infield (1988) reported that around Korup National Park, pangolin meat is highly favored, and Bobo and Kamgaing (2011) found that white-bellied pangolins were the second most harvested animal in villages northeast of Korup National Park. Both arboreal species (white- and black-bellied pangolin) are harvested by ethnic Mbo and Banyangi hunters in the Banyang-Mbo Wildlife Sanctuary (Willcox and Nambu, 2007). In Equatorial Guinea, pangolins made up an increasing proportion of the total carcasses at Central and Mundoasi markets between 1991 and 2003 (Kümpel, 2006) and between 2003 and 2010 (Gill, 2010). Giant pangolins, which do not occur on Bioko Island, are shipped to the Malabo market on Bioko from mainland Equatorial Guinea and Cameroon (Cronin et al., 2015; Hoffmann et al., 2015; Ingram et al., 2019; Fig. 15.3). In Gabon, the price of both arboreal and giant pangolins rose in sampled forest-gate villages, in Makokou (provincial town) and Libreville between 2002/03, and 2014 (Mambeya et al., 2018). These price increases

FIGURE 15.3 Giant pangolins (*S. gigantea*) are hunted and poached in West and Central Africa, and consumed or sold as bushmeat. *Photo credit: Stuart Nixon.*

were significant: in Libreville prices for giant pangolin rose by 211% and for arboreal pangolins by 73%, while inflation rose by 4.6% (Mambeya et al., 2018). This study also highlighted that Asian industry workers frequently request pangolins from hunters; this additional demand may explain price increases, or potentially the rarity of the species, and warrants further research.

There has been less research in CAR, Angola and Chad, but pangolins are hunted and consumed in these countries. Research in CAR has shown that Baka and non-Baka communities eat pangolins (Bahuchet, 1990; Hodgkinson, 2009), and that within the Bofi ethnic group, both Babingas-Bofis and Gbayas-Bofis catch pangolins for consumption (Lupo and Schmitt, 2002; Vanthomme, 2010). Giant pangolins are also regularly observed for sale in urban markets in CAR despite being a protected species (Fargeot, 2013). There is little available information on the use of pangolins as bushmeat in Angola and Chad. Svensson et al. (2014) reported that white-bellied pangolins are sold openly on roadside bushmeat markets in Angola, and Bräutigam et al. (1994) reported that local consumption of pangolins occurs in Chad and likely continues to take place.

Medicinal use

There are few documented ethnopharmacological uses of pangolins in Central Africa in the scientific literature. However, rural people near Korup National Park, Cameroon, report that white- and black-bellied pangolin scales are used to treat stomach disorders (Bobo and Ntum Wel, 2010). They are typically burnt and/or ground into a powder and mixed with palm oil or water which is then ingested to purge the stomach. Boki and Anyang people use the scales to treat coughs (Mouté, 2010). A report from Chad indicates that pangolin derivatives have been used to treat malaria (see Bräutigam et al., 1994).

Other uses

Pangolins have various associations and spiritual connections with people in Central Africa. This includes an association with fertility and attraction (see Chapter 12). However, not all associations are good omens. For the Mbuti people in eastern DRC, pregnant women are advised not to consume white-bellied or giant pangolins because it is believed that the animals "may cause a fatal bleeding during delivery" (Ichikawa, 1987). The Baka hunter-gatherers in Republic of Congo and Cameroon associate an illness with white-bellied pangolins, whereby infants are afflicted with an abdominal disorder if their parents consume the meat before birth or during the lactation period (Sato, 1998). In the Salonga-Lukenie-Sankuru landscape in the DRC, the giant pangolin is considered a totem animal, which if not shared will result in a death in the family (Abernethy et al., 2010). This species is also a totem for the Nkundu people, who hunt it under specific cultural norms; this includes the need to prepare the carcass at sacred sites (Steel et al., 2008).

Pangolin derivatives are also used as tools. In southwest Cameroon, the scales of arboreal pangolins are used as blades and their skins are used to make drums (Bobo et al., 2015).

East Africa

For the purposes of this chapter, East Africa encompasses Burundi, Ethiopia, Kenya, Malawi, Rwanda, Somalia, South Sudan, Tanzania, Uganda and Zambia. Three species of pangolin occur in the region, including Temminck's, giant and white-bellied, while the black-bellied species may occur marginally in Uganda (see Chapter 8).

Nutritional use

There has been little research on the use of pangolins in East Africa compared to West, Central and Southern Africa, but uses have been documented. In Tanzania, Banyamwezi hunters were recorded hunting Temminck's pangolin for bushmeat in the mid-1990s (Carpaneto and Fusari, 2000). In Uganda, Olupot et al. (2009) reported that in 2007–08, two Temminck's pangolins were killed in the Kafu Basin and Murchison Falls Conservation Area respectively, likely for bushmeat consumption.

Medicinal use

Ethnozoological use of pangolins in East Africa is limited. In Tanzania, scales from Temminck's pangolin have been used to reposition foetuses in pregnant women and to help expel the placenta following parturition (see Marshall, 1998). Crushed scales have also been used to treat nose-bleeds (Kingdon, 1974).

Other uses

Pangolins are sometimes associated with human attraction. In Uganda, women in the Buganda subnational kingdom buried the scales of giant pangolins under the doorsteps of their lover's homes, in the presence of a diviner, to influence them to fulfill their desires (Kingdon, 1974).

There is also a reported association between pangolins and rain. Walsh (1995/96) reported that the Sangu people of southwest Tanzania conducted ritual sacrifices of pangolins, which are thought to be associated with predicting rain, and thus the abundance of food, but such sacrifices were rare (Walsh, 1995/96; see Chapter 12). This is mirrored in other parts of Tanzania and Malawi where sighting a pangolin is a sign that rain will fall (Bräutigam et al., 1994; Mafongoya and Ajayi, 2017). Across several countries in East Africa, pangolins are also associated with good luck (see Chapter 12).

Pangolin derivatives are also used as protective charms. The scales of the giant pangolin are reported to have been mixed with tree bark by women in Uganda to neutralize evil spirits (Kingdon, 1974). Around Ruaha National Park and the Mbomipa Wildlife Management Area, Tanzania, Temminck's pangolin scales are used to ward off bad people and bad luck (Mbilinyi, 2014). In Malawi, the scales are reportedly used as protection against bad omens (Marshall, 1998). Bräutigam et al. (1994) report that smoke from burning pangolin scales has been used to ward off lions in Acholi, Uganda. They have also been used to repel wild animals in Tanzania (Marshall, 1998).

Southern Africa

Temminck's pangolin is the only species of pangolin to occur in Southern Africa, which for the purposes of this chapter includes Botswana, Mozambique, Namibia, South Africa, Swaziland, and Zimbabwe. This region has a multitude of ethnic tribal communities and there is a long history of utilizing Temminck's pangolin as a food source, to treat medical conditions, and in spiritual rituals.

Nutritional use

There are records of Temminck's pangolin having been hunted for bushmeat in South Africa (Jacobsen et al., 1991; Pietersen et al., 2016; van Aarde et al., 1990), Zimbabwe (Ansell, 1960; Coulson, 1989) and Botswana (Setlalekgomo, 2014). There are no traceable records for Namibia, but it seems likely that hunting and consumption has occurred in the country; it has occurred in Mozambique (D.W. Pietersen, pers. obs.). However, pangolins are consumed much less frequently than in West and Central Africa. This may be in part because of the fattiness of the meat (Bräutigam et al., 1994). Farm workers in the Kalahari region of South Africa are known to have eaten pangolins found as roadkill or that have been electrocuted on game farm fences, but apparently do not actively source them for food (Pietersen et al., 2014). The trade in bushmeat in Southern Africa has received little research attention (Hayward, 2009; Warchol and Johnson, 2009) and its extent and impact on Temminck's pangolin populations remains unknown.

Medicinal use

The use of traditional medicines is centuries old in Southern Africa. Historically, a specialist activity of rural herbalists and "Sangomas" (traditional healers who use natural plant and animal remedies), there is a thriving commercial trade in traditional medicine in Southern Africa (Cunningham and Zondi, 1991; Williams and Whiting, 2016), in part because the majority of rural people in Southern Africa consult traditional healers to treat medical conditions. Particularly "powerful" species are readily harvested over large areas, including neighboring countries, and sold in urban commercial markets (e.g., Faraday market, Johannesburg; Williams and Whiting, 2016). Pangolins are one of the most sought-after mammals in such markets in Southern Africa (Cunningham and Zondi, 1991), where their scales and bones can be found (Whiting et al., 2011), but availability of the species is low. Williams and Whiting (2016) posit this is because they are extremely difficult to source in the wild. Pangolin scales are also in demand in Zimbabwe, where they have long been regarded as a powerful medicine (Duri, 2017; Smithers, 1966). Pangolin body parts are infrequently seen in markets in Namibia and Mozambique, though Marshall (1998) reported that substantial demand exists in both countries. The specific body parts that are prescribed for particular ailments has received little attention. Setlalekgomo (2014) and Baiyewu et al. (2018) investigated the use of pangolins for traditional medicine in Botswana and South Africa respectively. Like in West Africa, body parts ranging from the liver and lungs, to the brain, blood, fat and scales are prescribed by traditional healers for ailments including arthritis, excessive menstrual bleeding, chicken pox, earache and diabetes (Table 15.3).

Other uses

Chance encounters with certain species are often associated with an omen or belief system in Southern Africa. Amongst the Shona people of Zimbabwe and the amaZulu of South Africa, it is considered a very good omen to come upon a pangolin, and typically the animal is captured and presented as a "valuable" gift to a chieftain, head of state or traditional healer (Coulson, 1989; Pietersen et al., 2014). By presenting a pangolin to a traditional healer, it is believed that person will receive protection and well-being (Challender and Hywood, 2012). Amongst the Venda and Tswana people of South Africa, special songs are sung and a sheep is slaughtered when a pangolin is sighted, and special treatment is bestowed upon the individual(s) who sighted the animal by the chief and tribesmen (Baiyewu et al., 2018).

Pangolins have been heavily associated with climatic phenomena, including rainfall (as in East Africa) and drought. It is widely believed in South African tribal culture that if the blood of a pangolin is spilled on the ground, it will

TABLE 15.3 Body parts of Temminck's pangolin (*S. temminckii*) prescribed for ailments by traditional healers in Botswana and South Africa.

Body part	Aliments treated
Scales	Body pains, arthritis, headache, back pain, swollen feet/legs, foot ache, excessive menstrual bleeding, menstrual cramps, illness in babies, stroke, chicken pox, epilepsy, heart disease, wounds, skin problems/dry skin, skin rash, sores, tuberculosis, tiredness, cancer, chest pains, cracked heels, diabetes, goiter, hypertension, persistent cough
Blood	Excessive menstrual bleeding, menstrual cramps, heart disease, nose bleeds, hypertension, chest pains, blood cleansing, for general health and well-being
Meat	Hearing problems, ear sores, earache, skin problems/dry skin, skin rash
Fat	Blood disease, nose bleed, skin problems/dry skin, skin rash, hearing problems, earache, blood cleansing
Heart, liver, intestines, lung	Nose bleeds, skin disease, internal parasites in children, asthma in children, to increase fertility
Claws	Aches and pains

*Source: Baiyewu, A.O., Boakye, M.K., Kotzé, A., Dalton, D.L., Jansen, R., 2018. Ethnozoological survey of the traditional uses of Temminck's pangolin (*Smutsia temminckii*) in South Africa. Soc. Anim. 26, 1–20; Setlalekgomo, M.R., 2014. Ethnozoological survey of the indigenous knowledge on the use of pangolins (*Manis sps [sic]*) in traditional medicine in Lentsweletau Extended Area in Botswana. J. Anim. Sci. Adv. 4 (6), 883–890.*

result in drought (Baiyewu et al., 2018; Niehaus, 1993). In the Venda and Tswana tribes, the animal is believed to "come flying down from the skies" during lightning and thunderstorms and the animal is often seen following such an event (Baiyewu et al., 2018). Bräutigam et al. (1994) report that increased harvesting of pangolins in Namibia was believed to be in response to increased demand for these rain-making tokens. In contrast, the amaZulu people in South Africa believe that if a pangolin is seen there will be a drought and to prevent this from happening the animal must be killed (Kyle, 2000). In Mozambique, Temminck's pangolin is associated with rain, and live pangolins can be a sign of abundance or famine (Bräutigam et al., 1994).

In Botswana, it is considered a bad omen if a pangolin is seen by a person and it does not curl up on spotting them, or if it is seen walking on two legs. It is believed that any pregnant woman who crosses its trail will give birth to a baby with scaly skin, and if a person treads on a pangolin's trail (i.e., spoor), they will develop cracked heels (Setlalekgomo, 2014).

Temminck's pangolin parts are used in a variety of spiritual remedies. Baiyewu et al. (2018) and Setlalekgomo (2014) examined such uses in South Africa and Botswana respectively (Table 15.4). Pangolin scales and blood have the greatest application by traditional healers in rural areas. Many of these spiritual associations relate to the bringing of good luck, conferring protection or power, and cleansing. Scales are often carried in a wallet to "protect" money, carried in a car to prevent accidents, or even swallowed to increase longevity or to avoid confessing to a wrong-doing. Blood is sometimes carried on a person in a vial and, with fat, is mixed with building materials used in the construction of homes or animal pens to ward off evil spirits (Baiyewu et al., 2018). In the KwaZulu-Natal Province of South Africa, the burning of scales

TABLE 15.4 Body parts of Temminck's pangolin (*S. temminckii*) prescribed for spiritual remedies by traditional healers in Botswana and South Africa.

Body part	Spiritual remedies
Scales	Prevention of lightning strikes, protection against severe weather, good luck, promotion, physical fortification, spiritual fortification, fortification in babies/children, protection against evil spirits/deeds, fire prevention, love charms, longevity, spiritual cleansing, to help chiefs stay in power, protection of cattle and livestock, cattle fertility, protection of crops against witchcraft
Blood	To help chiefs stay in power, spiritual and physical fortification, prevention of lightning strikes, protection against severe weather; cattle fertility, protection of cattle and livestock, protection against evil spirits, spiritual cleansing, blood cleansing, general health and well-being, to attract lovers, to attract customers
Fat	Protection against evil spirits or bad luck, ritualistic rites, spiritual cleansing, blood cleansing
Heart, liver, intestines, lung	To help chiefs stay in power, prevention of bewitching, protection against evil spirits/deeds
Head, brain, eyes, nose	To protect a cattle kraal (enclosure) and livestock against evil and predators, mind control, premonition
Claws	Good luck, fortification
Skin, carcass	Protection of cattle and livestock
Whole animal	To help chiefs stay in power, good luck, promotion, protection against evil spirits/deeds

*Source: Baiyewu, A.O., Boakye, M.K., Kotzé, A., Dalton, D.L., Jansen, R., 2018. Ethnozoological survey of the traditional uses of Temminck's pangolin (*Smutsia temminckii*) in South Africa. Soc. Anim. 26, 1–20; Setlalekgomo, M.R., 2014. Ethnozoological survey of the indigenous knowledge on the use of pangolins (*Manis *sps [sic]) in traditional medicine in Lentsweletau Extended Area in Botswana. J. Anim. Sci. Adv. 4 (6), 883–890.*

and inhalation of the smoke is believed to reduce hysteria (Cunningham and Zondi, 1991).

Impact of local and national use on pangolin populations

In West Africa, the consensus of research on the bushmeat trade and traditional medicinal use of pangolins (e.g., Akpona et al., 2008; Boakye et al., 2014, 2015; Soewu and Adekanola, 2011) is that these activities are driving population declines in the three tropical African species (see also Chapters 8–10). Opinions of respondents interviewed on the use of pangolins in bushmeat and traditional medicine trades in the last 30 years indicate that there has been a steady decline in pangolin numbers and distribution. Anadu et al. (1988) reported that bushmeat retailers in Bendel State (now Edo and Delta States), Nigeria, listed the white- and black-bellied pangolins among the species that were difficult to source (Anadu et al., 1988). Similarly, Sodeinde and Adedipe (1994) note that at the time of the research, hunters described white-bellied pangolins as rarer than in previous years. Traditional medicine practitioners interviewed by Soewu and Adekanola (2011) responded that pangolins had decreased in number and average body size over time.

In Central Africa, available evidence indicates that offtake for local use and trade is also likely unsustainable. Ingram et al. (2018) used local hunting data to estimate the exploitation of pangolin populations across the region.

They estimated that exploitation likely includes 0.4 million individuals annually, and based on a sample of 310 pangolins, 45% were sub-adults or juveniles. Moreover, it appears exploitation of pangolins in the region is increasing. Ingram et al. (2018) reported that the total annual catch of pangolins increased by ~150% from before (1975–1999) to post-2000 (2000–2014). The percentage of pangolin species in the catch also increased significantly from 0.04% in 1972 to 1.83% in 2014 (Ingram et al., 2018). Insights from local communities support the evidence that pangolins are becoming rarer. In Cameroon, local communities interviewed near Korup National Park in 2009 stated that giant pangolins were abundant 10 years prior to the survey, but in 2009 were very rare (Ngoufo et al., 2014). Although anecdotal, Infield (1988) noted that the species may have been extirpated from locations around Korup National Park, and Mouté (2010) reported that the species is no longer observed to the northeast of the park. Abugiche (2008) stated that the giant pangolin was feared locally extinct near 14 villages in the vicinity of Banyang-Mbo Wildlife Sanctuary in Cameroon. In Gabon, interviews with local Pouvi hunters in two villages revealed that the giant pangolin was either rare or had already disappeared from some areas (Schleicher, 2010).

There is little information on the impact of exploitation of East African pangolin populations. Trafficking of pangolin parts, mainly scales, in the region appears to be increasing based on seizure records, implicating, among other countries, Kenya and Uganda (see Chapter 16; Challender and Waterman, 2017; Heinrich et al., 2017). However, it is not known where pangolins are being sourced, or whether they are being targeted specifically for international trafficking or sourced as a by-product of bushmeat hunting.

Temminck's pangolin is highly sought-after in Southern Africa, and exploitation has had a negative impact on populations in parts of its range. In the KwaZulu-Natal Province of South Africa, the species is considered to be locally extinct as a direct result of overharvesting for traditional practices (see Chapter 11; Cunningham and Zondi, 1991; Pietersen et al., 2014, 2016). However, the species is elusive and rarely seen, which makes it challenging to determine the impact of use on populations elsewhere in the species' range (see Chapter 11). The low availability of the species in traditional markets in Southern Africa may be due to its scarcity, but further research on the status of populations is needed (Chapter 34).

Overall, evidence suggests that the number of pangolins being sold for bush meat and traditional medicine is likely unsustainable in many places in both the short- and long-term. Considering that hunting pressure is compounded by targeted exploitation for international trafficking (see Chapter 16), it is highly likely that pangolin populations are in decline across Africa (see Chapters 8–11). Further research is needed to quantify, qualify and monitor the impacts of local and national use on populations.

Conclusion

Use of pangolins across Africa follows similar patterns, but with some regional differences. The species are consumed as bushmeat, particularly in West and Central Africa, where there is substantial demand for all three tropical African species. The white-bellied pangolin is reportedly the most frequently used species and is most encountered in trade. The majority of rural African people still rely on traditional healers for their healthcare needs and pangolins are used for traditional medicinal purposes across Africa, potentially compounding exploitation for bushmeat. High human population growth forecasts in parts of Africa (e.g., DRC, Nigeria) suggest that the exploitative pressure on Africa's pangolin populations is unlikely to ease in the near future. There is, therefore, an urgent need for research to understand exploitation levels and their impact on populations to inform conservation planning and management.

References

Abernethy, K., Coad, L., Llambu, O., Makiloutila, F., Easton, J., Akiak, J., 2010. Wildlife Hunting, Consumption Trade in the Oshwe Sector of the Salonga-Lukenie-Sankuru Landscape, DRC. WWF CARPO, Kinshasa, Democratic Republic of Congo.

Abugiche, S.A., 2008. Impact of Hunting and Bushmeat Trade on Biodiversity Loss in Cameroon: A Case Study of the Banyang-Mbo Wildlife Sanctuary. Ph.D. Thesis, Brandenburg University of Technology, Germany.

Ajayi, S., 1978. Pattern of bushmeat production, preservation and marketing in West Africa. Niger. J. For. 8, 48–52.

Akani, G.C., Amadi, N., Eniang, E.A., Luiselli, L., Petrozzi, F., 2015. Are mammal communities occurring at a regional scale reliably represented in "hub" bushmeat markets? A case study with Bayelsa State (Niger Delta, Nigeria). Folia Zool. 64 (1), 79–86.

Akpona, H.A., Djagoun, C.A.M.S., Sinsin, B., 2008. Ecology and ethnozoology of the tree-cusped pangolin *Manis tricuspis* (Mammalia, Pholidota) in the Lama forest reserve, Benin. Mammalia 72 (3), 198–202.

Albrechtsen, L., Macdonald, D.W., Johnson, P.J., Castelo, R., Fa, J.R., 2007. Faunal loss from bushmeat hunting: empirical evidence and policy implications in Bioko Island. Environ. Sci. Policy 10 (7–8), 654–667.

Anadu, P.A., Elamah, P.O., Oates, J.F., 1988. The bushmeat trade in southwestern Nigeria: a case study. Human Ecol. 16 (2), 199–208.

Ansell, W.F.H., 1960. Mammals of Northern Rhodesia. The Government Printer, Lusaka.

Bahuchet, S., 1990. Food sharing among the pygmies of Central Africa. Afr. Study Monogr. 11 (1), 27–53.

Baiyewu, A.O., Boakye, M.K., Kotze, A., Dalton, D.L., Jansen, R., 2018. Ethnozoological survey of the traditional uses of Temminck's pangolin (*Smutsia temminckii*) in South Africa. Soc. Anim. 26, 1–20.

Bene, J.-C.K., Gamys, J., Dufour, S., 2013. A wealth of Wildlife Endangered in northern Nimba County, Liberia. Int. J. Innov. Appl. Stud. 2 (1), 314–323.

Boakye, M.K., Pietersen, D.W., Kotzé, A., Dalton, D.L., Jansen, R., 2014. Ethnomedicinal use of African pangolins by traditional medical practitioners in Sierra Leone. J. Ethnobiol. Ethnomed. 10, 76.

Boakye, M.K., Pietersen, D.W., Kotzé, A., Dalton, D.-L., Jansen, R., 2015. Knowledge and uses of African pangolins as a source of traditional medicine in Ghana. PLoS One 10 (1), e0117199.

Boakye, M.K., Pietersen, D.W., Kotzé, A., Dalton, D.L., Jansen, R., 2016. Unravelling the pangolin bushmeat commodity chain and the extent of trade in Ghana. Hum. Ecol. 44 (2), 257–264.

Bobo, K.S., Ntum Wel, C.B., 2010. Mammals and birds for cultural purposes and related conservation practices in the Korup area, Cameroon. Life Sci. Leaflets 9, 226–233.

Bobo, K.S., Kamgaing, T.O.W., 2011. *Etude chasse et contribution a l'evaluation de la durabilite des prelevements de Cephalophus monitcola* en peripherie nord-est du parc national de Korup (sud-ouest, Cameroun). Report for The Volkswagen Project, Dschang, Cameroon.

Bobo, K.S., Aghomo, F.F.M., Ntumwel, B.C., 2015. Wildlife use and the role of taboos in the conservation of wildlife around the Nkwende Hills Forest Reserve; southwest Cameroon. J. Ethnobiol. Ethnomed. 11, 2.

Bräutigam, A., Howes, J., Humphreys, T., Hutton, J., 1994. Recent information on the status and utilization of African pangolins. TRAFFIC Bull. 15 (1), 15–22.

Carpaneto, G.M., Fusari, A., 2000. Subsistence hunting and bushmeat exploitation in central-western Tanzania. Biodivers. Conserv. 9 (11), 1571–1585.

Challender, D.W.S., Hywood, L., 2012. African pangolins under increased pressure from poaching and intercontinental trade. TRAFFIC Bull. 24 (2), 53–55.

Challender, D., Waterman, C., 2017. Implementation of CITES Decisions 17.239 b) and 17.240 on Pangolins (*Manis* spp.), CITES SC69 Doc. 57 Annex. Available from <https://cites.org/sites/default/files/eng/com/sc/69/E-SC69-57-A.pdf>. [August 2, 2018].

Coulson, I., 1989. The pangolin (*Manis temminckii* Smuts, 1835) in Zimbabwe. Afr. J. Ecol. 27 (2), 149–155.

Cowlishaw, G., Mendelson, S., Rowcliffe, J.M., 2005. Evidence for post-depletion sustainability in a mature bushmeat market. J. Appl. Ecol. 42 (3), 460–468.

Cronin, D., Woloszynek, S., Morra, W.A., Honarvar, S., Linder, J.M., Gonder, M.K., et al., 2015. Long-Term urban market dynamics reveal increased bushmeat carcass volume despite economic growth and proactive environmental legislation on Bioko Island, Equatorial Guinea. PLoS One 10 (8), e0137470.

Cunningham, A.B., Zondi, A.S., 1991. Use of Animal Parts for Commercial Trade in Traditional Medicines. Working Paper No. 76. Institute for Natural Resources, University of Natal, South Africa.

Dethier, M., 1995. Projet ECOFAC-Composante Cameroun - Etude Chasse. AGRECO, Bruxelles, Belgium.

Djagoun, C.A., Akpona, H.A., Mensah, G.A., Nuttman, C., Sinsin, B., 2012. Wild mammals trade for zootherapeutic and mythic purposes in Benin (West Africa):

capitalizing species involved, provision sources, and implications for conservation. In: Nóbregra Alves, R.R., Rosa, I.L. (Eds.), Animals In Traditional Folk Medicine. Springer, Berlin, Heidelberg, pp. 367–381.

Dupain, J., Nackoney, J., Vargas, J.M., Johnson, P.J., Farfan, M.A., Bofaso, M., et al., 2012. Bushmeat characteristics vary with catchment conditions in a Congo market. Biol. Conserv. 146 (1), 32–40.

Duri, F.P.T., 2017. Development discourse and the legacies of pre-colonial Shona environmental jurisprudence: pangolins and political opportunism in independent Zimbabwe. In: Mawere, M. (Ed.), Underdevelopment, Development and the Future of Africa. Langaa Research and Publishing Common Initiative Group, Bamenda, Cameroon, pp. 435–460.

Fa, J.E., Seymour, S., Dupain, J., Amin, R., Albtrechtsen, L., Macdonald, D., 2006. Getting to grips with the magnitude of exploitation: bushmeat in the Cross-Sanaga region, Nigeria and Cameroon. Biol. Conserv. 129 (4), 497–510.

Fargeot, C., 2013. La chasse commerciale en Afrique central: une menace pour la biodiversite ou une activite economique durable? La cas de la Republique Centrafricaine. Ph.D. Thesis, L'Universite Paul Valery, Montpellier, France.

Gill, D.J.C., 2010. Drivers of Change in Hunter Offtake and Hunting Strategies in Sendje, Equatorial Guinea. M.Sc. Thesis, Imperial College London, UK.

Gonedelé Bi, S., Koné, I., Béné, J.C.K., Bitty, E.A., Yao, K. A., Kouassi, B.A., et al., 2017. Bushmeat hunting around a remnant coastal rainforest in Coté d'Ivoire. Oryx 51 (3), 418–427.

Greengrass, E., 2016. Commercial hunting to supply urban markets threatens mammalian diversity in Sapo National Park, Liberia. Oryx 50 (3), 397–404.

Hayward, M.W., 2009. Bushmeat hunting in Dwesa and Cwebe Nature Reserves, Eastern Cape, South Africa. South Afr. J. Wildlife Res. 39 (1), 70–84.

Heinrich, S., Wittman, T.A., Rosse, J.V., Shepherd, C.R., Challender, D.W.S., Cassey, P., 2017. The Global Trafficking of Pangolins: A Comprehensive Summary of Seizures and Trafficking Routes From 2010–2015. TRAFFIC, Southeast Asia Regional Office, Petaling Jaya, Selangor, Malaysia.

Hodgkinson, C., 2009. Tourists, Gorillas and Guns: Integrating Conservation and Development in the Central African Republic. Ph.D. Thesis, University College London, London, UK.

Hoffmann, M., Cronin, D.T., Hearn, G., Butynski, T.M., Do Linh San, E., 2015. A review of evidence for the presence of Two-spotted Palm Civet *Nandinia binotata* and four other small carnivores in Bioko, Equatorial Guinea. Small Carnivore Conserv. 52 & 53, 13–23.

Hoyt, R., 2004. Wild meat harvest and trade in Liberia: managing biodiversity, economic and social impacts. Overseas Development Institute Wildlife Policy Briefing Number 6. Overseas Development Institute, London, UK.

Ichikawa, M., 1987. Food restrictions of the Mbuti Pygmies, Eastern Zaire. Afr. Study Monogr. 6, 97–121.

Infield, M., 1988. Hunting, Trapping and Fishing in Villages Within and on the Periphery of the Korup National Park. WWF Report, UK.

Ingram, D.J., Coad, L., Abernethy, K.A., Maisels, F., Stokes, E.J., Bobo, K.S., et al., 2018. Assessing Africa-wide pangolin exploitation by scaling local data. Conserv. Lett. 11 (2), e12389.

Ingram, D.J., Cronin, D.T., Challender, D.W.S., Venditti, D. M., Gonder, M.K., 2019. Characterizing trafficking and trade of pangolins in the Gulf of Guinea. Glob. Ecol. Conserv. 17, e00576.

Jacobsen, N.H.G., Newbery, R.E., De Wet, M.J., Viljoen, P. C., Pietersen, E., 1991. A contribution of the ecology of the Steppe Pangolin *Manis temminckii* in the Transvaal. Zeitschrift für Säugetierkunde 56 (2), 94–100.

Kingdon, J., 1974. East African Mammals: An Atlas of Evolution in Africa, vol. 1. University of Chicago Press, Chicago.

Kümpel, N.F., 2006. Incentives for Sustainable Hunting of Bushmeat in Río Muni, Equatorial Guinea. Ph.D. Thesis, Imperial College London, UK.

Kyle, R., 2000. Some notes on the occurrence and conservation status of *Manis temminckii*, the pangolin, in Maputaland, KwaZulu/Natal. Koedoe 43, 97–98.

Lupo, K.D., Schmitt, D.N., 2002. Upper paleolithic net-hunting, small prey exploitation, and women's work effort: a view from the ethnographic and ethnoarchaeological record of the Congo Basin. J. Archeol. Method Theory 9 (2), 147–179.

MacKinney, L.C., 1946. Animal substances in materia medica. J. Hist. Med. Allied Sci. 1 (1), 149–170.

Mafongoya, P.L., Ajayi, O.C., 2017. Indigenous Knowledge Systems and Climate Change Management in Africa. CTA, Wageningen, The Netherlands.

Mambeya, M.M., Baker, F., Momboua, B.R., Pambo, A.F.K., Hega, M., Okouyi, V.J.O., et al., 2018. The emergence of a commercial trade in pangolins from Gabon. Afr. J. Ecol. 56 (3), 601–609.

Marshall, N.T., 1998. Searching for a Cure: Conservation of Medicinal Wildlife Resources in East and Southern Africa. TRAFFIC International, Cambridge, UK.

Mbete, R.A., 2012. Household Bushmeat Consumption in Brazzaville, the Congo. Ph.D. Thesis, University of Liege, Liege, Belgium.

Mbilinyi, S., 2014. Medicinal Use of Wild Animal Products by the Local Communities Around Ruaha National

Park and Mbomipa Wildlife Management Area. B.Sc. Thesis, Sokoine University of Agriculture, Morogoro, Tanzania.

Milner-Gulland, E.J., Bennett, E., the SCB 2002 Annual Meeting Wild Meat Group, 2003. Wild meat: the bigger picture. Trends Ecol. Evol. 18 (7), 351–357.

Mouté, A., 2010. Etat des lieux et perspectives de gestion durable de la chasse villageoise en peripherie nord-est du parc national de Korup, region du sud-ouest de Cameroun. M.Sc. Thesis, Universite de Dschang, Dschang, Cameroon.

Ngoufo, R., Yongyeh, N.K., Obioha, E.E., Bobo, K.S., Jimoh, S.O., Waltert, M., 2014. Social norms and cultural services – community belief system and use of wildlife products in the northern periphery of the Korup National Park, south-west Cameroon. Change Adapt. Socioecol. Syst. 1, 26–34.

Niehaus, I.A., 1993. Witch-hunting and political legitimacy: continuity and change in Green Valley, Lebowa, 1930-91. Africa: J. Int. Afr. Inst. 63 (4), 498–530.

Ntiamoa-Baidu, Y., 1987. West African wildlife: a resource in jeopardy. Unasylva 39 (2), 27–35.

Olupot, W., Mcneilage, A.J., Plumptre, A.J., 2009. An Analysis of Socioeconomics of Bushmeat Hunting at Major Hunting Sites in Uganda. Working Paper 38. Wildlife Conservation Society (WCS), Kampala, Uganda.

Ordaz-Nemeth, I., Arandjelovic, M., Boesch, L., Gatiso, T., Grimes, T., Kuehl, H.S., et al., 2017. The socio-economic drivers of bushmeat consumption during the West African Ebola crisis. PLoS Negl. Trop. Dis. 11 (3), e0005450.

Petrozzi, F., Amori, G., Franco, D., Gaubert, P., Pacini, N., Eniang, E.A., et al., 2016. Ecology of the bushmeat trade in West and Central Africa. Trop. Ecol. 57 (3), 547–559.

Pietersen, D.W., McKechnie, A.E., Jansen, R., 2014. A review of the anthropogenic threats faced by Temminck's ground pangolin, *Smutsia temminckii*, in southern Africa. South Afr. J. Wildlife Res. 44 (2), 167–178.

Pietersen, D., Jansen, R., Swart, J., Kotze, A., 2016. A conservation assessment of *Smutsia temminckii*. In: Child, M.F., Roxburgh, L., Do Linh San, E., Raimondo, D., Davies-Mostert, H.T. (Eds.), The Red List of Mammals of South Africa, Swaziland and Lesotho. South African National Biodiversity Institute and Endangered Wildlife Trust, South Africa.

Sato, H., 1998. Folk etiology among the Baka, a group of hunter-gatherers in the African rainforest. Afr. Study Monogr. Suppl. 25, 33–46.

Schleicher, J., 2010. The Sustainability of Bushmeat Hunting in Two Villages in Central Gabon. M.Sc. Thesis, University of Oxford, UK.

Setlalekgomo, M.R., 2014. Ethnozoological survey of the indigenous knowledge on the use of pangolins (*Manis* sps [sic]) in traditional medicine in Lentsweletau Extended Area in Botswana. J. Anim. Sci. Adv. 4 (6), 883–890.

Smithers, R.H.N., 1966. The Mammals of Rhodesia. Zambia and Malawi. Collins, London.

Sodeinde, O.A., Adedipe, S.R., 1994. Pangolins in south-west Nigeria – current status and prognosis. Oryx 28 (1), 43–50.

Soewu, D.A., Ayodele, I.A., 2009. Utilisation of pangolin (*Manis* sp. [sic]) in traditional Yorubic medicine in Ijebu Province, Ogun State, Nigeria. J. Ethnobiol. Ethnomed. 5, 39.

Soewu, D.A., Adekanola, T.A., 2011. Traditional-medical knowledge and perception of pangolins (*Manis* sps [sic]) among Awori People, Southwestern Nigeria. J. Ethnobiol. Ethnomed. 7, 25.

Steel, L., Colom, A., Maisels, F., Shapiro, A., 2008. The Scale and Dynamics of Wildlife Trade Originating in the South of the Salonga-Lukenie-Sankuru landscape. WWF, Democratic Republic of Congo.

Svensson, M., Bersacola, E., Bearder, S., 2014. Pangolins in Angolan bushmeat markets. News piece for the IUCN SSC Pangolin Specialist Group. Available from: <https://www.pangolinsg.org/2014/06/01/pangolins-in-angolan-bushmeat-markets-2/>. [November 3, 2018].

van Aarde, R.J., Richardson, P.R.K., Pietersen, E., 1990. Report on the Behavioural Ecology of the Cape Pangolin (*Manis temminckii*). Mammal Research Institute, University of Pretoria, South Africa.

van Vliet, N., Nebesse, C., Gambalemoke, S., Akaibe, D., Nasi, R., 2012. The bushmeat market in Kisangani, Democratic Republic of Congo: implications for conservation and food security. Oryx 46 (2), 196–203.

Vanthomme, H., 2010. L'exploitation durable de la faune dans un village forestier de la Republique Centrafricaine: une approche interdisciplinaire. Ph.D. Thesis, Museum National D'Histoire Naturelle, Paris, France.

Walsh, M.T., 1995/96. The ritual sacrifice of pangolins among the Sangu of south-west Tanzania. Bull. Int. Committ. Urgent Anthropol. Ethnol. Res. 37/38, 155–170.

Warchol, G., Johnson, B., 2009. Wildlife crime in the game reserves of South Africa: a research note. Int. J. Comp. Appl. Crim. Just. 33 (1), 143–154.

Whiting, M.J., Williams, V.L., Hibbitts, T.J., 2011. Animals traded for traditional medicine at the Faraday market in South Africa: species diversity and conservation implications. J. Zool. 284 (2), 84–96.

Willcox, A.S., Nambu, D.M., 2007. Wildlife hunting practices and bushmeat dynamics of the Banyangi and Mbo people of southwestern Cameroon. Biol. Conserv. 134 (2), 251–261.

Williams, V.L., Whiting, M.J., 2016. A picture of health? Animal use and the Faraday traditional medicine market, South Africa. J. Ethnopharmacol. 179, 265–273.

World Health Organization, 2013. World Health Organization Traditional Medicine Strategy 2014–2023. World Health Organization, Geneva, Switzerland.

Wright, J.H., Priston, N.E.C., 2010. Hunting and trapping in Lebialem Division, Cameroon: bushmeat harvesting practices and human reliance. Endanger. Sp. Res. 11 (1), 1–12.

Zeba, S., 1998. Community Wildlife Management in West Africa: A Regional Review. Evaluating Eden Series Working Paper No.9. International Institute for Environment and Development (IIED), London, UK.

CHAPTER

16

International trade and trafficking in pangolins, 1900–2019

Daniel W.S. Challender[1,2], *Sarah Heinrich*[2,3,4], *Chris R. Shepherd*[2,4] *and Lydia K.D. Katsis*[2]

[1]Department of Zoology and Oxford Martin School, University of Oxford, Oxford, United Kingdom [2]IUCN SSC Pangolin Specialist Group, ℅ Zoological Society of London, Regent's Park, London, United Kingdom [3]School of Biological Sciences, University of Adelaide, Adelaide, SA, Australia [4]Monitor Conservation Research Society, Big Lake Ranch, BC, Canada

OUTLINE

Introduction

Pangolins and their derivatives have long been in commercial, international trade. This can be traced back to at least the early 20th century, but likely took place earlier, and has continued in one form or another since, both legally and illegally, culminating in high volumes of pangolins and their parts being trafficked since the turn of the 21st century. This chapter examines international trade and trafficking in pangolins from 1900 to July 2019.

Pangolins
DOI: https://doi.org/10.1016/B978-0-12-815507-3.00016-2

It discusses historic trade and presents updated analyses of trade data from the Convention on International Trade in Endangered Species of Wild Fauna and Flora (CITES), and illegal trade data based on seizure records. It discusses spatial and temporal international trade and trafficking dynamics, its impact on populations, and contemporary drivers.

CITES trade data were downloaded from the CITES website (https://trade.cites.org/) for analysis on September 13, 2018, in the form of a comparative tabulation report for *Manis* spp. from 1975 to 2016 (the latest year for which complete data were available), including all parties, sources, purposes and trade terms. Data for *Manis* spp. were downloaded because the CITES taxonomic nomenclature reference for mammals (Wilson and Reeder, 2005) includes all pangolins in this genus. Results and discussion follow the taxonomy presented in Chapter 2, i.e., pangolins reside in three genera: *Manis*, *Phataginus* and *Smutsia* (see also Gaudin et al., 2009; Gaubert et al., 2018).

An updated version of the dataset presented in Challender et al. (2015), including African and Asian pangolins, was analyzed to characterize illegal international trade between 2000 and July 2019, and was chosen because it enabled analyses from as far back as 2000 and up to 2019. The conversion parameters used to estimate the number of animals in trade (e.g., from quantities of scales) are presented in Table 16.1. Estimating the number of each species of pangolin in illegal trade is difficult because most data sources do not record this information, rather only "pangolin" (i.e., Manidae spp.). Where this was the case conversion parameters for the Sunda pangolin (*Manis javanica*) were used (see Table 16.1). This is because accurate conversion parameters exist for this species, which is larger than the arboreal and semi-arboreal African pangolins, and smaller than the terrestrial African pangolins, the smallest and largest species of the Manidae respectively. However, this may under- or over-estimate the number of pangolins in any given seizure, and overall, depending on the species involved (Challender and Waterman, 2017). The same parameters were used to estimate the number of pangolins in historic trade (i.e., 1900–1970s) and trade reported to CITES where it involved quantities of scales.

Early-mid 20th century trade, 1900–1970s

Records indicate that commercial harvesting and international trade in pangolins took place throughout the early to mid-20th century. Dammerman (1929) reports that several tonnes of Sunda pangolin scales were exported from Java, Indonesia to China between 1925 and 1929, involving up to 10,000 animals annually, despite the species being legally protected (Nijman, 2015). Harrisson and Loh (1965) report that between 1958 and 1964 over 60 tonnes of scales were exported from Indonesian Borneo, via Sarawak in Malaysia, to Singapore and Hong Kong, now Hong Kong Special Administrative Region (hereafter "Hong Kong"), reportedly for re-export to China for use in Traditional Chinese Medicine (TCM) (Allen, 1938). Assuming the figures reported by Harrisson and Loh are accurate, and using contemporary conversion parameters (see Table 16.1), this trade from Indonesian Borneo likely involved more than 166,000 Sunda pangolins. Around the same time, Chinese pangolin (*M. pentadactyla*) populations in Taiwan, were in decline due to extensive hunting for the domestic leather industry, which involved a minimum estimate of 60,000 animals annually between the 1950s and 1970s (Anon, 1992). Consequently, and due to resulting population declines, Taiwan increasingly relied upon imports of pangolins from Southeast Asia for supply. An estimated

TABLE 16.1 Conversion parameters used to estimate number of pangolins in illegal trade. Parameters used presented only.

Species	Derivative		
	Individual (kg)	Scales (g)	Meat (kg)
Chinese pangolin		573.47[a]	
Sunda pangolin	4.96[b]	360.51[a]	4.59[c]
Philippine pangolin	4.96[b]	360.51[a]	4.59[c]
Indian pangolin		3400[d]	
White-bellied pangolin		301[e]	
Giant pangolin		3600[f]	
Manis spp.	4.96[b]	360.51[a]	4.59[c]
Phataginus spp.		301[g]	
Phataginus/Smutsia spp.		360.51[a]	
Manidae spp.	4.96[b]	360.51[a]	4.59[c]

[a]*From Zhou, Z.-M., Zhao, H., Zhang, Z.-X., Wang, Z.-H., Wang, H., 2012., Allometry of scales in Chinese pangolins (*Manis pentadactyla*) and Malayan pangolins (*Manis javanica*) and application in judicial expertise. Zool. Res. 33 (3), 271–275.*
[b]*Trimmed mean. Taken from Pantel, S., Anak, N.A., 2010. A preliminary assessment of Sunda pangolin trade in Sabah. TRAFFIC Southeast Asia, Petaling Jaya, Selangor, Malaysia.*
[c]*From Challender, D.W.S., Harrop, S.R., MacMillan, D.C., 2015. Understanding markets to conserve trade-threatened species in CITES. Biol. Conserv. 187, 249–259.*
[d]*From Mohapatra, R.K., Panda, S., Nair, M.V., Acharjyo, L.N., Challender, D.W.S., 2015. A note on the illegal trade and use of pangolin body parts in India. TRAFFIC Bull. 27 (1), 33–40.*
[e]*Calculated from parameters in Table 9.1 (Chapter 9).*
[f]*Data from the Tikki Hywood Foundation.*
[g]*As e.*

50,000–60,000 skins, most of which likely comprised Sunda pangolins, were imported annually throughout the 1970s from Cambodia, Lao People's Democratic Republic (Lao PDR), Indonesia, Malaysia, Myanmar, Vietnam and the Philippines, thereby also including the Philippine pangolin (*M. culionensis*; Anon, 1992). Although not recognized as a distinct species until 2005 (Gaubert and Antunes, 2005; see Chapter 7), trade in the Sunda pangolin originating from the Philippines must have involved the Philippine pangolin. Due to a hunting ban, combined with increasing labor costs and reported issues with international supply, the domestic pangolin leather industry in Taiwan closed in the 1980s (Anon, 1992). In addition to population declines in Taiwan, this trade likely contributed to declines in Sunda pangolin populations in Southeast Asia (see next section).

Late 20th century trade, 1975–2000

Global monitoring of international wildlife trade improved with the advent of CITES in 1975. At its inception, the Asian pangolin species recognized at the time, the Sunda, Chinese and Indian pangolin (*M. crassicaudata*) were included in Appendix II. Temminck's pangolin (*Smutsia temminckii*) was included in Appendix

I and the three remaining African species were included in Appendix III by Ghana in 1976. In 1995, all species of pangolin were included in Appendix II (see Chapter 19).

Trade reported to CITES between 1975 and 2000 involved an estimated 776,000 pangolins (Heinrich et al., 2016). This primarily comprised skins, an estimated 509,564 of which reportedly originated in, or were exported from, 11 of the 19 Asian pangolin range states (Fig. 16.1A). The majority of this trade involved the Sunda pangolin (87%; 442,966 of 509,564 skins), encompassing trade in the Philippine species, with much less trade involving the Chinese (11%; 53,874 skins) or Indian pangolin (2%; 10,555 skins; see Fig. 16.1B). However, it is known that pangolin skins in trade were misidentified and it is possible that this trade did not involve the Indian pangolin (Anon, 1999a,c; Broad et al., 1988).

These data indicate that a mean of 21,232 skins/year were traded internationally between 1977 and 2000 peaking at almost 60,000 skins in 1981 and notably in the year 2000, with nearly 74,000 skins traded (Fig. 16.1A and B). The latter peak is potentially due to a proposal to include Asian pangolins in CITES Appendix I at the 11th Conference of the Parties (CoP) meeting in the year 2000 (see Chapter 19). CITES listing proposals have been shown to stimulate trade (Rivalan et al., 2007). There was negligible reported trade in 1989 and declining volumes in the two preceding years, which has been attributed, in part, to the establishment of an import ban on pangolin skins from Thailand and Indonesia in the United States in 1987 (Anon, 1992; Nooren and Claridge, 2001). Although many range states were involved in this trade, the source shifted over time, reportedly due to dwindling availability of pangolins (Anon, 1992, 1999a), but also perhaps to avoid new prohibitions on imports to the United States, the foremost market for skins at the time (Heinrich et al., 2016; Nooren and Claridge, 2001). The main exporters were Indonesia and Thailand in the late 1970s and 1980s, Lao PDR in the 1990s, and Malaysia in the late 1990s (Fig. 16.1A). Most of this trade was for commercial purposes despite pangolins being protected species in key exporting states including Indonesia, Malaysia and Thailand. The vast majority of this trade was ultimately destined to the United States and Mexico, but with substantial re-exports from Japan and Singapore. Once in North America the skins were manufactured into leather goods including handbags, belts, wallets and boots for wholesale and retail (Fig. 16.2).

Scales and an array of other Asian pangolin derivatives were also traded internationally between 1975 and 2000 according to CITES trade data. This included nearly 17,000 kg of scales between 1994 and 2000, equivalent to an estimated 47,000 Sunda pangolins, which were exported from Malaysia and destined for use in TCM in China and Hong Kong. To a far lesser extent trade involved the Chinese pangolin and unidentified "*Manis* spp." It also involved comparably low volumes of live animals, meat, and other derivatives (Challender and Waterman, 2017; Heinrich et al., 2016).

By comparison, there was little international trade in African pangolins reported to CITES. It mainly involved live white-bellied (*Phataginus tricuspis*) and black-bellied (*P. tetradactyla*) pangolins, nearly 150 of which were imported to Japan and the United States for captive breeding and commercial purposes. It also involved small quantities of bodies and carvings among other derivatives (Challender and Waterman, 2017).

Although trade in Asian pangolins reported to CITES in this period was substantial, it was dwarfed by unreported and therefore seemingly illegal trade. Extracting data from the CITES Review of Significant Trade (RST) reports on pangolins (see Anon, 1992, 1999a,b, c), Challender et al. (2015) estimated that this trade involved, at minimum, an additional

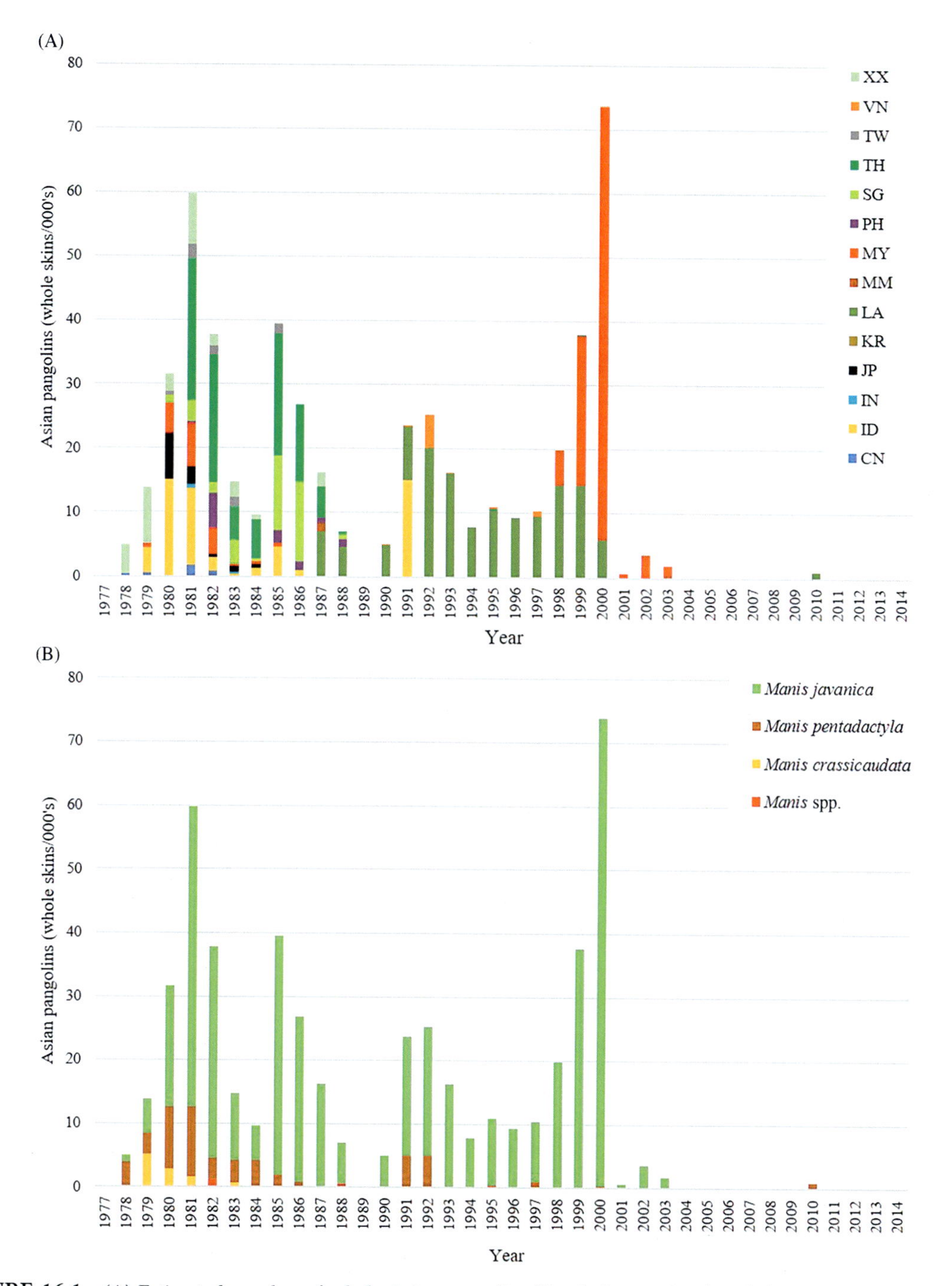

FIGURE 16.1 (A) Estimated number of whole Asian pangolin skins in international trade between 1977 and 2014 as reported by importers, by country of origin (or exporters if origin not reported). This includes trade reported as *Manis* spp. where it originated in or was exported from Asian pangolin range states. The last year for which trade in whole Asian pangolin skins was reported is 2014. Trade originated in or was exported from 11 Asian pangolin range states: China (CN), Indonesia (ID), India (IN), Lao PDR (LA), Myanmar (MM), Malaysia (MY), Philippines (PH), Singapore (SG), Thailand (TH), Taiwan (TW) and Vietnam (VN). JP=Japan, XX=Unknown. (B) Estimated number of whole Asian pangolin skins in international trade between 1977 and 2014 as reported by importers, by species. This includes trade in *Manis* spp. originating in or exported from Asian pangolin range states. *Source: CITES trade data.*

FIGURE 16.2 Pangolin leather products. *Photo credit: United States Fish and Wildlife Service.*

500,000–935,000 Asian pangolins. It took the form of imports of several tonnes or tens of tonnes of scales annually by Taiwan and South Korea in the 1980s and 1990s, and of similar volumes by China in the 1990s; China also imported tens of thousands of live pangolins mainly from Southeast Asia, most of which likely comprised the Chinese and Sunda species (Anon, 1992, 1999a,b,c; Broad et al., 1988; Li and Li, 1998; Wu et al., 2004; Wu and Ma, 2007). This trade was not formally reported to CITES, but its inclusion in the RST process meant it formed part of the evidence base that informed policy decisions during the 1990s (see Chapter 19).

A lack of knowledge of pangolin populations and how they behave (e.g., mortality and recruitment rates) has meant that determining the impact of international trade on populations, and disentangling it from local and domestic use, has proven challenging (Anon, 1992, 1999a,b). However, evidence indicates that international trade, legal and illegal, has played a major role in the decline and in some cases the commercial depletion of Asian pangolin populations. Through the CITES RST process between 1988 and 1999 there was increasing recognition that trade-driven harvesting had contributed to severe declines in populations of Chinese and Sunda pangolins, mainly in Southeast Asia (Anon, 1999a,b; Broad et al., 1988). Although anecdotal, this included estimates by villagers in different parts of Lao PDR in the 1990s that populations of the Sunda pangolin had declined in places by up to 99% of levels in the 1960s because of overexploitation (Duckworth et al., 1999). There is similar evidence from other parts of Southeast Asia. In various parts of Peninsular Malaysia, the indigenous "Orang Asli" reported that Sunda pangolins were abundant in the 1970s and 1980s, but populations declined severely up to the 2000s because of overharvesting (D. Challender, unpubl. data). One trading operation in Peninsular Malaysia reportedly harvested 100 tonnes of pangolins a month (equivalent to about 20,000 Sunda pangolins) for export during the 1990s, but ceased trading because populations had been all but commercially depleted (Anon, unpubl. data). In Vietnam, interviews conducted with hunters in three different provinces in northern and central parts of the country in 2007 revealed that populations of Chinese and Sunda pangolins declined in the 1990s and 2000s, predominantly due to overexploitation (Newton et al., 2008; P. Newton, pers. comm.). Similarly, in Cambodia, hunters have indicated that the Sunda pangolin has been extirpated from some areas entirely due to overharvesting, and the same scenario is apparent in parts of Myanmar and Thailand (see Chapter 6).

Pangolin populations also declined in East Asia in the late 20th century, and though not due to international trade, it is critical to understanding contemporary trade and trafficking dynamics. Between the 1960s and 1980s, an estimated 150,000–160,000 pangolins, were harvested annually in China (Zhang, 2009), predominantly or exclusively involving the Chinese pangolin (see Chapter 6 for discussion on the occurrence of the Sunda pangolin in China). This was for the consumption of meat, favored in southern China, with assumed health benefits (see Chapter 14), and the use of scales in TCM (Zhang, 2009). Such was the magnitude of this exploitation that it apparently led to the commercial extinction of pangolins in China by the mid-1990s (Anon, 1999b; Zhang, 2009). Wu et al. (2004) estimated that pangolin populations in China declined by up to 94% between the 1960s and early 2000s. Prior to the 1990s, China had essentially been self-sufficient in pangolins and their derivatives. However, due to shortages of domestic supply, by the early 1990s, it was importing large quantities of both live animals and scales from neighboring countries including Lao PDR, Myanmar and Vietnam (Anon, 1999b). Virtually all of this trade went unreported and was apparently illegal. Yet, international supplies reportedly collapsed in 1995 which led to the price of scales in southern China more than doubling between 1995 and 1996 (SATCM, 1996). As a consequence, reports suggest that there remained severe shortages of scales in China in the mid-late 1990s, which led to some TCM companies publically offering to purchase substantial quantities of scales (Anon, 1999b). This apparent shortage of scales, combined with persistent demand for pangolin meat, scales and products containing scales, means China plays a major role in contemporary pangolin trafficking.

Trade and trafficking, 2000–19

Trade reported to CITES (2001–16)

At CITES CoP11 in the year 2000, zero export quotas were established for commercial, international trade in wild-caught Asian pangolins based on concerns about the sustainability of trade and its impact on populations (see Chapter 19). Due to ongoing illegal trade in the interim period, at CoP17 in 2016, all eight species of pangolin were transferred from Appendix II to I, establishing an international ban on commercial trade in wild-caught pangolins globally, which entered into force on January 2, 2017.

Between 2001 and 2016, there were low levels of reported trade in Asian pangolins compared to pre-2000 (Fig. 16.1A,B). This mainly involved skins, with approximately 6000 Sunda pangolin skins traded between 2001 and 2003, and it appears that the introduction of zero export quotas caused the decline of this trade with little reported or illegal trade in skins since (Fig. 16.1A,B; Challender et al., 2015; Challender and Waterman, 2017; Heinrich et al., 2016). An exception is 1000 Chinese pangolin skins that were exported from Lao PDR to Mexico in 2010 that were reportedly ranched (see Challender and Waterman, 2017). Trade otherwise involved 3200 kg of scales that were exported from Malaysia to China and Hong Kong, and quantities of medicines, shoes and among other items, leather products (Challender and Waterman, 2017; Heinrich et al., 2016).

Quantities of African pangolins in international trade reported to CITES increased noticeably in the period 2001–2016. This mainly involved scales and live animals, and primarily the white-bellied and giant pangolin (*S. gigantea*). Accurately estimating trade volumes is problematic because reported quantities differ between importers and exporters. Based on importer records, trade involved an estimated 2510 kg of white-bellied pangolin scales (equivalent to an estimated ~8000 animals) between 2013 and 2016. These scales were exported from Democratic Republic of the Congo (DRC) and Congo primarily, and Togo, the vast majority of which, by weight (97%), were imported to China. Based on a combination of data from importers and exporters, trade also involved an estimated 11.3 tonnes of giant pangolin scales (equivalent to an estimated ~3100 animals). More than

90% of this trade was exported from Burundi to Hong Kong (57%, ~6500 kg) and from Uganda to China (36%, ~4000 kg) between 2014 and 2016. There is no evidence for the presence of the giant pangolin in Burundi (see Chapter 10) which suggests that these scales were harvested in another range country or were misidentified. A further 750 kg of unidentified African pangolin scales were exported from DRC to China in 2015.

Assuming trade reported to CITES represents actual trade volumes (as opposed to permits issued), trade post-2000 also involved ~1340 live African pangolins, most of which were exported to Asia. This included 650 white-bellied pangolins exported from Nigeria and Togo to China, Lao PDR and Vietnam between 2012 and 2015; 200 black-bellied pangolins exported from Nigeria to China in 2015; and 150 giant pangolins exported from Nigeria and Togo to China and Lao PDR between 2012 and 2015, all for captive breeding or commercial purposes. This trade is apparently associated with attempts to commercially captive breed or farm pangolins (see Chapter 32). Other trade in live African pangolins involved the white-bellied species, an estimated 287 of which were traded mainly for commercial purposes between 2001 and 2016; the majority (83%) of animals were exported from Togo, but also Benin and Cameroon. The United States imported 132 individuals and smaller numbers were imported by, among others, the Czech Republic, Italy, Japan, Malaysia, South Korea and the United Kingdom. Small numbers of black-bellied, giant and Temminck's pangolins were traded for the same purpose in this period. Other trade in African pangolins involved small quantities of skulls, skins and other derivatives (see Challender and Waterman, 2017).

Illegal international trade (2000–19)

Illegal wildlife trade is difficult to monitor, in part because it may be clandestine. Quantifying it is typically dependent on government or open source data (e.g., seizures reported in the media) or the use of innovative methods (e.g., Hinsley et al., 2017). Here, an updated dataset from Challender et al. (2015) was analyzed to characterize international pangolin trafficking, including selected government, non-governmental organization (NGO) and open source data. This dataset was chosen because it enabled an evaluation of pangolin trafficking across a near 20-year period (August 2000–July 2019). Species of pangolin in illegal trade were inferred where possible based on reported country or continent of origin (e.g., Africa) and species distribution. However, it was not possible to infer numbers of specific species involved in each seizure because reports (e.g., in the media) rarely record the species of pangolin involved (see Challender et al., 2015). Despite the utility of seizure data, inherent detection and reporting biases (see Underwood et al., 2013) mean that the results should not be interpreted as temporal trends or absolute trade volumes (Milliken et al., 2012). In addition, data on trafficking routes is only available for a subset of these data, and the results should be treated with caution due to potential inaccuracies (see Challender and Waterman, 2017; Heinrich et al., 2017).

International trafficking in pangolins and their derivatives between August 2000 and July 2019 involved an estimated equivalent of 895,000 animals (Fig. 16.3A). This is based on 1474 seizures and other records of illegal trade (e.g., trading syndicate logbooks; see Pantel and Anak, 2010). The actual number of pangolins trafficked likely exceeds this figure as only a portion of illegal trade is intercepted, but accounting for undetected and unreported illegal trade is challenging (Phelps and Webb, 2015) and beyond the scope of this chapter. It is possible that this figure over-estimates the equivalent number of pangolins seized, which may be the case if trafficking of African pangolin scales in particular mainly involved the

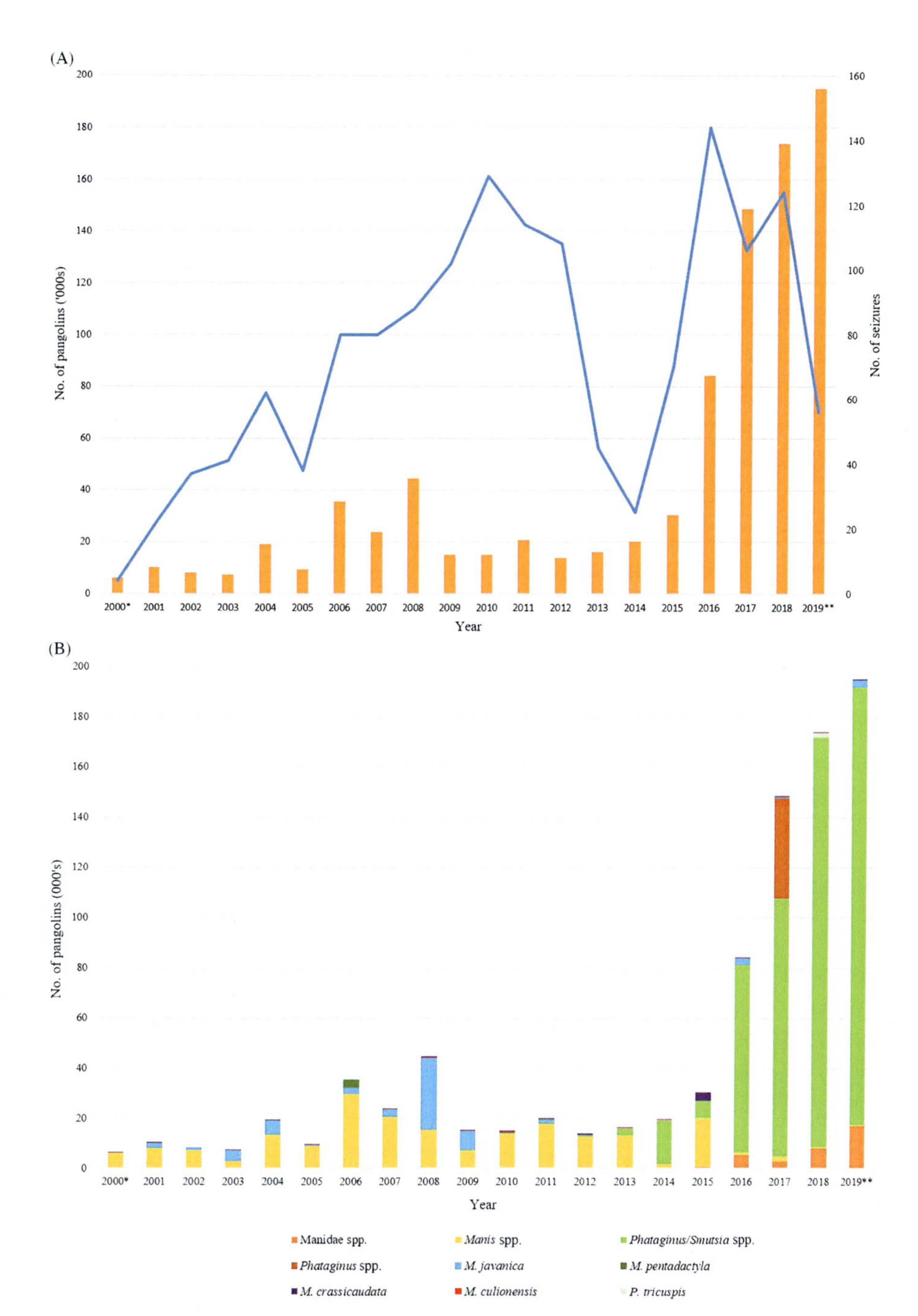

FIGURE 16.3 (A) Estimated number of pangolins illegally traded and number of seizures between August 2000 and July 2019. (B) Estimated number of pangolins illegally traded between August 2000 and July 2019 by species, genera, or presented as Manidae spp., inferred from available seizures data. *(A and B) Source: open source and pangolin range state government data. Trade volumes based on seizures and illegal trade records. For full methods see Challender, D.W.S., Harrop, S.R., MacMillan, D.C., 2015. Understanding markets to conserve trade-threatened species in CITES. Biol. Conserv. 187, 249–259. Orange bars=estimated no. of pangolins; blue line=no. of seizures. *Seizures from August–December 2000 only. **Seizures from January–July 2019 only.*

terrestrial species (e.g., giant pangolin) because family-level conversion parameters were used (see Table 16.1). However, it seems likely based on what is known about the abundance of the tropical African pangolins (see Chapters 9–11), that seizures involve a higher proportion of white-bellied pangolins the than other species, especially considering the indiscriminate nature of pangolin harvesting. More accurate recording and reporting of pangolins in illegal trade would enable more robust analyses (see Chapter 34).

International trafficking involved all eight species of pangolin (Fig. 16.3B). Historically, recorded illegal international trade has predominantly involved Asian pangolins, but alarmingly most trade in this period – in which it was possible to infer the genus or species in trade – comprised African pangolins, and primarily took place between 2016 and 2019 (Fig. 16.3B). This involved an estimated 585,000 African pangolins, 65% of overall trade by number of pangolins. It primarily involved trade in *Phataginus/Smutsia* spp. (93%, 544,000 of 585,000 animals), i.e., African pangolins or their derivatives that were not identified to species level. It involved an additional estimated equivalent of 39,000 *Phataginus* spp. (7%, 39,000 of 585,000; see Fig. 16.3B). There was little reported trade in specific species, which was limited to an estimated 1997 white-bellied, 118 giant, and 144 Temminck's pangolins (Fig. 16.3B). However, illegal trade in each of these species almost certainly involved higher numbers of animals, accounting for trade recorded at the genus and/or family level. Trafficking of the black-bellied pangolin was not recorded specifically (Fig. 16.3B), but is reflected in illegal trade inferred as involving *Phataginus/Smutsia* spp. and *Phataginus* spp. The presence of this species in intercontinental trafficking to Asia has been confirmed through both forensic means (Mwale et al., 2017) and visual inspection of seized scales (e.g., in Abidjan, M. Shirley, pers. comm.).

Illegal trade in Asian pangolins involved an estimated 275,000 animals (31% of overall trade), most of which (73%; 202,000 of 275,000) involved *Manis* spp. (i.e., it was not possible to infer the species involved). It did include an estimated 65,000 Sunda pangolins (7% of overall illegal trade), but much lower quantities of the Chinese (~3500), Indian (~3700) and Philippine (~700) pangolins. However, like for African pangolins, illegal trade in each species almost certainly involved higher numbers of individuals, reflected in trade in *Manis* spp. and/or Manidae spp. (Fig. 16.3B). Trade in Manidae spp. involved an estimated 35,000 pangolins.

International trafficking involved whole pangolins (i.e., live or dead) and meat, scales and skins, and involved more than 50 countries (see Fig. 16.4). Seizures took place in 40 countries, but 55 countries are implicated as a country of origin, export, transit and/or destination (Fig. 16.4). This broadly corresponds with previous analyses of contemporary trafficking between 2010 and 2015 in which 67 countries were implicated (Heinrich et al., 2017). The 55 countries included 17 of the 19 Asian, and 25 of the 36 African, range states (Fig. 16.4).

The vast majority of illegal trade by estimated number of animals (83%) involved scales, equivalent to an estimated 745,000 pangolins. Most of this trade (80%, 592,000 of 745,000 estimated pangolins) took place between 2016 and July 2019 (Fig. 16.3B), and by sea, with pangolin scales concealed in boxes or sacks in shipping containers and declared as, among other cargo, fish (Fig. 16.5). Virtually all trade in African pangolins involved scales. Assuming reported countries of origin or export are accurate, and based on available data, African pangolin scales were primarily exported from Nigeria (226,000 pangolins), DRC (44,000), Cameroon (34,000), Uganda (15,000; without reported destination), Congo (11,000) and Ghana (10,000). Côte d'Ivoire, Burkino Faso and Liberia were reported as the origin of scales from an estimated >8300 pangolins. Based on

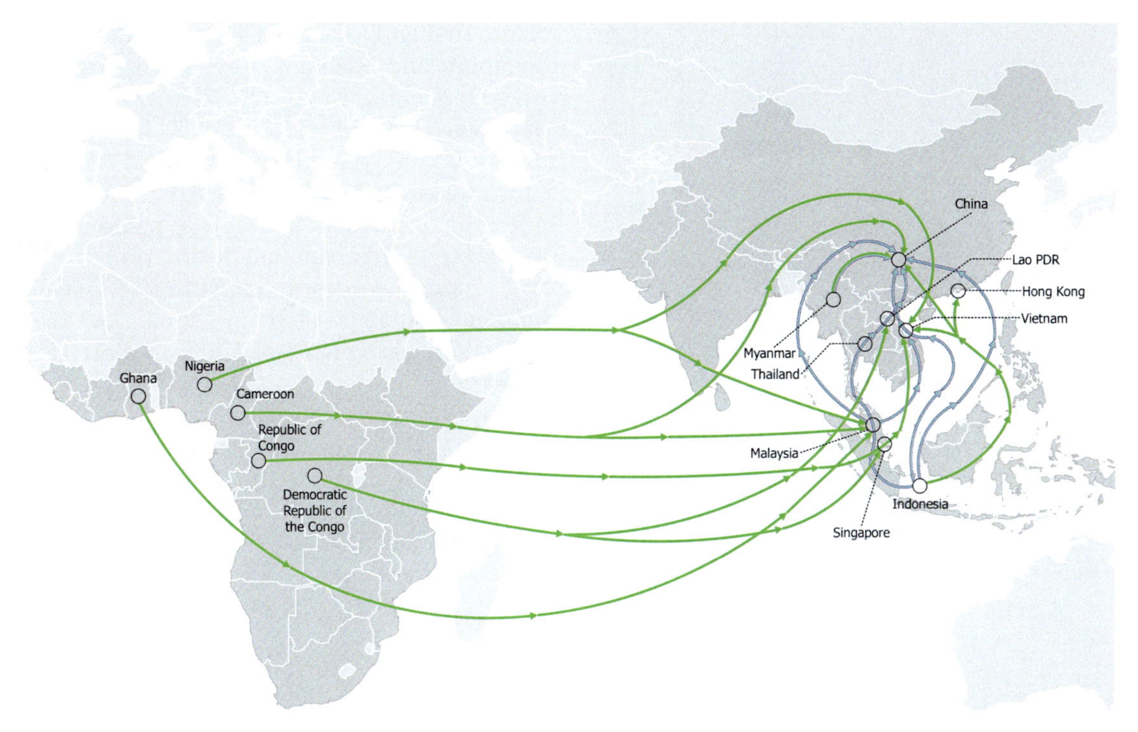

FIGURE 16.4 Pangolin range states in which trafficking is known to occur (shaded dark gray) and major trafficking routes between August 2000 and July 2019. Arrows indicate major trafficking routes based on available data for individual (i.e., live and dead) pangolins and meat (blue) and scales (green). Between 2000 and 2019 trafficking implicated the following countries: Angola, Australia, Austria, Belgium, Benin, Bhutan, Botswana, Burkino Faso, Cambodia, Cameroon, Central Africa Republic, China (including Hong Kong), Congo, Côte d'Ivoire, DRC, Ethiopia, France, Gabon, Ghana, Guinea, India, Indonesia, Kenya, Lao PDR, Liberia, Malawi, Malaysia, Morocco, Mozambique, Myanmar, North Korea, Namibia, Nepal, Nigeria, Pakistan, Philippines, Qatar, Sierra Leone, Singapore, South Africa, South Korea, South Sudan, Sri Lanka, Taiwan, Tanzania, Thailand, Turkey, United Arab Emirates, Uganda, United States, Vietnam, Zambia and Zimbabwe. *Map prepared by Helen O'Neill.*

available data this trade primarily transited through Central African Republic, Kenya, Malaysia, Singapore, Thailand, Turkey and Vietnam. Malaysia is notable as a well-known transit point for pangolin scales trafficked in Asia (Fig. 16.5; Krishnasamy and Shepherd, 2017). Malaysia was also reported as a destination, but it is unclear whether it actually comprises a destination or exists as a thoroughfare to East Asia. Assuming reported destinations are accurate, trafficked scales were primarily destined to China, as well as Vietnam and Lao PDR.

Trade in scales from Asian pangolins involved an estimated 128,000 animals, including all four species. For most of it (82%; 105,000 of 128,000 animals), it was not possible to infer the actual species involved. Where it was possible, most trade involved the Sunda pangolin (13%; 17,000 of 128,000 animals). Most of the trade in Asian pangolins originated in, or was exported from, Indonesia and Myanmar and destined for China, Hong Kong and Vietnam. However, countries of origin and export were diverse and included India, Malaysia, Nepal, Pakistan, the Philippines,

FIGURE 16.5 Pangolin scales seized in Malaysia en route from Africa to East Asia. *Photo credit: TRAFFIC.*

Singapore, Sri Lanka and Vietnam. Trade in scales in Manidae spp. involved an estimated equivalent of 31,000 pangolins and followed similar routes, both in terms of reported countries of origin and export, and destination.

Live and dead pangolins and meat accounted for a further 16% of trafficking by volume. It included an estimated 146,000 pangolins, primarily involving *Manis* spp. (66%; 97,000 of 146,000 animals) and the Sunda pangolin (30%; 44,000 of 146,000 animals). Trafficking in other species occurred at much lower volumes, but are likely included in the trafficking of *Manis* spp. Most of the trafficking either originated in or was exported from Indonesia and Malaysia, and to a lesser degree from Thailand, Lao PDR and Cambodia, Singapore and Vietnam. Key transit countries comprised Hong Kong and Vietnam, but also Malaysia, Thailand and Lao PDR. Large quantities of frozen pangolins (i.e., several tonnes at a time) were typically trafficked by sea from Indonesia to Vietnam and China, while smaller quantities of live animals were traded over land along routes comprising Malaysia-Thailand-Lao PDR-Vietnam-China. The main destination markets were China, followed by Vietnam.

Other trade involved an estimated 4450 skins, mainly involving the Sunda pangolin (3900 skins), but it is unclear whether this represents trade in skins only or trade in skins with the scales attached (i.e., skins being trafficked for the scales).

Impact of international trade and trafficking

Exploitation for international trade and trafficking has been a major cause of population decline in Asian pangolins, in particular for the Sunda and Chinese species. Although the Sunda pangolin has been in international commercial trade since at least the early 20th century, high volumes of both legal and illegal international trade in the 1950s to 1970s likely contributed to population declines: steep declines are documented for both species in parts of Southeast Asia from the 1980s to 2000s (see earlier discussion). In addition, Indigenous peoples groups in the Philippines report steep declines in populations of the Philippine pangolin between the 1980s and 2010s (Acosta and Schoppe, 2018). Due in part to international trade and trafficking, pangolins are increasingly rare, commercially extinct or apparently absent altogether at some sites in Southeast Asia (see Newton et al., 2008; Willcox et al., 2019). Where Asian pangolins do persist, they remain subject to exploitation, and it is likely that Singapore, Taiwan and smaller islands within the species ranges are the only locations where overexploitation is not a major threat.

International trafficking appears to be increasingly pervasive in Asia. Between 2000 and 2008, it largely involved Asian pangolins from Southeast Asia, typically trafficked to China and Vietnam (Challender et al., 2015; Pantel and Chin, 2009). However, between 2009 and 2019, 17 out of the 19 Asian pangolin range states were implicated in this trade (the exceptions were Bhutan and Brunei Darussalam). This included countries not prominent in illegal international pangolin trade historically, e.g., India, Nepal, Pakistan and Sri Lanka (Fig. 16.4). Evidence indicates that this is having a negative impact on populations in places. For example, population densities of the Indian pangolin in the Potohar Plateau region of Pakistan reportedly declined by 80% between 2010 and 2012 due to the killing of the animals and harvest of their scales for trafficking to China (Irshad et al., 2015; Mahmood et al., 2012; see Chapter 5). Crucially, there remains a need for the development of population monitoring methods to more rigorously quantify the impact of exploitation (see Chapters 34–35).

The most significant trend in international trade and trafficking of pangolins in the early 21st century has been the advent of intercontinental trade, both legal and illegal, from Africa to Asian markets since around 2008 (Challender and Hywood, 2012). This was not apparent in the early 2000s but it was suspected that there would be a shift in source to Africa following declines in Asian pangolins (Bräutigam et al., 1994). This shift is reflected in both the CITES trade data (see earlier discussion; Heinrich et al., 2016) and high levels of contemporary trafficking (Fig. 16.3B), which involves all four African species and at least 25 range countries (Fig. 16.4), and primarily scales destined to China.

Determining the impact of contemporary illegal international trade on populations of African pangolins is problematic because it is difficult to extricate it from local use. Pangolins have been consumed as bushmeat in West and Central Africa throughout human history and their body parts, including scales, are used in traditional medicines among other applications (see Chapter 15; Boakye et al., 2015; Soewu and Adekanola, 2011). Ingram et al. (2018) estimated that 0.4–2.7 million pangolins (most likely 0.4 million) are hunted annually for local use in Central Africa. However, evidence is starting to emerge which indicates that intercontinental trafficking is placing an additive pressure on populations and African pangolins are being targeted specifically for their scales for illicit export, and the meat consumed, sold as a by-product, or even discarded (CITES, 2019; S. Jones, pers. comm.). This appears to be the case in Cameroon and neighboring countries (O. Drori, pers. comm.), in Côte d'Ivoire (M. Shirley, pers. comm.) and Gabon, where trafficking routes appear distinct from the bushmeat trade (Mambeya et al., 2018). Elucidating these dynamics and how they vary both within and between range countries and temporally, is critical to understanding the impact of international trafficking, and overall exploitation, on African pangolin populations (Ingram et al., 2019).

Drivers of contemporary international trafficking

Overexploitation is the foremost threat to pangolins, and for most species international trafficking is the main threat (Chapters 4–11), the drivers of which are complex. Available evidence indicates that most pangolin products trafficked internationally are destined to China and Vietnam, where there is substantial demand for pangolin meat and scales and which, in part, is driving illegal trade

(Challender et al., 2019; see Chapters 14 and 22). This also applies to areas close to China e.g., the Myanmar-China border region (Nijman et al., 2016). Seizure data do indicate that other markets also exist (see Chaber et al., 2010; Heinrich et al., 2017). Regarding China, its position as a destination country can be traced back to the overexploitation of its own pangolin populations throughout the 1960s–1990s. Its size, combined with substantial demand, means that it plays a major role in contemporary international pangolin trafficking globally. However, an in-depth understanding of demand for pangolin products in both China and Vietnam is lacking (see Chapter 22). Available evidence indicates that demand for pangolin meat is characterized by conspicuous consumption and the desire for consumers to reinforce social status, while perceived rarity, illegality and high price are also important (Shairp et al., 2016). For instance, pangolin meat was offered for sale in restaurants in Ho Chi Minh City (HCMC), Vietnam, in 2018 for USD 300/kg (D. Challender, unpubl. data). Pangolin scales are in demand as an ingredient in TCM and traditional Vietnamese Medicine (TVM) to treat a range of ailments and prices are also high. In China the retail price of scales in 2016 was USD 800/kg, though scales are typically sold in much smaller quantities (e.g., 10 g).

Supply of pangolins and their parts trafficked internationally comes from rural parts of Asia and Africa. Historically, up to around 2008, supply to Asian markets was limited to Asia. Throughout the 1980s and 1990s (Anon, 1992; Anon 1999a,b), 2000s (Newton et al., 2008; Sopyan, 2009) and 2010s (e.g., Azhar et al., 2013; D'Cruze et al., 2018) pangolins have been collected, hunted or poached, and the animals and/or their parts entered into international trade, because there has been, and remains, a strong financial incentive to do so. The people who hunt and poach pangolins include poor and marginalized rural people living in absolute poverty in some cases (e.g., in parts of Malaysia and Indonesia), for which finding and selling one pangolin can equate to several months' income (D. Challender, unpubl. data). They also include those living in relative poverty, and hunting and trafficking of pangolins is used to earn extra income (Pantel and Chin, 2009) or support their livelihood aspirations (Challender and MacMillan, 2014). In 2004, one report indicated that individuals from one rural community in Vietnam were poaching and trafficking pangolins so that they could afford to buy a television (Anon, 2004). Rural dwellers (e.g., farmers in Indonesia) poach pangolins opportunistically (Sopyan, 2009), and plantation workers on monoculture plantations (e.g., in Malaysia and Indonesia) collect the animals for illegal trade (Azhar et al., 2013). In the 1980s and 1990s pangolins may have been kept and consumed locally, but evidence indicates that this is now typically foregone and the default response to finding a pangolin in most of Southeast Asia is to sell it illegally (MacMillan and Nguyen, 2014; Nuwer and Bell, 2014; G. Semiadi, pers. comm.). There is some variation to this dynamic with research in northeast India indicating that only pangolin scales are sold into illegal trade while the meat is consumed locally, but even selling the scales alone can equate to several months' income (D'Cruze et al., 2018).

The supply of pangolins for intercontinental trafficking from Africa are less well documented but there are parallels to Asia e.g., a high financial incentive to catch and sell the animals and/or their parts into trade. In West and Central Africa, pangolins are widely and openly traded in bushmeat markets in many places (e.g., Boakye et al., 2016). However, evidence suggests that in Gabon (Mambeya et al., 2018), Côte d'Ivoire (M. Shirley, pers. comm.) and Cameroon (O. Drori, pers. comm.) trafficking does not follow bushmeat supply routes. Rather, it is linked to the trafficking of other high value wildlife including elephant (Elephantidae) ivory (Mambeya et al., 2018), gorillas (*Gorilla* spp.), chimpanzees (*Pan troglodytes*), and leopard (*Panthera pardus*) skins, among other species (Ingram et al., 2019), and constitutes organized crime (UNODC, 2016).

Pangolin trafficking is known to be carried out by organized criminal groups in pursuit of financial profits. In Asia and Africa rural communities hunt or poach pangolins and store the animals and/or their scales at the household or village level, which are collected periodically by middlemen, and the animals and their derivatives may pass through various levels of traders or middlemen prior to export, and during transit to consumer markets (Pantel and Chin, 2009; Sopyan, 2009). Various criminal syndicates have been identified operating pangolin trafficking networks in Asia, and similar entities are likely responsible for intercontinental trafficking from Africa. Syndicates have been identified in Lao PDR (see Rademeyer, 2012), while Sopyan (2009) details that a syndicate operating in Palembang, Sumatra, in the mid-2000s exported 11 tonnes of Sunda pangolins (~2200 animals) a month to China and Vietnam. In Sabah, Malaysia, a syndicate operating between 2007 and 2009 poached ~22,000 Sunda pangolins for illegal export (Pantel and Anak, 2010). Evidence indicates that these groups operate by using legal businesses as a front for trafficking, by bribing officials to "look the other way" (Sopyan, 2009) and in some cases by using new trafficking routes and processing products to change their appearance (e.g., powdering scales; Heinrich et al., 2019; D. Challender, unpubl. data). Heinrich et al. (2017) identified a mean of 29 new pangolin trafficking routes a year between 2010 and 2015. Although seizures take place, available data suggest that arrests, prosecutions and conviction rates are low (Challender and Waterman, 2017). When traffickers are apprehended, they may also possess the means to secure their release. For example, in 2012, an individual caught trafficking pangolins was arrested in Thailand and charged and fined, and paid the USD 75,000 fine in full, in cash, within hours of arrest (see Challender and MacMillan, 2014). Zimbabwe is perhaps an exception, where pangolin seizures are increasing and perpetrators are often given the maximum jail sentence of nine years (Shepherd et al., 2017).

Addressing international pangolin trafficking

Pangolins have long been in international commercial trade, especially in Asia, but hunting and poaching for international trafficking now comprises a major threat to the continued survival of the species, perhaps with the exception of Temminck's pangolin (see Chapter 11). Historically, the pre-dominant approach to controlling international trade has been regulation of harvest and trade institutionalized through legislation. Pangolins are protected species in most range states, and have a long history in CITES (see Chapter 19), but measures adopted to date have been inadequate to prevent overexploitation.

Pangolin trafficking is complex, involving various actors and incentives, and takes place on a large geographic scale, and the problem is seemingly intractable. However, a range of conservation solutions exist which need to be implemented urgently to safeguard the species at the site level, to combat trafficking, and to address consumer demand. At the site level, there is a need for locally appropriate interventions to protect pangolins. In most cases, this will comprise a combination of law enforcement, which needs to be well resourced and effective, and local community engagement, for which partnerships should be built around mutually agreed theories of change (see Chapters 24 and 25). In addition, pangolin populations should be monitored in order to evaluate the effectiveness of these interventions (Chapter 35). Other interventions may also be needed to support site-based protection, such as innovative financing mechanisms (see Chapter 37) or pangolin tourism (see Chapter 38). Within, but also beyond, the site-level, and along international trafficking routes, capacitated and effective law enforcement based on good intelligence is needed in order to identify, disrupt and dismantle organized criminal networks smuggling pangolins and their parts, with efficient processing through judicial systems (see Chapters 17 and 18). Where countries are facing serious challenges implementing CITES, support is available from CITES and its parties (see Chapter 19). Finally, in key consumer markets, there is a need to reduce consumer demand for pangolin products, for example, by changing consumer behavior (see Chapter 22) in order to alleviate exploitation for international trafficking. These interventions are discussed in more detail in Section Three of this Volume.

References

Acosta, D., Schoppe, S., 2018. Proceedings of the stakeholder workshop on the Palawan Pangolin – Balintong, Bulwagang Princesa Tourist Inn, Puerto Princesa City, 17 February 2018. Katala Foundation Inc, Puerto Princesa City, Palawan, Philippines, pp. 1–10.

Allen, J.A., 1938. Chinese medicine and the pangolin. Nature 141, 72.

Anon, 1992. Review of Significant Trade in Animal Species included in CITES Appendix II, Detailed Review of 24 priority species, Indian, Malayan and Chinese pangolin. CITES, Geneva, Switzerland.

Anon, 1999a. Review of Significant Trade in Animal Species Included in CITES Appendix II, Detailed Review of 37 Species. *Manis javanica*. World Conservation Monitoring Centre, IUCN Species Survival Commission and TRAFFIC, Cambridge, UK.

Anon, 1999b. Review of Significant Trade in Animal Species Included in CITES Appendix II, Detailed Review of 37 Species. *Manis pentadactyla*. World Conservation Monitoring Centre, IUCN Species Survival Commission and TRAFFIC, Cambridge, UK.

Anon, 1999c. Review of Significant Trade in Animal Species Included in CITES Appendix II, Detailed Review of 37 Species. *Manis crassicaudata*. World Conservation Monitoring Centre, IUCN Species Survival Commission and TRAFFIC, Cambridge, UK.

Anon, 2004. Pangolins for televisions. Unpublished report. Hanoi, Vietnam.

Azhar, B., Lindenmayer, D., Wood, J., Fischer, J., Manning, A., McElhinny, C., et al., 2013. Contribution of illegal hunting, culling of pest species, road accidents and feral dogs to biodiversity loss in established oil-palm landscapes. Wildlife Res. 40 (1), 1–9.

Boakye, M.K., Pietersen, D.W., Kotzé, A., Dalton, D.-L., Jansen, R., 2015. Knowledge and uses of African pangolins as a source of traditional medicine in Ghana. PLoS One 10 (1), e0117199.

Boakye, M.K., Kotzé, A., Dalton, D.L., Jansen, R., 2016. Unravelling the pangolin bushmeat commodity chain and the extent of trade in Ghana. Hum. Ecol. 44 (2), 257–264.

Bräutigam, A., Howes, J., Humphreys, T., Hutton, J., 1994. Recent information on the status and utilization of African pangolins. TRAFFIC Bull. 15 (1), 15–22.

Broad, S., Luxmoore, R., Jenkins, M., 1988. Significant Trade in Wildlife, A Review of Selected Species in CITES Appendix II. IUCN Conservation Monitoring Centre, Cambridge, UK.

Chaber, A.S., Allebone-Webb, S., Lignereux, Y., Cunningham, A.A., Rowcliffe, J.M., 2010. The scale of illegal meat importation from Africa to Europe via Paris. Conserv. Lett. 3 (5), 317–323.

Challender, D.W.S., Hywood, L., 2012. African pangolins under increased pressure from poaching and intercontinental trade. TRAFFIC Bull. 24 (3), 53–55.

Challender, D.W.S., MacMillan, D.C., 2014. Poaching is more than an enforcement problem. Conserv. Lett. 7 (5), 484–494.

Challender, D.W.S., Harrop, S.R., MacMillan, D.C., 2015. Understanding markets to conserve trade-threatened species in CITES. Biol. Conserv. 187, 249–259.

Challender, D., Waterman, C., 2017. Implementation of CITES Decisions 17.239 b) and 17.240 on Pangolins (*Manis* spp.), CITES SC69 Doc. 57 Annex. Available from: <https://cites.org/sites/default/files/eng/com/sc/69/E-SC69-57-A.pdf>. [August 2, 2018].

Challender, D.W.S., t Sas-Rolfes, M., Ades, G., Chin, J.S.C., Sun, N.C.M., Chong, J.L., et al., 2019. Evaluating the feasibility of pangolin farming and its potential conservation impact. Glob. Ecol. Conserv. 20, e00714.

CITES, 2019. Wildlife Crime Enforcement Support in West and Central Africa. CITES CoP18 Doc. 34. Available from: https://cites.org/sites/default/files/eng/cop/18/doc/E-CoP18-034.pdf. [November 16, 2019].

Dammerman, K.W., 1929. Preservation of Wild Life and Nature Reserves in the Netherlands Indies. Proceedings of the 4th Pacific Science Congress, Java. Emmink, pp. 1–91.

D'Cruze, N., Singh, B., Mookerjee, A., Harrington, L.A., Macdonald, D.W., 2018. A socio-economic survey of pangolin hunting in Assam, Northeast India. Nat. Conserv. 30, 83–105.

Duckworth, J.W., Salter, R.E., Khounboline, K., 1999. Wildlife in Lao PDR: 1999 Status Report. IUCN, Wildlife Conservation Society, Centre for Protected Areas and Watershed Management, Vientiane, Lao PDR.

Gaubert, P., Antunes, A., 2005. Assessing the taxonomic status of the Palawan pangolin *Manis culionensis* (Pholidota) using discrete morphological characters. J. Mammal. 86 (6), 1068–1074.

Gaubert, P., Antunes, A., Meng, H., Miao, L., Peigné, S., Justy, F., et al., 2018. The complete phylogeny of pangolins: scaling up resources for the molecular tracing of the most trafficked mammals on Earth. J. Hered. 109 (4), 347–359.

Gaudin, T., Emry, R., Wible, J., 2009. The phylogeny of living and extinct pangolins (Mammalia, Pholidota) and associated taxa: a morphology based analysis. J. Mammal. Evol. 16 (4), 235–305.

Harrisson, T., Loh, C.Y., 1965. To scale a pangolin. Sarawak Museum J. 12, 415–418.

Heinrich, S., Wittmann, T.A., Prowse, T.A.A., Ross, J.V., Delean, S., Shepherd, C.R., et al., 2016. Where did all the pangolins go? International CITES trade in pangolin species. Glob. Ecol. Conserv. 8, 241–253.

Heinrich, S., Wittman, T.A., Rosse, J.V., Shepherd, C.R., Challender, D.W.S., Cassey, P., 2017. The Global Trafficking of Pangolins: A Comprehensive Summary of Seizures and Trafficking Routes From 2010–2015. TRAFFIC, Southeast Asia Regional Office, Petaling Jaya, Selangor, Malaysia.

Heinrich, S., Koehncke, A., Shepherd, C.R., 2019. The role of Germany in the illegal global pangolin trade. Glob. Ecol. Conserv. 20, e00736.

Hinsley, A., Nuno, A., Ridout, M., John St., F.A.V., Roberts, D.L., 2017. Estimating the extent of CITES noncompliance among traders and end-consumers: lessons from the global orchid trade. Conserv. Lett. 10 (5), 602–609.

Ingram, D.J., Coad, L., Abernethy, K.A., Maisels, F., Stokes, E.J., Bobo, K.S., et al., 2018. Assessing Africa-wide pangolin exploitation by scaling local data. Conserv. Lett. 11 (2), e12389.

Ingram, D.J., Cronin, D.T., Challender, D.W.S., Venditti, D. M., Gonder, M.K., 2019. Characterizing trafficking and trade of pangolins in the Gulf of Guinea. Glob. Ecol. Conserv. 17, e00576.

Irshad, N., Mahmood, T., Hussain, R., Nadeem, M.S., 2015. Distribution, abundance and diet of the Indian pangolin (*Manis crassicaudata*). Anim. Biol. 65 (1), 57–71.

Li, Y., Li, D., 1998. The dynamics of trade in live wildlife across the Guangxi border between China and Vietnam during 1993-1996 and its control strategies. Biodivers. Conserv. 7 (7), 895–914.

Krishnasamy, K., Shepherd, C.R., 2017. Seizures of African pangolin scales in Malaysia in 2017. TRAFFIC Bull. 29 (2), 53–55.

MacMillan, D.C., Nguyen, Q.A., 2014. Factors influencing the illegal harvest of wildlife by trapping and snaring among the Katu ethnic group in Vietnam. Oryx 48 (2), 304–312.

Mahmood, T., Hussain, R., Irshad, N., Akrim, F., Nadeem, M.S., 2012. Illegal mass killing of Indian Pangolin (*Manis crassicaudata*) in Potohar Region, Pakistan. Pak. J. Zool. 44 (5), 1457–1461.

Mambeya, M.M., Baker, F., Momboua, B.R., Pambo, A.F.K., Hega, M., Okouyi, V.J.O., et al., 2018. The emergence of a commercial trade in pangolins from Gabon. J. Afr. Ecol. 56 (3), 601–609.

Milliken, T., Burn, R.W., Underwood, F.M., Sangalakula, L., 2012. The Elephant Trade Information System (ETIS) and the Illicit Trade in Ivory: A report to the 16th meeting of the Conference of the Parties to CITES. TRAFFIC International, Cambridge, UK.

Mwale, M., Dalton, D.-L., Jansen, R., De Bruyn, M., Pietersen, D., Mokgokong, P.S., et al., 2017. Forensic application of DNA barcoding for identification of illegally traded African pangolin scales. Genome 60 (3), 272–284.

Newton, P., Nguyen, T.V., Roberton, S., Bell, D., 2008. Pangolins in peril: using local hunters' knowledge to conserve elusive species in Vietnam. Endanger. Sp. Res. 6, 41–53.

Nijman, V., 2015. Pangolin seizures data reported in the Indonesian media. TRAFFIC Bull. 27 (2), 44–46.

Nijman, V., Zhang, M.X., Shepherd, C.R., 2016. Pangolin trade in the Mong La wildlife market and the role of Myanmar in the smuggling of pangolins into China. Glob. Ecol. Conserv. 5, 118–126.

Nooren, H., Claridge, G., 2001. Wildlife Trade in Laos: The End of the Game. Netherlands Committee for IUCN, Amsterdam.

Nuwer, R., Bell, D., 2014. Identifying and quantifying the threats to biodiversity in the U Minh peat swamp forests of the Mekong Delta, Vietnam. Oryx 48 (1), 88–94.

Pantel, S., Chin, S.-Y., 2009. Pangolin capture and trade in Malaysia. In: Pantel, S., Chin, S.-Y. (Eds.), 2009. Proceedings of the Workshop on Trade and

Conservation of Pangolins Native to South and Southeast Asia, 30 June – 2 July 2008, Singapore Zoo, Singapore. TRAFFIC Southeast Asia, Petaling Jaya, Selangor, Malaysia, pp. 143–160.

Pantel, S., Anak, N.A., 2010. A preliminary assessment of Sunda pangolin trade in Sabah. TRAFFIC Southeast Asia, Petaling Jaya, Selangor, Malaysia.

Phelps, J., Webb, E.L., 2015. "Invisible" wildlife trades: Southeast Asia's undocumented illegal trade in wild ornamental plants. Biol. Conserv. 186, 296–305.

Rademeyer, J., 2012. Killing for Profit. Zebra Press, Cape Town.

Rivalan, P., Delmas, V., Angulo, E., Bull, L.S., Hall, R.J., Courchamp, F., et al., 2007. Can bans stimulate wildlife trade? Nature 447, 529–530.

SATCM, 1996. Guangxi Province: cross-border trade prices for pangolins rise further. Zhongyaocai (State Administration of Traditional Chinese Medicine) 19, 4.

Shairp, R., Veríssimo, D., Fraser, I., Challender, D.W.S., MacMillan, D.C., 2016. Understanding urban demand for wild meat in Vietnam: implications for conservation actions. PLoS One 11 (1), e0134787.

Shepherd, C.R., Connelly, E., Hywood, L., Cassey, P., 2017. Taking a stand against illegal wildlife trade: the Zimbabwean approach to pangolin conservation. Oryx 51 (2), 280–285.

Soewu, D.A., Adekanola, T.A., 2011. Traditional-medical knowledge and perceptions of pangolins (*Manis* sps [sic]) among the Awori people, Southwestern Nigeria. J. Ethnobiol. Ethnomed. 7 (1), 25.

Sopyan, E., 2009. Malayan Pangolin *Manis javanica* Trade in Sumatra, Indonesia. In: Pantel, S., Chin, S.-Y. (Eds.), 2009. Proceedings of the Workshop on Trade and Conservation of Pangolins Native to South and Southeast Asia, 30 June – 2 July 2008, Singapore Zoo, Singapore. TRAFFIC Southeast Asia, Petaling Jaya, Selangor, Malaysia, pp. 134–142.

Underwood, F.M., Burn, R.W., Milliken, T., 2013. Dissecting the illegal ivory trade: an analysis of ivory seizures data. PLoS One 8 (10), e76539.

UNODC, 2016. World Wildlife Crime Report: Trafficking in Protected Species. UNODC, Vienna, Austria.

Willcox, D., Nash, H., Trageser, S., Kim, H.-J., Hywood, L., Connelly, E., et al., 2019. Evaluating methods for detecting and monitoring pangolin (Pholidata: Manidae) populations. Glob. Ecol. Conserv. 17, e00539.

Wilson, D.E., Reeder, M., 2005. Mammal Species of the World, A Taxonomic and Geographic Reference, third ed. Johns Hopkins University Press, Baltimore.

Wu, S.B., Liu, N., Zhang, Y., Ma, G.Z., 2004. Assessment of threatened status of Chinese Pangolin (*Manis pentadactyla*). Chin. J. Appl. Environ. Biol. 10, 456–461. [In Chinese].

Wu, S.B., Ma, G.Z., 2007. The status and conservation of pangolins in China. TRAFFIC East Asia Newsl. 4, 1–5. [In Chinese].

Zhang, Y., 2009. Conservation and trade control of pangolins in China. In: Pantel, S., Chin, S.-Y. (Eds.), 2009. Proceedings of the Workshop on Trade and Conservation of Pangolins Native to South and Southeast Asia, 30 June – 2 July 2008, Singapore Zoo, Singapore. TRAFFIC Southeast Asia, Petaling Jaya, Selangor, Malaysia, pp. 66–74.

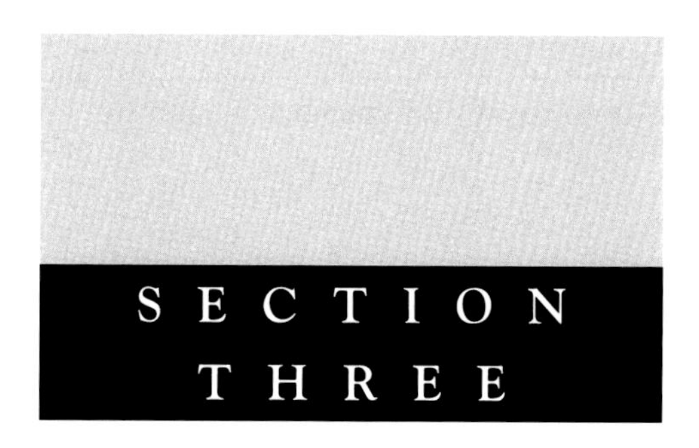

Conservation Solutions

Overview

Section Three focuses on conservation solutions. It is divided into five parts: law enforcement and regulation; awareness-raising and behavior change; site-based protection and local community engagement; ex situ conservation; and conservation planning, research and finance.

Part 1 starts with Chapter 17, which examines international law and assesses its application to pangolin conservation, before discussing the complexities of enforcing legislation, and arguing that pangolin trafficking ought to be treated as serious crime like other forms of syndicated, organized crime. This is followed by a front-line perspective of the practical challenges in enforcing such laws in Africa, and how these challenges may be overcome (Chapter 18). The chapter describes six essential elements of effective law enforcement, which can be applied even in resource-poor environments. Chapter 19 reviews the complex history of pangolins in CITES, the Convention on International Trade in Endangered Species of Wild Fauna and Flora, and evaluates whether the Convention has been effective at ensuring sustainability in international pangolin trade and future options for conserving pangolins using the convention. Part 1 concludes with a discussion of how forensic science can be used to support law enforcement and inform criminal investigations to combat illegal pangolin trade (Chapter 20).

Part 2 focuses on awareness raising and behavior change. Chapter 21 charts the rise of pangolins from virtual obscurity to icons of the illegal wildlife trade within the space of a decade. It identifies eight key events and activities that helped to raise the profile of the species between 2012 and 2018 and lessons that have been learnt which could be applied to other species. Chapter 22 highlights the importance of efforts to reduce demand for pangolin products in key markets through consumer behavior change. Demand-side measures to change consumer behavior are increasingly being implemented as a complement to traditional harvest and trade regulations discussed in Part 1.

Part 3 focuses on engaging varied stakeholders to conserve pangolin populations. Chapter 23 discusses the rationale for engaging with local communities and indigenous peoples, and the importance of doing so at the site level, which is critical to conserving pangolins. It introduces a theory of change with four pathways outlining how community-level actions can alter incentives for poaching. This theory of change was piloted in three community conservation areas in Kenya between 2016 and 2018, and Chapter 24 discusses this project and highlights lessons learnt on implementing the approach. This is followed by case studies on Nepal, Singapore and Cameroon (Chapters 25–27), each of which highlights why a collaborative and multi-faceted approach is crucial to protecting pangolin populations.

Part 4 focuses on ex situ conservation. Ex situ activities can complement the conservation of pangolin populations in the wild through the rehabilitation and release of individuals rescued from illegal trade or the maintenance of healthy captive populations as an insurance policy. Chapter 28 discusses the main challenges to keeping pangolins in captivity, and recent advances in pangolin husbandry that are enabling facilities to maintain healthy captive populations of some species. This is followed by a detailed review of pangolin veterinary health (Chapter 29), which also highlights key knowledge gaps that need addressing. Chapter 30 draws on the experiences of facilities in South Africa and Vietnam to discuss successes and challenges associated with the rescue, rehabilitation and release of pangolins confiscated from illegal trade. Chapter 31 discusses the current and potential future role of zoos in pangolin conservation. The final chapter in Part 4 (Chapter 32) introduces key variable factors and theoretical insights to consider when evaluating the potential impact of wildlife farming, and evaluates the potential of commercial captive breeding, or farming, of pangolins as a way to ensure a sustainable supply of derivatives (e.g., scales) for commercial markets.

Part 5 covers research and conservation priorities, and other approaches that could support pangolin conservation. Chapter 33 discusses the rationale and need for the development of participatory conservation strategies with associated monitoring and evaluation in order to guide successful conservation efforts, drawing on best practice from IUCN models and guidelines. Research priorities, including the urgent need to develop population monitoring methods for pangolins, informed by knowledge of pangolin ecology and biology, are discussed in the following two chapters (Chapters 34 and 35). The section then revisits conservation planning with a case study on the Chinese pangolin in Taiwan, which illustrates how a multi-stakeholder approach has worked in practice over a 13 year period (Chapter 36). The final two chapters explore the potential of impact investing and the use of pay-for results instruments to mobilize greater capital for pangolin conservation (Chapter 37), and the potential of ecotourism to help monitor and protect pangolins at the site level (Chapter 38).

The challenges of conserving pangolins are evident throughout this section but there is cause for optimism. There is greater awareness of pangolins than ever before, stronger political will to address poaching and trafficking, a greater understanding of the actions needed to conserve the species, and more funding opportunities to implement conservation action. A common theme throughout this section, and the case studies in particular, is that a multi-stakeholder approach is crucial to overcoming obstacles and implementing locally-appropriate conservation actions. The final section of this book expands on this and discusses challenges and opportunities for pangolin conservation and how existing actions can be scaled up to ensure pangolins are around for generations to come.

PART 1

Law Enforcement and Regulation

CHAPTER

17

Conserving pangolins through international and national regulation and effective law enforcement

Stuart R. Harrop
Kingston University, London, United Kingdom

OUTLINE

Introduction

Pangolins have gained almost iconic status in the last five or so years. This is almost entirely because of the threat they face from overexploitation for international trafficking, mainly to Asian markets (see Chapter 16). From a regulatory perspective, pangolins also have special status because of their complex relationship with humans. Consequently, the plight of pangolins illustrates almost all of the shortcomings of, and challenges to, local and global regulatory and enforcement regimes. Threatened with extinction from overexploitation, their natural habitat is also being lost through anthropogenic pressure, and the only significant limitation to the perfect storm that pangolins face is that they do not suffer high levels of persecution inflicted on many threatened, larger mammals from human-wildlife conflicts (though see Chapter 25).

The threats to pangolins are both national and international. However, in a globalized world, the significant drivers of pangolin extinction are increasingly cross-border issues: the proportion of pangolin offtake for

Pangolins
DOI: https://doi.org/10.1016/B978-0-12-815507-3.00017-4

illicit international trade appears to be increasing, which is driven in part by demand in East and Southeast Asian markets (see Chapter 16). Unfortunately, most international regulatory regimes are not geared to respond to rapidly changing issues with anything like matching speed. There is significant evidence that, in terms of the development of biodiversity preservation regulation, advances and priorities are not keeping up with developments in conservation science. Indeed, these regulatory developments are slowing rather than speeding up. Consequently, regulatory policies are failing to keep pace with the speed of the drivers of extinction (Harrop, 2013; Rands et al., 2010).

This chapter briefly explores the existing matrix of regulation and enforcement strategies that seek to protect pangolins. It also examines enforcement of regulations and other approaches that complement enforcement and thus have a bearing on effective regulatory protection.

Prior to the era of colonialism and conquest (roughly before the 15–16th centuries) communities living within the geographic ranges of pangolins—operating through slow evolution of practices and necessarily by trial and error—either failed or succeeded in applying sustainable practices to their hunting, gathering, fishing and other modes of exploitation of the natural environment. Failures to develop sustainable practices are consigned to history and pre-history. However, studies of traditional indigenous practices executed by rural communities that manifest sustainability, reveal a whole host of approaches to protect the natural world and secure the sustainable utilization of natural resources (Harrop, 2003). Bespoke strategies of regulation and enforcement comprise an important aspect of these approaches within traditional communities. They include: restrictions through endowing sites with a sacred nature to protect breeding grounds, taboos in time or in designated areas to restrict harvesting of specific animals, or among certain groups of people, the allocation of proprietary responsibility for seeds, animal products and knowledge to particular persons or groups of persons within a community and various modes of punishments for transgressions. These systems of customary regulation could operate effectively in a sparsely populated, pre-modern world (Harrop, 2003). However, depending on the level of power of the elders who administered such systems, these regimes were potentially vulnerable to migrations from other comparatively local communities bound by different traditions (Harrop, 2003).

European colonists brought a greater shock to most traditional systems of community regulation when they overlaid their own, alien regimes. Deploying the concept of *terra nullius* in some cases—an argument that is still contested (see Cavanagh, 2014)—they imposed overarching legal regimes that, in many cases, ignored the existence of subsidiary, traditional regulation or overlaid it with "higher" authority. These European regimes have left their mark. Indeed, in some instances colonial legal regimes remain a blueprint for regulatory approaches and national laws aiming to protect, *inter alia*, pangolins. In addition, the covering over of customary approaches to biodiversity preservation and allocation of rights to natural resources did not necessarily eradicate the old system or the persisting cultural memory. Moreover, when these systems were not integrated into a pluralistic legal approach, in some cases the concealment or attempted eradication of traditional regimes and rights fueled resentment and anger, which in turn created the potential to frustrate contemporary attempts to enforce laws to protect pangolins and other species (Colchester, 2004). This continues to drive anger and resentment and leads to the poaching of species directly (Massé and Lunstrum, 2016).

Many species were negatively affected by the dynamics of European colonization as farmers degraded natural habitats and replaced wildlife with domesticated animals. Furthermore, a whole new wave of natural resource users and recreational hunters carved into wild species populations and opened new national and global markets from the end of the 19th century onwards (Lindsey, 2008). In response, as some animal populations began to plummet, colonists began to impose early conservation regimes often merely to preserve their own sport hunting interests—as in Africa in relation to the larger mammals. However, when areas were allocated for conservation, local peoples, who often operated traditional practices on that land within their own customary laws, were largely dispossessed of their ancestral homes and removed elsewhere. There are of course some notable exceptions (e.g., the Maasai remained in Ngorongoro crater in Tanzania). The dispossession of lands exacerbated existing resentment and anger and was counterproductive to conservation aims in many instances (Brockington and Igoe, 2006). This resentment remains part of the contemporary challenge regarding conservation of various species, including pangolins (Massé and Lunstrum, 2016).

The contemporary legislative picture must be examined in a globalized world with rapid economic and social change where conservation is a cross-border and international concern. However, the historical picture is still relevant. Thus, consideration of both national laws and the relevant international legal instruments cannot ignore the customary regimes and persisting cultural memory that may be hidden under overarching legal regimes that have long since shed control. Without an understanding of these local dynamics, effective laws and enforcement to protect pangolins may never be achieved.

Pangolin protection in international law

International biodiversity preservation law has evolved in a disparate manner, consisting of a number of *ad hoc* instruments rather than an ideal collection of related and coordinated international laws that together deal comprehensively with all the threats to the natural world (Harrop, 2013). Further, some relevant international legal instruments create few, clear obligations to which member states should comply. When a state ratifies an international law instrument such as CITES (Convention on International Trade in Endangered Species of Wild Fauna and Flora) or the CBD (Convention on Biological Diversity), it undertakes, by that act, to implement the terms of the international law within its own regulatory system. However, if that instrument is designed generally rather than specifically and also watered down in its impact by qualifications, such that a member state is given enormous bandwidth in the manner it implements its provisions, then progress in dealing with the challenges might be slow or non-existent.

There are essentially two international instruments of relevance to pangolin conservation. The first, CITES, is by and large clearly and precisely drafted (a true lawyer's instrument; Harrop, 2013). By design, it focuses narrowly on a particular area; targeting sustainability in international wildlife trade, a crucial focus for pangolin conservation. However, CITES faces many challenges (see Chapter 19), not least because by design it overlooks the social, economic and cultural complexity of wildlife trade and there is little evidence that it has been effective (e.g., Challender et al., 2015a,b; 't Sas-Rolfes, 2000). Many species CITES seeks to protect and that are included in its Appendices are traded illegally in large volumes (Hinsley et al., 2018; UNODC, 2016). It must be emphasized that a skillfully designed legal text is only the first step, and many things can go wrong thereafter in application of the law and its enforcement. These include among many others: a fault in the conceptual design that is discovered long after the instrument comes into being; perverse impacts, for example the local extinction of species subject to policy decisions in CITES (see Leader-Williams, 2003); a failure of the global community to fully support the legal instrument with the capacity to execute its work; global priorities that override environmental considerations (Harrop and Pritchard, 2011); and the sheer complexity of forces that are negatively impacting species throughout the world.

The second instrument is the CBD. Although regarded as a pinnacle of achievement at the Earth Summit (the 1992 United Nations Conference on Environment and Development), this convention does not create many substantive obligations or lend itself to clear, specific and uniform implementation. Instead, its provisions are general and heavily qualified and expressed, often in such inappropriate non-normative language, that commentators have described it as comprising "soft diffuse obligations" (Braithwaite and Drahos, 2000). The original intention was to strengthen the CBD with subsidiary instruments known as protocols, but nothing has emanated from it in this respect to enhance the type of conservation strategy that would directly affect the status of pangolins. Instead, it has produced two strategic plans, both of which include non-binding targets. Thus, although the CBD —as a "hard law" international convention—requires the monitoring of the status of species and habitats (Article 7) and has limited and general requirements to establish protected areas and protect populations of species (Article 8), its qualified, "soft" and "diffuse" approach to designing obligations and its strategic trend towards expressly non-binding objectives is unlikely to induce a member state to increase its conservation efforts and priorities. This is especially so bearing in mind the profusion of other short-term obligations confronting all governments.

Other international laws

Two further areas of international regulation have relevance to pangolin conservation. The first, recognizing the need for multifaceted interventions to combat the threats facing pangolins, includes international instruments dealing with the rights of indigenous and traditional communities. Unlike the CBD, the provisions of which, seek to link indigenous peoples' rights directly to biodiversity conservation, these instruments focus on human rights from the indigenous rather than conservation perspective. They support the indigenous and local case to extend rights to these people in their ancestral territories, to secure their involvement in environmental planning, and to acknowledge and permit their traditional subsistence harvesting as exceptions to wider conservation prohibitions. Specifically, these instruments include ILO 169 (Convention concerning Indigenous and Tribal Peoples in Independent Countries, 1989) and the United Nations Declaration on the Rights of Indigenous Peoples 2007 (UN GA 61/295). The former has a limited number of ratifying countries and indeed none of the pangolin range states have ratified its text. Consequently, ILO 169 creates no direct obligations for these states. The UN Charter may not lend much assistance either because it has a contested status and may only constitute "soft" law thereby possessing no binding legal authority. Consequently, both of these instruments may only operate as persuasive policy rather than obligatory law. However, they are mentioned herein because, in theory, they could be leveraged to bring about the change necessary to build partnerships with indigenous peoples and local communities for pangolin conservation (see Chapters 23 and 24).

The subject of indigenous and traditional subsistence rights leads to the final convention in this context; the "CMS" (the Convention on the Conservation of Migratory Species of Wild Animals). This convention contains a specific provision permitting an exception to its prohibitions on taking endangered species where such harvesting is to "accommodate the needs of traditional subsistence users." This type of harvesting has been identified as a necessary exemption to legal prohibitions in order to encourage conservation incentives in local communities within the range states of threatened species including pangolins (Cooney et al., 2017). Unlike the CBD, the CMS contains comparatively comprehensive, clear and implementable provisions covering the whole spectrum of conservation measures specifically tailored for threatened species that fall into its definition of migratory species. It also contains guidelines for the creation of specific agreements to be entered into by range states to facilitate cooperation and coherent strategies across borders—a necessary requirement bearing in mind the cross-border challenges facing pangolins (see Chapter 16; Challender and Waterman, 2017). Similar to CITES, this convention lists species in Appendices. Appendix I contains protected "endangered" species and Appendix II comprises protected species that are not threatened but have, nevertheless, an "unfavorable conservation status." Specific provisions apply to each appendix and necessarily the approach to protecting Appendix II species is less stringent than the Appendix I species. Pangolins are not obvious species to be protected by a convention dealing with migratory species and are not included in the CMS Appendices. However, species of a relatively sedentary nature (e.g., gorillas, *Gorilla* spp.), in comparison to more obvious migratory animals, are listed and thereby protected by CMS. There is also a subsidiary "Agreement on the Conservation of Gorillas and Their Habitats", made under powers within the CMS text, which legally binds all ten gorilla range states to implement actions to conserve the species' and their habitats. Although gorillas are not, at first glance, obvious candidates for the epithet migratory species, they have been accepted in the CMS because its definition of "migratory species" emphasises the crossing of state borders by a species or a species population.

This definitional approach potentially invites into the protection of the CMS many other species that require greater international conservation attention. In the case of pangolins, not all of its range state borders are geographical barriers that would prevent migration from one state to another (indeed some of these state borders are merely lines drawn on a map). It is therefore possible that some populations or species of pangolin could fall within this definition and merit the protection of the CMS, which could be used to further catalyse conservation action.

National implementation of legislation

CITES has precise requirements for implementation, clear legal terms to guide implementation, and virtually all pangolin range states are members of CITES. Moreover, each member state is required to take measures to implement the convention, including enacting adequate implementing legislation, though only 101 parties (55%) have legislation that meets the requirements for implementation (CITES, 2019). The parties have also agreed on priority conservation actions for pangolins including effective law enforcement (see Chapter 19). This applies not only to pangolin range states, but to all CITES members (currently 182 countries plus the EU) because pangolin trafficking is a global issue (Heinrich et al., 2017). However, even with these measures in place, there are many challenges to implementation and effective law enforcement at the national and local level.

Pangolin range states regulate international trade in pangolins as do all parties to CITES, including those that are conduits for illegal trade (e.g., the EU) and destinations (Heinrich et al., 2017, 2019). Although all species of pangolin are included in Appendix I (see Chapter 19), there is some difference in the approach to implementation of CITES by range states. Implementing legislation in some range states includes all species of pangolin, but in others, it applies to native species only. In the latter case, this has the potential to create problems with the monitoring and enforcement of offences in that, given individual pangolins and their derivatives (e.g., scales) are similar looking to all but experts, there is scope for traffickers to claim that trade involves a non-native species, when really native species are being trafficked. Bearing in mind that different qualitative approaches to designing and implementing effective laws is inevitable, protection of pangolins within the specific requirements of CITES is generally in place nationally throughout most pangolin range states (Challender and Waterman, 2017).

Range states also control the harvest and domestic trade in pangolins through national legislation. This may or may not be the same legislation that implements CITES, and typically includes enforcement powers to confiscate, and provisions for storage and disposal of confiscated pangolins and their derivatives. Most range states have penalties in place for poaching and/or trafficking. These vary by country, ranging from fines of less than USD 10 in Côte d'Ivoire to USD 760,000 in South Africa and prison sentences up to a maximum of life imprisonment in China (Challender and Waterman, 2017). Suspended death sentences have been issued for pangolin trafficking offences in China (Anon, 2008).

In Zimbabwe, up to nine-year prison terms have been imposed apparently as part of a strengthened law enforcement response to pangolin harvesting and trade crimes since 2010, and there is a suggestion that high penalties may be acting as a deterrent (Shepherd et al., 2016). However, substantial fines or even prison sentences do not necessarily act to deter the illegal taking of wildlife (Leader-Williams and Milner-Gulland, 1993). Furthermore, effective deterrents differ from country to country, across different community and cultural backgrounds and from the local to the national and international context in the case of pangolin trafficking.

Consequently, the profusion of international soft and hard law, national implementation and unilateral national law thus far discussed only constitutes half the story. Without carefully and specifically designed enforcement approaches and sophisticated strategies throughout the trail of harvesting to consumer markets, from the local to the global, the law is impotent to mitigate the very real extinction threat to pangolins.

Enforcement of legal provisions

To be effective, laws to protect pangolins must be fully supported by associated sanctions and mechanisms for the administration of justice. If laws are unclear or too vague—or avoid language that creates obligations—this can rarely be the case. Thus, the implementation of the CBD text is less likely to be supported by effective enforcement than, for example, the implementation by member states of the clearer and obligatory CITES provisions and its underpinning national legislation. However, despite its comparatively clear portfolio of laws, CITES also faces challenges with the enforcement of its provisions. Enforcement difficulties in respect of pangolin harvesting and trafficking are extensive and include predominantly a lack of capacity across all areas. This includes equipment, biological expertise to identify pangolins and their parts and derivatives in illegal trade, and a lack of ability to monitor illegal trade that takes place online. Other issues include the inability to enforce regulations across long and complicated terrain on state borders "24 hours a day, seven days a week" and the difficulties faced in enforcing the law in remote, rural areas where high price inducements to local communities incentivize poaching (see Chapter 18; Challender and Waterman, 2017).

At the higher level, where a convention member state may not be fulfilling its obligations under international law, the mechanisms of enforcement and justice that may be taken for granted in a national context are for the most part impossible and there is a reliance on political pressure and often exposure of failings publicly by NGOs (non-governmental organizations) to achieve compliance (see Chapter 19). However, even when national laws are in place that implement international obligations and are bolstered by well-funded and directed enforcement and effective mechanisms for the application of justice, problems may persist. In addition, there are often unforeseen consequences which can flow from legislation. These are generated, not only because species are embedded in a complex matrix of ecological and environmental relationships which may be poorly understood by the designers of legislation (and conservation biologists), but also because the conservation of species requires a sophisticated understanding of the human relationship with particular species. This requires consideration of a wide range of cultural, social, economic and other local and global dynamics. The effects of some laws to protect species promulgated by colonists were so all-embracing that they also had the effect of removing subsistence rights in relation to traditional use of biodiversity within local communities. As discussed, this can create anger and resentment and undermine conservation efforts. Thus, one of the key components of an effective toolkit to protect pangolins requires local stakeholder involvement and conservation strategy buy-in (see Chapters 23 and 24). Without addressing local community issues within pangolin ranges, heavy-handed legal approaches can be blunt instruments that are limited in effectiveness at one end of the spectrum and may exacerbate the problem at the other.

A heavy-handed legal approach also creates scope for corruption by officials, particularly those that are underpaid, and can be a recipe for persecution and human rights violations. Consequently, corruption can be a further factor, in developing countries in particular, and can frustrate otherwise effective conservation strategies (Felbab-Brown, 2018). It can also exacerbate the problem, particularly where high value species, which includes pangolins in many range states, are concerned.

The effectiveness of conservation law requires that the law, along with the action that surrounds its enforcement, both act as a deterrent. The approach will necessarily differ according to the local context and from the point of harvest along illegal cross-border trade chains. Even at the lowest poaching level, high penalties are deterrents in certain contexts. Local poachers may be dissuaded by the potential imposition of a financial penalty or by a prison sentence for example. However, at the high end of sophisticated poaching and trafficking operations, which in themselves may be extremely expensive to organize and operate, an increase in detection rate, particularly for those in power, may act as a greater deterrent (Leader-Williams and Milner-Gulland, 1993).

Enforcement at the higher level of organized and syndicated illegal trade

Beyond the local level, where high value, coordinated crime involving pangolins takes place, financial punishment or short-term imprisonment deterrents may seem petty in comparison with the high level of potential profit. Such punishments may constitute no more than occasional and minor occupational hazards. Furthermore, local deterrent strategies beyond the imposition of penalties may be completely eclipsed. One of the major criticisms of CITES, and a reason for low rates of success, is the failure to fully understand market forces and the complex mechanisms and incentives surrounding harvest and use of wildlife (Challender et al., 2015a,b, 2019). Across and beyond range state borders, the smuggling of pangolins follows the trail of other high value wildlife products including ivory and timber, and drugs. At this higher level, such trafficking is entangled with organizations operating sophisticated money laundering schemes, narcotic cartels and other criminal syndicates (Ingram et al., 2019; Mambeya et al., 2018). International cooperation by states and with inter-governmental agencies such as Interpol and specialist investigators of complex financial activities is essential to match the nature of the criminal activity and the magnitude and complexity of the organization behind it (Nellemann et al., 2014). To some extent, this is occurring. Thus, pangolin range states in Africa and Asia have reported to CITES that they have worked with other states and Interpol to combat the illegal pangolin trade (Challender and Waterman, 2017).

Nevertheless, more can be done and attention is being directed to deploy "follow the money" approaches (Bawden, 2018) utilizing non-traditional legislation to counter wildlife trafficking. Discovering, following and shutting down the money trail behind large-scale illicit trade has the potential to stifle entire organizations from the highest level right down to the local actors. Raising finance is a major purpose of crime and is the key to the organization, success, power and expansion of global criminal syndicates. Money supports the networks and infrastructures and facilitates corruption so that officials can enable cross-border trafficking (Realuyo, 2017). Global financial systems enable rapid money transfers and the pressure for barrier-less trading has enabled a proliferation of money laundering strategies (Winer and Trifin, 2003). Winer and Trifin (2003) argue that, by using globally coordinated financial intelligence and banking mechanisms, states would have the facility to better detect and confiscate funds and thereby frustrate criminal syndicates that organize and operate internationally coordinated trafficking in wild species. Using these approaches, states could choke off the money required for the logistics of the illegal trade and shut down the lower echelon and political corruption that facilitates illegal harvesting and sale, and eases the passage of wildlife contraband across state borders (Winer and Trifin, 2003).

Haenlein and Keatinge (2017) have suggested a list of actions required to enable this aspect of combatting wildlife crime to be implemented. These include, among others:

- recognition by states that wildlife crime is a predicate offence to wider money laundering crimes and thus there is a need to promulgate national legal changes, where required, to combat this;
- capacity building of wildlife trade investigators to deal with the financial aspect of the overall criminal activity;
- use of alternative prosecution mechanisms such as those which relate to corruption and money laundering with their substantially higher penalties;
- inclusion of wildlife crime in national risk assessments at a higher level (since money laundering is often linked to terrorism and other organized crime affecting national security); and
- coordinated sharing of information of relevant financial information.

Conclusion

Pangolins illustrate many of the challenges to local, national and global regulatory and enforcement regimes. International laws that have a "diffuse" and over-general effect (e.g., the CBD) have limited utility to directly mitigate the threats that pangolins face directly. However, beyond CITES, the CMS, the Convention concerning Indigenous and Tribal Peoples in Independent Countries and the United Nations Declaration on the Rights of Indigenous Peoples offer means of bolstering existing efforts and catalyzing further conservation action for pangolins. Pangolins are well protected on paper in most range states, but a number of challenges need to be overcome for effective law enforcement in these countries. Moreover, clear laws and effectively coordinated national implementation is not the full story. Capacity-backed enforcement strategically geared to deter specific levels of criminal activity and sufficiently powerful to match the force of the criminals involved in commercial pangolin trafficking is needed. Equally, strategies to eradicate disincentives and to build in incentives are required to ensure that local communities support pangolin conservation, but without compromising their rights. All of these approaches are fundamental to a successful strategy to effective law enforcement, nationally and internationally.

References

Anon, 2008. China comes down on pangolin smugglers. Available from <https://www.reuters.com/article/environment-china-pangolins-dc-idUSPEK19856420080110>. [August 18, 2019].

Bawden, T., 2018. Duke of Cambridge urges governments to "follow the money" on wildlife crime', iNews the Essential Daily Briefing. Available from: <https://inews.co.uk/news/environment/duke-of-cambridge-urges-governments-to-follow-the-money-on-wildlife-crime/>. [January 9, 2018].

Braithwaite, J., Drahos, P., 2000. Global Business Regulation. Cambridge University Press, Cambridge.

Brockington, D., Igoe, J., 2006. Eviction for conservation: a global overview. Conserv. Soc. 4 (3), 424–470.

Cavanagh, E., 2014. Possession and dispossession in corporate New France, 1600–1663: Debunking a "Juridical History" and Revisiting Terra Nullius. Law Hist. Rev. 32 (1), 97–125.

Challender, D.W.S., Harrop, S.R., MacMillan, D.C., 2015a. Understanding markets to conserve trade-threatened species in CITES. Biol. Conserv. 187, 249–259.

Challender, D.W.S., Harrop, S.R., MacMillan, D.C., 2015b. Towards informed and multi-faceted wildlife trade interventions. Glob. Ecol. Conserv. 3, 129–148.

Challender, D., Waterman, C., 2017. Implementation of CITES Decisions 17.239 b) and 17.240 on Pangolins (*Manis* spp.), CITES SC69 Doc. 57 Annex. Available from: <https://cites.org/sites/default/files/eng/com/sc/69/E-SC69-57-A.pdf>. [August 2, 2018].

Challender, D.W.S., Hinsley, A., Milner-Gulland, E.J., 2019. Inadequacies in establishing CITES trade bans. Front. Ecol. Environ. 17 (4), 199–200.

CITES, 2019. National Laws for Implementation of the Convention. CITES CoP18 Doc. 26. Available from: <https://cites.org/sites/default/files/eng/cop/18/doc/E-CoP18-026-R1.pdf>. [March 15, 2019].

Colchester, M., 2004. Conservation policy and indigenous peoples. Cult. Surv. Quart. 28 (1), 17–23.

Cooney, R., Roe, D., Dublin, H., Phelps, J., Wilkie, D., Keane, A., et al., 2017. From poachers to protectors: engaging local communities in solutions to illegal wildlife trade. Conserv. Lett. 10 (3), 367–374.

Felbab-Brown, V., 2018. The threat of illicit economies and the complex relations with state and society. In: Comolli, V. (Ed.), Organized Crime and Illicit Trade: How to Respond to This Strategic Challenge in Old and New Domains. Palgrave Macmillan, pp. 1–21.

Haenlein, C., Keatinge, T., 2017. Follow the Money - Using Financial Investigation to Combat Wildlife Crime. Royal United Services Institute for Defence and Security Studies Occasional Paper. Royal United Services Institute, London, UK.

Harrop, S.R., 2003. Human diversity and the diversity of life: international regulation of the role of indigenous and rural communities in conservation. Malay. Law J. 4, xxxviii-lxxx.

Harrop, S.R., 2013. Biodiversity and conservation. In: Falkner, R. (Ed.), The Handbook of Global Climate and Environment Policy. Wiley-Blackwell, Oxford, pp. 37–53.

Harrop, S.R., Pritchard, D., 2011. A hard instrument goes soft: the implications of the Convention on Biological Diversity's current trajectory. Glob. Environ. Change 21 (2), 474–480.

Heinrich, S., Wittman, T.A., Ross, J.V., Shepherd, C.R., Challender, D.W.S., Cassey, P., 2017. The Global Trafficking of Pangolins: A Comprehensive Summary of Seizures and Trafficking Routes From 2010–2015. TRAFFIC, Southeast Asia Regional Office, Petaling Jaya, Selangor, Malaysia.

Heinrich, S., Koehncke, A., Shepherd, C.R., 2019. The role of Germany in the illegal global pangolin trade. Glob. Ecol. Conserv. 20, e00736.

Hinsley, A., de Boer, H.J., Fay, M.F., Gale, S.W., Gardiner, L.M., Gunasekara, R.S., et al., 2018. A review of the trade in orchids and its implications for conservation. Bot. J. Linn. Soc. 186 (4), 435–455.

Ingram, D.J., Cronin, D.T., Challender, D.W.S., Venditti, D. M., Gonder, M.K., 2019. Characterising trafficking and trade of pangolins in the Gulf of Guinea. Glob. Ecol. Conserv. 17, e00576.

Leader-Williams, N., 2003. Regulation and protection: successes and failures in rhinoceros conservation. In: Oldfield, S. (Ed.), The Trade in Wildlife. Regulation for Conservation. Earthscan, London, pp. 89–99.

Leader-Williams, N., Milner-Gulland, E.J., 1993. Policies for the enforcement of wildlife laws: the balance between detection and penalties in Luangwa Valley, Zambia. Conserv. Biol. 7 (3), 611–617.

Lindsey, P.A., 2008. Trophy hunting in sub-Saharan Africa: Economic Scale and Conservation Significance. Best Practices in Sustainable Hunting, 41–47.

Mambeya, M.M., Baker, F., Momboua, B.R., Pambo, A.F.K., Hega, M., Okouyi, V.J.O., et al., 2018. The emergence of a commercial trade in pangolins from Gabon. Afr. J. Ecol. 56 (3), 601–609.

Massé, F., Lunstrum, E., 2016. Accumulation by securitization: commercial poaching, neoliberal conservation, and the creation of new wildlife frontiers. Geoforum 69, 227–237.

Nellemann, C., Henriksen, R., Raxter, P., Ash, N., Mrema, E. (Eds.), 2014. The Environmental Crime Crisis – Threats to Sustainable Development From Illegal Exploitation and Trade in Wildlife and Forest Resources. UNEP Rapid Response Assessment. United Nations Environment Programme and GRID-Arendal, Nairobi and Arendal.

Rands, M.R.W., Adams, W.R., Bennun, L., Butchart, S.H. M., Clements, A., Coomes, D., et al., 2010. Biodiversity conservation: challenges beyond 2010. Science 329 (5997), 1298–1303.

Realuyo, C.B., 2017. "Following the Money Trail" to Combat Terrorism, Crime, and Corruption in the Americas. Available from: <https://www.wilsoncenter.org/sites/default/files/follow_the_money_final_0.pdf>. [January 9, 2019].

Shepherd, C.R., Connelly, E., Hywood, L., Cassey, P., 2016. Taking a stand against illegal wildlife trade: the Zimbabwean approach to pangolin conservation. Oryx 51 (2), 280–285.

't Sas-Rolfes, M., 2000. Assessing CITES: four case studies. In: Hutton, J., Dickson, B. (Eds.), Endangered Species Threatened Convention, The Past, Present and Future of CITES. Africa Resources Trust and Earthscan Publications Ltd, London, pp. 69–87.

UNODC, 2016. World Wildlife Crime Report: Trafficking in Protected Species. UNODC, Vienna, Austria.

Winer, J.M., Trifin, J.R., 2003. Follow the money: the finance of illicit resource extraction. In: Collier, P., Bannon, I. (Eds.), Natural Resources and Violent Conflicts: Options and Actions. World Bank, Washington, D.C., pp. 161–214.

CHAPTER

18

Combating the illegal pangolin trade - a law enforcement practitioner's perspective

Christian Plowman

Wildlife Conservation Society, Brazzaville, Republic of Congo

OUTLINE

Introduction

Pangolins are trafficked in high volumes and since 2008 there has been an apparent increase in the trafficking of African pangolins, mainly scales, to Asian markets (see Chapter 16). This is despite pangolins being protected by national legislation in nearly all range states, which in most cases means that exploitation and trade is prohibited, with some concessions for local use by indigenous peoples.

The ineffectiveness of law enforcement to adequately address pangolin trafficking (and that involving other species) has received some research attention. At the highest level there are political limitations to available funding for enforcement of national laws, which is in competition with infrastructure development, education, healthcare and other services

Pangolins
DOI: https://doi.org/10.1016/B978-0-12-815507-3.00018-6

provided by the state (Challender and MacMillan, 2016), and the implementation of wildlife legislation specifically has been, and is, frequently seen as a low priority (see Reeve, 2002). Corruption, collusion and nepotism are also known to undermine enforcement efforts (see Felbab-Brown, 2018; Reeve, 2002). Moreover, the drivers of wildlife trafficking are complex, ranging from relative and absolute poverty in some cases, to profit-seeking enterprises operating illegally, and strong social forces associated with consumer demand (Shairp et al., 2016). Ineffectiveness can in part be attributed to the fact that regulation rarely fails to adequately contend with these complex drivers (Challender and MacMillan, 2014; Roe et al., 2002). Practical constraints to effective law enforcement have historically included, and continue to include, a lack of training for front-line law enforcement personnel (e.g., customs officers), a shortage of resources (human and technical), and ineffective communication and collaboration between law enforcement agencies both within and between nation states (Challender and Waterman, 2017; Patel, 1996; Reeve, 2002). However, between the early 2000s and late 2010s, and in response to the contemporary poaching crisis, there has been substantial investment in "boots on the ground" law enforcement from the international donor community. This has included resources dedicated specifically to law enforcement training and capacity building. Between 2010 and 2016, more than USD 1.3 billion was spent on combating wildlife trafficking globally, an estimated 65% of which was invested in protected area management and intelligence-led law enforcement including international cooperation (World Bank, 2016).

How and why then is enforcement of wildlife and other applicable national legislation in pangolin range states and other (e.g., transit and destination) countries still not being implemented effectively? This chapter examines front-line practical issues for active law enforcement officers who are tasked with preventing and solving crime, and more specifically the poaching, unlawful trading and trafficking of pangolins and their parts. The chapter is informed by decades of policing and investigative experience in Europe, and observations and practical application of policing and enforcement skills in a range of African countries. Although mainly focused on Africa, this chapter, and the recommendations herein, is relevant more broadly to efforts to combat the trafficking of pangolins and other wildlife, particularly in austere, isolated or resource-poor environments.

A practitioner's perspective

Here, a "front-line" practitioner's perspective means the outlook of a law enforcement official, investigator, case-builder, or intelligence manager. These roles are an essential element in efforts to combat illegal wildlife trade (IWT). It is hoped that the front-line view presented here will complement existing strategies and tools designed to combat pangolin trafficking (e.g., Challender et al., 2014; see CITES Res. Conf. 17.10[1]). The focus of this chapter is on the initial stages of the investigative and intelligence process, and questions around legislative instruments, judicial processes, and international laws and conventions are not addressed (see Chapters 17 and 19).

The chapter starts by offering an appropriate definition of "law enforcement", in seeking to ensure it objectifies the essential elements of tackling the illegal trade in pangolins. It then posits the concept that there exist six elements of law enforcement activity that are essential to ensuring success in combating trafficking and disrupting criminal entities engaged in poaching and trafficking pangolins. It asserts that if these six elements are in place, the

[1] Resolution Conf. 17.10, Conservation of and trade in pangolins.

result should be a diminished illegal trade and heightened awareness amongst criminals and the public of the capabilities of law enforcement and the consequences of trafficking pangolins. It should be noted that pangolin trafficking is not an exclusive crime; it is often committed concurrently with other offences, including trafficking of other wildlife (Ingram et al., 2019; Mambeya et al., 2018). This means that the challenges and solutions discussed are applicable to a wide range of trafficked species.

What is "law enforcement?"

Simply, "law enforcement" means "application of the law." A "law enforcement authority" (hereafter referred to as an "LEA") is an agency, usually a government agency or division, which is charged with applying the law. In the context of combating wildlife trafficking in Africa, these agencies may comprise wildlife or forestry officials, police, customs, gendarmeries, prosecution and judicial departments, and military and paramilitary units. Often it is a combination of these bodies.

The application of the law can be broadly interpreted as preventing violations, and thereafter, identifying breaches of the law, and acquiring sufficient evidence in order that the perpetrator(s) can be brought before the appropriate judicial authorities and subjected to judicial proceedings, with the ultimate aim being prosecution. Some specific elements of applying the law, including attempting to achieve prosecution, may lend themselves to traditional constructs of what conservation law enforcement means to many people: namely fatigue-clad paramilitary rangers conducting dissuasive patrols in protected areas, technological monitoring of patrols and detentions, and/or inevitable celebratory social media posts picturing detained poachers with their contraband. However, law enforcement, should comprise all of the elements of a common investigative chain, with the end goal being a successful prosecution. These elements are intelligence, arrest, evidence, and prosecution. They form the basis of any investigation. Without each of these elements, the definition of "law enforcement" is not met and law enforcement goals are unlikely to be achieved.

From a universal viewpoint, there are a number of important requirements in order for the law to be applied correctly and to ensure it is effective. They are universal because they are not legislatively or jurisdictionally specific, and apply regardless of national or organizational capacity and resources. First, addressing illegal trade in pangolins should take place in partnership with other conservation efforts. There are multitude commonalities, especially regarding community engagement, protected area surveillance and species monitoring, which comprise essential elements of an intelligence picture for law enforcement. For example, electronic or aerial monitoring of species for conservation reasons can be used to identify potential poaching-for-profit hotspots, vulnerable trafficking points, or corroborate historical criminal intelligence about illegal activity (see Sandbrook, 2015 for discussion on the ethics of aerial monitoring). Second, developing intelligence at all levels—from rural communities and indigenous peoples to protected area and NGO (non-governmental organization) staff, and international organized crime—is key. This is often overlooked when stakeholders provide "traditional" law enforcement support, and may result in intelligence or information gaps, stifling the effectiveness of LEAs. Important to successful intelligence development is a local "partnership" approach with all stakeholders in the applicable area. Where there are issues around the sharing and use of data acquired for conservation purposes with law enforcement agencies and/or NGOs, they are typically ethical and can be addressed in many cases by virtue of implementing formal agreements or memoranda of understanding, e.g., by establishing robust and transparent information gathering or sharing protocols (see College of Policing, 2014). Third, law enforcement agents focusing on wildlife are often seen predominantly as officers from a local wildlife department, but pangolin trafficking is characteristically transnational, and occurs by road, air, and sea (Heinrich et al., 2017). Support to LEAs should therefore not be limited to agents working with a specific wildlife remit, but should include police, gendarmes, border and customs agents and ports authorities. Crucially, law enforcement support should *never* be limited to a specific protected area or a geographic region for the same reasons.

The six elements of effective law enforcement

Notwithstanding universally important requirements, there are six specific elements that would improve law enforcement in many African pangolin range states, enhancing efforts to combat pangolin trafficking. Often, support to LEAs focuses on only one of these aspects, but this rarely results in the scaling needed to effect organizational change or attitude-shift for the LEA concerned. Support to law enforcement should be focused on guiding best practice and changes in attitude and culture if needed, and every effort should be made to enable, rather than dictate, or do. Training of front-line law enforcement staff should be provided with an operational focus, ideally with a real-life enforcement activity planned or initiated as a result of the training. Ideally, training should be undertaken inhouse, by the law enforcement organization themselves. This ensures organizational standardization of training and methodologies, and elements of sustainability and longtermism regarding training provision. Where personnel from outside agencies are used they should have appropriate law enforcement expertise to enable the law to be applied to achieve prosecutions. This section examines each of the six elements and discusses how, in the real world, combatting illegal pangolin trade can be enhanced by considering each essential aspect.

1. Community engagement

A key challenge concerning enforcement of wildlife laws is the level and type of interaction between law enforcement officers and local community members. Often, those involved in pangolin hunting or poaching for sustenance are from marginalized communities and have little or no interaction with law enforcement officials. Any interactions they do have are usually negative (e.g., at a checkpoint or other "enforcement" incident). This does little to foster a positive relationship between the two parties, and can make it more likely that the communities will end up poaching on behalf of criminal facilitators, having little faith in, or respect for law enforcement. Negative interactions can also make it more likely that subsequent generations of communities will continue to perceive law enforcement negatively. Moreover, many communities have hunted pangolins (and other species) for food for decades, and countering this through dissuasive or intimidating tactics is unlikely to lead to a cessation of pangolin harvest in many cases.

In many Central African countries, the concept of "community policing" is almost nonexistent. This concept relies on the police policing "by consent" in the communities in which they serve, and the relationship is mutually beneficial. If law enforcement officers engaged with communities on a more educational, nonconfrontational level, and attempted to become part of the community, local community members may be more willing to assist them e.g., by sharing intelligence. This would yield benefits in many Central African pangolin range states. A project implemented by the German Office for International Cooperation (Deutsche Gesellschaft für Internationale Zusammenarbeit) between 2017 and mid-2018 saw a marked increase in "community interaction" with law enforcement at a border post on the Cameroonian/Gabonese border (D. Hauthoff, pers. comm.). Recognizing the community to be key in tackling wildlife trafficking, a village water pump was physically installed inside the border post, resulting in villagers being present at, and positively interacting with, law enforcement officers on duty. This type of initiative can pay dividends in terms of a community's willingness to address their own involvement in pangolin hunting, poaching and trafficking, but also their likelihood to seek assistance and/or provide information to law enforcement personnel. This is especially important as local and indigenous communities can become victims of predatory profiteers, facilitators and organizers, who take advantage of their traditional familial hunting skills and bush knowledge, to acquire wildlife for onward sale. These entities often provide them with a strong incentive to do so. In southern Cameroon for example, members of the Baka (a semi-nomadic indigenous group) are paid in whiskey to guide and accompany poaching missions.

Community focused activities or events that aim to counter wildlife trafficking should involve local law enforcement, to drive credibility and foster trust between law enforcement and the wider community. Trustworthy relationships can yield positive results and ensure that law enforcement officers are afforded the opportunity and autonomy to use their discretion and distinguish between sustenance-based hunting and criminally motivated, for-profit poaching. This would ensure they are appropriately targeting the latter, and not disenfranchising local communities with which they could be building relationships.

Training and education for LEA outreach officers, and for local conservationists who spend substantial periods in community areas is essential, and should benefit from the engagement of NGOs or other organizations that have fostered good relationships with local communities. Training should emphasize the importance of building a rapport with community members in order to obtain a picture of what sort of criminality (if any) takes place within local communities. This should include a clear definition of hunting or poaching for sustenance, and those individuals and groups who "poach-for-profit." The establishment of relationships between LEA personnel and communities can serve as a critical platform for information and intelligence sharing.

Even in the most complex and serious of criminal investigations, local law enforcement are usually the first port of call: their knowledge of, and relationships with, local communities are built over time and mean that they are often able to provide information on hunters, poachers, and traffickers, and provide insight into local criminal tactics and methods. Replicating something similar in isolated villages or rural small-town communities in pangolin range countries across Africa could have a positive impact on law enforcement (and NGO) capacity to intervene in pangolin trafficking. Local law enforcement should also have the opportunity to feed in information about trafficking activity to a relevant department, organization, contact, or agency. Many African LEAs do not have formal or centralized intelligence sharing structures. This is the case for wildlife-centric LEAs, and for front-line personnel in isolated, or resource-poor locations, the concept is non-existent. Personnel may have the capacity, or the organizational requirement, to pass any information on to their superiors or hierarchy. However, this does not always yield action due to factors including time constraints, resource issues, corruption or ineffectiveness, and in many areas, NGO partners are often seen (and act) as a de facto conduit for information sharing. An ongoing dialogue between community-based enforcement officers and a central point of contact in an LEA would afford them the opportunity to share relevant information. This can be as simple as having a telephone number for local enforcement officers to call, or a regular face-to-face visit from an NGO.

2. Improving the intelligence cycle

Intelligence is a much-vaunted concept and the term "intelligence cycle" is sometimes used within conservation circles with little understanding of what it is: essentially, common sense cyclical influential movement of information and subsequent action. It can be a very powerful tool for combatting pangolin trafficking. However, the majority of African pangolin range states have limited ability, knowledge and/or resources to manage intelligence effectively. For example, law enforcement authorities in Benin and Cameroon, which are responsible for tackling wildlife crime, lack access to basic tools such as records or databases, IT-based or otherwise. Although they may have the capacity to use field-based tools, such as the Spatial Monitoring and Reporting Tool, (SMART; see Chapter 27), this cannot replace the requirement to have dedicated intelligence capacity. Field based tools should also have the capacity to be evidentially viable in court, meaning that the data they receive and store (e.g., photographs, GPS information) should be capable of maintaining digital integrity and sustaining probity in a court of law, which could in theory mean the difference between a successful prosecution, or not.

LEAs and divisions within, and supporting NGOs, should ideally have dedicated "intelligence officers" or units. With a specific focus on intelligence, this would allow them to leverage the knowledge of "local" or "community" officers and personnel in field-based NGO communities, enabling them to evaluate and disseminate information accordingly. This position would ideally be the single point of contact for any enforcement response to intelligence and be trained in basic human source management (see next section) equipping them with the skills to do the job. Ideally, intelligence officers would also "debrief" arrestees, or persons of interest, in order to acquire information on trafficking methods, actors, geographies and chronologies of trafficking incidents. Finally, intelligence officers would need an ability to analyze intelligence to enable an understanding of criminal activity, which will warrant training.

3. Developing transparent human source management

The cultivation, recruitment and management of human sources ("informants") is complex, sensitive, and a highly specialized skill. In conservation circles, the effective construction and management of an "informant network" is frequently cited as an objective in tackling wildlife trafficking, especially in grant applications where it is often seen as an attractive prospect for donors. Given the complexities involved, it may be an ambitious aim to develop a locally based functioning human source system in many African pangolin range states. There is substantial risk, both ethically and organizationally, in attempting to do so without the requisite experience or expertise. This includes accounting for all elements of source handling: the need for specialist training; risks to sources/officers/handlers; payment amounts and systems; security of information (especially sources identities); legal aspects of source information being used in court (i.e., legislative measures to protect identities); establishment of covert units to handle sources and deploy associated covert techniques; establishment of NCND ("neither confirm nor deny") policies for sources in order to maintain their safety; rewarding sources transparently and ethically according to results; addressing issues of accountability and corruption in agencies through strict and transparent accounting and recording systems; and training of senior officials and managers in such matters to obtain buy-in, among others. Nevertheless, it is not impossible in the right circumstances and it is a scalable concept. A working template of a locally (or ecosystem) based unit or officer could be redeveloped and transplanted elsewhere once a network has been established and tangible enforcement goals (increased seizures, arrests and prosecutions) are being achieved.

In the European Union and United Kingdom, source management is subject to strict scrutiny, guidelines, protocols and laws. In the absence of any clarity around conducting source recruitment and management activities in particular countries, the UK codes of practice could offer guidance, with due regard to the European Convention on Human Rights as it pertains to covert law enforcement practices. The UK rules comprise comprehensive and auditable regulations concerning transparency, accountability and protection, and as such offer an appropriate benchmark.

Although there are many apparent obstacles to overcome, the effectiveness of a dedicated, credible and efficient human source capacity cannot be underestimated: they remain the definitive and ultimate intelligence-gathering tool.

4. Rapid response (action on intel)

In order to provide substantive support to the intelligence cycle concept, LEAs must be in a position to react swiftly to intelligence. Without reactive capability, the process often stalls, resulting in only minor disruption to criminal networks. The opportunity to collect further information and feed it back into the intelligence cycle is also lost. Capacity to react, for example, to human source information about pangolin traffickers would:

(1) provide the best chance of disrupting specific criminal activities;
(2) potentially allow for the arrest and processing by the LEA of the individuals involved; and
(3) afford law enforcement an opportunity to acquire requisite evidence and intelligence from arrestees, to be fed back into the intelligence cycle.

Feeding information back into the intelligence cycle should contribute to a deleterious effect on poaching in the long term, all other things being equal, ultimately resulting in the disruption of the more organized profiteers within criminal networks.

One of the greatest missed opportunities in combating pangolin trafficking in many African nations is the lack of intelligence acquisition from individuals hunting/poaching at a subsistence level or for small profit. Much can be gleaned from judicious and legitimate intelligence interviews, and correct methods of evidence seizure to identify opportunities to feed information back into the intelligence cycle. Such intelligence can be of great value in identifying networks, relationships and trafficking methods, furthering the disruptive potential of enforcement action.

While likely warranted in many instances, implementation of a dedicated response capacity requires careful deliberation. There are extensive budget considerations concerning the provision of vehicles and equipment, and training may be required (e.g., tactical and arrest training, detainee safety, human rights and community-focused communications).

5. Improving criminal investigations

Investigating criminal offences is the foundation of law enforcement, enabling the presentation of cases before a court so judicial processes can be completed. Maintenance of evidential integrity and chain of custody is vitally important to this process. This concept is often covered in law enforcement training, but it is difficult to monitor in implementation terms. When, for example, poachers are arrested LEA personnel should manage the evidence appropriately, including photographing weapons and other items, seizing and labeling them, and retaining them for forensic evaluation as applicable (including potential ballistic matching to other crimes). Where government agencies have little or no forensic capacity, forensic analyses may be facilitated under certain circumstances by various international agencies or NGOs. New tools are emerging for this purpose. For instance, in 2018 the University of Portsmouth in partnership with the Zoological Society of London (ZSL) and UK Border Force, acquired evidentially sound fingerprints from pangolin scales using forensic gel developed for this purpose. Ease of use in the field was an important consideration in development, and the gels can be deployed by law enforcement officers with limited forensic knowledge or training.

Traps and snares should be treated in the same way. Photography or drawings of items in situ are essential, and physical exhibits should be labeled and packaged appropriately using the best available resources. For example, agents patrolling the depths of the Congo Basin who happen upon a crime scene are unlikely to be in possession of tamper-evident self-seal bags of varying sizes to retain bullet casings, or weaponry, or personal effects left behind by suspects. However, they can minimize physical interaction and maximize evidential integrity by using other items available to them, including, for example, large leaves to package items of evidence.

LEAs should avoid photographing suspects holding contraband (e.g., bags of pangolin scales, dead pangolins). This is contrary to the suspect's moral right not to incriminate themselves (even if a confession has already been recorded, it is inappropriate). Relatedly, the posting of arrestees' pictures online is ethically controversial because it implies guilt prior to fair judicial processes having taken place. Effective defense lawyers could have the suspect acquitted using arguments of duress, forensic contamination, or inappropriate management of evidence. Agents should be *au fait* with the probative value of notes or pictures made relating to specific cases, using evidentially sound protocols. The same rules can be applied to locally used documents (for example, proces verbales).

Ideally, wildlife crime scenes should be treated as serious crime scenes. This means they should be preserved and secured for assessment, evaluation, and the removal of evidence. Consideration should be given to cordons, common approach paths, entry and exit routes, crime scene logs, sketch plans, videos and photographs, organic material samples, fluid and body (blood and tissue) samples (animal and human). The evidential chain should be maintained with integrity, ensuring minimal interference (if any) with physical evidence. The concept of a "chain of custody" should be in place, along with places for the appropriate storage of any forensic samples.

The ability to translate physical or forensic evidence into viable documents or images suitable for judicial proceedings is also essential. The much-vaunted case of Yang Fenglan (the so-called "Ivory Queen") in Tanzania, yielded a huge amount of press attention (e.g., Kriel and Duggan, 2015), with many organizations claiming involvement or success in her detention when she was arrested. The case against her was finally concluded in February 2019 following consistent adjournments, ostensibly due to procedural and paperwork inconsistencies. This lengthy case highlights the need for robust chains of custody to be in place.

Investigations need to be properly managed and implemented. The correct manner of interviewing arrestees in order to obtain more information about criminal activity is the single most missed opportunity in wildlife law enforcement at present. There are few (if any) examples of contemporary investigations where, for example, a low-level poacher or facilitator has been arrested, debriefed by appropriately trained personnel, and provided information leading to the investigation of transnational trafficking networks. It is easy for some LEAs to be satisfied with the arrest of low-level poachers or traffickers, but much more consideration should be given to the intelligence value of these individuals.

Acquisition of mobile phone, computer and financial data can provide a detailed understanding of criminal networks. This is not always conducted and is a further missed opportunity, and should be used as standard in the immediate future. To tackle international pangolin trafficking, which evidence indicates is highly organized (Chapter 16), LEAs should be perpetuating the intelligence cycle, using human sources effectively, and evaluating intelligence resulting from arrests to acquire further information to disrupt trafficking.

Evidential rules apply equally to the use of social media to sell pangolin parts. Activity conducted (even initially) online to obtain information, evidence or covertly engage with individuals or groups operating online are subject to the same processes as any other observational or documentary evidence.

6. Improving prosecution and conviction rates

Application of the law entails criminals being held to account for their actions, according to applicable legislation, including through prosecution and conviction. Prosecution and conviction rates could be improved in many African pangolin range countries in a number of ways. This includes conducting training for judiciaries, especially prosecutors, and investigators, eco-guards, police and customs officials, to ensure there is a sound understanding of the evidential and prosecutorial process. Notably, this applies to regional and local courts, which are often unaware of the protected status of wildlife, but are situated where poaching is rife. Such training will help ensure they are cognizant of the illegality of pangolin trafficking, including the fact that in many cases it comprises transnational organized crime. Notwithstanding local legislation, courts, judges and prosecutors should also be made aware that pangolins are afforded protection under international conventions, i.e., are included in CITES, the Convention on International Trade in Endangered Species of Wild Fauna and Flora, Appendix I (see Chapter 19). Similar to methods employed in drug trafficking cases, each case file relating to pangolin trafficking should contain an "impact statement" or a similar expert testimony, explaining the severity of trafficking and the impact on pangolin populations. It would also be useful to provide case studies or "stated cases" in respect of previous successful pangolin trafficking prosecutions. The combination of training, increased awareness by judiciaries and enhanced quality of professional cases presented before courts, should contribute to an increase in successful prosecution and conviction rates.

Judiciaries and governments should further be encouraged to enact legislation or protocols enabling lower level criminals to be considered for more lenient sentencing in exchange for substantive, actionable information about criminality higher up the criminal chain. This would enable LEAs to more readily pursue the higher echelons of organized criminal networks.

Finally, organizations that have ministerial, governmental or diplomatic influence, including some international NGOs, should leverage this influence, in conjunction with the activities discussed, to strengthen demands for appropriate sentencing guidelines. These should then be used appropriately for those involved in pangolin trafficking. This should include the use of laws pertaining to seizure of assets and finances (see Chapter 17), allowing organized pangolin traffickers to be prosecuted under legislation reflecting the seriousness of their crimes.

Conclusions

Effective application of the law is critical to addressing (i.e., reducing) the trafficking of pangolins and their parts. The six key elements of law enforcement highlighted here are not singular solutions; they need to be implemented synergistically to effect long-term change. It would be easy to understand criticism or skepticism of some of the concepts discussed, especially trying to apply them in fragile and conflict-affected states, where lawlessness and corruption can be an obstacle. However, all of these solutions are eminently adaptable to different circumstances and do not require overly sophisticated systems. They can be implemented using dedicated, enthusiastic, motivated front-line law enforcement personnel.

The biggest seizure of wildlife traded illegally in Benin was made in March 2018, at Cotonou airport, and comprised nearly 5 tonnes of pangolin scales (ZSL, 2018). The law enforcement officers were from a dedicated specialist anti-trafficking unit that had been provided with basic evidential training in 2017. This unit, whose general remit was anti-narcotics work, made it very clear that, without the training they had received in respect of recognizing behavioral traits of the alleged traffickers (who were arrested) and the scales themselves (the suspects claimed they were "dried fish scales") this major seizure would not have occurred. This example highlights the importance of providing law enforcement officers with the knowledge and capacity they need, both in terms of commodity and methodology, to target individuals, routes, and transport methods accordingly. There are various challenges to conserving pangolins. Only by ensuring that front-line law enforcement personnel right through to judiciaries have the resources and knowledge they need to make arrests, secure prosecutions and convictions, and feed intelligence back into the intelligence cycle, will it be possible to target the organized networks trafficking pangolin parts and reduce the threat from overexploitation.

References

Challender, D.W.S., MacMillan, D.C., 2014. Poaching is more than an enforcement problem. Conserv. Lett. 7 (5), 484–494.

Challender, D.W.S., MacMillan, D.C., 2016. Transnational environmental crime: more than an enforcement problem. In: Schaedla, W.H., Elliott, L. (Eds.), Handbook of Transnational Environmental Crime. Edward Elgar Publishing, Cheltenham and Northampton, Massachusetts, pp. 489–498.

Challender, D., Waterman, C., 2017. Implementation of CITES Decisions 17.239 b) and 17.240 on Pangolins (*Manis* spp.), CITES SC69 Doc. 57 Annex. Available from: <https://cites.org/sites/default/files/eng/com/sc/69/E-SC69-57-A.pdf>. [August 2, 2018].

Challender, D.W.S., Waterman, C., Baillie, J.E.M., 2014. Scaling Up Pangolin Conservation. IUCN SSC Pangolin Specialist Group, Zoological Society of London, London, UK.

College of Policing, 2014. Code of Ethics. A Code of Practice for the Principles and Standards of Professional Behaviour for the Policing Profession of England and Wales. Available from: <https://www.college.police.uk/What-we-do/Ethics/Ethics-home/Documents/Code_of_Ethics.pdf>. [June 18, 2019].

Felbab-Brown, V., 2018. The threat of illicit economies and the complex relations with state and society. In: Comolli, V. (Ed.), Organized Crime and Illicit Trade: How to Respond to This Strategic Challenge in Old and New Domains. Palgrave Macmillan, pp. 1–21.

Heinrich, S., Wittman, T.A., Ross, J.V., Shepherd, C.R., Challender, D.W.S., Cassey, P., 2017. The Global Trafficking of Pangolins: A Comprehensive Summary of Seizures and Trafficking Routes From 2010–2015.

TRAFFIC, Southeast Asia Regional Office, Petaling Jaya, Selangor, Malaysia.

Ingram, D.J., Cronin, D.T., Challender, D.W.S., Venditti, D. M., Gonder, M.K., 2019. Characterising trafficking and trade of pangolins in the Gulf of Guinea. Glob. Ecol. Conserv. 17, e00576.

Kriel, R., Duggan, B., 2015. 'Queen of Ivory' arrested in Tanzania. CNN. Available from: <https://edition.cnn.com/2015/10/09/africa/tanzania-elephant-ivory-queen-arrest/index.html>. [March 23, 2019].

Mambeya, M.M., Baker, F., Momboua, B.R., Pambo, A.F.K., Hega, M., Okouyi, V.J.O., et al., 2018. The emergence of a commercial trade in pangolins from Gabon. Afr. J. Ecol. 56 (3), 601–609.

Patel, P., 1996. The Convention on International Trade in Endangered Species: enforcement and the last unicorn. Houst. J. Int. Law 157–213.

Reeve, R., 2002. Policing International Trade in Endangered Species. The CITES Treaty and Compliance. The Royal Institute of International Affairs and Earthscan Publications Ltd, London.

Roe, D., Mulliken, T., Milledge, S., Mremi, J., Mosha, S., Greig-Gran, M., 2002. Making a killing or making a living? Wildlife trade, trade controls and rural livelihoods. Biodiversity and Livelihoods Issues, No.6. IIED, London, UK.

Sandbrook, C., 2015. The social implications of using drones for biodiversity conservation. Ambio 44 (Suppl. 4), S636–S657.

Shairp, R., Veríssimo, D., Fraser, I., Challender, D.W.S., MacMillan, D.C., 2016. Understanding urban demand for wild meat in Vietnam: implications for conservation actions. PLoS One 11 (1), e0134787.

World Bank, 2016. Analysis of International Funding to Tackle Illegal Wildlife Trade. The World Bank, Washington D.C.

ZSL, 2018. Anti-trafficking officials in Benin seize record haul of pangolin scales. Available from: <https://www.zsl.org/conservation/news/anti-trafficking-officials-in-benin-seize-record-haul-of-pangolin-scales>. [April 16, 2019].

CHAPTER

19

Addressing trade threats to pangolins in the Convention on International Trade in Endangered Species of Wild Fauna and Flora (CITES)

Daniel W.S. Challender[1,2] *and Colman O'Criodain*[3]

[1]Department of Zoology and Oxford Martin School, University of Oxford, Oxford, United Kingdom [2]IUCN SSC Pangolin Specialist Group, ℅ Zoological Society of London, Regent's Park, London, United Kingdom [3]WWF International, The Mvuli, Nairobi, Kenya

OUTLINE

Introduction

CITES, the Convention on International Trade in Endangered Species of Wild Fauna and Flora, is the primary multilateral environmental agreement for ensuring sustainability in international wildlife trade (CITES, 2019a; Wijnstekers, 2018). Born out of recognition of the need to better control international commerce in wildlife in the mid-late 20th century, it entered into force in 1975 (CITES, 2019a). CITES is implemented by 182 member

Pangolins
DOI: https://doi.org/10.1016/B978-0-12-815507-3.00019-8

countries (known as parties), plus the European Union which is a party in its own right, through national legislation and enforcement mechanisms and a system of permits and certificates (CITES, 2019b). States accede to the convention voluntarily but are required to take a number of measures to implement it, including enacting legislation to enforce its provisions and designating national Management and Scientific Authorities. These authorities are required to monitor and advise on levels of trade to ensure it is not biologically unsustainable, to verify the legal acquisition of specimens in trade, and to grant permits (e.g., import/export permits and re-export certificates) for trade (CITES, 2019c).

The principal tenet of CITES is trade controls, the extent of which is dependent upon the listing of species in the CITES Appendices. Species are typically listed in one of its three appendices with corresponding trade controls, based on an assessment of extinction risk evaluated against biological and trade criteria, the "listing criteria"[1]. Approximately 1000 species that are in international trade and are deemed to be threatened with extinction are listed in Appendix I. Commercial trade in these species is generally prohibited and otherwise permitted only in exceptional circumstances (CITES, 2019c). Where trade is allowed it requires not only the issuance of an export permit verifying that the specimen was legally obtained but also the issuance of a permit by the importing country. Most CITES-listed species (about 35,000) are included in Appendix II (CITES, 2019d). These are: (1) species that are deemed to be at risk of extinction in the future if trade is not regulated, and (2) species where trade must be regulated by virtue of their resemblance to species in the preceding category (i.e. "lookalike species"). Trade in these species is permitted subject to the issuance of an export permit. This is based on: (1) a legal acquisition finding, that is a determination that the specimen of the species in question was legally obtained, and (2) a non-detriment finding (NDF), that is a declaration by the exporting state that trade in specimens of a given species will not be detrimental to the survival of that species in the wild. About 200 species are listed in Appendix III whereby trade is regulated by one or more parties, but where the relevant party or parties have cooperation from other parties in preventing unsustainable trade (CITES, 2019d).

Parties can propose amendments to the appendices. The inclusion, deletion and transfer of species between Appendices I and II typically takes place at meetings of the Conference of the Parties (CoP). Amendments are dependent on consensus or at least a two-thirds majority vote of parties present and voting. CoPs, which take place every three years, are the convention's highest decision-making body, and governance of the convention in between CoPs falls to the Standing Committee.

To ensure parties are compliant with the convention, CITES uses a combination of "carrots" and "sticks" (Reeve, 2002). This includes technical assistance and capacity building (e.g., training and legal support) and conversely, the threat of, and/or establishment of trade sanctions for CITES-listed species where parties are noncompliant (Reeve, 2006; Sand, 2013). Sanctions take the form of a recommendation to parties by the Standing Committee to suspend trade in some or all CITES-listed species with the country in question. They may be applied, for example, where parties have failed to enact implementing legislation. As CITES has evolved, a number of additional mechanisms have been adopted to support compliance. These include the Review of Significant Trade (RST) process through which remedial recommendations are formulated for parties where trade in specific species in Appendix II is unsustainable or otherwise

[1] Resolution Conf.9.24 (Rev. CoP17), Criteria for amendment of Appendices I and II.

problematic, and a review process for trade in animal specimens reported as produced in captivity. Additionally, ad hoc measures have been taken for specific taxa or groups of species where non-compliance has concerned high levels of illegal trade. These include the adoption of CITES Resolutions and Decisions which direct parties, the CITES Secretariat and/or other CITES Committees and stakeholders to take specific actions. For example, urging parties to eliminate consumer demand for specific species or to destroy stockpiles of certain derivatives, the creation of bespoke illegal trade monitoring systems (e.g., ETIS, the Elephant Trade Information System), and high-level technical or political missions to problem countries (Challender et al., 2015a). With an increasing focus on escalating illegal trade in CITES-listed species, especially species listed in Appendix I, the CoP and the Standing Committee are spending more time on compliance issues and focussing in depth on countries of critical concern, e.g., Democratic Republic of the Congo (DRC), Guinea, Lao People's Democratic Republic (Lao PDR) and Nigeria (CITES, 2018).

Due to the high volumes of international trade in pangolins historically, including in the mid-late 20th century (see Chapter 16), they have been included in CITES since its inception and have a long and complex history in the convention. This chapter discusses how the convention has managed pangolins historically, whether or not it has been effective at ensuring sustainability in international pangolin trade, critically evaluates the impact of CITES decisions, and discusses means of furthering the conservation of pangolins in CITES.

A history of pangolins in CITES

At the inception of CITES, Asian pangolins (minus the Philippine pangolin [*Manis culionensis*], which wasn't described as a distinct species at the time) were listed in Appendix II and Temminck's pangolin (*Smutsia temminckii*) in Appendix I (Fig. 19.1). These listings were made in the absence of defined listing criteria, which were not developed until the first plenipotentiary CoP in 1976 (Huxley, 2000). Following debate at CoP8 in 1992 on the merits of Temminck's pangolin being included in Appendix I in the absence of listing criteria (CITES, 1994), Switzerland, as the depository government, proposed the transfer of this species from Appendix I to II, at CoP9 in 1994. It also proposed the inclusion of the giant pangolin (*S. gigantea*), white-bellied pangolin (*Phataginus tricuspis*) and black-bellied pangolin (*P. tetradactyla*) in Appendix II. The proposal was submitted at the request of the Chair of the Animals Committee, the CITES technical advisory committee for fauna, and at CoP9 the CoP agreed to a genus listing. Consequently, *Manis* spp. were included in Appendix II as of February 16, 1995. This included all extant species of pangolin since CITES uses Wilson and Reeder (2005) as its mammalian taxonomic reference, which considers each species to reside in the genus *Manis*.

Concerns over the sustainability of international trade in Asian pangolins led to their inclusion in the RST process multiple times. Between 1977 and the year 2000, international trade in pangolins reported to CITES involved more than an estimated 750,000 animals (Chapter 16; Heinrich et al., 2016). This mainly involved Asian pangolin skins and scales; more than 500,000 skins were traded primarily for commercial purposes, mostly from the Sunda pangolin (*Manis javanica*). Much of the skin trade was ultimately destined for Japan, the United States, and Mexico, for the manufacture and retail of leather goods (Anon, 1992). Despite reported trade being permitted by exporting parties, for which NDFs should have been made, concern over the sustainability of trade led to the inclusion of Asian pangolins in the RST process in its preliminary phase in 1988, in phase I in 1992, and phase IV in 1999 (Table 19.1).

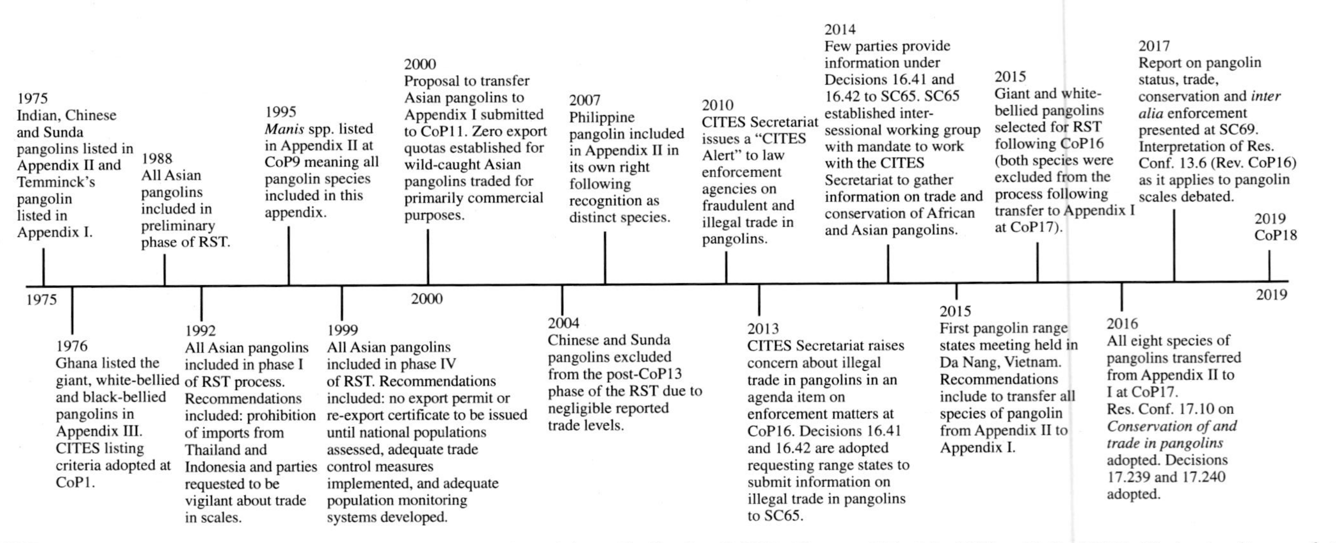

FIGURE 19.1 Selected pangolin and CITES timeline. *Adapted from Challender, D.W.S., Harrop, S.R., MacMillan, D.C., 2015b. Understanding markets to conserve trade-threatened species in CITES. Biol. Conserv. 187, 249–259.*

TABLE 19.1 Outcomes, recommendations and results for pangolins in the CITES Review of Significant Trade (RST) process.

Phase	Species included	Outcome	Recommendation (P=Primary, S=Secondary)	Party directed at	Result
Prelim (1988)	*M. javanica* *M. pentadactyla* *M. crassicaudata*	Possible problem			
I (1992)	*M. javanica* *M. pentadactyla* *M. crassicaudata*	Category C (current trade levels and/or conservation status insufficiently known). Primary and secondary recommendations made.	P: Prohibit imports of items originating in Indonesia and Thailand because of legislation protecting the species in both countries.	All parties	Notification 688 sent to parties. Secretariat satisfied.
			S: Parties involved in trade in oriental medicines, particularly Singapore, China (including Taiwan) and Hong Kong (now Hong Kong SAR) requested to be vigilant regarding trade in pangolin scales.	All parties trading in oriental medicine	Notification 688 sent to Parties. Secretariat satisfied.
			S: The Secretariat should advise non-party importing countries, in particular the Republic of Korea, that trade controls exist for these species and request that they co-operate in ensuring all specimens imported have been legally exported.	Non-party importing countries, particularly Republic of Korea	Secretariat wrote to Republic of Korea (non-party at the time).
			S: The Management Authority of Malaysia should advise the Secretariat of the protection status of pangolins, especially in Sabah and Sarawak.	Malaysia	Sunda pangolin protected in Malaysia. Secretariat satisfied.
			S: The Management Authority of Singapore should investigate the origin of pangolin scales imported, to verify the legality of exports and advise the Secretariat on outcomes of its inquiry.	Singapore	One importer prosecuted. Secretariat satisfied.
			The Management Authority of China should advise the Secretariat of the status of its research on *Manis pentadactyla*.	China	Field surveys and research on captive breeding planned in China. Secretariat satisfied.

(Continued)

TABLE 19.1 (Continued)

Phase	Species included	Outcome	Recommendation (P=Primary, S=Secondary)	Party directed at	Result
IV (1999)	*M. javanica*, *M. pentadactyla* *M. crassicaudata* *S. gigantea* *S. temminckii* *P. tricuspis* *P. tetradactyla*	*M. javanica*: Category d(i) – information indicates global population or population in a particular range state is being adversely affected by international trade. Primary recommendation made. *M. pentadactyla*: Category d(i)/d(ii) – information indicates global population or population in a particular range state is being adversely affected by international trade. Primary recommendation made. *M. crassicaudata, S. gigantea, S. temminckii, P. tricuspis, P. tetradactyla*: Category d(iii) – elimination from the review.	P: No export or re-export certificate should be issued, or accepted, for specimens of *M. javanica, M. pentadactyla*, and *M. crassicaudata* until the following actions have taken place and been reported on to the satisfaction of the Secretariat: • an assessment of the distribution and population status (including abundance) of the three species in all range states that authorize exports of specimens of these species; • the competent authority of Lao PDR and the Management Authorities of Singapore, Thailand, Cambodia, China, Malaysia, Vietnam, Myanmar and Indonesia have developed and implemented adequate control measures and inspection procedures to detect and intercept illegal shipments of specimens of all *Manis* spp.; and • the authorities of all range states wishing to trade in pangolins, their parts and derivatives have developed adequate, scientifically based population monitoring systems and measures to identify and regulate exports of legally obtained specimens.	All parties	CoP11 (2000) established zero export quotas for wild-caught specimens of *M. javanica, M. pentadactyla* and *M. crassicaudata*. SC45 (2001) agreed that if zero quotas were removed any range states wishing to trade in these species would need to satisfy the Secretariat that the 1999 recommendations had been implemented before exports took place.
Following CoP16 (2013)	*S. gigantea* *P. tricuspis*	Excluded from the process following transfer from Appendix II to I at CoP17.			

Adapted from Challender, D.W.S., Harrop, S.R., MacMillan, D.C., 2015b. Understanding markets to conserve trade-threatened species in CITES. Biol. Conserv. 187, 249–259.

Within each phase of the RST, a detailed review of each species' biology, threats and international trade was conducted. A lack of quantitative data on pangolin populations meant that it was difficult to determine the impact of exploitation on populations and to disentangle international trade from local (i.e., domestic) use. However, shifting trade patterns were observed, seemingly as populations were depleted in parts of Southeast Asia, and each review reported high collection pressure and hunting-driven population declines in many parts of the Chinese (*M. pentadactyla*) and Sunda pangolin ranges (Anon, 1992, 1999; Broad et al., 1988). High levels of unreported and seemingly illegal trade in pangolins were also documented, in particular involving live animals and scales, and in quantities that far surpassed trade levels reported to CITES (see Chapter 16). Similarly, the reviews documented that local communities in source areas for pangolins had very high incentives to poach and trade the animals due to the high prices they fetched, and that there was substantial demand for pangolin meat and scales in international markets (e.g., China). Time-bound remedial measures were introduced in phase I of the RST and a combination of short- (primary) and long-term (secondary) recommendations were prescribed in 1992 in an effort to ensure any future trade was sustainable (Table 19.1). Recommendations included the prohibition of imports of Asian pangolins and their parts from Indonesia and Thailand where the species were protected by law and, among other things, a request that parties involved in trade in oriental medicines exercise vigilance concerning trade in scales (Table 19.1).

Despite implementation of these recommendations to the satisfaction of the CITES Secretariat (see Table 19.1), the seven species of pangolin recognized at the time were included in phase IV of the RST owing to ongoing concerns about sustainability. The African species and the Indian pangolin (*M. crassicaudata*) were eliminated from the process on the basis that international trade was evidently not a problem (Table 19.1). For the Chinese and Sunda pangolins it was adjudged that populations were being adversely affected by international trade and a primary recommendation was made. It prescribed that no exports or re-exports of these species should be accepted by parties until range states had, among other things, assessed the distribution and population status (including abundance) of each species and developed scientifically based population monitoring methods (Table 19.1).

While still subject to phase IV of the RST process, Asian pangolins were the focus of a proposed transfer from Appendix II to I at CoP11 in the year 2000 (CITES, 2000a). Following debate at the meeting, and despite recognition by a number of parties and the CITES Secretariat that the Sunda pangolin met the criteria for inclusion in Appendix I (CITES, 2000b), the proposal was rejected. This was because the species were still in the RST process and it was considered premature to include them in Appendix I. Instead, the parties adopted zero export quotas for wild-caught Asian pangolins traded for commercial purposes, in effect, a proxy trade ban (Challender et al., 2015b), but one that did not require the issuance of import permits by importing countries.

Between the years 2001 and 2016 there was comparatively little reported trade in Asian pangolin skins, suggesting that the introduction of zero export quotas led to its near cessation (Chapter 16; Challender et al., 2015b). Trade involving Asian pangolins otherwise occurred at negligible levels compared to pre-2000 and involved a range of derivatives (Challender and Waterman, 2017). On the contrary, reported international trade in African pangolins increased after the year 2000 (see Chapter 16).

However, far exceeding all reported trade to CITES between the year 2001 and 2016 was illegal trade in pangolins and their meat and scales. Evidence from the RST process and other sources (e.g., government records, seizures reported in the media) indicate that there was an ongoing unreported, illicit trade in pangolins and their parts in Asia since at least the 1980s, which continued ostensibly unabated post-2000. Estimates suggest that illegal trade between August 2000 and July 2019 involved nearly 900,000 pangolins (see Chapter 16). This trade also involved all eight species, but particularly African pangolins, and almost exclusively scales, between 2016 and 2018 (Chapter 16). However, despite the advent of intercontinental trafficking of African pangolin scales to Asian markets in 2008 (Challender and Hywood, 2012), i.e., pangolin trafficking becoming an intercontinental issue and not one confined to Asia, pangolins received little concerted attention in CITES between 2001 and 2013 when they arguably needed it most.

After a virtual hiatus from CITES meetings, pangolins reappeared at CoP16 in 2013 under an agenda item on enforcement matters; the CITES Secretariat was concerned about frequent high volume seizures involving the animals. This ultimately led to a series of actions culminating in the transfer of the species from Appendix II to I at CoP17 in 2016. At CoP16, two Decisions were adopted which directed pangolin range states to submit information on the conservation and illegal trade in the animals to SC65 (the 65th meeting of the Standing Committee) and for this meeting to develop recommendations to address illegal trade and report to CoP17. At SC65, an inter-sessional working group was formed with a mandate to collect further information on the conservation of and trade in pangolins, and to draft recommendations for combatting illegal trade for consideration at SC66. By the time of the SC66 meeting (January 2016) the First Pangolin Range States meeting had been held in Da Nang, Vietnam, co-hosted by the Vietnamese and United States governments. This produced a series of recommendations, including that all pangolin species be transferred from Appendix II to I. Information submitted by parties to SC66 attested that there exists little knowledge of pangolin populations and/or that they are suspected to be declining. The SC66 meeting made a number of recommendations to CoP17 including that the parties adopt a Resolution (i.e., soft legislation elaborating on the provisions of the convention) urging parties to take a number of measures to combat trafficking and calling on all relevant stakeholders to do so.

Prior to CoP17, proposals to transfer all pangolin species from Appendix II to I were jointly submitted by a number of parties, including range states such as Nepal, the Philippines, Senegal, South Africa and Vietnam, and the United States. There was virtually unanimous support for the proposals at the meeting. One exception was Indonesia, which prevented consensus on the transfer of the Chinese and Sunda pangolins to Appendix I and forced a vote; however, it was the only country opposed to the amendment.

There was otherwise a groundswell of interest in pangolins at CoP17, characterized by the Egyptian delegation declaring in Committee I of the meeting that "CoP17 is the CoP for the pangolin." Each species of pangolin was transferred from Appendix II to I, a listing that came into effect on January 2, 2017. CoP17 also adopted a Resolution on pangolins[2], which urges parties and other stakeholders to take action to, among other things:

- ensure robust legislation is in place which provides deterrent penalties to address illegal trade in pangolins and their parts and ensure it is strictly enforced;
- undertake capacity-building activities which focus on detecting illegal trade;

[2] Resolution Conf. 17.10, Conservation of and trade in pangolins.

- ensure that where pangolin breeding facilities exist parts and derivatives do not enter into illegal trade;
- ensure that where stockpiles exist adequate controls measure are in place and stocks are secure;
- partner with local communities to ensure sustainable management of pangolin populations;
- implement measures to reduce demand for pangolin products traded illegally; and
- develop in situ monitoring, conservation and management programmes.

CoP17 also adopted two Decisions on pangolins; the first requested the CITES Secretariat to liaise with international law enforcement networks (e.g., the International Consortium on Combatting Wildlife Crime [ICCWC]) to ensure pangolins are prioritized within the work programs of its members, which in addition to the CITES Secretariat includes Interpol, the United Nations Office on Drugs and Crime (UNODC), the World Customs Organisation (WCO), and the World Bank. The second Decision called for a comprehensive report on pangolin status, trade, enforcement efforts, stockpiles, captive populations and demand reduction initiatives. Based on this report, which was presented at SC69 (see Challender and Waterman, 2017), multiple draft Decisions containing actions to support combatting illegal pangolin trade were proposed for adoption at CoP18 (CITES, 2017 see also CITES, 2019e). They included encouraging range states to implement urgent measures to conserve pangolins as articulated in Resolution Conf. 17.10, and for range states, intergovernmental organizations, aid agencies and non-governmental organizations (NGOs) that develop tools and guidance that could assist in implementing such measures to bring them to the attention of the Secretariat for sharing with the CITES parties. By the time of CoP18 (August 2019), additional draft Decisions were proposed for adoption by the CoP. They included directing the CITES Secretariat to work with experts to develop more robust conversion parameters to accurately estimate the number of pangolins in illegal trade, and to prepare a report on the status of pangolins, legal and illegal trade, stockpiles, and enforcement issues between CoP18 and CoP19. These Decisions were adopted at CoP18.

Additionally, pangolins were the cause of contention at SC69 (November 2017). Specifically, the issue of stockpiles of scales, and whether such stockpiles accumulated while the species were listed in Appendix II should be treated as specimens of species in Appendix I, thus prohibiting commercial trade, or Appendix II, which would allow such stockpiles to be traded commercially. There was heated debate at the meeting and polarized interpretations of guidance on this issue (specifically Resolution Conf. 13.6, Rev. CoP16[3]) between China, and Cameroon, Gabon, Nigeria, Senegal, the EU and the United States. The Standing Committee voted that, until CoP18, such stocks should be treated as specimens of species listed in Appendix I. This interpretation was agreed to by the parties at CoP18.

Has CITES been effective for pangolins?

The effectiveness of CITES has been and continues to be debated both in terms of species conservation and international collaboration (e.g., Bowman, 2013; 't Sas-Rolfes, 2000). Causally attributing international policy decisions made in CITES to the status of species in the wild, however, is difficult due to the myriad confounding factors affecting species status

[3] Resolution Conf. 13.6 (Rev. CoP16), Implementation of Article VII, paragraph 2, concerning 'pre-Convention specimens.'

(e.g., habitat loss, intrinsic and extrinsic biological factors; Martin, 2000). Yet, it is possible to make inferences about the impact of CITES decisions using changing trade dynamics and economic and other data. With hindsight, it is apparent that CITES has largely failed at ensuring sustainability in international trade in pangolins and their parts. This can be attributed to a few key factors including non-compliance among parties, especially range states, lack of appropriate ecological monitoring methods and knowledge of pangolin populations with which to inform decision-making and CITES processes, and decision-making in CITES not accounting for the economic reality of trade (Challender and MacMillan, 2014).

Despite exports of Asian pangolins being declared sustainable from the late 1970s to the late 1990s, i.e., parties presumably made NDFs and granted permits for trade, evidence suggests that such trade was in fact unsustainable and contributed to the decline of Chinese and Sunda pangolin populations, e.g., in Lao PDR, Malaysia and Thailand (Chapter 16; Nooren and Claridge, 2001). Some of this trade took place despite pangolins being protected species in exporting countries, including Indonesia, Malaysia and Thailand, meaning that they should have been protected from commercial exploitation. Effective ecological monitoring methods combined with greater knowledge of pangolin populations and how they behave (e.g., population estimates, recruitment rates, mortality rates) would have allowed more informed NDFs to be made and more sensitive estimates of offtake and management, which may have prevented instances of unsustainable trade. However, even with such knowledge, there is no guarantee that it would have improved sustainability and compliance with CITES. Problems with compliance among Asian pangolin range states during this time period were significant, and included lack of capacity, resources and personnel, ineffective communication between parties, the need for better training of customs and other front-line enforcement personnel, and the need for more robust legislation coupled with stronger enforcement (McFadden, 1987; Patel, 1996).

Asian pangolins entered the RST process on multiple occasions and recommendations were implemented. However, this did not prevent ongoing trade as well as concerns over sustainability. The failure here was that recommendations in the RST process were principally focused on regulatory measures and did not address the economics drivers of trade. The recommendations did nothing to address the very high financial incentives for local people in source countries, often the rural poor, to hunt or poach pangolins, or the demand for pangolins and their derivatives in key international markets (e.g., China). CITES has been criticized for failing to contend with the economic reality of wildlife trade (Challender and MacMillan, 2014; Challender et al., 2015a) and the outcomes for Asian pangolins may have been different had the RST process considered the drivers of poaching and trafficking in their real world context.

The establishment of zero export quotas for wild-caught Asian pangolins traded commercially in 2000 was both positive and negative. Combined with other measures, for example an import ban on pangolin skins imposed by the United States during the late 1980s, the introduction of zero quotas appears to have led to the near cessation of international trade in Asian pangolin skins, with little trade reported post-2000 (see Chapter 16). However, as a regulatory measure, the quotas did little to address the actual drivers of trade and high volumes of illicit trade in pangolins and their parts has taken place between 2000 and 2019. Zero export quotas otherwise ensured that any subsequent commercial trade that took place was illegal and in the hands of organized criminals, and thereby difficult to monitor, with the exception of seizure data, which has a number of problems and biases (Underwood et al., 2013).

It is too early to evaluate the long-term impact of the Appendix I listing. However, seizures of large quantities of pangolins and their derivatives have taken place throughout 2017–19, mainly in Africa and Asia, involving all eight pangolin species. While this is positive in the sense that seizures are being made (i.e., illegal trade is being intercepted), it suggests that there has been little immediate, positive impact regarding the exploitation of pangolin populations. There is also a risk that the Appendix I listing could have exacerbated exploitation rates, which commands research attention (see Chapter 34). Compliance also remains a problem, for example, fraudulent permits have been used in attempts to export commercial quantities of scales from Burundi, DRC and Nigeria to China, Hong Kong Special Administrative Region and Lao PDR (CITES, 2018), and pangolin range states remain subject to trade suspensions because of non-compliance (e.g., Guinea, Liberia).

What now for pangolins in CITES?

Pangolins are arguably at a crossroads in CITES, having recently been transferred to Appendix I. Pessimists may point to the fact that Appendix I listings have failed to solve the problems of many other species at risk of extinction from international trade. The most high profile example is, perhaps, the tiger (*Panthera tigris*), where efforts to halt the trade – including the imposition of a ban on bone trade in China, the main market, in 1993 – have not been successful (Novak, 1999; Nowell and Ling, 2007) and where numbers continued to decline until an apparent increase in 2016 (WWF, 2016). However, while such examples should serve as caution against naïve optimism, there are species where, in recent decades, an Appendix I listing provided legal and institutional support that enabled in situ conservation efforts to achieve some success. The greater one-horned rhinoceros (*Rhinoceros unicornis*) in Asia is a good example, the global population of which is increasing due to conservation interventions including strict protection (Talukdar et al., 2008). The fact that the CITES parties adopted a specific Resolution on pangolins and a suite of Decisions at CoP17 contrasts with the previous example, where Resolutions were adopted much later, and indicates that the CITES parties have learned from previous experience and are anticipating at least some of the implementation issues that may arise with pangolins.

Nevertheless, where there have been successes (e.g., greater one-horned rhinoceros), they have depended on political will at the national and provincial level, and in situ efforts at the landscape level in range states. It remains to be seen whether these are forthcoming across a sufficient number of pangolin range states to secure the future of all eight species.

In this regard, it is crucial that there is continued oversight of issues relating to pangolins in CITES and other relevant international fora. The adoption of Decisions 17.239 and 17.240 by CITES (Fig. 19.1) ensured that implementation of the Appendix I listing would continue to be scrutinized at SC69 and CoP18 and the adoption of further Decisions at CoP18 means there will be continued oversight up to CoP19.

Experiences for other species with high market value and high volumes of illicit trade dictate that ongoing scrutiny will be necessary long into the future. Based on previous experience, anticipated problems for pangolins may include:

- an increase in trade in pangolins purportedly coming from captivity;
- attempts to offload stockpiles that were acquired but not disposed of prior to the entry into force of the Appendix I listing;
- increasing sophistication of smuggling techniques; and
- greater problems with corruption facilitating illegal trade.

There are a number of ways that these issues could be addressed in CITES. The first is through the Standing Committee, which following adoption of Decisions at CoP18 has a mandate to focus on pangolins, including illegal trade, in the intersessional period between CoP18 and CoP19.

Future meetings of the Standing Committee will need to be able to identify instances where countries are facing genuine difficulties in implementation, such as lack of capacity or resources. The Standing Committee will also need to be able to identify countries where weak governance or lack of political will are the obstacles. Typically, these countries are identified through indicators such as:

1. cases where consignments are being seized in transit or destination countries, and where documentary or forensic evidence traces the problem back to range countries, or to other transit countries where few or no seizures have been recorded (see Underwood et al., 2013); and
2. consumer countries where pangolins and their products are openly on sale either despite a prohibition, which is not well enforced (e.g., Vietnam), or because of legislative gaps that allow the open sale of wildlife products without any verification of their origin.

The question that bedevils discussions in such cases is how best to gather the evidence that allows such countries to be identified. Inevitably, parties and non-party countries attending CITES meetings, insofar as they report at all, will inform the Standing Committee that they are doing everything that is required of them under the convention. In the absence of solid evidence to the contrary, their word as sovereign governments will be taken at face value. However, on occasion CITES has mandated independent scrutiny in certain cases to monitor trafficking trends and to identify which specific countries need to take further action. Such scrutiny has taken a number of forms.

Resolution Conf. 14.3 on CITES *compliance procedures*

This Resolution sets out general guidelines on compliance procedures. It provides *inter alia* that, when the CITES Secretariat receives information indicating that a party is not in compliance on a given issue, it should take up the matter in the first instance with that party and then, in the absence of an adequate response, with the Standing Committee. The Committee may ultimately decide to recommend a suspension of trade either in certain species or in all CITES-listed species with the country concerned. For example, Guinea has been subject to suspension of trade in all CITES-listed species because it was exporting gorillas (*Gorilla* spp.) certified as being captive-bred, despite the absence of a breeding facility in the country. However, while this Resolution is central to all compliance matters in CITES, when it is not used in conjunction with other mechanisms the information that is relied upon can be anecdotal or ad hoc, and the word of the country is often sufficient to close the discussions unless there is corroborating evidence. In addition, because of the punitive overtones of compliance procedures, they are not necessarily the best approach when the country's problems result from lack of resources, rather than lack of political will.

Article XIII of the convention

In many ways this Article lies at the heart of CITES provisions and underpins Resolution Conf. 14.3 on compliance procedures. It provides that when the Secretariat is satisfied that any species included in Appendix I or II is being adversely affected by international trade, or that the provisions of the convention are not being effectively implemented, it should take such information up with the party or parties concerned in the first instance. However, unlike the Resolution, which has been in force since 2007 and which merely codified pre-existing practice, the Secretariat only begun to invoke Article XIII in 2016. In 2018, Lao PDR, DRC, Guinea and – latterly – Japan and Nigeria were under scrutiny by the Standing Committee. However, it is only in the case of Lao PDR that this has led to the country being faced with specific time-bound requirements to improve its legislation, law enforcement and other aspects of its implementation of the convention.

Resolution Conf. 9.14 (Rev. CoP17) on *Conservation of and trade in African and Asian rhinoceroses*

Measures adopted for rhinos could offer promise for pangolins. At CoP13 in 2004, and in response to complaints from range states about the reporting burden under this Resolution, it was agreed that in future TRAFFIC and the IUCN Species Survival Commission (SSC) Rhino Specialist Groups would prepare a consolidated report on the status of and trade in all five rhino species, in consultation with range states. This has proven to be a much better way of analyzing trade and trafficking trends and identifying countries of concern, such as Vietnam, than the previous approach, which required the countries to report themselves. However, the mechanism has not been incorporated into other species-specific Resolutions. In most cases, where there has been independent reporting at all, it has been as the result of one-off reports mandated by Decisions of the CoPs, as in the case of pangolins for SC69.

The Elephant Trade Information System (ETIS)

The bespoke illegal trade reporting mechanism for elephants (Elephantidae spp.) could be an option for pangolins. In 1997, CoP10 adopted Resolution Conf. 10.10 on *Trade in elephant specimens*. This Resolution covers many aspects of elephant conservation and elephant ivory trade and it has been revised at nearly every subsequent CoP. Provisions of most relevance to this discussion are those mandating the creation of ETIS. The central component of ETIS is a database on seizures of elephant specimens that have been reported – either by governments or from other sources – anywhere in the world since 1989, and it is supported by a series of subsidiary databases that address biases in seizure data. Trends in illegal trade are also measured against events such as ivory sales, and against background economic variables. ETIS has enabled the identification of key countries whose role in ivory trafficking has contributed to the elephant poaching crises. This led CITES to develop the National Ivory Action Plan process, enshrined in Resolution Conf. 10.10 (Rev. CoP17) at CoP17, which serves as a mechanism to identify countries that warrant compliance measures under the convention, and those that might require additional capacity and support.

While ETIS is undoubtedly the most sophisticated mechanism available for monitoring illegal trade trends in any species, it does come with a cost, and is dependent on donor funding. Nevertheless, it has proved an invaluable tool for monitoring illegal trade trends, including identification of transit and consumer markets, and it offers a model that could be used for other key high value species where trafficking remains a major issue, including pangolins.

Illegal trade reports

At CoP17, two Resolutions were revised in order to facilitate systematic reporting of illegal trade events: Resolution Conf. 11.17 (Rev. CoP17) on *National reports*, and Resolution Conf. 11.3 (Rev. CoP17) on *Compliance and enforcement*. In addition, the meeting adopted Decision 17.121, mandating the CITES

Secretariat to engage with appropriate bodies, including the United Nations Environment-World Conservation Monitoring Centre (UN Environment-WCMC), which is the recipient of CITES parties' annual trade reports, and UNODC, which maintains a seizure database, known as WorldWISE, which underpinned UNODC's first world wildlife crime report (UNODC, 2016). The aim of this engagement is to establish a global framework to store and manage illegal trade data collected through the parties' annual illegal trade reports.

WorldWISE has proven useful in terms of comparing the quantities of different species in trade, including pangolins, by reference to their monetary value. However, while this database may be useful for inter-species comparison, most data comes from a relatively small number of countries. There is no correction of the data for enforcement effort and, for those countries that do not report seizures, it is not known whether this is because illegal trade is not encountered, or because seizures are not made, or simply because they are not reported. Nevertheless, it remains a useful mechanism that could inform decision makers on the magnitude of illegal trade in pangolins relative to other high value trades (e.g., ivory, rosewood [*Dalbergia* spp.]).

Pulling it all together

There are a number of mechanisms that CITES could use to both monitor illegal trade in pangolins and take steps to reduce it. For the most part these are existing mechanisms that simply have to be applied to pangolins. This is true of CITES compliance procedures, including Resolution Conf. 14.3 and Article XIII of the convention. It is also true of the illegal trade reports, where illegal trade data are already, via WorldWISE, contributing to an understanding of the nature and volume of such trade. On the other hand, imitating what has been done for species such as rhinoceroses, tigers or elephants will require the adoption of bespoke measures including, most likely, amendment of Resolution Conf. 17.10, and this will require additional funds for their implementation. Ultimately, the justification for such measures depends on the current conservation status of pangolins and their vulnerability to overexploitation for international trade. It is apparent from the rest of the material presented in this volume that there is substantial demand for pangolins, they are easy to catch, and increasingly threatened by overexploitation.

It is imperative that CITES, pangolin range states and the donor community have the best information on which to act, both in terms of assisting countries where there are resource and capacity gaps, and pressurizing countries where lack of political will is reflected in weak legislation, low conviction rates, and high levels of corruption associated with the illegal pangolin trade. Pangolins, perhaps more so than any other species, do need bespoke monitoring mechanisms under CITES. Such mechanisms should be in the nature of ongoing assessments rather than one or more occasional reports, and they should be sufficiently robust to yield the maximum amount of information on which to base decisions. ETIS is arguably the best model. Above all, however, it must not be assumed that the Appendix I listing has "saved pangolins." Rather, the listing, although historic, is only the start of a process that will need to continue long into the future. Moreover, given limitations to CITES in how it contends with the economic and social drivers of trade and trafficking, this listing will need to be accompanied by a suite of supporting interventions as outlined in this volume, ranging from local community engagement to consumer behavior change.

References

Anon, 1992. Review of Significant Trade in Animal Species included in CITES Appendix II, Detailed Review of 24 priority species, Indian, Malayan and Chinese pangolin. CITES, Geneva, Switzerland.

Anon, 1999. Review of Significant Trade in Animal Species included in CITES Appendix II, Detailed Review of 37 species. World Conservation Monitoring Centre, IUCN Species Survival Commission and TRAFFIC, Cambridge, UK.

Bowman, M., 2013. A tale of two CITES: divergent perspectives upon the effectiveness of the wildlife trade convention. Rev. Eur. Commun. Int. Environ. Law 22 (3), 228–238.

Broad, S., Luxmoore, R., Jenkins, M., 1988. Significant Trade in Wildlife, A Review of Selected Species in CITES Appendix II. IUCN Conservation Monitoring Centre, Cambridge, UK.

Challender, D.W.S., Hywood, L., 2012. African pangolins under increased pressure from poaching and intercontinental trade. TRAFFIC Bull. 24 (2), 53–55.

Challender, D.W.S., MacMillan, D.C., 2014. Poaching is more than an enforcement problem. Conserv. Lett. 7 (5), 484–494.

Challender, D.W.S., Harrop, S.R., MacMillan, D.C., 2015a. Towards informed and multi-faceted wildlife trade interventions. Glob. Ecol. Conserv. 3, 129–148.

Challender, D.W.S., Harrop, S.R., MacMillan, D.C., 2015b. Understanding markets to conserve trade-threatened species in CITES. Biol. Conserv. 187, 249–259.

Challender, D., Waterman, C., 2017. Implementation of CITES Decisions 17.239 b) and 17.240 on Pangolins (*Manis* spp.), CITES SC69 Doc. 57 Annex. Available from: <https://cites.org/sites/default/files/eng/com/sc/69/E-SC69-57-A.pdf>. [August 2, 2018].

CITES, 1994. Amendments to Appendices I and II of the Convention, Transfer from Appendix I to Appendix II of *Manis temminckii* and inclusion of *Manis gigantea*, *Manis tetradactyla* and *Manis tricuspis* in Appendix II. CITES, Geneva, Switzerland.

CITES, 2000a. Amendments to Appendices I and II of the Convention, Prop. 11.13 Transfer of *Manis crassicaudata*, *Manis pentadactyla*, *Manis javanica* from Appendix II to Appendix I. CITES, Geneva, Switzerland.

CITES, 2000b. Doc. 11.59.3, Consideration of proposals for amendment of Appendices I and II, Eleventh Meeting of the Conference of the Parties. CITES, Geneva, Switzerland.

CITES, 2017. SC69 Com. 9, Report of the Working Group on Pangolins (Manidae spp.). Available from: <https://cites.org/sites/default/files/eng/com/sc/69/com/E-SC69-Com-09.pdf>. [September 25, 2018].

CITES, 2018. SC70 Doc. 27.3.5, Application of Article XIII in Nigeria. Available from: <https://cites.org/sites/default/files/eng/com/sc/70/E-SC70-27-03-05.pdf>. [December 31, 2018].

CITES, 2019a. What is CITES? Available from: <https://www.cites.org/eng/disc/what.php>. [January 1, 2019].

CITES, 2019b. List of Parties to the Convention. Available from: <https://www.cites.org/eng/disc/parties/index.php>. [January 1, 2019].

CITES, 2019c. How CITES works. Available from: <https://www.cites.org/eng/disc/how.php>. [January 1, 2019].

CITES, 2019d. The CITES species. Available from: <https://www.cites.org/eng/disc/species.php>. [January 1, 2019].

CITES, 2019e. Pangolins (*Manis* spp.). CITES CoP18 Doc. 75. Available from: <https://cites.org/sites/default/files/eng/cop/18/doc/E-CoP18-075.pdf>. [September 22, 2019].

Heinrich, S., Wittmann, T.A., Prowse, T.A.A., Ross, J.V., Delean, S., Shepherd, C.R., et al., 2016. Where did all the pangolins go? International CITES trade in pangolins species. Glob. Ecol. Conserv. 8, 241–253.

Huxley, C., 2000. CITES: The vision. In: Hutton, J., Dickson, B. (Eds.), Endangered Species Threatened Convention, The Past, Present and Future of CITES. Africa Resources Trust and Earthscan Publications Ltd, London, pp. 3–12.

Martin, R.B., 2000. When CITES works and when it does not. In: Hutton, J., Dickson, B. (Eds.), Endangered Species Threatened Convention, The Past, Present and Future of CITES. Africa Resources Trust and Earthscan Publications Ltd, London, pp. 29–37.

McFadden, E., 1987. Asian compliance with CITES, problems and prospects. Boston Univ. Int. Law J. 5, 311–325.

Nooren, H., Claridge, G., 2001. Wildlife Trade in Laos: The End of the Game. Netherlands Committee for IUCN, Amsterdam.

Novak, R.M., 1999. *Panthera tigris* (tiger), Walker's Mammals of the World, sixth ed. Johns Hopkins University Press, Baltimore, pp. 825–828.

Nowell, K., Ling, X., 2007. Taming the tiger trade: China's markets for wild and captive tiger products since the 1993 domestic trade ban. TRAFFIC East Asia, Hong Kong, China.

Patel, P., 1996. The Convention on International Trade in Endangered Species: enforcement and the last unicorn. Houst. J. Int. Law 157–213.

Reeve, R., 2002. Policing International Trade in Endangered Species, The CITES Treaty and Compliance. The Royal Institute of International Affairs and Earthscan Publications Ltd, London.

Reeve, R., 2006. Wildlife trade, sanctions and compliance: lessons from the CITES regime. Int. Aff. 82 (5), 881–897.

Sand, P.H., 2013. Enforcing CITES: the rise and fall of trade sanctions. Rev. Eur. Commun. Int. Environ. Law 22 (3), 251–263.

Talukdar, B.K., Emslie, R., Bist, S.S., Choudhury, A., Ellis, S., Bonal, B.S., et al., 2008. *Rhinoceros unicornis. The IUCN Red List of Threatened Species* 2008: e.T19496A8928657. Available from: <https://www.iucnredlist.org/species/19496/8928657>. [December 31, 2018].

't Sas-Rolfes, M., 2000. Assessing CITES: four case studies. In: Hutton, J., Dickson, B. (Eds.), Endangered Species Threatened Convention, The Past, Present and Future of CITES. Africa Resources Trust and Earthscan Publications Ltd, London, pp. 69–87.

Underwood, F.M., Burn, R.W., Milliken, T., 2013. Dissecting the illegal ivory trade: an analysis of ivory seizures data. PLoS One 8 (10), e76539.

UNODC, 2016. World Wildlife Crime Report: Trafficking in Protected Species. UNODC, Vienna, Austria.

Wijnstekers, W., 2018. The Evolution of CITES, eleventh ed. International Council for Game and Wildlife Conservation, Budapest, Hungary.

Wilson, D.E., Reeder, M., 2005. Mammal Species of the World, A Taxonomic and Geographic Reference, third ed. Johns Hopkins University Press, Baltimore.

WWF, 2016. Global Wild Tiger Population Status. Available from: <http://tigers.panda.org/wp-content/uploads/Background-Document-Wild-Tiger-Status-2016.pdf>. [January 1, 2019].

CHAPTER

20

Understanding illegal trade in pangolins through forensics: applications in law enforcement

Antoinette Kotze[1,2], *Rob Ogden*[3,4], *Philippe Gaubert*[5,6], *Nick Ahlers*[7], *Gary Ades*[8], *Helen C. Nash*[9,10] *and Desire Lee Dalton*[1,11]

[1]National Zoological Garden, South African National Biodiversity Institute, Pretoria, South Africa [2]Genetics Department, University of the Free Sate, Bloemfontein, South Africa [3]TRACE Wildlife Forensics Network, Edinburgh, United Kingdom [4]Royal (Dick) School of Veterinary Studies and the Roslin Institute, University of Edinburgh, United Kingdom [5]Laboratoire Évolution & Diversité Biologique (EDB), Université de Toulouse Midi-Pyrénées, CNRS, IRD, UPS, Toulouse, France [6]CIIMAR, University of Porto, Matosinhos, Portugal [7]TRAFFIC, ℅ IUCN, Hatfield Gables, Pretoria, South Africa [8]Fauna Conservation Department, Kadoorie Farm & Botanic Garden, Hong Kong SAR, P.R. China [9]Department of Biological Sciences, National University of Singapore, Singapore [10]IUCN SSC Pangolin Specialist Group, ℅ Zoological Society of London, Regent's Park, London, United Kingdom [11]University of Venda, Thohoyandou, South Africa

OUTLINE

Pangolins
DOI: https://doi.org/10.1016/B978-0-12-815507-3.00020-4

Introduction

The global illegal trade in pangolins and their body parts has seemingly increased in both Africa and Asia in recent decades, to supply a growing demand, primarily for use as traditional medicines and wild meat (Chapters 14–16; Heinrich et al., 2016; Ingram et al., 2018). Analyses of pangolin seizures reveal an apparent, stark increase in the number of pangolin scales originating in Africa that are supplying traditional medicine markets in Asia, while scales and whole and eviscerated bodies are more frequently traded within Asia (Challender and Waterman, 2017; Heinrich et al., 2017).

Knowledge of illegal trade dynamics is important for informing law enforcement priorities and individual investigations, and desk-based studies of pangolin trafficking can both inform and are informed by scientific analysis. Seized pangolins and their body parts cannot usually be identified as anything more specific than "pangolin" by enforcement officers. This may be sufficient for prosecution in cases of international illegal trade where all pangolin species are protected (though see Chapter 17), but it limits investigations into range states and prevents the collection of potentially important information regarding the sources of seized pangolins. Conversely, from a forensic perspective, traditional trade studies guide the scientists' understanding of the species, transport routes and sample types to be encountered in pangolin trade control and allow forensic scientists to prepare for the types of investigative questions and evidential material they may face. Collaboration between the scientific community and trade monitoring bodies, including integration of their respective datasets, therefore represents an important aspect of illegal pangolin trade investigations.

Wildlife forensic analyses may be applied to investigations of poaching, illegal trade, verification of species that are captive-bred versus wild caught, and identification of protected species in traditional medicine trade. Different methods to support law enforcement have been developed to determine (1) species identification, (2) geographic origin, (3) individual identification, (4) kinship, (5) age, and (6) gender. This chapter reviews these methods, their potential relevance and application to pangolin trafficking, and how local and global efforts to apply them should be coordinated in order for them to be effective tools in combating pangolin trafficking.

Past, present and future methods

Species identification

Morphological methods such as osteology or microscopy can be used to identify species. These methods require specialist knowledge of comparative anatomy at macroscopic and microscopic scales (Bell, 2011). A study on the radiological anatomy and scales of Temminck's pangolin (*Smutsia temminckii*) indicated that patterns of scales on the body are associated with the underlying skeletal structure and thus may be used to identify the different pangolin species (Steyn, 2016). However, it is not always possible to identify different species of pangolin from scales alone as scales from different parts of the body differ in shape and size, and can vary with age and environmental variables. Microscopic analysis of animal material can be used to exclude possible pangolin species but it is generally not used for identification purposes due to overlapping ranges of measurements between species and due to limited access to reference data (Hillier and Bell, 2007).

Molecular genetic approaches to pangolin species identification rely on DNA barcoding using mitochondrial DNA (mtDNA) markers or species specific microsatellite markers. Targeted mitochondrial gene regions include Cytochrome c Oxidase I (COI), Cytochrome b (Cytb), control region (CR), 12 S and 16 S ribosomal RNA, which have all been validated in previous studies for forensic application (Balitzki-Korte et al., 2005; Dawnay et al., 2007; Gaubert et al., 2015). Using these approaches, a region of the mitochondrial genome is sequenced and compared with sequences in public online databases such as GenBank, the European Molecular Biology Laboratory (EMBL), the Barcode of Life Database (BOLD; Ratnasingham and Hebert, 2007), and/or expert-curated online identification tools such as ForCyt (Ahlers et al., 2017) and DNAbushmeat (Gaubert et al., 2015). Critical to successful analysis of any species identification in a forensic case is the construction of a DNA reference library. A case study demonstrating the utility of DNA barcoding in pangolins is provided in Box 20.1. The ForCyt initiative differs from other publicly accessible databases in that it is designed for wildlife forensic casework and only contains quality assured sequence data generated from validated voucher specimens.

At a broader level, collaboration with and critical input from relevant police services, prosecuting authorities and environmental enforcement agencies are essential to ensuring the impact of such forensic techniques. There is a need to raise awareness of DNA barcoding within law enforcement and prosecutor communities globally to promote the technology, which is currently being incorporated into actual investigations and court cases. Standard Operating Procedures (SOPs) and guideline documents are required, and staff involved in the collecting and analysis of samples need adequate training to ensure that the protocols are followed, reference libraries used and any subsequent investigations are legally defensible.

Geographic origin

Geographic source may be required for criminal investigations involving either live pangolins for potential repatriation, or dead animals or their parts and derivatives for understanding poaching hotspots and trade routes. There are two primary analytical approaches used for the geographic origin assignment of seized wildlife samples in trade: stable isotope analysis and molecular genetic (DNA) analysis. The geographic origin of biological materials, sometimes referred to as biogeolocation, can be determined using stable light isotopes (Hobson and Wassenaar, 2018; Oulhote et al., 2011), in a process which utilizes isotopic maps (or isoscapes) to assign a sample to its likely origin based on its isotopic profile. Measurements of stable isotopes have been used to trace origins of a range of wildlife products including African elephant ivory (van der Merwe et al., 1990; Vogel et al., 1990; Ziegler et al., 2016) and are widely used in analysis of traded foodstuffs. The technique is based on differences in elemental stable isotope ratios among geographic regions and may relate to underlying environmental differences (e.g., hydrogen, oxygen isotope ratios), vegetation type (e.g., nitrogen isotope ratios) or underlying geology (e.g., strontium isotope ratios). For stable isotope analysis to be useful in geographic origin assignment, it is usually necessary to generate a combined elemental profile for a sample and assign it to an isoscape map of multiple elements. One limitation to the approach is that despite having environmental isoscape maps available for many geographic regions, the way in which isotopes are incorporated into biological material may vary among species and among tissue types, so that separate isoscapes often need to be generated for each type of sample being used in the analysis. This limitation, combined with difficulties in accessing

BOX 20.1

Case Study: Forensic application of DNA barcoding for identification of illegally traded African pangolin scales (Mwale et al., 2017).

Between 2014 and 2015, 3.3 tonnes of pangolin scales were confiscated by the CITES (Convention on International Trade in Endangered Species of Wild Fauna and Flora) Management Authority in Hong Kong SAR. It was suspected that the consignment originated from Africa and a representative subsample of ten bags, each representing a different consignment with a scale net weight of 27 kg was sent to the forensic laboratory of the National Zoological Garden in South Africa for analysis (Fig. 20.1A–B). The contents were visually sorted into distinct scale types and provisionally assigned to a species by a taxonomic expert based on shape, coloration and morphology. Five samples per scale morph type were voucher reference specimens of three African pangolin species, Temminck's pangolin, black-bellied pangolin (*Phataginus tetradactyla*) and white-bellied pangolin (*P. tricuspis*), were analyzed. References were supplemented with sequences of the giant pangolin (*S. gigantea*), Sunda pangolin (*Manis javanica*) and Chinese pangolin (*M. pentadactyla*) retrieved from GenBank. The results indicated that the samples came from both African and Asian pangolins, and that confiscated scales may represent multiple pangolin species in one bag. This has laid the foundation for accurate identification of pangolin species in further forensic cases in South Africa.

(A)
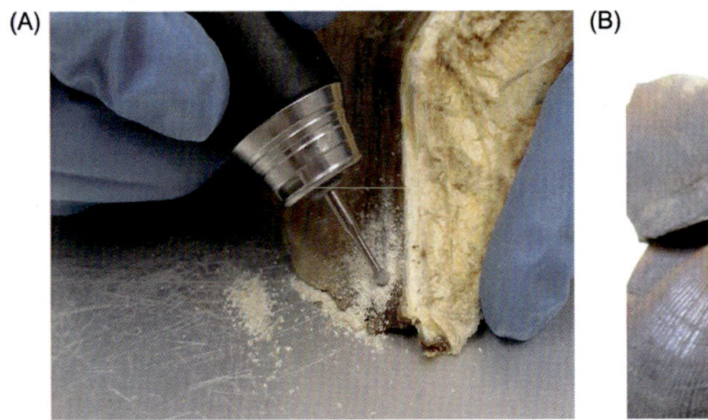
(B)

FIGURE 20.1 (A) Extraction of pangolin scale tissue and its DNA for forensic analysis. (B) Pangolin scales seized from illegal wildlife trade.

appropriate laboratory equipment and validating test methods means that isotopic profiling has seen limited use in forensic investigations (Meier-Augenstein et al., 2013) and although may have good potential for pangolin origin analysis, remains unproven to date.

Molecular genetic methods that rely on the analysis of mitochondrial DNA variation and nuclear DNA variation (microsatellite markers or single nucleotide polymorphisms [SNPs]) to assign individuals to a particular population provide an alternative option (Ogden and Linacre, 2015). These methodologies utilize genetic differentiation that has occurred as a result of gradual genetic separation between isolated populations in different geographic regions. For mitochondrial DNA, populations may show fixed discrete genetic differences, whereas for nuclear DNA, differences in the frequencies of genetic markers can be used to characterize this genetic structure. Accurate identification relies on the development of large genetic databases that are representative of the candidate source populations across the species' geographic distribution (Ogden et al., 2009; Wasser et al., 2015). Bayesian methods developed by Pritchard et al. (2000) and Falush et al. (2003) can be used for assignment of individuals to clusters. For example, the geographic origin of 46 rescued chimpanzees has been conducted using mtDNA sequences and microsatellite genotypes (Ghobrial et al., 2010). Specifically for pangolins, the examination of a series of mtDNA and nuclear genes have allowed for the identification of six geographic lineages within the white-bellied pangolin (Gaubert et al., 2016), a pattern that could help track the global trade of the species at a sub-regional scale.

Individual identification

Individual identification of animals or plants in wildlife forensics may be very useful when trying to link samples from different crime scenes or points along the supply chain. This is most common where the numbers of traded individuals are relatively low and the value of the information is considered high. For example, the use of individual DNA profiling to link horns back to individual rhinoceros carcasses through the RHODIS database (Harper et al., 2018), or linking pairs of elephant tusks being moved in separate shipments (Wasser et al., 2015). Such applications are not typically essential for securing a conviction and are usually only practicable in cases where the number of individuals trafficked is relatively low (tens to hundreds). For pangolins, often traded in many thousands of individuals (or hundreds of thousands of scales), the evidential or intelligence value of individual identification relative to the potential cost of the work is much reduced. Apart from the occasional presence of specific morphological features on live pangolins (e.g., easily identifiable scales or scale patterns), it is difficult to distinguish between individual animals. While photographic libraries of confiscated animals (live or dead) may prove useful reference materials, particularly to support evidential control processes, there is currently little immediate requirement or capacity for morphological identification of pangolins at an individual level.

The construction of DNA profiles can enable individualization, just as in humans. Individual DNA profiles may be utilized to regulate legal trade of species that are subject to quotas, or that can be used to determine the captive origin of an animal (e.g., where databases exist for captive stock). However, there are currently no techniques in place to use individual DNA profiling of pangolins for trade monitoring or forensic investigation, nor is there a requirement to do so.

Kinship investigations

Establishing levels of relatedness between animals in forensic investigations is generally employed in order to differentiate between captive-bred and wild caught animals (Ogden et al., 2009) and can currently only be achieved with DNA-based methods. The patterns of inheritance from parent to offspring allow DNA profiles to be used to verify family relationships. Parentage is refuted if all or part of the profile observed in an offspring individual is not present in its claimed parents. In addition to a DNA profiling system, kinship investigations usually rely on breeding records that allow parent-offspring trios to be identified for testing. A requirement for such breeding records should be included in breeding license regulations if any DNA testing system is to be effectively employed for enforcement. With the potential advent of pangolin farms (see Chapter 32), such a regulatory framework may be necessary to prevent the laundering of wild animals through captive breeding centers.

Age determination

In some wildlife crime investigations, it may be necessary to determine whether an individual was living at a certain point in time. For example, rhino horn or ivory collected prior to 1947 pre-dates laws prohibiting trade. The primary method of ageing recent biological material is a form of radio carbon dating, known as bomb-testing. During the early part of the 1950s, atmospheric nuclear weapons testing became common. This resulted in an artificial increase in the amounts of different carbon isotopes, particularly carbon 14 ($\delta^{14}C$), which had doubled in abundance by 1965. As such, biological samples that pre-date this period will be expected to have a lower ratio of $\delta^{14}C$ than more modern specimens. Determining the actual age of dead individuals may be achieved through analysis of morphological features, such as growth rings in otoliths in fish (Campana, 2001), and tooth cementum annulation (Wittwer-Backofen et al., 2004) in mammals, but for application to time-bound legislation these would need to be accompanied by reliable estimates of date of death. Determining the age of live animals largely relies upon external features which change predictably over time in a discriminate fashion, for example, pigmentation patterns on the ventral side of humpback whale (*Megaptera novaeangliae*) flukes; or artificial markings, such as unique numbered bands attached to the legs of birds (Sherley et al., 2014). Research on genetic age determination has also made significant progress, primarily in humans and suggests promise for forensic application (Jarman et al., 2015). However at present, investigations concerning the illegal trade in pangolins do not require knowledge of individual age or period of life; but this situation could change, for example if legal movement of pre-CITES convention pangolins were to create a loophole for the illegal trade in contemporary specimens, such as pangolin scales. Such horizon scanning would be required to allow time for the forensic community to develop and validate ageing methods for pangolins.

Gender determination

Gender determination via morphological and molecular genetic methods are available for many whole specimens or body parts of species in trade and may occasionally have application to law enforcement. For pangolins, sexing via direct observation of the genital region may be possible depending on species (e.g., easier in Temminck's pangolin to determine gender due to penis size) and if the full carcass is available for dead animals. Molecular gender determination can be conducted if only parts of the animal are available or when sex-specific characters are either absent or difficult to observe. Methods of DNA gender determination in pangolins are not currently available, but their development should be readily achievable if required. However, as with ageing, there is no immediate requirement for this type of information in support of law enforcement investigations.

Coordinating and managing wildlife forensics at global and local scales

The global capacity to undertake wildlife forensic analysis and the number of casework investigations performed are increasing year on year. Across Africa and Southeast Asia in particular, a significant increase in wildlife forensic capacity-building activities occurred in the decade from 2009 to 2019, resulting in the routine use of wildlife forensic evidence to support prosecutions in many countries. This process of international development is far from complete, with laboratories in different countries operating at different levels and offering a range of techniques related to variation in both national need and technical capability. The international nature of the illegal pangolin trade, both with respect to the number of range states across Africa and Asia and the subsequent movement of pangolins through non-range states, means that coordination of forensic activities also needs to operate at an international scale. There are two key aspects that should be considered in this respect: the availability and harmonization of pangolin identification techniques among countries; and the ability to implement these methods with sufficient forensic rigor to secure prosecutions and prevent the emergence of weak links in forensic capacity along trade routes.

Research and method development

Despite the high level of global attention given to pangolins due to their illegal trade, the group as a whole has remained remarkably under-studied. Although the species-level taxonomy of the African and Southeast Asian taxa has been described, relatively few phylogenetic studies on populations were conducted prior to 2010. This lack of data has prevented the rapid development of forensic genetic tools, requiring the implementation of biological baseline research before species identification and geographic origin tests could be designed to support pangolin trade investigations. Since 2010, several international groups have focused on the application of research on pangolin biology, ecology and evolutionary history, towards production of a number of validated traceability tools and forensic identification techniques.

Work at the University of Toulouse, France, by Phillippe Gaubert and colleagues, has focused on the evolutionary history of pangolins (Gaubert et al., 2018) and population genetics of African pangolins (Gaubert et al., 2016), distinguishing pangolins from different geographic regions based on genetic variation, and thus laying the foundations for traceability of seized pangolins to areas within broad species distributions. A complimentary approach is being employed by the South African National Biodiversity Institute (SANBI) research group at the National Zoological Garden, under Antoinette Kotze, where the wildlife forensics laboratory has extensive experience in performing a combination of species identification (Dalton and Kotze, 2011) and origin assignment (Mwale et al., 2017) to help investigate the illegal pangolin trade across southern Africa.

In East Asia, the University of Hong Kong's School of Biological Sciences recently established a forensic laboratory to support wildlife crime enforcement work including a focus on pangolin analysis. The Conservation Genetics laboratory at Kadoorie Farm & Botanic Garden (KFBG) is also providing an analytical function and has carried out preliminary investigations on pangolin scales seized in Hong Kong Special Administrative Region (hereafter "Hong Kong"), and donated to the center for scientific use (Zhang et al., 2015). KFBG has more recently focused on the illegal trade in African pangolins. Scales originating from

BOX 20.2

Case Study: Illegal trade in Sunda pangolins across insular Southeast Asia.

Sunda pangolins across insular Southeast Asia have been heavily poached for the illegal wildlife trade. For example in April 2015, five tons of frozen pangolins, 77 kg of pangolin scales, and 96 live pangolins were seized from a single haul in Medan, Indonesia (Nijman et al., 2016). Due to the large, transnational range of the species, even when it is possible via morphological features or other methods to identify a pangolin to species level, it is difficult to determine the national or local geographic origins of a seizure, unless informants have provided further details (although these can often be unreliable). The application of advanced genetic techniques was trialed across insular Southeast Asia to help provide information about the origin of 97 Sunda pangolins within several seizures. It included the use of geolocated reference samples from across the region to help assign the seized individuals to their likely geographic origin. The DNA analysis revealed three previously unrecognized genetic lineages of Sunda pangolins, from Borneo, Java and Singapore/Sumatra. For the seizure samples, it was possible to conclude that most of the pangolins had been captured from Borneo and exported to Java (Nash et al., 2018).

The project was a collaboration between The Indonesian Institute of Sciences (LIPI), the IUCN SSC Pangolin Specialist Group, the National University of Singapore (NUS), the University of Malaya (UM), the Universiti Malaysia Terengganu (UMT), and Wildlife Reserves Singapore (WRS), supported by funding from the Southeast Asian Regional Centre for Tropical Biology (SEAMEO BIOTROP DIPA) and other partners.

Africa seized in Hong Kong since 2009 are being analyzed in combination with statistical modeling to characterize the illegal trade and identify priority areas for targeted policy and enforcement efforts.

In Southeast Asia, research projects in Indonesia, Malaysia and Singapore have led to the development of novel methods for pangolin identification. The collaborative work on Sunda pangolin geolocation in Singapore (see Box 20.2) has demonstrated the feasibility of applying research and development to pangolin trade investigations in this arena, and the partnership approach employed across multiple government, academic and non-government organizations offers a powerful model for broader work within the region. At the same time, the Malaysian National Wildlife Forensics Laboratory at Perhilitan, Kuala Lumpur, has embarked on genomic analysis of pangolin species from around the region, resulting in novel insights into both the taxonomy of the Southeast Asian pangolins and the potential to generate DNA-based traceability tools for Malaysia and beyond.

Developing pangolin forensic capacity

The transfer and application of such research initiatives into the international wildlife forensic community represents the next significant step in the use of forensic science in investigations into illegal pangolin trade. For the results of pangolin sample analysis to be used successfully in law enforcement, it is necessary to raise awareness and build expertise from crime scene to courtroom. The use of biological material as forensic evidence requires seizures to be made following robust protocols that maintain both the evidential and biological integrity of the pangolin products confiscated. Laboratories analyzing such evidential samples need to operate under strict quality management systems using test methods previously validated to generate robust, reproducible, accurate results. Lastly, forensic reports must be generated and communicated to the legal profession, such that the prosecution, defence and judiciary alike are capable of evaluating and accepting the evidence placed before them.

These issues are not unique to pangolins and the international wildlife forensic community works to ensure that national enforcement agencies have access to forensic services that offer internationally recognized tests delivered within a common forensic framework. The Society for Wildlife Forensic Science (SWFS) sets international standards for the discipline. Long-term capacity building programmes supported by The United States Agency for International Development (USAID), the U.S State Department and the European Commission are all contributing to the dissemination and harmonization of best practice in this field, largely implemented through the work of technical specialist organizations such as TRACE Wildlife Forensics Network and the Netherlands Forensic Institute (NFI). A number of initiatives specifically relating to pangolin forensics are being developed within these over-arching programs, including field officer training in the identification of pangolin parts to support seizures, protocols for the sampling and collection of pangolin evidence for forensic analysis, laboratory training in DNA recovery from pangolin scales, and the production of guidelines for the coordination of pangolin seizure analysis among countries and continents (Fig. 20.2).

Conclusion

The development and application of wildlife forensic analysis is driven by the needs of the enforcement community. Enforcement agencies may be looking for intelligence level data that inform investigations concerning trade patterns and likely shipment routes, or they may need forensic evidence concerning the identity of a sample for use in legal proceedings. These requirements must guide the scientific community in how it can help to support the ongoing fight against the illegal pangolin trade. Within the field of wildlife forensics, only a subset of possible methods and applications are currently applicable to pangolin law enforcement; these focus primarily on species identification and geographic origin assignment. Species identification of whole specimens can be achieved using expert morphological examination, or more routine DNA sequencing analysis. Approaches for determining geographic origin are available for certain species in specific regions and work is underway to extend these capabilities to enable all pangolins to be traced back to their origin to at least some level of geographic resolution.

As with other species subject to wildlife forensic investigation, the initial development of analytical methods and their subsequent validation and application to legal casework crosses the boundary between academic research and forensic casework laboratory environments, requiring informed coordination between these two scientific communities. As the need for analytical data relating to the illegal trade in pangolins increases, the

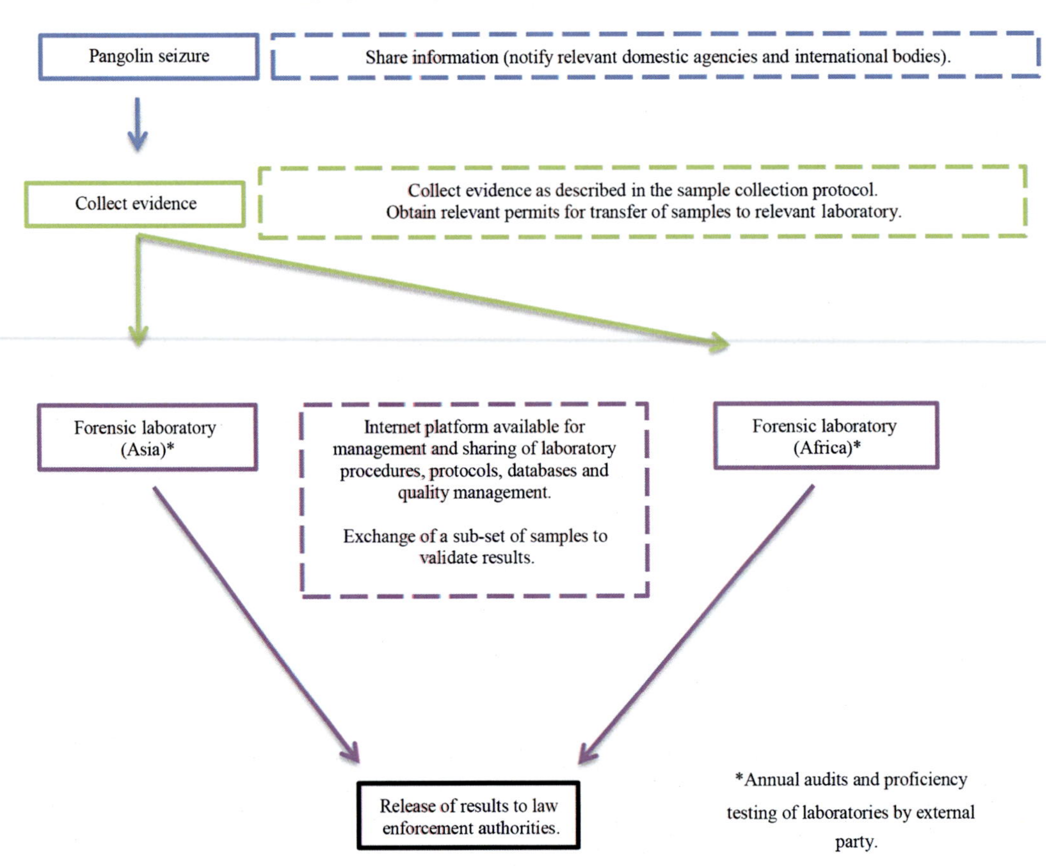

FIGURE 20.2 Proposed guidelines for wildlife forensic investigations involving pangolin confiscations. Excerpts from UNODC, 2014.

response from the scientific community is leading to increases in the availability of both pangolin-specific techniques and laboratories capable of implementing them for forensic application.

References

Ahlers, N., Creecy, J., Frankham, G., Johnson, R.N., Kotze, A., Linacre, A., et al., 2017. 'ForCyt' DNA database of wildlife species. Forensic Sci. Int.: Genet. Suppl. Ser. 6, e466–e468.

Balitzki-Korte, B., Anslinger, K., Bartsch, C., Rolf, B., 2005. Species identification by means of pyrosequencing the mitochondrial 12S rRNA gene. Int. J. Legal. Med. 119 (5), 291–294.

Bell, L.S., 2011. Forensic science in support of wildlife conservation efforts - morphological and chemical approaches (global trends). Forensic Sci. Rev. 23 (1), 29–35.

Campana, S.E., 2001. Accuracy, precision and quality control in age determination, including a review of the use and abuse of age validation methods. J. Fish. Biol. 59 (2), 197–242.

Challender, D., Waterman, C., 2017. Implementation of CITES Decisions 17.239 b) and 17.240 on Pangolins (*Manis* spp.), CITES SC69 Doc. 57 Annex. Available from: <https://cites.org/sites/default/files/eng/com/sc/69/E-SC69-57-A.pdf>. [March 22, 2018].

Dalton, D.L., Kotze, A., 2011. DNA barcoding as a tool for species identification in three forensic wildlife cases in South Africa. Forensic. Sci. Int. 207 (1–3), e51–e54.

Dawnay, N., Ogden, R., McEwing, R., Carvalho, G.R., Thorpe, R.S., 2007. Validation of the barcoding gene

COI for use in forensic genetic species identification. Forensic. Sci. Int. 173 (1), 1–6.

Falush, D., Stephens, M., Pritchard, J.K., 2003. Inference of population structure using multilocus genotype data: linked loci and correlated allele frequencies. Genetics 164 (4), 1567–1587.

Gaubert, P., Njiokou, F., Olayemi, A., Pagani, P., Dufour, S., Danquah, E., et al., 2015. Bushmeat genetics: setting up a reference framework for the DNA-typing of African forest bushmeat. Mol. Ecol. Resour. 15 (3), 633–651.

Gaubert, P., Njiokou, F., Ngua, G., Afiademanyo, K., Dufour, S., Malekani, J., et al., 2016. Phylogeography of the heavily poached African common pangolin (Pholidota, *Manis tricuspis*) reveals six cryptic lineages as traceable signatures of Pleistocene diversification. Mol. Ecol. 25 (23), 5975–5993.

Gaubert, P., Antunes, A., Meng, H., Miao, L., Peigné, S., Justy, F., et al., 2018. The complete phylogeny of pangolins: scaling up resources for the molecular tracing of the most trafficked mammals on Earth. J. Hered. 109 (4), 347–359.

Ghobrial, L., Lankester, F., Kiyang, J.A., Akih, A.E., De Vries, S., Fotso, R., et al., 2010. Tracing the origins of rescued chimpanzees reveals widespread chimpanzee hunting in Cameroon. BMC Ecol. 10 (1), 2.

Harper, C., Ludwig, A., Clarke, A., Makgopela, K., Yurchenko, A., Guthrie, A., et al., 2018. Robust forensic matching of confiscated horns to individual poached African rhinoceros. Curr. Biol. 28 (1), 13–14.

Heinrich, S., Wittmann, T.A., Prowse, T.A., Ross, J.V., Delean, S., Shepherd, C.R., et al., 2016. Where did all the pangolins go? International CITES trade in pangolin species. Glob. Ecol. Conserv. 8, 241–253.

Heinrich, S., Wittman, T.A., Ross, J.V., Shepherd, C.R., Challender, D.W.S., Cassey, P., 2017. The Global Trafficking of Pangolins: A Comprehensive Summary of Seizures and Trafficking Routes From 2010–2015. TRAFFIC, Southeast Asia Regional Office, Petaling Jaya, Selangor, Malaysia.

Hillier, M.L., Bell, L.S., 2007. Differentiating human bone from animal bone: a review of histological methods. J. Forensic. Sci. 52 (2), 249–263.

Hobson, K.A., Wassenaar, L.I., 2018. Tracking Animal Migration With Stable Isotopes, second ed. Academic Press, London.

Ingram, D.J., Coad, L., Abernethy, K.A., Maisels, F., Stokes, E.J., Bobo, K.S., et al., 2018. Assessing Africa-wide pangolin exploitation by scaling local data. Conserv. Lett. 11 (2), e12389.

Jarman, S.N., Polanowski, A.M., Faux, C.E., Robbins, J., Paoli-Iseppi, D., Bravington, M., et al., 2015. Molecular biomarkers for chronological age in animal ecology. Mol. Ecol. 24 (19), 4826–4847.

Meier-Augenstein, W., Hobson, K.A., Wassenaar, L.I., 2013. Critique: measuring hydrogen stable isotope abundance of proteins to infer origins of wildlife, food and people. Bioanalysis 5 (7), 751–767.

Mwale, M., Dalton, D.L., Jansen, R., De Bruyn, M., Pietersen, D., Mokgokong, P.S., et al., 2017. Forensic application of DNA barcoding for identification of illegally traded African pangolin scales. Genome 60 (3), 272–284.

Nash, H.C., Wirdateti, W., Low, G., Choo, S.W., Chong, J. L., Semiadi, G., et al., 2018. Conservation genomics reveals possible illegal trade routes and admixture across pangolin lineages in Southeast Asia. Conserv. Genet. 19 (5), 1083–1095.

Nijman, V., Zhang, M.X., Shepherd, C.R., 2016. Pangolin trade in the Mong La wildlife market and the role of Myanmar in the smuggling of pangolins into China. Glob. Ecol. Conserv. 5, 118–126.

Ogden, R., Dawnay, N., McEwing, R., 2009. Wildlife DNA forensics - bridging the gap between conservation genetics and law enforcement. Endanger. Sp. Res. 9 (3), 179–195.

Ogden, R., Linacre, A., 2015. Wildlife forensic science: a review of genetic geographic origin assignment. Forensic Sci. Int.: Genet. 18, 152–159.

Oulhote, Y., Le Bot, B., Poupon, J., Lucas, J.P., Mandin, C., Etchevers, A., et al., 2011. Identification of sources of lead exposure in French children by lead isotope analysis: a cross-sectional study. Environ. Health 10 (1), 75.

Pritchard, J.K., Stephens, M., Rosenberg, N.A., Donnelly, P., 2000. Association mapping in structured populations. Am. J. Hum. Genet. 67 (1), 170–181.

Ratnasingham, S., Hebert, P.D., 2007. BOLD: The barcode of life data system. Mol. Ecol. Notes 7 (3), 355–364. Available from: <http://www.barcodinglife.org>.

Sherley, R.B., Abadi, F., Ludynia, K., Barham, B.J., Clark, A.E., Altwegg, R., 2014. Age-specific survival and movement among major African Penguin Spheniscus demersus colonies. Ibis 156 (4), 716–728.

Steyn, S. 2016. The Radiological Anatomy and Scale Pattern of the Thoracic Limb of *Smutsia temminckii*. M.Sc. Thesis, University of Pretoria, Pretoria, South Africa.

UNODC, 2014. Guidelines for Forensic Methods and Procedures of Ivory Sampling and Analysis. United Nations Office of Drugs and Crime, United Nations, New York.

van der Merwe, N.J., Lee-Thorp, J.A., Thackeray, J.F., Hall-Martin, A., Kruger, F.J., Coetzee, H., et al., 1990. Source-area determination of elephant ivory by isotopic analysis. Nature 346 (6286), 744–746.

Vogel, J.C., Eglington, B., Auret, J.M., 1990. Isotope fingerprints in elephant bone and ivory. Nature 346 (6286), 747–749.

Wasser, S.K., Brown, L., Mailand, C., Mondol, S., Clark, W., Laurie, C., et al., 2015. Genetic assignment of large seizures of elephant ivory reveals Africa's major poaching hotspots. Science 349 (6243), 84–87.

Wittwer-Backofen, U., Gampe, J., Vaupel, J.W., 2004. Tooth cementum annulation for age estimation: results from a large known-age validation study. Am. J. Phys. Anthropol. 123 (2), 119–129.

Zhang, H., Miller, M.P., Yang, F., Chan, H.K., Gaubert, P., Ades, G., et al., 2015. Molecular tracing of confiscated pangolin scales for conservation and illegal trade monitoring in Southeast Asia. Glob. Ecol. Conserv. 4, 414–422.

Ziegler, S., Merker, S., Streit, B., Boner, M., Jacob, D.E., 2016. Towards understanding isotope variability in elephant ivory to establish isotopic profiling and source-area determination. Biol. Conserv. 197, 154–163.

PART 2

Awareness Raising and Behavior Change

CHAPTER

21

No longer a forgotten species: history, key events, and lessons learnt from the rise of pangolin awareness

Paul Thomson[1] *and Louise Fletcher*[2]

[1]Save Pangolins, ℅ Wildlife Conservation Network, San Francisco, CA, United States [2]IUCN SSC Pangolin Specialist Group, ℅ Zoological Society of London, Regent's Park, London, United Kingdom

OUTLINE

Introduction

In less than a decade, pangolins have increased in prominence, going from virtual obscurity to being widely recognized as an icon of the illegal wildlife trade (Harrington et al., 2018). Once rarely known outside of biodiversity conservation circles (and even within some such circles), and those consuming and trading them, pangolins have become an increasingly familiar species around the world. This is due in large part to a series of events and activities designed to raise awareness of the animals, the threats that they face, and the need for urgent conservation action.

Awareness raising is a means of placing issues of concern in the local, national and even global consciousness. It may also comprise a

Pangolins
DOI: https://doi.org/10.1016/B978-0-12-815507-3.00021-6

first step towards generating support for and eliciting behavioral change to address a particular issue (Schultz, 2010). Raising awareness is of strategic importance for pangolins, which face high levels of exploitation (Chapters 14–16), and have received little concerted conservation attention historically (Challender et al., 2012). Until around 2012, pangolins were virtually unknown to much of the public in regions such as North and South America and Europe, and even within some pangolin range states (Harrington et al., 2018; D. Hendrie, pers. comm.), and this lack of awareness likely hindered conservation action.

Within a short period of time, attention on pangolins and the threat from illegal trade has grown dramatically. Harrington et al. (2018) described increases in social media activity, frequency of news articles, and internet searches on pangolins over time. They recorded an almost 100-fold increase in Facebook activity (posts, comments, and likes) about pangolins between 2009 and 2016 and a nine-fold increase in the number of news articles published by the editorial media between 2005 and 2016.

This chapter discusses the movement to raise awareness of pangolins locally and globally, beginning with a brief history of early research and conservation efforts, and culminating in a global network of conservationists working to promote the species and their conservation. The chapter identifies eight key events and activities designed to increase global awareness of pangolins, and examines how they contributed to the growth in the profile of the species between 2012 and 2018.

However, despite increasing awareness pangolins remain relatively little known compared to traditionally iconic species (e.g., African lion [*Panthera leo*], rhinoceroses [Rhinocerotidae] and elephants [Elephantidae]) and continue to be threatened by overexploitation. There is a need for further awareness raising to generate additional support for, and ensure the successful conservation of pangolins. By examining key events and activities that have brought attention to these species, lessons learned can be applied to awareness raising efforts in the future both for pangolins and other seemingly "forgotten" species.

A movement begins

Centuries before contemporary efforts to raise the profile of pangolins, interest in the species was largely limited to naturalists, collectors and taxonomists (see Chapter 13). In the mid-late 20th century, the species were of interest to zoologists (e.g., Mohr, 1961) and research focused on the use of the species (Anadu et al., 1988), their ecology (e.g., Heath and Coulson, 1997; Pagès, 1975) and husbandry (e.g., Heath and Vanderlip, 1988) and subsequently on international trade (e.g., Wu and Ma, 2007). Despite the value of this knowledge generation, such research efforts were largely disparate and there was lack of an active international body or bodies coordinating priority research, thereby ensuring it focused on the most pressing questions of conservation concern, or garnering public awareness and support.

In the early 2000s, conservationists began noting the frequency of large seizures of pangolins and their meat and scales, especially in East and Southeast Asia (see Pantel and Chin, 2009). For instance, seizures of up to 17 tonnes of eviscerated pangolins and scales (the equivalent of nearly 3500 animals – see Chapter 16 for conversion parameters) took place, typically along trafficking routes from Indonesia and Malaysia to China and Vietnam (Pantel and Chin, 2009). This trade took place despite Asian pangolins being subject to zero export quotas in CITES (Convention on International Trade in Endangered Species of Wild Fauna and Flora; see Chapters 16 and 19). At the time, it was becoming apparent that illegal trade in pangolins was a serious conservation issue, but there were few targeted efforts in place to ensure there was a proportionate conservation response.

By the late 2000s, an emerging international pangolin conservation community began to form. It comprised individuals and organizations including the Carnivore and Pangolin Conservation Program (CPCP) in Vietnam (now Save Vietnam's Wildlife), the Tikki Hywood Foundation in Zimbabwe, TRAFFIC in Southeast Asia, Conservation International's Cambodia program, Education for Nature Vietnam (ENV), Taipei Zoo in Taiwan, and Prof. Wu Shi Bao and colleagues at South China Normal University, China. The formation of Save Pangolins in 2007 marked the beginning of significant events associated with raising awareness of pangolins and their threats (Harrington et al., 2018). The Save Pangolins website was the first ever dedicated solely to promoting pangolins and highlighting their threats. In 2008, TRAFFIC held a workshop on the trade and conservation of pangolins native to South and Southeast Asia. The aim was to convene government stakeholders, non-governmental organizations (NGOs), research institutions, zoos and rescue centers to share information and find solutions to tackle the illegal trade in pangolins (see Pantel and Chin, 2009). In 2011, the African Pangolin Working Group (APWG) was established to address increasing threats to pangolins in Africa, by coordinating a conservation response at the continental level. The International Union for Conservation of Nature Species Survival Commission (IUCN SSC) Pangolin Specialist Group (PSG) was re-established in 2012 in recognition of the need to catalyze conservation action for pangolins at the global level in light of the threats they face (Challender et al., 2012; see also Chapter 39). The PSG held a conservation conference in 2013 at Wildlife Reserves Singapore, the first event to convene stakeholders with expertise on both Asian and African pangolins and their conservation. The outputs from the meeting included the PSG's "Scaling up Pangolin Conservation" Action Plan (see Challender et al., 2014), which provided a broad strategy to address the conservation of all eight species globally.

From 2005–16, coverage of pangolins and their threats dramatically increased on social (e.g., like, shares, and comments) and in editorial (e.g., published articles) media, with peaks in public interest occurring with reports of various pangolin seizures (Harrington et al., 2018). For example, the seizure of 10 tonnes of pangolins, equivalent to an estimated 2,000 animals, in the Philippines in April 2013 (Harrington et al., 2018). It was during this period that awareness of pangolins and their plight grew beyond conservation NGOs, practitioners and academics, to governments, journalists and members of the public globally. For instance, pangolins were the centerpiece of a wildlife photography conference in Montier-en-Der, France in 2015, Europe's largest wildlife photography event; a seizure of 4000 frozen pangolins in Medan, Indonesia was widely publicized, with an associated image winning the Wildlife Photographer of the Year 2016 competition; pangolins were featured in a unique Angry Birds Friends game that was promoted by HRH The Duke of Cambridge; and a pangolin also made a prominent cameo in the animated Disney blockbuster film, The Jungle Book in 2016. The online movement of World Pangolin Day (WPD), which began in 2012, also became an annual event for conservation groups and interested citizens alike to raise awareness of pangolins and the threats they face, around the world.

The remainder of this chapter discuses eight key events and activities that had a substantial role in raising awareness of pangolins at a local level or global scale between 2012 and 2018. "Events" are defined as actions related to pangolins, their conservation, or illegal trade. Each of the events discussed here share the distinguishing characteristic that they were designed by conservationists, citizens, or other stakeholders to raise awareness of pangolins and their conservation in some form. For this reason, large seizures of pangolins were not included. The year 2012 was selected as the starting point for this exercise because it marked both the re-establishment of the PSG and the first WPD. The events discussed were identified through interviews with 21 pangolin conservation experts (both academics and practitioners) and media professionals. By design, the chapter is not exhaustive; reviewing all awareness raising activities at local to global levels for this period is beyond the scope of this chapter. Rather than attempting to capture every activity and actor, this chapter focuses on the greatest catalysts of awareness generation as identified through the aforementioned interviews.

The IUCN SSC Pangolin Specialist Group

Specialist Groups constitute the main working units of IUCN's SSC, providing the breadth of expertise and commitment to achieve, "*A just world that values and conserves nature through positive action to reduce the loss of diversity of life on earth*" (IUCN SSC, 2016). They are committed to achieving conservation, by drawing on the strength of their volunteer network of experts and proactively working to communicate conservation needs, design effective programs, build political will, and increase the financial and human resources needed to stimulate action to halt the decline in biodiversity.

In 2012, pangolins were the only group of mammals not represented by an SSC Specialist Group. Given the extent of illegal trade in the species, there was a clear need for a coordinated response at the global level to catalyze action and attention on the species: this niche was filled with the re-establishment of the PSG.

Similar to other Specialist Groups, the mandate of the PSG was broadly to convene relevant experts and conservation stakeholders and mobilize action to conserve pangolins. The PSG brought together individuals with expertise in ecology, social science, conservation policy, veterinary health, and genetics, among others, in order to coordinate the development of conservation strategies and direct research needed to determine and address the conservation needs of the species (Challender et al., 2012).

Between 2012 and 2018, the PSG raised international awareness of pangolins in conservation and policy circles by providing technical and scientific contributions to key meetings. This included organizing, with Wildlife Reserves Singapore, the first-ever international conservation conference focusing on Asian and African pangolins in 2013, and the first-ever regional conservation planning workshop for the Sunda pangolin (*Manis javanica*) in 2017 (IUCN SSC Pangolin Specialist Group et al., 2018). The PSG has also provided technical input and raised awareness of the threat of illegal trade at each relevant meeting of the CITES Animals and Standing Committees and Conferences of the Parties (CoP) since 2013, and the first pangolin range states meeting held in Vietnam in 2015.

Beyond conservation circles, the PSG helped raise public awareness by corresponding with the press and broadcast media, speaking as an authority on pangolins and their conservation. The group's leadership and members have frequently appeared in print and broadcast media (online, print, radio, and television), supplied fact-checking, illegal trade statistics, and participated in WPD activities.

Importantly, the PSG has provided a formalized means of information sharing among pangolin experts within and outside range states, allowing a level of coordination among experts and other stakeholders that was previously non-existent. As a formal body designed to catalyze conservation action for pangolins, the PSG has played an important role in helping garner attention to pangolins and their conservation through strategic efforts to coordinate and convene experts, organize technical workshops to deliver tools for conservation management, and assisting the media.

World Pangolin Day

World Pangolin Day is an annual citizen-led movement aimed at raising awareness of pangolins and building public support for their conservation around the world. The objectives of WPD are to highlight the threats facing pangolins, educate the public about them, and share work that the conservation community – local, national and international – is doing to conserve the species.

From humble beginnings in 2012 as an online campaign driven by a handful of individuals, WPD has grown to a point where thousands of individuals participate online (P. Thomson, unpubl. data), including high profile institutions such as CITES, IUCN, and United States Fish and Wildlife Service (USFWS). By 2018, Facebook users who "liked" the WPD page came from 47 countries ranging from the United States to Argentina, Kenya and Israel, indicating interest in pangolins among an international audience, extended well beyond range countries. The first WPD, in February 2012, correlates to the first apparent increase in monthly Google searches for "pangolins," and WPD in 2016 correlates with an increase in baseline Facebook posts (i.e., number of news articles or posts; see Harrington et al., 2018).

Beyond the internet, a growing number of physical events have taken place around the world for WPD between 2012 and 2018. Selected examples include the United States Agency for International Development (USAID) releasing a pangolin identification guide on WPD 2018 to support front line law enforcement officers. In Sabah, Malaysia, the Sandakan Airport unveiled an oversized sculpture of a pangolin in 2018 (Fig. 21.1). Fellows in the USFWS MENTOR-POP (Progress on

FIGURE 21.1 To coincide with World Pangolin Day 2018, Elisa Panjang of the Danau Girang Field Centre helped unveil a new "Pangolin Awareness Sculpture" at Sandakan Airport in Sabah, Malaysia. *Photo credit: Elisa Panjang.*

FIGURE 21.2 A youth football match was held in Cameroon's Kimbi Fungom National Park, along with other activities aimed at raising awareness among local residents for World Pangolin Day in 2018. *Photo credit: Jerry Kirensky Mbi.*

Pangolins) program have led "pangolin walks" in Cameroon, which included talks at a local school and a football match bringing together civil servants and young people (Fig. 21.2). In Nepal, KTK-BELT, a local biodiversity and outreach organization, held a pangolin awareness workshop for local residents, where they screened a pangolin documentary. Although these examples are selective, WPD has served as an umbrella campaign, under which awareness raising activities have been coordinated and executed, with the number of events around the world seemingly increasing year on year.

WPD started as an online movement created by citizens and has grown organically into a global series of online and real world events, with participation from citizens, NGOs, and even government, intergovernmental and U.N agencies, that have helped raise awareness of pangolins.

The world's most trafficked mammals

A very influential activity to increase the profile of pangolins was the declaration that they are "the most trafficked wild mammals in the world." The origins of the statement go back to 2011, when the PSG leadership discussed ways to propel pangolins into the public consciousness to garner conservation support. Extrapolation of seizure data available at the time suggested that between the year 2000 and 2013, an estimated one million pangolins may have been poached and trafficked globally (see IUCN SSC Pangolin Specialist Group, 2016). Comparison with available data for other taxa revealed that no other wild mammal had such high rates of illegal trade over the same period[1]; as such, pangolins were identified as the world's most trafficked wild mammals – a statement that would become a much-used tagline within the global press and media.

[1] It should be noted that there are taxonomic and geographic biases in the detection and reporting of seizures involving wildlife traded illegally, and that data is not collected for most species in illegal trade.

CNN, through publication of an article in April 2014 entitled "The Most Trafficked Mammal You've Never Heard Of" (Sutter, 2014), was the first major media outlet to use the statement. This appears to have triggered a groundswell of interest in pangolins from readers, other press outlets, and conservation groups. The viral nature of the internet played a critical role in spreading awareness by enabling individuals to share the article or react to it in creative ways. For example, the article inspired the creation of music videos on YouTube, a petition to Disney to include a pangolin character in a film, and a self-published children's book, among others (J. Sutter, pers. comm.). Perhaps the most impactful result of the CNN article was its headline. Following publication of the article, dozens of popular media articles and videos started using variations of the "the world's most trafficked wild mammal" statement, which has become synonymous with the species and seemingly made pangolins memorable to readers and viewers around the world.

Google Doodles

Google has played a role in raising awareness of pangolins. In 2015 and 2017, the tech company featured pangolins in two of its "Google Doodles", the graphical changes made to the Google logo on its landing page that are used to celebrate holidays, anniversaries, and the lives of famous artists, scientists, and pioneers.

For the 2015 Earth Day Doodle, users could take a quiz asking, "Which animal are you?" with one of the animals being a pangolin. The resulting page included information about the species and their conservation, and the ability for users to search the internet to learn more. Two years later, to coincide with Valentine's Day 2017, Google created a Doodle in the form of an interactive game, the main characters of which were pangolins. The game directed users to the World Wildlife Fund (WWF-US) website where they could find out more information about the species and their threats. The Doodle launched on mobile, web and the Google "app." in more than 60 countries on February 13–14 2017.

The Google Doodles led to a significant amount of online organic social conversation around pangolins. Google does not disclose statistics on the reach of its Doodles; however, Google Trends (GT), can be used to track levels of public awareness (Harrington et al., 2018; Proulx et al., 2014) by returning the number of Google searches made for a keyword in a given region over a defined period of time. Analysis of GT data shows that the 2017 Valentine's Day Doodle and the 2015 animal quiz Doodle correlated with the two highest peaks in Google searches for the term "pangolin" between 2004 and 2018 (Fig. 21.3). It should be acknowledged that this represents correlation and not necessarily causation. It is worth noting that the 2015 Doodle was released on April 22, 2015, one day before a sizeable seizure of pangolins occurred in Medan, Indonesia, that was covered by multiple news sources online. It is possible that the peak in searches in 2015 is the result of these two events combined.

While the 2017 Valentine's Day Doodle was active, more people visited the pangolin entry on Wikipedia (en.wikipedia.org) than on any other date between July 2015 (data before then not available) and the end of 2018 (Wikipedia, 2019). On the day the Doodle launched, the page received 101,472 page views (compared to a daily average of 3425), which was more than double the number of views of the previous record (August 26, 2016: 52,869 views).

During the two days the 2017 Doodle was live, WWF-US recorded almost 270,000 page visits and 305,000 page views of their landing page, which far exceeded all other pages on the website and more than 50% of all site traffic was on pangolin-related content. WWF-US received an undisclosed but reportedly

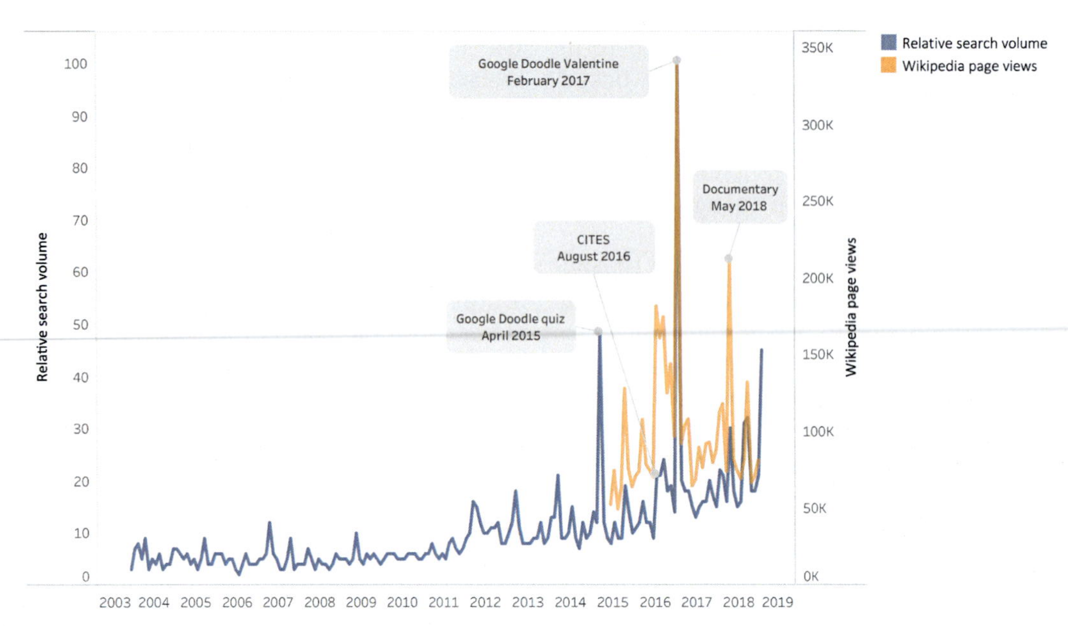

FIGURE 21.3 Monthly data from Google Trends and Wikipedia. Relative search volume shows Google search interest for the keyword "pangolin" between January 2004 and January 2019, relative to the highest point on the chart. Wikipedia page views is the number of visits to the pangolin Wikipedia page per month between July 2015 and January 2019 (data before July 2015 is not available). The peak in visits to Wikipedia and Google searches for the term "pangolin" both correlate with the Valentine's Day Google Doodle in February 2017.

substantial increase in pangolin-related donations because of the Doodle (D. Quigley, pers. comm.).

CITES CoP17

Pangolins have had a long history in CITES but the 17th meeting of the CoP (September–October 2016) proved a key moment for awareness raising, related to the transfer of all eight species from Appendix II to I.

Activities leading up to and during the CoP (see Fig. 21.4), and the response to the resulting international commercial trade ban ostensibly contributed to global awareness.

Strategic campaigns to raise awareness of pangolins and their threats, and generate support for uplisting of all eight species to Appendix I began months before CoP17. In 2015, a group of conservation NGOs petitioned the USFWS to protect seven pangolin species under the Endangered Species Act (ESA) (Temminck's pangolin [*Smutsia temminckii*] was previously listed under the ESA in 1976). A successful ESA listing would place tighter restrictions on import, export, take, interstate, and foreign commerce activities in the United States, in addition to the trade controls under CITES. The petition was a strategic move to generate awareness and support from the public and some policy-makers in the United States, as well as some pangolin conservation stakeholders internationally, and to use the listing as leverage to support the longer-term aim of uplisting pangolins to CITES Appendix I at CoP17.

In September 2016, the IUCN World Conservation Congress (WCC) took place in Hawaii, United States, where a number of pangolin-related events were held, some of

FIGURE 21.4 Members of the IUCN SSC Pangolin Specialist Group at the CITES CoP17 meeting in Johannesburg, South Africa. From left to right: Lisa Hywood, Jeff Flocken, Leanne Wicker, Dan Challender, Thai Van Nguyen, Keri Parker, and Darren Pietersen. *Photo credit: Frank Kohn/United States Fish & Wildlife Service.*

which actively promoted the transfer of pangolins to Appendix I as a solution to combatting illegal trade and conserving the species. Leading up to WCCs, IUCN Members and Commission members discuss and debate proposed resolutions that shape conservation and policy. A motion to urge all IUCN Members to support the proposed transfer from Appendix II to I was adopted in advance of the WCC. It happened in the context of associated media presence and growing hype around pangolins, using the tagline of "the world's most trafficked mammal." Although it is challenging to casually attribute these events to the near unanimous decision by the parties to CITES to transfer all species of pangolin from Appendix II to I at CoP17, they may, at least in part, have influenced the decision of some parties when voting on proposals to amend the appendices that involved pangolins.

The decisions to include pangolins in CITES Appendix I and the events leading up to them appear to have led to widespread attention around pangolins. August 2016 marked a peak in both monthly Google searches for "pangolin" and visits to the pangolin entry on Wikipedia, an indicator of the public's interest in the animals (Fig. 21.3). In addition, between August and October 2016, at least 58 international newspapers and online outlets published news of the uplisting (T. Shibaike, unpubl. data).

Documentaries

Compared to traditionally iconic species, pangolins are underrepresented in nature films and documentaries, perhaps because of the difficulty of finding wild pangolins, and low levels of public knowledge of their existence. Yet, there are two notable instances where documentaries about pangolins have likely contributed to an increase in public awareness of the species.

The first major contemporary appearance of pangolins in a documentary was in 2012 when the high profile naturalist Sir David Attenborough featured pangolins as one of ten animals he would most like to save from extinction. The segment appeared in "Attenborough's Ark: Natural World Special" which was broadcast on the BBC in the United Kingdom and watched by 4.57 million people, making it the most watched Natural World film shown that year. It has been broadcast again numerous times on the BBC and sold to many countries around the world (S. Greenwood, pers. comm.).

In 2018, a feature documentary on pangolins was broadcast on BBC2 in the United Kingdom and subsequently on Public Broadcast System (PBS) in the United States – the first time a documentary of this nature (i.e., focused solely on pangolins) had aired on networks of this size in these countries. The program, entitled "Pangolins: the world's most wanted animal" in the United Kingdom and "Nature: The World's Most Wanted Animal" in the United States, likely helped contribute to and reinforce increasing awareness of pangolins in these countries. A production of the BBC Natural History Unit, the documentary focused on the pangolin rehabilitation work of two organizations: the Rare and Endangered Species Trust in Namibia and Save Vietnam's Wildlife. The program was watched by 1.53 million viewers in the United Kingdom and 1.2 million viewers in the United States. The BBC Natural History Unit's dedicated webpage saw a significant increase in traffic in the two weeks following the broadcast, with 162,800 unique visits. The BBC Earth Twitter account was also the number one trending on Twitter following the broadcast of the show, for the first time in the show's history. More than two dozen media and news outlets in the United Kingdom and US also covered the program (V. Bromley and R. Davis, pers. comm.).

Earlier in 2018, the Discovery Channel made a documentary entitled, "Secrets of the Pangolin", which highlighted both field and captive care work of researchers and conservationists in Taiwan, as well as the story of a new-born pangolin. In the year of its launch, the show reached 200 million subscribers in Southeast Asia and has been used to raise awareness about pangolins among visitors, adults and children alike, through broadcasts at Taipei Zoo. Documentaries and nature films have the ability to draw widespread and sizeable viewership, and, as demonstrated with pangolins, can expose large numbers of viewers to lesser-known animals and their conservation issues.

Celebrity engagement

The engagement of celebrities can increase public willingness to engage with a conservation campaign (Duthie et al., 2017). While there are few studies that provide positive accounts of the role of celebrity-associated campaigns to change behavior related to a specific cause (Brockington and Henson, 2015), they remain a useful tool for raising awareness and generating interest in specific issues, including environmental concerns. Concerning pangolins, two notable campaigns involved celebrities in raising awareness of the species between 2012 and 2018. The first featured HRH The Duke of Cambridge in the United Kingdom, and the second featuring prominent television and film stars in China and Vietnam.

In 2014, Finnish gaming company Rovio Entertainment, makers of the global hit game Angry Birds Friends, collaborated with the conservation coalition United for Wildlife to create a pangolin edition of the game, "Roll with the pangolins." It was designed specifically to raise awareness of the species, and the designers chose pangolins because they wanted to start a conversation around pangolin conservation among an audience that had likely not heard of them (N. Doak, pers. comm.). The Angry Birds Friends game ran for a week in November 2014 and was played more than ten million times. The game was released with an accompanying video of HRH The Duke of Cambridge playing the game and discussing pangolins and the illegal wildlife trade. During the campaign, the video was watched more than 2.1 million times and had more than 61 million impressions on Twitter (United for Wildlife, 2014).

To raise awareness amongst public audiences in Asia, conservation group WildAid launched a pangolin campaign in May 2016. It was released with strategic media events featuring Angelababy, one of China's leading actresses and television personalities, and Jackie Chan, the global celebrity, actor, and martial artist. It was also one of the first pangolin campaigns to use celebrities within pangolin range states. Between May 2016 and the end of 2018, WildAid placed more than 100,000 video screens and billboards around subways, office blocks, government buildings and airports in more than 37 Chinese cities, and estimate that this resulted in more than 2.5 billion impressions. Impressions are defined as the number of times the content was potentially viewed (S. Blake, pers. comm.). The campaign's online videos collectively accumulated more than 200 million views between 2016 and 2018. In December 2017, a short online film featuring Angelababy recounting the story of a pangolin rescued from poachers in Namibia received ~40 million views on Weibo (a popular social media platform in China), including 25 million views on the first day (S. Blake, pers. comm.).

A global network of local champions

Connecting the aforementioned events, and permeating through awareness raising activities focused on pangolins between 2012 and 2018, is an eighth influential activity: the formation of a global network of pangolin champions. The strength of this network lies in the collective influence of individual efforts, both within and beyond pangolin range states, in raising awareness of pangolins through myriad activities carried out locally. These include running education and awareness programs, actively engaging local communities in conservation efforts, engaging with the media, fundraising initiatives, and pursuing policy changes.

Non-exhaustive examples include Thai Van Nguyen and Save Vietnam's Wildlife in Vietnam who have raised awareness of pangolins among school children and other visitors to their education center at Cuc Phuong National Park, and among government officials through the rescue and rehabilitation of pangolins in the illegal trade. In Hong Kong Special Administrative Region, Gary Ades of Kadoorie Farm and Botanic Gardens produced pangolin-related education products, exposing an estimated 450,000 visitors to pangolin conservation messaging between 2015 and 2018. In Sabah, Malaysia, Elisa Panjang played a key role in convincing the Sabah Wildlife Department to increase protection and penalties for hunting and trafficking the Sunda pangolin. In Nepal and India, Tulshi Laxmi Suwal, Ambika Khatiwada and Kumar Paudel, and Rajesh Mohapatra, respectively, have held round table discussions on conserving pangolins with multiple stakeholders ranging from local community members to government officials. In Pakistan, Tariq Mahmood's research on illegal pangolin trade has highlighted the threats to the Indian pangolin in the country, including through national news agencies. In Zimbabwe, Lisa Hywood was influential in encouraging the government to enact some of the world's toughest penalties for poaching and trafficking pangolins. In Cameroon, Fellows of the USFWS MENTOR-POP program made media appearances on television and radio programs to raise awareness of growing threats facing pangolins in Central Africa.

The network extends to individuals outside pangolin range states as well. In the United States, Paul Thomson (co-author of this chapter), Keri Parker, and colleagues created savepangolins.org in 2007, the first website dedicated solely to pangolins and raising awareness of their plight; it has since evolved into an NGO, Save Pangolins, which continues to raise awareness and funds for conservation activities in pangolin range countries. Conservation filmmaker Katie Schuler of Coral & Oak Productions made a powerful short documentary film tracking a poached pangolin from forest to dinner plate: it was viewed more than 50 million times in multiple countries between its launch in 2016 and the end of 2018. Educator Louise Fletcher (co-author of this chapter), has developed campaigns using pangolin art to inspire and educate school children about pangolins in range countries including, Indonesia, Nepal and South Africa.

From awareness to action

This chapter has identified and discussed eight key events and activities that have ostensibly contributed to raising awareness of pangolins at local, national and global levels and among important audiences including policy-makers, funders, and the global public between 2012 and 2018. There is an opportunity for conservationists to leverage this increased attention to acquire funding, influence policy, change consumer behavior, and take further action against overexploitation threatening pangolins.

Lessons can be drawn from these events to aid pangolin conservation in the future and efforts for other seemingly "forgotten" species. They include cost effective ways of raising awareness such as leveraging the power of free social networks including Facebook (e.g., for World Pangolin Day); and engaging businesses and corporations, especially technology companies, who may bear the cost of activities (e.g., Rovio Entertainment and the Angry Birds Friends game and Google's Doodles). The use of film and celebrities can also reach a large and widespread audience, especially when part of a coordinated campaign. Use of traditional methods (e.g., large press outlets such as CNN) and digital methods (e.g., online social media platforms) is important to raising awareness at a global level, but should be combined with local efforts by individuals and NGOs, for example acting as species champions. Finally, the events leading up to, during, and following CITES CoP17 suggest that specific policy events can generate much awareness and interest around a particular issue among policy-makers and other interested stakeholders, both editorial and social media, and potentially the public.

Awareness raising can play an important role in increasing the profile of species. Unlike more traditionally iconic species, there is an obvious need to raise awareness for species such as pangolins that are less well known, in order to generate interest that can mobilize funding and initiate strategic conservation initiatives. Although anecdotal, practitioners of pangolin conservation have indicated that increasing awareness of the species since 2012 has made funding more accessible, and that funds available for pangolin conservation have also increased over time. Pangolins are among priority species in terms of funding programs tackling illegal wildlife trade: as an example, the major funding programs to address this issue from the United States and United Kingdom governments both include pangolins as focal species.

There is a need for further awareness raising activities for pangolins to ensure they receive support in the future, especially in range states. Whether the cumulative efforts to raise awareness of the species discussed in this chapter, those not discussed and those in the future will be sufficient to catalyze conservation action at a scale and speed to reduce substantially the extinction risk to pangolins remains to be seen, but with so much interest and enthusiasm, as discussed in this chapter, there is real cause to be optimistic.

Acknowledgments

The authors wish to thank the following individuals who contributed to this chapter: Takumi Shibaike, University of Toronto; Dan Challender, Carly Waterman and Helen C. Nash; Lydia Katsis; John Sutter, CNN; Christina Vallianos and Steve Blake, WildAid; Perla Campos, Google; Diane Quigley, WWF-US; Naomi Doak, The Royal Foundation; Doug Hendrie, Education for Nature Vietnam; Harry Lawrence, Victoria Bromley, and Steve Greenwood, BBC; Nick Ching Min-Sun; Mark Hunt, Discovery Networks; Ryan Davis, KQED; Maria Diekmann, Rare and Endangered Species Trust; Keri Parker, Save Pangolins; Rhishja Cota-Larson, Annimaticus; Jeff Flocken, Humane Society International; and Gary Ades, Kadoorie Farm and Botanic Garden.

References

Anadu, P.A., Elamah, P.O., Oates, J.F., 1988. The bushmeat trade in southwestern Nigeria: a case study. Hum. Ecol. 16 (2), 199–208.

Brockington, D., Henson, S., 2015. Signifying the public: celebrity advocacy and post-democratic politics. Int. J. Cult. Stud. 18 (4), 431–448.

Challender, D.W.S., Baillie, J.E.M., Waterman, C., IUCN SSC Pangolin Specialist Group, 2012. Catalysing conservation action and raising the profile of pangolins - the IUCN SSC Pangolin Specialist Group (PangolinSG). Asian J. Conserv. Biol. 1 (2), 140–141.

Challender, D.W.S., Waterman, C., Baillie, J.E.M., 2014. Scaling up Pangolin Conservation. IUCN SSC Pangolin Specialist Group, Zoological Society of London, London, UK.

Duthie, E., Veríssimo, D., Keane, A., Knight, A.T., 2017. The effectiveness of celebrities in conservation marketing. PLoS One 12 (7), e0180027.

Harrington, L.A., D'Cruze, N.D., Macdonald, D.W., 2018. Rise to fame: events, media activity and public interest in pangolins and pangolin trade, 2005-2016. Nat. Conserv. 30, 107–133.

Heath, M.E., Vanderlip, S.L., 1988. Biology, husbandry, and veterinary care of captive Chinese pangolins (*Manis pentadactyla*). Zoo. Biol. 7 (4), 293–312.

Heath, M.E., Coulson, I.M., 1997. Home range size and distribution in a wild population of Cape pangolins, *Manis temminckii*, in north-west Zimbabwe. Afr. J. Ecol. 35 (2), 94–109.

IUCN Species Survival Commission, 2016. Terms of reference for Members of the IUCN Species Survival Commission 2017-2020. Available from: <www.iucn.org/sites/dev/files/tors_ssc_members_2017-2020_final_0.pdf>. [November 3, 2018].

IUCN SSC Pangolin Specialist Group, 2016. The status, trade and conservation of pangolins (*Manis* spp.). CITES CoP17 Inf. 59. Available from: <https://cites.org/sites/default/files/eng/cop/17/InfDocs/E-CoP17-Inf-59.pdf>. [April 12, 2019].

IUCN SSC Pangolin Specialist Group, IUCN SSC Asian Species Action Partnership, Wildlife Reserves Singapore, IUCN SSC Conservation Planning Specialist Group, 2018. Regional Sunda Pangolin (*Manis javanica*) Conservation Strategy 2018-2028. IUCN SSC Pangolin Specialist Group, Zoological Society of London, London, UK.

Mohr, E., 1961. Schuppentiere. Neue Brehm-Bucherei. A. Ziemsen Verlag, Wittenberg Lutherstadt.

Pagès, E., 1975. Etude eco-ethologique de *Manis tricuspis* par radio-tracking. Mammalia 39, 613–641.

Pantel, S., Chin, S.-Y. (Eds.), 2009. Proceedings of the Workshop on Trade and Conservation of Pangolins Native to South and Southeast Asia, 30 June – 2 July 2008, Singapore Zoo, Singapore, TRAFFIC Southeast Asia, Petaling Jaya, Selangor, Malaysia.

Proulx, R., Massicotte, P., Pépino, M., 2014. Googling Trends in Conservation Biology. Conserv. Biol. 28 (1), 44–51.

Schultz, P.W., 2010. Making energy conservation the norm. In: Ehrhardt-Martinez, K., Laitner, J. (Eds.), People-Centered Initiatives for Increasing Energy Savings. American Council for an Energy Efficient Economy, pp. 251–262.

Sutter, J.D., 2014. The most trafficked mammal you've never heard of. Change the list, CCN.com. Available from: <edition.cnn.com/interactive/2014/04/opinion/sutter-change-the-list-pangolin-trafficking/>. [July 27, 2018].

United for Wildlife, 2014. Play Angry Birds Roll with the Pangolins Campaign. Available from: www.unitedforwildlife.org/#!/2014/12/play-angry-birds-roll-with-the-pangolins/. [July 28, 2018].

Wikipedia, 2019. Page views analysis for the term pangolin. Available from: <https://en.wikipedia.org>. [January 6, 2019].

Wu, S.B., Ma, G.Z., 2007. The status and conservation of pangolins in China. TRAFFIC East Asia Newsl. 4, 1–5. [In Chinese].

CHAPTER

22

Changing consumer behavior for pangolin products

Gayle Burgess[1,2], *Alegria Olmedo*[3,4], *Diogo Veríssimo*[5,6] *and Carly Waterman*[7,8]

[1]Institution of Environmental Sciences, London, United Kingdom [2]Society for the Environment, Coventry, United Kingdom [3]Department of Zoology, University of Oxford, Oxford, United Kingdom [4]People for Pangolins, London, United Kingdom [5]Department of Zoology and Oxford Martin School, University of Oxford, Oxford, United Kingdom [6]Institute for Conservation Research - San Diego Zoo, Escondido, CA, United States [7]Conservation and Policy, Zoological Society of London, Regent's Park, London, United Kingdom [8]IUCN SSC Pangolin Specialist Group, ℅ Zoological Society of London, Regent's Park, London, United Kingdom

OUTLINE

Pangolins
DOI: https://doi.org/10.1016/B978-0-12-815507-3.00022-8

Introduction

Addressing the illegal trade in wildlife products requires a multi-faceted approach (Burgess, 2016; Challender et al., 2015). Regulatory measures, harvest and trade controls and appropriate enforcement thereof, in source countries, along trafficking routes and in major end-use markets, are essential (see Chapters 17 and 18). Interventions that enable community engagement should also ensure benefits arise from wildlife protection and/or legal, sustainable trade (see Chapter 23: Cooney et al., 2017). However, these measures can be, and are, undermined by illegal markets that are impossible to contain whilst demand persists (Nijman, 2010; Veríssimo et al., 2012).

Complementary efforts to effectively address demand for wildlife products are therefore increasingly recognized as critical in tackling illegal wildlife trade (TRAFFIC, 2016; Veríssimo and Wan, 2019). Economic theory, and some situational crime prevention models, assert that reducing demand for wildlife products among end consumers (Gore, 2011), and thereby average market price, reduces incentives for market actors, including international criminal networks, to engage in poaching and illicit trade. This should lead to less poaching, and all other things being equal, allow overexploited wild populations to recover (Challender and MacMillan, 2014).

Within this context "demand reduction" has been recognized as a specific mechanism through which to address illegal trade in wildlife. The term and associated approaches have been recognized since the 1970s (e.g., Arthur and Wilson, 1979), but have gained substantial traction since around 2012, with examples of demand reduction actions agreed by governments in the 2014 and 2018[1] London Declarations and associated statements.

Another notable example of demand reduction commitment to action was the UN General Assembly Resolution (69/314) on "Tackling Illicit Trafficking in Wildlife." It also features in key regional agreements such as those via the Forum on China-Africa Cooperation (FOCAC) and Asia-Pacific Economic Cooperation (APEC). These agreements and declarations were reinforced through adoption of measures to address demand at the 17th meeting of the Conference of the Parties (CoP) to CITES, the Convention on International Trade in Endangered Species of Wild Fauna and Flora (CITES CoP17). Resolution Conf. 17.4[2] requires parties to take action to reduce demand for illegally sourced wildlife products through evidence-based behavior change strategies. Demand reduction

[1] London Conference on the Illegal Wildlife Trade (October 2018) Declaration.

[2] Resolution Conf. 17.4, Demand reduction strategies to combat illegal trade in CITES-listed species.

action was also prescribed for pangolins specifically, in CITES Res. Conf. 17.10[3] and in conservation strategies developed for the species.

In this chapter, the focus is on the behavior change component of demand reduction actions, which is recognized as complementary to measures imposing societal control (e.g., regulations/legislation, enforcement, and retailers removing products from sale; Burgess, 2016). The chapter first provides some background and context regarding consumer demand for pangolins, before discussing challenges and considerations relating to pangolin demand reduction efforts. It then summarizes insights in relation to consumer demand in Asia, and discusses opportunities to use these insights to reduce demand for pangolin products through behavior change.

Background and context regarding consumer demand for pangolins

Consumer demand for pangolins has been recorded throughout most, if not all, of their range countries in Asia (Chapter 14) and Africa (Chapter 15; Ingram et al., 2018). Pangolins can represent an important source of protein and their parts are used for a variety of medicinal, ornamental and ritualistic purposes (Chapters 12, 14–16). Such demand is hard to characterize given the multiple motivations for use, various consumer groups and the variety of derivatives that are used and consumed. This complexity is compounded by variations evident across major markets, due to, for example, disruptive marketing by suppliers and fads and trends in consumer markets.

That said, much of the contemporary illegal international trade is driven by demand in Asian countries, especially China, but also Vietnam, and Myanmar (Chapter 16; Heinrich et al., 2017; Nijman et al., 2016). Demand for pangolin products has also been shown to exist in non-range countries, such as the United States, some European countries, and Japan (Heinrich et al., 2016, 2017). Most demand reduction research and interventions that aim to address the consumption of illegally traded wildlife, including pangolin products, have focused on Asia, particularly China and Vietnam (Veríssimo and Wan, 2019). This is due to the size of these markets for illegal wildlife products in general, combined with phenomenal rates of economic growth in recent decades (Nijman, 2010). More is understood about the nature of demand in these countries as a result and, for this reason, the rest of this chapter focuses primarily on these locations. However, it should be noted that further research into demand for pangolin products in other countries is needed to understand the impact of these markets on wild populations.

Challenges and considerations regarding demand reduction efforts

Before detailed consideration of the research evidence, it is worth recognizing that addressing the consumer demand driving illegal wildlife trade is challenging. Demand reduction campaigns often call into question long-held beliefs or a culturally-ingrained behavior on the basis of their impact on wildlife populations. Some of the challenges of these campaigns are, therefore, ethical in nature. Demand reduction campaigns focus on voluntary behavior change and as such do not aim to limit the freedom of choice of their target audience.

Disciplines such as social marketing should be used to benefit both the target audience and

[3] Resolution Conf. 17.10, Conservation of and trade in pangolins.

society at large. In strict terms, this means that they should not be designed to benefit biodiversity at the expense of the target audience. Still, these efforts do often have as a premise not only that the target audience is not behaving in its own best interest, but that conservationists know better how the target audience ought to behave instead. This has led to behavior change efforts being criticized for being paternalistic and even manipulative (Andreasen, 2002). In the context of biodiversity conservation the situation can be further complicated when those promoting the importance of wildlife are based in countries far from species' range states or areas where target audiences reside, and thus accusations of neo-colonialism arise due to complex historical power imbalances.

It is thus important to recognize that while many demand reduction campaigns take their social license from laws and regulation enacted by a national government, there are numerous contexts where the assumption that those delivering demand reduction interventions work primarily for the benefit of the people they serve, does not hold. This means that to ensure a degree of legitimacy, conservationists engaging in demand reduction efforts should build broader partnerships with civil society in the countries they work and ensure there is ample evidence to support addressing the priorities they are focusing on. Moreover, the ethics of influencing behavior should also bear in mind the potential unintended consequences that interventions may have (Cho and Salmon, 2007). While in some conservation contexts, a failed intervention represents a simple loss of time and resources, in the case of the behavioral sciences, unintended consequences can mean that hastily implemented interventions can worsen the threats they were meant to mitigate (Pfeiffer, 2004).

Historically, efforts have lacked an in-depth and culturally sensitive understanding of both demand and the most appropriate interventions with which to change consumer behavior (Greenfield and Veríssimo, 2019; Olmedo et al., 2017). Many of those delivering communications to persuade different types of purchasing preferences and consumer choice, have tended to focus on achieving breadth, rather than depth (Burgess et al., 2018). In terms of evaluation, impact measurement has often been described in terms of message "reach" rather than "resonance" (Burgess et al., 2018; Veríssimo and Wan, 2019), and on celebrities engaged as messengers, irrespective of their anticipated influence or measurable impact on key buyer, user and intender groups (e.g., Duthie et al., 2017).

Further, the links between information provision, awareness and actual behavior change are in most cases often tenuous at best (Burgess, 2016; Veríssimo et al., 2012). Behaviors are not just influenced by knowledge, but are shaped by a variety of forces, such as social context, political and cultural forces, as well as the degree to which the source of information is trusted (Burgess et al., 2018). Changes in knowledge may eventually lead to changes in attitudes and practice, but even where attitudes are supportive of, for example, biodiversity conservation and animal welfare concerns, this still may not result in changes in behavior (e.g., Lane and Potter, 2007).

Case study evidence and relevant context in this regard, is available from rhino (Rhinocerotidae) horn demand reduction efforts in Vietnam. Much international attention and donor support has been focused on combatting this illegal market. Interventions have thus enjoyed the strongest opportunities to demonstrate not just message outreach, but also best practice in behavior change approaches.

Olmedo et al. (2017) developed a framework against which to evaluate the design of behavior change interventions aimed at reducing consumption of rhino horn in Vietnam. This included clearly defined, measurable objectives, explicit theories of change, supported by research, with built-in monitoring and evaluation, thus enabling evaluation of outcomes, learning, and adaptive management. Findings, however, included that very few interventions featured all requisite steps (Olmedo et al., 2017). Instead they tended to focus on communications, raising awareness of legal constraints and highlighting the cruelty and threat to rhinos, including of extinction. With some audiences seeking to acquire such products specifically because they are illegal or rare (e.g., Burgess, 2016; Kennaugh, 2016), such messaging risks being counterproductive at best. This illustrates the imperative of understanding in-depth what is driving demand before designing approaches to reduce it. Within this context the following section aims to summarize what is understood about demand for pangolin products in two major markets for products trafficked internationally.

Summary of insight into consumer demand in Asia

Asia has been the focus of the majority of systematic research into demand for pangolin products. This effort has focused mostly on the larger urban areas of China and Vietnam, associated to some extent with the role these urban populations play in the broader demand for wildlife (Nijman, 2010; Zhang and Yin, 2014). Thus, through a primary lens on these geographies, subsequent information is split into knowledge around consumption of each product type. Key gaps and limitations are discussed first.

An associated note on nomenclature. The primary lens applied here is on demand reduction actions aiming to change behavior. Consumers are thus considered all those who buy, gift, or use products, or intend to do so. Market research is that identifying price and other details around the availability and sale of products in the marketplace; consumer research is that revealing insights into the socio-economic and psycho-demographics of consumer knowledge, attitudes, values and behaviors.

Gaps and limitations

One key limitation to understanding the nature of demand for pangolin products is the lack of access to an adequate quantity or quality of consumer research. The authors are aware of nine surveys specifically focused on the consumption of pangolin products. Many remain either internal to the institutions that conducted them or have only been summarized as part of broader reports or press releases, which frequently omit information needed to interpret results. As a result, this chapter reports on the full results of two of those surveys. A related limitation is the dearth of information on the demographic and psychographic profile of consumers of pangolin products, beyond the multi-species studies carried out by Shairp et al. (2016) in Vietnam and USAID in both China (2018a) and Vietnam (2018b).

The scope of several surveys carried out in China and Vietnam had the potential for overlap. This raises the question of whether organizations communicated prior to, or during, the survey design process or research phase. Lack of collaboration can hinder the possibility of surveying consumer populations systematically across countries of interest. It also limits the opportunities for coordinated efforts at the implementation stage.

Lastly, due to differences in methodology, comparison across the results of these surveys is not possible. Many used methods unlikely to yield a representative sample and sample sizes implying large confidence intervals. This makes the resulting estimates harder to interpret meaningfully. Agreeing on common standards for conducting consumer research across various stakeholders would ensure findings can be used in a complimentary manner, and could provide deeper insights on consumption patterns and motivations. It would also ensure interventions resulting from these findings are based on best practice as recommended by CITES (e.g., in Resolution Conf. 17.4).

Within this frame, the following insights represent the state of knowledge regarding demand for pangolin products in China and Vietnam. Further research is required to ensure that any diversification of illegal trade is identified before it proliferates, and that demand reduction action can be undertaken.

China

A survey of higher income households in Beijing, Shanghai, Guangzhou, Hangzhou, Nanning and Kunming, found that 10% of a total of 3000 respondents report having consumed pangolin products in the past (WildAid, 2016). A different survey of 1800 respondents in six cities (five of which overlapped with the previous study) found that 7% had purchased pangolin products in the last year (USAID, 2018a). According to this second survey, cities with the highest prevalence of consumption were Beijing and Shanghai, and retail stores and Traditional Chinese Medicine (TCM) pharmacies across all cities, were the locations where most consumers reported purchasing pangolin products (USAID, 2018a).

Pangolin meat

Research which surveyed 1037 randomly selected adult residents in Hong Kong Special Administrative Region (hereafter "Hong Kong SAR"), estimated that only 0.1% had consumed pangolin meat in the past year; however, the majority of these individuals had consumed this product in mainland China (Humane Society International, Hong Kong University, 2015). Another survey conducted in 2016, where 1892 residents of ten large, medium and small Chinese cities were interviewed, found that 9% of respondents reported eating pangolin meat in the previous year (Horizon China, 2016). WildAid (2016) found around a quarter of respondents had consumed pangolin meat and had done so mostly because it was regarded as an "expensive status symbol" and "exotic wild animal." A previous study of 969 people in six cities found that 0.6% of respondents reported having eaten pangolin in the last year (Wasser and Bei Jiao, 2010).

Pangolin scales

Results of a survey showed that 78% of self-declared consumers have purchased scales in the last year; these are used to cure illnesses and contribute to overall well-being (USAID, 2018a; Fig. 22.1). Meanwhile, Wasser and Bei Jiao (2010) found that 1.4% of 969 respondents reported consuming pangolin as a tonic or as medicine but did not specify which product. A different survey indicates that 14% of urban Chinese residents have used pangolin products for medicinal purposes in the previous year (Horizon China, 2016).

WildAid's research in China suggests that scales can be one of many ingredients in

FIGURE 22.1 Seized pangolin scales. *Photo credit: Linh Bao Nguyen.*

prescription medicines; medicines containing pangolin scales are one of the most consumed pangolin products, with about two thirds of self-reported consumers having consumed them. These medicines are believed to treat an array of illnesses including rheumatism, skin diseases, swelling and pus, asthma, promoting lactation and treating cancer, among others (see Chapter 14; Nash et al., 2016; WildAid, 2016). Similarly, Yu and Hong (2016) describe scales as a common ingredient in medicines to treat symptoms associated with breast cancer and lymphoma. Pangolin scales are listed in the official Pharmacopeia in China where they are prescribed to promote lactation, improve circulation and treat skin diseases (Chinese Pharmacopeia Commission, 2015).

Although the sale and trade of some products derived from pangolin species has been prohibited by China's National Forestry and Grasslands Administration (NFGA), there is a legal trade in scales for medicinal use. This allows scales from government stockpiles to be used in the manufacture of around 70 patented traditional Chinese medicines by about 200 pharmaceutical companies and, once certified, sold by about 700 hospitals in China (China Biodiversity Conservation and Green Development Foundation, 2016). However, evidence indicates that unlicensed hospitals still prescribe and use pangolin scales, which are also readily available in unregistered pharmacies (Xu, 2009; Xu et al., 2016). Further research is needed to determine whether the existing legal trade creates social acceptability around the consumption of pangolin products, further encouraging demand.

WildAid's survey found 70% of respondents reported they believe that pangolin scales have medicinal properties (WildAid, 2016). In contrast, only 39% of Hong Kong SAR residents surveyed believed scales had medicinal benefits and only 0.2% had used scales for medicinal purposes (Humane Society International, Hong Kong University, 2015).

Interestingly, 57% of respondents from research conducted by USAID (2018a) in mainland China suggested that synthetic products similar to those from pangolins could substitute for scales used for medicinal purposes, while 17% said that nothing could perform the same function.

Although it is not possible to compare research findings across studies due to different methodologies employed, overall all studies found a higher prevalence of consumption of pangolin products for medicinal purposes, when compared to other use types and motivations. This finding has however to be interpreted in the light that medicinal use most commonly entails the use of a much smaller part of the animal than other uses.

Consumers

The only information regarding the beliefs, attitudes, values and other socio- and psycho-demographics of pangolin consumers in China is the study conducted by GlobeScan for USAID (2018a). This research found that 61% of self-declared pangolin consumers (n = 126) were men. The majority of consumers, both men and women, were in the 31–40 year old range and in a middle-income category. Among 67% of those who purchased pangolin products in the last 12 months, there was a high social acceptability for buying or owning such products.

Pangolin wine

Pangolin wine (i.e., body parts steeped in rice liquor) has been reported to be consumed by a "large proportion of consumers" in China (WildAid, 2016), but there is little information on the prevalence of consumption among populations surveyed, or the profile of the consumers or the motivations driving their consumption.

FIGURE 22.2 Pangolin scale carved into a hair comb. *Photo credit: Xiao Yu/TRAFFIC.*

Ornamental use

There is some evidence that pangolin scales are available online as engraved pieces (for example, as hairgrips; Fig. 22.2), and also used for ornamental purposes in China (Soewu and Sodeinde, 2015) but there is very limited information and further consumer research will be key to understanding it.

Vietnam

In Vietnam, consumer research conducted in 2015 found that 4% of all respondents (815 high income individuals from Hanoi, Ho Chi Minh City [HCMC] and Da Nang) reported purchasing pangolin products. Within this 4%, wine and meat were the most consumed products, with about half having used each of these (WildAid, 2016). In a survey of wildlife consumption in general, Do et al. (2011) found that less than 10% of adult respondents in HCMC (total sample size: 4062) and just over 10% of children surveyed (total sample size: 3562) reported consuming pangolin products. In contrast, a consumer survey of randomly selected adult residents of HCMC suggested that only 0.3% had consumed products derived from pangolins (Education for Nature Vietnam, 2016).

A 2018 survey was conducted with 1400 participants via mobile phone in five cities in Vietnam: Hanoi, Hai Phong, Da Nang, Can Tho and HCMC. This research included only respondents who were over 18 years old and earning a minimum of VND 10,000,000 monthly before tax. Results show 10% of respondents across the five cities report purchasing pangolin products at some point in their life; 6% having done so in the last year and 5% in the last six months. Most purchases are made physically in-country from private individuals (USAID, 2018b). The difference in results is substantial but may be due to the differing methodologies employed in each study. However, it is not possible to evaluate this hypothesis due to the lack of detail about the studies undertaken.

At 93% and 90% respectively, the vast majority of respondents were aware that selling pangolin products and purchasing them is illegal (WildAid, 2016). The consumption prevalence, despite known illegality, suggests law enforcement is currently not an effective deterrent in Vietnam. With the adoption of a new Penal Code, as of January 1, 2018 some conservationists expect that laws relating to possession of illegal wildlife products will feature more prominently in the public consciousness and be enforced more thoroughly. USAID's research (2018b) found that despite 52% of respondents who have purchased pangolin products in the last year having heard about the revision in the Penal Code, and 86% of the same group agreeing with these revisions, 60% of these buyers are likely to purchase pangolin products again in the future.

Pangolin meat

Do et al. (2011) found that pangolin meat was the most consumed pangolin product in HCMC (less than 10% of all respondents); similarly, Venkataraman (2007) found that just

FIGURE 22.3 Pangolin meat being cooked. *Photo credit: Linh Bao Nguyen.*

over 10% of 2000 respondents surveyed in Hanoi, consumed pangolin meat (Fig. 22.3). These findings echo other results that found 10% of respondents had consumed pangolin meat, although most of them had done so more than five years ago (WildAid, 2016). Research conducted in 2018 found that 12% of those who have purchased pangolin products across five cities have purchased pangolin meat (USAID, 2018b).

The most common motivations for consumption of pangolin meat in Vietnam are related to the rareness of pangolins, the perceived medicinal properties of the meat and the prestige that comes with purchasing a very expensive item (WildAid, 2016). These findings align with Shairp et al.'s (2016) research which determined that the high price and rarity of pangolin meat, particularly of wild-sourced pangolins, are attractive attributes for those who wish to display their social status in urban centers in Vietnam. Individuals identified as pangolin meat consumers included wealthy business elites and government officials. These findings support those of Do et al. (2011) relating to the general consumption of wildlife in Vietnam, although there is the need for more up to date research.

Pangolin scales

Research conducted by WildAid (2016), found that pangolin scales were consumed by 41% of self-declared pangolin consumers. Although prescription medicines containing pangolin scales are used in Vietnam, unlike China only a quarter of pangolin consumers claimed they purchase them. These results support findings from Do et al.'s research (2011) in which pangolin products used as medicine and pangolin wine were some of the products most consumed. Similarly, results from USAID's research show that scales and powdered scales are the products being purchased the most. Out of those who admitted having purchased pangolin products, 37% have purchased scales and 31% have purchased powder (USAID, 2018b). Scales are believed to treat several ailments including cancer, chronic varicella, malaria, chills, rheumatism, menstruation stagnation, breast feeding problems, styptic and scrofula (Vo, 1998).

In contrast with China and Hong Kong SAR, WildAid's research determined that only 8% of respondents in Vietnam believe pangolin scales have medicinal benefits. However, 64% of all respondents are aware of the "curative" properties of this product but are unsure whether the claims are true (WildAid, 2016). Additionally, research has found that pangolin products are believed to increase the efficacy of other medicinal ingredients (USAID, 2018b).

Pangolin wine

WildAid (2016) reported pangoln wine was consumed, but no research has been conducted on this product specifically. Research conducted by USAID (2018b) found that only 1% of those who have purchased pangolin products reported purchasing pangolin wine (Fig. 22.4).

Ornamental use

USAID's research (2018b) found that out of the total number of respondents who have

FIGURE 22.4 Pangolin wine in Vietnam. *Photo credit: Save Vietnam's Wildlife.*

purchased pangolin products, 6% have purchased sculptures/statues.

Consumers

Research conducted under USAID's Saving Species project is the only research thus far that shows insights on who pangolin buyers are. This research found that 64% of those who have purchased pangolin products in the last year are men with a higher income and who are on average 35.8 years old. The top three purchase occasions of pangolin products are: to promote wellness, treat an illness and to gift the product to someone else (USAID, 2018b).

Opportunities to reduce demand through behavior change

Types of behavior to change

Available consumer research reveals demand for pangolin products in China and Vietnam is driven by, amongst other motivations, perceptions and beliefs that scales can treat illness or promote wellness; and that meat can demonstrate status (whilst also providing nutrition). These two "clusters" of motivations, i.e., medicinal and emotional, are also apparent for other types of illegally traded wildlife - including rhino horn, tiger (*Panthera tigris*) products and elephant (Elephantidae) ivory (e.g., Burgess, 2016; Burgess et al., 2018).

Within this broad classification, the occurrence and frequency of product purchase and use (i.e., "ingestion"), becomes important in determining the relevant behavioral science theories, models and frameworks to apply for demand reduction. Additional considerations relate to the covert or overt nature of the practice, and the social proof required to validate them. Consideration also needs to be given as to whether the primary target for behavior change initiatives should be the person ingesting the product, or rather those procuring or prescribing it.

It is also important to note that the strategies and approaches suggested largely span the "social mobilization" and "behavioral change communications" realms of Social and Behavioral Change Communications (SBCC: per Clark et al., 2017). Social marketing (Kotler and Zaltman, 1971) is considered a cross cutting approach that can be applied to either realm.

Although there are a diverse range of reasons for scale consumption, the examples in this section will focus on medicinal and "emotional" use.

Medicinal motivations

In relation to the "medicinal" cluster, while exceptions are evident, a generalization is that products used to promote wellbeing or treat chronic conditions need replenishing over time and thus, their purchase and use has the potential to be more routine or habitual. This is distinct in behavioral terms, from occasional or "one-off" practices.

"Habitual" behaviors tend to be more "sticky", less malleable and more challenging to change. They may also be subject to more heuristics and biases (e.g., Caputo, 2013), illustrate "System 1" thinking (Kahneman, 2012) and an associated diminished sense of agency and process of cognition. In simple terms, the behavior has the potential to be more of a repeated pattern than active choice, as compared to more deliberative types of decision-making. This creates challenges for empathy-based demand reduction campaigns, which are also subject to compassion fatigue (e.g., Kinnick et al., 1996) and concepts such as the Social Cognitive Theory of Morality (Bandura, 1991). The latter considers the influences on what is a moral thought or action, and thus on an individual's self-regulation or control in relation to, e.g., purchase choice and practice.

An additional complicating factor is that the behaviors and actors for "purchase" and "use" are distinct and may require very different strategies for engagement and influence (Cialdini, 1984). By way of illustration, in Vietnam, the grandmother of new-born babies may both supply and recommend the new mother consume pangolin scales to promote lactation (Thomas-Walters, 2017). Elders typically are beyond reproach, and so the treatment is consumed. A corollary from China, is that the healthcare professional prescribing traditional medicine treatments may be above questioning by the patient receiving the prescription. Much the same as in "Western" medicine systems, it would be highly irregular to challenge the judgment, knowledge or decision of the professional recommending the treatment. This reinforces that careful consideration is required around the target of the demand reduction intervention and the manner through which change can be most effectively influenced.

"Emotional" motivations

In relation to the "emotional" cluster (whereby consumption choices largely fulfil a need for hedonistic pleasures or social gain), while exceptions are once again evident, a generalization is that eating meat to demonstrate status is, by contrast to habitual use, a more occasional or one-off practice. Inherently the experience of ingestion should be an infrequent, elusive or exclusive one, if it is to adequately exemplify the "rarity" attribute sought after.

Consumption of meat would thus, for this type of motivation, typically be conspicuous or overt, i.e., the behavior requires observation and may also be subject to social proof (Aronson et al., 2005) to ensure peer acknowledgment and status "validation." This suggests that whether the context for eating pangolin meat is a corporate, political or social one, the person ingesting the dish may not always be the same as the person who ordered it. The target for demand reduction campaigns therefore also, in this motivational cluster, requires careful consideration.

In general, with both medicinal and emotional motivations, more in-depth research is required, around the frequency and occasion of use. Clear insight should be acquired in relation to the underlying knowledge, values, attitudes and beliefs that pangolin product consumers have. This would be in addition to understanding their definitions of quality; how they identify the best product to acquire, whether they research it extensively first or rather go to trusted sources; their reaction to price; motivators and inhibitors to purchase, intention and use; any catalyst or "gateway" behaviors governing pangolin product consumption more broadly; the channels of influence generally and triggers for ingestion. All aspects are important in determining the best theoretical foundation to employ, through initiatives aiming to change behavior and reduce demand (see Fig. 22.5).

FIGURE 22.5 A pangolin carved into a fruit pip by master carvers in China, illustrating a behaviorally informed approach to persuade an alternative consumption choice by ornamental collectors in China. The artist Liu Baodong, is a Master Carver, and keen to show how sustainable materials can fulfil the need for an exclusive product while protecting pangolins. *Photo credit: Liu Baodong.*

Existing experience

At least five non-governmental organizations (NGOs) and one inter-governmental organization (IGO) are either designing, developing or starting to deliver, "demand reduction" initiatives intended to reach potential pangolin product consumers in China and Vietnam. The majority seem to follow a mass media distribution or "spray and pray" approach; their aim is maximum coverage rather than that targeted to specific channels of influence. Only one initiative is understood to have any substantial behavioral change theoretical foundation (Olmedo et al., 2017; Veríssimo and Wan, 2019), and none are at a sufficient level of implementation for success factors and lessons learned to be deduced i.e., there is no evaluation of impact yet on either buyer, user or intender groups. The strategic approaches summarized in the sections that follow therefore, are based largely on theories applied, and experience gained, in other fields, or taxa consumed for similar motivations in the same markets.

It is also important to note that while a range of approaches to reduce demand are discussed, this should not erode emphasis on the importance of effective legal deterrents that are appropriately enforced. Whether the target behavior is associated with product purchase or use, or in the medicinal or emotional cluster, effective societal controls, enforcement thereof and the removal of products from sale, will always be critical in complementing messaging to shape motivation, to ensure enduring behavior change. This is important in itself, but also because the perception of prosecution can be an important influence on individual intention and action (Schneider, 2008; USAID, 2018b).

Multiplicity of models

While behavioral science is replete with theories, frameworks, principles, concepts and models of change that could have application to this issue, a limited number are introduced here. This is both for clarity and the purpose of illustration, but also because those selected are considered amongst the most notable, for their potential application in relation to pangolin demand reduction initiatives. The narrative focuses on the mixed methods typically required around messaging, messengers and mechanisms for communication activities and distribution. It is difficult to disaggregate the models that can guide campaign structure and design, from the concepts underpinning the models. This is not discussed further here therefore, but additional resources include

Darnton (2008), Michie and Johnston (2012), and Change Wildlife Consumers (2019).

Specific behavior change theories relevant to reducing "medicinal" demand

In relation to more routine or habitual behaviors, such as those that might surround medicinal motivations, Lewin's "Unfreeze-Change-Refreeze" Model, later coined "Change Theory" (Lewin, 1947), is of particular relevance.

The central premise of this cornerstone in behavioral theory, is that by raising the level of cognition an undesirable behavior can be "frozen", and new more socially responsible/environmentally desirable behavior identified and recommended to replace it. The new behavior is then "refrozen." Subsequent research has suggested that the overarching nature of "keystone habits" (Duhigg, 2012) should also be considered. These may have an important influence on the context within which the target behavior is conducted. Disrupting any subconscious practices around it should also be a priority. An additional consideration is that interventions delivered during a "Key Life Change" (e.g., changing jobs, moving home, getting married, or –of relevance to pangolin scale users– having a baby) may be especially impactful, as automatic patterns of behavior are naturally disrupted through such fundamental changes in routine and day to day decision-making (Duhigg, 2012).

From this composite set of insights, a recommendation for demand reduction professionals aiming to target, for example, the use of pangolin scales to promote lactation, could thus be to engage those developing e.g., "Take Home" bags from hospitals or maternity clinics, or others providing advice through medical referral services to new mums, as "messengers" in initiatives aiming to reduce demand for the medicinal use of pangolin scales. Such professionals may prove especially powerful in unfreezing, changing, and refreezing desired behaviors.

Alternative products providing an equivalent function for the mother, would need to be identified alongside this to increase the probability of success. Research is required in order for suitable alternatives to be identified (see also Broad and Burgess, 2016). Stories from those who have successfully challenged conventional wisdom handed down through generations regarding how best to stimulate milk production, should also be sought at the "change" stage, to reinforce what is possible and increase the "agency" the target audience perceives regarding their ability to make a more environmentally desirable choice.

Of further note regarding communicating about alternative products, is that those typically issuing messaging around pangolin scale purchase and use, need to be those with the most credibility to comment on suitable medical treatments. Conservation NGOs or government departments with a natural resource or wildlife protection mandate, are not healthcare professionals, and thus may offer either no or low credibility and an equivalent ability to influence.

Engaging messengers who can identify and prescribe alternative treatments to promote wellness/reduce illness, is therefore key. Ministries of Health, Traditional Medicine Colleges, training institutions, research bodies and universities, alongside proponents of a sustainable industry with a respected reputation, such as the World Federation of Chinese Medicine Societies (WFCMS), may be crucial and should be engaged.

Regular reinforcement, recognition and reward of the new "desirable" behavior are also important. Messaging that promotes a positive rather than aims to negate a negative, would also be implicit, and in line with evidence about the approaches to messaging that achieve greatest affect (e.g., Keller and Lehmann, 2008; Schaffner et al., 2015).

Specific behavior change theories relevant to reducing "emotional" demand

In relation to overt or conspicuous consumption practices, such as those involved in the status driven behavior of consuming pangolin meat, validation is required by a respected peer group, i.e., "social proof." Social Network Theory (Lin, 1999), is thus of particular note for this type of demand. The central premise of this behavioral theory, is the insight it provides into pathways of influence in social networks. Attributes such as "Nodes" and "Ties" are the subject of much research in "social network analysis," and evidence is therefore rife around dynamics such as node types offering the most connections into, and social proof for, selected target audiences, alongside the ties that may be strongest in ensuring the uptake of new behaviors. Dunbar's Number (Dunbar, 1992) sheds light on the most influential circles surrounding the target individual – 15 close family members and friends are generally deemed to be the most impactful in changing an individual's habits and decisions.

Several influential personality types are identified as relevant to behavior change in more populist texts such as Roger's Diffusion of Innovations model (Rogers, 1983) and Gladwell's Maven's Connectors and Salespeople (Gladwell, 2000). In Roger's model, "Innovators" (estimated to be 2.5% of any populace or group) are the first to adopt a new practice and thus generally most open to change. This then shifts to increasing momentum/shaping more substantial change via "connectors" engaging "early adopters"; and subsequently by "salespeople" convincing the "early majority." Although this set of insights focuses on message distribution, due to the validation required through the overt set of behaviors in conspicuous consumption, it represents an essential mechanism to consider, in ensuring enduring and meaningful change.

Relevant models to both medicinal and emotional motivations

The "Stages of Change" model (Prochaska and Diclemente, 1983) originated from the application of behavior change in relation to public health practices, and is one of the few that recognizes the "journey" people proceed through when moving from a socially or environmentally "undesirable" to "desirable" behavior. Relapse, retention and refinement are important parts of the process; in simple terms, redirecting those consuming pangolin scales to treat illness/promote wellness, or consuming meat to display status, is anticipated to happen through a series of stages, and messaging and messengers need to be engaged differently at each stage.

The "Socio-Ecological Model" (SEM: Bronfenbrenner, 1979), originally the *Ecological Framework for Human Development*, is also relevant and overarching. Similar to Vlek's "Needs-Opportunity-Abilites" model (Vlek et al., 1997), SEM essentially recognizes the various influences on deliberations around the pros and cons of adopting a new choice, amongst broader decision-making processes. The layers distinguish inherent (i.e., internal) factors from those more inter-personal, institutional, or at the community or society level. Influences on behavioral intention and action, and the barriers, facilitators, inhibiting or enabling factors involved, are thus implied as distinct for each layer. Each should therefore be considered in demand reduction initiative design, planning and execution.

Conclusion

This chapter has sought to summarize the imperatives, opportunities and potential mechanisms for using behavioral science to reduce demand for illegally traded pangolin products. A central premise for the chapter is that a multifaceted approach is critical. That is, one that employs social science interventions such as behavior change, alongside others more legislative in nature.

While there have been a number of studies aiming to increase insight into what drives consumer demand for pangolins in Asia, so as to guide, shape and inform efforts to dissuade it, few such studies have been published in full in the public domain, and further, few have focused on psycho-graphic and socio-demographic elements driving purchase and use.

Thus, while it is clear pangolin scales are consumed in many markets for a perception of efficacy in treating illness and promoting wellness; and pangolin meat is consumed for an association with demonstrating status, substantial knowledge gaps unfortunately remain around various aspects that might help to shape and inform truly transformative efforts to change both these types, and other types, of consumer choice.

Examples of research priorities therefore, aside from those identified for the geographies aforementioned, include various elements necessary to conduct behavioral journey mapping and a stages of change analysis – such as purchase pathway triggers and drivers, inhibitors and facilitators, and catalyst and gateway behaviors (e.g., Austin et al., 2011).

Strategic prioritization of demand reduction efforts could also occur if comparative analyses were conducted around which use types and motivations were the most common amongst consumers, and perhaps the most "destructive" in terms of product. Scales are often identified in seizures but they feature in small amounts typically in traditional medicine treatments – so are pangolins poached mostly for meat with scales a profitable by-product? Or vice-versa? (see Chapter 16).

Such data and insight would arise from research designed to deliver a more fine-grained level of resolution than is available currently. Once such knowledge is secured, the development of segmentation models identifying priority behaviors to target and the consumer clusters most willing and able to act, might occur. While these approaches to behavior change are quite commonplace in other fields of application (e.g., Defra, 2008), they have yet to be brought to bear for the benefit of the heavily trafficked yet enigmatic pangolins. This chapter has aimed to help inform action to address such issues.

Acknowledgment

The lead author thanks IUCN SSC Pangolin Specialist Group member and TRAFFIC Senior Director for Asia, James Compton, for his review and inputs into early drafts of this chapter.

References

Andreasen, A.R., 2002. Marketing social marketing in the social change marketplace. J. Public Policy Market. 21 (3), 3–13.

Aronson, E., Wilson, T.D., Akert, A.M., 2005. Social Psychology, fifth ed. Prentice Hall, Upper Saddle River, New Jersey.

Arthur, L., Wilson, W., 1979. Assessing the demand for wildlife resources: a first step. Wildlife Soc. Bull. (1973-2006) 7 (1), 30–34.

Austin, A., Cox, J., Barnett, J., Thomas, C., 2011. Exploring Catalyst Behaviours: Full Report. A Report to the Department for Environment, Food and Rural Affairs. Brook Lyndhurst for Defra, London.

Bandura, A., 1991. Social cognitive theory of moral thought and action. In: Kurtines, W.M., Gewirtz, J.L. (Eds.), Handbook of Moral Behavior and Development, vol. 1. Erlbaum, Hillsdale, New Jersey, pp. 45–103.

Broad, S., Burgess, G., 2016. Synthetic biology, product substitution and the battle against illegal wildlife trade. TRAFFIC Bull. 28 (1), 23–28.

Bronfenbrenner, U., 1979. The Ecology of Human Development. Experiments by Nature and Design. Harvard University Press, Cambridge.

Burgess, G., 2016. Powers of persuasion: conservation communications, behavioural change and reducing demand for illegal wildlife trade. TRAFFIC Bull. 28 (2), 65–73.

Burgess, G., Zain, S., Milner-Gulland, E.J., Eisingerich, A., Sharif, V., Ibbett, H., et al., 2018. Reducing Demand for Illegal Wildlife Products: Research Analysis on

Strategies to Change Illegal Wildlife Product Consumer Behaviour. Available from: <https://www.traffic.org/site/assets/files/11081/demand_reduction_research_report.pdf>. [April 10, 2019].

Caputo, A., 2013. A literature review of cognitive biases in negotiation processes. Int. J. Conflict Manage. 24 (4), 374–398.

Challender, D.W.S., MacMillan, D.C., 2014. Poaching is more than an enforcement problem. Conserv. Lett. 7 (5), 484–494.

Challender, D.W.S., Harrop, S.R., MacMillan, D.C., 2015. Understanding markets to conserve trade-threatened species in CITES. Biol. Conserv. 187, 249–259.

China Biodiversity Conservation and Green Development Foundation, 2016. An Overview of Pangolin Data: When Will the Over-Exploitation of the Pangolin End? Available from: <http://www.cbcgdf.org/English/NewsShow/5011/6145.html>. [March 19, 2019].

Chinese Pharmacopeia Commission, 2015. Pharmacopoeia of the People's Republic of China 2015. Medical Science and Technology Press, Beijing, China.

Cho, H., Salmon, C.T., 2007. Unintended effects of health communication campaigns. J. Commun. 57 (2), 293–317.

Cialdini, R.B., 1984. Influence: The Psychology of Persuasion. Harper Business.

Clark, C.J., Spencer, R.A., Shrestha, B., Ferguson, G., Oakes, M., Gupta, J., 2017. Evaluating a multicomponent social behaviour change communication strategy to reduce intimate partner violence among married couples: study protocol for a cluster randomized trial in Nepal. BMC Public Health 17, 75.

Cooney, R., Roe, D., Dublin, H., Phelps, J., Wilkie, D., Keane, A., et al., 2017. From poachers to protectors: engaging local communities in solutions to illegal wildlife trade. Conserv. Lett. 10 (3), 367–374.

Darnton, A., 2008. GSR Behaviour Change Knowledge Review Reference Report: An Overview of Behaviour Change Models and Their Uses. Available from: <https://assets.publishing.service.gov.uk/government/uploads/system/uploads/attachment_data/file/498065/Behaviour_change_reference_report_tcm6-9697.pdf>. [March 19, 2019].

Defra, 2008. A Framework for Pro-Environment Behaviours. Available from: <https://www.gov.uk/government/publications/a-framework-for-pro-environmental-behaviours>. [April 10, 2019].

Do, H.T.T., Bui, M.H., Hoang, H.D., Do, H.T.H., 2011. Consumption of Wild Animal Products in Ho Chi Minh City, Vietnam – Results of Resident and Student Survey. Wildlife At Risk, Ho Chi Minh City, Vietnam.

Duhigg, C., 2012. The Power of Habit: Why We Do What We Do and How to Change. Random House Books, London.

Dunbar, R., 1992. Neocortex size as a constraint on group size in primates. J. Hum. Evol. 22 (6), 469–493.

Duthie, E., Veríssimo, D., Keane, A., Knight, A.T., 2017. The effectiveness of celebrities in conservation marketing. PLoS One 12 (7), e0180027.

Education for Nature Vietnam (ENV), 2016. Pangolin Consumer Crime in Vietnam: The Results of ENV Surveys and Enforcement Campaigns, 2011-2015. Available from: <http://envietnam.org/images/News_Resources/Publication/jan-28-2016-pangolin-TCM-survey-results.pdf>. [March 19, 2019].

Gladwell, M., 2000. The Tipping Point – How Little Things Make a Big Difference. Little, Brown and Company, Boston.

Gore, M., 2011. The science of conservation crime. Conserv. Biol. 25 (4), 659–661.

Greenfield, S., Veríssimo, D., 2019. To what extent is social marketing used in demand reduction campaigns for illegal wildlife products? Insights from elephant ivory and rhino horn. Soc. Market. Quart. 25 (1), 40–54.

Heinrich, S., Wittmann, T.A., Prowse, T.A., Ross, J.V., Delean, S., Shepherd, C.R., et al., 2016. Where did all the pangolins go? International CITES trade in pangolin species. Glob. Ecol. Conserv. 8, 241–253.

Heinrich, S., Wittman, T.A., Ross, J.V., Shepherd, C.R., Challender, D.W.S., Cassey, P., 2017. The Global Trafficking of Pangolins: A Comprehensive Summary of Seizures and Trafficking Routes From 2010–2015. TRAFFIC, Southeast Asia Regional Office, Petaling Jaya, Selangor, Malaysia.

Horizon China, 2016. Report on the Survey on the Attitude of the Chinese Public Towards the Consumption of Pangolins and Their Products. Report Presented to AITA Foundation and Humane Society International.

Humane Society International, Hong Kong University, 2015. Survey on Pangolin Consumption Trends in Hong Kong. Available from: <https://www.scribd.com/document/274722859/Hong-Kong-Pangolin-Survey-2015-HKU-HSI>. [March 19, 2019].

Ingram, D.J., Coad, L., Abernethy, K.A., Maisels, F., Stokes, E.E., Bobo, K.S., et al., 2018. Assessing Africa-wide pangolin exploitation by scaling local data. Conserv. Lett. 11 (2), e12389.

Kahneman, D., 2012. Thinking Fast and Slow. Penguin, United Kingdom.

Keller, P.A., Lehmann, D.R., 2008. Designing effective health communications: a meta-analysis. J. Public Policy Market. 27 (2), 117–130.

Kennaugh, A., 2016. Rhino rage: What is driving illegal consumer demand for rhino horn? NRDC. Available from: <https://www.savetherhino.org/assets/0002/8719/Rhino_rage_report_by_Alex_Kennaugh_Dec_2016.pdf>. [March 19, 2019].

Kinnick, K.N., Krugman, D.M., Cameron, G.T., 1996. Compassion fatigue: communication and burnout toward social problems. J. Mass Commun. Quart. 73 (3), 687–707.

Kotler, P., Zaltman, G., 1971. Social marketing: an approach to planned social change. J. Market. 35, 3–12.

Lane, B., Potter, S., 2007. The adoption of cleaner vehicles in the UK: exploring the consumer attitude-action gap. J. Cleaner Prod. 15 (11–12), 1085–1092.

Lewin, K., 1947. Frontiers in group dynamics: concept, method and reality in social science; social equilibria and social change. Hum. Relat. 1 (1), 5–41.

Lin, N., 1999. Building a network theory of social capital. Connections 22, 28–51.

Michie, S., Johnston, M., 2012. Theories and techniques of behaviour change: developing a cumulative science of behaviour change. Health Psychol. Rev. 6 (1), 1–6.

Nash, H.C., Wong, M.H., Turvey, S.T., 2016. Using local ecological knowledge to determine status and threats of the Critically Endangered Chinese pangolin (*Manis pentadactyla*) in Hainan, China. Biol. Conserv. 196, 189–195.

Nijman, V., 2010. An overview of international wildlife trade from Southeast Asia. Biodivers. Conserv. 19 (4), 1101–1114.

Nijman, V., Zhang, M.X., Shepherd, C.R., 2016. Pangolin trade in the Mong La wildlife market and the role of Myanmar in the smuggling of pangolins into China. Glob. Ecol. Conserv. 5, 118–126.

Olmedo, A., Sharif, V., Milner-Gulland, E.J., 2017. Evaluating the design of behavior change interventions: a case study of rhino horn in Vietnam. Conserv. Lett. 11 (1), e12365.

Pfeiffer, J., 2004. Condom social marketing, Pentecostalism, and structural adjustment in Mozambique: a clash of AIDS prevention messages. Med. Anthropol. Quart. 18 (1), 77–103.

Prochaska, J.O., Diclemente, C.C., 1983. Toward a comprehensive model of change. In: Miller, W.R., Heather, N. (Eds.), Treating Addictive Behaviors. Applied Clinical Psychology, vol. 13. Springer, Boston.

Rogers, E.M., 1983. Diffusion of Innovations. Free Press, New York.

Schaffner, D., Demarmels, S., Juettner, U., 2015. Promoting biodiversity: do consumers prefer feelings, facts, advice or appeals? J. Consum. Market. 32 (4), 266–277.

Schneider, J.L., 2008. Reducing the illicit trade in endangered wildlife: the market reduction approach. J. Contemp. Crim. Just. 24 (3), 274–295.

Shairp, R., Veríssimo, D., Fraser, I., Challender, D.W.S., MacMillan, D.C., 2016. Understanding urban demand for wild meat in Vietnam: implications for conservation actions. PLoS One 11 (1), e0134787.

Thomas-Walters, L.A., 2017. Mapping Motivations. Combatting Consumption of Illegal Wildlife Trade in Vietnam. Available from: <https://www.traffic.org/site/assets/files/4313/understanding-motivations-summary-final-web.pdf>. [March 19, 2019].

TRAFFIC, 2016. Changing Behaviour to Reduce Demand for Illegal Wildlife Products: Workshop Proceedings. Available from: <http://www.traffic.org/general-reports/traffic_pub_gen108.pdf>. [March 19, 2019].

Soewu, D.A., Sodeinde, O.A., 2015. Utilization of pangolins in Africa: fuelling factors, diversity of uses and sustainability. Int. J. Biodivers. Conserv. 7 (1), 1–10.

USAID (2018a). Research Study on Consumer Demand for Elephant, Pangolin, Rhino and Tiger parts and products in China. USAID Wildlife Asia. Available at: <http://www.usaidwildlifeasia.org/resources/reports/usaid_china_wildlife-demand-reduction_english_presentation_june12_2018_final.pdf/view>. [March 19, 2019].

USAID (2018b). Research Study on Consumer Demand for Elephant, Rhino and Pangolin Parts and Products in Vietnam. Available at: <https://www.usaidwildlifeasia.org/resources/reports/ussv-quant-report-saving-elephants-pangolins-and-rhinos-20181105.pdf>. [March 19, 2019].

Venkataraman, B., 2007. A Matter of Attitude: The Consumption of Wild Animal Products in Ha Noi, Vietnam. TRAFFIC Southeast Asia, Greater Mekong Programme, Ha Noi, Vietnam.

Veríssimo, D., Challender, D.W.S., Nijman, V., 2012. Wildlife trade in Asia: start with the consumer. Asian J. Conserv. Biol. 1 (2), 49–50.

Veríssimo, D., Wan, A.K.Y., 2019. Characterizing efforts to reduce consumer demand for wildlife products. Conserv. Biol. 33 (3), 623–633.

Vlek, C., Jager, W., Steg, L., 1997. Modellen en strategieën voor gedragsverandering ter beheersing van collectieverisico's. Nederlands Tijdschrift voor de Psychologie 52, 174–191.

Vo, V.C., 1998. Dictionary of Vietnamese Medicinal Fauna and Minerals, Medicine. Ha Noi Health Publishing House, Hanoi. [In Vietnamese].

Wasser, R.M., Bei Jiao, P., 2010. Understanding the Motivations: The First Step Toward Influencing China's Unsustainable Wildlife Consumption. TRAFFIC East Asia report.

WildAid, 2016. Pangolins On the Brink. Available from: <http://wildaid.org/wp-content/uploads/2017/09/WildAid-Pangolins-on-the-Brink.pdf>. [March 19, 2019].

Xu, L., 2009. The pangolin trade in China. In: Pantel, S., Chin, S.-Y. (Eds.), Proceedings of the Workshop on Trade and Conservation of Pangolins Native to South and Southeast Asia, 30 June – 2 July 2008, Singapore Zoo, Singapore. TRAFFIC Southeast Asia, Petaling Jaya, Selangor, Malaysia, pp. 189–193.

Xu, L., Guan, J., Lau, W., Xiao, Y., 2016. An Overview of Pangolin Trade in China. TRAFFIC Briefing Report. TRAFFIC, Cambridge, UK, pp. 1–10.

Yu, R., Hong, H., 2016. Cancer Management With Chinese Medicine - Prevention and Complementary Treatments. World Scientific Publishing, New Jersey.

Zhang, L., Yin, F., 2014. Wildlife consumption and conservation awareness in China: a long way to go. Biodivers. Conserv. 23 (9), 2371–2381.

PART 3

Site-Based Protection and Local Community Engagement

CHAPTER

23

Engaging local communities in responses to illegal trade in pangolins: who, why and how?

Rosie Cooney[1,2] and Daniel W.S. Challender[3,4]

[1]IUCN CEESP/SSC Sustainable Use and Livelihoods Specialist Group, Gland, Switzerland [2]Fenner School of Environment and Society, Australian National University, Canberra, ACT, Australia [3]Department of Zoology and Oxford Martin School, University of Oxford, Oxford, United Kingdom [4]IUCN SSC Pangolin Specialist Group, ℅ Zoological Society of London, Regent's Park, London, United Kingdom

OUTLINE

Introduction

For decades it has been recognized that conserving biodiversity is best done with the support and engagement of the indigenous peoples and local communities living with wildlife (Brown, 2002; CBD, 2004; Ghimire and Pimbert, 1997; IUCN, 1994, 1996; WCED, 1987). Where conservation doesn't gain local support, it frequently fails to leverage the energy, input and knowledge of locally influential actors, and where it arouses hostility or resentment it faces an uphill battle against opposing forces to achieve its aims (Cooney et al., 2018;

DOI: https://doi.org/10.1016/B978-0-12-815507-3.00023-X

Duffy et al., 2016; Twinamatsiko et al., 2014). This recognition has been the foundation of decades of efforts on community-based conservation and natural resource management. While outcomes are mixed, there have been notable success stories including various geographies and species (e.g., Anderson and Mehta, 2013; Brooks et al., 2012; Frisina and Tareen, 2009; Hulme and Murphree, 2001; Roe et al., 2009). Specific to pangolins, there are myriad local community-focused projects in Africa and Asia, including in but not limited to, Bangladesh (see Trageser et al., 2017), India, Malaysia, Nepal (Chapter 25), Singapore (Chapter 26) and Cameroon (Chapter 27).

This chapter draws on an extensive process of consultation and case study exploration undertaken between 2015 and 2017 by IUCN's Sustainable Use and Livelihoods Specialist Group (SULi) and several regional offices, the International Institute for Environment and Development (IIED), TRAFFIC and other partners, in addition to published literature. This "Beyond Enforcement" initiative was prompted by the observation by some (e.g., Cooney et al., 2016a,b; 2017) that the flurry of global policy attention on illegal wildlife trade (IWT) that began around 2013 framed the contemporary "poaching crisis" predominantly as a law enforcement problem, at least in terms of action required in "source" countries, especially in sub-Saharan Africa (Challender and MacMillan, 2014; Duffy, 2014; Lunstrum, 2014), and as distinct from consumer markets.

However, recognition of the role and importance of indigenous peoples and local communities (IPLCs) in combating IWT has been strengthened in official documentation. For example, the London Declaration (2014), signed by Heads of State and senior government ministers at the first inter-governmental conference on IWT, includes a commitment to "Work with, and include local communities in, establishing monitoring and law enforcement networks in areas surrounding wildlife." The Kasane Statement (2015), highlighted the need to "Promote the retention of benefits from wildlife resources by local people where they have traditional and/or legal rights over these resources." This was echoed at the third and fourth inter-governmental conferences on IWT in Hanoi (2016) and London (2018) respectively (see also Chapter 24). Greater voice for local communities in international and national policy-making is also the subject of discussion in the Convention on International Trade in Endangered Species of Wild Fauna and Flora (CITES; see CITES, 2019).

This chapter is about the people who live with pangolins – the indigenous peoples and local communities who live in and around the protected areas, habitats and landscapes where pangolins live (hereafter "communities"). It argues that top-down enforcement is at best only one part of solving the problem of pangolin poaching and trafficking, and at worst can exacerbate the problem. It reinforces an approach to pangolin conservation that views communities as (in most contexts) essential partners, and sets out insights drawn from the "Beyond Enforcement" initiative and beyond, on effective approaches to supporting and engaging communities in conservation. In particular, it draws on a framework developed through this work for understanding community-level incentives for engaging in IWT vs. supporting conservation (Cooney et al., 2016a,b; 2017), and a "Theory of Change" that articulates a set of pathways for action at community level to reduce poaching (Biggs et al., 2016). Although the work that primarily informs this chapter has not focused on pangolins *per se*, it has involved examination of a very broad array of poaching contexts across regions and taxa, including pangolins, and can offer relevant insights for conservation of the species. The chapter draws out lessons learnt and interventions likely to be of most relevance to pangolin conservation, based on the specific characteristics of this most charming of taxa.

Why communities?

Why is understanding and engaging with communities essential in reducing poaching of pangolins for IWT? The narrative of inhumane, greedy and ruthless poachers exploiting pangolins for profit and justifying increasingly harsh and punitive responses is a compelling one. At the most basic level, laws that specify that a wild species should enjoy protection from exploitation are being broken, with demonstrably negative impacts on pangolin populations (Chapters 14–16). It is tempting to therefore reach for law enforcement as the obvious remedy. However, as with many other societal arenas of regulation, relying on law enforcement as the sole or major response is simplistic: it risks failing to engage with and counter underlying drivers of rule-breaking behavior, alienating key partners, and has other drawbacks including limited political will and funding (see Chapter 17; Challender and MacMillan, 2014, 2016) and challenges associated with providing a law enforcement presence in remote areas (e.g., border regions, monoculture plantations; Challender and Waterman, 2017). Further, the potential for excessive or misdirected use of force to counter poaching can and has led to abuses, human rights violations and serious social hardship (IUCN-SULI, IIED, CEED, TRAFFIC, 2015).

There are doubtless some circumstances where the engagement of local communities is not important or unrealistic in combating pangolin poaching – particularly where poaching takes place in remote areas far from settled communities and is carried out by "outsiders" without any significant local participation in hunting or trade, and without local support. In most cases, however, communities and community-based approaches are important for compelling reasons.

First, pangolins inhabit lands that communities live on, use and/or manage. Approximately 25% of lands outside Antarctica are owned, managed or influenced by Indigenous people (Garnett et al., 2018; see also Molnar et al., 2004), and around two-thirds of these lands have not been intensively developed – a much higher percentage than non-Indigenous lands. Indeed, 40% of the lands formally listed by state governments as being managed for conservation are indigenous lands. As a practical reality, pangolins in these vast landscapes can only be most effectively conserved with the leadership, support and buy-in of communities.

Second, community members are often involved in hunting or poaching pangolins themselves – they may actively target pangolins, or opportunistically take them in the course of other activities (e.g., collecting, forestry or agriculture), and/or be involved in trade and trafficking. They may also contribute to poaching less directly by providing poachers with information, food, or accommodation. The motivations for hunting and poaching can be highly diverse within and between communities and contexts, but can include meeting subsistence needs, gaining income or status, pursuing traditional practices of cultural significance, as well as being driven by contemporary or historical injustices linked with conservation (Duffy, 2010; Duffy et al., 2016; Harrison et al., 2015). Some of these drivers can be targeted by appropriate interventions to shift poaching patterns.

Third, and consequently, addressing poaching will typically affect community members and the way they use species and landscapes for food, shelter, income and other livelihood needs, and whose cultures, traditions and wellbeing may be deeply embedded in these practices. Addressing hunting and poaching can therefore have significant implications for the rights and livelihoods of IPLCs. In the current climate of increasingly militarized conservation approaches in some places (the "war on poaching"; Bűscher and Ramutsindela, 2016; Lunstrum, 2014) the potential for severe abuses and human rights violations is real. Poorly directed or heavy-handed efforts can involve harassment, impose unjustified restrictions on people's use of wildlife resources, infringe rights, and undermine the benefits that local people can gain from conservation and wildlife protection (Büscher and Ramutsindela, 2016; Corry, 2015; Duffy, 2014; Lunstrum, 2014). Those that have received media attention include the anti-poaching initiative Operation Tokomeza in Tanzania in 2013 (Makoye, 2014), treatment of indigenous Baka people in Cameroon (Survival International, 2014), and shootings of local villagers as suspected poachers in Kaziranga National Park in India (Rowlatt, 2017). Human rights abuses are unacceptable in themselves. Government and non-governmental organization (NGO)-led conservation has a long and often brutal history of pursuing conservation at the expense of communities, from evictions of Native Americans from Yellowstone (Poirier and Ostergren, 2002) to ongoing resettlements of communities from Indian tiger reserves (Dash and Behera, 2018; Torri, 2011). Fairness demands conservation should not be pursued at the expense of some of the world's most marginalized and poor people. Harsh approaches can also be counter-productive in conservation terms. When people feel unjustly treated in the name of conservation and wildlife protection, anger and resentment can itself be a significant driver of poaching (Hübschle, 2017; Massé et al., 2017; Twinamatsiko et al., 2014). Approaches that respect community rights, priorities and livelihoods and support and engage communities as partners in conservation on the other hand, offer a more just, inclusive and robust way to secure wildlife populations.

Fourth, communities living with wildlife and dependent on natural resources typically hold extensive indigenous and local knowledge of direct value and importance for conservation – this includes knowledge of pangolin presence and distribution, habitat use, interactions with other species, and perceptions of abundance (e.g., Nash et al., 2016; Newton et al., 2008). Working with communities on a basis of reciprocity and trust can enable this knowledge to be harnessed and mobilized together with mainstream scientific knowledge for more effective conservation interventions.

Finally, wherever communities are living near wildlife, their support for and cooperation with enforcement authorities is likely to have a major impact on the effectiveness of law enforcement efforts. Effective enforcement relies critically on good intelligence, which primarily comes from local people. This is a well-known conclusion of research in policing generally – as relevant in urban contexts as in rural areas (Wilkie et al., 2016). Efforts to prevent crime and enforce laws are in general most effective when there is cooperation between citizens and authorities (Hawdon and Ryan, 2011). Local people are best placed to provide critical intelligence on who is poaching and where. The effectiveness of enforcement is seriously compromised without good relations with communities and their willingness to work with and provide intelligence to anti-poaching authorities (see Chapter 18).

Recognizing that community support for conservation and conservation authorities can be essential leads to the question of what conditions are likely to create and enable that support. Here, both the Beyond Enforcement work and decades of literature on community wildlife management provides some clear and consistent lessons.

How can conservation interventions support and engage communities?

As highlighted above, the evolution of intergovernmental statements includes increased emphasis on the roles and rights of local communities (Cooney et al., 2018). However, there is little evidence that such calls and commitments are being translated into practice. One reason for this may be that governments, donors and project designers lack a detailed understanding of how to deliver community-based interventions on the ground (IUCN-SULI, IIED, CEED, TRAFFIC, 2015). There is no blueprint approach and no detailed, clear understanding of how to engage a wide range of relevant communities, which are diverse, heterogeneous and often complex. Socio-economic, political, legal, environmental, and historical factors influence the nature of interactions between communities and wildlife and wildlife authorities, and their perceptions of and attitudes towards IWT (Biggs et al., 2016; Cooney et al., 2016a,b). All these factors will influence the types of community engagement interventions that are likely to be effective.

Here, insights on community engagement in addressing poaching of pangolins and related IWT are framed by drawing attention to the incentives facing individual community members. Cooney et al. (2016a,b; 2017) highlight that both supporting conservation and supporting IWT will typically involve both costs and benefits (tangible and intangible), and that it is reasonable to posit that in general conservation will need to deliver net positive benefits to individuals to enjoy wide support (see Fig. 23.1). Building on this framework, Biggs et al. (2016) developed a theory of change for how community level approaches on the ground can shift incentives facing individual community members and shape their decisions on whether to hunt and poach pangolins or cooperate with poachers versus protecting pangolins and/or cooperating with conservation and enforcement authorities (Fig. 23.2). These pathways are introduced and discussed below.

Strengthening the disincentives for illegal behavior

This is the most widely emphasized response to poaching and IWT, involving increasing the costs to people of engaging in it (Pathway A in Fig. 23.2) through tightening restrictions on harvest and trade; strengthening legal frameworks; increasing the likelihood of detection, arrest and conviction; and/or

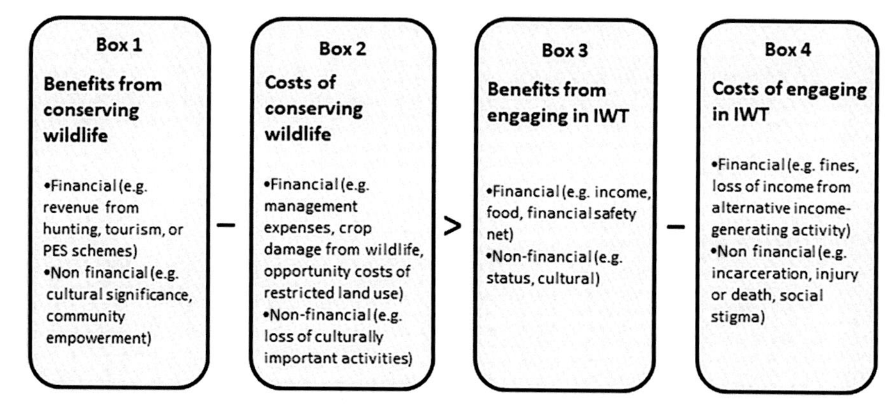

FIGURE 23.1 Conceptual framework for exploring the conditions likely to be required for local wildlife conservation in the context of IWT. Wildlife is more likely to be conserved where net benefits (financial and non-financial) to individuals in local communities are greater than net benefits of engaging in IWT. PES – Payment for Ecosystem Services. *Reproduced from Cooney, R., Roe, D., Dublin, H., Phelps, J., Wilkie, D., Keane, A., et al., 2016. From poachers to protectors: engaging local communities in solutions to illegal wildlife trade. Conserv. Lett. 10 (3), 367–374.*

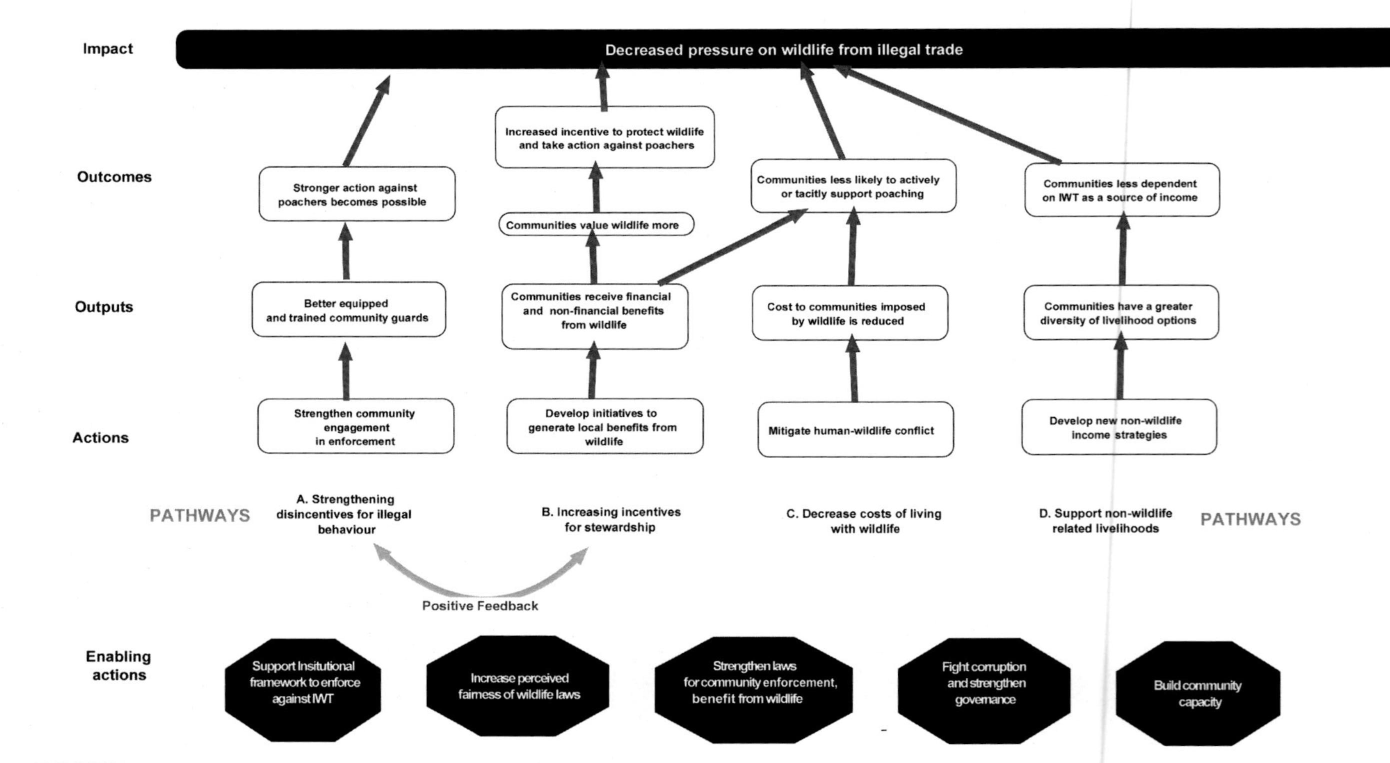

FIGURE 23.2 A simplified theory of change for community-based actions against illegal wildlife trade. There is positive feedback between pathways A and B because communities with increased incentives for stewardship will have more resources to combat poaching and will be more likely to do so. *Reproduced from Biggs, D., Cooney, R., Roe, D., Dublin, H.T., Allan J.R., Challender, D.W.S., et al., 2016. Developing a theory of change for a community-based response to illegal wildlife trade. Conserv. Biol. 31 (1), 5–12.*

increasing sanctions and penalties (Duffy, 2014; St. John et al., 2015). However, these disincentives can also be strengthened by empowering and partnering with communities as motivated and active agents in law enforcement (Lotter and Clark, 2014; Naidoo et al., 2016; Roe et al., 2015). In some cases, motivated communities are protecting their lands and wild species themselves from outside incursions, with little or no assistance from centralized authorities (Blomberg, 2018; Eaton, 2016). Elsewhere, community members may work cooperatively with enforcement authorities as scouts, informants and guides in joint patrols or through information sharing (Lotter and Clark, 2014; Wilkie et al., 2016). Community members can act as "eyes and ears" of enforcement authorities, and mechanisms can be established to enable people to easily, anonymously and safely report information, increasingly through mobile technologies. Further, social and informal sanctions ("peer pressure") can exert a powerful influence against behavior considered unacceptable.

However, such motivation and engagement depends on the attitudes of community members toward wildlife and conservation authorities. This will be strongest where people feel a strong sense of ownership or stewardship over wildlife – where they are protecting "their" wildlife (Wilkie et al., 2016). Conversely, it will be damaged or undermined by conservation approaches that marginalize, disrespect or alienate communities (IUCN-SULI, IIED, CEED, TRAFFIC, 2015), or where corruption among enforcement authorities is evident.

In the context of well-armed and organized poaching by "outsiders", there may be considerable risks in community members taking on an anti-poaching role, either alone or in concert with official authorities (IUCN-SULI, IIED, CEED, TRAFFIC, 2015). Where community members are employed or mobilized by external organizations (governments or NGOs), the risks they are exposed to must be clearly understood, properly rewarded and carefully managed (Wilkie et al., 2016). Further, communities need effective and responsive backup from formal armed authorities with the power of arrest in order to protect their own security and their own natural assets.

Trust in police and the legal system, recognition of wildlife laws as legitimate and fair, and guaranteed personal security are critical in motivating people to play an active role in enhancing enforcement and increasing disincentives for poaching and complicity in poaching. These themes are returned to below.

Increasing incentives for conservation

This pathway is about enhancing the benefits that communities gain from conservation, in order to incentivize their active stewardship and protection. This lesson has a long history in community conservation and natural resource management, but needs stating clearly in the context of IWT, where it may be lost in an over-emphasis on coercive enforcement. Field conservation interventions often include a community engagement component that is essentially a one-way process of "education", without providing any meaningful reasons for local communities to invest time and resources in conservation (e.g., Nilsson et al., 2016). Communities need realistic and locally relevant incentives to engage constructively with law enforcement authorities and support conservation.

This is of particular relevance in the context of high value species including pangolins, where involvement in poaching and IWT can be very lucrative. Since at least the 1980s the high price of pangolins, especially in Asia, has been an incentive for local communities to poach and illegally trade in the animals (see Chapter 16). Finding a pangolin has been referred to as a "lottery win" in various parts of Southeast Asia due the high price that can be fetched by selling just one animal, which can be sold for the equivalent of several months wages in many rural areas (D. Challender, unpubl. data). Returns from illicit activity must be countered by robust incentives for protection stemming from high social or economic value, particularly where people are struggling to meet urgent subsistence needs. The literature is split as to whether monetary or non-monetary; financial or non-financial; tangible or intangible benefits, are key to providing conservation incentives that are meaningful to communities. In other cases, research suggests that the benefits needed to motivate pro-conservation behavior must be tangible and monetary (Mazambani and Dembetembe, 2010; Musavengane and Simatele, 2016). However, others have noted that intangible social incentives can be more important in community-based conservation, including equity and empowerment (Berkes, 2004; Horwich and Lyon, 2007). Cultural factors such as "pride in place" can be powerful drivers of conservation (Govan et al., 2006). For some communities, conserving wild species may be intrinsic to their culture, identity and spiritual values, and maintaining culture and identity will provide a powerful motivation. Broadly, however, it is likely that people who are struggling to meet basic necessities of life are more likely to require tangible economic (if not financial) benefits to offset the returns from poaching, particularly where the likelihood of facing penalties is low.

There are a wide range of ways that benefits from conservation for communities can be enhanced, such as strengthening community ownership rights and/or capacity to use, manage and benefit from wildlife (for subsistence, cultural and/or commercial purposes), participating in Payments for Ecosystem Services (PES) schemes, gaining a share of protected area gate fees or similar revenue flows, securing jobs as community guards or in nature-based tourism enterprises, or strengthening cooperation and communication with conservation or wildlife management agencies (IUCN-SULI, IIED, CEED, TRAFFIC, 2015; Roe et al., 2015).

Such benefits can be powerful in motivating communities to be active and committed conservation actors against poaching and IWT, as evidenced in conservancies in Namibia (Naidoo et al., 2016) and Kenya (Blackburn et al., 2016). Effectiveness of different interventions will vary according to local context: for example, benefits from tourism are only feasible where certain conditions are met, such as political stability, tourism infrastructure and scenic landscapes (Naidoo et al., 2016).

Empowerment of communities through rights to access, use, and decision-making about natural resources are particularly important in predicting where community-based conservation approaches are successful. A number of studies have found that these rights to exert control over wild lands and their conservation and management are more important than benefits themselves and their distribution (Lokina and Robinson, 2008; Waylen et al., 2010). Devolution of such rights, however, battles against vested interests of more powerful public and private stakeholders and is strongly resisted by central governments (Nelson, 2010). In Africa there has been little devolution of rights, at least for forest land: here governments still own 98% of forest land (see Rights and Resources Initiative [RRI], 2012 in Anderson and Mehta, 2013). In Asia, the situation is somewhat better – governments control about 68% of forested land, while individuals and firms own about 24% and the rest is owned by or designated for IPLCs. Only in Latin America has a significant shift occurred in tenure – governments control about 36% while IPLCs control about 39%.

It is important to note that strengthening community rights to manage and use lands and resources does not automatically imply opening up legal use of pangolin, which are protected species in most range states in both

Asia and Africa. There are numerous factors that make a managed sustainable use of wild pangolins a challenging concept: the very high level of threat faced by most populations, the difficulty of monitoring (see Chapter 35), and paucity of knowledge of populations, meaning achieving sustainability would be challenging. There is also a prohibition on international commercial trade in wild-caught pangolins (see Chapter 19). Further, as Freese (2012) highlighted, exceptionally high resource values (as for pangolins) raises potential challenges for community management and use, including: (1) difficulty in enforcing community property rights when poachers have enormous incentives to take high risks; (2) powerful vested interests over-riding weaker community rights; and (3) uncertainty over the future value of the resource resulting in a tendency to maximize immediate harvest while the price is high (and to prevent poachers [e.g., from outside an area] capturing the value instead).

What benefits are appropriate and effective will vary from site to site and community to community - what is important is that these reflect the priorities of the community itself, not the priorities of those implementing conservation interventions and projects. These benefits need to be freely chosen: engaging communities in a meaningful way needs to go well beyond, for example, providing jobs that are controlled and funded from external sources, and need to be linked to greater empowerment, voice, rights, and ownership or stewardship over wildlife (Cooney et al., 2018; Duffy et al., 2016).

Benefits of course are not a panacea, and can go wrong in various ways. The distribution and sharing of such benefits can involve serious pitfalls, and has been the downfall of many initiatives. Elite capture is a constant threat that can undermine the potential engagement of a community as a whole (Spiteri and Nepalz, 2006). The distribution of benefits must be supported by and perceived as equitable by the community at large (Hartley and Hunter, 1997). They must also be perceived by the community as connected to and conditional on avoiding poaching and protecting pangolins and their habitat, or they will have little effect (see "*Supporting alternative (non-wildlife) livelihoods*", below). If the intention is that community benefits incentivize pro-conservation behavior, whether that is refraining from poaching, carrying out patrols, sharing information, or something else, the benefits need to be conditional upon that behavior and broadly commensurate with the varying contributions made by different people (Chevallier, 2016; Roe et al., 2000). Finally, the benefits individuals receive may be inadequate to offset the costs they incur from wildlife or from conservation (Child, 1995; Gibson and Marks, 1995).

Decreasing the costs of living with wildlife

This pathway to reducing poaching may not appear immediately applicable to pangolins, as unlike e.g., elephants (Elephantidae) or the tiger (*Panthera tigris*), they do not impose direct costs on communities in terms of threatening human security or livelihoods. However, a broader understanding of this pathway is about reducing the costs that conservation itself imposes on local people. Exclusionary and coercive forms of conservation impose massive social costs – removal of livelihood options, curtailment of culturally important practices, displacement, marginalization of indigenous knowledge, as well as the social costs of large scale incarceration and penalization of poachers. Indigenous people and local community members are often "low-hanging fruit" for enforcement agents tasked with apprehending offenders. They are typically the "foot soldiers" of poaching and trade chains, gaining minimal benefits, often having little knowledge of the global conservation context or the status of species at larger scales, and with little power. More powerful players further up trade chains may be too powerful or well-connected to be targeted by enforcement agents, even when their involvement is well-known. Massé et al. (2017) details the trauma and disruption in communities in Mozambique adjacent to rhinoceros conservation areas, with large numbers of widows and orphans left behind after the killing of scores of young men suspected of poaching. Likewise, escalated anti-poaching activity in Sabah, Malaysia led to the targeting of indigenous communities. This resulted in families losing breadwinners and having to sell land, their sole livelihood asset, to meet the cost of fines (Cooney et al., 2016a), and plunging people into a downward spiral of increasing poverty, further increasing reliance on returns from poaching. Illegal activity can be exacerbated by anger and resentment against conservation authorities – particularly when communities see corruption among enforcement agents and lack of enforcement against powerful criminal interests, and this can stymie more cooperative efforts. Approaches based on respect, justice and appropriate targeting of enforcement effort on those involved in trafficking further along in illegal trade chains can reduce tensions and provide a more conductive environment for developing robust and cooperative conservation strategies.

Supporting alternative (non-wildlife) livelihoods

Community engagement in anti-poaching initiatives through providing "alternative livelihoods" for local communities is popular in wildlife conservation. Here these approaches are understood as livelihoods not based on (legal or illegal) use of wild resources, but other activities such as small-scale farming and retail enterprises. Note that livelihoods based on sustainable use of wild species (for example, through hunting and tourism) would fall into the first pathway of "strengthening incentives for conservation" (Pathway B in Fig. 23.2).

However, there is a lack of evidence for the effectiveness of alternative livelihoods in delivering conservation outcomes. Projects are often poorly-designed, lack sufficient monitoring and evaluation and/or are poorly documented in the literature (Roe et al., 2009, 2015; Wicander and Coad, 2014; Wright et al., 2016). The most commonly cited rationale for these initiatives is that they provide an alternative source of income to IWT, so reduce people's dependence on it, and also reduce the time available for illegal hunting or harvesting. Yet, project designers often ignore the cultural and economic drivers of wildlife use (for example, introducing domestic sources of protein and assuming people will prefer them), inadequately consider the economic viability of new enterprises (such as marketing honey or crafts) and inappropriately target beneficiaries (e.g., not targeting those likely to be involved in poaching [Roe et al., 2015]).

Most importantly, the benefits people gain from alternative livelihoods interventions are rarely conditional on conservation outcomes, and so no link is made between the two. That is to say, where people can gain the benefits of the alternative livelihoods without reducing poaching, the intervention is unlikely to seriously impact poaching. Alternative livelihood benefits may not replace, but simply supplement benefits from poaching and IWT (Wright et al., 2016). Indeed in some cases they can even undermine conservation in the long term, if livelihoods such as agriculture are so profitable they incentivize land use change and thereby undermine conservation. While there are some positive examples where alternative livelihoods form part of a package of interventions to tackle IWT (see Lotter and Clark, 2014) or where alternative income opportunities incentivize "reformed poachers associations" (see Harrison et al., 2015), this approach should only be adopted with considerable care and caution.

It's not what you do, it's the way that you do it: co-creating approaches built on equality and trust

Faced with the above, policy designers and project implementers focused on pangolins may deliberate over what strategies and approaches are likely to deliver their objectives in particular areas. However, a strong and consistent message from research and from the Beyond Enforcement dialogues is that there is indeed no blueprint, and that what is more important than the specific strategies used is *how* these were decided upon – and *who* was involved in those decisions (Cooney et al., 2016a,b; 2017, 2018; IUCN-SULI, IIED, CEED, TRAFFIC, 2015). Success in site level interventions against poaching and IWT typically relies critically on the approach taken and relationships. Local motivation for and ownership of conservation interventions is an important part of their success (see Chapter 24).

It is important for anti-poaching initiatives involving local communities to move slowly, build trust, and develop reciprocal and cooperative relationships with communities, not something that can easily be achieved within the typical 2–5 year timeframe of many interventions. Strategies should be co-created with communities, not imposed from the outside by governments, NGOs or international policy deliberations. Communities need to be involved in framing poaching problems and determining what strategies are chosen to address them, not just engendering a culture of passive reliance on an externally provided financial benefit. Livelihood options and ways to benefit from wildlife need to be chosen by community members themselves according to cultural and socio-economic values, not imposed by external actors. This applies also to enforcement interventions – these will be more effective where they are "co-created" i.e., communities have a say in the setting of rules and penalties for breaking them, where traditional authorities are respected, and relations of trust between enforcement authorities and communities have been built. Interventions against poaching and IWT can help communities reach their own goals (e.g., security, livestock protection, food security, resource management, maintaining culture and traditional knowledge) but understanding these priorities requires trust to be gained through seeking input and listening to communities at every stage of planning and implementation.

A particular insight from the IWT context is that the "theories of change" or beliefs about cause and effect held by community members (on one hand) and project designers/implementers (on the other) can be substantially different, and this can pose a major stumbling block to implementation of projects and interventions. In an approach called "First Line of Defence (FLoD)" (see Chapter 24), the basic theory of change set out above (Fig. 23.2) has been field-tested and refined with communities in Kenya, and highlights the mismatches in practice and the problems this poses for implementation. Involving communities from the start in designing and planning interventions to combat poaching and IWT should improve prospects for success, and should be applied to pangolins.

Conclusions

The lessons and insights underlying this chapter have been repeatedly learnt and stressed over decades of research and practice. What is evident, however, is that these lessons have, for the most part, not been acted upon. The contemporary crisis of poaching and IWT involving pangolins and many other species has sparked "crisis-mode" responses in conservation: this has meant an over-reliance on simplistic, punitive and exclusionary measures in some places (e.g., sub-Saharan Africa) that offer little hope of effectively tackling the problem longer-term, while imposing harsh costs on the communities that live with wildlife. These approaches can backfire in conservation terms – driving disenfranchisement, resentment and anger, undermining the potential for collaborative approaches, and indeed driving increased community involvement in poaching for IWT. Wherever the behaviors and decisions of communities living close to wildlife affect (or could affect) patterns of pangolin hunting, poaching and trafficking, either through direct engagement in it, support to poachers, or providing intelligence and cooperation to enforcement agencies, the cooperation and support of communities for pangolin conservation will be critical.

Tackling pangolin hunting and poaching by engaging and supporting communities can enable, motivate and empower community members to protect wild species and lands. Encouragingly, numerous local community focused pangolin conservation projects are being implemented in Africa and Asia but there is scope for much more scale in order to ensure there is a proportionate response to the threats pangolins face. These interventions can enhance community rights to manage and benefit from wildlife resources and boost their sense of ownership and stewardship, and reduce the costs they face from living with wildlife or with conservation interventions. Cooperative models of enforcement can mobilize the energies, knowledge and capacities of community members and be powerful and effective. Enforcement is critical in community-based approaches, but crucially, it is enforcement that upholds and protects the rights of communities and their members, rather than potentially undermining them.

An enduring challenge is not simply to draw out more and more lessons and best practices, but to find ways to ensure they are taken into account in the design and implementation of new policies and projects. Nelson (2010) highlighted that, at least in an African context, endless technical evidence can be (and has been) produced on what works, what doesn't, and why – but this will never counter the powerful interests that underpin the way policy decisions are made in the real world. Real reforms are only likely to happen when communities themselves are able to demand rights and hold policy makers to account (Nelson, 2010). For this they need to be organized and mobilized, and need democratic mechanisms for representation and accountability.

Communities need greater voice in the decision-making and policy development that affects them, including on anti-poaching. If pangolins are to persist alongside the communities who control or influence most of their habitat, the rest of the conservation community needs to help these communities exercise their voice and listen to what they have to say.

References

Anderson, J., Mehta, S., 2013. A Global Assessment of Community Based Natural Resource Management: Addressing the Critical Challenges of the Rural Sector. USAID, Washington, D.C., USA. Available from: <https://rmportal.net/library/content/global-assessment-cbnrm-challenges-rural-sector/view>. [December 11, 2018].

Berkes, F., 2004. Rethinking community-based conservation. Conserv. Biol. 18 (3), 621–630.

Biggs, D., Cooney, R., Roe, D., Dublin, H.T., Allan, J.R., Challender, D.W.S., et al., 2016. Developing a theory of change for a community-based response to illegal wildlife trade. Conserv. Biol. 31 (1), 5–12.

Blackburn, S., Hopcraft, G.C., Ogutu, J.O., Matthiopoulos, J., Frank, L., 2016. Human-wildlife conflict, benefit sharing and the survival of lions in pastoralist community-based conservancies. J. Appl. Ecol. 53 (4), 1195–1205.

Blomberg, M., 2018. Meet the 'vigilante' grandfathers protecting indigenous forest life in Cambodia. Al Jazeera. Available from: <https://www.aljazeera.com/indepth/features/cambodia-indigenous-vigilante-grandfathers-

protect-forest-life-181115215336028.html>. [December 11, 2018].

Brooks, J.S., Waylen, K.S., Borgerhoff Mulder, M., 2012. How national context, project design, and local community characteristics influence success in community-based conservation projects. Proc. Natl. Acad. Sci. 109 (52), 21265–21270.

Brown, K., 2002. Innovations for conservation and development. Geogr. J. 168 (1), 6–17.

Büscher, B., Ramutsindela, M., 2016. Green violence: rhino poaching and the war to save Southern Africa's peace parks. Afr. Aff. 115 (458), 1–22.

CBD, 2004. Programme of Work on Protected Areas, approved at COP 7, Kuala Lumpur, February 2004. Available from: <https://www.cbd.int/doc/publications/pa-text-en.pdf>. [February 4, 2019].

Challender, D.W.S., MacMillan, D.C., 2014. Poaching is more than an enforcement problem. Conserv. Lett. 7 (5), 484–494.

Challender, D.W.S., MacMillan, D.C., 2016. Transnational environmental crime: more than an enforcement problem. In: Schaedla, W.H., Elliott, L. (Eds.), Handbook of Transnational Environmental Crime. Edward Elgar Publishing, Cheltenham and Northampton, Massachusetts, pp. 489–498.

Challender, D., Waterman, C., 2017. Implementation of CITES Decisions 17.239 b) and 17.240 on Pangolins (*Manis* spp.), CITES SC69 Doc. 57 Annex. Available from: <https://cites.org/sites/default/files/eng/com/sc/69/E-SC69-57-A.pdf>. [December 2, 2018].

Chevallier, R., 2016. The State of Community-Based Natural Resource Management in Southern Africa-Assessing Progress and Looking Ahead. Occasional Paper 240. South African Institute of International Affairs, Johannesburg, South Africa.

Child, B., 1995. The practice and principles of community-based wildlife management in Zimbabwe: the CAMPFIRE Programme. Biodivers. Conserv. 5 (3), 369–398.

CITES, 2019. Participatory Mechanism for Rural Communities, CITES CoP18 Doc. 17.3. Available from: <https://cites.org/sites/default/files/eng/cop/18/doc/E-CoP18-017-03.pdf>. [April 14, 2019].

Cooney R., Brunner J., Roe D., Compton J., Laurenson J., 2016a. Workshop Proceedings: Beyond Enforcement: Engaging Communities in Combating Illegal Wildlife Trade. A Regional Workshop for Southeast Asia, with a focus on the Lower Mekong Basin. Published by IUCN SULi. Available at: <https://www.iucn.org/commissions/commission-environmental-economic-and-social-policy/our-work/specialist-group-sustainable-use-and-livelihoods-suli/communities-and-illegal-wildlife-trade/beyond-enforcement-initiative>. [December 11, 2018].

Cooney R., Roe D., Melisch R., Dublin H., Dinsi S., 2016b. Workshop Proceedings: Beyond Enforcement: Involving Indigenous Peoples and Local Communities in Combating Illegal Wildlife Trade. Regional Workshop for West and Central Africa. Published by IUCN SULi. Available at: <https://www.iucn.org/commissions/commission-environmental-economic-and-social-policy/our-work/specialist-group-sustainable-use-and-livelihoods-suli/communities-and-illegal-wildlife-trade/beyond-enforcement-initiative>. [December 11, 2018].

Cooney, R., Roe, D., Dublin, H., Phelps, J., Wilkie, D., Keane, A., et al., 2017. From poachers to protectors: engaging local communities in solutions to illegal wildlife trade. Conserv. Lett. 10 (3), 367–374.

Cooney, R., Roe, D., Dublin, H., Booker, F., 2018. Wild Lives, Wild Livelihoods: Engaging Communities in Sustainable Wildlife Management and Combating Illegal Wildlife Trade. United Nations Environment Program, Nairobi, Kenya. Available at: <http://wedocs.unep.org/bitstream/handle/20.500.11822/22864/WLWL_Report_web.pdf>. [December 11, 2018].

Corry, S., 2015. When Conservationists Militarize, Who's the Real Poacher? Truthout, August 9, 2015.

Dash, M., Behera, B., 2018. Biodiversity conservation, relocation and socio-economic consequences: a case study of Similipal Tiger Reserve. India Land Use Policy 78, 327–337.

Duffy, R., 2010. NatureCrime: How We're Getting Conservation Wrong. Yale University Press, New Haven, Connecticut.

Duffy, R., 2014. Waging a war to save biodiversity. Int. Aff. 90 (4), 819–834.

Duffy, R., St. John, F.A.V., Büscher, B., Brockington, D., 2016. Toward a new understanding of the links between poverty and illegal wildlife hunting. Conserv. Biol. 30 (1), 14–22.

Eaton, S., 2016. These Indian Women Said They Could Protect Their Local Forests Better Than the Men in Their Village. The Men Agreed. PRI, The World. Public Radio International. Available from: <https://www.pri.org/stories/2016-03-21/women-are-india-s-fiercest-forest-protectors>. [December 11, 2018].

Freese, C., 2012. Wild Species as Commodities: Managing Markets and Ecosystems for Sustainability. Island Press, Washington, D.C.

Frisina, M.R., Tareen, S.N.A., 2009. Exploitation prevents extinction: case study of endangered Himalayan sheep and goats. In: Dickson, B., Hutton, J., Adams, W.M. (Eds.), Recreational Hunting, Conservation and Rural Livelihoods. Blackwell Publishing Ltd, Chichester, pp. 141–156.

Garnett, S.T., Burgess, N.D., Fa, J.E., Fernandez-Llamázares, Á., Molnár, Z., Robinson, C.J., et al., 2018. A spatial overview of the global importance of indigenous lands for conservation. Nat. Sustain. 1 (7), 369–374.

Ghimire, K.B., Pimbert, M.P. (Eds.), 1997. Social Change and Conservation: Environmental Politics and Impacts of National Parks and Protected Areas. Earthscan, London.

Gibson, C.C., Marks, S.A., 1995. Transforming rural hunters into conservationists: an assessment of community-based wildlife management programs in Africa. World Dev. 23 (6), 941–957.

Govan, H., Tawake, A., Tabanukawai, K., 2006. Community-based marine resource management in the South Pacific. Parks 16 (1), 63–67.

Harrison, M., Roe, D., Baker, J., Mwedde, G., Travers, H., Plumptre, A., et al., 2015. Wildlife Crime: A Review of the Evidence on Drivers and Impacts in Uganda. IIED, London, UK.

Hartley, D., Hunter, N., 1997. Community wildlife management: turning theory into practice. Paper prepared for the DFID Natural Resource Advisors Conference 6–10 July 1997. Sparsholt College, Winchester, UK.

Hawdon, J., Ryan, J., 2011. Neighborhood organizations and resident assistance to police. Sociol. Forum 26 (4), 897–920.

Horwich, R.H., Lyon, J., 2007. Community conservation: practitioners' answer to critics. Oryx 41 (3), 376–385.

Hulme, D., Murphree, M. (Eds.), 2001. African Wildlife and Livelihoods: The Promise and Performance of Community Conservation. James Currey, Oxford.

Húbschle, A.M., 2017. The social economy of rhino poaching: of economic freedom fighters, professional hunters and marginalized local people. Curr. Sociol. 65 (3), 427–447.

IUCN, 1994. The Importance of Community-Based Approaches. Resolution 023 of the 19th General Assembly of IUCN, Buenos Aires, 1994. Available from: <https://portals.iucn.org/library/sites/library/files/resrecfiles/GA_19_RES_023_The_Importance_of_Community_based_Ap.pdf>. [December 11, 2018].

IUCN, 1996. Collaborative Management for Conservation. Resolution 042 of the First World Conservation Congress of IUCN, Montreal, 1996. Available from: <https://portals.iucn.org/library/sites/library/files/resrecfiles/WCC1_REC_042_COLLABORATIVE_MANAGEMENT_FOR_CONSERVA.pdf>. [December 11, 2018].

IUCN-SULI, IIED, CEED, TRAFFIC, 2015. Beyond enforcement: communities, governance, incentives and sustainable use in combating wildlife crime. Symposium Report, 26–28 February 2015, Glenburn Lodge, Muldersdrift, South Africa. IIED, London, UK.

Lokina, R.B., Robinson, E.J., 2008. Determinants of Successful Participatory Forest Management in Tanzania. The Environment for Development Initiative, Tanzania.

Lunstrum, E., 2014. Green militarization: anti-poaching efforts and the spatial contours of Kruger National Park. Ann. Assoc. Am. Geogr. 104, 816–832.

Lotter, W., Clark, K., 2014. Community involvement and joint operations aid effective anti-poaching in Tanzania. Parks 20 (1), 19–28.

Makoye, K., 2014. Anti-Poaching Operation Spreads Terror in Tanzania. Available from: >http://www.ipsnews.net/2014/01/anti-poaching-operation-spread-terror-tanzania/>. [February 4, 2019].

Massé, F., Gardiner, A., Lubilo, R., Themba, M.N., 2017. Inclusive anti-poaching? Exploring the potential and challenges of community-based anti-poaching. South Africa Crime Quart. 60, 19–27.

Mazambani, D., Dembetembe, P., 2010. Community Based Natural Resource Management Stocktaking Assessment: Zimbabwe Profile. USAID, Washington, D.C.

Molnar, A., Scherr, S., Khare, A., 2004. Who Conserves the World's Forests? Community-Driven Strategies to Protect Forests and Respect Rights. Forest Trends, Washington D.C.

Musavengane, R., Simatele, D., 2016. Significance of social capital in collaborative management of natural resources in Sub-Saharan African rural communities: a qualitative meta-analysis. South African Geogr. J. 99 (3), 1–16.

Naidoo, R., Weaver, C.L., Diggle, R.W., Matongo, G., Stuart-Hill, G., Thouless, C., 2016. Complementary benefits of tourism and hunting to communal conservancies in Namibia. Conserv. Biol. 30 (3), 628–638.

Nelson, F., 2010. Community Rights, Conservation and Contested Land: The Politics of Natural Resource Governance in Africa. Routledge, London.

Newton, P., Nguyen, T.V., Roberton, S., Bell, D., 2008. Pangolins in peril: using local hunters' knowledge to conserve elusive species in Vietnam. Endanger. Sp. Res. 6 (1), 41–53.

Nash, H.C., Wong, M.H.G., Turvey, S.T., 2016. Using local ecological knowledge to determine status and threats of the Critically Endangered Chinese pangolin (*Manis pentadactyla*) in Hainan, China. Biol. Conserv. 196, 189–195.

Nilsson, D., Baxter, G., Buler, J.R.A., McAlpine, C.A., 2016. How do community-based conservation programs in developing countries change human behavior? A realist synthesis. Biol. Conserv. 200, 93–103.

Poirier, R., Ostergren, D., 2002. Evicting people from nature: indigenous land rights and National Parks in Australia, Russia, and the United States. Nat. Resour. J. 42 (2), 331–351.

Roe, D., Nelson, F., Sandbrook, S. (Eds.), 2009. Community Management of Natural Resources in Africa: Impacts, Experiences, and Future Directions. Natural Resources No. 18, International Institute for Environment and Development, London, UK.

Roe, D., Mayers, J., Grieg-Gran, M., Kothari, A., Fabricius, C., Hughes, R., 2000. Evaluating Eden - Exploring the Myths and Realities of Community Based Wildlife Management. Evaluating Eden Series Overview. IIED, London, UK.

Roe, D., Booker, F., Day, M., Zhou, W., Allebone-Webb, S., Hill, N.A.O., et al., 2015. Are alternative livelihood projects effective at reducing local threats to specified elements of biodiversity and/or improving or maintaining the conservation status of those elements? Environ. Evid. 4, 22.

Rowlatt, J., 2017. Kaziranga: The Park That Shoots People to Protect Rhinos. BBC News. Available from: <https://www.bbc.co.uk/news/world-south-asia-38909512>. [December 2, 2018].

Spiteri, A., Nepalz, S.K., 2006. Incentive-based conservation programs in developing countries: a review of some key issues and suggestions for improvements. Environ. Manage. 37 (1), 1–14.

St. John, F.A.V., Mai, C.H., Pei, K.J.C., 2015. Evaluating deterrents of illegal behavior in conservation: carnivore killing in rural Taiwan. Biol. Conserv. 189, 86–94.

Survival International, 2014. Hunters or Poachers? Survival, the Baka and WWF. Available from: <https://www.survivalinternational.org/campaigns/wwf.com>. [February 4, 2019].

Torri, M.C., 2011. Conservation, relocation and the social consequences of conservation policies in protected areas: case study of the Sariska Tiger Reserve, India. Conserv. Soc. 9 (1), 54–64.

Trageser, S.J., Ghose, A., Faisal, M., Mro, P., Mro, P., Rahman, S.C., 2017. Pangolin distribution and conservation status in Bangladesh. PLoS One 12 (4), e0175450.

Twinamatsiko, M., Baker, J., Harrison, M., Shirkhorshidi, M., Bitariho, R., Wieland, M., et al., 2014. Linking Conservation, Equity and Poverty Alleviation: Understanding Profiles and Motivations of Resource Users and Local Perceptions of Governance at Bwindi Impenetrable National Park, Uganda. IIED Research Report. IIED, London, UK. Available from: <http://pubs.iied.org/14630IIED>. [December 11, 2018].

Waylen, K.A., Fischer, A., McGowan, P.J.K., Thirgood, S.J., Milner-Gulland, E.J., 2010. The effect of local cultural context on community-based conservation interventions: evaluating ecological, economic, attitudinal and behavioural outcomes. Environ. Evid. 1–36.

WCED, 1987. Our Common Future. World Commission on Environment and Development. Oxford University Press, Oxford.

Wicander, S., Coad, C., 2014. Learning Our Lessons: A Review of Alternative Livelihood Projects in Central Africa. Environmental Change Institute, University of Oxford, UK and IUCN, Gland, Switzerland. Available from: <http://cmsdata.iucn.org/downloads/english_-version.pdf/>. [February 4, 2019].

Wilkie, D., Painter, M., Jacob, I., 2016. Rewards and Risks Associated With Community Engagement in Anti-Poaching and Anti-Trafficking. Biodiversity Technical Brief. United States Agency for International Development, Washington D.C. Available from: <https://pdf.usaid.gov/pdf_docs/PA00M3R9.pdf>. [February 4, 2019].

Wright, J.H., Hill, N.A., Roe, D., Rowcliffe, J.M., Kumpul, N.F., Day, M., et al., 2016. Reframing the concept of alternative livelihoods. Conserv. Biol. 30 (1), 7–13.

CHAPTER

24

Exploring community beliefs to reduce illegal wildlife trade using a theory of change approach

Diane Skinner[1], *Holly Dublin*[2], *Leo Niskanen*[3], *Dilys Roe*[4] *and Akshay Vishwanath*[3]

[1]IUCN CEESP/SSC Sustainable Use and Livelihoods Specialist Group, Harare, Zimbabwe [2]IUCN CEESP/SSC Sustainable Use and Livelihoods Specialist Group, Nairobi, Kenya [3]IUCN Eastern and Southern Africa Regional Office, Nairobi, Kenya [4]International Institute for Environment and Development, London, United Kingdom

OUTLINE

Introduction

Illegal wildlife trade (IWT) has been growing in the period 2009–2019 and international attention to and funding for this challenge has also increased (Challender and MacMillan, 2014; World Bank, 2016). A number of high-value species are the target of this trade, including pangolins (Manidae), elephants (Elephantidae), rhinos (Rhinocerotidae), a variety of big cats (e.g., tiger [*Panthera tigris*]), as well as many trees, medicinal, aromatic and ornamental plants (e.g., orchids [Orchidaceae]), birds, reptiles and fish (Rosen and Smith, 2010). Pangolins in particular are subject to high levels of poaching and trafficking, both in Africa and Asia (Heinrich et al., 2017).

Pangolins
DOI: https://doi.org/10.1016/B978-0-12-815507-3.00024-1

Much of the focus of both national and international responses, funding allocation and implementation of interventions has been on law enforcement from source to destination (World Bank, 2016). Despite this focus, there has also been a strong recognition that the long-term survival of wildlife, and in particular the success of efforts to combat IWT, depends to a large extent on local communities that live with these species (Cooney et al., 2018). Recognition of the important role of local communities in tackling IWT is reflected in many intergovernmental statements and commitments (see Table 24.1).

Despite this increased policy attention, there has been a lack of practical guidance on *how* to most effectively partner with local communities (see Chapter 23). Part of the problem is that there is no blueprint approach. Communities are diverse. Socio-economic, political, legal and environmental factors influence the nature of their interactions with wildlife and hence there are different perceptions of, and attitudes towards IWT (Biggs et al., 2015). This influences the types of community engagement interventions that are likely to be effective. Furthermore, communities themselves are rarely consulted on what they think about IWT and how best to tackle it, while project developers are often driven by their own strong beliefs and assumptions which may prove to be inconsistent with community beliefs (Roe et al., 2018).

Local Communities: First Line of Defence against Illegal Wildlife Trade (FLoD) – developing a methodology

Building on the "Beyond Enforcement" initiative (see Chapter 23), in 2016, the IUCN Sustainable Use and Livelihoods Specialist Group (IUCN-SULi), International Institute for Environment and Development (IIED), IUCN Eastern and Southern Africa Regional Office (ESARO), and the IUCN Species Survival Commission (SSC) African Elephant Specialist Group (AfESG) secured funding from the UK government's Illegal Wildlife Trade Challenge Fund to field test the straw-model ToC in the "Local Communities: First Line of Defence against Illegal Wildlife Trade" initiative (FLoD). While the Beyond Enforcement ToC was intentionally generic, the FLoD initiative sought to test the ToC's integrity in the field, refine it and develop methods by which the ToC could be used at the project level to design improved interventions in engaging communities against IWT.

Working with a number of local partners in Kenya (Big Life Foundation, Cottar's Safari Services, South Rift Landowners Association, and the Kenya Wildlife Conservancies Association), the project team identified three pilot sites in southern Kenya: Olderkesi Conservancy adjacent to Masai Mara National Reserve, the Shompole-Olkiramatian Group Ranches, and the Kilitome Conservancy adjacent to Amboseli National Park. The focus of the project was on working with the local communities to understand the context of wildlife crime in the area, and the success or failure of interventions to combat it. While the focus of the project was the African elephant (*Loxodonta africana*), many species that are targeted for the high-value illegal wildlife trade are present in the pilot sites, including Temminck's pangolin (*Smutsia temminckii*).

The project adopted a "participatory action research" approach (Rowe et al., 2013), a research methodology that emphasizes participation and action with the target communities. Whereas conventional research tends to be characterized by detached researchers and research "subjects," participatory action research seeks to understand issues through a process of collaboration, reflection and stakeholder engagement (Rowe et al., 2013).

An inception workshop was held in Nairobi in May 2016 at which the Beyond Enforcement ToC (see Fig. 23.2) and associated assumptions were discussed with project partners and refined following their feedback. The ToC, including the overall pathway titles,

TABLE 24.1 International policy commitments on the role of communities in combatting IWT.

Policy declaration	Relevant statement
African Elephant Summit (2013)[a]	"Engage communities living with elephants as active partners in their conservation."
London Declaration (2014)[b]	"Increase capacity of local communities to pursue sustainable livelihood opportunities and eradicate poverty."
	"Work with, and include local communities in, establishing monitoring and law enforcement networks in areas surrounding wildlife."
Kasane Declaration (2015)[c]	"Promote the retention of benefits from wildlife resources by local people where they have traditional and/or legal rights over these resources. We will strengthen policy and legislative frameworks needed to achieve this, reinforce the voice of local people as key stakeholders and implement measures which balance the need to tackle the illegal wildlife trade with the needs of communities, including the sustainable use of wildlife."
Brazzaville Declaration (2015)[d]	"Recognize the rights and increase the participation of indigenous peoples and local communities in the planning, management and use of wildlife through sustainable use and alternative livelihoods and strengthen their ability to combat wildlife crime."
UN General Assembly adopted Resolution 69/314 on Tackling Illicit Trafficking In Wildlife (2015)[e]	"Strongly encourages Member States to support, including through bilateral cooperation, the development of sustainable and alternative livelihoods for communities affected by illicit trafficking in wildlife and its adverse impacts, with the full engagement of the communities in and adjacent to wildlife habitats as active partners in conservation and sustainable use, enhancing the rights and capacity of the members of such communities to manage and benefit from wildlife and wilderness."
Sustainable Development Goal 15 (2015)[f]	"Enhance global support for efforts to combat poaching and trafficking of protected species, including by increasing the capacity of local communities to pursue sustainable livelihood opportunities."
Hanoi Statement on Illegal Wildlife Trade (2016)[g]	"Recognize the importance of supporting and engaging communities living with wildlife as active partners in conservation, through reducing human-wildlife conflict and supporting community efforts to advance their rights and capacity to manage and benefit from wildlife and their habitats; and developing collaborative models of enforcement."
	"The active participation of local people is critical to effective monitoring and law enforcement as well as sustainable socio-economic development."
UN General Assembly adopted Resolution 71/326 on Tackling Illicit Trafficking In Wildlife (2017)[h]	"Encourages Member States to increase the capacity of local communities to pursue sustainable livelihood opportunities, including from their local wildlife resources."

(Continued)

TABLE 24.1 (Continued)

Policy declaration	Relevant statement
	"Strongly encourages Member States to enhance their support, including through transnational and regional cooperation, for the development of sustainable and, as appropriate, alternative livelihoods for communities affected by illicit trafficking in wildlife and its adverse impacts, with the full engagement of the communities in and adjacent to wildlife habitats as active partners in conservation and sustainable use, enhancing the rights and capacity of the members of such communities to manage and benefit from wildlife and wilderness."
London Declaration (2018)[i]	"Recognize the essential engagement role and rights of local communities and indigenous people to ensure a sustainable solution to addressing the illegal wildlife trade."

[a]*Government of Botswana, IUCN, 2014. African Elephant Summit, Gaborone, Botswana, 2–4 December 2013: Summary record.*
[b]*London Conference on the Illegal Wildlife Trade, 2014. Declaration.*
[c]*Kasane Statement on the Illegal Wildlife Trade, 2015.*
[d]*African Union, 2015. African Strategy on Combating Illegal Exploitation and Illegal Trade in Wild Fauna and Flora in Africa (No. EX.CL/Dec.873–897(XXVII)). African Union, Johannesburg.*
[e]*Albania, 84 other UN Member States, 2015. Tackling illicit trafficking in wildlife.*
[f]*UN General Assembly, 2015. Transforming our world: the 2030 Agenda for Sustainable Development (Resolution adopted by the General Assembly No. A/RES/70/1). UN General Assembly, New York.*
[g]*Hanoi Conference on Illegal Wildlife Trade, 2016. Hanoi Statement on Illegal Wildlife Trade.*
[h]*Australia, 15 other UN Member States, 2017. Tackling illicit trafficking in wildlife.*
[i]*London Conference on the Illegal Wildlife Trade, 2018. Declaration.*

interventions, results, outputs and outcomes were refined based on feedback from stakeholders. This refined version of the ToC, for the purposes of the project in Kenya now called the "Baseline ToC" (see Fig. 24.1), along with a set of refined assumptions, was then used as the starting point for the testing process.

The first step in the FLoD process was to understand the logic behind the design of a particular IWT intervention. An in-depth interview with the IWT project designers and/or implementers helped to interrogate and articulate the expected causal results chain and, critically, the key *assumptions* – explicit and implicit – that underpin them. This logic was then taken to the community targeted by the IWT intervention to explore their perspectives, and in particular whether the assumptions were valid. This process included a series of age and gender-disaggregated focus group meetings, which allowed the airing and sharing of a diverse range of viewpoints from within a single community.

Finally, communities and project designers/implementers were convened to hear the results of the consultations, explore differences in perspectives, and to provide a forum for the reconciliation of differences, and identification of potential new or changed interventions.

The process has been documented in detail in case studies (see Niskanen et al., 2018) and in the FLoD guidance for implementation (see Skinner et al., 2018). The full set of case studies, policy briefs, guidance and tools are available at https://www.iucn.org/flod.

Lessons learned

Key lessons were learned throughout the testing process. First, an initial scoping visit to the

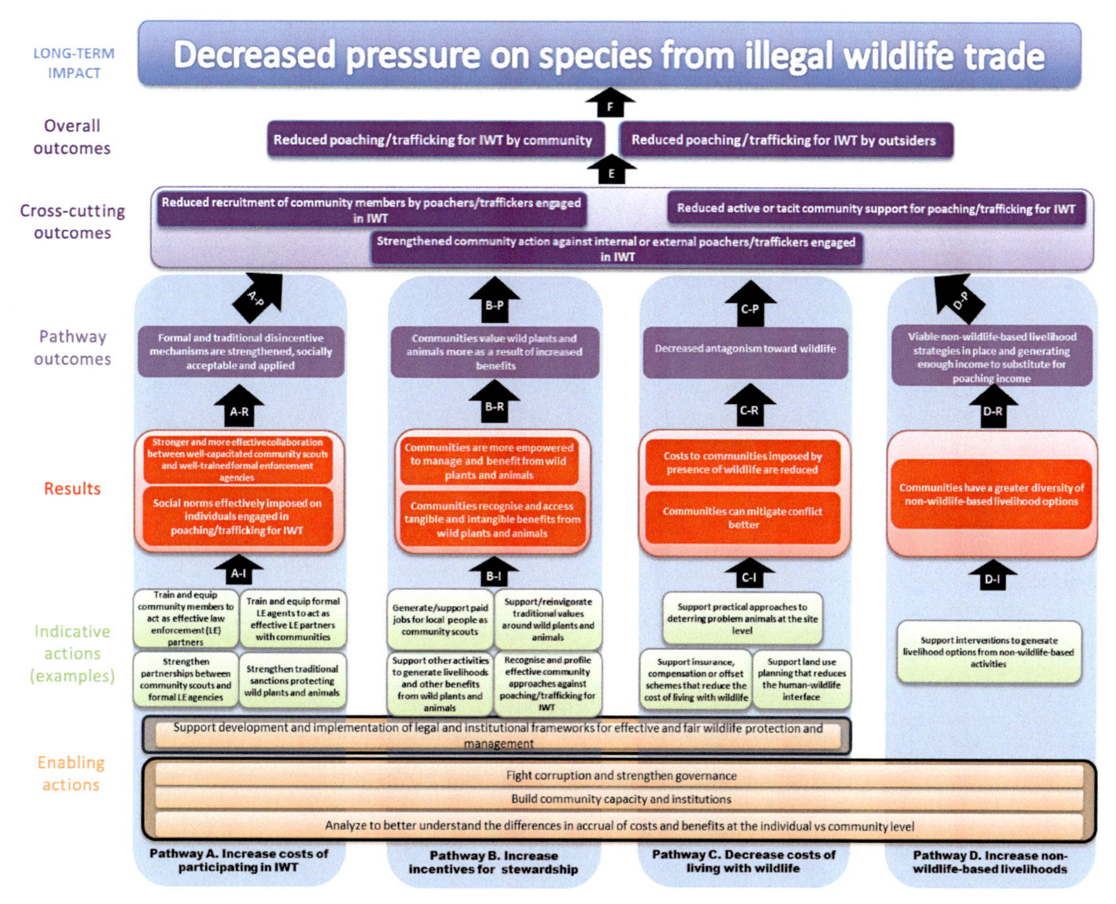

FIGURE 24.1 Baseline Theory of Change, updated for inclusion in the First Line of Defence Guidance. *Source: Skinner et al. (2018).*

case study area and an inception workshop are critical to: (1) explain the process to the project designer/implementer and to local partners; (2) collect the necessary background information about the community, wildlife, and poaching in the area; (3) identify the key stakeholders and define the target community; (4) define the geographical boundaries of the study site; and (5) gain insights into the extent of poaching (Fig. 24.2).

Second, key stakeholder interviews provide essential independent triangulation to help validate the assumptions behind both the community and the project designer ToCs, as well as confirm information about institutions (both formal and informal), economic activities, and poaching dynamics.

Third, community focus group discussions (Fig. 24.3) require expert facilitation, by someone who is perceived by the communities as independent and not representing any particular interest or point of view. Facilitation tools must be well understood and relevant to participants.

Fourth, it is critical to translate complex concepts into local languages through the participation of a skilled interpreter and to manage overly domineering voices (e.g., traditional leaders and other members of the community elite) allowing all voices to be heard.

FIGURE 24.2 Women in Shompole-Olkirimatian debate the most important pathways to reduce illegal wildlife trade. *Photo credit: IUCN/Akshay Vishwanath.*

FIGURE 24.3 A working group discusses the findings from FLoD at the feedback workshop in Shompole-Olkirimatian. *Photo credit: IUCN/Akshay Vishwanath.*

Finally, a process of continual feedback and verification is a core part of the process. The process is relatively complex and the resources to take the process to completion should be secured before beginning.

Lessons emerging from the southern Kenya pilot sites have demonstrated that the FLoD initiative is an effective way of bringing community voice out around the issue of combatting IWT in different contexts. Subtleties are important and by articulating each community's views on which anti-IWT strategies are more effective, and why others are not, interventions can be improved (or designed from scratch) to directly respond to community needs and viewpoints, and are therefore likely to result in higher success rates.

The overall structure and logic flow of the Baseline ToC for engaging communities in combatting IWT appeared to be valid based on our case studies. The level of emphasis given to each pathway differed between the different communities, and the challenges faced in each area were different. For example, human-wildlife conflict, resulting in revenge killing of wildlife, seemed more of an issue for the Kilitome community than for the Olderkesi or Shompole-Olkiramatian communities.

Despite having similar cultural and economic characteristics there were clear differences of opinion and belief systems on key issues among different gender and age groups within communities in different areas. This may have significant implications for the sustainability of some IWT interventions that may rely more heavily on the cooperation of one gender or age group. For example, the views of the youth were sometimes at odds with those of the elders in terms of the future vision for their communities, with a greater emphasis for the youth on activities that were cash-based. The women had very different views as well, with much greater attention paid to benefits that were related to education of children. Therefore, projects that engage with all sectors of a community to better understand its dynamics, beliefs and attitudes, may be more likely to have successful outcomes.

The long-term visions for the pastoral Maasai communities extended beyond the specific goal of reducing pressure from poaching to the much broader goal of securing intact ecosystems for sustainable wildlife-based land use that is also beneficial to their core livelihood: livestock. This more holistic focus may be partially explained by the fact that none of the conservancies were experiencing high levels of poaching pressure at the time of the study. However, it was also clear that the project designers/implementers and the communities were fully aware of the fact that while poaching needed to be dealt with as a matter of priority, ultimately the future of wildlife depended on whether it would remain possible to secure enough land to accommodate wildlife in human and livestock dominated landscapes. The larger and more complex problem of preventing habitat loss requires different strategies, investments and a broader range of interventions than combatting poaching and wildlife trafficking.

While poaching for IWT may not have been a pressing challenge at the time in the three communities, the persistent problem of human-wildlife conflict, the major promises made through legislative reform and the weak response to this issue by the Kenyan government had become a highly politicized issue that was causing widespread resentment. This was contributing to revenge killing of elephants and other wildlife, and potentially also causing some communities to turn a blind eye to poaching by outsiders. Strong objections were heard from communities about the fact that the Kenyan government's responses to IWT and retaliation to wildlife killings were stronger and swifter than their responses to human deaths, injuries and other losses caused by wildlife.

The benefits from tourism were considered absolutely critical in all sites. However, there was a mismatch between the community expectations and what tourism could realistically be expected to generate for the local communities, even in prime wildlife areas where all the necessary conditions for a viable tourism operation were in place. This was considered of particular concern in light of the growing human populations in these areas and the fickle nature and volatility of the international tourism markets worldwide and in Kenya (as evidenced, for example by the dramatic decline in tourist numbers following the Ebola outbreak in 2014). This also underscored the importance of ensuring transparency in how large revenues from tourism actually are, as well as how they are generated and distributed. This was seen as a critical pre-condition for building and maintaining trust between communities and tourism operators.

Overall, revenues from wildlife were not seen as sufficient, and viable incomes from other means were seen as critical to stave off future engagement in IWT. Other non-wildlife related revenue streams, such as livestock-keeping and agriculture can help reduce dependency on tourism and may reduce incentives to poach. Some livelihood strategies, however, may not be compatible with wildlife-based land use in the long term, a fact raised by the communities.

The generation of revenue to local communities from multiple land uses while hoping to accommodate wildlife that require large areas was considered contingent on effective land use planning and governance at the landscape level. These conditions were being met in two of the three sites. Management of conflict and coexistence through land zoning – traditional or contemporary – was raised by all stakeholders as critical to the success of conservancies.

Local communities in all sites recognized the importance of law enforcement, provided it was in partnership with the communities. For example, a well trained and equipped local community scout program supported by a local community informer network can be a formidable first line of defence in the fight against IWT. The sustainability of such programs was, however, questionable as long as they remained dependent on external funding sources and the risk to individual informants was high.

There seemed to be consensus that stiff penalties, introduced in Kenya in the The Wildlife Conservation and Management Act (2013), were having a deterrent effect against poaching and wildlife trafficking. The penalties, although extremely punitive, were generally perceived as fair. Both social sanctions and social pressures were believed to effectively reinforce the government penalties and help deter poaching.

Conclusions

While communities are sometimes consulted on or engaged in the planning of externally-driven IWT (or broader conservation) projects, this is often quite tokenistic and could more accurately be described as "informing" than consulting. Moreover, it often involves the local chiefs, elders or key decision-makers (usually men) rather than the wider community. In contrast, the FLoD approach attempts more representative community consultations – involving young and old people, women and men – and facilitates interaction, feedback, and the reconciliation of differences both amongst different segments of the community and between project designers/implementers and the communities they target.

A ToC approach such as FLoD can help enhance stakeholders' understanding of: (1) implicit ToCs of both communities and project designers; (2) differences within communities and between communities and designers; and (3) reasons for success or failure of particular components of a project. It can also provide useful lessons for other projects (existing and new).

The FLoD approach can provide an excellent entry point for communities and project designers to engage in true dialogue on fundamental issues of mutual concern. Although initially aimed at addressing the issue of IWT, experience has shown that the FLoD approach can also help to unearth issues and solutions that have broader applicability to community based natural resource management.

Communities can be an effective first line of defence against the illegal wildlife trade (Cooney et al., 2017). However, wildlife management needs to become a viable and more competitive land-use option for local people; to achieve this effective land-use planning and governance is required at a landscape level, at large-scale, across diverse land uses, including in and around protected areas. Equally, benefits from wildlife are required, and the cost of living with wildlife must be reduced or managed. Individual IWT projects can be very appealing, promising direct action and on-the-ground results in the short term. However, in the long-term, creating an environment of good governance, enlightened policy and strong partnership that fosters and supports communities to be active participants in conservation and wildlife-based land use is likely to provide far greater hope for the future of Africa's wildlife, including pangolins.

References

Biggs, D., Cooney, R., Roe, D., Dublin, H., Allan, J., Challender, D., Skinner, D., 2015. Engaging Local Communities in Tackling Illegal Wildlife Trade: Can a 'Theory of Change' help? (IIED Discussion Paper). IIED, London, UK.

Challender, D.W.S., MacMillan, D.C., 2014. Poaching is more than an enforcement problem. Conserv. Lett. 7 (5), 484–494.

Cooney, R., Roe, D., Dublin, H., Phelps, J., Wilkie, D., Keane, A., et al., 2017. From poachers to protectors: engaging local communities in solutions to illegal wildlife trade. Conserv. Lett. 10 (3), 367–374.

Cooney, R., Roe, D., Dublin, H., Booker, F., 2018. Wild Life, Wild Livelihoods: Involving Communities in Sustainable Wildlife Management and Combatting the Illegal Wildlife Trade. United Nations Environment Porgramme, Nairobi, Kenya.

Heinrich, S., Wittman, T.A., Ross, J.V., Shepherd, C.R., Challender, D.W.S., Cassey, P., 2017. The Global Trafficking of Pangolins: A Comprehensive Summary of Seizures and Trafficking Routes From 2010–2015. TRAFFIC, Southeast Asia Regional Office, Petaling Jaya, Selangor, Malaysia.

Niskanen, L., Roe, D., Rowe, W., Dublin, H., Skinner, D., 2018. Strengthening Local Community Engagement in Combatting Illegal Wildlife Trade: Case Studies From Kenya. IUCN, Nairobi, Kenya.

Roe, D., Dublin, H., Niskanen, L., Skinner, D., Vishwanath, A., 2018. Local Communities: The Overlooked First Line of Defence for Wildlife (No. 17455IIED), IIED Briefing Papers. IIED, London, UK.

Rosen, G.E., Smith, K.F., 2010. Summarizing the evidence on the international trade in illegal wildlife. Ecohealth 7 (1), 24–32.

Rowe, W.E., Graf, M., Agger-Gupta, N., Piggot-Irvine, E., Harris, B., 2013. Action Research Engagement: Creating the Foundation for Organizational Change, Monograph. Action Learning, Action Research Association Inc, Victoria, BC.

Skinner, D., Dublin, H., Niskanen, L., Roe, D., Vishwanath, A., 2018. Local Communities: First Line of Defence Against Illegal Wildlife Trade (FLoD). Guidance for Implementing the FLoD Methodology. IIED and IUCN, London, UK, and Gland, Switzerland.

World Bank, 2016. Analysis of International Funding to Tackle Illegal Wildlife Trade. The World Bank, Washington D.C.

CHAPTER

25

Community conservation in Nepal – opportunities and challenges for pangolin conservation

Ambika P. Khatiwada[1], Tulshi Laxmi Suwal[2,3], Wendy Wright[4], Dilys Roe[5], Prativa Kaspal[6], Sanjan Thapa[2] and Kumar Paudel[7]

[1]National Trust for Nature Conservation, Lalitpur, Nepal [2]Small Mammals Conservation and Research Foundation (SMCRF), Kathmandu, Nepal [3]Department of Tropical Agriculture and International Cooperation, National Pingtung University of Science and Technology, Pingtung, Taiwan [4]School of Health and Life Sciences, Federation University Australia, Gippsland, VIC, Australia [5]International Institute for Environment and Development, London, United Kingdom [6]Women for Conservation/ Bhaktapur Multiple Campus, Tribhuvan University, Kirtipur, Nepal [7]Greenhood Nepal, New Baneshwor, Kathmandu, Nepal

OUTLINE

Pangolins
DOI: https://doi.org/10.1016/B978-0-12-815507-3.00025-3

Introduction

Two species of pangolin occur in Nepal, the Chinese (*Manis pentadactyla*) and the Indian pangolin (*M. crassicaudata*). Hodgson (1836) was the first to publish reports of the Chinese pangolin and its habitat in the lower and central regions of the country. Corbet and Hill (1992) reported that the distribution of Chinese pangolins includes eastern Nepal. It was not until the 1990s that Suwal and Verheugt (1995) reported the presence of the Indian pangolin in Chitwan National Park, Parsa National Park and the districts of Bara, Parsa and Chitwan. However, both species are protected in Nepal under the National Parks and Wildlife Conservation Act (1973) signifying that the protection of pangolins in the country has long been recognized as a priority.

Nepal is 885 km long and 193 km wide, stretching from east to west in a roughly trapezoidal shape and covering an area of 147,181 km^2. It is landlocked by China to the north and otherwise by India. The range in elevation within the country is vast: from 60 to 8848 m above sea level (asl). Nepal is one of 47 countries classified as "least developed" by the United Nations (United Nations, 2018). However, it is rich in natural resources. Water availability and forest cover is higher than twice the South Asian per capita average (World Bank, 2018). It is the 31st most biodiverse country in the world and the 10th most biodiverse country in Asia (Joshi et al., 2017).

Nepal is commonly divided into five physiographic (Fig. 25.1) and six bioclimatic zones. The Terai (lowland) comprises about 14% of the country's land and the rest is mountainous. The mountainous area is typically described as comprising the Siwalik, hill, middle mountain and the high mountain areas. 33% of the mountainous area is perennially covered by

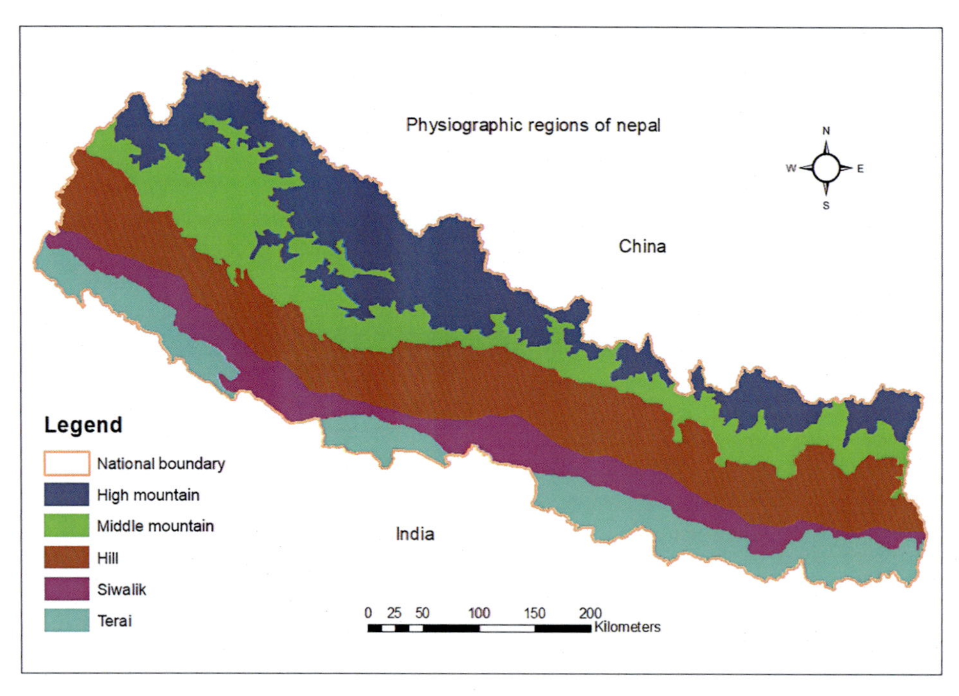

FIGURE 25.1 Physiographic regions of Nepal. *Source: Department of Survey, Government of Nepal.*

snow; meaning only 67% of Nepal's land area is suitable for human settlement (Baral and Bhatta, 2005).

Pangolins are mainly recorded in the Terai, Siwalik, hill and middle mountains (Gurung, 1996; Jnawali et al., 2011), which include tropical, sub-tropical and temperate climates. Presence data from camera trap surveys conducted by Nepal's National Trust for Nature Conservation (NTNC), sightings from experts and citizen scientists, and records of confiscated and rescued animals indicate that the Chinese pangolin occurs in multiple conservation areas and national parks (NPs) (Fig. 25.2). These include the lower elevation belt of the Kangchenjunga Conservation Area, and Makalubarun, Sagarmatha, Chitwan and Parsa NPs, Gaurishankar Conservation Area, Shivapuri-Nagarjun NP, Annapurna Conservation Area and the Salyan district in western Nepal. Similar sources show records of the Indian pangolin in Suklaphanta, Bardia, Banke and Parsa NPs and Kailali and Surkhet districts. These data suggest that Chinese pangolins are distributed across eastern, central and mid-western Nepal, and Indian pangolins are found in Nepal's western regions, at lower elevations (Fig. 25.2). However, it also suggests that the two species are sympatric (e.g., in Parsa NP). This may be the case in other parts of the Terai,

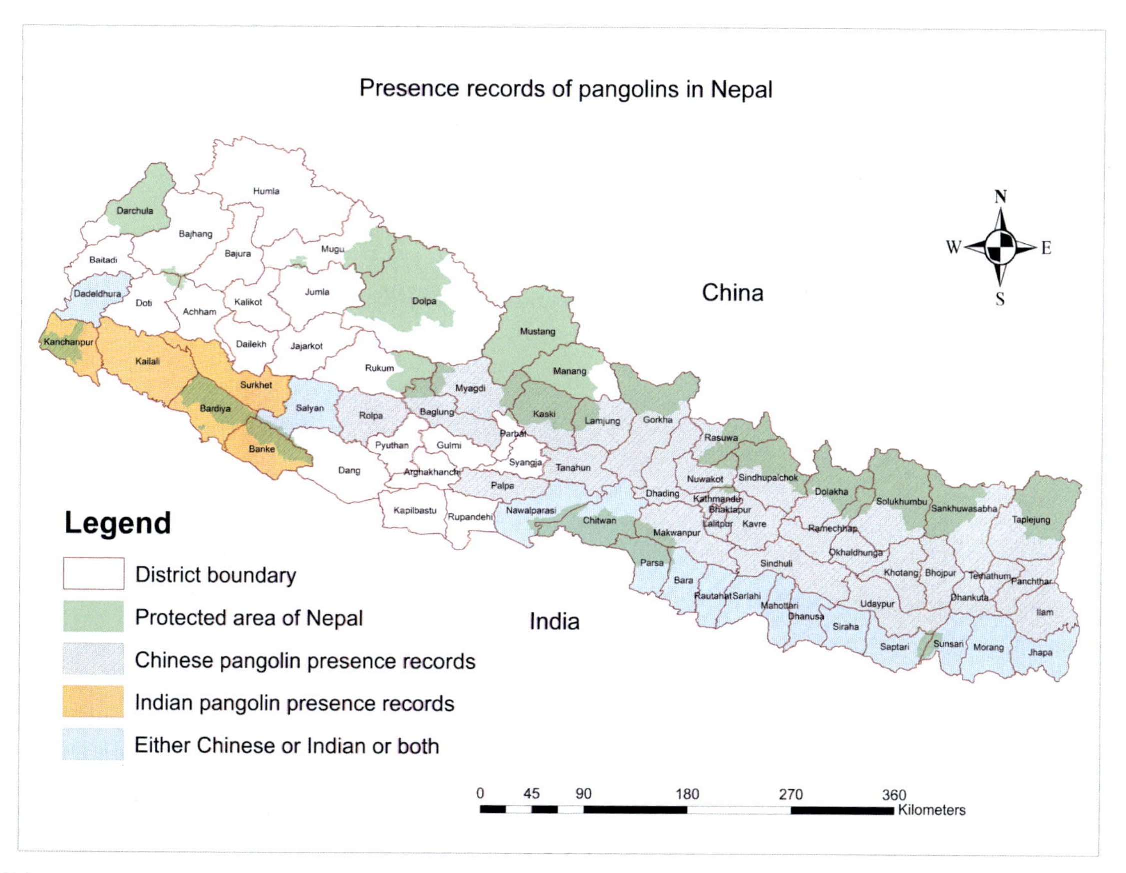

FIGURE 25.2 Presence records of Chinese and Indian pangolins in Nepal. *Source: Department of Survey, Department of National Parks and Wildlife Conservation, National Trust for Nature Conservation, Government of Nepal.*

Siwaliks and middle mountains, but research is needed to verify whether this is the case.

The history of conservation and forest management in Nepal

Conservation in Nepal began in the 18th century. King Surendra Shah (1847–1881) promulgated a law that forbade the killing of wild animals by the public (Upreti, 2017). However, from the mid-1800s until the late 1950s, it was relatively common for members of the elite ruling classes to visit the Terai regions to hunt tiger (*Panthera tigris*) and greater one-horned rhinoceros (*Rhinoceros unicornis*) (Gurung, 1980; Smythies, 1942). In 1964, a sanctuary was declared to protect rhinoceroses and their habitat due to their declining numbers (Bhatt, 2003; Mishra, 1982).

Contemporary history and scientific management of Protected Areas (PAs) in Nepal began in 1973 when King Birendra approved the National Parks and Wildlife Conservation Bill, establishing Chitwan National Park (Bhattarai et al., 2017; Upreti, 2017). The Department of National Parks and Wildlife Conservation (DNPWC), which is responsible for overall management of PAs, was subsequently founded in 1980. The King Mahendra Trust for Nature Conservation (now the National Trust for Nature Conservation [NTNC]) was established by the National Trust for Nature Conservation Act 1982 as an autonomous, not-for-profit organization to meet the growing requirements of field based wildlife conservation in Nepal. There are now 12 NPs, six conservation areas (CAs), one wildlife reserve (WR), one hunting reserve (HR) and 13 buffer zones (BZs) in the country, covering approximately 34,000 km^2 (23%) of the total land area (Table 25.1; DNPWC, 2018). Buffer zones are multiple-use zones, which surround protected areas, and may include villages, community forests, and croplands.

Nepal has experienced several major phases in the management of its protected areas. Initially, "command and control approaches" (1970s–80s; Baral, 2005); also known as "fine and fence" (Bhattarai et al., 2017) served to protect wildlife for the elite hunting parties, and excluded local people and prevented their access to forest resources. In 1973, species-oriented conservation approaches (species-level conservation) became a focus; this progressed to ecosystem-level conservation in the early 1980s. It was not until the introduction of an Integrated Conservation and Development (ICD) approach, initiated in the 1990s, that local people were included as key participants in protected area management. ICD has taken many different forms in different countries, but in Nepal it has entailed the introduction of the BZ concept and the establishment of CAs. CAs and BZs are category VI protected areas using IUCN criteria, and support local communities within their boundaries. In these areas, local people are granted access to forest products (e.g., firewood, timber, grasses and non-timber forest products [NTFPs]) based on operational plans authorized and approved by community forest managers.

Pangolins in Nepal are not restricted to the country's PA network. Substantial additional habitat exists in forested areas that are managed for purposes other than conservation. Forests and other wooded land occupy 66,100 km^2 (6.61 million ha); 44.7% of the total area of the country. Of the total forested area, 49,300 km^2 (4.93 million ha) or 82.7% lies outside of PAs (DFRS, 2016). Pangolin conservation in Nepal must take account of these areas.

Nepal is well known for its very successful system of community forest management, which is characterized by the decentralization of authority to local people to conserve, manage and use forest resources on a sustainable basis (Gilmour and Fisher, 1991; Government of Nepal, 1988, 2015; Hobley and Malla, 1996). The country is arguably more progressive in

TABLE 25.1 Summary of PAs in Nepal. CP=Chinese pangolin, IP=Indian pangolin.

PA name	PA declaration date	Core area (km^2)	BZ declaration date	BZ area (km^2)[a]	Physiographic zones(s)	Pangolin presence
Chitwan NP	1973	952.63	1996	729.37	Inner terai (low land)	CP, IP
Bardiya NP	1976	968	1996	507	Terai to inner terai	IP
Banke NP	2010	550	2010	343	Terai to inner terai	IP
Khaptad NP	1984	225	2006	216	High mountain	–
Langtang NP	1976	1710	1998	420	Middle mountain to high mountain	CP
Makalu Barun NP	1991	1500	1999	830	Middle mountain to high mountain	CP
Parsa NP	1984	627.39	2005	285.3	Terai to inner terai	CP, IP
Rara NP	1976	106	2006	198	High mountain	–
Sagarmatha NP	1976	1148	2002	275	High mountain	CP
Shey-Phoksundo NP	1984	3555	1998	1349	High mountain	–
Shivapuri-Nagarjun NP	2002	159	2015	118.61	Middle mountain	CP
Suklaphanta NP	1976	305	2004	243.5	Terai (low land)	IP
Koshitappu WR	1976	175	2004	173	Terai (low land)	–
Dhorpatan HR	1987	1325	Proposed	750	High mountain	–
Annapurna CA	1992	7629	–	–	Middle mountain to high mountain	CP
Api Nampa CA	2010	1903	–	–	Middle mountain to high mountain	–
Krishnasaar CA	2009	16.95	–	–	Terai (low land)	–
Gaurishankar CA	2010	2179	–	–	Middle mountain to high mountain	CP
Kangchenjunga CA	1997	2035	–	–	Middle mountain to high mountain	CP
Manaslu CA	1998	1663	–	–	Middle mountain to high mountain	–
Total Core Area		28,731.97	Total BZ Area	5687.78		

[a]*The proposed Dhorpatan HR is not included in the total BZ area.*

terms of its management of production forests than it is for its PAs.

Community-Based Organizations (CBOs), such as buffer zone committees and conservation area management committees, have been established to ensure participatory decision making for conservation and development interventions. About 30–50% of the revenue generated from PAs is invested in community development via BZ management councils (DNPWC, 2017). This approach has encouraged communities to participate in conservation activities and to view wildlife and their habitats as valuable community assets (Bajracharya and Dahal, 2008; Bhattarai et al., 2017). However, human-wildlife conflict threatens the success of community-based conservation in Nepal (Acharya et al., 2016; Lamichhane et al., 2018) and education, mitigation, and compensation measures are recognized as important responses to this issue.

As of January 2019, 19,361 Community Forest User Groups (CFUG), involving 2.46 million households (about 45% of Nepal's population) manage forested areas totaling 18,135 km^2 across the country (Central Bureau of Statistics, 2012; Department of Forestry, 2018). Similarly, 875,021 households (16.1% of the population [Central Bureau of Statistics, 2017]) are engaged in the management of 428.35 km^2 of leasehold forest and 576.63 km^2 of collaborative forest. Originally introduced in the 1970s, with the objective of addressing unsustainable forest utilization, and to protect timber resources, the remit of many CFUGs has extended to include the protection of wildlife. This presents an opportunity to build on the existing participatory approach to conservation in Nepal's production forests to conserve pangolins.

Challenges to pangolin conservation in Nepal

Few studies have been conducted on pangolins and their distribution within Nepal's protected area network (including NP, WR, CA and BZs). However, information on pangolins in human-dominated landscapes outside of the PA network is more readily available. Recent surveys have confirmed the presence of pangolins in 44 of Nepal's 77 districts (Ministry of Forest and Soil Conservation, 2016). Their known distribution extends from lowland tropical regions (the Terai) to temperate regions, which occur at higher elevations, up to 3000 m asl in eastern Nepal (Fig. 25.1; Khatiwada, 2016). The wide distribution of pangolins outside protected areas raises concerns about the vulnerability of the animals to threats including interactions with humans (hunting, poaching and illegal trade, persecution); deforestation; and infrastructure development, including road construction and hydropower projects that destroy and fragment habitat. Additional challenges include limited levels of awareness among local communities concerning the importance and conservation needs of the species; weak and under-resourced law enforcement; and excessive use of chemical fertilizers and pesticides in agricultural practices. The latter may adversely affect pangolins through loss of prey availability and, potentially, bioaccumulation. Research in Taiwan suggests that long-term exposure to toxins in the environment may be linked with a high prevalence of hepatic and respiratory lesions in Chinese pangolins (Khatri-Chhetri et al., 2016; Sun et al., 2019).

As a result of these threats, there is concern in Nepal that pangolins are now locally extinct in several areas where they were previously known to occur (e.g., the districts of Bhaktapur, Kavrepalanchowk and Sindhupalchowk; Kaspal et al., 2016). The biggest challenges to conserving pangolins in the country are a lack of population and ecological data with which to target conservation actions, and a poor understanding of the extent and impact of hunting and poaching pangolins for local use and the illegal wildlife trade. Limited funding for conservation is also a problem in Nepal.

Absence of ecological data

A national pangolin survey was conducted in 2016 and several small-scale pangolin conservation projects were implemented between 2012 and 2018. The national survey was a first step in determining the distribution of pangolins in Nepal. It was carried out in several districts and depended in large part on reports of pangolin presence from local community members rather than systematic ecological surveys (see Ministry of Forest and Soil Conservation, 2016). Small-scale projects aiming to collect basic ecological information on the Chinese pangolin and raise awareness of the species among local communities have also been implemented in several locations (see Fig. 25.3). However, key ecological data regarding pangolins in Nepal, and a strategic research approach to address the knowledge gaps, are lacking. These are needed to develop effective and appropriate conservation programs. Data on species distribution, habitat requirements, occupancy and population densities, potential strongholds, the extent to which the two species are sympatric, feeding ecology, burrow use and genetic information are all required. Baseline information and ongoing monitoring for all of these parameters is important in order to inform conservation and management.

Social and cultural information about local traditions relating to pangolins, attitudes to

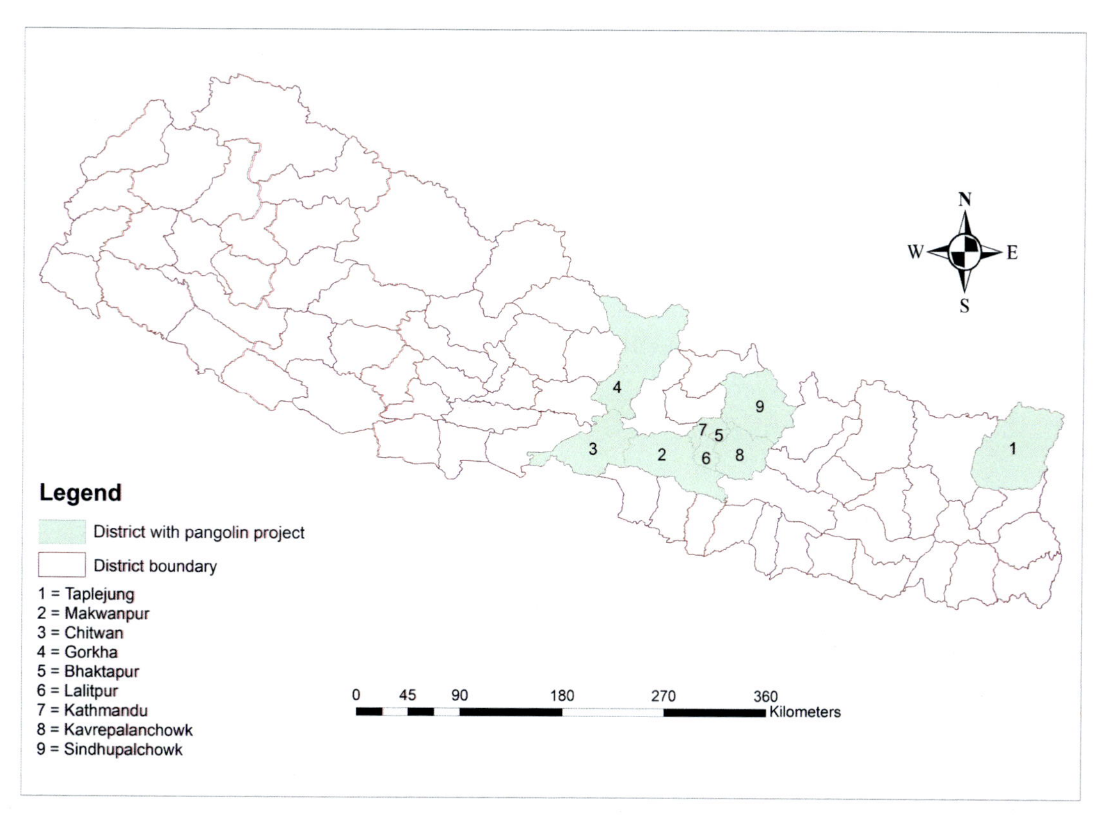

FIGURE 25.3 Districts in which community-based pangolin conservation projects have been implemented. *Source: Department of Survey, Department of National Parks and Wildlife Conservation, National Trust for Nature Conservation, Government of Nepal.*

pangolin conservation and the extent and type of local ecological knowledge regarding the species would also improve understanding of the conservation context for pangolins in Nepal. In addition, a better understanding of the key threats is required. Climate induced threats (forest fires, droughts, flooding, landslides, climate change) and anthropogenic threats (poaching, illegal trade, impact of chemical fertilizers on agricultural crops and prey species, impact of infrastructure development e.g., hydropower, roads) are likely both important influences on pangolin abundance. There is also a need to collect detailed rescue and release information in order to better understand pangolin presence and the threats the species face. The strategic and effective design and implementation of pangolin conservation projects in Nepal depends on the acquisition of such data and on the successful engagement with various stakeholders, but in particular, local communities.

Poaching and trafficking of pangolins in Nepal

Hunting and trade in pangolins and their body parts (e.g., scales) is illegal in Nepal. The National Parks and Wildlife Conservation Act (1973 – specifically Section 26.2) stipulates that any person who kills or injures a pangolin shall be punished with a fine ranging from NPR 40,000 (ca. USD 400) to NPR 75,000 (ca. USD 750) and/or face a prison sentence ranging from one to 10 years. In addition, the Government of Nepal recently ratified the Convention on International Trade in Endangered Species (CITES) Act 2018. This act has provisions for punishments associated with breaches of the Act. Fines ranging from NPR 500,000 (ca. USD 5000) to NPR 1000,000 (ca. USD 10,000) and/or a prison sentence of five to 15 years may be imposed on any person who keeps, uses, grows, breeds, sells, purchases, transfers or obtains pangolins or their parts without permission. Despite these strong legal provisions, poaching and trafficking of pangolins is frequently observed (Katuwal et al., 2013; Paudel, 2015). At least 46 seizures of pangolins or their scales were made in Nepal between May 2010 and July 2018, involving 785 kg of scales (Fig. 25.4).

Nepal's community approach to pangolin conservation offers opportunities to safeguard the species (see subsequent discussion), and there is a growing level of interest in pangolin conservation projects among local communities. However, local people remain involved in

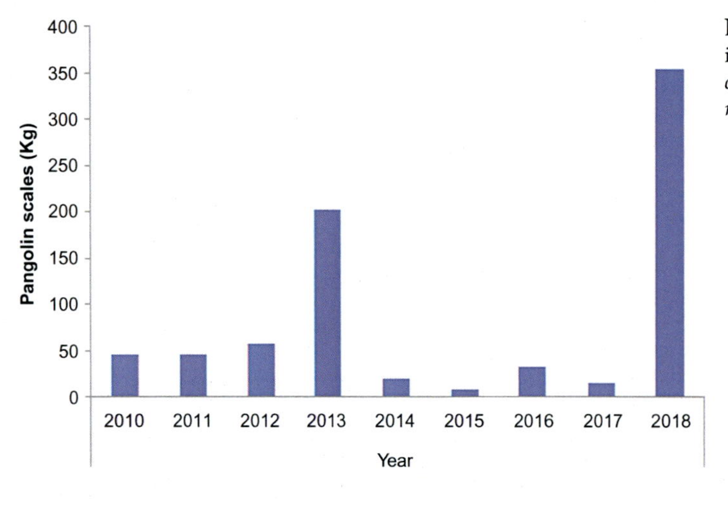

FIGURE 25.4 Pangolin scales confiscated in Nepal between 2010 and July 2018. *Data compiled by A. Khatiwada from online news reports in Nepal.*

poaching and trafficking activities. Poaching is likely perceived as a low risk/high reward endeavor in rural areas in Nepal. The drivers of poaching are largely related to the financial rewards available, although there is anecdotal evidence that pangolin meat is still eaten in some communities. A lack of awareness of the laws and regulations protecting pangolins and of the importance of conservation are also common.

Two important approaches, of the many possible, could help to prevent poaching at the local level. First, the development of a reward and sanction mechanism through locally agreed rules and regulations; and second, the introduction of income generating activities or skills training to facilitate alternative livelihood options. The former could help to reward people who contribute to pangolin conservation and punish those who are involved in poaching; and the latter could reduce local peoples' reliance on forest resources and may motivate and empower local communities for long-term pangolin conservation. There are limitations to these approaches including that the latter may merely provide additive income as opposed to an actual alternative livelihood. Such programs need to be carefully planned with local communities (see Chapters 23 and 24).

Human–pangolin conflict

Human–pangolin conflict occurs in the hilly and middle mountain areas of Nepal (see Fig. 25.1), particularly in terraced agricultural land. There is anecdotal evidence of a high density of pangolin burrows in such areas. Local people claim that the burrows interfere with crop production and degrade the aesthetic value of the terraces. There are also unsubstantiated claims that clusters of burrows lead to landslides during the rainy season. Local people sometimes act on these perceptions, capturing and killing pangolins found on farmland. Such incidents have been reported in various districts, including Solukhumbu, Taplejung and Ramechhap in northeastern Nepal. Here, local people consume the meat of the animals they kill and sell the scales to middlemen who trade them illegally.

Even where pangolin burrows do not cause such problems, a commonly held understanding is that pangolins bring bad luck to those who encounter them, which is sufficient motivation for members of rural communities to persecute the animals. Nepal's wildlife conservation authorities and agencies have considerable experience in mitigating human-wildlife conflict (involving other species such as the tiger and rhinoceroses). The importance of changing community attitudes through awareness raising and education programs has been key to the success of such programs; and is likely to be important to reduce human-pangolin conflict.

Limited resources

There are limited global resources for the conservation of wildlife. Nepal obtains much of its funding for conservation from international donors and aid agencies. A substantial proportion of these resources are directed to the conservation of iconic megafauna including elephants (Elephantidae) and rhinoceroses. Although more than two dozen studies (e.g., Bhandari, 2013; Dhakal, 2016; Kaspal, 2016; Khatiwada, 2014; Khadgi, 2016; Paudel, 2015; Sapkota, 2016; Suwal, 2011; Tripathi, 2015) have been conducted on pangolins in Nepal, scarce resources prevent these efforts translating into concerted and strategic conservation actions. Much of the existing effort is reliant on a small number of researchers, conservationists, university scholars and CFUGs who are championing pangolin conservation, often relying on their own financial support or using short term funding grants to build a better understanding of these species. This lack of a strategic approach means that potential synergies between projects are not recognized or acted upon and reduces the effectiveness and efficiency of each individual project. This piecemeal approach is not sustainable.

Community support for pangolin conservation is present in some regions. For example, the Taudolchhap CFUG in Bhaktapur has been conserving rescued and resident Chinese pangolins in their community forest as part of a pangolin conservation program established in 2012. This has been recognized by the Government of Nepal's Department of Forests and Soil Conservation in the form of a cash prize from the local District Forest Office (DFO). Many other local communities are willing to safeguard pangolins and want to be involved in pangolin research and conservation activities, including establishing pangolin sanctuaries and rescue and research centers. However, they wish to be paid fairly for their contribution. The Taudolchhap CFUG has established some innovative approaches to ensure that they are remunerated for their conservation activities. For instance, they request a fee from researchers who wish to study pangolins in their community forest and from visitors who want to access the forest for a chance to see a pangolin.

Opportunities for pangolin conservation in Nepal – the promise of community conservation

Nepal has an enviable reputation for curbing poaching within its protected areas (Acharya, 2016; WWF, 2018). The organizations involved in Nepal's anti-poaching activities range from government to grass-roots organizations and include the Nepalese Government's Ministry of Forests and Environment (MoFE), the Nepalese Army, provincial and local level government authorities, non-governmental organizations (NGOs), and CBOs. Community-Based Anti-Poaching Units (CBAPUs) are the key CBOs in this context. They have played a crucial role in the achievement of six zero poaching years (i.e., no poaching incidents relating to three key megafauna species: greater one-horned rhinoceros, tiger and Asian elephant (*Elephas maximus*) in a 12-month period) since 2012. Recent institutional and legislative changes allowing strict enforcement of laws and effective involvement of local communities have been integral to these achievements (Aryal et al., 2017). Local communities have adopted and supported the concept of CBAPUs, the main role of which is to provide timely and accurate intelligence regarding poaching activities to enforcement agencies. The successful contributions of local communities towards Nepal's achievements in wildlife conservation are an important starting point for further developing and refining community-led pangolin conservation in forested areas (in human dominated landscapes) and around PAs.

Outside protected areas (including in forest estates) there are no organized or dedicated law enforcement mechanisms for wildlife protection. Furthermore, the emphasis in community forests has traditionally been on the protection and production of key forest resources including the collection of NTFPs. Conservation in general, and pangolin conservation in particular, has not been a priority (and indeed often runs counter to the forestry production/direct benefit focus of CFUGs). In some areas, however, community conservation initiatives are beginning to emerge (Fig. 25.3). Several local communities have successfully developed and introduced strategies to enforce laws relating to the protection of key wildlife species, including pangolins, in community forests, for example, in Makwanpur district (Raniban CF).

Importantly, most pangolin conservation projects in Nepal have focused on the Chinese pangolin. They have comprised various outreach programs intended to raise public awareness of pangolin conservation, and included the holding of workshops, radio shows, school presentations and the production of products portraying pangolin conservation messages (e.g., posters, booklets, t-shirts, cups and badges). A book entitled *'Saalak'* (the Nepali word for pangolin) was published and widely distributed in 2016 (Kaspal et al., 2016). The book links effective and successful conservation of pangolins with opportunities for economic well-being via nature-based tourism and is intended to alter peoples' perceptions about the value of the animals in their natural habitats. It is hoped that successful community-based pangolin conservation efforts that have been initiated can be scaled up and replicated across the country.

On a broader scale, a range of stakeholders are discussing the interlinking of existing protected areas in Nepal via biological corridors, building on existing and successful landscape-scale conservation programs such as the Terai Arc Landscape and Chitwan-Annapurna Landscape. The inclusion of existing forests outside of the protected area network should provide significant opportunities for pangolin conservation at the landscape level, including possibilities for on-the-ground conservation projects and local community engagement through improving and diversifying livelihoods.

Community-based pangolin conservation initiatives in Nepal

Despite the challenges described above, there are several examples of successful community-based pangolin conservation initiatives in Nepal, where local communities are engaged and active in protecting pangolins. These initiatives have been implemented in various districts in Nepal (see Fig. 25.3; Kaspal, 2016; Khatiwada, 2014; NTNC, 2018; SMCRF, 2018; ZSL, 2018).

Taplejung, Makwanpur, Chitwan and Gorkha

In 2012, NTNC, in collaboration with the Zoological Society of London (ZSL), initiated a project in Nangkholyang and Dokhu villages in the Taplejung District in eastern Nepal (Fig. 25.3). The aim was to support conservation activities to raise awareness and control illegal trade at the local level. A pangolin conservation committee was established.

Prior to the project, villagers who came across a pangolin would most likely have killed it. The meat is considered a delicacy and may have been eaten or sold. The scales would have been sold into illegal trade. There is evidence of a change in attitudes following the implementation of the project. It is now more common for local people who come across a live pangolin in the fields or on a road, to bring it back to the village and to the attention of the conservation committee members. Committee members use the opportunity to gather a wider group of people to talk about the pangolin and explain the law protecting it before releasing it back into the wild. Between 2013 and 2016, six pangolins were found by local community members and brought to committee members for safe release into appropriate habitat. Many members (approximately 45 people) of local communities were actively involved in organizing patrolling events, community meetings and raising awareness among villagers between 2012 and 2014, but this has since dropped to 2–3 individuals. This is because financial and logistical support and technical backup waned following the end of the project period. With so few people now involved, project activities have all but ceased. The situation is unlikely to improve until continued support (financial, logistical, and technical) is available. This could occur if local government takes ownership or if communities begin to benefit directly from their involvement in pangolin conservation (e.g., via nature-based tourism).

On World Pangolin Day 2018, two community managed pangolin conservation areas (CMPCAs), the Chuchchekhola CFUG and Situ BZCFUG, covering an area of approximately 7 km^2, were established in Makwanpur district – close to Chitwan and Parsa NPs. These new CMPCAs are the result of a collaboration between ZSL, DNPWC, NTNC, Himalayan Nature (HN), Mithila Wildlife Trust (MWT) and CFUGs. Associated community engagement programs, including camera trapping surveys, were conducted in 2016 and 2018 in seven CFUGs within these CMPCAs. One of the CFUGs, Raniban CFUG, initiated a pangolin conservation program that resulted in a community-led declaration of a pangolin core area within the community forest. This was an example of increased ownership and participation in pangolin conservation by a CMPCA within a community forest.

The NTNC has been working with four CFUGs in the Chitwan district and four in the Gorkha district within a single program between 2016 and 2019. The acquisition of ecological data from camera traps is a key feature of this project. Camera trap surveys were conducted in 2016 and 2018 and confirmed the presence of Chinese pangolins in three of the eight community forests surveyed (one in Chitwan and two in Gorkha). All eight community forests surveyed are considered potential pangolin habitat.

The NTNC also supports CFUGs to implement outreach programs and trains them to monitor pangolins and to protect them from poaching and illegal trade. These projects are developing towards the project goals of community-based and community-led pangolin conservation.

Bhaktapur, Lalitpur, Kathmandu, Kavrepalanchowk and Sindhupalchowk

The Small Mammals Conservation and Research Foundation (SMCRF), Women for Conservation (WC), Greenhood Nepal (GHN), Natural Heritage Nepal (NHN) and Nepal Rural Development and Environment Protection Council (NRDEPC) are collaboratively implementing community-based pangolin conservation projects in Bhaktapur, Lalitpur, Kathmandu, Kavrepalanchowk and Sindhupalchowk. The goal of these projects is to educate local community members about pangolins and to engage them in community-based pangolin conservation efforts.

Project activities have included: outreach, such as conservation education programs in local schools, essay or artwork competitions among school students, educational public speaking events, round table discussions, radio broadcasts, workshops, and the production and distribution of posters, books, t-shirts and caps (SMCRF, 2018); community capacity building programs, including training in ecological survey and monitoring of pangolins and collection of local ecological knowledge, use of GPS and camera traps and the formation and mobilization of local CBAPUs; and alternative livelihoods and income generating activities (e.g., cooperatives, micro-loans, and homestay promotion).

Traditional ecological knowledge held by community members, and their constant presence "on the ground", provide valuable sources of information to improve knowledge and understanding of pangolin populations and distribution. In addition, the involvement of local people in patrolling forest areas – an activity that is otherwise lacking outside the protected area network – reduces the willingness of other community members to participate in pangolin poaching. Community members who are engaged in pangolin conservation educate their peers about the laws relating to the killing of pangolins, and inform their neighbors, relatives and friends who may have been knowingly engaged in illegal activities (Roe, 2015). Importantly, community members engaged in conservation also discourage outsiders from coming into the villages in search of pangolins or scales because of a higher risk of being reported to the relevant authorities.

The future of pangolin conservation in Nepal – opportunities and challenges

There are several characteristics of conservation in Nepal that bode well for pangolins. The majority of the Nepalese population is either Hindu or Buddhist. Both religions have strong teachings relating to respect for wildlife. People living in Nepal's rural communities typically accept the premise underpinning species conservation – that the protection and conservation of wildlife and forest habitats is beneficial. This is partly because of their religious values and partly because they understand that, they are dependent on natural resources (e.g., NTFPs) for their livelihoods. These resources are therefore highly valued. Although some rural people believe that seeing a pangolin is bad luck and that killing the animal can prevent associated problems for themselves and their families, others consider pangolins helpful because they feed on termites and ants, thereby regulating populations of these insects. Typically, community members have shown keen interest during the early stages of community-projects focused on pangolins because they receive new information and can take part in capacity building training and workshops. The authors of this chapter have observed social rewards in the form of recognition from peers and, the media, and project organizers, and some participants have reported that this has increased their sense of self-worth and self-efficacy.

Despite being protected species in Nepal since the 1970s, the active conservation of pangolins has not been a priority in the country. Charismatic, flagship megafauna such as rhinoceroses, tigers and even snow leopards (*Panthera uncia*) have received much of the spotlight, and much of the available funding. This is unsurprising as the protected area network in Nepal was established to protect such species. However, the threat of extinction of Nepal's two pangolin species, and the alarmingly high levels of illegal trade (see also Chapter 16), has resulted in a higher priority for pangolin conservation by the Nepalese government. The national pangolin survey, completed in 2016, was preceded by development of a protocol for monitoring pangolins, and a Pangolin Conservation Action Plan for Nepal (2018–22; see DNPWC, 2018 and Department of Forestry, 2018). One DFO (Kavrepalanchowk) has also prepared a pangolin management plan for community forests within its jurisdiction (Kaspal, 2017).

Several pangolin conservation projects are being implemented in Nepal and there are impressive conservation success stories, where communities have taken ownership and developed capacity to conserve pangolins within their jurisdiction, whether that be a community forest or on privately owned land. Yet, there are challenges to overcome to secure the conservation of pangolins in Nepal, including limited funds, lack of ecological data and the growing threat of poaching and illegal trade within Nepal. Nevertheless, with the necessary resources and collaboration, Nepal is well placed to play a leading role in pangolin conservation.

References

Acharya, K.P., 2016. A Walk to Zero Poaching for Rhinos in Nepal. Department of National Parks and Wildlife Conservation, Kathmandu, Nepal.

Acharya, K.P., Paudel, P.K., Neupane, P.R., Kohl, M., 2016. Human-wildlife conflicts in Nepal: patterns of human fatalities and injuries caused by large mammals. PLoS One 11 (9), e0161717.

Aryal, A., Acharya, K.P., Shrestha, U.B., Dhakal, M., Raubenhiemer, D., Wright, W., 2017. Global lessons from successful rhinoceros conservation in Nepal. Conserv. Biol. 31 (6), 1494–1497.

Bajracharya, S.B., Dahal, N. (Eds.), 2008. Shifting Paradigms in Protected Area Management. National Trust for Nature Conservation, Kathmandu, Nepal.

Baral, N., 2005. Resources Use and Conservation Attitudes of Local People in the Western Terai Landscape, Nepal. M.Sc. Thesis, Florida International University, Miami, FL, United States.

Baral, T.N., Bhatta, G.P., 2005. Maximizing Benefits of Space Technology for Nepalese Society, Journal 5 (Published in 2063 B.S.), Nepalese Journal of Geoinformatics. Published by Ministry of Land Reform and Management, Government of Nepal.

Bhandari, N., 2013. Distribution, Habitat Utilization and Threats Assessment of Chinese Pangolin (*Manis pentadactyla* Linnaeus, 1758) in Nagarjun Forest of Shivapuri Nagarjun National Park. M.Sc. Thesis, Tribhuvan University, Kathmandu, Nepal.

Bhatt, N., 2003. Kings as wardens and wardens as kings: Post-Rana ties between Nepali royalty and National Park staff. Conserv. Soc. 2003 (1), 247–268.

Bhattarai, B.R., Wright, W., Poudel, B.S., Aryal, A., Yadav, B.P., Wagle, R., 2017. Shifting paradigms for Nepal's protected areas: history, challenges and relationships. J. Mountain Sci. 14 (5), 964–979.

Central Bureau of Statistics (CBS), 2012. National Population and Housing Census 2011 (National Report). NPHC 2011, 01. Central Bureau of Statistics, Kathmandu, Nepal.

Central Bureau of Statistics (CBS), 2017. A Compendium of National Statistical System of Nepal. Central Bureau of Statistics, Kathmandu, Nepal.

Corbet, G.B., Hill, J.E., 1992. The Mammals of the Indomalayan Region: A Systematic Review. Oxford University Press, Oxford.

Department of Forestry, 2018. Community Forestry. Available from: <http://dof.gov.np/dof_community_forest_division/community_forestry_dof>. [June 13, 2018].

Department of Forest Research and Survey (DFRS), 2016. Forest Resource Assessment Nepal. Kathmandu, Nepal.

Department of National Parks and Wildlife Conservation (DNPWC), 2017. Management Effectiveness Evaluation of Selected Protected Areas of Nepal. Ministry of Forests and Soil Conservation, Department of National Parks and Wildlife Conservation. Kathmandu, Nepal.

Department of National Parks and Wildlife Conservation (DNPWC), 2018. Protected Areas of Nepal. Available from: <http://www.dnpwc.gov.np/>. [June 9, 2018].

Dhakal, S., 2016. Distribution and Conservation Status of Chinese Pangolin in Palungtaar Municipality of Gorkha District, Western Nepal. Tribhuvan University, Kathmandu, Nepal.

Gilmour, D.A., Fisher, R.J., 1991. Villagers, Forests, and Foresters: The Philosophy, Process, and Practice of Community Forestry in Nepal. Sahayogi Press, Kathmandu.

Government of Nepal, 1988. Master Plan for the Forestry Sector, Nepal. Kathmandu, Nepal.

Government of Nepal, 2015. Forest Policy. Government of Nepal, Kathmandu.

Gurung, H., 1980. Vignettes of Nepal. Sajha Prakashan, Kathmandu, Nepal.

Gurung, J.B., 1996. A pangolin survey in Royal Nagarjung Forest in Kathmandu, Nepal. Tiger Paper 23 (2), 29–32.

Hobley, M., Malla, Y., 1996. From Forests to Forestry - The Three Ages of Forestry in Nepal: Privatization, Nationalization, and Populism. In: Hobley, M. (Ed.), Participatory Forestry: The Process of Change in India and Nepal. Rural Development Forestry Network, Overseas Development Institute, London, pp. 65–92.

Hodgson, B.H., 1836. Synoptical description of sundry new animals, enumerated in the catalogue of Nepalese mammals. J. Asiat. Soc. Bengal 5 (52), 231–238.

Joshi, B.K., Acharya, A.K., Gauchan, D., Chaudray, P. (Eds.), 2017. The State of Nepal's Biodiversity for Food and Agriculture. Ministry of Agricultural Development, Kathmandu, Nepal.

Jnawali, S.R., Baral, H.S., Lee, S., Acharya, K.P., Upadhyay, G.P., Pandey, M., et al., 2011. The Status of Nepal's Mammals: The National Red List Series. Department of National Parks and Wildlife Conservation, Kathmandu, Nepal.

Kaspal P., Shah, K.B., Baral. H.S., 2016. Saalak. Himalayan Nature, Kathmandu, Nepal.

Kaspal, P., 2016. Scaling up the Chinese Pangolin Conservation Through Education and Community-Based Monitoring Programs to Combat Pangolin Trade in Sindhupalchowk District, Nepal, Natural Heritage Nepal. A report submitted to the Ocean Park Conservation Foundation, Hong Kong.

Kaspal, P., 2017. Pangolin Conservation Management Plan for the Community Forests of Nepal. Natural Heritage Nepal. Submitted to District Forest Office, Kavrepalanchowk, Nepal.

Katuwal, H.B., Neupane, K.R., Adhikari, D., Thapa, S., 2013. Pangolin Trade, Ethnic Importance and Its Conservation in Eastern Nepal. Small Mammals Conservation and Research Foundation and WWF-Nepal, Kathmandu, Nepal.

Khadgi, B., 2016. Distribution, Conservation Status and Habitat Analysis of Chinese Pangolin (*Manis pentadactyla*) in Maimajuwa VDC, Ilam, Eastern Nepal. School of Environmental Science and Management (SchEMS), Pokhara University, Kathmandu, Nepal.

Khatiwada, A.P., 2014. Conservation of Chinese Pangolin (*Manis pentadactyla*) in the Eastern Himalayas of Nepal. Unpublished final report to Zoological Society of London, London, UK.

Khatiwada, A.P., 2016. A Survival Blueprint for the Chinese Pangolin, *Manis pentadactyla*. Zoological Society of London, London, UK.

Khatri-Chhetri, R., Chang, T.C., Khatri-Chhetri, N., Huang, Y.-L., Pei, K.J.-C., Wu, H.-Y., 2016. A retrospective study of pathological findings in endangered Formosan pangolins (*Manis pentadactyla pentadactyla*) from southern Taiwan. Taiwan Vet. J. 43 (1), 55–64.

Lamichhane, B.R., Persoon, G.A., Leirs, H., Poudel, S., Subedi, N., Pokheral, C.P., et al., 2018. Spatio-temporal patterns of attacks on human and economic losses from wildlife in Chitwan National Park, Nepal. PLoS One 13 (4), e0195373.

Ministry of Forest and Soil Conservation, 2016. National Pangolin Survey (NPS) Final Report. Government of Nepal, Kathmandu, Nepal.

Mishra, H.R., 1982. Balancing human needs and conservation in Nepal's Royal Chitwan Park. Ambio 11 (5), 246–251.

National Trust for Nature Conservation (NTNC), 2018. Annual Performance Report, Hariyo Ban Program, National Trust for Nature Conservation 2018. Kathmandu, Nepal.

Paudel, K., 2015. Assessing Illegal Wildlife Trade in Araniko-Trail, Nepal. M.Sc. Thesis, Pokhara University, Kathmandu, Nepal.

Roe, D., (Ed.), 2015. Conservation Crime and Communities: Case Studies of Efforts to Engage Local Communities in Tackling Illegal Wildlife Trade. IIED, London.

Sapkota, R., 2016. Habitat Preference and Burrowing Habits of Chinese Pangolin. M.Sc. Thesis, Tribhuvan University, Kirtipur, Nepal.

Small Mammals Conservation and Research Foundation (SMCRF), 2018. Educating and Empowering Local Community for Pangolin Conservation in Kathmandu Valley. Small Mammals Conservation Research Foundation, Kathmandu, Nepal.

Smythies, E., 1942. Big Game Shooting in Nepal. Thacker Spink, Calcutta.

Sun, N.C.-M., Arora, B., Lin, J.-S., Lin, W.-C., Chi, M.-J., Chen, C.-C., et al., 2019. Mortality and morbidity in wild Taiwanese pangolin (*Manis pentadactyla pentadactyla*). PLoS One 14 (2), e0198230.

Suwal, T.L., 2011. Status, Distribution, Behaviour and Conservation of Pangolins in Private and Community Forest of Balthali in Kavrepalanchowk, Nepal. M.Sc. Thesis, Tribhuvan University, Kathmandu, Nepal.

Suwal, R., Verheugt, Y.J.M., 1995. Enumeration of Mammals of Nepal. Biodiversity Profiles Project Publication No. 6. Department of National Parks and Wildlife Conservation, Ministry of Forest and Soil Conservation. His Majesty's Government, Kathmandu, Nepal.

Tripathi, A., 2015. Distribution and Conservation Status of Chinese Pangolin (*Manis pentadactyla*) in Prithivinarayan municipality, Gorkha, Western Nepal. B.Sc. Thesis, Tribhuvan University, Kathmandu, Nepal.

United Nations, 2018. UN list of Least Developed Countries. Available at: <http://unctad.org/en/pages/aldc/Least%20Developed%20Countries/UN-list-of-Least-Developed-Countries.aspx>. [September 25, 2018].

Upreti, B.N., 2017. Early Days of Conservation in Nepal: A Collection of Papers and Views Since 1970. Nepal Biodiversity Research Society, Kathmandu, Nepal.

World Bank, 2018. Nepal Systematic Country Diagnostic. Available at: <http://documents.worldbank.org/curated/en/361961519398424670/pdf/Nepal-SCD-Feb1-02202018.pdf>. [April 3, 2019].

WWF, 2018. How Nepal Achieved Zero Poaching. Available from: <http://tigers.panda.org/news/achieve-zero-poaching/>. [July 22, 2018].

ZSL, 2018. Launching the World's First Community Managed Pangolin Conservation Areas. Available from: <https://www.zsl.org/blogs/asia-conservation-programme/launching-the-world%E2%80%99s-first-community-managed-pangolin>. [July 22, 2018].

CHAPTER

26

The Sunda pangolin in Singapore: a multi-stakeholder approach to research and conservation

Helen C. Nash[1,2], Paige B. Lee[3], Norman T-L Lim[2,4], Sonja Luz[3], Chenny Li[5], Yi Fei Chung[5], Annette Olsson[6], Anbarasi Boopal[7], Bee Choo Ng Strange[8] and Madhu Rao[9]

[1]Department of Biological Sciences, National University of Singapore, Singapore [2]IUCN SSC Pangolin Specialist Group, ℅ Zoological Society of London, Regent's Park, London, United Kingdom [3]Department of Conservation, Research and Veterinary Services, Wildlife Reserves Singapore, Singapore [4]National Institute of Education, Nanyang Technological University, Singapore [5]Conservation Division, National Parks Board, Singapore [6]Conservation International, Singapore [7]ACRES Wildlife Rescue Centre (AWRC), Singapore [8]Vertebrate Study Group, Nature Society Singapore, Singapore [9]Wildlife Conservation Society, Singapore

OUTLINE

Pangolins
DOI: https://doi.org/10.1016/B978-0-12-815507-3.00026-5

Introduction

The Sunda pangolin (*Manis javanica*) is the only species of pangolin native to Singapore. It is distributed across the country, including offshore islands such as Pulau Ubin and Pulau Tekong (Fig. 26.1). Singapore is a unique environment for pangolins, being a small island city state and, hence, heavily urbanized. Sunda pangolins in Singapore occur in highly developed sites, such as housing estates, schools and universities, as well as green spaces and parks with natural vegetation more consistent with the rest of their geographic range.

Singapore is the only range state where poaching of pangolins is not known to be a threat domestically. Possible reasons for the cessation of poaching of the species, which has occurred historically, include strong law enforcement (National Parks Board (NParks), 2019), socio-economic factors, including relatively high per capita income (Department of Statistics, Singapore, 2019) compared to other countries, and good public access to westernized healthcare (Ministry of Health, Singapore, 2019). Nevertheless, government agencies and other stakeholders recognize that the safety and conservation of pangolins in Singapore should not be taken for granted. This recognition drives continued attention and ongoing actions to maintain and enhance pangolin populations in Singapore.

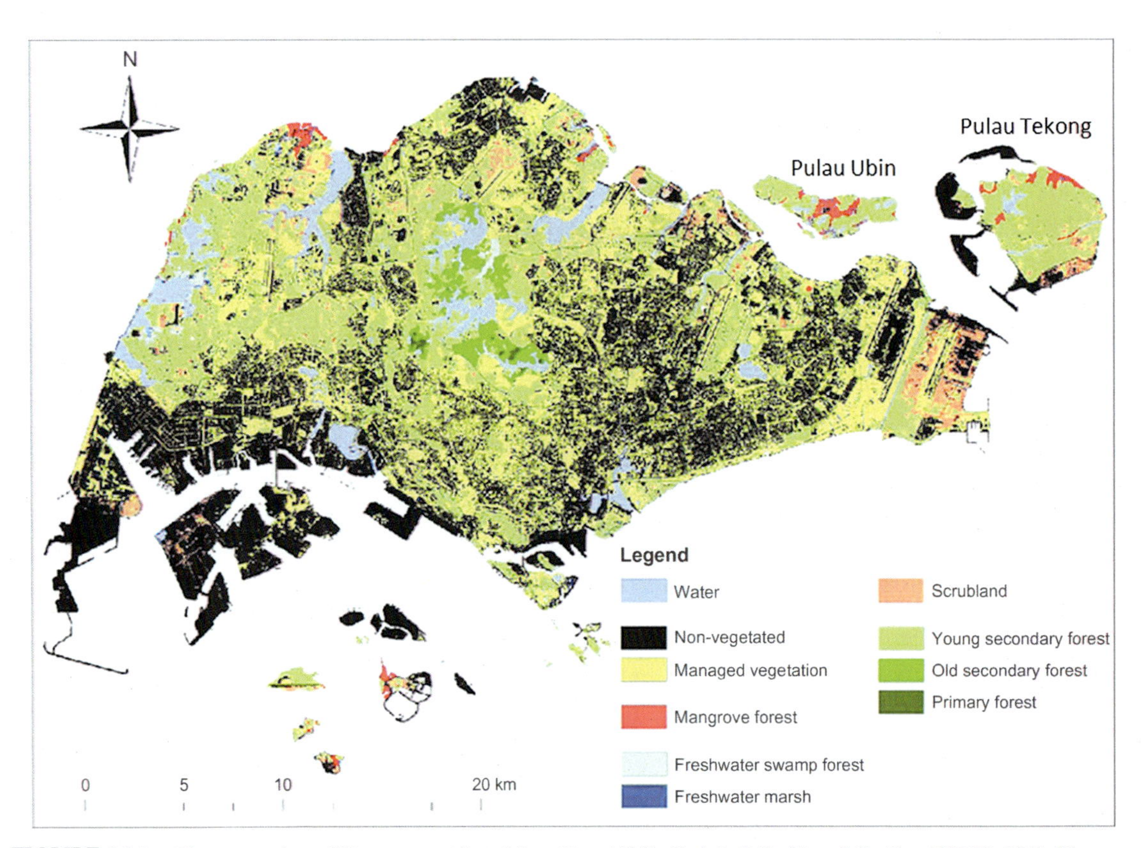

FIGURE 26.1 The vegetation of Singapore. *Adapted from Yee, A.T.K., Corlett, R.T., Liew, S.C., Tan, H.T.W., 2011. The vegetation of Singapore -an updated map. Gardens' Bull. Singapore 63 (1 & 2), 205–212.*

This chapter summarizes knowledge of pangolins in Singapore, reviews research and conservation actions that have been undertaken for the species, and outlines future plans for securing Sunda pangolin populations in the country. Throughout, it emphasizes the benefits of having a national-level working group, the Singapore Pangolin Working Group (SPWG), to foster collaboration between stakeholders and drive forward pangolin conservation.

Population

Singapore is only approximately 720 km^2 and as pangolins are considered rare across the country, it has been suspected that the population is small. In 2019, the SPWG facilitated data sharing across agencies and stakeholders, to enable the Singapore population of Sunda pangolins to be estimated for the first time. The estimate of 1046 pangolins ($575 < N_1 < 1604$; D. Fung Yu En, unpubl. data) was derived from a variety of mark-recapture estimators and environmental niche modeling (D. Fung Yu En, pers. comm.). Data sources included camera trap surveys undertaken by Singapore National Parks Board (NParks; Chung et al., 2016) and a National University of Singapore (NUS) student within nature reserves, camera traps on private land deployed by Mandai Park Holdings, logged callouts by local charity, Animal Concerns Research and Education Society (ACRES), microchip data from Wildlife Reserves Singapore (WRS) on rescued and released pangolins, and other radio telemetry and pangolin tracking data. Difficulties in uniquely identifying individuals from photos, and the existence of pangolins within urban areas and beyond the camera trap grids used complicated the process, but the combinatorial approach of using a variety of mark-recapture estimators and environmental niche modeling helped to address these issues. It would have been extremely difficult to achieve a robust population estimate prior to the establishment of the SPWG, which was created in 2014 specifically to improve coordination between research, conservation and outreach efforts for pangolins in Singapore.

Habitat use

Singapore is heavily urbanized, and some of the resident pangolins appear to have modified their behavior to survive across the fragmented patches of natural habitat. Natural habitat comprises primary forests of lowland dipterocarp, coastal hill and freshwater swamp species, as well as secondary forests, Adinandra belukar, grasslands, gardens and urban parkland (NParks, 2014). In addition to natural habitats, pangolins can be found at highly developed sites, for example, housing estates, schools, universities, and roadsides. They have been documented to use drains and culverts to cross beneath roads, and have been observed using urban structures such as pipes to pass beneath buildings (Nash et al., 2018a). They have also been known to explore construction sites, and have been found in noisy or brightly lit areas. Whether these behaviors are widespread in pangolins across Singapore or just representative of a minority of highly urban individuals is not known. Other individuals in Singapore occupy forests, grasslands, and other green spaces, and exhibit characteristics that are more commonly documented for Sunda pangolins across their geographic range, including sheltering in forested areas and making use of large trees for natal dens (see Chapter 6).

Sunda pangolins use landscapes and structures that occur within areas managed by a wide variety of stakeholders, including NParks, the Ministry of Defence (MINDEF), Public Utilities Board (PUB), and Land Transport Authority (LTA). Green spaces with natural vegetation and mature trees are mostly confined to four nature reserves with peripheral green buffer zones in the center of the country (see Fig. 26.1), and several regional parks, managed by NParks. Pangolin presence has also been recorded in military zones, such as the Western Catchment Area, which have restricted entry.

Ecology

Little is known about Sunda pangolins in Singapore. For instance, their social structure, how they interact with conspecifics and other species, and the fragmentation of populations and associated gene flow. Yet, they are known to use a wide range of structures for shelter including burrows, trees, ferns and grasses, and urban structures (e.g., pipes), among others. Preliminary research suggests that diet varies with habitat type, and particular species such as weaver ants (*Oecophylla smaragdina*) might be preferential prey (A. Srivathsan, pers. comm; see Chapter 6). The species is semi-arboreal, spending time foraging for subterranean, terrestrial, and arboreal prey. Predation on ants and termites likely contributes to the control of local pests (i.e., termites), but there is little knowledge of the quantities of prey consumed.

Little is known about home ranges of Sunda pangolins in Singapore (Lim, 2007; Lim and Ng, 2008a; Nash, 2018). The species is typically shy, nocturnal and rarely seen, although some individuals, especially sub-adult males, are more commonly seen in public areas (ACRES, unpubl. data). This is likely because they are dispersing and in search of a home range; for example, a young male has been spotted at a busy urban crossroads during daylight (Nash, 2018). The species appears resilient and able to adapt to new habitats if threats such as poaching and roadkill can be minimized (Chapter 6). However, there are reports that female pangolins prefer large trees in which to make natal dens (Lim and Ng, 2008a). These dens are typically well-hidden in dense shrubbery and sometimes pangolins select den sites with other obstacles such as creepers and tree roots to protect the den entrance. This suggests the presence of mature forests with large trees (>50 cm DBH) are important for the reproduction of Sunda pangolins and their conservation.

Pangolins in Singapore have few natural predators because large carnivores have been extirpated from the country. Reticulated pythons (*Malayopython reticulatus*) do predate on the species (Lim and Ng, 2008b) and juveniles are at higher risk due to their smaller size. However, overall predation rates are thought to be low.

Threats

As noted in the introduction, poaching by humans is believed to be minimal, if any does still occur. Feral dogs pose some threat to pangolins, and they have been seen to chase individuals and cause injuries such as severe scratches and tail bites. One pangolin was found with only a stump of a tail, possibly due to a dog attack, although alternative causes such as mechanical damage by construction machinery are feasible.

Road-related mortality is likely the major threat to pangolins in Singapore. Chua et al. (2017) recorded 59 road kills between 1995 and 2017. ACRES reported that between June 2014 and March 2018 it received 107 phone calls about pangolins, 44% (47 calls) related to pangolins considered to be at risk of harm on roads. Other reasons for call-outs included that pangolins had been found within residential housing blocks, under cars, in private gardens, at swimming pools, inside schools, universities, temples and military buildings, and other public places. A few nuisance reports have also been received about pangolins, for example they have been found digging unwanted holes through wooden doors or in gardens. Typically, local people seem pleased and excited if they are lucky enough to encounter a pangolin, and photos or videos of sightings can be quickly shared via social media and local nature networks.

ACRES is able to rapidly respond to most call-outs, often within an hour, and after assessing the situation personnel are licensed to judge how best to respond using established protocols (ACRES, 2018). All pangolins are transported to Singapore Zoo within WRS for veterinary attention. Following rehabilitation, NParks arranges translocation to a safe and confidential natural location.

Legislative protection and law enforcement

In the early 1900s, pangolins were legally hunted in Singapore during authorized times of the year, but not on private property or other restricted areas (Tan and Tan, 2013). Pangolin soup was a distinguished dish for birthday and other celebrations. Some elderly Singaporeans recall celebrations at restaurants in the city's Chinatown where they ate pangolin (H.C. Nash, unpubl. data). Traditional Chinese Medicine (TCM) products also included pangolin parts, for example, tablets, ointments and creams containing powdered pangolin scales were recommended for a range of ailments, including, skin conditions, improving lactation, and heart conditions among others (Y.F. Chung, unpubl. data).

As awareness about the need to protect pangolins and other wildlife grew, the Singaporean government decided to end the use of native pangolins in Singapore and protect other wildlife through enactment of The Wild Animals and Birds Act (1965). Singapore also became a signatory to CITES, the Convention on International Trade in Endangered Species of Wild Fauna and Flora in 1986 and introduced the Endangered Species Act (2006) which further restricted trade in pangolins, including the unlicensed sale or use of pangolin products. The culture and state-society relations of Singapore is such that national legislation is strictly enforced and there are tough penalties for illicit acts. This, together with the country's small size and higher-than-average standard of living, perhaps explains the relative ease in which animal protection legislation was introduced and is enforced across Singapore. Globally, only a few countries have successfully minimized poaching rates, Singapore being a positive example. It is now illegal to trade, consume, or use any pangolin product within Singapore without a permit issued by NParks, which acts as the country's CITES Management and Scientific Authority. NParks also spot-checks TCM shops and other commercial traders and monitors stockpiles of pangolin products. Under the Endangered Species (Import and Export) Act, the maximum penalty for illegal trade (import, export and re-export) of wildlife listed in CITES is a fine of SGD 50,000 per specimen (not exceeding an aggregate of SGD 500,000) and/or imprisonment of up to two years. In addition, members of the public are only allowed to use designated trails in nature reserves, and are not allowed to enter these sites at night. NParks officers routinely check for snares, traps and suspicious behavior, and police patrols monitor sites frequently.

Meanwhile, illegal wildlife trade likely continues to pass undetected through Singapore's busy ports: not every shipping vessel can realistically be inspected. As of early 2019, the last known seizure of live pangolins was in 2003. This comprised 20 Sunda pangolins trafficked from Indonesia. Scales have also been seized; for example, in December 2015, 324 kg of pangolin scales trafficked from Nigeria were seized at Changi Airfreight Centre. The scales were en route to Lao People's Democratic Republic (Lao PDR) via Singapore. On April 3, 2019, Singapore Customs and Immigration Checkpoint Authority (ICA) and NParks seized 12.9 tonnes of pangolin scales at Pasir Panjang Export Inspection Station, and another 12.7

tonnes of scales on April 8, 2019 (NParks, pers. comm.). On both occasions the scales were seized from shipping containers on route to Vietnam from Nigeria. Efforts have been made to minimize illegal trade and prevent trafficking via Singapore, including improved whistle-blowing, reporting and enforcement (NParks, pers. comm.). However, in the face of rampant illegal wildlife trade globally involving highly organized criminal networks (Chapter 16), illicit shipments can be difficult to detect. Given that pangolins are under heavy poaching pressure in neighboring countries such as Indonesia and Malaysia, it is imperative that Singapore law enforcement and customs agencies remain alert to the possibility of persons and networks targeting pangolins in Singapore or using Singapore as a transit point.

Pioneering research and conservation efforts

Pangolin-focused conservation activities in Singapore started in the 1990s by NParks, WRS and local nature enthusiasts, with the objective of rescuing and rehabilitating pangolins found in hazardous urban estates where there is a risk of harm from roadkill and other hazards (e.g., construction machinery). The rehabilitated pangolins were released into suitable habitats around the island. This rescue, rehabilitation and release initiative continues.

Ex situ conservation research has been ongoing since 2005 at WRS, including Singapore Zoo and the Night Safari, which was globally the first organization to successfully breed and raise Sunda pangolins under human care (Fig. 26.2). Between 2012 and 2018, WRS received 103 rescued Singaporean Sunda pangolins. Of these, seven had to be euthanized on welfare grounds and seven were dead on arrival; the rest were rehabilitated, microchipped and released. Animal behavior experts at WRS have used feeding challenges to investigate the spatial memory of Sunda pangolins and concluded that the species can learn where to find reliable sources of food (P. B. Lee, 2018, unpubl. data). The diet in captivity has been assessed and improved, with input from animal nutritionists (see Chapter 28; Cabana and Tay, 2019). WRS has decades of husbandry experience which has led to successful conservation breeding, and rehabilitation of rescued Sunda pangolins.

In situ research on Sunda pangolin ecology and behavior was pioneered in 2005–06 on Pulau Tekong (Lim, 2007). It included an investigation of habitat preference, prey preference, natural behavior and baseline natural history. Wild pangolins were captured by hand and tracked using radio telemetry and infra-red triggered camera traps. The project ran between September 2005 and November 2006 and 22 pangolins were captured. The radio tag drop-off rate was high, 80% within two weeks, but four adult males were successfully tracked for seven days (see Chapter 6).

The Singapore Pangolin Working Group

Established in 2014, the SPWG includes a broad range of stakeholders that advise on research and conservation, engage with developers and policy makers, and raise public awareness of pangolins and the threats they face. The SPWG was created to improve coordination between conservation, research and outreach efforts for pangolins in Singapore. It is chaired by WRS and comprises varied stakeholders including ACRES, Conservation International (CI), Lee Kong Chian Natural History Museum (LKCNHM), NParks, Nature Society (Singapore; [NSS]), local NGO The Pangolin Story, Wildlife Conservation Society (WCS), and academics and other researchers. The SPWG strategizes how best to conserve pangolins and implement conservation action and management plans. Activities are guided

FIGURE 26.2 A captive-bred juvenile Sunda pangolin at Wildlife Reserves Singapore. *Photo credit: Wildlife Reserves Singapore/David Tan.*

through biannual working group meetings and ongoing communication between participants. Both in situ and ex situ research projects are facilitated, as well as community outreach, education, media work, and pangolin rescue, rehabilitation and release.

The SPWG has been instrumental in driving forwards research and conservation action for pangolins in Singapore. It has forged strong relationships between diverse members, fostered greater collaboration and coordination of research methodologies, and enabled long-term conservation plans to be developed and adaptively managed. The generation of the first robust population estimate for pangolins in Singapore is one example of a significant research output that came about as a direct result of the SPWG facilitating standardization of methods, data sharing and coordination between government agencies, research institutes and NGOs.

Ongoing conservation efforts

Since August 2014, the SPWG has coordinated research and conservation projects across Singapore. Projects have focused primarily on increasing understanding of Sunda pangolins in Singapore. In situ research projects aim to address critical research questions on population size, home range and habitat preferences. Research tools are wide-ranging and include the use of camera traps, tracking methods, sightings and roadkill reports, profiling and monitoring of rescued pangolins, and genetic research. Ongoing efforts are now discussed.

In situ research on population size, distribution and habitat selection

Camera trap surveys

Camera trap grids have been used by NParks to investigate the distribution of pangolins, estimate population size, and understand habitat selection. Camera traps have been deployed in nature reserves and at the "Eco-Link@BKE", an overhead wildlife bridge that connects reserves across a six-lane highway. Repeat surveys to facilitate an understanding of population trends through time are being conducted. Pangolins have been detected on the bridge and in reserves. In Singapore, over 100 camera traps spaced 50 m apart in 500 × 500 m grids were deployed across several different habitat types including primary and secondary forest and more open shrub and grassland. Analyses using generalized linear mixed models (GLMM) and generalized least squares (GLS) suggest that pangolin presence did not vary significantly with forest type, vegetation density (normalized difference vegetation index [NDVI]), or other habitat variables across the grids (Y.F. Chung, unpubl. data). Better knowledge of home range sizes is needed to produce occupancy models using camera trap data. This could potentially be addressed through other tracking methods.

Tracking and monitoring of translocated pangolins

Singapore is a good testing site for new technology due to the large number of pangolins that are rescued and released annually, relative accessibility of pangolin habitat, and minimal poaching, if any. In 2015, pangolin rescue, rehabilitation and release were identified as priority actions for Singapore by the SPWG. Two approaches have been piloted for post-release monitoring: radio telemetry and microchips. Radio telemetry has been conducted since March 2016 to further understand pangolin ecology and behavior (Fig. 26.3), and comprises post-release monitoring of rescued and translocated animals. Preliminary results indicate statistically significant differences between sub-adult and adult post-release behavior, with sub-adults seeming to disperse over large distances per night (over 3 km reported), unlike adults, which moved less per night and may be territorial (see Chapter 6; Nash, 2018). The finding that sub-adult pangolins can disperse across large distances, likely in search of new territory, is a novel result in Singapore. Urban sites are frequently used by both adult and sub-adult pangolins, including drains, culverts and roadside verges (Nash, 2018). Such extensive use of urban sites had also not previously been documented for Sunda pangolins (Chapter 6). Ten pangolins have so far been tracked; however, the radio tag drop off rate was high, with the majority lasting less than three weeks. Only two males were observed for a longer period, but this was only because each was surprisingly re-found over one month later (without tags) and then re-tagged and re-released (Nash, 2018). Due to these problems, the post-release survival rate of pangolins in Singapore is not well understood but only two of the ten animals have been confirmed dead.

The second in situ tracking method involves microchips. Each rescued and released pangolin seen by the veterinarians at Singapore Zoo is tagged with a subcutaneous Radio-frequency Identification (RFID) microchip, so that if re-found dead or alive it can be identified via a Trovan tag reader. As of 2018, only four pangolins have been re-found and re-released, a small percentage of the total tagged.

FIGURE 26.3 Post-release monitoring of a Sunda pangolin in Singapore. *Photo credit: Wildlife Reserves Singapore/ David Tan.*

As technology advances, Global Positioning System (GPS) tagging options are also becoming more feasible for pangolins. Due to the nocturnal behavior of Sunda pangolins, radio-tracking is often difficult and at times impossible due to the low transmission range of the radio tags – sometimes less than 50 m in dense habitat or if the animal is underground. Use of GPS tags would significantly improve understanding of pangolin ecology and behavior, and potentially improve post-release monitoring.

Monitoring data on sightings and roadkill

Other collections of pangolin data include sightings and roadkill reports across a variety of databases, for example, by ACRES, LKCNHM, NUS, NParks Biodiversity and Environment Database System (BIOME) and SGBioAtlas, and the National Environment Agency (NEA), which have enabled roadkill locations and hotspots to be mapped (Ong, 2017; Lee et al., 2018). Other existing data, particularly sightings records, are being collated. The SPWG is useful to bring together varied data sources from different organizations for a national-level summary of results. SPWG members also promote the existence of such databases to the public wherever possible, such as through public talks about pangolins or printed advisory leaflets, to encourage citizen science and reporting of pangolin sightings. ACRES has observed an increase in pangolin call-outs and reporting following national media coverage of their work online and on air (television and radio) (ACRES, unpubl. data).

Diet

Further research is required to better understand dietary preferences of Sunda pangolins in Singapore, including the potential role of pangolins in pest control by predating on ants and termites. Metagenomic studies are helping to improve understanding of these subjects (see *Genetics*). The SPWG has played an important role in linking field staff, veterinarians, laboratory scientists and bioinformaticians together for this research.

Ex situ research

Profiling and monitoring of rescued pangolins

A comprehensive system of profiling and monitoring of each rescued pangolin taken to WRS is in place. Each pangolin is microchipped and profiled, and morphometric data collected. Samples of blood hormones, DNA, sperm and stools are taken, and an ultrasound is performed to check for pregnancy or internal injury. The profiles and microchips provide a comprehensive basis for wide-ranging conservation research topics, including captive breeding, rehabilitation, survival, and disease screening. The biannual SPWG meetings, and regular communication between members, helps Singapore Zoo to adapt their profiling and monitoring protocols as and when needed for new research objectives or other agendas for pangolin conservation.

Genetics

Genetic analysis of Singaporean pangolins has been conducted at NUS and Nanyang Technological University (NTU). The SPWG has been essential to link the varied researchers and stakeholders together to implement this research, to secure funding from SPWG partners, and obtain relevant sample collection and research permits.

NTU researchers have sequenced several whole pangolin genomes, both nuclear and mitochondrial, as a basis for further study of local population genetics. This research is important to help understand the genetic health and viability of Singapore's pangolin population.

Researchers at NUS have used metagenomic techniques to compare the microbiomes of wild and captive pangolins and characterize the diet of wild pangolins. This was achieved through investigating the gut contents of dead pangolins and stool samples of live animals (A. Srivathsan, pers. comm.). A lack of prey reference markers in public databases limited the dietary analysis, and the pangolin's own DNA can also mask prey DNA; but useful findings included confirmation that weaver ants are a preferred prey species, and further research is ongoing (A. Srivathsan, 2019, pers. comm.).

Mitochondrial and RADseq markers have also been used at NUS to compare Singaporean pangolins with other regional pangolin lineages to better understand genetic differentiation and population structure. Preliminary results suggest that Singaporean pangolins are likely to be genetically similar to North Sumatran Sunda pangolins (Nash et al., 2018b).

Conservation outreach and education

Past and current outreach and education events relating to pangolins have been organized intermittently by local NGOs, such as NSS and The Pangolin Story, and NParks, NUS and WRS since 2005. These include lectures, craft events, dance productions, storytelling, banner displays, festival booths, games, gifts, pamphlets, national advisories, and production of other educational materials.

The Singapore Zoo and Night Safari have a strong focus in their outreach programs and conservation interpretation on Singaporean biodiversity, including pangolins. One aim of outreach in Singapore has been to discourage consumption of illegally trafficked wildlife through initiatives such as the year-long "You Buy, They Die" campaign run by WRS and TRAFFIC in 2015. While consumption seems unlikely in Singapore, Singaporean residents traveling overseas to countries with major markets for pangolin products may have more opportunities to use or consume pangolin products.

In celebration of World Pangolin Day in February 2018, NParks and The Pangolin Story, organized a morning of activities at Windsor Nature Park to help children and adults learn more about native Sunda pangolins and conservation efforts in Singapore. Activities included coloring scales to complete a pangolin banner (Fig. 26.4), taking part in a word search along the Hanguana Trail to spot several handcrafted wooden pangolin cut-outs, story-telling (Fig. 26.5), art workshops, and a guided walk. The event was part of several broader conservation education programs run by NParks, including "I heART Nature" children's workshops, Art in Nature, and nature appreciation walks. Although the educational impact of the event was not formally assessed, informal questioning of some participants indicated that the event had increased their knowledge of pangolins. Several participants were inspired to make a donation to support pangolin conservation.

FIGURE 26.4 Coloring scales to complete a pangolin banner at Windsor Nature Park 2018, Singapore. *Photo credit: NParks.*

Awareness-raising initiatives such as these have aimed to instill a sense of pride in the fact that Sunda pangolins occur in Singapore. The impact of these conservation outreach initiatives is not fully known, and would benefit from evaluation. The SPWG plans to conduct baseline surveys of peoples' attitudes and beliefs about pangolins in Singapore. The results will inform the design of future conservation outreach and education projects, which will incorporate formal monitoring and evaluation to assess impact.

Since the formation of the SPWG in 2014, pangolins in Singapore have received increasing media attention and they have been the focus of a growing number of outreach activities (SPWG, 2019, unpubl. data). This has included television documentaries, news programs and articles, magazine features, story books, brochures, museum exhibits, artwork, and branding for an eco-film festival, among others, aimed at a broad age range of people from mixed backgrounds. Although the impact of these activities has not been evaluated, there is anecdotal evidence that people are doing more of the desired behaviors such as reporting pangolin sightings (ACRES, unpubl. data). However, it is likely that many Singaporeans are still unfamiliar with pangolins, and those who do recognize the species are not necessarily aware of its presence in Singapore or what they can do to help protect them. The SPWG intends on addressing this and will initiate monitoring and evaluation of the impact of these activities.

Policy

National legislation to protect pangolins in Singapore is strong relative to some other

FIGURE 26.5 Storytelling at Windsor Nature Park 2018, Singapore. *Photo credit: NParks.*

range states. However, conflicts of interest between protecting natural areas and development, which can result in the loss of pangolin habitat for housing or infrastructure, are inevitable in such a densely populated country. In general, efforts are made to conduct environmental studies and impact assessments, and mitigate any negative impacts on pangolins where possible. Between 2013 and 2019, several developers conducted phased construction using wildlife shepherding techniques to help move pangolins and other wildlife into safe areas. This is a voluntary process in Singapore, which some developers have chosen to implement, encouraged by local conservation groups, SPWG, the public, environment-related government agencies, and other interested parties, in keeping with the national policy agenda of making Singapore a "city in a garden." The SPWG engages with developers whenever contacted directly or indirectly through its membership to help encourage mitigation and inspire innovative approaches which might facilitate the coexistence of people and pangolins. One possibility is the construction of subterranean wildlife crossings for pangolins and other wildlife.

National Conservation Strategy and Action Plan 2018–2030

In July 2017, WRS hosted and funded the first National Conservation Planning Workshop for Sunda pangolins in Singapore (Fig. 26.6). The need for a national strategy had been discussed at SPWG meetings, and the resultant national workshop was co-organized by WRS and NParks, in collaboration with the IUCN Species Survival Commission (SSC) Asian Species Action Partnership (ASAP) and the IUCN SSC

FIGURE 26.6 The Singapore Pangolin Working Group and other participants at a conservation planning workshop hosted by WRS. *Photo credit: Wildlife Reserves Singapore/David Tan.*

Pangolin (PSG) and Conservation Planning Specialist Groups (CPSG). Participants of the workshop were from local NGOs including ACRES, NSS and The Pangolin Story, government agencies, including the Housing and Development Board (HDB), PUB, and Urban Redevelopment Authority (URA), and academics and researchers from tertiary institutions. The outcome of the workshop was Singapore's National Conservation Strategy and Action Plan for Sunda pangolins (see Lee et al., 2018). The SPWG is charged with catalyzing its implementation, monitoring and evaluation, and adaptive management of the action plan. Communication and inter-agency collaboration was identified as a key priority, as was information gathering and sharing; increasing sustainability and connectivity between habitats; ensuring sensitive urban planning and associated policies; and rescue, rehabilitation and release of pangolins (Lee et al., 2018).

Reflections on the work of the SPWG

Although the threats facing Sunda pangolins in Singapore differ in some ways to those in other range states, many of the challenges and characteristics of the SPWG are relevant to these countries.

The establishment of a national-level working group has been instrumental in driving forward pangolin conservation in Singapore.

This model could be applied to other range states, particularly where a lack of co-operation and/or direct competition between organizations, is constraining research and conservation at the national level.

A national level working group helps stakeholders move away from a situation where research and conservation is implemented by a small number of individuals accessing short-term funding, toward a coordinated approach whereby larger-scale, multi-stakeholder projects can be designed, implemented, monitored and evaluated.

Regular communication between stakeholders and biannual face-to-face meetings have enabled strong teamwork and productive outcomes for pangolin conservation. They provide opportunities for potential synergies between projects to be recognized and acted upon, increasing the effectiveness and efficiency of each project.

An important factor in the success of pangolin conservation in Singapore has been good morale between stakeholders, and having a diverse team of working group members who are engaged around a similar cause. This has helped to motivate stakeholders and other agencies and people in Singapore.

A key benefit of a national-level working group is the ability to secure large-scale funding through collaborative approaches to donors. The research and conservation projects conducted by various SPWG members received substantial technical and financial support from WRS and the Wildlife Reserves Singapore Conservation Fund (WRSCF), as well as government support through NParks and other agencies. Financial stability and support from government agencies likely contribute to the productivity of the SPWG. However, many pangolin conservation projects in Singapore are reliant on unpaid volunteers and retaining such enthusiastic helpers has been assisted by the positive dynamic of the SPWG.

Much progress has been made for pangolins in Singapore, but there remains a long way to go to fully secure their population. The Singapore National Conservation Strategy and Action Plan for Sunda pangolins is a clear guide for ongoing and future conservation action, and reflective, adaptive management by the SPWG will be essential to secure the conservation of the species.

References

ACRES, 2018. Animal Concerns Research and Education Society. Available at: <http://acres.org.sg>. [March 26, 2019].

Cabana, F., Tay, C., 2019. The addition of soil and chitin into Sunda pangolin (*Manis javanica*) diets affect digestibility, faecal scoring, mean retention time and body weight. Zoo Biol. Early View.

Chung, Y.F., Lim, N.T.-L., Shunari, M., Wang, D.J., Chan, S.K.L., 2016. Record of the Malayan porcupine, *Hystrix brachyura* (Mammalia: Rodentia: Hystricidae) in Singapore. Nat. Singapore 9, 63–68.

Department of Statistics, Singapore, 2019. Household Income. Available at: <https://www.singstat.gov.sg/find-data/search-by-theme/households/household-income/latest-data>. [April 12, 2019].

Lee, P.B., Chung, Y.F., Nash, H.C., Lim, N.T.-L., Chan, S.K. L., Luz, S., et al., 2018. Sunda Pangolin (*Manis javanica*) National Conservation Strategy and Action Plan: Scaling Up Pangolin Conservation in Singapore. Singapore Pangolin Working Group, Singapore.

Lim, N.T.-L., 2007. Autecology of the Sunda Pangolin (*Manis javanica*) in Singapore. M.Sc. Thesis, National University of Singapore, Singapore.

Lim, N.T.-L., Ng, P.K.L., 2008a. Home range, activity cycle and natal den usage of a female Sunda pangolin *Manis javanica* (Mammalia: Pholidota) in Singapore. Endanger. Sp. Res. 4, 233–240.

Lim, N.T-L., Ng, P.K.L., 2008b. Predation of *Manis javanica* by *Python reticulatus* in Singapore. Hamadryad 32 (1), 62–65.

Ministry of Health, Singapore, 2019. Singapore's Healthcare System. Available at: <https://www.moh.gov.sg/our-healthcare-system>. [April 12, 2019].

Nash, H.C., 2018. The Ecology, Genetics and Conservation of Pangolins. Ph.D. Thesis, National University of Singapore, Singapore.

Nash, H.C., Lee, P., Low, M.R., 2018a. Rescue, rehabilitation and release of Sunda pangolins (*Manis javanica*) in Singapore. In: Soorae, P.S., (Ed.), Global Re-Introduction Perspectives. Case-Studies From Around the Globe. IUCN/SSC Re-introduction Specialist Group, Abu Dhabi, UAE.

Nash, H.C., Wirdateti, Low, G., Choo, S.W., Chong, J.L., Semiadi, G., et al., 2018b. Conservation genomics reveals possible illegal trade routes and admixture across pangolin lineages in Southeast Asia. Conserv. Genet. 19 (5), 1083–1095.

National Parks Board (NParks), 2014. Terrestrial Ecosystems. Available at: <https://www.nparks.gov.sg/biodiversity/our-ecosystems/terrestrial>. [June 11, 2019].

National Parks Board (NParks), 2019. Do's and Don'ts - Animal Advisory for Pangolins. Available at: <https://www.nparks.gov.sg/gardens-parks-and-nature/dos-and-donts/animal-advisories/pangolins>. [April 12, 2019].

Ong, S.Y., 2017. The Identification, Characterisation and Management of Mammal Roadkill Hotspots on Mainland Singapore. B.Sc. Thesis, National University of Singapore, Singapore.

Tan, M.B.N., Tan, H.T.W., 2013. The Laws Relating to Biodiversity in Singapore. Raffles Museum of Biodiversity Research. National University of Singapore, Singapore.

CHAPTER

27

Holistic approaches to protecting a pangolin stronghold in Central Africa

Andrew Fowler

Conservation and Policy, Zoological Society of London, Regent's Park, London, United Kingdom

OUTLINE

Introduction

Three species of pangolin occur in the Republic of Cameroon (hereafter "Cameroon"), the giant pangolin (*Smutsia gigantea*), white-bellied (*Phataginus tricuspis*) and black-bellied pangolin (*P. tetradactyla*). The first two species are listed as Endangered, and the black-bellied pangolin as Vulnerable on The IUCN Red List of Threatened Species (see Chapters 8–10). Within Cameroon, the giant pangolin is totally protected under the 1994 Cameroon Forest Act, meaning all forms of hunting and trade in its body parts are prohibited by law (Challender and Waterman, 2017). White- and black-bellied pangolins were partially protected (i.e., could be harvested with a hunting or collection permit) until 2017, when they were listed as fully protected species in the above legislation. All exploitation of pangolins in Cameroon is prohibited.

Poaching and trafficking of pangolins in Cameroon occurs despite the protection afforded to them under Cameroonian law (Ingram et al., 2019). It is largely unchecked and is driven by demand for pangolins and their derivatives in two main markets: (1) local markets, where the meat is consumed, and (2)

Pangolins
DOI: https://doi.org/10.1016/B978-0-12-815507-3.00027-7

markets in Asia to which pangolin scales are trafficked internationally (Ingram et al., 2019). Local markets can be further broken down into consumption by local communities and indigenous peoples living adjacent to pangolin habitats, and the commercial bushmeat trade supplying larger urban centers (Furnell, 2019).

This chapter presents a case study on pangolin conservation in Cameroon. It focuses primarily on activities being undertaken in the Dja Biosphere Reserve (DBR), a World Heritage Site containing the three tropical African pangolin species and important populations of forest elephants (*Loxodonta cyclotis*) and great apes. In the DBR, a holistic approach is being taken to conservation which includes improving protected area management capacity, providing support to front-line law enforcement agents and the judiciary, creating community surveillance networks, and engaging local communities in conservation efforts.

Background to Cameroon

Located in Central Africa, Cameroon is often described as "Africa in Miniature" because it contains a diversity of cultures, ecosystems and geological features that represent those found across the African continent (Mbenda et al., 2014). These range from a semi-arid landscape in the north, through the Congo Basin rainforest to marine and freshwater habitats, including mangroves along the coast. Cameroon is bordered by Nigeria to the west, Chad to the northeast, the Central African Republic to the east, and Equatorial Guinea, Gabon and the Republic of the Congo to the south. The >22 million hectares of tropical forest covering the south of the country are a vital part of the Congo Basin forest ecosystem (De Wasseige et al., 2009). These forests provide a source of livelihoods for local communities and indigenous peoples, and habitat for over 8000 plant species, about 900 species of birds and over 300 species of mammal, including the western lowland gorilla (*Gorilla gorilla gorilla*) and central African Chimpanzee (*Pan troglodytes troglodytes*; Republic of Cameroon, 2012).

The majority of the population practice subsistence farming and live in poverty. This is despite a high degree of economic development and increasing exploitation of natural resources in the form of plantations, logging concessions and hydro-electric power stations. A rapidly expanding infrastructure is opening up new areas and threatening wildlife. Many of the contracts for developing this infrastructure are taken by East Asian companies, which has led to strengthened links with East Asia, and a sizeable population of Asian ex-patriates in the country (Nordtveit, 2011). This has resulted in increased access to East Asian markets for uncertified wood and the facilitation of wildlife trafficking (Clarke and Babic, 2016).

Several factors make effective and long-term conservation interventions in Cameroon challenging, and these are relevant to efforts to disrupt the illegal trade in pangolins. The Cameroonian government agency tasked with managing protected areas and protected species is the Ministry of Forests and Wildlife (MINFOF). MINFOF, in common with many governments in Central Africa, lacks sufficient resources, both financial and in terms of human capacity, to effectively fulfil its mandate. There is also widespread corruption at all levels of the Cameroonian government, which ranks 152 out of 180 on Transparency International's corruption perception index (Transparency International, 2019).

The result of a lack of resources and poor governance is a virtual breakdown of effective judicial procedures. This is particularly acute when dealing with the Illegal Wildlife Trade (IWT). Many cases are not pursued from arrest to conviction. Even with those that are, there is often no follow-up to pursue the collection of penalties and impose the ruling of the court. It is important that MINFOF is supported in the application of laws protecting pangolins and other wildlife. This includes support from mandated law enforcement agencies including the gendarmerie, customs and airport security as well as related Ministries. Non-governmental organization (NGO) partners are helping to address existing capacity gaps through training on procedures and criminal process, the mentoring of key personnel to increase commitment and competence, and the funding of the legal processes. The costs of transporting witnesses to court hearings and arranging for adequate legal representation are often borne by conservation NGOs, without the assistance of which, many trials would not result in a verdict and sentencing. There has been a concerted effort by the NGO conservation sector to ensure better follow-up of cases in terms of tracking the penalties and damages that are owed to MINFOF by those previously convicted and sentenced but who have never been pursued for payment. However, obtaining basic data can be very difficult.

Pangolins in Cameroon

The relative impact of, and interactions between, local consumption (primarily bushmeat) and international trafficking (primarily scales) of pangolins are largely unknown. Surveys of wild meat markets and consumers in the Central, East and southern regions of Cameroon, undertaken between February 2017 and August 2018, found that pangolins are the fourth most available species (Furnell, 2019). There was apparently no local market for scales prior to the surge in demand from Asia in the last decade. Furnell (2019) reports that scales and live pangolins were rarely found in markets in Yaoundé, but were readily available in smaller urban centers (e.g., Djoum and Abong Mbang), suggesting that traders in the capital are increasingly aware of the sensitivity of openly trading in pangolins.

TABLE 27.1 Year, quantity and reported destination of major seizures of pangolin scales made in Cameroon (2013–19).

Year	Quantity (kg)	Destination
2014	1500	
2015	214	
2016	680	Malaysia
2016	200	
2017	4898	China
2017	1050	Nigeria
2018	1000	China
2018	718	
2019	2000	

Cameroon is a key exporter of pangolin scales traded illegally to Asian markets (Challender and Waterman, 2017; CITES, 2019; see Table 27.1). This includes the harvesting of scales specifically for illicit export (see CITES, 2019). Mambeya et al. (2018) recorded a large increase in the monetary value and demand for pangolins in Gabon between the early 2000s and 2014, noting that trafficking in Gabon followed forest export routes used for trafficking elephant ivory rather than public transport routes. This aligns with the experiences of law enforcement agencies in Cameroon, where trade in ivory has concentrated on the Gabonese border area in the northern Tri-National Dja-Odzala-Minkébé (TRIDOM protected area) landscape (Wasser et al., 2015).

In addition to the factors discussed in the previous section, effective conservation of pangolins in Cameroon is constrained by a lack of accurate data and information on the distribution, ecology and population dynamics of remaining populations. Small-scale research projects have been conducted (e.g., Bruce et al., 2018), and the presence of pangolins has been recorded in multiple protected areas, but there is limited knowledge of pangolin ecology, distribution and reproductive parameters. Baseline information and ongoing monitoring of pangolin populations is essential for informing management and conservation. Due to the difficulty in reliably identifying pangolin signs, camera trap data from comprehensive coverage of protected areas is required. A large-scale camera trap project has been underway in the DBR since 2017 (see next section).

Significant support is needed to effectively protect remaining pangolin populations in Cameroon. Government law enforcement agencies such as MINFOF and customs officials at ports and airports need increased resources to control hunting and poaching and subsequent transport of pangolins and their derivatives. The remainder of this chapter focuses on activities in the DBR, and concludes with an assessment of the success of these activities and recommendations for future activities.

The Dja Biosphere Reserve

The Dja Faunal Reserve (DFR), in southern Cameroon (Fig. 27.1), was created in 1950 and became a UNESCO Man and the Biosphere Reserve (DBR) in 1981. The Faunal Reserve was designated as a UNESCO World Heritage Site in 1987. The DFR covers 5260 km^2 of near intact rainforest, and notable rocky outcrops in the north and west accompanied by patches of savanna habitat. The reserve is part of the TRIDOM conservation landscape and holds populations of several key large mammals, including forest elephant, forest buffalo (*Syncerus caffer nanus*), western lowland gorilla, central African chimpanzee, bongo (*Tragelaphus eurycerus*) and the three tropical African pangolin species (Dupain et al., 2004).

Local communities around the DBR consist of a mix of ethnicities, including local indigenous Baka and Bantu (Muchaal and Ngandjui, 2001). Communities typically live along main roads passing through the landscape that link Cameroon to Gabon and the Republic of Congo. No communities are permanently resident in the reserve. Shot guns are the main tool used for hunting. Snares are also used extensively (Muchaal and Ngandjui, 2001; Wright and Priston, 2010). There is no accurate census of the human population, but estimates suggest village inhabitants around the DBR number 19,500, with up to 30,000 in the wider landscape (Ngatcha, 2019). In 2005, the total population was estimated at 129,059 inhabitants, primarily belonging to six ethnic groups, including two semi-nomadic, the Baka and the Kaka. Traditional agriculture remains the primary means of subsistence with bushmeat the main source of protein (BUCREP, 2019).

Conservation action in the DBR

SMART

The Spatial Monitoring and Reporting Tool (SMART) offers a systematic approach to patrol planning, data recording, reporting and feedback. The system is designed to maximize existing patrol activity by making the recording of data easy and reliable, using pre-set data entry protocols on an electronic Personal Digital Assistant (PDA) or smartphone. Reporting templates can be set up and patrol reports generated as required to suit different levels of management need (Critchlow et al., 2016). The SMART data model, constituting the entire range of species and signs that can be encountered in a given protected area, is tailored to the needs of the local conservation service. This is done in consultation with relevant authorities, such as eco-guards and protected area management.

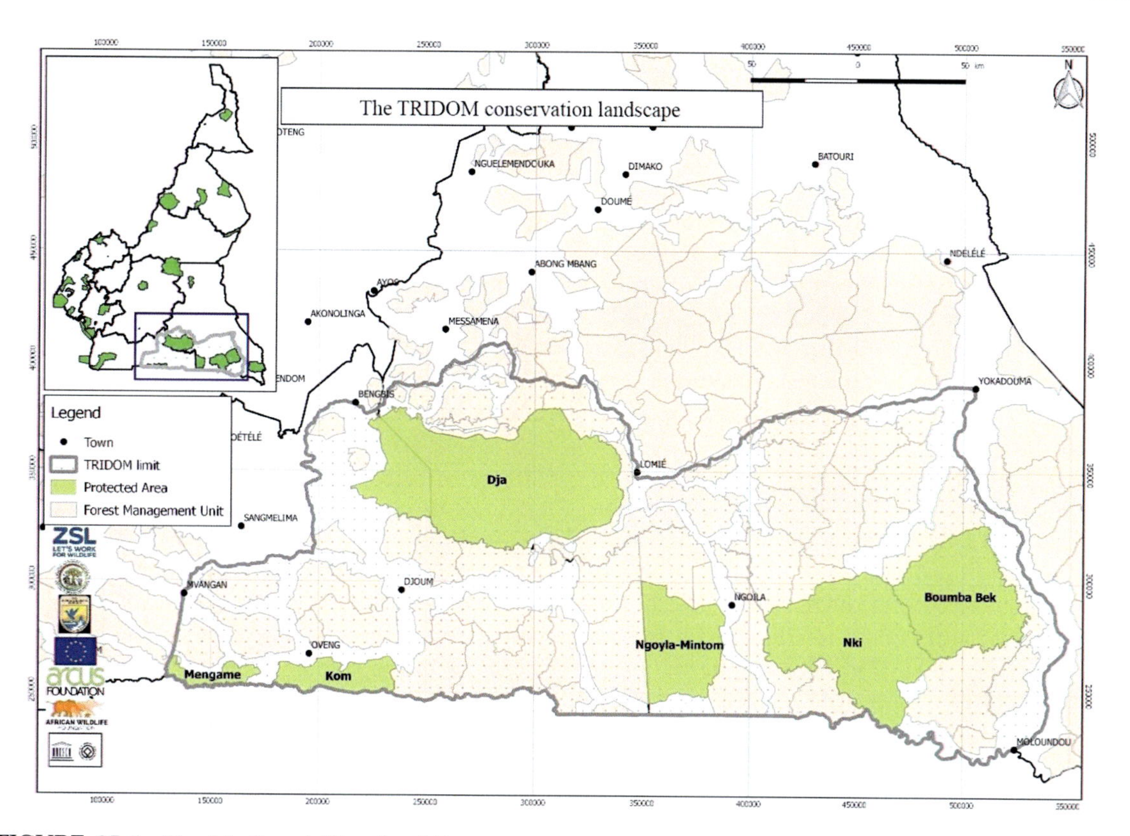

FIGURE 27.1 The Dja Faunal [Biosphere] Reserve situated in the northern section of the Tri-National Dja-Odzala-Minkébé (TRIDOM) conservation landscape. The landscape encompasses large expanses of natural forest, some of which are managed as forestry concessions and community forests. Several major protected areas occur in the landscape. *Map adapted from Global Forest Watch; spatial data from MINFOF/WRI, https://cmr.forest-atlas.org.*

SMART was introduced to the DBR in 2014, and has since been adopted by patrol teams in all four management sectors (northern, eastern, western and southern). SMART requires considerable and ongoing investment to cover the training of personnel, purchase of appropriate equipment and costs associated with ensuring that continuous patrol coverage is maintained.

Data are collected on signs of human activity, including snares, hunting camps and encounters resulting in arrest or caution, using PDA devices (more recently lower-cost smart phones have been successfully used), and downloaded to laptop computers for analysis and reporting. The SMART patrol management system enables data from previous patrols, such as identified hotspots of illegal human activity and areas of interest for particular species, to be used to inform the planning of future patrols. This allows for a more adaptive management approach than that previously implemented. The system is designed for ease of use in terms of producing accurate maps and generating reports, according to prepared templates, for individual patrols as well as monthly, quarterly and annual reports. SMART also enables individual eco-guard performance to be tracked easily and thus assists management in remote areas where direct oversight is difficult.

The SMART approach enables protected area managers to track trends in signs of human activity within the reserve, together with direct sightings and indirect signs of wildlife recorded during patrols. The system enables managers to identify gaps in patrol coverage through an improved understanding of patrol effort. The system can be configured to provide specific geographical patrol targets that must be met by passing within a designated distance from the pre-established location. Patrol intensity is measured by calculating the number of 5 km^2 grids passed through during a patrol.

Table 27.2 shows the total distance of patrols covered by MINFOF eco-guards in the years 2014–18. The patrol distance increased over time from around 600 km in 2014 to over 12,000 km in 2017, and decreased to 11,247 km in 2018. The main reason for this decrease was the transfer of 80 MINFOF eco-guards from the Conservation Service and a corresponding influx of new personnel. This required a period of training to bring the patrol teams up to standard, and in the first year, patrol distance decreased as a result.

The patrol coverage achieved and recorded in the DBR using SMART in 2016 and 2017 is shown in Fig. 27.2. Patrol coverage increased markedly from 2016 to 2017, with more of the reserve covered, and more grids cells entered. In 2016 a total of 9692 km was covered during MINFOF eco-guard SMART patrolling, with a total of 650 indirect pangolin signs recorded, including footprints. This increased to a total of 12,422 km in 2017, with a total of 878 signs, and the encounter rate remained the same at around 0.07/km (see Table 27.2). It should be noted that indirect pangolin signs are not easily identified with any degree of accuracy since there are a number of other animal species which produce similar digging and scratching signs. The results of SMART patrols are thus reported with this caveat.

SMART has been accepted by MINFOF at the central government level, and SMART focal points have been appointed for ten different regions and sub-regions, and a national focal person appointed to oversee data management and reporting. At the local level, the DBR Conservation Service has readily adopted SMART, and eco-guards have reacted favorably to the adoption of the system, despite some initial hesitancy. The Zoological Society of London (ZSL), in collaboration with other conservation partners, has arranged training for MINFOF personnel at all levels and a national strategy has been discussed.

There is general consensus among MINFOF and partners that the SMART approach has merit for improving the management of protected areas within Cameroon. However,

TABLE 27.2 Total distance covered by MINFOF eco-guard patrols from 2014 to 2018 and corresponding observations identified as pangolin signs.

Year	Distance (km)	Total signs identified as pangolin	Number of observations identified as pangolin sign per km
2014	609.29	7	0.01
2015	956.00	30	0.03
2016	9691.71	650	0.06
2017	12,421.62	878	0.07
2018	11,246.90	283	0.02

(A)

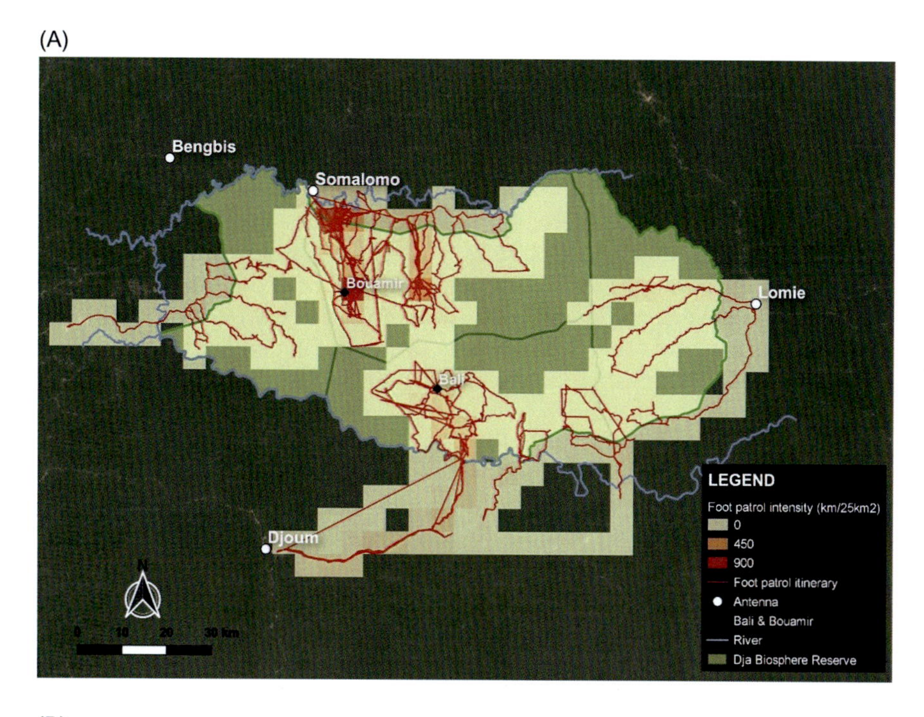

(B)

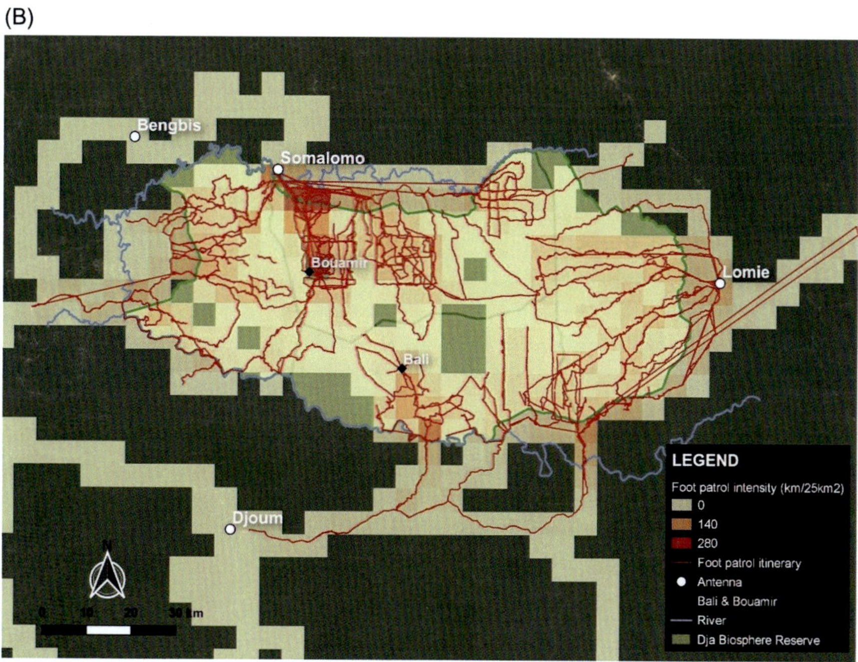

FIGURE 27.2 Patrol coverage by MINFOF eco-guards in the Dja Biosphere Reserve during 2016 (A) and 2017 (B). Patrol routes are indicated in red.

projects tend to be stand-alone with a one-to-one relationship between a donor and implementing agency (typically an NGO). This means there is often little incentive for close collaboration and harmonization of patrol strategies and SMART data-models in different areas. Cross area comparisons are therefore rendered difficult, preventing effective regional and national conservation strategies from being implemented. An exception is a collaboration between ZSL and the African Wildlife Foundation (AWF), which are implementing a European Union funded project (ECOFAC 6) to jointly support MINFOF in effective management of the DBR. To achieve this, the two organizations and MINFOF have agreed on a unified central SMART data-model which is being used for patrolling in all sectors.

A central problem to this kind of investment in a system that requires intensive training and oversight is the periodic re-deployment of MINFOF eco-guards. This affects all protected areas in Cameroon, and has been outside the influence of conservation partners working in protected areas, who invest significant funds in training eco-guards. In late 2017, 80% of the eco-guards based in the DBR were transferred to other locations, and replaced with personnel either new to the service or previously operating in other posts. While this may increase the overall capacity of eco-guards, it disrupts attempts to implement specific management approaches, such as SMART, in specific locations. Training needs must be quickly identified and provided as appropriate after each re-deployment. The issue of re-deployment has been raised with MINFOF by a number of different donors and remains problematic. Eco-guards may be posted to remote locations with a minimum of training and preparation, which often leads to low morale, an inability to perform their duties effectively, and perhaps consequent mistreatment and abuse of local communities. Following discussions with donors and NGO partners, MINFOF is considering a system of local recruitment of eco-guards, which might act as a "feeder" into the national eco-guard recruitment process, allowing qualified candidates from communities adjacent to protected areas to be appointed as community eco-guards.

Despite a marked increase in patrol effort on the part of eco-guards throughout the DBR, poaching and trafficking is still a major threat to wildlife. It is unclear what direct effect patrolling and seizures are having on the actual rate of illegal activity in the DBR. It is a large and remote area and it is unlikely that sufficient financial and human resources will ever be available to achieve the level of direct patrolling by eco-guards required to significantly reduce these threats, particularly given the potential financial rewards available from the relatively unrestricted illegal trade in Cameroon. In the context of the rural, subsistence-based farming economy in which most of the human population exists, the attraction of relatively easy and substantial financial rewards from IWT, including pangolins, are difficult to exclude. Many of the poachers and middle men involved in the trade of pangolin scales (and ivory) do not come from the communities surrounding the protected area. However, local communities are involved in supplying human labor as porters, expertise as guides or in providing information on the area, and coercion is often used to force local inhabitants to serve these functions.

It is imperative that any long term, sustainable solution to the problem of pangolin hunting and poaching, and broader IWT, in large-scale landscapes such as the DBR involve significant and lasting incentives to local populations. These are needed to reduce their direct reliance on exploitation of natural resources close by, where it is unsustainable, and to empower them to resist the often coercive demands of poaching groups to participate in illegal trade. To achieve this, it is necessary to work directly with local communities to conceive and implement effective solutions (see Chapters 24 and 25). The provision of material benefits, tailored to local needs, which contribute to increasing their well-being, and economic and food security will likely need to be a part of any successful conservation solution.

Local community engagement

Community engagement work around the DBR has included a variety of approaches, ranging from supporting community-based natural resource management to information gathering and intelligence-led protection. Forest peoples, originally defined as those who access the tropical forest for subsistence purposes (Bailey et al., 1992), often depend on the resources the forest provides for food, shelter, and income, and rural communities around the DBR have highlighted concerns over outsiders unsustainably depleting "their wildlife." They have expressed a desire to take action against criminals they see as a threat to their security and livelihoods.

Communities' mandates over the use of natural resources in proximity to their villages can be strengthened through the creation of Community Surveillance Networks (CSNs). CSNs have been created in eight communities in areas adjacent to the DBR and are considered to have a high potential impact on reducing poaching and trafficking. The network operates anonymously and all communication is through a phone donated to the communities with credit provided on a monthly basis. In the DBR, the CSN intelligence phone is managed by ZSL and calls are logged in a secure database.

Given the sensitivity of the information potentially being exchanged, it is paramount that anonymity is maintained and information concerning the identities of informants is restricted to an individual staff member. ZSL follows detailed protocols on correct processes to maintain sound chains of intelligence handling. Pre-set rewards for valuable information are paid when information is provided. Due to the difficulty in effectively mobilizing law enforcement personnel and the challenges of ensuring that judicial processes are effectively followed-up, it is not possible to make payments for information contingent upon arrests being made or successful prosecutions. Therefore, payment is made on receipt of credible information. ZSL receives on average around five credible calls per month on the CSN phone.

Sustainable livelihood initiatives have been established in 12 communities, including a mix of Bantu (6) and Baka (6). The activities are carefully monitored in terms of the participation of often marginalized groups, including women and indigenous peoples. Income Generating Activities (IGAs) are primarily concerned with supplying improved crop varieties, development of plant nurseries and poultry rearing. Other communities have suggested developing honey production as a potential IGA. Activities such as chicken rearing supply communities with substantial quantities of alternative protein, and are intended to reduce the amount of protein that is removed from the forest. As members of local communities spend more time on IGAs, they have less time to dedicate to hunting. The next step is to identify the "redoubtable" hunters and encourage them to enter the IGAs.

Village Savings and Loan Associations (VSLAs) are another way to engage communities in an activity that brings them potential livelihood and well-being benefits (Allen, 2006; Ksoll et al., 2016). They are being used in the DBR to enhance the value chain of Non-Timber Forest Products (NTFP). VSLAs are a micro-finance scheme that initially brings non-conservation related benefits to communities. The presence of these benefits can then be used to introduce ideas of sustainable natural resource use through the introduction of projects relevant to ensuring a clean water supply and other environmentally linked factors. There are currently 16 active VSLAs in communities around the DBR. The VSLAs do not have an immediate conservation impact, but bring benefits to local communities and are a way of introducing concepts around conservation and sustainable management of natural resources.

Another benefit of VSLAs is that the CSNs can be connected to them to encourage information to be passed on to law enforcement authorities to protect communities' forests against outsiders. As noted, outsiders enter and take large quantities of meat, including pangolins (Furnell, 2019) for urban markets, destroying resources and preventing effective sustainable management by communities living adjacent to the forest.

UCL-ExCiteS

The University College London's Extreme Citizen Science (ExCiteS) group (UCL-Excites) and ZSL are supporting communities around the DBR to monitor and report on illegal activities of concern to them. Communities are supported to integrate their traditional ecological knowledge and cultural values with new technology (ExCiteS Sapelli) to monitor and report on natural resource use, wildlife crime and law enforcement actions. Sapelli is an icon based data collection tool for non-literate communities; the structure of the tool and its icons are tailored to the specific context of the project. The communities lead the process of project design and implementation, producing a tool which remains community owned, thereby promoting both community empowerment and a local data collection mechanism. The project promotes data sovereignty through co-designing community protocols which are centered on local ownership of the data (Lewis, 2015). The tool has previously been successfully piloted with forest peoples in the Central African Republic and Northern Congo (Vitos et al., 2017), where it was used to report sites of illegal logging and IWT. Project partners are providing ongoing support and technical advice to communities in the DBR via consultations, workshops, village meetings and skills sessions, during which the evolutionary and ecological significance of pangolins are highlighted. Together, the ExCiteS Sapelli project, CSNs, VSLAs and IGAs constitute a suite of complementary initiatives that local communities can use to enhance their security and livelihoods, and help to protect pangolins and other threatened species.

Private sector engagement

Engagement of private sector actors, including logging concessions, rubber plantations, hydro-electric power suppliers and trophy hunting concessions, in the area around the DBR, have resulted in significant gains for several species of wildlife, including the creation of corridors for great ape and elephant dispersal. The creation of wildlife management plans for logging concessions, including the commitment of concessionaires not to open permanent roads and to supply protein alternatives to bushmeat for their employees in the form of canteens, for example, have established best practice standards in several concessions, including Pallisco and Rougier (Asanga et al., 2018). These commitments can help reduce pressure on pangolins and other threatened wildlife in the region.

Wildlife monitoring

ZSL has undertaken camera trap surveys in several locations within the DBR since 2017. The camera traps have revealed important insights into the presence of key species, including leopard (*Panthera pardus*), golden cat (*Caracal aurata*), central African chimpanzee, western lowland gorilla, and forest elephant, in addition to pangolins. Pangolins have been encountered more frequently as distance to the reserve boundary increased, which is likely due to increased rates of hunting and disturbance by humans nearer the periphery of the DBR.

Two camera trap grids, each comprising 40 cameras, were deployed in the northern and eastern sectors of the DBR in 2018. Cameras were placed at distances of 2 km apart in 6 × 7 grids, with 4 cameras placed closer to the interior of the reserve. The camera traps were deployed for 100 days between January and May 2018 in both sectors. The camera traps were active at the same time as a full faunal inventory was underway in the reserve to allow a comparison of species-specific metrics gathered by both survey methods.

Camera traps were not targeted specifically at pangolins. However, both white-bellied and giant pangolins were recorded in the northern sector of the DBR (white-bellied; 23 events providing 174 images: giant pangolin; 10 events providing 72 images). Black-bellied pangolins were not recorded, likely because arboreal camera traps were not set.

The results provided some insights into pangolin ecology. All giant pangolins detected (10 occasions) were solitary adults, and 90% of them were located more than 8 km from the reserve boundary. A nocturnal pattern was detected, with activities peaking at 0400, though one event occurred at 0700 (see also Chapter 10). White-bellied pangolins were recorded 23 times, and the timing of the camera trap records suggests an entirely nocturnal activity pattern: activity peaked at midnight to 0100. The distribution records suggests that this species is more frequently encountered in the interior of the DBR.

Notably, previous surveys have revealed that local knowledge is somewhat unreliable in terms of accurately identifying pangolin field signs in tropical forest (Bruce et al., 2018). There is a tendency on the part of guides to categorise ambiguous indirect signs of animal activity, such as scratch marks and signs of digging, as having been made by pangolins because the accompanying researchers are interested in the species, without any way of verification. Consequently, during the wildlife inventory, indirect signs were not attributed to pangolins, only sightings of animals. This resulted in low encounter rates of 0.001/km for black- and white-bellied pangolins, representing total records of one and two instances respectively. No direct sightings of giant pangolins were recorded.

Conclusions and recommendations

To successfully achieve long term conservation at the landscape level, it is necessary to combine law enforcement actions (e.g., patrolling inside protected areas, the interzones between them, and forest concessions) with local community engagement and private sector actors. Addressing one of these in isolation, even effectively, will have only limited impact. Arrests and seizures are unlikely to change behavior away from hunting, and may merely displace hunting to areas that are not patrolled. Communities need to be actively engaged in sustainably managing natural resources, with income generating activities developed to alleviate pressure on wildlife populations threatened by overexploitation. Private sector actors such as logging concessions need to provide subsidized alternative sources of protein through canteens and warehoused goods to alleviate the pressure created on wildlife by their staff and families. Similarly, these actors ought to contribute to efforts to effectively control hunting within their concessions by creating wildlife monitoring units and collaborating with MINFOF in facilitating law enforcement patrols.

Increasing the personnel capacity in MINFOF and other law enforcement agencies in Cameroon is of paramount importance. This should continue and be accompanied by a law enforcement program targeting pangolin and wildlife traffickers along known routes. There have been some signs of success in combatting the illegal trade in southern Cameroon in recent years; for example, MINFOF eco-guards from the Dja Conservation Service seized 216 elephant tusks in December 2017, the largest ever ivory seizure Cameroon. It is suspected pangolin body parts are trafficked along similar routes to ivory, and though seizures in themselves are not a measure of success, confiscations such as the above show promise for the future.

Similarly, an understanding of the local, regional and national market for bushmeat is crucial to inform knowledge of the illegal trade in pangolins. Bushmeat market, consumer surveys and household surveys are needed to provide a better understanding of trade dynamics and how this affects populations. While there is a major need for natural protein in the form of small mammals such as porcupine and duiker, to feed local communities, there is very little reliance on protected species for survival. There is, therefore, potential to reduce consumption of pangolins through well designed and locally appropriate behavior change campaigns.

Efforts to reduce consumption and trafficking of pangolins through law enforcement actions and senitization to the threats facing pangolins should be undertaken alongside initiatives designed to ensure conditions remain suitable for populations to continue to survive in the wild. Conservation interventions are increasingly being planned and implemented at a landscape level, rather than concentrating on single site or single species interventions. This approach is in response to the fact that issues such as forest clearance, subsistence agriculture, bushmeat hunting and poaching, particularly pangolins and forest elephants, are landscape-wide, encompassing the current and future needs of significant human populations in areas often immediately adjacent to protected areas. It also considers the fact that significant populations of pangolins and other key medium and large-bodied mammal species exist outside of protected areas, in the interzones, often including logging concessions, commercial plantations and small-scale farming plots.

Conservation planning at the landscape scale is also required to effectively manage the needs of expanding human populations. Meeting the aspirations of local communities while ensuring a future for pangolins and other species is a major challenge. In Cameroon, several large-scale efforts have been made; for example, the KfW, a German Development Bank is funding a Programme for the Sustainable Management of Natural Resources (PSMNR) in southwestern Cameroon, which has attempted to create a co-management approach. This has involved creating Conservation-Development Agreements with local communities who gain benefits such as improved varieties of cocoa, in exchange for certain conservation-related outcomes. For example, no hunting in forests adjacent to villages, which are often enclaved inside protected areas. The emphasis is on collaborative solutions being generated through consultation with local communities at project inception (Fouth et al., 2017; see Chapters 23 and 24). A similar approach is being implemented in the TRIDOM landscape: initial engagement in VSLAs is followed by the implementation of IGAs, which are selected and developed with community members, and tailored to their needs.

The effectiveness of pangolin conservation in Cameroon will be measured by the status of its wild pangolin populations. The species are challenging to survey, and standardized monitoring protocols are still in development (Chapter 35). In the DBR, preliminary studies of pangolins through wildlife inventories and camera trap surveys have yielded interesting but limited results. Future work will hopefully advance knowledge of populations in the country enabling long-term trends to be measured and appropriate conservation actions taken.

References

Allen, H., 2006. Village savings and loans associations – sustainable and cost-effective rural finance. Small Enterprise Dev. 17 (1), 61–68.

Asanga, C., De Ornellas, P., Dethier, M., Fankem, O., Grange, S., Ngo Bata, M., et al., 2018. Boite a outils pour la prise en compte de la faune dans les forets de production du bassin du Congo. Zoological Society of London, Royaume-Uni.

Bailey, R.C., Bahuchet, S., Hewlett, B., 1992. Development in the Central African rainforest: concern for forest peoples. In: Cleaver, K.M. (Ed.), Conservation de la Forêt Dense en Afrique Centrale Et de L'Ouest. World Bank Publications, pp. 202–211.

BUCREP, 2019. Central Bureau of the Census and Population Studies. Available from: <http://www.bucrep.cm/index.php/en/recensements/3eme-rgph/20-3eme-rgph/presentation/57-population-en-chiffre>. [June 30, 2019].

Bruce, T., Kamta, R., Mbobda, R.B.T., Kanto, S.T., Djibrilla, D., Moses, I., et al., 2018. Locating giant ground pangolins (*Smutsia gigantea*) using camera traps on burrows in the Dja Biosphere Reserve, Cameroon. Trop. Conserv. Sci. 11, 1–5.

Challender, D., Waterman, C., 2017. Implementation of CITES Decisions 17.239 b) and 17.240 on Pangolins (*Manis* spp.), CITES SC69 Doc. 57 Annex. Available from: <https://cites.org/sites/default/files/eng/com/sc/69/E-SC69-57-A.pdf>. [April 18, 2019].

CITES, 2019. Wildlife Crime Enforcement Support in West and Central Africa. CITES CoP18 Doc. 34. Available from: <https://cites.org/sites/default/files/eng/cop/18/doc/E-CoP18-034.pdf>. [April 22, 2019].

Clarke, A., Babic, A., 2016. Wildlife trafficking trends in sub-Saharan Africa. OECD, Illicit Trade: Converging Criminal Networks, OECD Reviews of Risk Management Policies. OECD Publishing, Paris, France.

Critchlow, R., Plumptre, A., Alidria, B., Nsubuga, M., Driciru, M., Rwetsiba, A., et al., 2016. Improving law-enforcement effectiveness and efficiency in protected areas using ranger-collected monitoring data. Conserv. Lett. 10 (5), 572–580.

De Wasseige, C., Devers, D., de Merken, P., Eba'a Atyi, R., Nasi, R., Mayaux, P. (Eds.), 2009. Les forêts du Bassin du Congo: état des forêts 2008. EU Publications Office, Brussels, Belgium.

Dupain, J., Guislain, P., Nguenang, G., De Vleeschouwer, K., Van Elsacker, L., 2004. High chimpanzee and gorilla densities in a non-protected area on the northern periphery of the Dja Faunal Reserve Cameroon. Oryx 38 (2), 209–216.

Fouth, D., Nkolo, M., Scholte, P., 2017. Analysis of Protected or Conservation Areas Governance Models, Practical experiences of GIZ projects/programmes in Africa. Available from: <https://www.snrd-africa.net/wp-content/uploads/2018/02/1801Doc_CapitalisationGouvernanceAP_vf_eng-2.pdf>. [March 03, 2019].

Furnell, S., 2019. Analysis of wild meat markets and consumers in the Central, East and South regions of Cameroon: with a focus on pangolins. Zoological Society of London, London, UK.

Ingram, D.J., Cronin, D.T., Challender, D.W.S., Venditti, D.M., Gonder, M.K., 2019. Characterizing trafficking and trade of pangolins in the Gulf of Guinea. Glob. Ecol. Conserv. 17, e00576.

Ksoll, C., Bie Lilleør, H., Helth Lønborg, J., Dahl Rasmussen, O., 2016. Impact of village savings and loan associations: evidence from a cluster randomized trial. J. Dev. Econ. 120, 70–85.

Lewis, J., 2015. Where goods are free but knowledge costs. Hunter-gatherer ritual economics in Western Central Africa. Hunter Gatherer Res. 1 (1), 1–27.

Mambeya, M.M., Baker, F., Momboua, B.R., Pambo, A.F.K., Hega, M., Okouyi, V.J.O., et al., 2018. The emergence of a commercial trade in pangolins from Gabon. Afr. J. Ecol. 56 (3), 601–609.

Mbenda, H.G.N., Awasthi, G., Singh, P.K., Gouado, I., Das, A., 2014. Does malaria epidemiology project Cameroon as 'Africa as miniature'? J. Biosci. 39 (4), 727–738.

Muchaal, P.K., Ngandjui, G., 2001. Impact of village hunting on wildlife populations in the Western Dja Reserve, Cameroon. Conserv. Biol. 13 (2), 385–396.

Ngatcha, L., 2019. Contribution a la preservation de la biodiversite par la mise on oeuvre des activites generatrices de revenus (AGRs) au profit des populations riveraines de la reserve de la biosphere du Dja. Memoire presente en vue de l'obtention du Diplome de Master Professional en Sciences Forestieres. Universite de Yaoundé 1, Cameroon.

Nordtveit, B.H., 2011. An emerging donor in education and development: a case study of China in Cameroon. Int. J. Educ. Dev. 31 (2), 99–108.

Republic of Cameroon, 2012. National Biodiversity Strategy and Action Plan – Version II –MINEPDED, Yaoundé, Cameroon.

Transparency International, 2019. Cameroon. Available from: <https://www.transparency.org/country/CMR>. [August 10, 2019].

Vitos, M., Altenbuchner, J., Stevens, M., Conquest, G., Lewis, J., Haklay, M., 2017. Supporting Collaboration with Non-Literate Forest Communities in the Congo-Basin. In: Proceedings of the 2017 ACM Conference on Computer Supported Cooperative Work and Social Computing, pp. 1576–1590.

Wasser, S.K., Brown, L., Mailand, C., Mondol, S., Clark, W., Laurie, C., et al., 2015. Genetic assignment of large seizures of elephant ivory reveals Africa's major poaching hotspots. Science 349 (6243), 84–87.

Wright, J., Priston, E.C., 2010. Hunting and trapping in Lebialem Division, Cameroon: bushmeat harvesting practices and human reliance. Endanger. Sp. Res. 11, 1–12.

PART 4

Ex Situ Conservation

CHAPTER

28

Captive husbandry of pangolins: lessons and challenges

Leanne Vivian Wicker[1], *Francis Cabana*[2], *Jason Shih-Chien Chin*[3], *Jessica Jimerson*[4], *Flora Hsuan-Yi Lo*[3], *Karin Lourens*[5], *Rajesh Kumar Mohapatra*[6], *Amy Roberts*[7] *and Shibao Wu*[8]

[1]Australian Wildlife Health Centre, Healesville Sanctuary, Zoos Victoria, Healesville, VIC, Australia [2]Wildlife Reserves Singapore, Singapore [3]Taipei Zoo, Taipei, Taiwan [4]Save Vietnam's Wildlife, Cuc Phuong National Park, Ninh Binh Province, Vietnam [5]Johannesburg Wildlife Veterinary Hospital, Johannesburg, South Africa [6]Nandankanan Zoological Park, Bhubaneswar, India [7]Chicago Zoological Society/Brookfield Zoo, Brookfield, IL, United States [8]School of Life Science, South China Normal University, Guangzhou, P.R. China

OUTLINE

Pangolins
DOI: https://doi.org/10.1016/B978-0-12-815507-3.00028-9

Introduction

The high level of indiscriminate exploitation and range of unique ecological and physiological characteristics, including a low reproductive rate, make wild pangolin populations particularly vulnerable to extinction (Gaubert, 2011; Sodeinde and Adedipe, 1994). As such, there is an urgent need to consider ex situ conservation as part of a holistic approach to conserving pangolins (see Chapter 31). However, despite a long history of keeping pangolins in captivity, with the first known captive records arising from a Buddhist monastery in Burma in 1859 (Yang et al., 2007), captive pangolins have historically faced high mortality rates, and maintaining healthy captive populations over the long term is challenging (Chin and Tsao, 2015; Mohapatra and Panda, 2014a).

As the illegal trade in pangolins continues to escalate, the need to share and expand knowledge of the ex situ care requirements for the rescue, rehabilitation and conservation breeding of pangolins is great. A number of existing pangolin rescue centers in range countries, including, but not limited to, those in Cambodia, China, India, Lao People's Democratic Republic (Lao PDR), Singapore, Vietnam, South Africa and Zimbabwe, assess and rehabilitate confiscated pangolins prior to their release back to the wild, or transfer them to conservation breeding programs in captivity (Chin and Yang, 2009; Clark et al., 2009; Mohapatra, 2016; Nash et al., 2018; Sun et al., 2019; Zhang et al., 2017), and this practice has contributed significantly to the existing knowledge base. However, there remain insufficient appropriately resourced rescue centers to care for pangolins confiscated from the illegal wildlife trade in range states in both Africa and Asia given the extent of illegal trade (see Chapter 16).

There is also a resurgence of interest in establishing sustainable zoo-based captive populations, outside the natural geographic range of pangolins (see Chapter 31). There is a growing number of white-bellied pangolins (*Phataginus tricuspis*) in the United States (Aitken-Palmer et al., 2017; Anon, 2017), a group of four Chinese pangolins (*Manis pentadactyla*) at Zoo Leipzig, Germany (R. Holland, pers. comm.), and a Chinese pangolin at Ueno Zoo, Tokyo. A number of zoos in pangolin range states, including Wildlife Reserves Singapore (Vijayan et al., 2009), Taipei Zoo (Chin et al., 2015) and the Pangolin Conservation and Breeding Center (PCBC) at Nandankanan Zoological Park, India (Mohapatra and Panda, 2014a), have contributed significantly to knowledge of pangolin husbandry, nutrition and reproduction, with over sixty years of collective experience in maintaining captive pangolin populations.

This chapter presents a summary of the husbandry of pangolins in zoos and rescue centers worldwide, including descriptions of appropriate quarantine and long term care enclosures, a brief introduction to diet, a discussion of enrichment, training and welfare of captive pangolins, an outline of successful conservation breeding strategies, and highlights the main challenges, successes and lessons learnt in recent decades. It is hoped that this information will facilitate improved outcomes for captive pangolins, increasing the contribution of ex situ programs to conservation of the Manidae worldwide.

Husbandry of pangolins worldwide

Pangolins may be kept in captivity temporarily, such as for the veterinary care of injured wild animals, or the quarantine, health assessment and rehabilitation of confiscated pangolins prior to release back to the wild; or over the long term, such as for inclusion in an ex situ conservation breeding program or in a zoological institution. The size and style of pangolin enclosures, the building materials they are constructed from, and the approach to diet, enrichment and husbandry varies widely depending on whether it is intended for short or long term animal care, and the species of pangolin in question.

Enclosure design and captive environment

Building materials and design considerations

Pangolins are incredibly agile and strong, with the long claws of the forelimbs designed to enable the animal to climb trees, dig burrows or pull apart strong termite mounds, depending on species behavioral and dietary preferences. Climbing and digging skills, and the overall strength of the pangolin must be considered in the design and construction of transport boxes and captive enclosures. Where concrete is used, it should be laid sufficiently thickly and at high enough cement to sand ratio to prevent pangolins from digging to escape. Soft materials, such as caulk or silicone sealant, can be easily picked away by the animals and are not recommended for sealing seams or cracks (Anon, 2017). All light fixtures, wires, plumbing and other sensitive equipment should remain out of reach of the animals, and, if placed outside of the enclosure, at a safe distance given the long reach of a pangolin's tongue (Anon, 2017). A further consideration is the considerable financial value placed on pangolins within the illegal wildlife trade, placing captive pangolins at risk of being stolen where enclosure design and animal management do not include adequate security measures. For all of these reasons, the captive husbandry routine must include regular enclosure checks to look for evidence of breakage and potential areas of weakness.

Pangolins have also frequently been known to injure themselves in poorly designed or inappropriately constructed captive environments. Significant trauma, including dislocation, laceration or accidental amputation of tail tips, digits and claws; and skin lacerations, particularly around the face and forelimbs, are seen in pangolins which climb and pull on mesh wire enclosure netting in a manner that can be abnormally repetitive (Challender et al., 2012). Pangolins have also become stuck, damaging their scales in the process, in narrow spaces found in poorly designed enclosures. All enclosure furnishings should be placed so as to allow the animal to pass completely and, preferably, turn around, as their scales can get caught and damaged trying to back out of tight spaces (Anon, 2017). Significant trauma to the foot pads and tail tips, and scrapes to the bridge and tip of the nose have also been seen in pangolins housed in rough concrete floored enclosures.

Pangolins in captivity appear to be extremely sensitive to rapid changes in the thermal environment. As such, close monitoring and management of the thermal environment is important in maintaining a healthy population of pangolins, with the onset of respiratory disease, including pneumonia, commonly attributed to a rapid drop in the environmental temperature (Chin and Tsao, 2015; Mohapatra and Panda, 2014a). Ensuring appropriate protection from wind, rain and sun; and the provision of an appropriate thermal range for the species must be considered when designing the captive environment. In the Pangolin Research Base for Artificial Rescue and Conservation Breeding of South China Normal University (PRB-SCNU), where rescued Sunda pangolins (*M. javanica*) are cared for outside the species' natural geographic range, under floor heating and a humidifier are included in enclosures to ensure the temperature remains in the range of 18–30 °C, and humidity remains above 60% year round (Zhang et al., 2017). Where pangolins are captive within their natural geographic

TABLE 28.1 Thermal environment required to maintain healthy pangolins in captivity.

Species	Temperature (°C)	Humidity (%)
Sunda pangolin[a,b]	18–30	84–100
Chinese pangolin[c,d]	24–26 (min. 10, max. 28)	Above 80
	18–24 (Chinese pangolin in Taiwan)	70 (Chinese pangolin in Taiwan)
Indian pangolin[e,f]	15.5–34.5 (winter)	23–89.5 (winter)
	20–40 (summer)	22.5–98 (summer)
Philippine pangolin	Unknown	Unknown
White-bellied pangolin[g]	25.6–27.8 (min. 23.8, max 29.4)	35–55
Black-bellied pangolin	Unknown	Unknown
Giant pangolin	Unknown	Unknown
Temminck's pangolin[h]	18–35	~59

Data from the following sources:
[a]*Pangolin Research Base for Artificial Rescue and Conservation Breeding of South China Normal University (PRB-SCNU), China.*
[b]*Wildlife Reserves Singapore, Singapore.*
[c]*Taipei Zoo, Taiwan.*
[d]*Zoo Leipzig, Germany.*
[e]*Pangolin Conservation and Breeding Center (PCBC), Nandankanan Zoological Park, India.*
[f]*Mohapatra, R.K. Panda, S., Nair, M.V., 2014. Architecture and microclimate of burrow systems of Indian pangolins in captivity. Indian Zoo Year Book. vol. VIII, 12–24.*
[g]*Brookfield Zoo/Chicago Zoological Society, United States.*
[h]*Johannesburg Wildlife Veterinary Hospital, South Africa.*

range maintaining an appropriate thermal environment is less complicated. However, the use of humidity and temperature sensors for environmental monitoring, with the judicious use of heating, cooling and humidifiers as appropriate, is recommended. Table 28.1 presents the appropriate thermal range required for keeping different species of pangolin healthy in captivity.

Hospital enclosure for sick or injured pangolins

Pangolins undergoing intensive veterinary care can be safely held in strongly built hospital crates or small, concrete based enclosures. Sunda and Temminck's pangolins (*Smutsia temminckii*) have been kept in wooden crates for a period of hospitalization, with those used for Temminck's pangolin strengthened by an internal steel frame. The interior of the hospital crate or enclosure must be large enough to safely provide ad lib access to water and an appropriate food bowl (for those species which will consume an artificial captive diet) and can be lined with towels, fleece blankets, yoga matting or natural bedding materials.

Heat mats or other forms of active patient warming are often used within hospital crates and enclosures to ensure environmental temperatures remain stable. However, care must be taken as pangolins have been known to sustain iatrogenic thermal burns by burrowing between a heat mat and the wall of a wooden bed box, particularly when animals are debilitated (Nguyen et al., 2010).

Small hospital enclosures are useful for those animals requiring cage rest and confinement to facilitate medical treatment and healing, but are not suitable for long term care of pangolins in captivity.

Quarantine or short-term holding enclosure

Quarantine housing should be constructed of materials which are easily cleaned and disinfected between inhabitants in order to maintain good biosecurity standards (Vogelnest, 2008). Concrete floored and walled enclosures are commonly used for this purpose, as they are easy to clean and disinfect between inhabitants.

However, to provide some shelter, but ensure adequate ventilation, outdoor enclosures incorporating at least a partial solid roof with the remainder made of chain-link mesh netting are commonly utilized (Fig. 28.1A).

Animal welfare and the success of wildlife reintroduction programs are both supported by a captive environment which enables the maintenance of behavioral diversity (Rabin, 2003). While quarantine enclosures are frequently smaller than those designed for long term captivity, pangolins should still be provided with sufficient space to move around and display some natural behaviors, including climbing and bathing. This is particularly important where pangolins are being held for quarantine prior to release back to the wild.

Water for bathing can be provided within the constraints of a quarantine enclosure by incorporating a small pond depression within the concrete flooring of each enclosure. Utilizing intelligent design, including suitable drainage within the pond and the ability to empty and refill the pond from outside the enclosure to minimize stress, ensures that these features are easily cleaned while improving the quarantine environment for the animals in care (Fig. 28.1B).

FIGURE 28.1 (A) Quarantine enclosures at Cuc Phuong National Park, Vietnam. The partially solid roof provides adequate shelter from sun and rain, while still allowing ventilation and UV access to the enclosure. The solid walls provide a visual barrier from pangolins in adjacent enclosures. Branches allow pangolins the opportunity to climb. *Photo credit: Leanne Wicker, Carnivore and Pangolin Conservation Program/Save Vietnam's Wildlife.* (B) The interior of a quarantine enclosure showing an inbuilt pond to allow bathing. *Photo credit: Leanne Wicker, Carnivore and Pangolin Conservation Program/Save Vietnam's Wildlife.*

Simple nest boxes, pipes or hollow logs should be provided in quarantine enclosures to provide a quiet, dark space for pangolins to retreat and rest. Where pipes or hollow logs are used, larger pangolins may curl up inside taking up most of the internal diameter. It can then be incredibly difficult to observe or remove pangolins for examination. Where more than one pangolin is housed within a single quarantine enclosure, there should be sufficient sleeping spaces provided so that there is at least one retreat location for each individual.

Long-term captive enclosures

In order to improve welfare of animals in long term captive care, larger enclosures, with substrate and furnishings chosen to mimic the natural environment and sufficient size to allow pangolins to display natural climbing, digging and bathing behavior are recommended, and have been found to improve health and wellbeing (Mohapatra, 2016; Rabin, 2003).

Most captive institutions holding pangolins over the long term utilize natural substrates for enclosure flooring. A deep soil layer provides the opportunity to dig burrows or bury feces, a natural behavior in some species, including Chinese and Indian pangolins (Nguyen et al., 2010), and some researchers have confirmed that the incidence of abnormal repetitive behaviors decreases while time spent engaged in exploratory behaviors increases when a soil substrate is provided (Challender et al., 2012; Mohapatra, 2016). A soft, naturalistic substrate also prevents abrasive lesions which can develop on the soft pads of pangolin feet when animals are housed in concrete floored enclosures. However, where a deep soil layer is included in enclosure design an escape proof base must also be laid deep in to the soil substrate to prevent animals digging out of their enclosure. Given the propensity for self-trauma with meshed or chain-linked netting, a concrete base, with adequate drainage, overlaid with a deep soil substrate is recommended. Care must also be taken to ensure that feces and old food is removed daily to maintain a hygienic captive environment. It is also worth noting that the ability to closely monitor animal health and welfare can be compromised where animals dig themselves into deep soil burrows, as individuals may not be easily sighted or retrieved from these burrows. In the Sunda pangolin, retreat to a deep burrow can be particularly common in females that are pregnant or have recently given birth. In Taiwan, Taipei Zoo have investigated substrate type and depth preferences of the terrestrial Chinese pangolin, concluding that a soft soil mixed with leaf litter, which is not too wet to become sticky and not too dry that the burrows dug by pangolins simply cave in, at a depth of 50–100 cm, provides sufficient natural substrate to allow natural digging behavior to be displayed while still facilitating careful management and monitoring of pangolins in captivity.

Solid partitions are recommended between adjacent enclosures housing pangolins, particularly where two males are housed side by side. This prevents excessive climbing and "clawing" on the wire mesh, behavior previously seen in male Sunda pangolins immediately following the introduction of a new male in an adjacent enclosure. Believed to be related to stress, this behavior led to significant self-trauma, and eventually death, in one of the male pangolins (Challender et al., 2012).

Long term enclosure furnishings must provide the pangolin with opportunities to shelter, exercise, investigate and display natural behaviors. Wooden logs and branches, felled tree trunks with intact branches, thick hanging ropes and large live plants are used to allow arboreal species the opportunity to climb and use their prehensile tail. In India, the PCBC include an earthen mound of $2 \times 2 \times 1$ m dimension to allow pangolins to dig a deep burrow. Pangolins are known to move into the deep burrows during day time, facilitating thermoregulation and the maintenance of their normal circadian rhythm (Mohapatra, 2016).

Pangolins are naturally good swimmers (Yang et al., 2007), and some species, including Sunda pangolins, are known to defecate in water (Challender et al., 2012; Zhang et al., 2017). Since they are not known to exhibit grooming behavior, provision of water in small pools large enough for the animal to immerse itself also provides an opportunity for bathing. For some species, including the Indian pangolin (see Fig. 28.2), the provision of a water bath large enough for the pangolin to completely submerge itself is considered essential for thermoregulation during the hot summer months (Mohapatra and Panda, 2014b). In Vietnam, captive Sunda pangolins have frequently been observed bathing in ponds when the environmental temperature was greater than 30 °C. Bathing has also been observed following feeding on live ants from a harvested tree ant nest (*Crematogaster* sp.). This may be to cool down, to relieve itchiness caused by ant bites or to remove ants from underneath the scales (Nguyen et al., 2010).

Most organizations caring for pangolins in captivity offer a range of options for pangolins to sleep in and retreat to. Concrete, wooden or thick plastic boxes, which can be placed on the ground, raised off the ground or set below the enclosure surface, are commonly provided. Subterranean nest boxes, which can be accessed by pangolins via a concrete pipe which opens up at ground level inside the enclosure, and by keepers via a small door which opens into a sunken walkway (Nguyen et al., 2010), offer the additional benefit of increasing the thermal range in the pangolin's captive environment, as the temperature in the nest box is generally cooler than the ambient temperature. Plastic or ceramic water pipes, large hollow tree stumps and a deep clay/soil substrate for digging also provide opportunities for shelter (Fig. 28.3) and sleeping as well as allowing pangolins some choice and control over their thermal environment (Mohapatra, 2016; Zhang et al., 2017).

FIGURE 28.2 Indian pangolin bathing in a pond in its enclosure at the Nandankanan Zoological Park, India. *Photo credit: Rajesh Kumar Mohapatra.*

FIGURE 28.3 A captive Indian pangolin mother and young in a hollow tree stump in their enclosure at the Nandankanan Zoological Park, India. *Photo credit: Rajesh Kumar Mohapatra.*

Social grouping

While research into wild social behavior of pangolins is lacking, it is generally accepted that, aside from during mating and rearing of young, pangolins are solitary in nature (Challender, 2009; Mohapatra and Panda, 2014a; Richer et al., 1997). Most organizations successfully keeping pangolins alive in captivity over the long term house pangolins individually, aside from during mating introductions or maternal care (Mohapatra and Panda, 2014a). Pangolins are reported to have been housed in pairs consisting of one male and one female (Nguyen et al., 2010; van Ee, 1966), two females (Challender et al., 2012; Nguyen et al., 2010) or in small groups (Nguyen et al., 2010; Wilson, 1994). However, males housed together show behavioral signs attributable to stress, including aggression and fighting (Mohapatra and Panda, 2014a; Nguyen et al., 2010). Indian pangolins appear to be most settled in captivity when housed alone, apart from during socialization and mating, or during the period of maternal care of young (Mohapatra and Panda, 2014b).

Behavior monitoring, welfare and environmental enrichment

Since pangolins are nocturnal in nature (with the exception of the black-bellied pangolin [*P. tetradactyla* – see Chapter 8]), behavioral observations of captive animals must occur during night-time hours. Infra-red enabled video cameras mounted in enclosures can provide an accurate understanding of activity patterns and time budgets, without the interference or disturbance which can be caused by the presence of human observers. A number of zoos, including Taipei Zoo, Taiwan and Chicago Zoological Society/Brookfield Zoo, United States (hereafter "Brookfield Zoo"), utilize cameras embedded within the roof of the pangolin's wooden nest boxes to monitor the health and behavior of animals at rest, particularly during breeding, or where females are with young.

There is scant literature on the captive behavior in pangolins. However, observational research in both Indian (Mohapatra, 2016; Mohapatra and Panda, 2014a,b) and Sunda pangolins (Challender et al., 2012) confirm activity in captivity is nocturnal, and is influenced by thermal and light cycle changes between seasons.

A small number of abnormal repetitive behaviors have been reported. These include "clawing" on enclosure wiring in Sunda pangolins (Challender et al., 2012), a behavior believed to be related to stress; and pacing in a repetitive route, including a figure of eight or circular pattern, which has been reported for Chinese, Sunda and Indian pangolins (Challender et al., 2012; Mohapatra and Panda, 2014b; Zhang et al., 2017).

The provision of appropriate environmental enrichment reduces the incidence of abnormal repetitive behaviors and improves the welfare of captive wild animals (Mason et al., 2007). At least one study has confirmed that this is also the case for pangolins (Mohapatra, 2016). Enrichment options used for captive pangolins include rearrangement of enclosure furnishings to interfere with pacing behavior; offering rotting, termite infested logs, abundant leaf and mulch litter substrate; hanging ropes from the enclosure roof and branches to provide an alternative climbing structure; scattering of frozen ant eggs and larvae to redirect clawing behavior; and utilization of novel scents (Anon, 2017; Challender et al., 2012; Mohapatra and Panda, 2014b; Nguyen et al., 2010). Keepers at Wildlife Reserves Singapore (WRS) create papier mache "nests" into which they put live weaver ants (*Oecophylla* spp.), and these are provided to pangolins as food based enrichment. Entire weaver ant nests, harvested from wild ant colonies in Vietnam, are provided intermittently as both nutrition and enrichment to Sunda pangolins residing in Save Vietnam's Wildlife's (SVW) rescue center (Nguyen et al., 2010). In some zoos, including Zoo Leipzig in Germany and Brookfield Zoo, United States, novel items such as timed feeders and challenging puzzle feeders have led to an increase in foraging time and exercise.

In South Africa and Zimbabwe, Temminck's pangolins undergoing rehabilitation prior to release back to the wild are frequently taken outside to allow natural foraging for up to 8 hours per day. In addition to providing an important source of nutrition, this serves as an effective enrichment program. While walking, pangolins engage in a range of normal pangolin behaviors, including sniffing and digging for food, sleeping in between feeding and rolling in dung. Temminck's pangolins can become accustomed to one handler, and are very relaxed when taken by the handler on walks, foraging in close proximity to them or falling asleep inside termite mounds. Stressed animals may try to escape and, when approached, will flatten themselves close to the ground. This behavior disappears once they are used to a handler, but will return in the presence of strangers or loud noises.

While there are very few reports of training or conditioning in pangolins, a small number of zoos, including Zoo Leipzig and Brookfield Zoo have conditioned female pangolins to allow for conscious ultrasound examination to assess reproductive status and monitor pregnancy. Brookfield Zoo have habituated one pangolin for voluntary nail trims using food as a reinforcer.

As focus turns towards maintenance of sustainable captive populations of pangolins, it is vital that attention is given to providing a captive environment that will support the captive health and wellbeing of pangolins.

Captive nutrition

Pangolins are myrmecophagous mammals. Like most other insect-eating species, providing an artificial captive diet which is not only nutritionally complete, but also palatable to the animals, has historically been one of the major challenges to keeping healthy pangolin populations alive in captivity (see Yang et al., 2007). Only rare records of historic diets are published in the available literature. However, unsuccessful diets trialed with Asian pangolin species in captivity have included a variety of food ingredients aimed at sustaining carnivores, including porridge and minced meat, dry dog chow and milk, evaporated milk and pabulum (a suspension or solution of nutrients in a state suitable for absorption), commercial wet feline diets, beef heart, chopped mealworms (*Tenebrio molitor*) and cockroaches (Blattodea), beef bouillon and even baby food (Yang et al., 2007). Gruel type diets had some success at maintaining Chinese pangolins after a slow adaptation process. Insectivore pelleted feeds have been trialled with Chinese and Sunda pangolins, but none have been found to be palatable. Some pangolin young have been observed eating items such as canned cat food or sour cream while others have refused to ingest all food items, showing a strong individual preference.

Taipei Zoo has pioneered the most successful artificial Chinese pangolin diet to date, transitioning from a diet of silkworm (*Bombyx mori*), low fat milk powder, yeast, coconut powder, bread, bee larvae (*Apis* spp.), rice, sweet potato and egg yolk to one of bee larvae, mealworm, egg yolks, apples, coconut powder, vitamin supplements and soil (Yang et al., 2007). Strange colouration and consistency of fecal matter were the strongest motivators for this change, and this amended diet resulted in significant improvements to fecal consistency, shape and color as well as reproductive success (Chin et al., 2009). Successful diets for

Sunda pangolins at WRS are largely insect-based and have supported long lived individuals and breeding (Cabana et al., 2017). The diet was originally beef-based but removing the sinews proved very time consuming and the fatty acid profiles of vertebrate meat was very different to those found in ants and termites.

Unsuccessful artificial diets given to African pangolins are similar to those given to Asian pangolins. Those reported have included a semi-liquid "soup" of ground meat, bran or other cereal and milk with a small quantity of ants, eggs and formic acid (Menzies, 1966). Standard diets at rescue centers historically included maise meal with honey, progessing to cereals and meat products but with little success. White-bellied pangolins kept in US zoos since 2016 are fed a gel-like substance comprising black soldier fly larvae (*Hermetia illucens*), blue bottle fly larvae (*Calliphora vomitoria*), mealworms, silkworm pupae, house crickets (*Acheta domesticus*) and agar (J. Watts, pers. comm.), however, even this reportedly palatable diet required a slow transition from pure ants and ant eggs (Lombardi, 2018). A white-bellied pangolin was reportedly maintained on a commercially available insectivorous bird feed (comprising grains, insect parts and sweetened with honey) for 13 months (Menzies, 1963). There also appears to be variation in food preference between individual pangolins, with reports that the individual kept at the Bronx Zoo, and one of two kept at San Diego Zoo had no problems accepting different artificial diets; however, a single male had to be tube fed for its whole life (M. Shlegal, E. Dierenfeld, pers. comm.). No palatable artificial diet has been reported for rescued Temminck's pangolins.

Given the difficulties in development of a successful, artificial diet, many rescue centers within pangolin range states offer a natural diet of ants, and their eggs and larvae. This is particularly important for recently confiscated pangolins which often fail to recognize artificial diets as food. Both Sunda and Indian pangolins will accept red weaver ants (*Oecophylla smaragdina*; Mohapatra and Panda, 2014a; Nguyen et al., 2010), and Sunda pangolins also consume black heart-shaped ants (*Crematogaster* sp.; Nguyen et al., 2010). Adult Indian pangolins consume around 600 g of red weaver ants per day (~5% of their body weight), while young are provided with 10%. Finding a sustainable supply of weaver ant nests for captive pangolins can present a significant challenge. Where supplies are short and in times of need, boiled poultry eggs mixed with milk powder have been provided as alternate food for Indian pangolins (Mohapatra and Panda, 2014a).

In Zimbabwe and South Africa, Temminck's pangolins undergoing rehabilitation prior to release are taken on supervised foraging walks, as there have been no successful artificial diets developed for this species. This provides an element of enrichment while in captivity, and it has been suggested improves survival rates and the outcomes of rehabilitation (see Chapter 30).

There has been much debate over successful diets for pangolins (Cabana et al., 2017; Yang et al., 2007). A successful diet is one which is both accepted and able to keep an animal alive over the different stages of its life. Acceptance of artificial diets in pangolins remains a significant challenge, and none of those used within rescue centes and zoos are fully palatable to all individuals. All require a slow adaptation process, using ants, ant eggs and larvae as an initial diet, followed by slowly incorporating the artificial mixture in increasing proportions. Without the addition of ants, ant eggs and larvae, acceptance of most diets designed for pangolins has historically been poor and unreliable between individuals.

Diet components

Chitin

The exoskeleton of insects is a sizeable component of a wild pangolin's diet, and its addition to captive diets has proven beneficial by increasing digestability (Chin et al., 2009). Increased apparent organic matter digestion rate has been reported in Chinese pangolins, possibly due to slowing down the food passage rate (Chin et al., 2009). Diets low in insect matter may benefit from the addition of chitin for a number of possible reasons. Chitin may act as a preobiotic, either feeding beneficial gut microbes directly, or result in endogenous chitin digestion. Trials with Sunda pangolins where chitin comprised up to 10% of the diet support this theory (Cabana and Tay, 2019). Chitin may also support digestion of organic matter, crude protein and fiber without impacting the palatability of the diet (see Cabana and Tay, 2019).

Soil

Soil has a significant impact on the fecal consistency of giant anteaters (*Myrmecophaga tridactyla*), with the addition of up to 40% of captive diets consisting of soil leading to better formed feces (Clark et al., 2016). This is expected for myrmecophagous species in captivity, which would naturally ingest soil when breaking into ant nests and termites mounds and predating on the species. Accordingly, the finding that 60–70% of the stomach content of rescued Sunda pangolins at WRS was sand is not surprising (WRS, unpubl. data). Supplying soil in captive diets may provide a number of benefits. First, it may increase the digestibility of organic matter if it helps to "masticate" the food in the stomach. Second, the ingestion of organic matter may increase fiber digestion, which may increase short chain fatty acid production and be beneficial for gut health. Third, it may help to dilute the energy density of the overall diet, allowing pangolins to consume more food while more easily controlling their weight (Cabana and Tay, 2019). This is important because pangolins in captivity can develop obesity problems (see Chapter 29). Finally, soil may provide beneficial environmental bacteria which may have a probiotic-like effect. However, adding soil to the diet may not be as simple as described, because the change in texture, smell and taste may impact palatability in some pangolins. As such, a slow transition, gradually increasing the amount of soil and chitin, may be required.

Nutrient requirements

Nutrient requirements of wild animals are largely unknown (Cheeke and Dierenfeld, 2010). As a result, diets for wildlife are generally based on a "model" species for which there is existing knowledge of nutritional needs. Choice of model species will depend on phylogenetic relationships, similarity in feeding ecology and/or digestive morphology and physiology. For pangolins, the giant anteater would appear suitable based on similarities in feeding ecology (Redford, 1985). The nutritional requirements for the giant anteater are generally based on the nutritional model of the domestic dog (*Canis lupus familiaris*). Cabana et al. (2017) reviewed successful pangolin diets at institions maintaining the species in captivity and the more successful diets fell within nutrient ranges similar to canines (see Table 28.2). It would therefore be appropriate to use the nutritional requirements of domestic canines, combined with insights from giant anteaters, as a basis for developing future captive diets for pangolins.

While there is evidently sufficient knowledge of pangolin nutrition to maintain several species in captivity (Cabana et al., 2017) and in some cases breed (e.g., Chinese and Sunda pangolins; see Chapter 36; Cabana et al., 2019), this is not the case for all species (Cabana et al., 2017), and knowledge of pangolin nutrition is far from complete and warrants further research.

Reproduction in captivity

Given the challenges that have resulted in poor survival of pangolins in captivity, it is

TABLE 28.2 Nutrient recommendations for pangolin diets in captivity (dry matter).

Nutrient (unit)	Range
Crude Fat (%)[a]	19–31
Crude Protein (%)[a]	32–53
ADF (Acid detergent fiber) (%)[a]	5–16
Ca (%)[a]	0.4–1.3
P (%)[a]	0.25–0.80
Vitamin A (IU A/kg)[b]	5000–25,000
Vitamin D (IU A/kg)[b]	500–5000
Na (%)[b]	0.06
K (%)[b]	0.6
Fe (mg/kg)[b]	80–200
Cu (mg/kg)[b]	7.3–250
Zn (mg/kg)[b]	120–1000
Se (mg/kg)[b]	0.11–2.0
Vitamin E (mg/kg)[b]	50–1000

[a]*Data from Cabana, F., Plowman, A., Nguyen, V.T., Chin, S.-C., Sung-Lin, W., Lo, H.-Y., et al., 2017. Feeding Asian pangolins: An assessment of current diets fed in institutions worldwide. Zoo Biol. 36 (4), 298–305.*
[b]*Data from National Research Council, 2006. Nutrient Requirements of Dogs and Cats. National Research Council, National Academies Press, Washington, DC.*

not surprising that efforts to develop captive breeding for conservation in all species have been largely unsuccessful (Yang et al., 2007). Difficulties associated with developing a nutritionally complete diet to support health and reproduction, high levels of morbidity and mortality due to a poor understanding of health needs and their apparent susceptibility to stress induced immune suppression, and a lack of knowledge of the reproductive behavior of pangolins, particularly as it pertains to mate choice, maternal care and weaning management, have all contributed to this lack of success (Hua et al., 2015; Pattnaik, 2008; Sun et al., 2018; Yang et al., 2007).

There are irregular reports of small numbers of captive births of most pangolin species over the past 150 years (see Hua et al., 2015; Yang et al., 2007), including births of white-bellied pangolins since 2016 in the United States (Lombardi, 2018). The institutions that have seen most success in captive breeding include Taipei Zoo, Taiwan, the PCBC at Nandankanan Zoological Park, India, the PRB-SCNU, China and WRS, Singapore which have all spent considerable time and resources developing an adequate diet, enclosures that are large and complex enough to allow a range of natural behaviors to be displayed, careful management of breeding introductions and close monitoring of maternal care (Chin et al., 2009; Hua et al., 2015; Mohapatra and Panda, 2014a; Yang et al., 2007). It is possible that a number of records of captive births, particularly involving sick and injured wild pangolins and those confiscated from the illegal trade and transferred to rescue centers, relate to animals that were conceived in the wild, rather than being a true representation of "captive breeding." In addition, many captive born pangolins have not survived beyond their first year (Yang et al., 2007), and there are only a few records of third generation captive born pangolins, most notably Chinese pangolins at Taipei Zoo, Taiwan (Chapter 36; Hua et al., 2015).

There appears to be considerable variation in the reproductive traits of different species. Research has suggested that the Chinese pangolin has a defined breeding season (Zhang et al., 2016), while other species are understood to breed year round (Lim and Ng, 2008; Zhang et al., 2015). Births of Indian pangolins in captivity have been reported in all months except for May and June (Mohapatra and Panda, 2014a).

Gestation periods and time to weaning reported in the literature also vary widely, both between and within species (see Tables 28.3 and 28.4), but in general, the gestation period is long. In addition, while some older reports suggest that twin births are possible (Payne et al., 1998; Prater, 1971), contemporary research

TABLE 28.3 Gestation periods reported for pangolins.

Species	Gestation period
Sunda pangolin	Varied reports: 176–188 d (Zhang et al., 2015), < 168 d (Nguyen et al., 2010)
Chinese pangolin	Varied reports: 180–225 d (Zhang et al., 2016), 318–372 d[a] (see also Chapter 4)
Indian pangolin	Varied reports: 165 d[b], 251 d (Mohapatra et al., 2018; see also Chapter 5)
Philippine pangolin	Unknown (see Chapter 7)
White-bellied pangolin	140–150 d[c], 186 d[d], 209 d[e] (see also Chapter 9)
Black-bellied pangolin	140 d[c,f]
Giant pangolin	Unknown
Temminck's pangolin	105–140 d[g,h]

[a]*Chin, S.-C., Lien, C.-Y., Chan, Y.-T., Chen, C.-L., Yang, Y.-C., Yeh, L.-S., 2011. Monitoring the gestation period of rescued Formosan pangolin (*Manis pentadactyla*) with progesterone radioimmunoassay. Zoo Biol. 31 (4), 479–489.*
[b]*Panda, S., Mishra, S., Mishra, A.K., Mohapatra, S.N., 2010. Nandankanan Faunal Diversity. Nandankanan Biological Park, Forest and Environment Department, Government of Odisha, India.*
[c]*Pagès, E., 1972. Comportement maternal et développement due jeune chez un pangolin arboricole (*M. tricuspis*). Biol. Gabon. 8, 63–120.*
[d]*Menzies, J.I., 1971. The birth in captivity of a tree pangolin (*Manis tricuspis *Rafinesque) and observations on its development. Niger. J. Sci. 5, 77–84.*
[e]*Kersey, D., Guilfoyle, C., Aitken-Palmer, C., 2018. Reproductive hormone monitoring of the tree pangolin (*Phataginus tricuspis*). Chicago International Symposium on Pangolin Care and Conservation, Brookfield Zoo, Chicago, IL, 23–25 August 2018.*
[f]*Pagès, E., 1972, Comportement maternel et développement du jeune chez un Pangolin arboricole (*M. tricuspis*). Biol. Gabon. 8, 63–120.*
[g]*van Ee, C., 1966. A note on breeding the Cape pangolin *Manis temminckii* at Bloemfontein zoo. Int. Zoo Yearb. 6 (1), 163–164.*
[h]*D.W. Pietersen, unpubl. data.*

TABLE 28.4 Time to weaning reported for pangolins.

Species	Time to weaning
Sunda pangolin	Varied reports: 107–112 d[a], 90–120 d (Lim and Ng, 2008)
Chinese pangolin	Varied reports: 113 d[b], 157 d (Sun et al., 2018; see Chapter 4)
Indian pangolin	150–240 d[c]
Philippine pangolin	Unknown (see Chapter 7)
White-bellied pangolin	90–180 d[d,e]/at next parturition (see Chapter 9)
Black-bellied pangolin	At next parturition[d]
Giant pangolin	At next parturition (Chapter 10)
Temminck's pangolin	135–360 d (see Chapter 11)

[a]*Wildlife Reserves Singapore, unpubl. data.*
[b]*Masui, M., 1967. Birth of a Chinese pangolin *Manis pentadactyla* at Ueno Zoo, Tokyo. Int. Zoo Yearb. 7 (1), 114–116.*
[c]*Mohapatra, R.K., Panda, S., 2014. Husbandry, behaviour and conservation breeding of Indian pangolin. Folia Zool. 63 (2), 73–80.*
[d]*Kingdon, J., 1971. East Africa Mammals. At Atlas of Evolution in Africa, vol. I, Primates, Hyraxes, Pangolins, Protoungulates, Sirenians. Academic Press, London.*
[e]*Pagès, E., 1972. Comportement maternal et développement due jeune chez un pangolin arboricole (*M. tricuspis*). Biol. Gabon. 8, 63–120.*

(Lim and Ng, 2008; Mohapatra and Panda, 2014a; Mohapatra et al., 2018; Zhang et al., 2015) and captive records from some of the more successful conservation programs, including the centers managed by SVW and the PCBC, and the zoo-based population managed by Brookfield Zoo, report only single births. An exception is a trade confiscated Sunda pangolin that died soon after rescue from illegal trade at SVW, and on necropsy a twin pregnancy was confirmed, one fetus on the left uterine horn and one on the right (N.D.H. Nguyen, pers. comm.). Since adult females are known to adopt young (Yang et al., 2007), and given multiple orphans will ride on the back of any nearby adult pangolin (Nguyen et al., 2010), it is possible that some anecdotal reports of twins arise from confiscated females observed to be carrying two (or more) infants. The variation seen in published reports on reproductive parameters for pangolins underscores the need for formal, and focused, research in this area.

Hand-rearing orphaned pangolins

To optimize the chances that a hand-reared wild animal will enjoy good health, appropriate growth rate, and be successfully released into the wild with behavior which enables it to assimilate into wild populations, it is important that the hand-rearing process mimics, as far as possible, the natural situation (McCracken, 2008). This relies on a good understanding of the normal growth and development of wild young, maternal care, diet changes with growth, age at weaning, and other environmental variables. Unfortunately, there is scant knowledge of the normal reproductive behavior and weaning of pangolins (Challender, 2009). As such, hand-rearing of orphaned pangolins is based on a poor understanding of the growth of pangolins of the same age in the wild. Regardless, this remains an important aspect of the management of pangolins in captivity, and Chinese, Sunda and Indian pangolins have been successfully hand-reared (Mohapatra, 2013; Nguyen et al., 2010).

Pangolins may require hand-rearing for a number of reasons. Where a pangolin has been born in captivity, the mother may fail to provide adequate maternal care necessitating supplementary feeding, or complete hand-rearing, by carers, or the mother may have died or be severely injured, for example, if rescued from illegal trade. Where orphan pangolins are confiscated from the illegal trade, how or when they were separated from their mother may remain unknown. The need to intervene to hand-rear orphaned pangolins is indicated by small size and body length and an individual's refusal to consume an adult diet of ants and ant eggs or larvae (Mohapatra, 2013; Nguyen et al., 2010).

Like all orphaned animals, pangolin young must be stabilized before any food is offered. Since contact with their mother and appropriate maternal care ensures that juvenile wild animals are appropriately hydrated and at normal body temperature, orphaned young may be hypothermic and/or dehydrated on presentation (Gage, 2008). Prior to the introduction of replacement milk formula, hydration can be achieved using oral, intravenous or subcutaneous electrolytes and fluid therapy (Gage, 2008). Successful milk formulas used for pangolins include kitten milk replacer for Sunda and Chinese pangolins (Jaffar et al., 2018; Nguyen et al., 2010) and puppy milk replacer for Indian pangolin (Mohapatra and Panda, 2014a). To reduce the risk of gastrointestinal upset, milk is introduced slowly, being made up at 25% strength initially, and gradually increasing to 100% over the first 24–72 hours (Nguyen et al., 2010). Teat sizes used successfully have included those sold commercially for kittens or macropod marsupials. Volume of food to be fed is dependent on the age of the animal and its development stage. Since growth charts for pangolins aren't available, daily monitoring of weight, assessment of body condition and demeanor can help to guide carers. A successfully hand-reared Sunda pangolin, which was 700 g on arriving in captive care, was initially fed 20% of body weight in volume of milk, divided into four equal meals daily (Nguyen et al., 2010). Even where orphans are hand-reared successfully, growth rates rarely match those of young raised naturally (Cabana et al., 2019; Sun et al., 2018), although the addition of an extra protein source, particularly casein, during the early part of lactation, with the protein to fat ratio slowly decreasing during the milk feeding period, may be helpful in achieving a desried growth rate (Cabana et al., 2019).

Housing of pangolin young depends on age at presentation and species. A hand-reared Indian pangolin was housed in a box of $60 \times 40 \times 30$ cm with towels for sleeping and shelter (Mohapatra, 2013). While this was successful initially, the individual began to show behaviors interpreted as stereotypic, necessitating the provision of a larger enclosure (Mohapatra, 2013). In Vietnam, one orphaned pangolin was confiscated with a group of juveniles at, or close to, weaning age. The orphan was housed in a crèche like situation with the other juveniles, being removed from the group enclosure only for bottle feeds as required. This eliminated the need to provide additional heat in the nest box and exposed the animal to the adult diet during the entire hand-rearing period, facilitating weaning from milk feeds to adult diet (Nguyen et al., 2010). As with all orphaned animals, strict hygiene and careful husbandry is required, as these individuals are more susceptible to infectious diseases than those being raised naturally by their mothers (Nguyen et al., 2010).

Successful weaning of hand-reared pangolins remains challenging and mortality rates in captivity remain high. A deeper understanding of the behavioral ecology of maternal care in all species of pangolins remains an urgent need.

Conclusion

As unsustainable, illegal trade continues to threaten wild pangolin populations in Asia and Africa, efforts to develop evidence based protocols for the care and conservation breeding of pangolins in captivity have increased around the world. Inroads have been made into understanding the nutritional requirements of pangolins, and there are reliable reports of artificial diets being both accepted by pangolins and having a positive impact on pangolin health and reproductive success in captivity. Acknowledging the frequency with which pangolins can seriously injure themselves on, or escape from, poorly designed and constructed enclosures, and an improved awareness of their sensitivity to fluctuations in both environmental temperatures and humidity, have all lead to significant developments in the design and construction of enclosures for captive care of pangolins of different species and different ages. There is also increased recognition of the significant consequences of chronic and acute stress in pangolins, leading to a significantly greater focus on pangolin captive welfare, and the provision of enrichment to those pangolins which remain in captivity over the long term.

However, provision of a captive diet which is nutritionally complete as well as readily accepted as food by pangolins remains a significant challenge, as is a working understanding of the reproductive behavioral ecology of wild pangolins so that such knowledge can be applied in ex situ situations to improve the success of captive breeding for conservation purposes. In addition, while the importance of captive welfare is increasingly acknowledged, very few captive institutions have developed training or conditioning programs to facilitate handling and captive management of pangolins in a manner which offers free choice and control to individuals.

There have been great advances in the general approach to pangolin husbandry in recent decades, but much of the focus has been on the Sunda, Chinese, Indian, and increasingly white-bellied pangolin. There is less knowledge of the long term captive husbandry needs of Temminck's pangolin and a complete lack of information on the captive management of the Phillipine, giant and black-bellied pangolin.

References

Aitken-Palmer, C., Sturgeon, G.L., Bergmann, J., Knightly, F., Johnson, J.G., Ivančić, M., et al., 2017. Enhancing conservation through veterinary care of the white-bellied tree pangolin (*Manis tricuspis*). 49th AAZV Annual Conference Proceedings. American Association of Zoo Veterinarians, Texas, United States.

Anon, 2017. Brookfield Zoo: White-Bellied Tree Pangolin Standards of Care. Chicago, IL, United States.

Cabana, F., Tay, C., 2019. The addition of soil and chitin into Sunda pangolin (*Manis javanica*) diets affect digestibility, faecal scoring, mean retention time and body weight. Zoo Biol. Early View.

Cabana, F., Plowman, A., Nguyen, V.T., Chin, S.-C., Sung-Lin, W., Lo, H.-Y., et al., 2017. Feeding Asian pangolins: an assessment of current diets fed in institutions worldwide. Zoo Biol. 36 (4), 298–305.

Cabana, F., Tay, C., Arif, I., 2019. Comparison of growth rates of hand-reared and mother-reared Sunda pangolin (*Manis javanica*) pups at the Night Safari (Singapore). J. Zoo Aquarium Res. 77 (1), 44–49.

Challender, D., 2009. Asian Pangolins: how behavioural research can contribute to their conservation. In: Pantel, S., Chin, S.-Y. (Eds.), Proceedings of the Workshop on Trade and Conservation of Pangolins Native to South and Southeast Asia, 30 June – 2 July 2008, Singapore Zoo, Singapore. TRAFFIC Southeast Asia, Petaling Jaya, Selangor, Malaysia, pp. 95–102.

Challender, D.W.S., Nguyen, V.T., Jones, M., May, L., 2012. Time-budgets and activity patterns of captive Sunda pangolins (*Manis javanica*). Zoo Biol. 31 (2), 206–218.

Cheeke, P.R., Dierenfeld, E.S., 2010. Comparative Animal Nutrition and Metabolism. Cambridge University Press, Cambridge.

Chin, J.S.-C., Tsao, E.H., 2015. Pholidota. In: Miller, R.E., Fowler, M.E. (Eds.), Fowler's Zoo and Wild Animal Medicine, vol. 8. Saunders, St. Louis, pp. 369–375.

Chin, S.-C., Yang, C.-W., 2009. Formosan pangolin rescue, rehabilitation and conservation. In: Pantel, S., Chin, S.-Y. (Eds.), Proceedings of the Workshop on Trade and Conservation of Pangolins Native to South and Southeast Asia, 30 June – 2 July 2008, Singapore Zoo, Singapore. TRAFFIC Southeast Asia, Petaling Jaya, Selangor, Malaysia, pp. 108–110.

Chin, S.-C., Yang, C.-W., Lien, C., Chen, C., Guo, J., Wang, H., et al., 2009. The effect of soil addition to the diet formula on the digestive function of Formosan pangolin (*Manis pentadactyla pentadactyla*). Third meeting of the Asian Society of Zoo and Wildlife Medicine 18–19 August 2009, Seoul National University, Seoul, Republic of Korea.

Chin, S.-C., Lien, C.-Y., Chan, Y., Chen, C.-L., Yang, Y.-C., Yeh, L.-S., 2015. Hematologic and serum biochemical parameters of apparently healthy rescued formosan pangolins (*Manis pentadactyla pentadactyla*). J. Zoo Wildlife Med. 46 (1), 68–76.

Clark, L., Nguyen, T.V., Phuong, T. Q., 2009. A long way from home: the health status of Asian pangolins confiscated from the illegal wildlife trade in Viet Nam. In: Pantel, S., Chin, S.-Y. (Eds.), Proceedings of the Workshop on Trade and Conservation of Pangolins Native to South and Southeast Asia, 30 June – 2 July 2008, Singapore Zoo, Singapore. TRAFFIC Southeast Asia, Petaling Jaya, Selangor, Malaysia, pp. 111–118.

Clark, A., Silva-Fletcher, A., Fox, M., Kreuzer, M., Clauss, M., 2016. Survey of feeding practices, body condition and faeces consistency in captive ant-eating mammals in the UK. J. Zoo Aquarium Res. 4 (4), 183–195.

Gage, L.J., 2008. Hand-Rearing Wild and Domestic Mammals. Wiley-Blackwell, Oxford.

Gaubert, P., 2011. Family Manidae. In: Wilson, D.E., Mittermeier, R.A. (Eds.), Handbook of the Mammals of the World, vol. 2. Hoofed Mammals. Lynx Edicions, Barcelona, pp. 82–103.

Hua, L., Gong, S., Wang, F., Li, W., Ge, Y., Li, X., et al., 2015. Captive breeding of pangolins: current status, problems and future prospects. ZooKeys 507, 99–114.

Jaffar, R., Kurniawan, A., Maguire, R., Anwar, A., Cabana, F., 2018. WRS Husbandry Manual for the Sunda Pangolin (*Manis javanica*), first ed. Wildlife Reserves Singapore Group, Singapore.

Lim, N.T.L., Ng, P.K., 2008. Home range, activity cycle and natal den usage of a female Sunda pangolin *Manis javanica* (Mammalia: Pholidota) in Singapore. Endanger. Sp. Res. 4, 233–240.

Lombardi, L., 2018. U.S. zoos learn how to keep captive pangolins alive, helping wild ones. Available from: <https://news.mongabay.com/2018/01/u-s-zoos-learn-how-to-keep-captive-pangolins-alive-helping-wild-ones/>. [January 15, 2019].

Mason, G., Clubb, R., Latham, N., Vickery, S., 2007. Why and how should we use environmental enrichment to tackle stereotypic behaviour? Appl. Anim. Behav. Sci. 102 (3–4), 163–188.

McCracken, H., 2008. Veterinary aspects of hand-rearing orphaned marsupials. In: Voglenest, L., Woods, R. (Eds.), Medicine of Australian Mammals. CSIRO Publishing, Clayton South, Victoria, Australia, pp. 13–37.

Menzies, J., 1963. Feeding pangolins (*Manis* spp [sic]) in captivity. Int. Zoo Yearb. 4 (1), 126–128.

Menzies, J., 1966. A note on the nutrition of the tree pangolin *Manis tricuspis* in captivity. Int. Zoo Yearb. 6 (1), 71-71.

Mohapatra, R.K., 2013. Hand-rearing of rescued Indian pangolin (*Manis crassicaudata*) at Nandankanan Zoological Park, Odisha. In: Acharjyo, L.N., Panda, S. (Eds.), Indian Zoo Year Book. Indian Zoo Directors Association and Central Zoo Authority, New Delhi, India, pp. 17–25.

Mohapatra, R., 2016. Studies on Some Biological Aspects of Indian Pangolin (*Manis crassicaudata* Gray, 1827). Ph.D. Thesis, Utkal University, Bhubaneswar, India.

Mohapatra, R.K., Panda, S., 2014a. Husbandry, behaviour and conservation breeding of Indian pangolin. Folia Zool. 63 (2), 73–80.

Mohapatra, R.K., Panda, S., 2014b. Behavioural descriptions of Indian pangolins (*Manis crassicaudata*) in Captivity. Int. J. Zool. 795062.

Mohapatra, R.K., Panda, S., Sahu, S.K., 2018. On the gestation period of Indian pangolins (*Manis crassicaudata*) in captivity. Biodivers. Int. J. 2 (6), 559–560.

Nash, H.C., Lee, P., Low, M.R., 2018. Rescue, rehabilitation and release of Sunda pangolins (*Manis javanica*) in Singapore. In: Soorae, P.S. (Ed.), Global Re-Introduction Perspectives. Case-Studies From Around the Globe. IUCN/SSC Re-introduction Specialist Group, Abu Dhabi, UAE.

Nguyen, V.T., Clark, L., Tran, Q., 2010. Management Guidelines for Sunda Pangolin (*Manis javanica*), first ed. Carnivore and Pangolin Conservation Program, Cuc Phuong National Park, Vietnam.

Pattnaik, A., 2008. Enclosure design and enrichment key to the successful conservation breeding of Indian pangolin (*Manis crassicaudata*) in captivity. Indian Zoo Year Book V, 91–102.

Payne, J., Francis, C.M., Phillipps, K., 1998. Field Guide to the Mammals of Borneo, third ed. Sabah Society, Kota Kinabalu, Malaysia.

Prater, S.H., 1971. The Book of Indian Mammals, third ed. Bombay Natural History Society, Bombay.

Rabin, L., 2003. Maintaining behavioural diversity in captivity for conservation: natural behaviour management. Anim. Welfare 12 (1), 85–94.

Redford, K.H., 1985. Feeding and food preference in captive and wild giant anteaters (*Myrmecophaga tridactyla*). J. Zool. 205 (4), 559–572.

Richer, R., Coulson, I., Heath, M., 1997. Foraging behaviour and ecology of the Cape pangolin (*Manis temminckii*) in north-western Zimbabwe. Afr. J. Ecol. 35 (4), 361–369.

Sodeinde, O.A., Adedipe, S.R., 1994. Pangolins in south-west Nigeria - current status and prognosis. Oryx 28 (1), 43–50.

Sun, N.C.-M., Sompud, J., Pei, K.J.-C., 2018. Nursing period, behavior development, and growth pattern of a newborn Formosan pangolin (*Manis pentadactyla pentadactyla*) in the Wild. Trop. Conserv. Sci. 11, 1–6.

Sun, N.C.-M., Arora, B., Lin, J.-S., Lin, W.-C., Chi, M.-J., Chen, C.-C., et al., 2019. Mortality and morbidity in wild Taiwanese pangolin (*Manis pentadactyla pentadactyla*). PLoS One 14 (2), e0212960.

van Ee, C., 1966. A note on breeding the Cape pangolin Manis temniincki at Bloemfontein zoo. Int. Zoo Yearb. 6 (1), 163–164.

Vijayan, M., Leong, C., Ling, D., 2009. Captive Management of Malayan Pangolins *Manis javanica* in the Night Safari. In: Pantel, S., Chin, S.-Y. (Eds.), Proceedings of the Workshop on Trade and Conservation of Pangolins Native to South and Southeast Asia, 30 June – 2 July 2008, Singapore Zoo, Singapore. TRAFFIC Southeast Asia, Petaling Jaya, Selangor, Malaysia, pp. 119–130.

Vogelnest, L., 2008. Veterinary considerations for the rescue, treatment, rehabilitation and release of wildlife. In: Vogelnest, L., Woods, R. (Eds.), Medicine of Australian Mammals. CSIRO, Publishing, pp. 1–12.

Wilson, A.E., 1994. Husbandry of pangolins *Manis* spp. Int. Zoo Yearb. 33 (1), 248–251.

Yang, C.-W., Chen, S., Chang, C.-Y., Lin, M.F., Block, E., Lorentsen, R., et al., 2007. History and dietary husbandry of pangolins in captivity. Zoo Biol. 26 (3), 223–230.

Zhang, F., Wu, S., Yang, L., Zhang, L., Sun, R., Li, S., 2015. Reproductive parameters of the Sunda pangolin, *Manis javanica*. Folia Zool. 64 (2), 129–135.

Zhang, F., Wu, S., Zou, C., Wang, Q., Li, S., Sun, R., 2016. A note on captive breeding and reproductive parameters of the Chinese pangolin, *Manis pentadactyla* Linnaeus, 1758. ZooKeys 618, 129–144.

Zhang, F., Yu, J., Wu, S., Li, S., Zou, C., Wang, Q., et al., 2017. Keeping and breeding the rescued Sunda pangolins (*Manis javanica*) in captivity. Zoo Biol. 36 (6), 387–396.

CHAPTER

29

Veterinary health of pangolins

Leanne Vivian Wicker[1], *Karin Lourens*[2] *and Lam Kim Hai*[3]

[1]Australian Wildlife Health Centre, Healesville Sanctuary, Zoos Victoria, Healesville, VIC, Australia
[2]Johannesburg Wildlife Veterinary Hospital, Johannesburg, South Africa [3]Save Vietnam's Wildlife, Cuc Phuong National Park, Ninh Binh Province, Vietnam

OUTLINE

Pangolins
DOI: https://doi.org/10.1016/B978-0-12-815507-3.00029-0

Introduction

Pangolins have been kept in captivity for over 150 years (Yang et al., 2007). However, despite rare reports of longevity of more than 20 years, the vast majority have died within the first six months, and overall captive survival rates are low (Chin and Tsao, 2015; Mohapatra and Panda, 2014; Wilson, 1994; Yang et al., 2007; Zhang et al., 2017). A poor understanding of their highly specialized nutritional needs, reproductive and husbandry requirements, and a general maladaptation to captive stress have all contributed to high morbidity and mortality rates in pangolins (Clark et al., 2009; Hua et al., 2015; Perera et al., 2017).

As wild populations decline there is an urgent need to improve the health and welfare of captive pangolins, as part of a holistic conservation response (see Chapter 31). While normal physiology and common health concerns of pangolins remain poorly studied compared to many more well-known species (Langan, 2014), the challenges faced by rescue centers and conservation breeding programs in pangolin range states, and increased interest in maintaining pangolins in zoos, have led to a rapid proliferation of health information for pangolins. This chapter presents a review of the infectious organisms and commonly seen health issues reported for pangolins in zoos and rescue centers worldwide, normal physiological variables (including blood biochemistry and hematology), examination techniques, diagnostic approach and a comprehensive formulary of medical treatment and anesthesia regimens used in pangolins.

Animal restraint

Physical restraint

Pangolins in good health are extremely strong, and their normal behavioral response to stress is to curl up into a tight ball, with their tail covering their head. It is not possible to uncurl a healthy, or very stressed, pangolin which is determined to remain curled up, and attempts to do so can result in injury to the pangolin (Langan, 2014; Robinson, 1999). While they are generally docile animals, caution is advised given their strength, sharp edged scales and long, strong claws.

Captive pangolins can become conditioned to handling for veterinary procedures such as weighing, physical examination, tube feeding (Wilson, 1994) and ultrasound (C. Aitkin-Palmer, pers. comm.); and juveniles or habituated animals may uncurl without the need for sedation or chemical restraint (Chin and Tsao, 2015). Chinese (*Manis pentadactyla*) and Sunda pangolins (*M. javanica*) accustomed to human contact may be restrained for physical examination by an experienced handler holding the base of the tail while gently shaking the animal and supporting its body weight by placing its forelimbs on the ground until it uncurls (Nguyen et al., 2010). This is not an appropriate method of manual restraint for extremely stressed pangolins, or for the heavier species: Temminck's pangolin (*Smutsia temminckii*) and giant pangolin (*S. gigantea*).

Animals undergoing frequent handling for treatment over a prolonged period may begin to anticipate a procedure, curling up tightly to resist efforts to medicate or examine wounds. Temminck's pangolins will curl up with their scales clamped down so tightly following subcutaneous or intramuscular injection that it is difficult to remove the needle. As a result, sedation or anesthesia is recommended for many veterinary procedures.

Sedation and chemical restraint

Sedation and general anesthesia in healthy pangolins is often uncomplicated, and pre-anesthetic fasting does not appear to be required. A wide range of anesthetic and sedative drugs and approaches have been used (see Table 29.1).

Sedation to facilitate veterinary procedures, or to calm stressed animals in captivity, has been achieved in Sunda, Chinese and Temminck's pangolins using diazepam given via intramuscular or subcutaneous injection. A combination of butorphanol and medetomidine hydrochloride has also been used for Temminck's pangolin, however only a very light plane of sedation was achieved at the dose rate used (see Table 29.1), and was insufficient to obtain radiographs or pass a stomach tube.

While a range of injectable anesthesia regimens have been used, including Zoletil, a commercially available dissociative anesthetic drug combination of tiletamine and zolazepam, to allow for short procedures (Chin and Tsao, 2015), inhalational anesthesia using isoflurane delivered in oxygen is the most commonly used protocol for surgical anesthesia (Chin and Tsao, 2015; Khatri-Chhetri et al., 2015; Nguyen et al., 2010). Pangolins may be induced by administration of isoflurane via a face mask, providing the animal is able to be uncurled, or placed into a sealed anesthetic induction box. Given the elongated face, small opening to the oral cavity and caudally placed larynx, intubation is extremely challenging in pangolins and has not been reported (Chin and Tsao, 2015) necessitating maintenance of anesthesia by delivery of isoflurane and oxygen using a face mask (see Fig. 29.1A–D).

Historically, ketamine hydrochloride was reported to provide safe anesthesia in Chinese pangolins (Heath and Vanderlip, 1988) and black-bellied pangolins (*Phataginus tetradactyla*) (Robinson, 1983). However, muscle tone persists even at high doses, salivation is common and duration of action is short (10–20 min) so this regimen is no longer recommended (Jaffar et al., 2018; Langan, 2014)

Anesthetic monitoring

Monitoring during anesthesia is similar to that used for other small mammals. At a minimum, heart rate, respiratory rate and core body temperature should be monitored (see Box 29.1).

Heart rate and respiratory rate are easily measured during anesthesia using visual observations, stethoscope or patient monitors (Fig. 29.1B). A pulse oxymeter probe may be placed on the cloacal folds, nipple or the skin folds just in front of the hindlimbs. Femoral pulses may be palpated in some species, however are not palpable in Temminck's pangolin. Use of electrocardiogram (ECG; see Fig. 29.1A, B) to monitor electrical activity of the heart and detect cardiac abnormalities has also been used (A. Grioni, pers. comm.).

Physical examination

To assess for external or behavioral evidence of disease, examination should begin with a distant observation of the animal's demeanor, ambulation, respiratory effort and responsiveness to normal stimuli. A healthy pangolin feels warm and dry to touch, and responds to stressful stimuli by contracting further into a curled up shape (Nguyen et al., 2010). Reducing stress and noise may encourage the animal to uncurl and begin to investigate its surroundings. When uncurled, pangolins frequently sniff their environment and the people handling them, and may attempt to grasp and curl around objects in their reach (Nguyen et al., 2010).

Sick or injured animals may be more weakly, or only partially, curled up, however they should still respond to touch or sound stimuli by curling, albeit with less vigor than a healthy individual.

Severely debilitated pangolins lie uncurled in lateral or ventral recumbency, and make

TABLE 29.1 Therapeutic drugs reported for treatment of pangolins. This table presents the results of a review of the published literature, reports from the gray literature, medical records of rescue centers and zoos and personal communication from veterinarians employed in institutions holding pangolins. Definitions used within the table: PO – orally, IM – intramuscular, SC – subcutaneous, IV – intravenous, TOP – topically, SID – once per day, BID – twice per day, TID – three times per day, QID – four times per day, EOD – Every second day, IU – international units.

Drug name	Dose/route/frequency	Species	Comments	Literature source
Antimicrobial agents				
Amikacin sulfate	4.4 mg/kg loading dose then 2.2 mg/kg SC BID	*P. tricuspis*	Used to treat infections caused by susceptible strains of gram negative bacteria	
Amoxicillin[a]	15 mg/kg PO SID up to 14 days	*P. tricuspis*	Used to treat range of bacterial infections	
Amoxicillin (long acting formulation)[b]	8 mg/kg IM EOD, 4–6 doses	*M. javanica*	Used in stressed animals to reduce handling frequency; without further research, the efficacy of this long acting formulation remains unknown	
Amoxicillin clavulanic acid[a,b,c,d,h]	10–20 mg/kg PO, IM SID-BID	*M. javanica* *M. pentadactyla* *S. temminckii* *P. tricuspis*	Used to treat range of infections, including dermatitis beneath scales and other skin wounds	Nguyen et al. (2010)
Ampicillin[e]	20–40 mg/kg IV, 3–4 times per day	*P. tricuspis*	Used to treat range of bacterial infections	
Benzylpenicillin[a,b]	20 mg/kg IV, IM, 3–7 days	*M. javanica*	Used to treat respiratory infections	
	50,000 IU SC, SID, 10 days	*P. tricuspis*		
Ceftazidime	20 mg/kg IM, BID	*P. tricuspis*	Used to treat range of bacterial infections	
Ceftiofur[a]	4.4 mg/kg, SC, SID, 10 days	*P. tricuspis*	Used to treat range of bacterial infections	
Ceftiofur crystalline free acid (long acting preparation)[b,e]	6–8 mg/kg SC every 3–5 days	*M. javanica* *P. tricuspis*	Used in stressed animals to reduce handling frequency; without further research, the efficacy of this long acting formulation remains unknown	Lam (2018)
Ceftriaxone[a]	50 mg/kg IM, BID, 7 days	*P. tricuspis*	Used to treat range of bacterial infections	
Cephalexin[b,f]	15 mg/kg PO	*M. javanica*	Used to treat respiratory, skin and bone infections	

(*Continued*)

TABLE 29.1 (Continued)

Drug name	Dose/route/frequency	Species	Comments	Literature source
Ciprofloxacin[a,h]	5–15 mg/kg, PO, BID, 14 days	*P. tricuspis*	Used to treat range of bacterial infections	
Clindamycin[b,h]	10 mg/kg, PO, BID, 7–10 days	*M. javanica*	Used to treat infections of the bone, tail or limb amputations	
Enrofloxacin[a,b,c,d,e,f]	5–15 mg/kg PO SID or 2.5–5 mg/kg IV	*M. javanica* *S. temminckii* *P. tricuspis*	Used for infections where anaerobic bacteria are not suspected	Jaffar et al. (2018)
Fluconazole[a]	5 mg/kg PO BID 5 days	*P. tricuspis*	Used to treat oral candidiasis	
Itraconazole[a]	5 mg/kg PO SID, 4 weeks	*P. tricuspis*	Used to treat oral candidiasis	
Lincomycin hydrochloride	0.3 ml/kg	*M. pentadactyla*	Used to treat serious bacterial infections which have failed to respond to other broad spectrum antibiotics	Zhang et al. (2017)
Marbofloxacin[f,h]	2–8 mg/kg PO SID, 14 days	*M. javanica*	Used to treat range of bacterial infections	Jaffar et al. (2018)
Metronidazole[a,b,e,h]	15–25 mg/kg PO BID, 7 days	*M. javanica* *P. tricuspis*	Used to treat bacterial and protozoal infections	
Nystatin[a]	50,000–150,000 units PO BID	*P. tricuspis*	Used to treat oral candidiasis	
Toltrazuril[b]	2.5–5 mg/kg PO SID, 3days	*M. javanica*	Used in treatment of coccidiosis	Clark et al. (2010), Jaffar et al. (2018), Lam (2018), Nguyen et al. (2010)
Trimethroprim sulfamethoxazole[f]	30 mg/kg PO SID, 7 days	*M. javanica*	Used in treatment of coccidiosis	Jaffar et al. (2018)
Antiparasitic agents				
Abamectin	400 mg/kg PO SID for 3 days	*M. pentadactyla*	Used to treat broad range of internal and external parasites	Zhang et al. (2017)
Albendazole[b]	20 mg/kg PO	*M. crassicaudata*	Used to treat gastrointestinal nematode infections	Lam (2018), Mohapatra and Panda (2014)
Ivermectin 1% injection[a,b,d,e,f,g]	0.2–0.4 mg/kg PO, SC	*M. pentadactyla* *M. javanica* *P. tricuspis*	Used to treat broad range of internal and external parasites	Chin et al. (2015), Clark et al. (2010), Jaffar et al. (2018), Lam (2018), Nguyen et al. (2010)

(Continued)

TABLE 29.1 (Continued)

Drug name	Dose/route/frequency	Species	Comments	Literature source
Fipronil spray[a]	7–10 mg/kg TOP	*M. tricuspis*	Topical treatment used where tick burdens are heavy	
Imidocarb[c]	6 mg/kg once (two treatments 24 hours apart may be required)	*S. temminckii*	Used to treat prioplasmosis, presumed babesia. May cause salivation which can be controlled using low dose of Atropine IM	
Levamisole[b]	5 mg/kg PO	*M. javanica*	NOT recommended – caused significant hypersalivation in pangolins	
Lindane dust 0.5%[g]	TOP	*M. pentadactyla*	Has been used to treat ticks in pangolins, however caution is advised due to its toxic effects in mammals	Chin and Tsao (2015)
Niclosamide[g]	157 mg/kg	*M. pentadactyla*	Used to treat gastrointestinal cestode infections	Chin et al. (2015)
Piperazine[g]	88–110 ml/kg	*M. pentadactyla*	Used to treat gastrointestinal nematode infections	Chin et al. (2015)
Pyrantel[a]	5 mg/kg PO once per week for 4 weeks	*P. tricuspis*	Used to treat gastrointestinal nematode infections	
Praziquantel[a,b,e]	7–25 mg/kg PO SID, 3 days	*M. javanica* *P. tricuspis*	Used to treat gastrointestinal cestode infections	Lam (2018)
Thiabendazole	59 mg/kg in food, once	*M. pentadactyla*	Used to treat gastrointestinal roundworm infections	Heath and Vanderlip (1988)
Analgesia				
Buprenorphine[c,e]	0.01–0.02 mg/kg IM, IV every 8 hours	*S. temminckii* *P. tricuspis*	Opiate analgesia	
Butorphanol[d,f]	0.05–0.09 mg/kg SC, IM, IV 0.2 mg/kg IM, SC, IV	*M. javanica* *M. pentadactyla*	Opiate analgesia. Also has light sedative properties in *M. javanica*	Jaffar et al. (2018)
Carprofen[f]	4 mg/kg SID	*M. javanica*	Non-steroidal anti-inflammatory drug. Use with care, given propensity to gastric ulceration. Ensure well hydrated prior to administration	Jaffar et al. (2018)
Fentanyl[c]	25 μg/h (patch dosage) TOP (skin)	*S. temminckii*	Potent opiate analgesia. Placed on ventral chest or abdomen, covered with Elastoplast with superglue to attach securely to skin	

(*Continued*)

TABLE 29.1 (Continued)

Drug name	Dose/route/frequency	Species	Comments	Literature source
Meloxicam[b,c,f,h]	0.2 mg/kg SC/PO SID once, then 0.1 mg/kg SC/PO SID	*M. javanica* *S. temminckii* *P. tricuspis*	Non-steroidal anti-inflammatory drug. Use with care, given propensity to gastric ulceration. Ensure well hydrated prior to administration	Jaffar et al. (2018)
Tramadol hydrochloride[b]	1–3 mg/kg PO, IM BID	*M. javanica*	Synthetic opiate analgesia	
Anesthesia and sedation				
Azaparone[c]	0.5 mg/kg deep IM	*S. temminckii*	Light, prolonged sedation, used to reduce stress in newly captive animals to allow tube feeding and reduce stress based behaviors	
Diazepam[b,c]	0.5–1 mg/kg IM 1–3 mg/kg PO	*M. javanica* *S. temminckii*	Useful sedation to allow non-invasive veterinary procedures and to reduce stress in captivity	Lam (2018)
Ketamine HCL	22–25 mg/kg	*M. pentadactyla*, *P. tetradactyla*	No longer recommended as a sole agent due to hyper-salivation and muscle rigidity	Heath and Vanderlip (1988), Robinson (1983)
Ketamine/medetomidine[c]	Ketamine 2 mg/kg, Medetomidine 0.3 mg/kg, both IM, Reversal Atipamezole 0.05 mg/g IM	*M. javanica*	Commonly used for examination of healthy animals	Jaffar et al. (2018)
Ketamine/midazolam[f]	Ketamine 5 mg/kg, Midazolam 0.05 mg/kg both IM, Reversal Flumazenine 0.01 mg/kg IM	*M. javanica*	Useful for short procedures requiring anesthesia	Jaffar et al. (2018)
Medetomidine[c]	0.05–0.1 mg/kg	*S. temminckii*	Good for sedation for short, non-invasive procedures	
Midazolam[f]	0.05–1 mg/kg IM, IV	*M. javanica*	Sedative effect, short procedures only (10–20 min)	Jaffar et al. (2018)
Tiletamine/zolazepam[g]	3–5 mg/kg	*M. pentadactyla*	Useful for short procedures requiring anesthesia	Chin and Tsao (2015)
Behavior modifying agents				
Trazodone[b]	2–5 mg/kg PO SID-TID	*M. javanica*	Used to reduce short term stress in captive animals	
Haloperidol[b]	0.5–1 mg/kg PO SID, 3–5 days	*M. javanica*	Used to reduce anxiety and stress in captive animals	

(Continued)

TABLE 29.1 (Continued)

Drug name	Dose/route/frequency	Species	Comments	Literature source
Gastrointestinal agents				
Bismuth subsalicylate[h]	8.7 mg/kg PO TID	*P. tricuspis*	Used to treat range of gastrointestinal disorders, including diarrhea and excess stomach acid production	
Capromorelin[h]	2.88 mg/kg PO SID	*P. tricuspis*	Used as an appetite stimulant	
Cimetidine[b,g]	5–10 mg/kg PO BID	*M. pentadactyla* *M. javanica*	Histamine H receptor antagonist that inhibits stomach acid production	Chin and Tsao (2015), Lam (2018)
Famotidine[a,e,h]	0.15–0.5 mg/kg SC, PO SID	*P. tricuspis*	Histamine H receptor antagonist that inhibits stomach acid production	
Maropitant citrate[e]	1–2 mg/kg PO, SC	*P. tricuspis*	Anti-nausea medication, used to control vomiting	
Metoclopramide[a]	0.5 mg/kg IM as required	*P. tricuspis*	Stimulates movement and contraction of the stomach, use with caution if gastrointestinal blockage is a possibility	
Omeprazole[a,e]	2 mg/kg PO SID,	*P. tricuspis*	Decreases stomach acid production, used to treat gastrointestinal ulcers	
Ondansetron[e]	1 mg/kg PO BID	*P. tricuspis*	Anti-nausea medication, used to control vomiting	
Ranitidine[a,b,f,g]	2–3.5 mg/kg PO BID	*M. pentadactyla* *M. javanica* *P. tricuspis*	Can be placed on top of a small amount of food prior to offering main meal	Chin and Tsao (2015), Jaffar et al. (2018), Nguyen et al. (2010)
Dioctahedral smectite[b]	75–150 mg/kg PO QID	*M. javanica*	Used prophylactically in first 7 days after arrival to reduce incidence of stress related gastric ulceration	
Sucralfate[a,b,e,f,g]	50–100 mg/kg PO BID at least 1 hour prior to feeding	*M. pentadactyla* *M. javanica* *P. tricuspis*	Lines the gastric mucosa, binding to ulcers to protect them from acid. Efficacy improved if administered at least 30 minutes prior to feeding. Can be added to a small amount of food prior to offering main meal	Chin et al. (2015), Jaffar et al. (2018)

(*Continued*)

TABLE 29.1 (Continued)

Drug name	Dose/route/frequency	Species	Comments	Literature source
Vitamins and minerals				
Ascorbic acid (vitamin C)[h]	22 mg/kg PO BID	*P. tricuspis*	Used to assist severely debilitated individuals	
Cyanocobalamin (vitamin B12)[h]	500 mcg SC	*P. tricuspis*	Used to treat anemia and to assist severely debilitated individuals	
Ferric Hydroxide[h]	10 mg/kg IM once	*P. tricuspis*	Used to treat anemia	
Phytonadione (vitamin K)[h]	1.5 mg/kg IM SID	*P. tricuspis*	Used in case of bleeding disorders, based on the supplementation of vitamin K in giant anteaters; further research required to substantiate its use in this manner	
Emergency drugs				
Atropine[b,g]	0.04–0.05 mg/kg SC, IM or slow IV	*M. pentadactyla* *M. javanica*	Used to treat some nerve and pesticide poisonings, can elevate heart rate during anesthesia. Successfully used to reduce excessive secretions as a side effect to use of Levamisole in *M. javanica*	Chin and Tsao (2015)
Dopram[b]	1–2 mg/kg IV	*M. javanica*	Used to stimulate respiration during anesthesia	
Topical ophthalmological drugs				
Ciprofloxacin 0.3% solution[a]	One drop TOP BID	*P. tricuspis*	Used to treat bacterial infections of the eye and periocular tissues	
Gentamicin sulfate ointment[a]	One drop TOP BID	*P. tricuspis*	Used to treat bacterial infections of the eye and periocular tissues	
Tricin eye and ear ointment (Zinc bacitracin 500 IU/g, Neomycin sulfate 5 mg/g, Polymyxin B sulfate 10,000 IU/g)[b]	One drop TOP BID - QID	*M. javanica*	Used to treat corneal ulceration	Nguyen et al. (2010)

Institutions reporting use of therapeutic agent: [a]San Diego Zoo, United States, [b]Save Vietnam's Wildlife, Vietnam, [c]Johannesburg Wildlife Veterinary Hospital, South Africa, [d]Kadoorie Farm and Botanic Garden, Hong Kong SAR, [e]Brookfield Zoo/Chicago Zoological Society, United States, [f]Wildlife Reserves Singapore, Singapore, [g]Taipei Zoo, Taiwan; [h]Gladys Porter Zoo, United States.

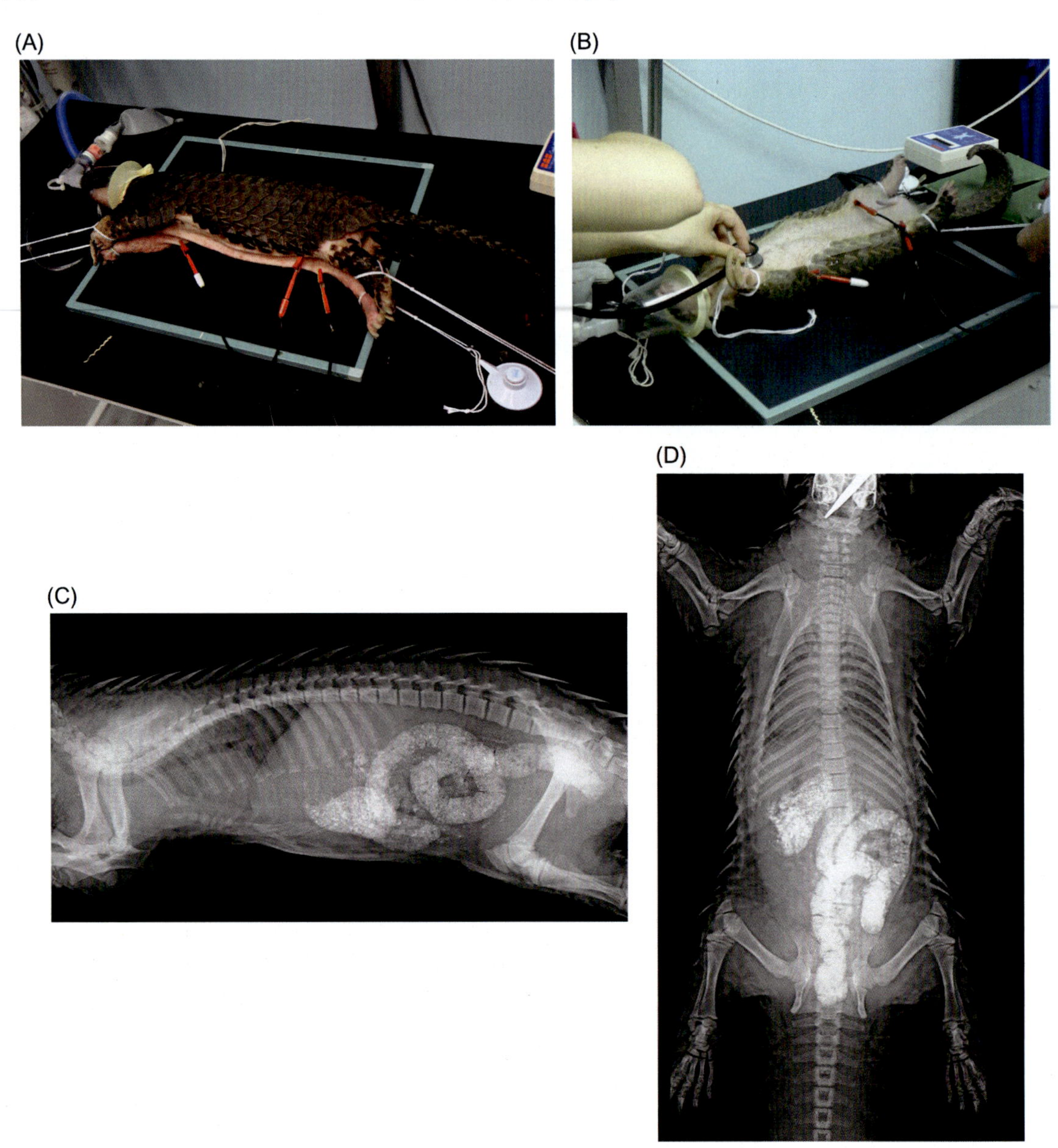

FIGURE 29.1 (A–D) A male Chinese pangolin anesthetized using inhalational isoflurane delivered in oxygen via a face mask, in order to obtain diagnostic radiographs. (A) The pangolin is positioned in lateral recumbency. Note the placement of electrocardiography (ECG) clips to monitor the electrical conductivity of the heart. In (B) the pangolin is positioned in dorsal recumbency, and in addition to the ECG clips, a stethoscope is being used to auscultate heart and lung sounds to measure heart rate and respiratory rate during anesthesia. (C) A lateral radiograph showing normal thoracic, abdominal and skeletal anatomy. The scales over the dorsal body wall are clearly visible. There is a significant volume of feces in the distal gastrointestinal tract. (D) A normal ventrodorsal radiograph of the same individual, taken while the animal was in dorsal recumbency, as shown in (B). *Photo credit: Kadoorie Farm and Botanic Garden (KFBG).*

BOX 29.1

Physiological variables for pangolins

Respiratory rate (measured in breaths per minute)

M. pentadactyla 14–53 (Heath and Vanderlip, 1988)

M. javanica 12–16 (asleep) and 60–100 (awake and active; Nguyen et al., 2010)

S. temminckii 19–30 (asleep; K. Lourens, unpubl. data)

P. tricuspis 20–50 (C. Singleton, pers. comm.)

Heart rate (measured in beats per minute)

M. pentadactyla 80–86 (Heath and Vanderlip, 1988)

M. javanica 80–200 (Nguyen et al., 2010)

S. temminckii 65–80 (asleep; K. Lourens, unpubl. data)

P. tricuspis 80–140 (C. Singleton, pers. comm.)

Core body temperature (measured via rectal thermometer)

M. pentadactyla 32.2–35.2 °C (Heath, 1987; Chin and Tsao, 2015)

M. javanica 32.2–35.2 °C (Nguyen et al., 2010)

S. temminckii 32–34 °C (K. Lourens, unpubl. data)

P. tricuspis 31.1–33.9 °C (C. Singleton, pers. comm.)

little or no attempt to respond as expected to stressful stimuli, remaining uncurled during handling. Very sick pangolins feel slightly cold and clammy to touch, and the skin around their mouth, nares and ventral abdomen may appear pale and gray in color (Nguyen et al., 2010). Pangolins that present in this way have a guarded prognosis for survival.

When an animal is tightly curled up, only the scales and the skin beneath scales of the lateral and dorsal body wall and tail can be examined. It is not possible to examine the face, the limbs, the ventral aspect of the tail or the ventral neck, thorax and abdomen, and sedation or light anesthesia may be required.

During physical examination, the following should be noted:

Physiological variables

Respiratory rate, heart rate and core body temperature can vary widely depending on the activity, stress level and health status at the time of measurement, and whether the animal is conscious or affected by chemical sedation or anesthesia. Core body temperature in pangolins is lower than most eutherian mammals. Box 29.1 lists respiratory rate, heart rate and core body temperature measured via rectal thermometer for a range of pangolin species.

Weight and body condition score

Pangolins in good body condition are well muscled over the vertebrae, scapular spines and pelvis, and the transition from scaled skin to scale free skin at the ventrolateral abdominal wall is smooth.

Pangolins rapidly lose weight and body condition when severely debilitated (Nguyen et al., 2010). Animals in poor body condition are visibly sunken over the pelvis, vertebrae and scapular spines, and these bony prominences can be palpated even beneath scales. Scaled skin appears to hang down over the scale free skin at the ventrolateral abdominal wall.

Obesity can become a concern in pangolins reliably consuming an artificial diet. Fat bulges can be seen around the neck, there is a rounded appearance to the dorsal midline and over the rump, and a bulge may be seen at the transition from scaled skin to scale free skin at the ventrolateral abdominal wall.

Hydration status

The presence of scales can make assessment of hydration status challenging. A degree of skin tenting can be found in normally hydrated animals. Clinically dehydrated pangolins are generally quiet and lethargic with sunken eyes, poor skin turgor on tenting of the pre-femoral skin, and dry mucous membranes around the eyes and inside the mouth. Evaluation of clinical signs alongside the blood parameters hematocrit and total serum protein will provide the clinician with a reasonable assessment of hydration status.

Scales and integument

Healthy wild pangolins frequently have damaged edges to their scales, which grow continuously throughout life. Damage to, or exposure of, the scale bed, however, is indicative of traumatic injury (see Fig. 29.2). Confiscated pangolins frequently present with severe traumatic skin wounds. It is important to check beneath scales for evidence of dermatitis or external parasites, including ticks (see *Infectious organisms and pangolins*). Severely debilitated pangolins sometimes present with non-healing, superficial ulcerative lesions of unknown etiology around the mouth, nares and eyes and on the foot pads (Nguyen et al., 2010).

Eyes and nares

Inflammation of the periocular or nasal tissues and mucopurulent ocular or nasal discharge are indicators of ill health. Healthy pangolins may have moist tear staining around the eyes and a small volume of serous nasal discharge can be a clinically normal finding.

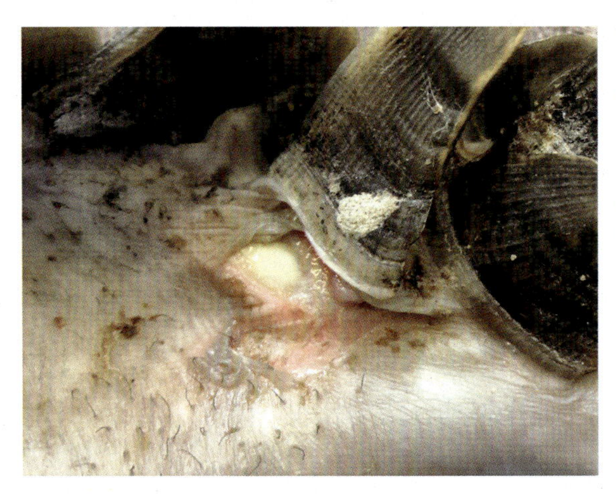

FIGURE 29.2 Skin lesion at the base of a scale in a Temminck's pangolin confiscated from the illegal wildlife trade. The lesion is at the junction between the scaled and scale-less skin, on the ventrolateral abdominal wall. The scale is slightly lifted, and there is a necrotic central area. This individual responded well to antibiotics and topical treatment. *Photo credit: Karin Lourens, Johannesburg Wildlife Veterinary Hospital.*

Thoracic auscultation

Given the reported susceptibility of pangolins to respiratory disease, thoracic auscultation is important. The ability to auscultate accurately is diminished by the scraping of scales on the bell of the stethoscope, and by sniffing sounds which mask respiratory and heart sounds. Approaching auscultation from the un-scaled skin of the ventral thorax provides a more accurate result. However, this is challenging in a conscious, stressed pangolin, as many attempt to curl around the stethoscope (Nguyen et al., 2010).

Oral cavity

Pangolins lack teeth, so dental examination is not relevant. Debilitated pangolins may have ulcerative lesions on the tongue and oral mucosa, but the narrow opening to the oral cavity makes these very difficult to examine. An endoscope can be used to examine inside the oral cavity of a quiet, conditioned or chemically restrained pangolin. The animals sometimes voluntarily extend their long, saliva covered tongues, enabling examination. Given their significant salivary glands, pangolins may have considerable sticky saliva around the lips and at the entrance to the oral cavity, which should not be mistaken for evidence of respiratory disease.

Musculoskeletal exam

The strong scales of pangolins can mask muscular and skeletal injuries, particularly along the vertebral column. Observation of ambulation, use of the limbs and prehensile tail for climbing and careful palpation along the appendicular and axial skeleton all contribute to the musculoskeletal exam. Imaging is necessary if musculoskeletal injury is suspected (see *Diagnostics*; Fig. 29.1A–D).

Sexing pangolins

It is difficult to accurately sex a completely curled up individual. Once an animal is uncurled, the external genitalia can be examined. Male pangolins have inguinal testicles and a penis just cranial to the anus. Female pangolins have a small vulva just cranial to the anus. Both sexes have a single cranial pair of nipples in the axillary region. During lactation, females develop palpable mammary glands associated with the nipple (Nguyen et al., 2010).

Diagnostics

Blood biochemistry and hematology

While conscious blood draws are possible, a light plane of anesthesia is often required to allow safe and reliable venous access for blood collection or intravenous injection (Nguyen et al., 2010).

A number of locations for venepuncture have been described; however, while not visible or palpable, the coccygeal vein on the ventral midline of the tail is most commonly accessed (Fig. 29.3). A 21–23 gauge needle is advanced at an angle of ~45°, in a dorsocranial direction, at the point where two ventral tail scales meet at the midline (Chin and Tsao, 2015; Heath and Vanderlip, 1988; Nguyen et al., 2010). In anesthetized Sunda pangolins blood is also routinely collected from the cephalic vein.

Blood biochemistry and hematology references have been published for Chinese, Sunda, white-bellied (*P. tricuspis*) and Indian pangolins (Tables 29.2 and 29.3). Not all published ranges have been statistically validated, and some represent very small sample sizes and include unhealthy animals. Sex does not appear to significantly affect these parameters. Further research, investigating the effect of seasonality and age, and encompassing all

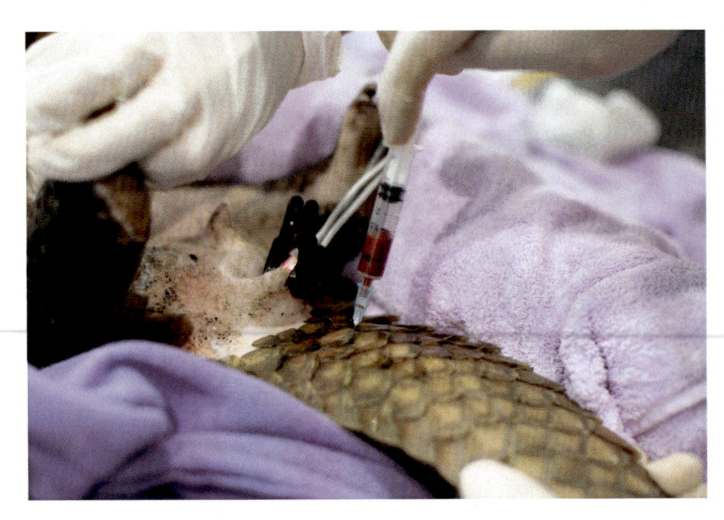

FIGURE 29.3 An anesthetized Sunda pangolin has blood collected from the coccygeal vein on the ventral midline of the tail. The pangolin is also receiving intravenous fluids into the coccygeal vein using a 24-gauge butterfly needle and giving set. *Photo credit: Save Vietnam's Wildlife.*

species of pangolin with statistically significant sample sizes of healthy individuals, is required (see also Chapter 34).

Urinalysis and coprological examination

Urine is preferably collected via cystocentesis from anesthetized pangolins, using aseptic technique. Urinalysis includes microscopic examination to check for bacteria, cells, casts and crystals, measurement of urine specific gravity and the use of commercially available urine dipsticks to check for protein, bilirubin, urobilinogen, glucose, ketones, and blood, and to indicate the urine's pH. Published normal urinalysis values are lacking for all species of pangolin.

Since the prevalence of gastrointestinal parasitism is extremely high in recently captive pangolins, and given severe parasite infection in debilitated animals can be fatal, coprological examination is imperative, particularly for those individuals showing signs of maladaptation to captivity. This includes fecal flotation to assess for the presence of gastrointestinal worm eggs or coccidian oocytes, gram stained fecal smear to examine microbial diversity and wet preparation to assess for motile protozoan. Whether or not treatment is required depends on the number of parasite eggs present, gross examination of the feces, and a clinical examination of the pangolin, but is strongly recommended when watery or bloody diarrhea occurs (see Table 29.1).

Fecal microbial culture has also been utilized in the diagnostic work up of debilitated pangolins, and a number of bacterial species of significance for both human and animal health have been isolated (Table 29.4).

Imaging

Imaging is strongly recommended for the diagnostic investigation of sick or injured pangolins. Ultrasonography and radiography are the most commonly used imaging modalities in pangolins. A minimum of two radiographic views (ventrodorsal or dorsoventral and lateral) is required to identify the location of lesions, with additional views taken as required (Fig. 29.1A–D). The scales are visible but do not reduce the ability to obtain diagnostic radiographs.

TABLE 29.2 Plasma biochemistry data published for pangolins.

		M. pentadactyla[a,b]			*M. pentadactyla*[c]			*M. crassicaudata*[a,d,e]			*M. javanica*[f]			*P. tricuspis*[g]		
		n	Mean (SD)	Range	n	Mean (SD)	Range	n	Mean	Range	n	Mean (SD)	Range	n	Mean	Range
Total protein	g/l	51	74.60 (±0.08)	52–96	99	61.75 (±6.25)	49.5–74	1	71.7	–	51	73.8 (±9.3)	50.00–93.00	10	59.6	–
Albumin	g/l	50	36.60 (±5.2)	27–45	100	35.18 (±3.79)	27–42.95	–	–	–	51	39 (±9.2)	27.00–63.00	–	–	–
Globulin	g/l	–	–	–	–	–	–	–	–	–	51	34.9 (±12.9)	11.00–66.00	–	–	–
AST	U/l	32	23.91 (±11.31)	4–49	99	24.4 (±15.43)	11–87	–	–	–	–	–	–	–	–	–
ALT	U/l	51	156.43 (±99.81)	46–528	100	154.86 (±81.98)	48.05–395.83	–	–	–	47	140.1 (±87.8)	71.00–569.00	–	–	–
Total bilirubin	μmol/l	42	7.18 (±7.18)	1.71–30.78	100	10.52 (±6.57)	3.42–30.69	–	–	–	51	9.8 (±3.6)	6.00–22.00	–	–	–
Urea nitrogen	mmol/l	51	11.41 (±4.17)	5.89–31.06	100	12.96 (±3.84)	7.38–23.73	1	31.26	–	49	9.64 (±4.47)	3.70–20.60	10	5.85	–
Creatinine	μmol/l	51	33.59 (±18.56)	8.84–114.92	100	20.51 (±9.39)	8.84–48.4	1	22.1	–	43	37.2 (±25.7)	4.00–104.00	10	66.3	–
Uric acid	μmol/l	19	30.93 (±22.6)	11.90–107.06	100	47.72 (±15.18)	23.6–91.3	–	–	–	–	–	–	–	–	–
Glucose	mmol/l	48	3.25 (±2.14)	1.89–9.99	99	5.04 (±1.38)	2.3–8.63	–	–	–	–	–	–	–	–	–
ALP	U/l	48	209.06 (±142.71)	42–623	–	–	–	–	–	–	44	482.5 (±192.3)	156.00–903.00	–	–	–
Cholesterol	mmol/l	50	5.61 (±2.93)	2.69–11.03	–	–	–	1	113.71	–	–	–	–	–	–	–
Triglycerides	mmol/l	17	1.44 (±0.87)	0.24–3.56	–	–	–	–	–	–	–	–	–	–	–	–
Amylase	U/l	22	280 (±105.14)	148–538	99	201.88 (±99.31)	50.5–475	–	–	–	49	351.4 (±104.7)	114.00–653.00	–	–	–
Lipase	U/l	–	–	–	99	50.94 (±13.96)	25–86.5	–	–	–	–	–	–	–	–	–
Calcium	mmol/l	41	2.66 (±0.25)	2.05–3.1	100	2.53 (±0.3)	1.96–3.1	–	–	–	51	2.46 (±0.16)	1.96–2.78	–	–	–
Phosphorus	mmol/l	32	1.79 (±0.34)	1.32–2.36	99	1.97 (±0.4)	1.18–2.84	–	–	–	51	2.47 (±0.51)	1.52–4.23	–	–	–
Sodium	mmol/l	21	148.86 (±3.24)	144–156	99	137.04 (±6.31)	124.66–149.42	–	–	–	51	144.4 (±4.7)	135.00–160.00	10	142.6	–
Potassium	mmol/l	21	4.94 (±0.62)	4–5.9	100	4.5 (±0.75)	3.41–6.64	–	–	–	50	4.59 (±0.56)	3.70–6.20	10	5.6	–
Chloride	mmol/l	21	101.9 (±2.81)	95–107	100	92.46 (±6.18)	80.05–104.43	–	–	–	–	–	–	10	109	–

[a]Values converted to SI units to enable comparison across all reported ranges.
[b]*Data from Chin, S.C., et al., 2015. Hematologic and serum biochemical parameters of apparently healthy rescued Formosan pangolins* (Manis Pentadactyla Pentadactyla). *J. Zoo Wildlife Med. 46 (1), 68–76.* [c]*Data from Khatri-Chhetri, R., et al., 2015. Reference intervals for hematology, serum biochemistry, and basic clinical findings in free-ranging Chinese Pangolin* (Manis pentadactyla) *from Taiwan. Vet. Clin. Pathol. 44 (3), 380–390.* [d]*The single individual sampled was clinically unwell at the time blood was collected. All other ranges represent clinically healthy pangolins.* [e]*Data from Mohapatra, R.K., Prafulla, M.K., Panda, S., 2014. Haematological, biochemical and cytomorphometric analysis of an Indian Pangolin. Int. Res. J. Biol. Sci. 3 (8), 77–81.* [f]*Data from Ahmad, A.A., Samsuddin, S., Oh, S.J.W.Y., Martinez-Perez, P., Rasedee, A., 2018. Hematological and serum biochemical parameters of rescued Sunda pangolins* (Manis javanica) *in Singapore. J. Vet. Med. Sci. 80 (12), 1867–1874.* [g]*Data from Oyewele, J.O., Ogunsanmi, A.O., Ozegbe, P., 1998. Plasma electrolyte, enzyme protein and metabolite levels in the adult African white-bellied pangolin* (Manis tricuspis). *Trop. Vet. 16 (1), 73–79.*

TABLE 29.3 Hematology data published for pangolins.

		M. pentadactyla[a,b]			*M. pentadactyla*[c]			*M. crassicaudata*[a,d,e]			*M. javanica*[f]			*P. tricuspis*[g]		
		n	Mean (SD)	Range	n	Mean (SD)	Range	n	Mean	Range	n	Mean (SD)	Range	n	Mean (SD)	Range
PCV	%	50	39.09 (±6.63)	23.5–55.3	100	37 (±0.08)	0.18–0.53	1	20.2	–	51	41.26 (±6.61)	25–55	10	40.40 (±4.95)	–
Hemoglobin	g/l	50	142.4 (±23.4)	83–186	96	130.85 (±25.12)	81.83–179.88	1	75	–	51	140.20 (±26.8)	61–194	10	100.1 (±14.4)	–
RBC	x10^{12}/l	50	5.67 (±1.03)	3.5–8.6	99	5.47 (±1.13)	2.66–7.73	1	2.8	–	51	6.6 (±1.6)	1.92–9.65	10	4.19 (±0.68)	–
MCV	fl	–	–	–	100	68.94 (±6.45)	58.5–83.59	1	72.14	–	51	65.0 (±3.4)	56–75	10	97.75 (±14.35)	–
MCH	Pg	50	25.24 (±1.97)	20.1–28.9	100	23.38 (±2.69)	18.36–31.61	1	26.78	–	49	20.99 (±1.85)	17.3–29.5	10	24.13 (±3.43)	–
MCHC	g/l	50	34.46 (±1.17)	31.3–38.6	100	339.7 (±41.80)	254.4–463.2	1	371.2	–	50	322.90 (±22.00)	289–426	10	248.4 (±24.6)	
Platelets	x10^{9}/l	–	–	–	100	233.52 (±108.96)	63.8–530.93	–	–	–	–	–	–	–	–	–
WBC count	x10^{9}/l	–	–	–	96	5.46 (±1.59)	2.34–8.58	1	25.9	–	51	7.82 (±3.13)	1.86–17.86	10	4.8 (±2.09)	–
Lymphocytes	x10^{9}/l	–	–	–	–	–	–	–	–	–	49	1.29 (±0.69)	0.3–3.0	10	2.22 (±1.01)	–
Lymphocytes	%	–	–	–	–	–	–	1	59	–	–	–	–	10	46.9 (±9.61)	–
Monocytes	x10^{9}/l	–	–	–	–	–	–	–	–	–	51	0.43 (±0.41)	0.01–2.5	10	0.10 (±0.11)	–
Monocytes	%	–	–	–	–	–	–	1	12	–	–	–	–	10	2.7 (±2.79)	–
Neutrophils	x10^{9}/l	–	–	–	–	–	–	–	–	–	50	5.7 (±2.85)	1.29–13.96	10	2.44 (±1.29)	–
Neutrophils	%	–	–	–	–	–	–	1	18	–	–	–	–	10	49.3 (±11.71)	–
Eosinophils	x10^{9}/l	–	–	–	–	–	–	–	–	–	50	0.14 (±0.19)	0–0.97	10	0.04 (±0.04)	–
Eosinophils	%	–	–	–	–	–	–	1	11	–	–	–	–	10	0.90 (±0.99)	–
Basophils	x10^{9}/l	–	–	–	–	–	–	–	–	–	50	0.01 (±0.02)	0–0.08	10	0.01 (±0.03)	–
Basophils	%	–	–	–	–	–	–	–	–	–	–	–	–	10	0.2 (±0.63)	–

[a]Values converted to SI units to enable comparison across all reported ranges.
[b]*Data from Chin, S.C., et al., 2015. Hematologic and serum biochemical parameters of apparently healthy rescued Formosan pangolins* (Manis Pentadactyla Pentadactyla). *J. Zoo Wildlife Med. 46 (1), 68–76.* [c]*Data from Khatri-Chhetri, R., et al., 2015. Reference intervals for hematology, serum biochemistry, and basic clinical findings in free-ranging Chinese Pangolin* (Manis pentadactyla) *from Taiwan. Vet. Clin. Pathol. 44 (3), 380–390.* [d]*The single individual sampled was clinically unwell at the time blood was collected. All other ranges represent clinically healthy pangolins.* [e]*Data from Mohapatra, R.K., Prafulla, M.K., Panda, S., 2014. Haematological, biochemical and cytomorphometric analysis of an Indian Pangolin. Int. Res. J. Biol. Sci. 3 (8), 77–81.* [f]*Data from Ahmad, A.A., Samsuddin, S., Oh, S.J.W. Y., Martinez-Perez, P., Rasedee, A., 2018. Hematological and serum biochemical parameters of rescued Sunda pangolins* (Manis javanica) *in Singapore. J. Vet. Med. Sci. 80 (12), 1867–1874.* [g]*Data from Oyewele, J.O., Ogunsanmi, A.O., Ozegbe, P., 1998. Plasma electrolyte, enzyme protein and metabolite levels in the adult African white-bellied pangolin* (Manis tricuspis). *Trop. Vet. 16 (1), 73–79.*

TABLE 29.4 Organisms known for pangolins. This presents the results of a review of the published literature, reports in the gray literature, medical records of rescue centers and zoos and personal communication from veterinarians working in institutions that hold pangolins in captivity.

Pathogen	Host species	Country	Clinical signs in pangolins	Literature source
Bacteria				
Anaplasma pangolinii	*M. javanica*	Malaysia	Unknown clinical significance	Koh et al. (2016)[a]
Escherichia coli	*M. pentadactyla*	United States	Isolated from debilitated pangolins	Heath and Vanderlip (1988), Narayanan et al. (1977)
	M. crassicaudata	India		
Proteus vulgaris	*M. pentadactyla*	United States	Isolated from debilitated pangolins	Heath and Vanderlip (1988)
Streptococcus faecalis	*M. crassicaudata*	India	Isolated from debilitated pangolins	Narayanan et al. (1977)
Staphylococcus sp.	*M. crassicaudata*	India	Isolated from debilitated pangolins	Narayanan et al. (1977)
Pseudomonas aeruginosa	*M. crassicaudata*	India	Isolated from debilitated pangolins	Narayanan et al. (1977)
P. fluoroscens	*M. pentadactyla*	United States	Isolated from debilitated pangolins	Heath and Vanderlip (1988)
Klebsiella pneumoniae	*M. pentadactyla*	United States	Isolated from debilitated pangolins	Heath and Vanderlip (1988)
Mycoplasma sp.	*M. javanica*	Malaysia	Unknown clinical significance	Jamnah et al. (2014)[b]
Protozoan				
Babesia spp.	*M. javanica*	Thailand	Unknown clinical significance	Sukmak et al. (2018)[c]
Toxoplasma gondii	*M. crassicaudata*	India	Unknown clinical significance	Kegaruka and Willaert (1971)[d]
Trypanosoma brucei non-gambiense	*P. tetradactyla*	Cameroon	Unknown clinical significance	Herder (2002)[f], Njiokou et al. (2004)[g], Njiokou et al. (2006)[h]
	P. tricuspis			
T. vivax	*P. tetradactyla*	Cameroon	Unknown clinical significance	Herder (2002)[f], Njiokou et al. (2004)[g]
	P. tricuspis			
Eimeria spp.[i]	*M. javanica*	Vietnam	Low burden in healthy animals, severe burden in debilitated animals can be fatal	Narayanan et al. (1977)
E. tenggilingi	*M. javanica*	Malaysia	Unknown clinical significance	Else and Colley (1976)[j]
E. nkaka	*P. tricuspis*	Angola	Unknown clinical significance	Jirků et al. (2013)[k]
Plasmodium sp.	*P. tricuspis*	Gabon	Unknown clinical significance	Coatney and Roudabush (1936)[l]
P. tyrio	*M. pentadactyla*	India	Unknown clinical significance	Coatney and Roudabush (1936)[l], Perkins and Schaer (2016)[m]
Piroplasm (species not identified, but likely *Babesia* spp.)	*S. temminckii*	United Kingdom, South Africa	Unknown clinical significance	Rewell (1950)[n]

(*Continued*)

TABLE 29.4 (Continued)

Pathogen	Host species	Country	Clinical signs in pangolins	Literature source
Viruses				
Canine Distemper Virus (CDV)	*M. pentadactyla*	Taiwan	CDV lesions have been found in lungs, digestive tract and brain of clinically unwell pangolins, may be associated with respiratory disease	Chin and Tsao (2015)
Endogenous retrovirus (MPERV1)	*M. pentadactyla*	China	Unknown clinical significance	Zhuo and Feschotte (2015)[o]
Ticks				
Amblyomma sp. (includes those formerly reported as *Aponoma* sp.)	*Manis* spp.	Vietnam	Low burden common in clinically healthy animals, severe burden may affect debilitated individuals	Toumanoff and Maillard (1957)[p], Ugiagbe and Awharitoma (2015)[q], Orhierhor et al. (2017)[r]
	P. tricuspis	Nigeria		
	S. temminckii			
A. compressum[s]	*P. tricuspis*	Central African Republic	Low burden common in clinically healthy animals, severe burden may affect debilitated individuals	Rahm (1956)[t], Pourrut et al. (2004)[u], Ntiamoa-Baidu et al. (2007a[v], b[w]), Mediannikov et al. (2012a,b), Uilenberg et al. (2013)[x]
	P. tetradactyla	Côte d'Ivoire , Central African Republic		
	S. gigantea	Democratic Republic of the Congo		
	S. temminckii	Ghana, Liberia, Gabon, United States		
A. clypeolatum	*M. crassicaudata*	Sri Lanka	Unknown clinical significance	Liyanaarachchi et al. (2013, 2015[y], 2016)
A. gerviasi (formerly *Aponomma gervaisi*)	*M. crassicaudata*	India	Unknown clinical significance	Pillai and George (1997)[z], Mohapatra and Panda (2014)
A. javanense[s]	*M. culionensis*	Philippines	Unknown clinical significiance	see Chapter 7
	M. javanica	Cambodia, China, Malaysia, Taiwan, Thailand, Vietnam	Low burden common in clinically healthy animals, severe burden may affect debilitated individuals	Liyanaarachchi et al. (2013, 2016), Hoogstraal (1971)[aa], Kollars Jr and Sithiprasasna (200)[bb], Parola et al. (2003)[cc], Yang et al. (2010), Hassan et al. (2013)[dd]
	M. crassicaudata	Sri Lanka		
A. testudinarium	*M. pentadactyla*	Taiwan, Sri Lanka	Low burden common in clinically healthy animals, severe burden may affect debilitated individuals	Chin and Tsao (2015), Liyanaarachchi et al. (2013, 2015[y], 2016), Khatri-Chhetri et al. (2016)
	M. crassicaudata			
Haemaphysalis formosensis	*M. pentadactyla*	Taiwan	Unknown clinical significance	Khatri-Chhetri et al. (2016)
H. hystricis	*M. pentadactyla*	Taiwan	Unknown clinical significance	Khatri-Chhetri et al. (2016)
H. parmata	*P. tricuspis*	Ghana	Unknown clinical significance	Ntiamoa-Baidu et al. (2007b[w])

Ixodes rasus	*S. temminckii*	Central African Republic	Unknown clinical significance	Uilenberg et al. (2013)[x]
	P. tetradactyla	Gabon		see Chapter 8
Ornithodoros moubata	*S. temminckii*	Mozambique	Unknown clinical significance	Dias (1954[ee], 1963[ff])
O. compactus	*S. temminckii*	South Africa	Unknown clinical significance	Jacobsen et al. (1991)[gg]
Rhipicephalus sp.	*P. tricuspis*	Nigeria	Unknown clinical significance	Orhierhor et al. (2017)[r]
R. haemaphysaloides	*M. crassicaudata*	Sri Lanka	Unknown clinical significance	Liyanaarachchi et al. (2015)[y]
R. longus	*S. temminckii*	Central African Republic	Unknown clinical significance	Uilenberg et al. (2013)[x]
R. muhsamae	*P. tricuspis*	Central African Republic	Unknown clinical significance	Uilenberg et al. (2013)[x]
	S. temminckii			
R. theileri	*S. temminckii*	South Africa	Unknown clinical significance	Jacobsen et al. (1991)[gg]
Mites				
Manitherionyssus heterotarsus	*S. temminckii*	South Africa	Unknown clinical significance	Jacobsen et al. (1991)[gg]
Manisicola africanus	*S. temminckii*	South Africa	Unknown clinical significance	Lawrence (2009)[hh]
Sarcoptiform mite (species not identified)	*M. pentadactyla*	Taiwan	Severe skin irritation, consistent with mange mite infection	Khatri-Chhetri et al. (2017)
Nematodes				
Ancylostoma spp.	*M. pentadactyla*	Central African Republic, Taiwan, United States	Unknown clinical significance	Chin and Tsao (2015), Uilenberg et al. (2013)[x]
	P. tricuspis			
	S. gigantea			
Brugia malayi	*M. javanica*	Malaysia	Unknown clinical significance	Laing et al. (1960)[ii], Wilson (1961)[jj]
B. pahangi	*M. javanica*	Malaysia	Unknown clinical significance	Laing et al. (1960)[ii], Wilson (1961)[jj]
Capillaria spp.	*M. pentadactyla*	Taiwan	Unknown clinical significance	Chin and Tsao (2015)
Cylicospirura spp.	*M. pentadactyla*	United States	Unknown clinical significance	Heath and Vanderlip (1988)
Chenospirura kwangiungensis	*M. pentadactyla*	China	Unknown clinical significance	Kou et al. (1958)[kk]
Dipetalonema fausti	*M. pentadactyla*	China	Unknown clinical significance	Esslinger (1966)[ll]
Habronema hamospiculatum	*M. pentadactyla*	Sri Lanka	Unknown clinical significance	Baylis (1931)[mm]
	S. temminckii	South Africa		
Leipernema leiperi	*M. pentadactyla*	India	Unknown clinical significance	Singh (2009)[nn]

(*Continued*)

TABLE 29.4 (Continued)

Pathogen	Host species	Country	Clinical signs in pangolins	Literature source
Manistrongylus meyeri	*M. pentadactyla*	Taiwan	Unknown clinical significance	Cameron et al. (1960)[oo]
M. manidis	*P. tricuspis*	Democratic Republic of the Congo	Unknown clinical significance	Baer (1959)[pp]
Microfilaria lukakae	*P. tricuspis*	Angola	Unknown clinical significance	Pais Caeiro (1959)[qq]
M. lundae	*P. tricuspis*	Angola	Unknown clinical significance	Pais Caeiro (1959)[qq]
M. nobrei	*P. tricuspis*	Angola	Unknown clinical significance	Pais Caeiro (1959)[qq]
M. vilhenae	*P. tricuspis*	Angola	Unknown clinical significance	Pais Caeiro (1959)[qq]
Necator americanus	*M. javanica*	Indonesia	Unknown clinical significance	Baylis (1933)[rr], Cameron et al. (1960)[oo], Khatri-Chhetri et al. (2017)
	M. pentadactyla	Taiwan		
Parastrongyloides sp.	*P. tricuspis*	Nigeria	Unknown clinical significance	Orhierhor et al. (2017)[r]
Pholidostrongylus armatus	*P. tricuspis*	Democratic Republic of the Congo	Unknown clinical significance	Baer (1959)[pp]
Strongyloides spp.	*M. pentadactyla*	Taiwan	Unknown clinical significance	Chin and Tsao (2015), Khatri-Chhetri et al. (2017)
Strongyle type[s]	*M. pentadactyla*	United States	Unknown clinical significance	Heath and Vanderlip (1988), Ugiagbe and Awharitoma (2015)[q]
	S. temminckii	Nigeria		
	M. javanica	Vietnam		
Trichochenia meyeri	*M. crassicaudata*	India	Unknown clinical significance	Naidu and Naidu (1981)[ss]
Cestodes				
Echinococcus sp.	*M. crassicaudata*	India	Unknown clinical significance	Rao et al. (1972)[tt]
Metadavainea sp.	*P. tricuspis*	Nigeria	Unknown clinical significance	Orhierhor et al. (2017)[r]
M. aelleni	*P. tricuspis*	Côte d'Ivoire	Unknown clinical significance	Rahm (1956)[t]
	P. tetradactyla			
Raillietina rahmi	*P. tricuspis*	Côte d'Ivoire	Unknown clinical significance	Rahm (1956)[t]
	P. tetradactyla			
R. anoplocephaloides	*P. tricuspis*	Côte d'Ivoire	Unknown clinical significance	Rahm (1956)[t]
	P. tetradactyla			
Oochoristica sp.	*S. temminckii*	Nigeria	Unknown clinical significance	Ugiagbe and Awharitoma (2015)[q]

Acanthocephalan				
Nephridiacanthus gerberi	*S. gigantea*	Democratic Republic of the Congo	Unknown clinical significance	Baer (1959)[pp]
Macracanthorhyncus sp.	*P. tricuspis*	Nigeria	Unknown clinical significance.	see Chapter 9
Oncicola sp.	*P. tricuspis*	Nigeria	Unknown clinical significance.	see Chapter 9
Pentastome				
Armillifer sp.	*S. temminckii*	Nigeria	Unknown clinical significance	Ugiagbe and Awharitoma (2015)[q]
Unidentified pentastome[uu]	*P. tricuspis*	United States	Unknown clinical significance	

Institutions reporting organism: [e]Johannesburg Wildlife Veterinary Hospital, South Africa, [s]Save Vietnam's Wildlife, Vietnam, [uu]San Diego Zoo, United States.

References: [a]*From Koh, F.X., Kho, K.L., Panchadcharam, C., Sitamand, F.T., Tay, S.T., 2016. Molecular detection of Anaplasma spp. in pangolins (*Manis javanica*) and wild boars (*Sus scrofa*) in Peninsular Malaysia. Vet. Parasitol. 227, 73–76.* [b]*Jamnah, O., Faizal, H., Chandrawathani, P., Premaalatha, B., Erwanas, A.I., Rozita, L., Ramlan, M., 2014. Eperythrozoonosis (*Mycoplasma *sp.) in Malaysian pangolin. Malay. J. Vet. Res. 5 (1), 65–69.* [c]*Sukmak, M., Yodsheewan, R., Sangkharak, B., Kaolim, N., Ploypan, R., Soda, N., Wajjwalku, W., 2018. Molecular detection of Babesia spp. from confiscated Sunda pangolin (*Manis javanica*) in Thailand. Proceedings of the 11th International Conference of Asian Society of Conservation Medicine. One Health in Asia Pacific, Wildlife Disease Association Australasia (WDAA), Udayana University, Bali, Indonesia, 28–30 October, p. 45.* [d]*Kageruka, P., Willaert, E., 1971. Toxoplasma gondii (Nicolle and Manceaux 1908) isolated from Goura cristata Pallas and Manis crassicaudata Geoffroy. Acta Zoologica et Pathologica Antverpiensia 52, 3–10.* [f]*Herder, S., 2002. Identification of trypanosomes in wild animals from Southern Cameroon using the polymerase chain reaction (PCR). Parasite 9 (4), 345–349.* [g]*Njiokou, F., Simo, G., Nkinin, S., Laveissière, C., Herder, S., 2004. Infection rate of Trypanosoma brucei sl, T. vivax, T. congolense "forest type", and T. simiae in small wild vertebrates in south Cameroon. Acta Trop. 92 (2), 139–146.* [h]*Njiokou, F., Laveissière, C., Simo, G., Nkinin, S., Grébaut, P., Cuny, G., Herder, S., 2006. Wild fauna as a probable animal reservoir for Trypanosoma brucei gambiense in Cameroon. Infect., Genet. Evol. 6 (2), 147–153.* [j]*Else, J.G., Colley, F.C., 1976. Eimeria tenggilingi sp. n. from the scaly anteater* Manis javanica *Desmarest in Malaysia. J. Eukaryotic Microbiol. 23 (4), 487–488.* [k]*Jirků, M., Kvičerová, J., Modrý, D., Hypša, V., 2013. Evolutionary plasticity in coccidia – striking morphological similarity of unrelated coccidia (Apicomplexa) from related hosts: Eimeria spp. from African and Asian Pangolins (Mammalia: Pholidota). Protist 164 (4), 470–481.* [l]*Coatney, G.R., Roudabush, R. L., 1936. A catalog and host-index of the Genus Plasmodium. J. Parasitol. 22 (4), 338–353.* [m]*Perkins, S.L., Schaer, J., 2016. A Modern Menagerie of Mammalian Malaria. Trends Parasitol. 32 (10), 772–782.* [n]*Rewell, R.E., 1950. Report of the Society's Pathologist for the year 1949. Proc. Zool. Soc. Lond. 120 (3), 485–495.* [o]*Zhuo, X., Feschotte, C., 2015. Cross-species transmission and differential fate of an endogenous retrovirus in three mammal lineages. PLoS Pathog. 11 (10), 1–23.* [p]*Toumanoff, C., Maillard, D., 1957. A new tick of the genus Aponomma occurring on the pangolin in South Vietnam. Bulletin de la Société de pathologie exotique et de ses filiales 50 (5), 700–703.* [q]*Ugiagbe, N., Awharitoma, A., 2015. Parasitic infections in African pangolin (*Manis temminckii*) from Edo State, southern Nigeria. Zoologist (The) 13, 17–21.* [r]*Orhierhor, M., Okaka, C.E., Okonkwo, V.O., 2017. A survey of the parasites of the African white-bellied pangolin,* Phataginus tricuspis*, in Benin City, Edo State, Nigeria. Niger. J. Parasitol. 38 (2), 266.* [t]*Rahm, U., 1956. Notes on Pangolins of the Ivory Coast. J. Mammal. 37 (4), 531–537.* [u]*Pourrut, X., Emane, K.A., Camicas, J-L., Leroy, E., Gonzalez, J.P., 2011. Contribution to the knowledge of ticks (Acarina: Ixodidae) in Gabon. Acarologia 51 (4), 465–471.* [v]*Ntiamoa-Baidu, Y., Carr-Saunders, C., Matthews, B.E., Preston, P.M., Walker, A.R., 2007. An updated list of the ticks of Ghana and an assessment of the distribution of the ticks of Ghanaian wild mammals in different vegetation zones. Bull. Entomol. Res. 94 (3), 245–260.* [w]*Ntiamoa-Baidu, Y., Carr-Saunders, C., Trueman, B.E., Preston, P.M., Walker, A.R., 2007. Ticks associated with wild mammals in Ghana. Bull. Entomol. Res. 95 (3), 205–219.* [x]*Uilenberg, G., Estrada-Peña, A., Thal, J., 2013. Ticks of the Central African Republic. Exp. Appl. Acarol. 60 (1), 1–40.* [y]*Liyanaarachchi, D. R., Rajakaruna, R.S., Dikkumbura, A.W., Rajapakse, R.P.V.J., 2015. Ticks infesting wild and domestic animals and humans of Sri Lanka with new host records. Acta Trop. 142, 64–70.* [z]*Pillai, K.M., George, P.O., 1997. Tick infestation in an Indian pangolin (*Manis crassicaudata*). Indian Vet. J. 74 (1), 71–72.* [aa]*Hoogstraal, H., 1971. Identity, hosts, and distribution of Haemaphysalis (Rhipistoma) canestrinii (Supino) (Resurrected), the Postulated Asian Progenitor of the African Leachi Complex (Ixodoidea: Ixodidae). J. Parasitol. 57 (1), 161–172.* [bb]*Kollars Jr, T.M., Sithiprasasna, R., 2000. New host and distribution record of* Amblyomma javanense *(Acari: Ixodidae) in Thailand. J. Med. Entomol. 37 (4), 640–640.* [cc]*Parola, P., Cornet, J.P., Sanogo, Y.O., Miller, R.S., Van Thien, H., Gonzalez, J.P., et al., 2003. Detection of* Ehrlichia *spp.,* Anaplasma *spp.,* Rickettsia *spp., and other eubacteria in ticks from the Thai-Myanmar border and Vietnam. J. Clin. Microbiol. 41 (4), 1600–1608.* [dd]*Hassan, M., Sulaiman, M.H., Lian, C.J., 2013. The prevalence and intensity of Amblyomma javanense infestation on Malayan Pangolins (*Manis javanica *Desmarest) from Peninsular Malaysia. Acta Trop. 126 (2), 142–145.* [ee]*Dias, J.A.T.S., 1954. Alternative Hosts of* O. moubata *In Mozambique. Anais do Instituto de Medicina Tropical 11 (3/4), 635–639.* [ff]*Dias, J.T.S., 1963. The importance of wart-hog (*Phacochoerus aethiopicus*) in the epidemiology of the relapsing fever or tick fever in Mozambique. South African J. Sci. 59 (12), 573–574.* [gg]*Jacobsen, N., Newbery, R., De Wet, M., Viljoen, P., Pietersen, E., 1991. A contribution of the ecology of the Steppe pangolin* Manis temminckii *in the Transvaal. Zeitschrift für Säugetierkunde 56 (2), 94–100.* [hh]*Lawrence, R.F., 2009. A new mite from the South African pangolin. Parasitology 31(4), 451–457.* [ii]*Laing, A., Edeson, J., Wharton, R., 1960. Studies on filariasis in Malaya: the vertebrate hosts of Brugia malayi and B. pahangi. Ann. Trop. Med. Parasitol. 54 (1), 92–99.* [jj]*Wilson, T., 1961. Filariasis in Malaya—a general review. Trans. R. Soc. Trop. Med. Hyg. 55 (2), 107–129.* [kk]*Kou, C. C., 1958. Studies on parasitic nematodes of mammals from Canton. I. Some new species from* Paradoxurus minor exitus *Schwarz,* Paguma larvata larvata *(Hamilton Smith), and* Manís pentadactyla aurita *Hodgson. Acta Zool. Sin. 10 (1), 60–72.* [ll]*Esslinger, J.H., 1966. Dipetalonema Fausti sp. n. (Filarioidea: Onchocercidae), a Filarial Parasite of the Scaly Anteater, Manis pentadactyla L. (Pholidota), from China. J. Parasitol. 52 (3), 494–497.* [mm]*Baylis, H.A., 1931. XXIII.—On a Nematode parasite of pangolins. Ann. Mag. Nat. Hist. 8 (44), 191–194.* [nn]*Singh, S.N., 2009. On a new nematode* Leipernema leiperi *n.g., n.sp. (Strongyloididae), parasitic in the pangolin* Manis pentadactyla *from Hyderabad, India. J. Helminthol. 50 (4), 267–274.* [oo]*Cameron, T.W.M., Myers, B.J., 1960. Manistrongylus meyeri (Travassos, 1937) Gen. Nov., and Necator americanus from the pangolin. Can. J. Zool. 38 (4), 781–786.* [pp]*Baer, J.G., 1959. Helminthes parasites. Exploration des Parcs Nationaux du Congo Belge: Mission, Baer, J.G., Gerber, W. (1958).* [qq]*Pais Caeiro, V., 1959. Quatro especies de microfilarias do Phataginus Manis tricuspis (Rafinesque). An Escola Superior Medicina Veterinária Lisbon 2, 83–94.* [rr]*Baylis, H.A., 1933. XLIII.—On some parasitic worms from Java, with remarks on the Acanthocephalan genus* Pallisentis*. Ann. Mag. Nat. Hist. 12 (70), 443–449.* [ss]*Naidu, K.V., Naidu, K.A., 1981.* Trichochenia meyeri *(Travassos, 1937) Naidu KV and Naidu KA comb. nov.(Nematoda:* Trichostrongylidae Leiper*, 1912) from Pangolin in South India. Proc. Anim. Sci.-Indian Acad. Sci. 90 (6), 615–618.* [tt]*Rao, A., Misra, S., Acharjyo, L., 1972. Pulmonary hydatidosis in captive animals at Nandankanan Zoo. Indian Vet. J. 49 (8), 842–843.*

Health issues and infectious disease

This section outlines the common health issues seen in pangolins recently captured from the wild, rescued or confiscated from the illegal wildlife trade, and those that have lived in long-term captivity in rescue centers and zoos worldwide.

Infectious organisms and pangolins

An inventory of potential pathogens is of fundamental importance in the management of species (Munson, 1991). It supports the development of sensible preventative medicine and husbandry protocols to maintain healthy captive populations (Hope and Deem, 2006), provides a baseline from which to plan disease surveillance (Lonsdorf et al., 2006), and can indicate the role of infectious diseases in species conservation (Leendertz et al., 2006). Given an increasing awareness of the transfer of pathogens between wild animals, humans and livestock (Jones et al., 2008), this information also allows a more thorough understanding of the risk of direct contact with species to both human and animal health (Travis et al., 2006).

There are a large number of infectious organisms reported for pangolins (Table 29.4). This includes eight genera of bacteria, five genera of protozoan (and one unidentified piroplasm), two viruses, five genera of tick, two genera of mite (and one unidentified sarcoptid mite), 15 genera of nematode, four genera of cestode, three genera of acanthocephalan and two pentostomes reported for the eight species of pangolin. This list will undoubtedly evolve with further pangolin health research, and as molecular diagnostic capacity becomes more common. Inaccuracies are also likely to be included here, given the traditional reliance on morphological identification to identify parasites (Nadler and De Leon, 2011), and due to on-going revision of taxonomy, particularly in some of the lesser known taxa.

Most studies included here (which builds on Mohapatra et al., 2016) utilized very small sample sizes and gave no reference to the clinical health status of infected pangolins. As such, the significance for pangolin health is known for only a few organisms (see Table 29.4). Three published papers describe pangolins as being in poor condition (Heath and Vanderlip, 1988; Narayanan et al., 1977; Yang et al., 2010). However, researchers failed to indicate whether the organism isolated was the cause of debilitation or a coincidental finding. One paper describes a severe, mange like skin disease in a Chinese pangolin in Taiwan (Khatri-Chhetri et al., 2017) due to infestation of an unidentified sarcoptid mite (see Fig. 29.4; R. Khatri-Chhetri, pers. comm.).

Ticks frequently accumulate on the skin beneath the scales and around the eyes. While many species have been reported (Table 29.4), two from the genus *Amblyomma* – *A. compressum* in Africa and *A. javanense* in Asia – are found almost exclusively on pangolins, and all life cycle stages have been found on pangolins (Kolonin, 2007). While healthy wild pangolins carry small tick burdens, confiscated and rescued pangolins often have significant tick burdens, possibly due to stress related immunosuppression. This may contribute to malaise and anemia, and manual removal of ticks or treatment with a topical acaricide is recommended (Table 29.1).

Gastrointestinal parasites, including a wide range of nematodes, cestodes and protozoan, have been found on necropspy and fecal assessment of pangolins. Coccidiosis, leading to watery diarrhea, melena and death in advanced cases, has been described in rescue center and zoo reports (Clark et al., 2009; Clark et al., 2010; Jaffar et al., 2018). Coccidian oocysts (*Eimeria* spp.) are found in healthy

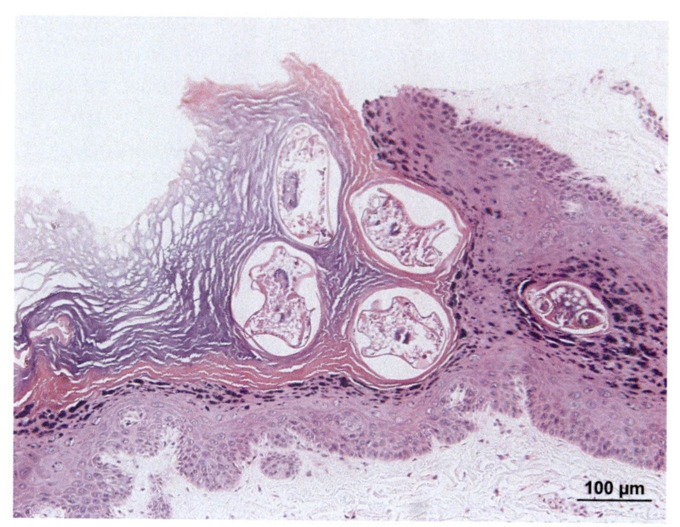

FIGURE 29.4 Histopathology of a section of skin from a Chinese pangolin with a severe mite infection (hematoxylin and eosin stain). This high powered view of hyperkeratotic, non-haired skin shows multiple unidentified sarcoptiform mites in cross section within the thickened epidermis. Clinically, this pangolin was emaciated with a severe dermatitis. *Photo credit: Rupak Khatri-Chhetri.*

wild pangolin feces (Jaffar et al., 2018). The belief that coccidia may be normal flora in pangolins is supported by the finding that many infected individuals are healthy, with the development of clinically significant coccidiosis influenced by the immune status and general health of the host.

A number of studies have suggested a role for pangolins as maintenance hosts for some significant vector-borne, zoonotic diseases. In one study, *Rickettsia conorii* subspecies *capsia, Ehrlichia* sp. and *Anaplasma* sp. – rickettsial bacteria which cause spotted fever, ehrlichiosis and anaplasmosis respectively in people, were isolated from *Haemaphysalis hystricis* ticks removed from free ranging Chinese pangolins in Taiwan (Khatri-Chhetri et al., 2016). *Anaplasma* sp. has also been isolated from *Amblyomma javanense* ticks removed from Sunda pangolins along the Thai-Myanmar border and Vietnam (Parola et al., 2003). Half of all *A. compressum* ticks removed from a giant pangolin and 10% of the ticks removed from one white-bellied pangolin in bushmeat markets in Democratic Republic of the Congo and Liberia respectively were positive for *Rickettsia africae*, the causative agent of African tick-bite fever in people (Mediannikov et al., 2012a; Mediannikov et al., 2012b).

The risk of transmission and spillover of pathogens from their wildlife host to humans and domestic animals is increased with close proximity (Cantlay et al., 2017; Smith et al., 2017). Encroachment into previously wild spaces and the global trade in wild animals are anthropogenic changes which facilitate unnatural proximity, transboundary disease movement and pathogen spillover between species (Daszak et al., 2001). The isolation of spotted fever rickettsial bacteria from *A. javanense* ticks removed from peri-urban pangolins in Sri Lanka may suggest a role for pangolins in maintenance or transmission of this pathogen (Liyanaarachchi et al., 2013; Liyanaarachchi et al., 2016). As pangolins are traded globally within the illegal wildlife trade (Chapter 16) they serve as a vector for any pathogens they carry, as evidenced by the isolation of five zoonotic bacteria – *Listeria monocytogenes, Staphylococcus aureus, Streptococcus* sp., *Enterobacter* sp. and *Klebsiella oxytoca* – from frozen African pangolin meat confiscated on arrival at a French airport (Chaber and Cunningham, 2016).

Infectious organisms are reported in Table 29.4 for all eight species of pangolin. This is significant for a family often described as being "poorly studied" in terms of health. However, only a single species is reported for the Philippine pangolin, and only two reported for giant pangolin; and while the zoonotic potential of many of these is known, their significance for the health of their pangolin hosts remains unclear. As such, further understanding of the breadth of infectious pathogens seen in pangolins, and their impact on pangolin health, should be the focus of future research.

Health status in confiscated and rescued pangolins

Animals confiscated from the illegal wildlife trade are generally in very poor health due to the harsh conditions in which they are hunted, stored and trafficked (Bell et al., 2004; Wicker et al., 2016). This is true for trade confiscated pangolins, many of which have been without food and water for prolonged periods of time (Clark et al., 2009; Sun et al., 2019), with severe debilitation, malnutrition and dehydration common diagnoses on arrival in captivity. The vast majority (82.9%) of pangolins rescued by one center in Taiwan were very unhealthy on arrival, with almost one quarter (23%) diagnosed as malnourished (Sun et al., 2019). Emaciation has also been reported on necropsy of confiscated pangolins in India (Mohapatra and Panda, 2014), Taiwan (Khatri-Chhetri et al., 2017) and northern Vietnam (Clark et al., 2009). Zhang et al. (2017) describe Sunda pangolins confiscated outside of their natural range in China as debilitated and weak, and commonly suffering from hypothermia. In South Africa, all Temminck's pangolins presented to a wildlife veterinary hospital displayed evidence of malnutrition, dehydration and weakness, being unable to hold their tail in the normal elevated position when walking. In some of the sickest individuals, ascites and pleural effusion secondary to hypoproteinemia has been seen.

Severely debilitated pangolins must be provided with intensive supportive therapy to survive, and it is vital that rescue centers and zoos receiving rescued and confiscated pangolins are appropriately resourced, with a trained veterinary team capable of providing the care pangolins require to survive.

Respiratory disease

Severe respiratory disease, including pneumonia, is one of the most widely reported causes of morbidity and mortality in captive pangolins (Chin and Tsao, 2015; Khatri-Chhetri et al., 2017; Mohapatra and Panda, 2014; Sun et al., 2019). Clinical signs include mucopurulent nasal and ocular discharge, shivering, dyspnea, dysphagia and, in severe cases, ataxia (Chin and Tsao, 2015). It is important to note that effusive serous nasal and ocular discharge can commonly be seen in healthy pangolins, and this may be more prominent with acute stress. Coughing has not been seen as a clinical sign in pangolins with respiratory disease. Pneumonia is believed to be secondary to stress related immune suppression, and is commonly seen when ambient temperature drops or fluctuates, and where pangolins face malnutrition due to either an inappropriate diet or refusal to consume artificial captive diets (Zhang et al., 2017). Elevated ammonia levels due to poor hygiene in overcrowded captive conditions may also contribute to the development of respiratory disease.

Diagnosis and treatment of respiratory disease in pangolins is based on the normal small animal veterinary approach, and should include a thorough clinical examination, imaging, blood biochemistry and hematology. Treatment is frequently unsuccessful, particularly where individuals are severely affected and significantly immunosuppressed however some success has been achieved with Temmincks pangolin with

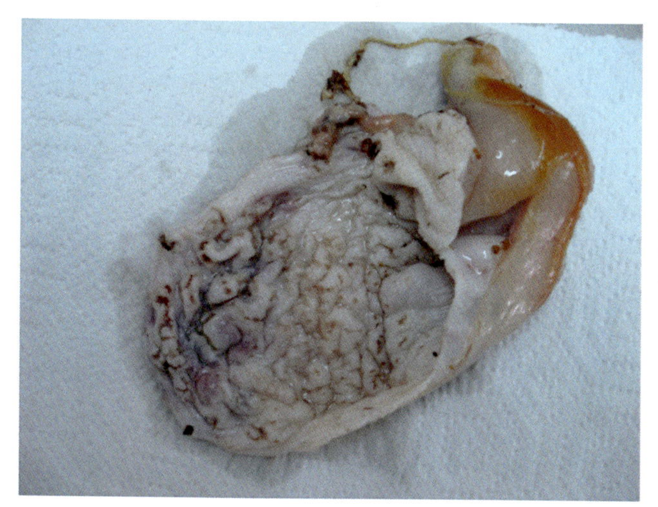

FIGURE 29.5 Severe gastrointestinal ulceration seen on necropsy of a Sunda pangolin after confiscation from the illegal wildlife trade. This individual was severely debilitated, with poor appetite and blood visible in the feces prior to death. While gastroprotectant agents, including ranitidine and sucralfate, were used this animal failed to improve with medical treatment and was euthanized. *Photo credit: Leanne Wicker, Carnivore and Pangolin Conservation Program (CPCP), Vietnam.*

nebulisation in saline three times per day and oxygen supplementation into the hospital enclosure.

Gastrointestinal disorders

Hemorrhagic, ulcerative lesions of the mucosa of the digestive system (tongue, esophageal mucosa, gastric mucosa and intestinal mucosa), associated with mild to severe gastritis, are also commonly seen at necropsy of recently captive pangolins (Chin and Tsao, 2015; Chin et al., 2006; Clark et al., 2009; Jaffar et al., 2018; Mohapatra and Panda, 2014; Sun et al., 2019; Yang et al., 2007). Ulceration ranges from mild, pinpoint lesions to severe, disseminated lesions that cover a vast area of the gastric mucosa (Fig. 29.5). Clinically affected pangolins are quiet, with a poor to absent appetite. The feces may contain blood. The integument and mucous membranes are generally very pale, and the skin may feel cool to touch. Anemia is often noted on blood evaluation.

No definitive cause has been found for these lesions on gross and histological examination of necropsy tissues. However, maladaptation to transport and captive stress, and inadequate nutrition affecting the gastrointestinal microbiome are believed to contribute to their development (Chin and Tsao, 2015; Clark et al., 2009; Jaffar et al., 2018). Inappropriate captive diet leading to abnormally high stomach acid levels has been linked to acute gastritis in the short-beaked echidna (*Tachyglossus aculeatus*), an Australian marsupial with a very similar, myrmecophagous diet as that consumed by pangolins (M. Shaw, pers. comm.). Further research on the gastric physiology and microbiome of pangolins, and their contribution to digestion, is warranted. However, given gastrointestinal ulceration is less commonly seen in pangolins which have settled into captivity, and are reliably consuming an appropriate captive diet, acute and chronic stress is likely to play a significant role in the development of ulcers.

The use of prophylactic gastro-protectants is recommended in highly stressed individuals; particularly those new to captivity (see Table 29.1), and the addition of soil to the captive diet may reduce the incidence of gastric ulceration and enteritis (Chin and Tsao, 2015). A potent, long-acting butyrophenone neuroleptic, Haloperidol, has also been used to calm particularly stressed individuals in an attempt to prevent their occurrence (J. Jimerson, pers. comm.).

Given the propensity of pangolins to develop gastric ulcers, non-steroidal anti-inflammatory drugs should be used with caution in pangolins, and only once hydration has been restored.

Gastrointestinal impaction due to the ingestion of straw or wood shaving bedding has been reported in Sunda (Zhang et al., 2017) and Chinese pangolins (J. Chin, pers. comm.). Care should be taken in choosing bedding material for captive pangolins.

Disorders of the integument

Skin abrasions and moist, ulcerative dermatitis beneath the scales (likely due to the poor hygiene and dirty conditions in which pangolins are trafficked) are also seen (Clark et al., 2009; Khatri-Chhetri et al., 2017; Perera et al., 2017). This is compounded by the common practice of trafficking pangolins in netting sacks, tied tightly in a curled up position, inside which they become covered in their own feces and urine (Fig. 29.6).

Daily bathing using one of a variety of topical solutions (including dilute povidine iodine, dilute chlorhexidine solution, an Epsom salt bath and sterile lactated ringers or NaCl 0.9%), followed by application of topical antimicrobial creams, such as sulfa silvadiazene cream, is efficacious. More complicated cases may require parenteral antibiosis (see Table 29.1; Clark et al., 2009; Perera et al., 2017).

FIGURE 29.6 Two Sunda pangolins confiscated from the illegal wildlife trade in Vietnam. Both pangolins are curled up tightly within individual netting sacks. Since pangolins in illegal trade are generally sold per kilogram, heavier animals obtain a higher price. The yellow, viscous substance visible on the ground is cornflour mixed with water, which is used to increase the weight of the animals. Pangolins are extremely stressed by this process, and are frequently injured during force feeding. Yellow colored diarrhea tinged with blood is a common finding in newly confiscated pangolins in Vietnam. *Photo credit: Leanne Wicker, Carnivore and Pangolin Conservation Program (CPCP), Vietnam.*

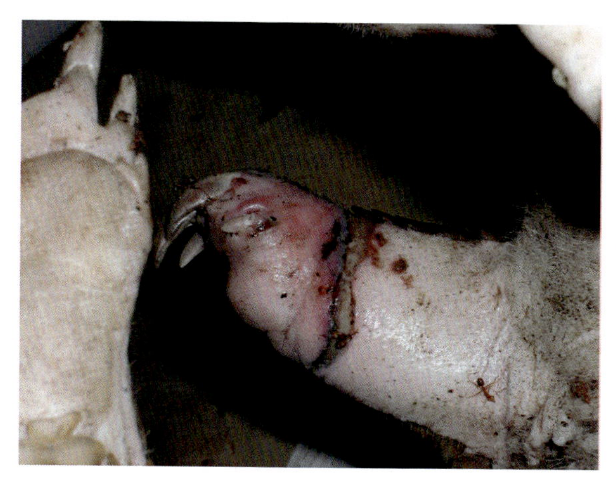

FIGURE 29.7 Severe, encircling snare trap wound to the right forelimb in a confiscated Sunda pangolin in Vietnam. The wound was treated with parenteral antibiosis (amoxycillin clavulanic acid for ten days), daily bathing in dilute chlorhexidine solution, followed by application of silver sulphadiazine cream and bandaged with a non-adherent, absorbent primary layer and a cohesive bandage secondary layer. The wound eventually healed completely. *Photo credit: Leanne Wicker, Carnivore and Pangolin Conservation Program (CPCP), Vietnam.*

Trauma

Hunting snares, traps, guns or knives cause a range of injuries (Fig. 29.7) including fractures, traumatic loss of a limb or the tail, lacerations or encircling lesions around the neck or body (Clark et al., 2009; Jaffar et al., 2018; Khatri-Chhetri et al., 2017; Sun et al., 2019). Bite wounds from hunting dogs and other predators vary in severity, but include punctures and lacerations to the limbs and tail which, if left untreated, can lead to sepsis and death (Clark et al., 2009; Jaffar et al., 2018). Fractures of the appendicular skeleton and pelvic girdle, with concurrent soft tissue trauma, are commonly encountered due to motor vehicle accidents (Jaffar et al., 2018).

Trauma is not limited to newly captive pangolins. Repetitive, abnormal behavior (such as clawing at enclosure wire), believed to be a response to stress, is documented as a cause of traumatic lesions in captive pangolins (Challender et al., 2012), and lesions to the tail tip, claws and skin of the face and forearms are seen as a result of this behavior.

Diagnosis and treatment is based on the general approach in small animal medicine, and should include a physical examination with close attention paid to the musculoskeletal exam, neurological examination and imaging. Choice of antibiotics should be guided by bacterial culture and sensitivity results where available (Jaffar et al., 2018). Surgical intervention may be warranted and topical dressings and bandaging help facilitate wound healing. Improved healing has been reported using laser therapy (L. Hai, unpubl. data). In many cases, lesions are so severe that humane euthanasia is the most appropriate option.

Opthalmologic disorders

Corneal ulceration, hypopyon, inflammation and edema of the periocular tissues have been reported. Where pangolins are confiscated from the illegal wildlife trade these are likely the result of rough handling and poor hygiene during hunting and transport. Diagnosis and treatment follows the approach utilized in small animal veterinary medicine. Surgical enucleation may be required, however, topical antimicrobial ophthalmic ointments have successfully treated uncomplicated cases (Clark et al., 2009).

Organ dysfunction

With increasing availability of physiological reference ranges for pangolins and establishment of captive pangolin populations in zoos, knowlegde of the underlying causes and progression of organ dysfunction in pangolins will develop. Hepatic pathology with no clear underlying cause has been described in Indian, Sunda, Chinese and white-bellied pangolins on the basis of elevated liver enzymes, enlarged or abnormally colored liver on gross necropsy and histopathological findings of inflammation of the liver (Chin and Tsao, 2015; Khatri-Chhetri et al., 2017; Zhang et al., 2017). Hepatic lipidosis has been seen in captive white-bellied pangolins (C. Aitkin-Palmer, pers. comm.).

Undifferentiated renal disease has also been diagnosed in Indian, Sunda and white-bellied pangolins on the basis of the abnormal appearance of kidneys on necropsy, and on histopathological descriptions of nephritis (Chin and Tsao, 2015; Khatri-Chhetri et al., 2017; Zhang et al., 2017).

Therapeutics

Drug regimens used previously in pangolins are reported in Table 29.1. There are no published studies on the pharmacokinetics, pharmacodynamics and efficacy of therapeutic agents in pangolins. As a result, the general approach and principles of medicating pangolins are extrapolated from those applied to small domestic animals. Given the lower core body temperature and unusual digestive physiology of pangolins (Nisa et al., 2005), it is likely that gastrointestinal uptake and drug metabolism differs to domestic dogs and cats. However, in the absence of dedicated research, treatment of pangolins remains based on published small animal dose rates, modified according to personal experience and anecdotal evidence of response to treatment in captive pangolins.

Oral medications can be administered in a small volume of food if the animal is reliably self-feeding, and gently moving a pangolin so that it wakes up enables delivery multiple times during the day in this manner despite the primarily nocturnal behavior of the animals. Where pangolins are not reliably self-feeding, drugs have been delivered via a stomach tube placed under sedation or anesthesia, or conscious if the animal is debilitated or conditioned to handling.

Parenteral medications can be delivered into the muscles of the lumbar, sacral and coccygeal spine even with the pangolin in the curled up position. If delivering medication subcutaneously, the scales must be gently lifted and the needle advanced at an angle through the very thick skin into the subcutaneous space. Given significant tension over the dorsal midline, particularly when the animal is curled up, this is more easily achieved by approaching from the lateral body walls.

Fluid therapy

Pangolins confiscated from illegal trade are moderately to severely dehydrated on presentation, depending on the time since their capture from the wild (Clark et al., 2009). Rehydration via provision of fluids is one of the most important therapeutic responses in the immediate period post confiscation.

Fluids can be delivered orally, subcutaneously, intravenously or intraperitoneally, depending on the level of debilitation and dehydration seen (Jaffar et al., 2018). Choice of fluids depends on clinical findings, and is based on the approach used for small domestic mammals. Subcutaneous fluids (Fig. 29.8) are generally administered into the pre-femoral space of the ventrolateral abdominal wall (Jaffar et al., 2018). Intravenous fluids can be

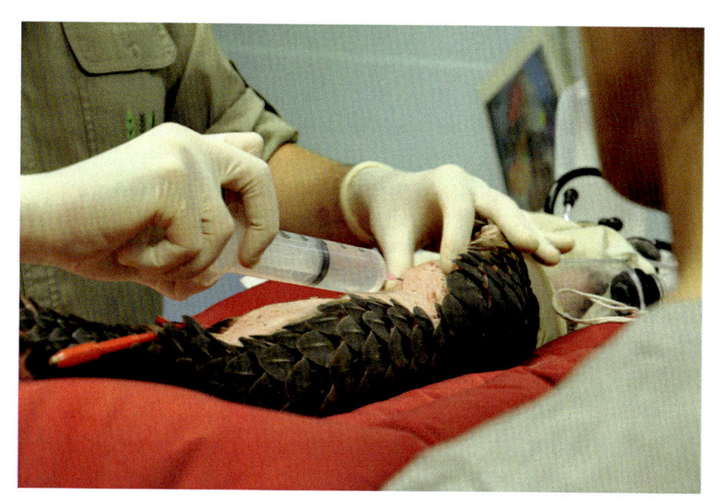

FIGURE 29.8 Administration of fluids to the subcutaneous space ventrally in an anesthetized Chinese pangolin. The red material on which the animal is laying is a warm air blanket, delivering warmed air beneath and around the animal during anesthesia to maintain normothermia. *Photo credit: Kadoorie Farm and Botanic Garden (KFBG).*

administered to a sedated, anesthetized or debilitated pangolin as a bolus into the coccygeal or cephalic vein, or via an indwelling catheter placed into the cephalic vein. Stabilization of the limb by use of a thermoplastic splint and three layer bandage has successfully enabled ongoing fluid administration to a Sunda pangolin (Jaffar et al., 2018).

Supplementary feeding

As myrmecophagous mammals with a highly specialized wild diet, the provision of adequate nutrition remains a significant challenge to the maintenance of healthy pangolins in captivity (Cabana et al., 2017; Yang et al., 2007). It can be difficult, or to date impossible in some species, including Temminck's pangolin, to transition wild pangolins onto an artificial diet. Recently captive pangolins show a strong, species specific, preference for particular ant or termite species, and many rescue centers spend considerable resources obtaining live or frozen ants to feed pangolins on arrival in captivity (Challender et al., 2012). Given the impracticality of providing live ants and termites to large numbers of animals, particularly when pangolins are transferred to institutions outside of their range states (Challender et al., 2012), there is a need to further develop appropriate artificial diets that are accepted by pangolins, provide adequate nutrition and can be sustainably sourced (Cabana et al., 2017).

Emergency nutritional support may be required for severely debilitated individuals. Stressed or debilitated pangolins may refuse to eat voluntarily (Nguyen et al., 2010), and malnutrition is responsible for significant morbidity and mortality in pangolins (Cabana et al., 2017; Lin et al., 2015). In these cases, delivery of nutrition via a tube may be initiated, and can be life-saving. Appropriately sized, lubricated nasogastric tubes are advanced into the stomach via the oral cavity in conscious, sedated or lightly anesthetized pangolins. Passing feeding tubes into the stomach via the nares has been attempted in Temminck's pangolin, however, in all cases the animal has dislodged the tube as strength returns over the first 48 hours; and in one case emergency nutrition was provided to a Temminck's pangolin via placement of a PEG (percutaneous endoscopic gastrostomy) tube, a method which involves placing a feeding tube directly into the stomach via the skin and abdominal wall, guided by endoscopy (K. Lourens, unpubl. data).

A range of nutritional support products have been used, including commercially available carnivore recovery, intensive care and diabetic diets, commercially available powdered insectivore diets, commercial tinned cat food mixed with water to form a slurry, a mix of frozen ants eggs and larvae blended to form a smooth, viscous paste and the artificial diet created, and published, by Taipei Zoo (see Lin et al., 2015).

While nutritional support via tube feeding is felt to be vital for some pangolins, and has been required for prolonged periods due to individual refusal to accept artificial diets (Aitken-Palmer et al., 2017), the procedure is not without risk. While the damaging metabolic and physiological disturbances caused by overly rapid reinstitution of nutrition are well described in the human medical literature, there are fewer reports in the veterinary literature (Brenner et al., 2011). It is possible that "re-feeding syndrome" contributes to the sudden deterioration and death seen in some pangolins which have initially responded well to supportive therapy. A number of more practical considerations exist: care must be taken to ensure the tube remains in the esophagus, rather than entering the blind ended sac which surrounds the tongue; regurgitation, predisposing the individual to aspiration pneumonia, is reported following tube withdrawal; prolonged tube feeding can predispose pangolins to oral candidiasis; and dependence on tube feeding can develop, requiring a gradual transitioning process to train the pangolin to reliably consume food from a bowl.

Conclusions

While historically poor survivability in captivity has hampered efforts to maintain pangolins in zoos, successfully rehabilitate pangolins following confiscation from the illegal trade and establish conservation breeding programs to supplement declining wild populations, this chapter has highlighted an upsurge in understanding of pangolin health and veterinary care requirements.

The general approach to physical examination, disease investigation and treatment follows that utilized in small domestic mammal medicine, bearing in mind some important features of pangolins which make them unique, including the lack of dentition, their unusual diet and digestive physiology, their lower core body temperature and the presence of scales over most of their body. Knowledge of a large number of infectious organisms isolated from the eight species of pangolin worldwide, and understanding of the common non-infectious health conditions seen provides an excellent baseline for those caring for pangolins in captivity. This assists in the diagnosis and treatment of debilitated pangolins and helps to ensure that captive institutions are adequately resourced to better ensure the health of the animals in their care. The publication of some physiological reference ranges, including blood biochemistry and hematology for a small number of species, has also provided a significant advantage in the investigation of health and treatment of sick and injured pangolins. It is envisaged that further work in this area will broaden the physiological parameters studied, the species for which this information is available, and that larger sample sizes of healthy pangolins over different age and reproductive stages will improve their diagnostic value.

A number of areas for health research are conspicuous in their absence. These include an imperative need to develop clinical pathology and post-mortem diagnostic capacity to understand more thoroughly the underlying causes and progression of disease in pangolins, and pharmacological research to improve treatment options and outcomes for sick and injured animals. This is particularly vital for pangolins confiscated from the illegal wildlife trade where mortality rates, although improving, remain high. Development of safe and effective protocols for the emergency care of severely debilitated pangolins, and ensuring that institutions caring for pangolins worldwide have the capacity to implement these protocols, is needed.

While the breadth of knowledge presented here is remarkable, it is focused on Sunda, Chinese and Indian pangolins in Asia and on Temminck's and white-bellied pangolins in Africa. There is little veterinary health information available for the Philippine, giant and black-bellied pangolins. While many of the lessons learned may be safely applied to improve the health and welfare of the lesser-studied species, there remains a need for veterinary health research dedicated to these species.

Acknowledgments

The authors gratefully acknowledge the assistance from many within the global pangolin conservation community. In particular, Gary Ades and Dr. Alessandro Grioni, Kadoorie Farm and Botanic Gardens (KFBG), Hong Kong SAR; Dr. Cora Singleton and Dr. Ilse Stalis, San Diego Zoo, United States; Dr. Copper Aitken-Palmer, Chicago Zoological Society – Brookfield Zoo, United States; Dr. Shangzhe Xie, Wildlife Reserves Singapore; Dr Rajesh K. Mohapatra, Nandankanan Zoological Park, India; Jessica Jimerson and Dr. Nguyen Ngoc Duyen Huong, Save Vietnam's Wildlife, Vietnam; Dr. Judy St. Leger, Seaworld Parks, United States; Dr. Thomas W deMaar, Gladys Porter Zoo, United States; Dr. Bonnie Raphael; Dr. Paolo Martelli, Ocean Park Conservation Foundation, Hong Kong SAR; Dr. Christa van Wessem, Paignton Zoo Environmental Park, UK; Rupak Khatri-Chhetri, National Pingtung University of Science and Technology, Pingtung, Taiwan; Emeritus Professor William Ian Beveridge, Veterinary Biosciences, The University of Melbourne, Australia and Dr. Elizabeth Dobson, Specialist Veterinary Pathologist, Australia for their contributions of knowledge, experience and data.

References

Aitken-Palmer, C., Sturgeon, G.L., Bergmann, J., Knightly, F., Johnson, J.G., Ivančić, M., et al., 2017. Enhancing conservation through veterinary care of the white-bellied tree pangolin (*Manis tricuspis*). 9th AAZV Annual Conference Proceedings. American Association of Zoo Veterinarians, Texas, United States.

Bell, D.J., Roberton, S.I., Hunter, P.R., 2004. Animal origins of SARS Coronavirus: possible links with the international trade in small carnivores. Philos. Trans. R. Soc. B: Biol. Sci. 359 (1447), 1107–1114.

Brenner, K., Kukanich, K.S., Smee, N.M., 2011. Refeeding syndrome in a cat with hepatic lipidosis. J. Feline Med. Surg. 8 (13), 614–617.

Cabana, F., Plowman, A., Nguyen, V.T., Chin, S.-C., Sung-Lin, W., Lo, H.-Y., et al., 2017. Feeding Asian pangolins: an assessment of current diets fed in institutions worldwide. Zoo Biol. 36 (4), 298–305.

Cantlay, J.C., Ingram, D.J., Meredith, A.L., 2017. A review of zoonotic infection risks associated with the wild meat trade in Malaysia. EcoHealth 14 (2), 361–388.

Chaber, A.-L., Cunningham, A., 2016. Public health risks from illegally imported African bushmeat and smoked fish. EcoHealth 13 (1), 135–138.

Challender, D.W.S., Nguyen, V.T., Jones, M., May, L., 2012. Time-budgets and activity patterns of captive Sunda pangolins (*Manis javanica*). Zoo Biol. 31 (2), 206–218.

Chin, J.S.-C., Tsao, E.H., 2015. Pholidota. In: Miller, R.E., Fowler, M.E. (Eds.), Fowler's Zoo and Wild Animal Medicine, vol. 8. Saunders, St. Louis, pp. 369–375.

Chin, S.C., Liu, C.H., Guo, J.C., Chen, S.Y., Yeh, L.S., 2006. A 10-year review of autopsy of rescued Formosan pangolin (*Manis pentadactyla pentadactyla*) in Taipei Zoo. 2nd Meeting of the Asian Society of Conservation Medicine, Asian Zoo and Wildlife Medicine Conference, Chulalongkorn University, Bangkok, Thailand.

Chin, S.-C., Lien, C.-Y., Chan, Y., Chen, C.-L., Yang, Y.-C., Yeh, L.-S., 2015. Hematologic and serum biochemical parameters of apparently healthy rescued Formosan pangolins (*Manis Pentadactyla Pentadactyla*). J. Zoo Wildlife Med. 46 (1), 68–76.

Clark, L., Van Thai, N., Phuong, T.Q., 2009. A long way from home: the health status of Asian pangolins confiscated from the illegal wildlife trade in Viet Nam. In: Pantel, S., Chin, S.-Y. (Eds.), Proceedings of the Workshop on Trade and Conservation of Pangolins Native to South and Southeast Asia, 30 June – 2 July 2008, Singapore Zoo, Singapore. TRAFFIC Southeast Asia, Petaling Jaya, Selangor, Malaysia, pp. 111–118.

Clark, L.V., Nguyen, T.V., Tran, Q.P., Higgins, D.P., 2010. A Retrospective Review of Morbidity and Mortality in Pangolins Confiscated From the Illegal Wildlife Trade in Vietnam. American Association of Zoo Veterinarians, South Padre Island, Texas, United States.

Daszak, P., Cunningham, A.A., Hyatt, A.D., 2001. Anthropogenic environmental change and the emergence of infectious diseases in wildlife. Acta Trop. 78 (2), 103–116.

Heath, M.E., 1987. Twenty-four-hour variations in activity, core temperature, metabolic rate, respiratory quotient in captive Chinese pangolins. Zoo Biol. 6 (1), 1–10.

Heath, M.E., Vanderlip, S.L., 1988. Biology, husbandry, and veterinary care of captive Chinese pangolins (*Manis pentadactyla*). Zoo Biol. 7 (4), 293–312.

Hope, K., Deem, S.L., 2006. Retrospective study of morbidity and mortality of captive jaguars (*Panthera onca*) in North America: 1982 to 2002. Zoo Biol. 25 (6), 501–512.

Hua, L., Gong, S., Wang, F., Li, W., Ge, Y., Li, X., et al., 2015. Captive breeding of pangolins: current status, problems and future prospects. ZooKeys 507, 99–114.

Jaffar, R., Kurniawan, A., Maguire, R., Anwar, A., Cabana, F., 2018. WRS Husbandry Manual for the Sunda Pangolin (*Manis javanica*), first ed. Wildlife Reserves Singapore Group, Singapore.

Jones, K.E., Patel, N.G., Levy, M.A., Storeygard, A., Balk, D., Gittleman, J.L., et al., 2008. Global trends in emerging infectious diseases. Nature 451 (7181), 990–993.

Khatri-Chhetri, R., Sun, C.M., Wu, H.Y., Pei, K.J.C., 2015. Reference intervals for hematology, serum biochemistry, and basic clinical findings in free-ranging Chinese Pangolin (*Manis pentadactyla*) from Taiwan. Vet. Clin. Pathol. 44 (3), 380–390.

Khatri-Chhetri, R., Wang, H.-C., Chen, C.-C., Shih, H.-C., Liao, H.-C., Sun, C.-M., et al., 2016. Surveillance of ticks and associated pathogens in free-ranging Formosan pangolins (*Manis pentadactyla pentadactyla*). Ticks and Tick-borne Dis. 7 (6), 1238–1244.

Khatri-Chhetri, R., Chang, T.-C., Khatri-Chhetri, N., Huang, Y.-L., Pei, K.J.-C., Wu, H.-Y., 2017. A retrospective study of pathological findings in endangered Formosan pangolins (*Manis pentadactyla pentadactyla*) from Southeastern Taiwan. Taiwan Vet. J. 43 (1), 55–64.

Kolonin, G.V., 2007. Mammals as hosts of Ixodid ticks (Acarina, Ixodidae). Entomol. Rev. 87 (4), 401–412.

Lam, H.K., 2018. Evaluating and monitoring health of confiscated pangolins: veterinary responses at rescue sites. Back to the Wild - Training Workshop. USAID, Save Vietnam's Wildlife, Khao Yai National Park, Thailand.

Langan, J.N., 2014. Tubulidentata and Pholidata. In: West, G., Heard, D.J., Caulkett, N. (Eds.), Zoo Animal and Wildlife Immobilization and Anesthesia, second ed. Wiley Blackwell, Oxford, pp. 539–542.

Leendertz, F.H., Pauli, G., Maetz-Rensing, K., Boardman, W., Nunn, C., Ellerbrok, H., et al., 2006. Pathogens as drivers of population declines: the importance of systematic monitoring in great apes and other threatened mammals. Biol. Conserv. 131 (2), 325–337.

Lin, M.F., Chang, C.Y., Yang, C.W., Dierenfeld, E.S., 2015. Aspects of digestive anatomy, feed intake and digestion in the Chinese pangolin (*Manis pentadactyla*) at Taipei zoo. Zoo Biol. 34 (3), 262–270.

Liyanaarachchi, D., Rajapakse, R., Dilrukshi, P., 2013. Tick Vectors of Spotted Fever Rickettsia in Sri Lanka, 17. Book of abstracts of the Peradeniya University Research Sessions, Sri Lanka, p. 146.

Liyanaarachchi, D., Rajakaruna, R., Rajapakse, R., 2016. Spotted fever group rickettsia in ticks infesting humans, wild and domesticated animals of Sri Lanka: one health approach. Ceylon J. Sci. (Biol. Sci.) 44 (2), 67–74.

Lonsdorf, E.V., Travis, D., Pusey, A.E., Goodall, J., 2006. Using retrospective health data from the Gombe chimpanzee study to inform future monitoring efforts. Am. J. Primatol. 68 (9), 897–908.

Mediannikov, O., Davoust, B., Socolovschi, C., Tshilolo, L., Raoult, D., Parola, P., 2012a. Spotted fever group rickettsiae in ticks and fleas from the Democratic Republic of the Congo. Ticks and Tick-borne Dis. 3 (5–6), 371–373.

Mediannikov, O., Diatta, G., Zolia, Y., Balde, M.C., Kohar, H., Trape, J.-F., et al., 2012b. Tick-borne rickettsiae in Guinea and Liberia. Ticks and Tick-borne Dis. 3 (1), 43–48.

Mohapatra, R.K., Panda, S., 2014. Husbandry, behaviour and conservation breeding of Indian pangolin. Folia Zool. 63 (2), 73–80.

Mohapatra, R.K., Panda, S., Nair, M.V., Acharjyo, L.N., 2016. Check list of parasites and bacteria recorded from pangolin (*Manis* sp.). J. Parasit. Dis. 40 (4), 1109–1115.

Munson, L., 1991. Strategies for integrating pathology into single species conservation programs. J. Zoo Wildlife Med. 22 (2), 165–168.

Nadler, S.A., De Leon, G.P.-P., 2011. Integrating molecular and morphological approaches for characterizing parasite cryptic species: implications for parasitology. Parasitology 138 (13), 1688–1709.

Narayanan, S., Kirchheimer, W., Bedi, B., 1977. Some bacteria isolated from the Indian pangolin (*Manis crassicaudata*) Geoffroy. Indian Vet. J. 54 (9), 692–988.

Nguyen, T.V., Clark, L., Tran, P.Q., 2010. Management Guidelines for Sunda pangolin (*Manis javanica*), first ed. Carnivore and Pangolin Conservation Program, Cuc Phuong National Park, Vietnam.

Nisa, C., Kitamura, N., Sasaki, M., Agungpriyono, S., Choliq, C., Budipitojo, T., et al., 2005. Immunohistochemical study on the distribution and relative frequency of endocrine cells in the stomach of the Malayan pangolin, *Manis javanica*. Anat., Histol. Embryol. 34 (6), 373–378.

Parola, P., Cornet, J.-P., Sanogo, Y.O., Miller, R.S., Van Thien, H., Gonzalez, J.-P., et al., 2003. Detection of Ehrlichia spp., Anaplasma spp., Rickettsia spp., and other eubacteria in ticks from the Thai-Myanmar border and Vietnam. J. Clin. Microbiol. 41 (4), 1600–1608.

Perera, P.K.P., Karawita, K.V.D.H.R., Pabasara, M.G.T., 2017. Pangolins (*Manis crassicaudata*) in Sri Lanka: a review of current knowledge, threats and research priorities. J. Trop. For. Environ. 7 (1), 1–14.

Robinson, P., 1983. The use of ketamine in restraint of a black-bellied pangolin (*Manis tetradactyla*). J. Zoo Anim. Med. 14 (1), 19–23.

Robinson, P.T., 1999. Pholidota (Pangolins). In: Miller, R.E., Fowler, M.E. (Eds.), Fowler's Zoo and Wild Animal Medicine, vol. 8. Saunders, St. Louis, pp. 407–410.

Smith, K.M., Machalaba, C.M., Jones, H., Caceres, P., Popovic, M., Olival, K.J., et al., 2017. Wildlife hosts for OIE-Listed diseases: considerations regarding global wildlife trade and host–pathogen relationships. Vet. Med. Sci. 3 (2), 71–81.

Sun, N.C.-M., Arora, B., Lin, J.-S., Lin, W.-C., Chi, M.-J., Chen, C.-C., et al., 2019. Mortality and morbidity in wild Taiwanese pangolin (*Manis pentadactyla pentadactyla*). PLoS One 14 (2), e0212960.

Travis, D.A., Hungerford, L., Engel, G.A., Jones-Engel, L., 2006. Disease risk analysis: a tool for primate conservation planning and decision-making. Am. J. Primatol. 68 (9), 855–867.

Wicker, L.V., Canfield, P.J., Higgins, D.P., 2016. Potential pathogens reported in species of the family Viverridae and their implications for human and animal health. Zoonoses Public Health 64 (2), 75–93.

Wilson, A.E., 1994. Husbandry of pangolins Manis spp. Int. Zoo Yearb. 33 (1), 248–251.

Yang, C.W., Chen, S., Chang, C.Y., Lin, M.F., Block, E., Lorentsen, R., et al., 2007. History and dietary husbandry of pangolins in captivity. Zoo Biol. 26 (3), 223–230.

Yang, L., Su, C., Zhang, F., Wu, S., Ma, G., 2010. Age structure and parasites of Malayan pangolin (*Manis javanica*). J. Econ. Anim. 14 (1), 22–25. [In Chinese].

Zhang, F., Yu, J., Wu, S., Li, S., Zou, C., Wang, Q., et al., 2017. Keeping and breeding the rescued Sunda pangolins (*Manis javanica*) in captivity. Zoo Biol. 36 (6), 387–396.

CHAPTER

30

The rescue, rehabilitation and release of pangolins

Nicci Wright[1,2] and Jessica Jimerson[3]

[1]Humane Society International – Africa, Johannesburg, South Africa [2]African Pangolin Working Group, Johannesburg, South Africa [3]Save Vietnam's Wildlife, Cuc Phuong National Park, Ninh Binh Province, Vietnam

OUTLINE

Introduction

In part due to their inclusion in Appendix I of CITES (Convention on International Trade in Endangered Species of Wild Fauna and Flora), in 2016 (see Chapter 19), there is increased awareness of pangolins among national natural resource management, conservation, and law enforcement agencies in pangolin range states. Pangolins are protected species in most range countries, though the degree of protection varies between species and country. There is also variation in how effectively laws are implemented and a number of challenges to law enforcement concerning pangolins have been identified (Challender and Waterman, 2017). Nonetheless, seizures of pangolins and their parts are made frequently both in Africa and Asia (Chapter 16; Heinrich et al., 2017). This includes the recovery of live animals, which, in a number of range states, are taken to rescue centers for health assessments and rehabilitation, before being released back in to the wild. This process has

Pangolins

DOI: https://doi.org/10.1016/B978-0-12-815507-3.00030-7

allowed the development of expertise on pangolin rehabilitation, veterinary health and captive care. The aim of this chapter is to document and contrast experiences of rehabilitating pangolins in different countries and under different circumstances, with different species. It draws primarily on experiences in South Africa and Vietnam. The chapter starts by contextualizing rescue, rehabilitation and release efforts in the two countries before discussing a number of case studies on pangolins recovered from illegal trade.

Rescue

In Southern Africa, there are two main causes of Temminck's pangolin (*Smutsia temminckii*) needing rescue and rehabilitation. First, pangolins are poached and trafficked, typically individual animals, and on detection by law enforcement agencies (e.g., through sting operations) are transported to the closest appropriate veterinary hospital or wildlife rehabilitation facility with the necessary expertise to deal with pangolins (Fig. 30.1). Ideally, in these scenarios forensic sampling of the rescued animal will be conducted to inform judicial proceedings (see Chapter 20). Second, Temminck's pangolins are electrocuted on electric fences surrounding game farms and reserves (Beck, 2008; Pietersen et al., 2014; see Chapter 11). The animals inadvertently curl around the bottom strand of electrified wire and most pangolins do not survive, though a small proportion do. Some animals may be immediately released on detection by landowners with no knowledge of their survival. In addition, in Namibia and the Northern Cape, South Africa, estimates suggest that up to 280 pangolins are killed annually by vehicles while crossing roads (Pietersen et al., 2016). Pangolins that survive electrocution and traffic collisions also form part of rescue efforts.

In contrast, the situation is very different in Vietnam, principally because of the number of animals that are seized. Vietnam, as well as being a consumer market for pangolin products (see Chapter 22), also serves as a thoroughfare for pangolins trafficked from others parts of Southeast Asia, largely to China (Chapter 16; Nguyen, 2009). Save Vietnam's Wildlife (SVW), an established Vietnamese non-governmental organization (NGO), frequently liaises with the government in order to rescue pangolins from illegal trade. Seizures vary in size from one to more than 200 pangolins, primarily involving the Sunda pangolin (*Manis javanica*) but also the Chinese pangolin (*M. pentadactyla*). The number of animals in any given seizure greatly impacts the quality of care provided at the rescue site and the way animals are triaged. A typical seizure in Vietnam entails releasing the animals from the confinement of small netted bags, in

FIGURE 30.1 An adult Temminck's pangolin (*S. temminckii*) rescued from illegal trade in South Africa. *Photo credit: African Pangolin Working Group.*

which they have been kept and trafficked for an indeterminate period (see also Chapter 29). They are usually covered in their own feces and urine. In addition, many pangolins have been force-fed a mixture of cornmeal and water to increase their weight and thereby their financial value to traffickers. Animals that have been subject to this procedure have yellow, liquid diarrhea, and often display inappetance. This technique traumatizes the esophagus through forceful entry of the feeding tube and causes gastritis. Where large numbers of animals are seized the strongest individuals are prioritized for care. This selective process is used to stabilize individuals in preparation for transport to the nearest rescue center.

Rehabilitation

Rehabilitating pangolins for release back into the wild is complex, and not possible in all cases. Techniques applied to the captive care of a pangolin depends on the species concerned, location, access to veterinary and husbandry resources, and experience level of the rehabilitator. In Southern Africa, pangolin rehabilitation efforts range from professionally trained and equipped rehabilitation personnel with modern facilities to game farmers and local conservation officers in remote areas with little knowledge and few resources. Frequently, advice and treatment options are relayed over the phone to assist whoever is in possession of the animal.

Initial stabilization of rescued animals is crucial regardless of geography. Rescued pangolins in both South Africa and Vietnam are placed in quarantine enclosures of a suitable temperature, and provided with bedding (e.g., blankets), food and water. Unless the animal is critically compromised, they are left alone to de-stress. Caution is needed because handling pangolins can cause them stress (see Chapter 29).

Restoring an animal to optimal health begins with a thorough physical exam. At SVW, pangolin diagnostic tools are limited. Animals are treated preventatively with dewormers, gastroprotectants and a broad spectrum of antibiotics if any wounds or disease symptoms are detected on initial assessment or within the first week of quarantine. Dehydrated and anorexic pangolins undergo fluid therapy with added vitamins for a number of days in the initial quarantine period in an attempt to normalize hydration status.

Pangolins are inspected for wounds, infections, abnormal parasite loads (both external and fecal), eye infections and respiratory issues (see also Chapter 29). Where wounds are found they are cleaned and may be treated with oral antibiotics in Sunda and Chinese pangolins. Fractures are also checked for and if suspected confirmed via further physical examination and radiographs if available. Whether the fracture is bandaged or results in amputation, depends on the animal's ability to locomote and survive in the wild. Euthanasia is sometimes necessary if there are no options for life long captive management and the animal has a poor prognosis for survival in the wild. Neurological issues are common with animals that have been clubbed or hit by a motor vehicle (e.g., in Southern Africa).

Blood samples are taken and full blood counts and serum chemistry are important tools to measure the animal's condition. Most important are the blood glucose and albumin levels, which are also used for monitoring purposes (see Chapter 29). Blood smears are also done and attention is paid to thrombocyte number and morphology. Many pangolins display signs of poor clotting factors at SVW. A fecal floatation test and direct wet smear are also carried out to assess which parasites are present and to determine the parasite load (see Chapter 29).

Pangolins seized from illegal trade are typically malnourished and underweight (Fig. 30.2) and caution is exercised to avoid re-feeding syndrome. This consists of metabolic disturbance when nutrition is incorrectly given to a

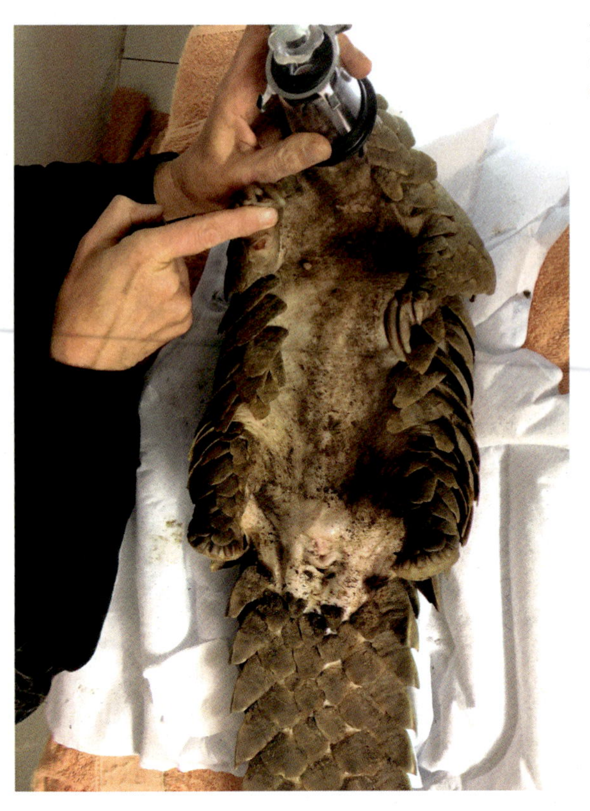

FIGURE 30.2 An emaciated Temminck's pangolin (*S. temminckii*) following rescue from illegal trade. *Photo credit: African Pangolin Working Group.*

system which has been severely malnourished. The volume of food consumed each night is monitored, and if an animal is not eating, an alternative food option is offered: foraged live ants. For Temminck's pangolin this can be accomplished by allowing the animal concerned to forage naturally. In Vietnam, for Chinese and Sunda pangolins, ant nests are collected from local forests and offered in the animal's enclosure, while ant eggs manufactured for human consumption are also offered and may be mixed with farmed silkworms (*Bombyx mori*).

In South Africa, once a Temminck's pangolin is deemed fit enough, the animal is taken on foraging excursions to consume natural prey (see Richer et al., 1997). This species tends not to feed in captivity and does not consume a captive diet. For this reason, Temminck's pangolins under rehabilitation require daily walking in order to forage. This is not the case for the Sunda or Chinese pangolin.

Housing is provided that meets the pangolins' basic needs. This includes a strongly built enclosure to prevent escape, a resting place that simulates a burrow, with good ventilation, and potentially a heat source depending on the species and season, and age of the animal (see Chapter 28). For semi-arboreal species (e.g., Sunda pangolin), enclosure furniture is provided in the form of native plants, tree trunks and a network of tree branches so the animals concerned can exhibit natural climbing behavior. For animals that cannot be released and need long-term captive care, perhaps because they are being hand-reared or are in protractive recovery, the husbandry needs are somewhat different (see Chapter 28).

Release

Release back into the wild is the final step in the rescue, rehabilitation and release process. In Southern Africa, increasing experience is indicating that Temminck's pangolins have a higher probability of survival if they are subject to a "soft release" (R. Jansen, pers. comm.). This also applies to hand-reared pangolins. A soft release entails habituating the animal to the new environment while under close observation, i.e., animals remain at the rehabilitation facility but are walked daily in the new environment by a minder. In South Africa, soft releases take place over a five-day period. The animal is weighed before and after every feeding session in order to record weight gain. Experience with young white- (*Phataginus tricuspis*) and black-bellied pangolins (*P. tetradactyla*) weighing less than 1 kg at the Sangha Pangolin Project in Central African Republic suggests that they should have a soft-release with a dedicated minder (A. Kriel, pers. comm.); the local Baka monitor both species.

In South Africa, all pangolins are tagged on release and monitored with telemetry equipment. They are monitored twice a day for three weeks, then twice a week for three months, and thereafter weekly for a period of 12 months. This is to generate knowledge of dispersal behavior and survival rates. Release sites are chosen carefully. Key considerations include whether a site is free from poaching, or within game reserves, national parks or conservancy areas suitable for release, and the extent of monitoring by anti-poaching units (Pietersen et al., 2016). The presence of electric fences and vicinity of such fences to release sites is also considered. Encouragingly, some reserves in Southern Africa are mitigating the impact of electric fences thereby making it safer for smaller species including pangolins (Beck, 2008; Pietersen et al., 2014). This mitigation is done by either switching off the electric current to the trip wire or increasing the height of it to prevent smaller species being electrocuted.

In Vietnam, SVW aims to rehabilitate and release pangolins as efficiently as possible. After the quarantine period, clinically stable pangolins are listed as release candidates. The veterinary team performs visual exams with attention to ambulation, stress level, activity, body condition score, sleep habits, and weight. Currently, disease screening is done individually based on presentation of clinical symptoms. "Stable" animals are not screened for any pathogens prior to release. Pangolins are notorious for sustaining captive-related health issues and decisions on suitability for release are often made with a long-term view. For example, a pangolin may have minor a wound on its footpads, tail or nose and a decision has to be made on whether to attempt to heal such a wound to prevent further infection or leave it untreated to avoid causing the animal stress. If the animal is otherwise in good condition physically, eating well, producing formed feces regularly and maintaining weight, the protocol is for release as soon as possible to minimize risk of further captive related issues. Keeping these pangolins in captivity longer in an attempt to resolve superficial issues often ends in mortality. Many individuals will remain stable for weeks but a variable such as drop in environmental temperature can result in sudden inappetance. Once ready for release, pangolins are microchipped and dewormed.

SVW primarily rehabilitates and releases Sunda pangolins. This involves transporting animals to one of two national parks (270 km and ~1500 km away from the conservation center respectively) within the species' geographic distribution. The field team considers the fact that SVW has been releasing pangolins in these parks for several years, and an assumption is made that stable populations must exist where animals have been released previously. This is considered when planning releases. Finite protected natural habitat for wildlife in

Vietnam is a major limiting factor for the release of pangolins. The journey to the release site is very stressful on the animals, especially given the poor road surfaces in parts of Vietnam. The animals are transported in wooden boxes and are fed overnight, given a blanket and water in their transport boxes, and fed again before they are released. Typically, ~25 pangolins are released at a time due to transport limitations. Animals are released in separate locations to avoid potential stress between individuals given the solitary nature of the species.

Releases, as far as possible, follow best practice guidance (see IUCN/SSC, 2013).

Prior to release, sites are surveyed to confirm suitable food availability, habitat and ecological structures (e.g., tree hollows for resting). Workshops are also held with local communities surrounding release sites to generate support for pangolin conservation. Release sites must be able to accommodate new pangolins, as released animals will need to establish a home range. However, a key challenge to the release of Sunda pangolins in Vietnam is identifying that release sites can accommodate additional pangolins. This is due, in part, to the lack of standardized monitoring methods for pangolins (see Chapter 35) with which to determine that this is the case. It is also because pangolin populations in many parts of Vietnam have been heavily depleted meaning very extensive survey effort would be needed to confirm sufficiently low densities, or theoretically absence, to warrant release (see Willcox et al., 2019). Consequently, there is typically a lack of detailed knowledge of resident pangolin populations. In practical terms, the need to determine population status is weighed up against the need to release urgently, in some cases, up to 200 pangolins in a very short period, in order to give the animals the best chance of survival. Whenever possible, camera traps are set up at release sites and GPS trackers are attached to selected pangolins to monitor movement and survival rates, but a practical challenge is monitoring post-release behavior when so many pangolins are concerned.

Case studies

A badly compromised Temminck's pangolin

A 13 kg adult male Temminck's pangolin was received at the Johannesburg Wildlife Veterinary Hospital in April 2018, following confiscation from a sting operation. The animal was severely dehydrated and had a marked limp of the left hindlimb, and a wound on the outside of the same leg. It was strongly suspected that the animal had been tied up in captivity or had been caught in a snare. The pangolin was anesthetized with isoflurane gas to facilitate intravenous fluid administration, and given a full clinical examination. Radiographs revealed a left midshaft tibia and fibula fracture. The animal was stabilized and given pain medication and antibiotic therapy. After liaison with a veterinary surgical specialist, a plate and screws were used to repair the leg. The animal was allowed to walk following surgery, as is standard for all small animals after this type of fracture repair. For the first 24 hours, the pangolin was fed through a tube placed into the stomach through the mouth. The animal was lightly sedated because unrolling an adult Temminck's pangolin is practically impossible.

After two days, the animal was given the opportunity forage naturally. For the following 10 days, the animal fed well, consuming up to 600 g of pugnacious ants per night, and maintaining body weight. However, the wound on the left leg dehisced and it was determined that due to the muscle strength of the animal, the metal plate had moved. This caused a small piece of bone to protrude through the outside of the leg, underneath the scales, and caused a severe wound infection that led to the wound opening. The wound was cleaned and flushed under sedation on a daily basis. Attempts were made to bandage the wound but it was almost impossible due to the animal's physiology. The pangolin was also tranquilized to limit movement using Valium, haloperidol, midazolam, but on this animal, these drugs lasted only a few hours at a time. Antibiotics and pain medication were also administered allowing the animal to forage for short bouts which appeared to have a calming effect. However, after two weeks, a follow-up radiograph revealed that the plate had sheared off the bone. A further round of surgery was needed to re-plate the leg, and a locking plate was used. However, due to the damage to the bone and the concurrent infection, the prognosis was guarded. The pangolin needed to be kept as still as possible post-operatively. Despite sedation, this proved impossible and within 48 hours of the second surgery, the locking plate had been broken and the tibia, dorsal to the plate, shattered. The decision was made to euthanize the animal.

Rehabilitation and release of a Temminck's pangolin

South African Environmental Law Enforcement Officers confiscated a male Temminck's pangolin, weighing 11.3 kg, in the Chinese quarter of Johannesburg in January 2018. The animal was taken to Johannesburg Wildlife Veterinary Hospital for treatment and rehabilitation.

The pangolin was heavily contaminated with motor oil as a result of being transported in the spare wheel hub of a vehicle. The animal was washed with a mild detergent in warm water to prevent ingestion/poisoning. Isoflurane was then used to sedate the animal enabling the belly and inner scales to be cleaned. A low dose of Valium was given, intramuscularly, to lower stress levels; blood glucose levels were checked and blood tests were also run including a blood count and serum chemistry analysis.

This pangolin's physical condition was generally good, especially considering the animal had endured an extended period without food or water. The animal was taken foraging every night for 11 nights, spending up to three hours foraging for prey each night. When not foraging, the pangolin was housed in a custom made sleeping box with a heating pad and water tray. The pangolin maintained weight and was evidently ingesting an adequate quantity of ants and termites. The animal was anaesthetized to facilitate chain of custody sampling process for the BioBank situated at the National Zoological Gardens. This involves taking blood and scale samples, weight and measurements.

In early February 2018, the pangolin was fitted with a telemetry unit and was released into a protected area consisting of typical bushveld habitat. The animal was observed on release and was located on a daily and weekly basis thereafter. After several months, the pangolin had moved approximately 10 km, and remained in a particular area, suggesting that the animal had established a home range.

Sunda pangolin caught in a snare trap

A single, adult female Sunda pangolin was rescued in Lang Son Province, Vietnam and admitted to SVW in October 2018. On arrival, the animal weighed 4 kg and was visually bright, alert and active and did not appear stressed. Physical exam findings included a series of malodorous, deep, mildly necrotic assumed snare-trap wounds slicing diagonally around the ventral aspect of the chest cutting through the muscle layer and reaching around to the lateral aspects of the chest (Fig. 30.3). The wounds varied in depth (0.5–1 cm) and purulent discharge was excreting from the deepest wounds. Upon further examination, the animal was noted to be mildly dehydrated and pale. There was also evidence of diarrhea with loose fecal material observed around the rectum. Several ticks were also removed. The decision

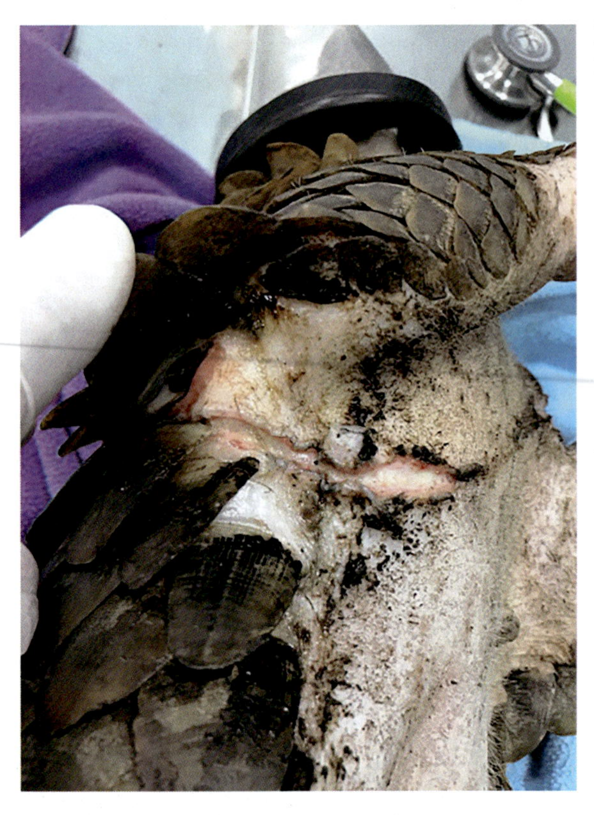

FIGURE 30.3 A female Sunda pangolin (*M. javanica*) presenting snare wounds. *Photo credit: Jess Jimerson.*

was made to anesthetize the animal for further investigation. Subsequent investigation confirmed that necrotic tissue was also present under the lateral scales, and given the depth of infection and limited mobility in the affected region the prognosis was guarded.

The entire wound was flushed with 0.9% saline and dilute iodine to remove organic material, and areas of necrosis were debrided. In total, seven scales were also removed. A small amount of saline soaked gauze was applied to the wound on the lateral aspect of the body topped with thick pads of dry gauze in an attempt to remove any residual contamination within the wounds without drying it out. Tegaderm was applied and wrapped around the body with vetwrap to keep the bandage in place and prevent further contamination (Fig. 30.4). The animal was kept in a wooden box for the remainder of treatment to minimize mobility. The animal was fed on thawed frozen ant eggs (200 g/daily) and treated using a combination of Clavamox, meloxicam, cemitidine and ivermectin.

On changing of the bandages the following day, the pangolin's abdomen was distended and firm and an ultrasound confirmed pregnancy. However, two days later the pangolin was found with dark blood around the vulva and on ultrasound no neonate heartbeat was detected and it was concluded that the animal had a stillbirth.

Four days later, and following two bandage changes under anesthetic, including additional cleaning of the wound due to the presence of purulent discharge, the animal was hydrated and in good condition and its food intake increased to 250 g/daily. However, upon further examination in subsequent days, ulcers

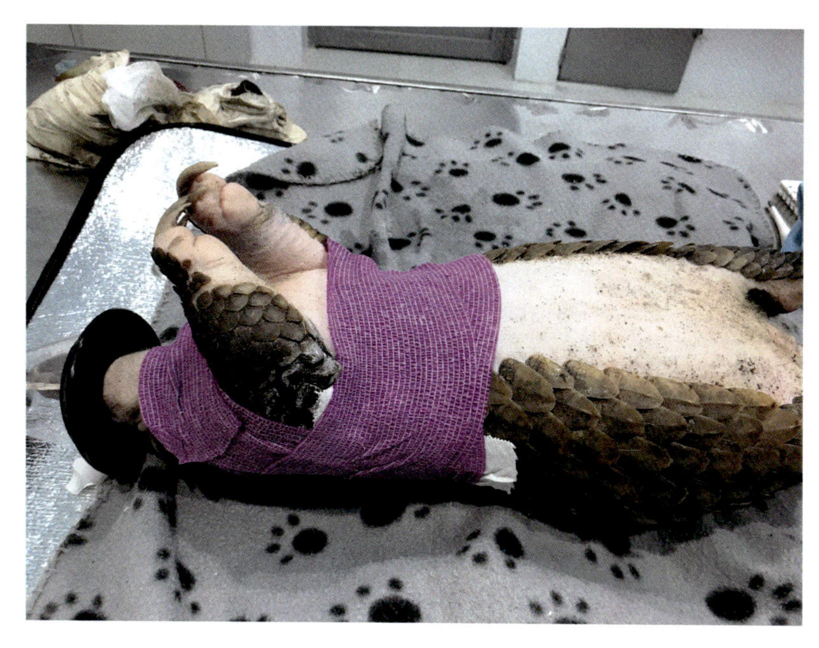

FIGURE 30.4 Female Sunda pangolin (*M. javanica*) under anesthetic following surgery to clean wounds from a snare. *Photo credit: Jess Jimerson.*

were found on the rear footpads measuring 0.2–0.3 cm. They were cleaned with chlorhexidine, rinsed with 0.9% saline, and silver sulfadiazine (SSD) cream was applied.

Within a few days, the pangolin's weight had increased to 4.5 kg and the chest wound was healing well. The animal was anaesthetized when needed in the following week to change the bandages on the chest wound and the footpad ulcers were repeatedly cleaned and treated with SSD cream. By the end of October 2018, the animal had increased in weight to 4.7 kg and all wounds had healed. A minor respiratory problem caused by poor ventilation due to limiting the movement of the animal was resolved within 72 hours by moving the pangolin to larger quarantine enclosure. By early November, the pangolin was listed for release back in to the wild.

Conclusions

Pangolins are seized with regularity in range states in Africa and Asia. Experience gained in the rescue, rehabilitation and release of pangolins in South Africa and Vietnam since the early 2000s means that for animals that are rescued, subject to their health status, there is a chance of rehabilitation and release back in to the wild. The rehabilitation and release of African pangolins in particular, is a relatively new phenomenon and there is little documented knowledge of best practice. The observations and data collected during rehabilitation and release therefore provide new and critical insights to further understanding of these species. Challenges remain to the successful release of pangolins back in to the wild, which in Vietnam include determining the status of resident pangolin populations prior to release and monitoring of animals post-release, given the numbers of pangolins successfully rescued and readied for release. The rescue, rehabilitation and release of pangolins makes an important contribution to the conservation of the species by offering a lifeline to rescued animals. Critical to success is working relationships between stakeholders, including government agencies (e.g., law enforcement) and rescue centers and veterinary hospitals, such that pangolins that are seized receive the care they need in a timely fashion and have the best chance of survival.

References

Beck, A., 2008. Electric Fence Induced Mortality in South Africa. M.Sc. Thesis, University of the Witwatersrand, Johannesburg, South Africa.

Challender, D., Waterman, C., 2017. Implementation of CITES Decisions 17.239 b) and 17.240 on Pangolins (*Manis* spp.), CITES SC69 Doc. 57 Annex. Available from <https://cites.org/sites/default/files/eng/com/sc/69/E-SC69-57-A.pdf>. [February 2, 2018].

Heinrich, S., Wittmann, T.A., Ross, J.V., Shepherd, C.R., Challender, D.W.S., Cassey, P., 2017. The Global Trafficking of Pangolins: A Comprehensive Summary of Seizures and Trafficking Routes From 2010–2015. TRAFFIC, Southeast Asia Regional Office, Petaling Jaya, Selangor, Malaysia.

IUCN SSC (IUCN Species Survival Commission), 2013. Guidelines for Reintroductions and Other Conservation Translocations. Version 1.0. Gland. IUCN Species Survival Commission, Switzerland.

Nguyen, T.V.A., 2009. ENV wildlife crime unit's efforts to combat illegal wildlife trade in Vietnam. In: Pantel, S., Chin, S.-Y. (Eds.), Proceedings of the Workshop on Trade and Conservation of Pangolins Native to South and Southeast Asia, 30 June – 2 July 2008, Singapore Zoo, Singapore. TRAFFIC Southeast Asia, Petaling Jaya, Selangor, Malaysia, pp. 169–171.

Pietersen, D.W., McKechnie, A.E., Jansen, R., 2014. A review of the anthropogenic threats faced by Temminck's ground pangolin, *Smutsia temminckii* in southern Africa. South Afr. J. Wildlife Res. 44 (2), 167–178.

Pietersen, D.W., Jansen, R., Swart, J., Kotze, A., 2016. A conservation assessment of *Smutsia temminckii*. In: Child, M.F., Roxburgh, L., Do Linh San, E., Raimondo, D., Davies-Mostert, H.T. (Eds.), The Red List of Mammals of South Africa, Swaziland and Lesotho. South African National Biodiversity Institute and Endangered Wildlife Trust, South Africa.

Richer, R.A., Coulson, I.M., Heath, M.E., 1997. Foraging behavior and ecology of the Cape pangolin (*Manis temminckii*) in north-western Zimbabwe. Afr. J. Ecol. 35 (4), 361–369.

Willcox, D., Nash, H., Trageser, S., Kim, H.J., Hywood, L., Connelly, E., et al., 2019. Evaluating methods for detecting and monitoring pangolin (Pholidota: Manidae) populations. Glob. Ecol. Conserv. 17, e00539.

CHAPTER

31

Zoo engagement in pangolin conservation: contributions, opportunities, challenges, and the way forward

Keri Parker[1,2] *and Sonja Luz*[2,3,4]

[1]Save Pangolins, ℅ Wildlife Conservation Network, San Francisco, CA, United States [2]IUCN SSC Pangolin Specialist Group, ℅ Zoological Society of London, Regent's Park, London, United Kingdom [3]Department of Conservation, Research and Veterinary Services, Wildlife Reserves Singapore, Singapore [4]IUCN SSC Conservation Planning Specialist Group, Apple Valley, MN, United States

OUTLINE

Introduction

Wildlife enthusiasts periodically ask where they can see pangolins at a zoo. Their desire to see pangolins is often intertwined with a desire to help the species: "Where can I see pangolins? How can I help?" Sometimes the answer is easy. Visitors to Singapore's Night Safari, for example, have the opportunity to see a pangolin, and can rest assured that their entrance fee will support the local and international conservation and research programs of Wildlife Reserves Singapore (WRS). Other conservation zoos, responsive to increasing

Pangolins
DOI: https://doi.org/10.1016/B978-0-12-815507-3.00031-9

interest from the public and keen to assist, reach out to the IUCN SSC Pangolin Specialist Group (PSG) from time to time seeking guidance on obtaining pangolins for their captive collections: "How can we obtain pangolins, so that we can contribute to their conservation?"

It is easy to understand why conservation zoos are eager to exhibit pangolins, which are deserving of public awareness. While this is a benevolent intent, the answers to what role captive populations can and should play to further the conservation of pangolins are far from straightforward. This chapter discusses opportunities for zoos to benefit pangolin conservation, including examples of contributions that conservation zoos have made, examines the challenges and risks that impede these opportunities, and provides recommendations for zoo engagement in pangolin conservation.

Zoo engagement in conservation

First, a definition of a "conservation zoo." While a zoo can be defined as any facility that displays live animals for viewing by the public (Gray, 2017), a conservation zoo is a center for species conservation. They are, to borrow from the World Association of Zoos and Aquariums (WAZA), "organizations that save populations of species in the wild, while delivering the highest standards of care and welfare for their resident animals, and providing exceptional, behavior-changing, guest experiences" (Barongi et al., 2015). A reasonable credential that signals that a zoo is striving to uphold these standards is accreditation, which covers all aspects of zoo operations (Gray, 2017), in particular, the standards put forward by the Association of Zoos and Aquariums (AZA), the European Association of Zoos and Aquaria (EAZA), and the Australia-based Zoo and Aquarium Association (ZAA). Ideally, visitors to a conservation zoo can expect to encounter animals that are receiving optimal care, and that their visit will not only be filled with wholesome, fun recreation, it will contribute to impactful conservation outcomes and they will learn how they can personally help conservation (Barongi et al., 2015).

Collectively, international conservation zoo networks have tremendous opportunities to influence positive change for wildlife, not least by raising awareness and environmental literacy among everyday people and contributing financially to wildlife conservation. A WAZA survey of national and regional zoo and aquarium associations determined that the world zoo and aquarium community reportedly spends roughly USD350 million on wildlife conservation annually (Gusset and Dick, 2011), a number that WAZA and accrediting bodies seek to increase (Cress, 2018). For people in many parts of the world, especially city dwellers, zoos provide an important opportunity to encounter wildlife and experience nature, and in fact may be the only places to experience diverse live animals from around the world and to develop pro-environmental behaviors (Grajal et al., 2018). An estimated 700 million zoo visits take place at about 400 WAZA member facilities each year (Gusset and Dick, 2011). Conservation zoos strive to create an experience for visitors that will ignite a life-long love affair with wildlife and engage them in conservation. Recent research indicates that progressive zoo environmental education programs with real animals can support biodiversity conservation when they empower visitors to develop pro-environmental behaviors (Grajal et al., 2018).

Ex situ management activities, where animals are maintained in artificial conditions, such as those conducted by zoos, when aligned and integrated with in situ conservation goals, can make important contributions to species conservation. Opportunities for ex situ management activities to benefit species conservation are summarized in the *IUCN SSC Guidelines on the Use of Ex Situ Management for*

Species Conservation (hereafter IUCN Ex Situ Guidelines or Guidelines). Ex situ management activities have the potential to address the causes of primary threats to species, such as through conservation research, conservation training, or conservation education activities. They can offset the effect of threats such as through head start programs to address high juvenile mortality; buy time for conservation communities to tackle catastrophic threats through the establishment of ex situ rescue or insurance populations; and support population restoration through reintroduction (IUCN SSC, 2014).

Zoo leadership in pangolin conservation

How have these opportunities been explored for pangolins? Separate and distinct from zoos, conservation organizations such as Tikki Hywood Foundation, Zimbabwe and Save Vietnam's Wildlife (SVW) have championed the development of rescue, rehabilitation, and release protocols through experience gained across decades. Meanwhile, conservation zoos are multifaceted institutions with the capacity to host and initiate myriad ex situ management activities including rescue and rehabilitation, and have shown impactful leadership in both in situ and ex situ pangolin conservation. While not an exhaustive list, several examples are highlighted here.

Taipei Zoo and Wildlife Reserves Singapore (WRS) have operated rehabilitation and release programs for hundreds of Chinese (*Manis pentadactyla*) and Sunda pangolins (*M. javanica*) respectively, rescued from illegal trade and human-wildlife conflicts, such as traffic accidents, for more than a decade. Both zoos have cautiously worked with non-releasable pangolins to develop educational exhibits for the public and behind-the-scenes husbandry, captive breeding, and research programs. They have sponsored and conducted extensive in situ and ex situ research on dozens of topics ranging from hematologic and serum biochemical parameters (Khatri-Chhetri et al., 2017) and gestation periods (Chin et al., 2011), to methods for attaching radio transmitters to pangolins for post-release monitoring.

Taipei Zoo convened in 2004 the first Population and Habitat Viability Assessment (PHVA) workshop for the Chinese pangolin in Taiwan in collaboration with the IUCN SSC Conservation Planning Group (CPSG), formerly known as the Conservation Breeding Specialist Group (CBSG). This resulted in the first ever pangolin conservation action plan (see Chao et al., 2005). In 2017, the zoo hosted a follow-up workshop to review progress and formulate new recommendations (see Chapter 36). Meanwhile, WRS has played a particular role in sponsoring, organizing and hosting opportunities for pangolin experts to meet and collaborate in person. These include the first ever workshop on the trade and conservation of pangolins in 2008, led by TRAFFIC; the first international PSG conference in 2013; and both Singapore and Regional Sunda Pangolin Conservation Planning Workshops in 2017 (see Chapter 33).

The National Zoological Gardens of South Africa (NZG) conducts extensive education and science awareness activities on Temminck's pangolin (*Smutsia temminckii*). NGZ scientists successfully sequenced the entire mitochondrial genome of all four African pangolin species in order to inform policy and contribute to management decisions for the species. The NGZ also serves as the appointed forensics service provider for the South African Police Department. It has funded and organized capacity building initiatives such as training for Zambia National Parks and Wildlife on the use of molecular techniques to geo-reference African pangolins. The NZG does not hold any pangolins in captivity, and instead works closely with an approved rehabilitation center (A. Kotze, pers. comm.).

Examples of conservation zoos outside the native range of pangolins that have championed the species' conservation, despite not hosting live exhibits, include the Zoological Society of London (ZSL), UK, Australia's Taronga Zoo, and Houston Zoo in the United States. ZSL has hosted the PSG since its re-formation in 2012, providing fiscal sponsorship, support to the group's part-time Program Officer, and hosting the group's website, as well as undertaking pangolin conservation projects in Cameroon, China, Nepal, the Philippines and Thailand. In 2016, Taronga Zoo launched a Legacy Species campaign, dedicating their next decade to the conservation of ten critical species including the Sunda pangolin. This includes providing advice to SVW to enhance the well-being of rehabilitated pangolins and supporting their work to develop a disease risk strategy for pangolin releases (Taronga Zoo, 2019). Houston Zoo has provided financial and technical support to SVW since 2007, and to ecological research and capacity building endeavors in Sabah, Malaysia (P. Riger, pers. comm.).

Opportunities for increased zoo engagement in pangolin conservation

Conservation zoos have shown leadership in pangolin conservation, yet more is possible. The collective investment of international zoo networks, when organized and united with in situ stakeholders, can be a powerful force for species recovery. International conservation programs for the golden lion tamarin (*Leontopithecus rosalia*), giant panda (*Ailuropoda melanoleuca*), and Tasmanian devil (*Sarcophilus harrisii*) are notable examples of what can be achieved when international coalitions of diverse stakeholders work together to break through conservation barriers to secure species from extinction.

Golden lion tamarin conservation stakeholders pioneered collaborative conservation methods, beginning with the 1972 conference "Saving the Lion Marmoset," which included status reviews and the development of collective conservation recommendations, including a captive management and research program (Rylands et al., 2002). Following breakthroughs to understand and overcome factors that contributed to poor reproductive success in captivity, hundreds of golden lion tamarins were bred by dozens of zoos working collaboratively to ensure maximum genetic diversity of the captive population. Ultimately the wild population, once fewer than 200, was bolstered through reintroduction efforts, and by 2018 comprised ~3200 animals. The 500 golden lion tamarins managed in 150 zoos globally provide an insurance policy against extinction, critical for this species which is vulnerable to disease (e.g., yellow fever). Sustained investment from international conservation zoo networks remains critical to supporting Brazil's Associação Mico-Leão-Dourado's (AMLD) work to connect and protect fragmented golden lion tamarin forest habitat and secure the species from the threats of disease outbreaks, deforestation and illegal trade (Mickelburg and Ballou, 2013; Perez, 2018).

China's giant panda conservation program is another important example. The Chinese Association of Zoological Gardens (CAZG) initiated a biomedical survey for the giant panda in 1998 in order to address chronically poor reproductive success. Inspired by a CBSG workshop and pioneering biomedical surveys for cheetah (*Acinonyx jubatus*; Wildt et al., 2006), Sumatran tigers (*Panthera tigris sumatrae*; Tilson et al., 2001), and South China tigers (*P. t. amoyensis*; Traylor-Holzer et al., 2010), an international and interdisciplinary team of specialists from seven institutions with expertise in veterinary medicine, reproductive physiology, endocrinology, animal behavior, genetics, nutrition, and pathology evaluated 61 giant pandas in order to uncover barriers to well-being and reproduction. The survey uncovered significant new information that had a direct impact on improving giant panda husbandry and exponentially increased knowledge about the biology of the species (Wildt et al., 2006). Over time, and under the leadership of a Chinese scientific management committee, stakeholders collaborated to transform the global captive panda population into a demographically healthy population that can serve both as an insurance population against catastrophic decline in the wild and as a source for reintroduction or population reinforcement (Traylor-Holzer and Ballou, 2016). By 2018 the ex situ population included 548 individuals in 93 institutions worldwide with a well-balanced age and sex structure (K. Traylor-Holzer, pers. comm.). International zoo investment has contributed to improvements in the wild population, which by 2015 was estimated to have increased by 17% over a decade, and also supported expansion of the giant panda reserve system to 67 protected areas (Parker, 2005; Traylor Holzer and Ballou, 2016).

Building on these lessons, collaborative stakeholder efforts sparked hope for the Tasmanian devil, at risk of extinction due to a fast-spreading disease known as Devil Facial Tumor Disease (DFTD). Beginning in 2003, periodic government-organized workshops in collaboration with CPSG brought together field researchers, ecologists, disease specialists, non-governmental organizations (NGOs), zoo community leaders, cultural leaders, and policy makers to collectively plan multifaceted conservation initiatives. From breakthroughs in understanding and managing DFTD to the revamping of the captive population into a progressive, adaptively-managed insurance program that is actively contributing to the species' re-establishment in the wild, the initiative is designed to not only insure the species against extinction, but ultimately provide for full ecological recovery. As of 2014 the insurance program was comprised of approximately 500 individuals, with around 75% held in intensively managed zoos and other captive facilities, 20% in free-ranging enclosures and 5% released to a protected island site. Inclusion of diverse stakeholders from the beginning, including zoo industry leaders, as well as periodic opportunities to meet under the neutral and independent facilitation of CPSG have empowered significant and sustained investments from zoos (Lees et al., 2013).

Challenges and risks that impede zoo engagement in pangolin conservation

Given the contributions zoos have made to wildlife conservation, it is easy to see why international conservation zoo networks are keen to contribute to ex situ management of pangolins. However, various factors impede these opportunities. Zoo visitors may be disappointed to learn that pangolin exhibits in zoos are few, especially outside of pangolins' native ranges. This is because survival rates of pangolins brought into captivity have historically been poor, and despite advancements in pangolin husbandry in recent years, maintaining them continues to be challenging (see Chapter 28). Infrequent captive breeding successes have largely taken place in institutions within the native range of pangolins. Sensitive to these risks, WRS and Taipei Zoo have only engaged in captive breeding with animals deemed non-releasable following rescue and subsequent rehabilitation, although this may pre-dispose the captive population to lower survival and reproductive success.

To complicate matters further, zoos have tough critics, and public opinion wrestles with the morality of holding wild animals in captivity, and challenges the ethics and relevance of zoos (Grajal et al., 2018). Zoos may be perceived to have a conflict of interest between their conservation role and commercial status (McGowan et al., 2016). Carefully crafted stories of conservation successes may be superseded in the media by anti-zoo rhetoric (Maynard, 2018). Moreover, conservation zoos are not the norm. Of the approximately 16,000 zoos and aquariums worldwide, only about 400 are members of WAZA (Cress, 2018), and fewer than 1000 are accredited by AZA, EAZA, and ZAA (AZA, 2018; EAZA, 2018; ZAA, 2018). WAZA is not an accrediting body, but it develops progressive strategies that outline best practices, and it requires members to agree to a code of ethics and animal welfare (Barongi et al., 2015; Cress, 2018; WAZA, 2003). By 2023, WAZA will require that all of its institutional members are accredited by one of the regional zoo and aquarium associations (Cress, 2018).

Certain criticisms of zoos are not unfounded and highlight the importance of conservation zoos adhering to the highest animal care and welfare standards. And, it is important to recognize that not all threatened species benefit from ex situ management (IUCN SSC, 2014; McGowan et al., 2016). Snyder et al. (1996) highlight significant concerns with poorly considered captive breeding initiatives. These range from the need to continually source animals from the wild for unsustainable captive populations, to high costs, domestication, preemption of other recovery techniques, and poor success of reintroductions. Further concerns include disease outbreaks stemming from exposing wild animals to foreign habitats and species, and the challenges of maintaining administrative continuity in order to maintain a species in perpetuity. Regardless, conservation initiatives sometimes elect, in the face of catastrophic threats, to initiate captive breeding before a thorough examination of risks and benefits. Bowkett (2009) cautions that failure to acknowledge the "lessons from the past concerning how and when to employ captive breeding in species conservation risks the failure of recovery programs and, ultimately, the loss of species."

Appropriately sourcing pangolins for zoological purposes is particularly difficult. Pangolins are victims of such comprehensive exploitative pressure that even well-meaning conservation zoos may inadvertently contribute to overexploitation unless they are hosting or collaborating with a rescue and rehabilitation program for pangolins. This is a recognized challenge that zoos face for species targeted by wildlife traffickers (Cress, 2018; WAZA, 2014). Worst of all, unscrupulous entities may masquerade as "zoos" in order to launder wildlife through legal mechanisms (Beastall and Shepherd, 2013; Nijman and Shepherd, 2009; WAZA, 2014), similar to concerns about commercial wildlife farming entities (Brooks et al., 2010; Bulte and Damania, 2005). The WAZA Code of Ethics and Animal Welfare recognizes that sourcing animals from the wild may be appropriate under certain conditions (WAZA, 2003). However, on the defense against anti-zoo interests convinced that there is no such thing as a good zoo—that animals are "better off dead than captive-bred" (Barongi, 2018), zoos may become secretive in their acquisition practices, a phenomenon that may jeopardize open communication and effective collaboration with other stakeholders.

A schism in species conservation planning may emerge when ex situ facilities and other stakeholders do not collaborate sufficiently in the development of species management plans (Traylor-Holzer et al., 2018). Insufficient collaboration between ex situ facilities, the field community and other stakeholders may exacerbate inherent risks: ex situ populations may not be appropriately structured to recover the species, and conversely, imperiled species may not have the opportunity to benefit from ex situ management when it is needed. Ashe (2018) describes a bias among environmental and academic communities regarding the practice of conservation breeding, which may have contributed to the loss of ex situ management opportunities for zoos to intervene in the imminent extinction of the vaquita porpoise (*Phocoena sinus*), the bycatch victim of relentless totoaba (*Totoaba macdonaldi*) poaching and trafficking. This bias may exacerbate the preexisting likelihood of tension between ex situ and in situ conservation practitioners. Ultimately, the schism may develop into a conflict between conservation stakeholders.

Conflicts within conservation, where stakeholders clash over objectives and parties may be perceived as acting at the expense of each other's interests, are a recognized challenge (Madden and McQuinn, 2014; Redpath et al., 2013). Redpath et al. (2013) warn of the challenges of preventing such disagreements from devolving into damaging conflicts, as well as the significant investments required to manage and minimize destructive conflicts when they do emerge. Some stakeholders in the diverse pangolin conservation community are predisposed towards conflict around what, if any role captive populations should play in pangolin conservation. This became evident in 2016 when a group of zoos in the United States, collectively known as the Pangolin Consortium, imported wild-sourced white-bellied pangolins (*Phataginus tricuspis*) from Togo into the United States for the purposes of research and captive breeding (Pangolin Consortium, 2019). The actions of these zoos drew public criticism from certain environmental NGOs and pangolin experts in range states (Cassidy, 2017; Hywood, 2018; Pepper, 2017). Management of this conflict required significant time and resources for pangolin conservation stakeholders. The Pangolin Consortium subsequently committed to ceasing pangolin imports (Hywood, 2018). It has commenced a grant-making initiative and is retroactively applying the IUCN Ex Situ Guidelines, such as by convening a symposium on opportunities for ex situ management (Pangolin Consortium, 2019).

Capping off these challenges, conservation zoos with financial and in-kind resources that could benefit pangolins, conscious of potential sensitivities and controversies, may be reluctant to get involved in pangolin conservation at all, and pangolin experts employed by zoos and governments may find themselves caught in the middle.

The way forward

Conservation zoos have shown leadership in pangolin conservation, and conservation zoo networks have a proven track record of contributing to species recovery when they collaborate effectively with the field community and other stakeholders. To ensure pangolins benefit from ex situ management activities that align with and support in situ conservation efforts and goals, it is paramount that plans to undertake ex situ activities with live pangolins are developed collectively through decision making processes that involve all important stakeholders. Given historically poor survival rates of pangolins in captivity, ongoing challenges of maintaining the species in captivity, risks of poorly considered captive breeding programs, sensitivities around sourcing pangolins for zoological purposes, and pre-disposition of some pangolin conservation stakeholders to disagree around what, if any, role captive populations should play in pangolin conservation, transparency and collaboration with all important stakeholders is essential to ensure synergy of ex situ management activities with in situ conservation goals.

Byers et al. (2013) and Traylor-Holzer et al. (2018) describe an integrated One Plan Approach to species conservation, wherein species management plans are developed jointly by "all responsible parties to produce a single, comprehensive plan" for species, with the ultimate goal of supporting its conservation in the wild. The One Plan Approach incorporates the IUCN Ex Situ Guidelines, which articulate a five-step decision making process to consider opportunities for ex situ activities and ensure that the tools of ex situ conservation complement, and do not undermine field conservation (Byers et al., 2013; IUCN SSC, 2014; Traylor-Holzer et al., 2018). This structured, scientific and comprehensive assessment process, which involves all of the important in situ and ex situ stakeholders, enables species conservation communities to develop an understanding of the potential benefits, costs, risks and feasibility of ex situ management for a species. Opportunities for ex situ activities are neither automatically dismissed nor included in this process, rather they are systematically evaluated. The five steps involve (1) compiling a status review of the species, including a threat analysis; (2) defining the potential role(s) that ex situ management may play in the overall conservation of the species; (3) determining the characteristics of the ex situ population needed to fulfill the conservation role(s); and (4) defining the resources and expertise needed for the ex situ management program to meet its role(s) and appraising the feasibility and risks involved. Only after this evaluation is completed is the fifth and final step of making an informed and transparent decision taken. The decision to include ex situ management in the conservation strategy for a species should be determined by weighing the potential conservation benefit to the species against the likelihood of success and overall costs and risks of the proposed ex situ program, as well as alternative conservation actions or inaction (IUCN SSC, 2014).

The IUCN Ex Situ Guidelines can be applied to existing captive populations (IUCN SSC, 2014). This is appropriate for pangolins already held in ex situ facilities, which should be included in the initial status review of the species. However, the ideal process is to proactively apply this five-step decision process before acquiring pangolins for zoological purposes except in cases of emergency acquisition (e.g., the accommodation of rescued pangolins). Following the five-step decision-making process within the Guidelines in order will ensure that ex situ management activities are aligned with conservation goals. It will be critical to achieve collective agreement before taking high-risk actions such as removing pangolins from the wild for zoological purposes. Engaging in a collaborative process with all important stakeholders through the One Plan Approach will minimize the risk of costly disagreements around the role of captive populations in conservation. Zoological facilities skipping these steps risk fostering the perception that they are taking unilateral action to obtain pangolins in order to satisfy personal or institutional agendas rather than for bona fide conservation purposes.

Face-to-face engagement, such as through a facilitated multi-stakeholder workshop, is recommended to support this process. While the costs of bringing together international stakeholders in person may seem prohibitive, the benefits outweigh the initial investment as face-to-face contact facilitates the development of rapport and mutual cooperation, which can significantly improve the chances for more satisfactory outcomes even when participants have different motives (Drolet and Morris, 2000; Emerson et al., 2009). Factors that optimize the chances for positive workshop outcomes, especially where preventing or minimizing conflict is a concern, include facilitation from a neutral third-party that is responsible for guiding participants through a structured, participatory, deliberative process (Madden and McQuinn, 2014; Redpath et al., 2013). When diverse groups of wildlife conservation stakeholders, including zoos, have come together under the leadership of strong, neutral third party facilitation, and together followed a structured, deliberative planning process to strive for a shared goal, species conservation communities have overcome conflicts, developed plans collectively and formed collaborative partnerships whose outcomes have surpassed their initial goals and expectations (CBSG, 2017).

The PSG has commenced a collaborative process with the CPSG and other stakeholders to develop conservation strategies for the eight pangolin species, with the ultimate aim of one consolidated global strategy for pangolin conservation. This process includes conservation planning workshops where appropriate and feasible (see Chapter 33). Thus far, four workshops have taken place: a Southeast Asia regional workshop for the Sunda pangolin (IUCN SSC Pangolin Specialist Group et al., 2018); a national workshop for the Sunda pangolin in Singapore (Lee et al., 2018); a national workshop for the Philippine pangolin in the Philippines (IUCN SSC Pangolin Specialist Group, 2018); and a PHVA workshop for the Chinese pangolin in Taiwan, which reviewed progress since the analogous 2004 PHVA workshop (see Chapter 36). A range of interventions were identified at these workshops to address priorities such as combatting trafficking, building capacity for effective law enforcement, and engagement with local communities (see Chapters 33 and 36; Lee et al., 2018). Consensus has not yet been reached on what role rescued pangolins should play in conservation (IUCN SSC Pangolin Specialist Group, 2018). An Ex Situ Conservation Needs Assessment Workshop was therefore held in Thailand in 2019 to provide in situ and ex situ Sunda pangolin conservation stakeholders with an opportunity to together undertake the five-step decision making process of the IUCN Ex Situ Guidelines, and set the stage for similar evaluations for other pangolin species.

Likewise, future conservation planning endeavors for pangolins should include an evaluation of ex situ management opportunities through the lens of the IUCN Ex Situ Guidelines. To ensure recommendations are sustainable and acceptable to in situ and ex situ stakeholders, both should participate in the process under the framework of the One Plan Approach and commit to following the steps of the IUCN Ex Situ Guidelines. Application of the five-step decision making process may or may not ultimately result in recommendations for captive breeding, ambassador animal exhibits, husbandry research, or other ex situ management activities. The Regional Conservation Strategy developed for the Sunda pangolin recommends the formation of a PSG task force to consider matters with captive pangolins (see IUCN SSC Pangolin Specialist Group, 2018). A logical action item for that task force would be to assess the feasibility of developing a centralized global studbook or similar organizing system so that registered captive pangolin information could be managed for conservation purposes.

Acknowledging pre-existing conflicts and incorporating supportive measures into conservation planning processes can empower pangolin conservation stakeholders to move beyond entrenched conflicts, where they exist, to collaborative engagement with conservation zoos. CPSG workshops are designed to provide a supportive environment to surface and address underlying conflicts, and CPSG facilitators are trained to support participants in navigating that process (CBSG, 2017). Appropriate pre-planning prior to each multi-stakeholder process is critical to increase the probability of long-lasting outcomes. The principles and techniques of conservation conflict transformation (CCT) are important tools available to support workshop planners in creating the conditions of a collaborative environment and designing processes that support the participation of diverse stakeholders (Madden and McQuinn, 2014). The Center for Conservation Peace Building offers resources and training that can build the capacity of workshop conveners and pangolin conservation practitioners in developing conservation conflict transformation skills (CCPB, 2018).

While conservation strategies are under development for pangolins, there are many ways conservation zoos can make a difference for pangolins and contribute to their conservation. The Zoological Society of London, Houston Zoo, and Taronga Zoo provide excellent examples of how conservation zoos can have a substantive impact on pangolin conservation without having pangolins in their collections. Pangolin conservation endeavors require ongoing investment, and myriad opportunities to support field conservation exist. Contributing to field programs not only provides much-needed funding, but also can serve to build bridges with the field community to strengthen collaboration. Zoos can also help to raise awareness of the threats pangolins face, for example by participating in World Pangolin Day (see Chapter 21). Further, Grajal et al. (2018) offer recommendations for developing educational strategies that can encourage the adoption of pro-environmental behaviors. ZACC (Zoos and Aquariums Committing to Conservation) (2018) recommend actions that zoos can take to combat wildlife trafficking. The WAZA Conservation Strategy offers an overview of best practices in zoo operations and guidance on how to strive for conservation outcomes, including recommendations for influencing behavior change for conservation (Barongi et al., 2015).

Finally, a critical action that conservation zoos, governments, other decision makers and donors can undertake is to support the development of integrated conservation strategies for pangolins. Once conservation strategies are developed, conservation zoos will have the opportunity to support and participate in their implementation. Recommendations that result from the collective thinking of all important stakeholders within the framework of the One Plan Approach can provide a road map to ongoing positive zoo engagement in pangolin conservation.

Acknowledgments

Thanks to Dr. David Wildt (Smithsonian Conservation Biology Institute), Lou Ann Dietz and Dr. James Dietz (Save the Golden Lion Tamarin) for their advice and encouragement in the early stages of our process. Dr. Peter Riger (Houston Zoo), Dr. Antoinette Kotze (South African National Biodiversity Institute (SANBI)), Dr. Leanne Wicker (Zoos Victoria), Dr. Dan Challender (IUCN SSC Pangolin Specialist Group), Paul Thomson (Save Pangolins), Flora Hsuan Yi Lo and Dr. Jason Shih-Chien Chin (Taipei Zoo) generously shared information. Dr. Kathy Traylor-Holzer (Conservation Planning Specialist Group) provided substantive and helpful review and feedback.

References

Ashe, D., 2018. The Immeasurable Distance Between Late and Too Late: Learning From the Ex Situ Attempt for the Vaquita. IUCN SSC Quarterly Report, June 2018, pp. 24–25. Available from: <https://www.iucn.org/sites/dev/files/media-uploads/2018/09/iucn_ssc_quarterly_report_jun2018_web.pdf>. [October 28, 2018].

AZA (Association of Zoos and Aquariums), 2018. Currently Accredited Zoos and Aquariums. Available from: <https://www.aza.org/current-accreditation-list>. [September 21, 2018].

Barongi, R., 2018. Committing to conservation: can zoos and aquariums deliver on their promise? In: Minteer, B.A., Maienschein, J., Collins, J.P. (Eds.), The Ark and Beyond: The Evolution of Zoo and Aquarium Conservation. University of Chicago Press, Chicago and London, pp. 108–121.

Barongi, R., Fisken, F.A., Parker, M., Gusset, M. (Eds.), 2015. Committing to Conservation: The World Zoo and Aquarium Conservation Strategy. WAZA Executive Office, Gland.

Beastall, C., Shepherd, C., 2013. Trade in 'captive bred' echidnas. TRAFFIC Bull. 25 (1), 16–17.

Bowkett, A., 2009. Recent captive-breeding proposals and the return of the ark concept to global species conservation. Conserv. Biol. 23 (3), 773–776.

Brooks, E.G., Roberton, S.I., Bell, D.J., 2010. The conservation impact of commercial wildlife farming of porcupines in Vietnam. Biol. Conserv. 143 (11), 2808–2814.

Bulte, E.H., Damania, R., 2005. An economic assessment of wildlife farming and conservation. Conserv. Biol. 19 (4), 1222–1233.

Byers, O., Lees, C., Wilcken, J., Schwitzer, C., 2013. The one plan approach: the philosophy and implementation of CBSGs approach to integrated species conservation planning. WAZA Mag. 14, 2–5.

Cassidy, R., 2017. Pangolin Conservation Won't Be Achieved in American Zoos. Pittsburgh Post-Gazette. Available from: <https://www.post-gazette.com/opinion/letters/2017/06/01/Pangolin-conservation-won-t-be-achieved-in-American-zoos/stories/201706010017>. [September 21, 2018].

CBSG (Conservation Breeding Specialist Group), 2017. Second Nature: Changing the Future of Endangered Species. IUCN SSC Conservation Breeding Specialist Group, St. Paul, Minnesota.

CCPB (Center for Conservation Peace Building), 2018. Center for Conservation Peace Building. Available from: <https://cpeace.ngo/>. [September 21, 2018].

Chao, J.-T., Tsao, E.H., Traylor-Holzer, K., Reed, D., Leus, K. (Eds.), 2005. Formosan Pangolin Population and Habitat Viability Assessment: Final Report. IUCN SSC Conservation Breeding Specialist Group, Apple Valley, Minnesota.

Chin, S.-C., Lien, C.-Y., Chan, Y.-T., Chen, C.-L., Yang, Y.-C., Yeh, L.-S., 2011. Monitoring the gestation period of rescued Formosan pangolin (*Manis pentadactyla pentadactyla*) with progesterone immunoassay. Zoo Biol. 31 (4), 479–489.

Cress, D., 2018. Keynote Speaker. Zoos and Aquariums Committing to Conservation Conference, 26 January, Jacksonville, Florida.

Drolet, A.L., Morris, M.W., 2000. Rapport in conflict resolution: accounting for how face-to-face contact fosters mutual cooperation in mixed-motive conflicts. J. Exp. Soc. Psychol. 36 (1), 26–50.

EAZA (European Association of Zoos and Aquariums), 2018. Accreditation. Available from: <https://www.eaza.net/members/accreditation/>. [September 21, 2018].

Emerson, K., Orr, P.J., Keyes, D.L., McKnight, K.M., 2009. Environmental conflict resolution: evaluating performance outcomes and contributing factors. Conflict Resolution Quart. 27 (1), 27–64.

Grajal, A., Luebke, J.F., Kelly, L.A.D., 2018. Why zoos have animals: exploring the complex pathyway from experiencing animals to pro-environmental behaviors. In: Minteer, B.A., Maienschein, J., Collins, J.P. (Eds.), The Ark and Beyond: The Evolution of Zoo and Aquarium Conservation. University of Chicago Press, Chicago and London, pp. 192–203.

Gray, J., 2017. Zoo Ethics: The Challenges of Compassionate Conservation. CSIRO Publishing, Ithaca and London.

Gusset, M., Dick, G., 2011. The global reach of zoos and aquariums in visitor numbers and conservation expenditures. Zoo Biol. 30 (5), 566–569.

Hywood, L., 2018. Response to Mongabay pangolin article. Available from: <https://news.mongabay.com/2018/01/u-s-zoos-learn-how-to-keep-captive-pangolins-alive-helping-wild-ones/>. [September 21, 2018].

IUCN SSC (IUCN Species Survival Commission), 2014. Guidelines on the Use of Ex Situ Management for Species Conservation. Version 2.0. IUCN Species Survival Commission, Gland, Switzerland.

IUCN SSC Pangolin Specialist Group, 2018. Scaling up Palawan Pangolin Conservation – Developing the First National Conservation Strategy for the Species. Available from: <https://www.pangolinsg.org/2018/04/23/scaling-up-palawan-pangolin-conservation-developing-the-first-national-conservation-strategy-for-the-species/>. [September 21, 2018].

IUCN SSC Pangolin Specialist Group, IUCN SSC Asian Species Action Partnership, Wildlife Reserves Singapore, IUCN SSC Conservation Planning Specialist Group, 2018. Regional Sunda Pangolin (*Manis javanica*) Conservation Strategy 2018-2028. IUCN SSC Pangolin Specialist Group, Zoological Society of London, London, UK.

Khatri-Chhetri, R., Chang, T.-C., Khatri-Chhetri, N., Huang, Y.L., Pei, K.J.C., Wu, H.Y., 2017. A retrospective study of pathological findings in endangered Taiwanese pangolins (*Manis pentadactyla pentadactyla*) From Southeastern Taiwan. Taiwan Vet. J. 43 (1), 55–64.

Lee P.B., Chung Y.F., Nash H.C., Lim N.T-L., Chan S.K.L., Luz S., et al., 2018. Sunda Pangolin (*Manis javanica*) National Conservation Strategy and Action Plan: Scaling Up Pangolin Conservation in Singapore. Singapore Pangolin Working Group, Singapore.

Lees, C., Andrews, P., Sharman, A., Byers, O., 2013. Saving the devil; one species, one plan. WAZA Mag. 14, 37–40.

Madden, F., McQuinn, B., 2014. Conservation's blind spot: the case for conflict transformation in wildlife conservation. Biol. Conserv. 178, 97–106.

Maynard, L., 2018. Media framing of zoos and aquaria: from conservation to animal rights. Environ. Commun. 12 (2), 177–190.

McGowan, P.J., Traylor-Holzer, K., Leus, K., 2016. IUCN guidelines for determining when and how ex situ management should be used in species conservation. Conserv. Lett. 10 (3), 361–366.

Mickelburg, J., Ballou, J.D., 2013. The golden lion tamarin conservation programme's one plan approach. WAZA Mag. 14, 2–5.

Nijman, V., Shepherd, C.R., 2009. Wildlife trade from ASEAN to the EU: issues with the trade in captive-bred reptiles from Indonesia. TRAFFIC Europe, Brussels, Belgium.

Pangolin Consortium, 2019. Care and Conservation Through Collaboration: The Pangolin Consortium Story. Available from: <http://zaa.org/members/membonly/images/pangolin.pdf>. [January 19, 2019].

Parker, K., 2005. State of the Panda Policy 2005: An Overview of the United States Policy on Giant Panda Import Permits and A Review and Analysis of Conservation Projects in China Sponsored by US Zoos. Unpublished Scholarly Paper, University of Maryland, Maryland, United States.

Pepper, E., 2017. Zoos Take a Step Backward in Pangolin Conservation, Scientific American. Available from: <https://blogs.scientificamerican.com/observations/zoos-take-a-step-backward-in-pangolin-conservation/>. [September 21, 2018].

Perez, L.P., 2018. Golden Lion Tamarin Conservation Program: history and challenges, the strategic role of zoos. Zoos and Aquariums Committing to Conservation Conference, 24 January, Jacksonville, Florida.

Redpath, S.M., Young, J., Evely, A., Adams, W.M., Sutherland, W.J., Whitehouse, A., et al., 2013. Understanding and managing conservation conflicts. Trends Ecol. Evol. 28 (2), 100–109.

Rylands, A.B., Mallinson, J.J.C., Kleiman, D.G., Coimbra-Filho, A.F., Mittermeier, R.A., Damara, R.A., et al., 2002. A history of lion tamarin research and conservation. In: Kleiman, D.G., Rylands, A.B. (Eds.), Lion Tamarins: Biology and Conservation. Smithsonian Institution Press, Washington D.C. and London, pp. 3–41.

Snyder, N.F., Derrickson, S.R., Beissinger, S.R., Wiley, J.W., Smith, T.B., Toone, W.D., et al., 1996. Limitations of captive breeding in endangered species recovery. Conserv. Biol. 10 (2), 338–348.

Taronga Zoo, 2019. Our Legacy Commitment. Available from: <https://taronga.org.au/conservation-and-science/our-legacy-commitment>. [January 6, 2019].

Tilson, R., Traylor-Holzer, K., Brady, G., Armstrong, D., Byers, O., Nyhus, P., 2001. Training, transferring technology, and linking in situ and ex situ tiger conservation in Indonesia. In: Conway, W., Hutchins, M., Souza, M., Kapetanekos, Y., Paul, E. (Eds.), AZA Field Conservation Resource Guide. Zoo Atlanta, Atlanta, pp. 245–255.

Traylor-Holzer, K., Ballou, J.D., 2016. Is conservation really black and white? WAZA News 16 (1), 2–7.

Traylor-Holzer, K., Xie, Z., Yin, Y., 2010. The struggle to save the last South China tigers. In: Tilson, R., Nyhus, P. (Eds.), Tigers of the World, second ed. Academic Press, London, Burlington, San Diego, pp. 457–461.

Traylor-Holzer, K., Leus, K., Byers, O., 2018. Integrating ex situ management options as part of a one plan approach to species conservation. In: Minteer, B.A., Maienschein, J., Collins, J.P. (Eds.), The Ark and Beyond: The Evolution of Zoo and Aquarium Conservation. University of Chicago Press, Chicago and London, pp. 129–141.

WAZA (World Association of Zoos and Aquariums), 2003. Code of Ethics and Animal Welfare. Available from: <http://www.waza.org/files/webcontent/1.public_site/5.conservation/code_of_ethics_and_animal_welfare/Code%20of%20Ethics_EN.pdf>. [September 21, 2018].

WAZA (World Association of Zoos and Aquariums), 2014. Resolution 69.1: Legal, Sustainable and Ethical Sourcing of Animals. Available from: <https://aboutzoos.info/images/stories/files/WAZA_resolution_69-1_animal-sourcing.pdf>. [September 21, 2018].

Wildt, D.E., Zhang, A., Zhang, H., Xie, Z., Janssen, D., Ellis, S., 2006. The giant panda biomedical survey: how it began and the value of people working together across cultures and disciplines. In: Wildt, D.E., Zhang, A., Zhang, H., Janssen, D.L., Ellis, S. (Eds.), Giant Pandas: Biology, Veterinary Medicine, and Management. Cambridge University Press, New York, pp. 17–36.

ZAA (Zoo Aquarium Association), 2018. Membership. Available from: <https://www.zooaquarium.org.au/index.php/membership/>. [September 21, 2018].

ZACC (Zoos and Aquariums Committing to Conservation), 2018. ZACC Conference 2018. Available from: <https://zaccjax.weebly.com/>. [October 30, 2018].

CHAPTER

32

Evaluating the impact of pangolin farming on conservation

Michael 't Sas-Rolfes[1] *and Daniel W.S. Challender*[2]

[1]School of Geography and the Environment and Oxford Martin School, University of Oxford, Oxford, United Kingdom [2]Department of Zoology and Oxford Martin School, University of Oxford, Oxford, United Kingdom

OUTLINE

Introduction

Pangolins have been exploited for consumptive use throughout human history (Chapters 14–16). Although national legislation in most range states prohibits commercial exploitation of the species, they are hunted or harvested live (both legally and illegally), consumed locally (Ingram et al., 2018), and trafficked internationally in high volumes, mainly to China and Vietnam (Chapter 16). Following increasing concern about the impact of ongoing exploitation on populations, including the emergence of intercontinental trafficking of African pangolin parts to Asia around 2008 (Challender and Hywood, 2012; Heinrich et al., 2016), all pangolin species were included in Appendix I of CITES, the Convention on International Trade in Endangered Species of Wild Fauna and Flora, at CoP17 in 2016. Accordingly, international commercial trade in wild pangolins and their derivatives is prohibited (Chapter 19).

However, there is widespread evidence of continued, and possibly even increasing, demand

Pangolins
DOI: https://doi.org/10.1016/B978-0-12-815507-3.00032-0

for pangolins among individual consumers in various countries in Africa and Asia (Boakye et al., 2015; Ingram et al., 2018; Shairp et al., 2016). Furthermore, the governments of some countries appear deeply committed to the use of pangolins and their parts. As an example, China has taken measures to regulate domestic trade and use of pangolin scales through a certification system established in 2007 (Zhang, 2009). Pangolins also have an important role as a protein source in both Africa and Asia, and contribute to local livelihoods through income generation, be it legal or illegal (Boakye et al., 2016; D'Cruze et al., 2018).

Conversely, a number of actors oppose the commercial trade and consumptive use of pangolins or their derivatives. This includes some pangolin range states and non-range parties to CITES, which are concerned about high levels of poaching and trafficking and are keen to avoid the commercial overexploitation of the species. It also includes actors concerned that any permitted trade in stockpiles of derivatives (e.g., scales) that range states have acquired, would facilitate the "laundering" of scales from wild-caught animals declared as having come from stockpiles. Additionally, a suite of mainly northern and western, primarily protectionist and animal welfare non-governmental organizations (NGOs) appear philosophically opposed to any consumptive use of pangolins and their derivatives for any purpose in any context.

However, theory, economic and otherwise, suggests that where there is demand for commodities, there will be actors willing to provide a supply, generally motivated by the prospect of earning a financial profit (Marshall, 2009). This is true even if such activity is illegal, in which case actors typically respond to the probability of incurring a punishment (Becker, 1968) in the context of prevailing social norms and relevant regulatory frameworks. Illegal activity thus persists wherever there is no consensus on the legitimacy of legal status, at a level determined by the extent to which the profits of such activities exceed their perceived potential costs. The persistence of illegal activity has been observed with myriad banned wildlife products and other commodities, notably drugs and alcohol, but also regular consumer goods when subject to unpopular regulations which has stimulated organized crime (e.g., Glenny, 2008).

Although regulation is typically used to control actors and their access to natural resources, it has limitations (e.g., resource constraints, political limitations; Challender and MacMillan, 2016) while demand reduction and behavior change have not yet been proven as an effective means of conservation intervention (Veríssimo and Wan, 2019; see Chapter 22). Moreover, at the global level, where there is a demand for wildlife products, including those listed in CITES, the solution being implemented for many species is artificial propagation or captive breeding (Harfoot et al., 2018). Whereas for a number of species this approach appears to have a limited (e.g., African lion [*Panthera leo*]; Williams et al., 2017), or even a largely positive (e.g., crocodilians; Hutton and Webb, 2003) effect on wild populations, for others, this is contested (e.g., bears; Dutton et al., 2011) or clearly not the case (e.g., xaté palm [*Chamaedorea ernesti-augusti*]; Williams et al., 2014). Furthermore, commercial captive breeding raises other concerns such as genetic effects, especially with leakage into wild populations or reintroductions (e.g., Araki et al., 2007).

This chapter discusses whether pangolin farming may help or hinder the conservation of wild pangolins. The discussion is placed in the context of certain plausible scenarios that may emerge in the next few decades. Broadly speaking these are (1) that farming pangolins is possible and taken to scale (e.g., in China) through the application of bio-technological and other solutions to current captive breeding challenges; (2) that there is a worldwide ban on all sales of pangolin products; and (3) that behavior change campaigns are successful leading to material changes in consumer preferences. These scenarios are defined somewhat loosely and are distinct from more precise technical configurations used in conventional scenario planning. The remainder of the chapter discusses how pangolin farming might evolve in relation to key changing variables, i.e., factors that require consideration when assessing the potential impact of farming on wild animal populations, and then reviews the basic theory of economic incentives and the theoretical underpinning for understanding potential responses to farming. Finally, it considers what the application of this theory in relation to the key factors suggests might transpire for pangolin farming and its impact on wild populations in the future.

Will pangolin farming help or hinder the conservation of wild pangolins? Key variables to consider

Determining whether farming pangolins will help or hinder the conservation of the species in the wild requires consideration of how farming might evolve in relation to key variable factors. These include:

- The extent to which farming is facilitated or suppressed by domestic legislation. Pangolins are protected by national legislation in most range states. This prohibits commercial exploitation of wild populations but not captive breeding and farming, which is encouraged in some range states as a legitimate means of conservation (e.g., China; Harris, 2009). However, this is not a binary consideration because farming for commercial production may be outlawed, but captive breeding may not be, resulting in an ambiguous outcome. This is currently the case with captive tigers (*Panthera tigris*) in China.
- The extent to which domestic trade is allowed or suppressed through laws. Again, this is not a simple binary consideration: trade may be allowed but heavily controlled and/or taxed, thereby creating incentives for parallel illegal activity.
- The extent of law enforcement and related ability to launder wild products through a farming system. Not only do laws matter, but so does the extent to which they are enforced, from monitoring through to effective conviction and punishment of offenders. This in turn depends on both the extent to which they are socially accepted and the economic affordability of effective enforcement.
- Consumer preferences for pangolin products. These may vary and/or change over time, both quantitatively and qualitatively. For example, relative preferences between products from farmed and wild pangolins may change over time.
- The impact of stockpile policy. Accumulated stockpiles of pangolin scales and other derivatives and the sale or destruction thereof may also indirectly influence the relative economic competitiveness of farming and wild harvesting.
- The impact of changing technology. New technologies may change the relative economic competitiveness of farming versus wild harvesting over time, for instance owing to bio-technological solutions.

Theory of wildlife harvesting, legal supply and illegal trade

To address the question of how pangolin farming might help or hinder the conservation of wild populations, established theory of wildlife harvesting and farming provides some guidance. Following seminal articles by Hotelling (1931) and Gordon (1954), the discipline of resource economics developed theoretical tools for understanding the effects of commercial exploitation of natural resources and consequent implications for conservation. This early work identified the crucial role of enforced legal rights to property and access in regulating exploitation of a common-pool resource, an issue that was highlighted to scientists by Hardin (1968) in his "Tragedy of the Commons" article. As Ostrom (1990) and others have pointed out, Hardin's article describes an unregulated open access problem rather than communal ownership and there are many instances of well-regulated resource management under collective arrangements. Nonetheless, when considering the sustainability of wildlife harvesting regimes, it is vitally important to understand the specific institutional arrangements under which management of the species in question takes place, given that institutions have a non-trivial effect on harvesting incentives (Barrett et al., 2001; 't Sas-Rolfes, 2017). In this context, "institutions" are defined as "the humanely devised constraints that structure political, economic and social interaction" (North, 1991).

When analyzing the relative incentives for harvesting and farming it is therefore important to identify the specified legal rights to property and associated access to natural resources of different actors, and to understand what arrangements and rules are in place for the management of given species in the wild across all relevant jurisdictions. Understanding this institutional landscape for pangolins is important because different conditions will provide different actors with incentives and disincentives to take part in, for instance, pangolin farming or pangolin poaching, trafficking and associated activities. The extent to which there are differences between countries will also influence harvesting and trade dynamics. As a general rule, countries that create and enforce stronger property rights (whether collective or individual) over wild pangolins would be expected to perform better at conserving them than those that continue to treat them as open access resources.

With a strong focus on fisheries, Clark (1973, 2010) developed the field of resource economics further, specifically to examine the overexploitation of biological resources under open access conditions. This work, combined with insights from the economics of crime (Cook, 1977; Ehrlich, 1973) formed a basis for modeling the incentives for illegal wildlife harvesting of terrestrial species (e.g., Milner-Gulland and Leader-Williams, 1992). Subsequent refinements to Clark's work (Barbier and Schulz, 1997; Swanson, 1994) included habitat conservation and management considerations for large terrestrial species. It differentiated these cases from open access fisheries in important ways and suggested a potential role for appropriately regulated sustainable legal harvesting as a means to compete with and crowd out illegal harvesting.

Although pangolins are terrestrial species, it is unclear to what extent (if any) proceeds from wild harvesting might be re-invested in management and protection, even though there are implied costs of protecting both pangolins and their habitat. There have also been no formal attempts at bio-economic modeling of wild pangolin harvesting to determine the extent to which sustainable harvests are feasible and under what conditions. Such research could provide useful indicators of the relative feasibility of controlled legal wild harvesting versus illegal harvesting, both of which could then be compared with the economic feasibility of farming under various different scenarios. If such scenarios do not allow for proceeds of legal wild harvesting to be reinvested in conservation, then the principal mechanism by which such harvesting would impact illegal trade is by direct competition in the consumer marketplace. In other words, sales of legal pangolin products would impact purchases of illegal products to the extent that they might offer better value for money (i.e., a combination of perceived better quality and lower selling prices) to consumers.

Consistent with this competitive approach, the notion that a legal supply of a commercially valuable wild harvested resource may exert downward pressure on black market prices, and thereby reduce incentives for illegal harvesting, follows elementary economic principles and provides the basis of a "supply-side approach" to conservation, as proposed by Bergstrom (1990). Simply interpreted, this approach implies that introducing additional stocks of a wildlife product into the market – whether from legal harvesting, stockpiles (legal or illegal), or farming – will benefit wild populations of given species. However, this apparent justification for lifting trade bans and promoting legal trade in species products is subject to certain caveats. These are highlighted by Fischer (2004) who describes four theoretical conditions, the presence of which may undermine conservation objectives when legalizing a previously illegal trade in an endangered species product. These four conditions can be labeled as (1) existence of parallel markets, (2) stigma effects, (3) laundering potential, and (4) countervailing illegal supply cost effects.

The parallel markets condition would exist if, for example, newly introduced farmed pangolin products were not purchased by existing consumers of wild pangolin meat and scales (and associated products), instead finding their way to a completely new set of consumers. A stigma effect would be said to exist if an existing ban was accepted by consumers as a credible indication that pangolin products had no benefits, health or otherwise; under this condition, lifting a ban might signal to consumers that these products are newly worthy of purchase. The existence of stigma effects for pangolin products remains unproven. Laundering is defined as the (typically unlawful) introduction of illegally obtained products into legal supply chains and markets, such that the legal status of the products is changed, with the illegal sources of such products typically concealed from end users. Laundering is considered a pervasive problem for many illegally harvested wildlife products for which farming is permitted (see Hinsley et al., 2017; Hinsley and Roberts, 2018). Finally, in the case of products for which some legal trade is allowed, there is an argument that the existence of this legal market enables outright smuggling and illegal consumer purchases (in addition to possible laundering) by complicating law enforcement. This argument holds that smuggling and other transactions costs are effectively lowered for products that are completely illegal.

Fischer's four conditions highlight issues that must be considered (and ideally empirically determined) when evaluating the merits of trade policy and designing appropriate mitigation measures (e.g., traceability mechanisms). Furthermore, when considering the extent to which farmed products may displace the demand for wild products, it is crucial to understand the extent to which consumers are willing to substitute the two types of product for one another and at what price. Although consumers may claim to prefer wild-harvested products to ones from farmed animals, such preferences are also shaped by differences in price between, and access to, the two products. Many consumers are likely to buy supposedly inferior farmed products if they are cheaper or more readily available. The extent to which such substitution actually takes place is poorly researched for terrestrial wildlife products, including pangolins, but has been examined in some detail for seafood products (see Brayden et al., 2018). The choice is further complicated when the farmed products are certified as genuine, legal, and sustainable, and the competing "wild" products are of unknown provenance, and therefore potentially fake or illegally and unsustainably sourced.

In the case of pangolins, global policy is moving in the direction of prohibition, not the opposite, in attempts to stigmatize consumption and trade, and penalize those continuing to trade internationally and within certain countries. It is unclear as to how these moves toward prohibition might affect market prices for illegally harvested pangolin products. Whether they will drop or increase will depend at least in part on the persistence of consumer demand for all pangolin products. If consumer demand persists and prices remain high, adding a legal supply from farmed products could suppress market prices for wild-harvested products. However, applying Fischer's analysis to the situation, farmed products might simply supply a separate parallel market. Furthermore, the continued existence of some legal markets may also undermine attempts to stigmatize products, provide opportunities for laundering illegally harvested pangolin products, and complicate domestic law enforcement efforts in consumer countries for products already within country borders.

Recent theoretical work by economists identifies other very specific circumstances under which wildlife farming could be problematic for species threatened by commercial exploitation (Bulte and Damania, 2005; Mason et al., 2012), but these are special cases that simply highlight the need for more careful analysis of industry structures when setting policy. Supplementing the work of theoretical economists, there has been an increasing interest in the topic of wildlife farming for conservation among practitioners and scientists (Mockrin et al., 2005; Phelps et al., 2014; Tensen, 2016). However, a gap in understanding between conservation scientists and economists remains evident and attempts to provide generic policy decision-making frameworks tend to oversimplify issues and fall short. Cooney et al. (2015) highlight the high level of complexity and variability associated more broadly with questions relating to wildlife trade and emphasize the importance of examining the specific characteristics of each traded species across all geographies and along entire trade chains.

Evaluating pangolin farming on current evidence

Will pangolin farming help or hinder conservation of the eight species in the wild, now and in the future? To address this question, this section draws upon the key variable factors and theoretical propositions discussed and considers them in the light of current knowledge relating to pangolins (based on both quantitative and qualitative data gleaned from relevant research). The question must be addressed within a context of varying and even conflicting actions, i.e., different actors currently pursuing different policies and outcomes, such as: the scaling up of farming in China and possibly other countries; a universal ban on sale of pangolin products; and behavior change campaigns that strive for reduction (if not elimination) of consumer demand.

This question can be considered with only limited knowledge of attempts to farm pangolins: in China, potentially in Lao PDR and Vietnam, and it is known that a farm existed in Uganda until 2016 and another in Mozambique until 2018 (Challender et al., 2019). Critically, however, there has been no known commercial captive breeding success for any species of pangolin. Despite attempts at captive breeding historically, most animals in captivity have died after short time periods (e.g., two years) and there has been very limited breeding success (see Chapter 28; Hua et al., 2015; Yang et al., 2007). However, it is recognized that the aim of investors in pangolin farming is to ensure that commercial breeding is successful in the near future, and that consideration of the potential impacts of farming must therefore account for if it were possible.

As noted previously, pangolins are protected species in most range states and national legislation typically prohibits harvest and trade of the animals (Challender and Waterman, 2017), though there are exceptions. These include Gabon, where white-bellied (*Phataginus tricuspis*) and black-bellied pangolins (*P. tetradactyla*) can be harvested and traded within a defined season (see Chapters 8 and 9). It also includes China, which permits use of certified scales and medicines containing scales. The certification system covers the manufacture of patented Traditional Chinese Medicine (TCM) including scales, and certified scales and medicines including scales can only be sold through 716 designated hospitals in China (China Biodiversity Conservation and Green Development Foundation, 2016). Commercial captive breeding of wildlife to supply markets is encouraged in national policy in China (Harris, 2009).

Most pangolins occur on state-owned land in Africa and Asia, where local people typically lack land or resource tenure, or it is limited (e.g., to certain common game or bushmeat species and plant resources). Exceptions include parts of Southern (e.g., South Africa) and perhaps East Africa. This is unlikely to change in the near future. Despite regulatory frameworks precluding access to pangolins, enforcement is undermined by a number of factors, including lack of human and technical capacity and resources (Bennett, 2011; Challender and MacMillan, 2016). Moreover, although there has been little research on the economic incentives of harvesting pangolins specifically, it is clear that the high financial value of the animals provides a strong incentive for poaching and trafficking pangolins (see Chapter 16). This results in high levels of illegal harvest, local use and consumption (Ingram et al., 2018), and trafficking (Cheng et al., 2017). There is typically little involvement of local people in formal management, and little to no proceeds from wild harvesting are applied to management and protection of the species.

Critical to understanding the potential impact of farming is understanding the role of pangolin product consumers and factors driving their decisions. Consumer demand is at least a partial driving force of pangolin poaching (and attempts to farm pangolins). Consumers are willing to pay money to meet their demands. Suppliers of pangolin products are motivated to earn money. Hence, there is a market.

A consumer can decide whether to buy products legally or illegally. Some consumers will care about legality and not buy an illegal product at any price. Others might not care about legality, but still care about price. Most consumers are likely to be price-sensitive and as a general rule, consumers will buy less of a product at higher prices and more at lower prices (an exception to this is high status goods, but this is a special case[1] and does not apply in aggregate to the entire market because of the law of resource scarcity).

A consumer might care about the nature of the product, i.e., whether it is wild or farmed (e.g., for health reasons); whether the animal that produced it suffered, and may prefer – and be willing to pay for – certain attributes. However, these preferences are not necessarily binary, nor are they static. As an example, consider consumers of salmon – some consumers might only eat wild-caught salmon, irrespective of price; others might only eat farmed salmon. However, many, if not most, consumers may be willing to eat either, depending on the price. Tastes for pangolin products are likely to vary too, depending on both qualitative characteristics and price. Existing research suggests a potential preference for wild pangolin meat (Drury, 2009; Shairp et al., 2016) and wild sourced products in TCM (Liu et al., 2016), and by extension, scales from wild pangolins. It also appears that scales from different pangolin species are substitutes, while a number of alternatives products exist that could comprise scale substitutes; however,

[1] In certain instances, as products become scarce, they may acquire additional value to certain consumers even if their prices increase: in fact, such price increases may even result in greater demand among these elite purchasers – this is known as a Veblen or snob effect (Chen, 2016).

there is less research on consumer preferences or among TCM practitioners (see Challender et al., 2019). The same applies to substitutes for pangolin meat in key consumer markets.

Prices are determined at least in part by the cost of production and supply. No-one can afford to continually sell a product for less than it cost them to obtain it, so the costs of supply underpin the price. What are the factors that determine the cost of pangolin supply? In the case of legal supply, it is simply the cost of harvesting (in the case of wild), or the cost of farming, plus the cost of transporting the product to the consumer. In the case of illegal supply, there are additional transaction costs such as concealment and the possible payment of bribes, in addition to a "penalty-risk premium." Economic crime analysts determine a hypothetical penalty-risk premium as a product of the probability of being apprehended, arrested and incurring a penalty by the (discounted) value of that penalty (Milner-Gulland and Leader-Williams, 1992).

Either legal or illegal suppliers may have cost advantages in certain cases, depending on circumstances that may vary considerably across space and time. Even at a particular time, there will be great variability: because of the laws of diminishing returns (and associated varying marginal costs), farming costs will drop with increasing volumes and illegal harvesting costs will increase as a species becomes rarer. Accurate costs of production at farms that are known to exist in China are unknown. However, existing research suggests that rearing costs are high and it is likely cheaper to source wild pangolins and their parts and transport them to local and international markets (Challender et al., 2019). More research is needed on the economic incentives associated with various forms of pangolin harvesting and how these might possibly change in the future.

The existence and management of stockpiles is also a potentially important factor in influencing market prices of pangolin products and, therefore, incentives to harvest or farm them. Since 2009, China has released an average of 26.6 tonnes of scales on to a legal market in China each year from government stockpiles (China Biodiversity Conservation and Green Development Foundation, 2016). Use and trade of these scales and associated products falls under China's certification system. Private stockpiles may also exist in China. Some African nations also possess stockpiles and have expressed a wish to trade in scales purportedly acquired before pangolins were listed in CITES Appendix I. These include, for example, the Democratic Republic of the Congo, which reported in 2017 to possess 22 tonnes of scales (CITES, 2017).

Although discussed in some depth in relation to other species such as elephants (Elephantidae; see Kremer and Morcom, 2000; 't Sas-Rolfes et al., 2014), the links between stockpile policy and conservation impacts remain contested and this topic warrants further research in relation to the markets for pangolin products. Yet, economic theory generally predicts that, as long as a Chinese consumer market persists and domestic trade remains legal, supplying stockpiled pangolin scales to the market can go some way toward satisfying demand and suppressing future scarcity-driven price increases. It is possible, if not likely, that in China, and under a scenario where commercial captive breeding of pangolins was feasible, that locally farmed pangolin scales would be integrated in to the existing stockpile system, thus adding to the supply of the already legal market.

It seems likely that technological advances are being sought to overcome the existing barriers to pangolin reproduction in captivity. Challenges include addressing the high mortality rates caused by inadequate captive diets and stress-induced immune suppression, and a general lack of knowledge regarding pangolin reproductive biology, especially female reproductive cycles and weaning (see Chapter 28; Challender et al., 2019; Hua et al., 2015). If these challenges could be overcome, resultant economies of scale could potentially reduce marginal costs of production, theoretically making at least some pangolin farming financially viable.

There is much uncertainty over the conservation impact of pangolin farming, current and future. Existing farms may be laundering wild-sourced scales as captive-bred, thereby facilitating and further incentivizing poaching and trafficking in Asia and Africa. Such farms may be supported by speculators, banking on being able to trade or launder scales in the future. Existing harvesting incentives are also poorly understood, as are preferences of consumers and TCM practitioners, which mediate both demand for scales and the substitutability of pangolin products. Current unresolved disputes over stockpile policies add a further measure of uncertainly to this landscape.

Appropriate future research may assist in reducing at least some of this uncertainty and, crucially, inform policy and decision-making at local, national and international levels (e.g., CITES) to benefit pangolin conservation. Issues deserving specific attention include:

(1) the economic incentives that drive harvesting in key source areas in Asia and Africa;
(2) the nature of demand (e.g., cross-price elasticity of demand) and substitutability of pangolin products in key consumer markets (China, Vietnam);
(3) the costs of farmed supply, including marginal costs of production, and comparison with costs of sourcing wild pangolins; and
(4) how legal and illegal markets may interact, including the extent of any laundering, and how farmed supply may impact on stockpile policies and *vice versa*.

Conclusion

Pangolins are threatened by overexploitation, and farming is being attempted in some countries in order to ensure a sustainable source of pangolin derivatives, principally scales, for consumptive use. The impact that this may have on the conservation of wild populations deserves evaluation. Established theory, economic and otherwise can guide such evaluation. Key variable factors to consider include legislation (particularly whether it permits farming or not) and the extent to which it is enforced, whether laundering may occur, consumer preferences for pangolin products and potential substitutes, how farming may affect stockpile policies, and how technology may alter the economic competitiveness of farming. Pangolins have not been bred successfully in captivity for commercial purposes and the immediate impact of pangolin farming on the conservation of wild pangolins is unclear. Yet, it is possible that the existence of farms is incentivizing poaching and trafficking in Africa and Asia. To better understand the potential impact of farming, further research is needed as outlined. This should take account of how the identified key variable factors may change under various scenarios such as scaled-up farming activity, more comprehensive prohibition of trade, or significant changes in consumer preferences due to behavior change campaigns. The longer-term future for pangolin conservation, trade, and farming remains highly uncertain.

References

Araki, H., Cooper, B., Blouin, M.S., 2007. Genetic effects of captive breeding cause a rapid, cumulative fitness decline in the wild. Science 318 (5847), 100–103.

Barbier, E.B., Schulz, C.-E., 1997. Wildlife, biodiversity and trade. Environ. Dev. Econ. 2 (2), 145–172.

Barrett, C.B., Brandon, K., Gibson, C., Gjertsen, H., 2001. Conserving tropical biodiversity amid weak institutions. Bioscience 51 (6), 497–502.

Becker, G.S., 1968. Crime and punishment: an economic approach. J. Polit. Econ. 76 (2), 176–177.

Bennett, 2011. Another inconvenient truth: the failure of enforcement systems to save charismatic species. Oryx 45 (4), 476–479.

Bergstrom, T., 1990. On the economics of crime and confiscation. J. Econ. Perspect. 4 (3), 171–178.

Boakye, M.K., Pietersen, D.W., Kotzé, A., Dalton, D.-L., Jansen, R., 2015. Knowledge and uses of African pangolins as a source of traditional medicine in Ghana. PLoS One 10 (1), e0117199.

Boakye, M.K., Kotzé, A., Dalton, D.L., Jansen, R., 2016. Unravelling the pangolin bushmeat commodity chain and the extent of trade in Ghana. Hum. Ecol. 44 (2), 257–264.

Brayden, W.C., Noblet, C.L., Evans, K.S., Rickard, L., 2018. Consumer preferences for seafood attributes of wild-harvested and farm-raised products. Aquacult. Econ. Manage. 22 (3), 362–382.

Bulte, E.H., Damania, R., 2005. An economic assessment of wildlife farming and conservation. Conserv. Biol. 19 (4), 1222–1233.

Challender, D.W.S., Hywood, L., 2012. African pangolins under increased pressure from poaching and intercontinental trade. TRAFFIC Bull. 24 (2), 53–55.

Challender, D.W.S., MacMillan, D.C., 2016. Transnational environmental crime: more than an enforcement problem. In: Schaedla, W.H., Elliott, L. (Eds.), Handbook of Transnational Environmental Crime. Edward Elgar Publishing, Cheltenham and Northampton, Massachusetts, pp. 489–498.

Challender, D., Waterman, C., 2017. Implementation of CITES Decisions 17.239 b) and 17.240 on Pangolins (*Manis* spp.), CITES SC69 Doc. 57 Annex. Available from: <https://cites.org/sites/default/files/eng/com/sc/69/E-SC69-57-A.pdf>. [August 2, 2018].

Challender, D.W.S., 't Sas-Rolfes, M., Ades, G., Chin, J.S.C., Sun, N.C.-M., Chong, J.L., et al., 2019. Evaluating the feasibility of pangolin farming and its potential conservation impact. Glob. Ecol. Conserv. 20, e00714.

Chen, F., 2016. Poachers and snobs: demand for rarity and the effects of antipoaching policies. Conserv. Lett. 9 (1), 65–69.

Cheng, W., Xing, S., Bonebrake, T.C., 2017. Recent pangolin seizures in China reveal priority areas for intervention. Conserv. Lett. 10 (6), 757–764.

China Biodiversity Conservation and Green Development Foundation, 2016. An Overview of Pangolin Data: When Will the Over-Exploitation of the Pangolin End? Available from: <http://www.cbcgdf.org/English/NewsShow/5011/6145.htm>. [March 19, 2019].

CITES, 2017. SC69 Doc. 29.2.2, Application of Article XIII in the Democratic Republic of the Congo. Available from: <https://cites.org/sites/default/files/eng/com/sc/69/E-SC69-29-02-02.pdf>. [March 3, 2019].

Clark, C.W., 1973. The economics of overexploitation. Science 181 (4100), 630–634.

Clark, C.W., 2010. Mathematical Bioeconomics: The Mathematics of Conservation. John Wiley & Sons, Hoboken, New Jersey.

Cook, P.J., 1977. Punishment and crime: a critique of current findings concerning the preventive effects of punishment. Law Contemp. Probl. 41 (1), 164–204.

Cooney, R., Kasterine, A., MacMillan, D., Milledge, S., Nossal, K., Roe, D., et al., 2015. The Trade in Wildlife: A Framework to Improve Biodiversity and Livelihood Outcomes. International Trade Centre, Geneva, Switzerland.

D'Cruze, N., Singh, B., Mookerjee, A., Harrington, L.A., Macdonald, D.W., 2018. A socio-economic survey of pangolin hunting in Assam, Northeast India. Nat. Conserv. 30, 83–105.

Drury, R., 2009. Reducing urban demand for wild animals in Vietnam: examining the potential of wildlife farming as a conservation tool. Conserv. Lett. 2 (6), 263–270.

Dutton, A., Hepburn, C., Macdonald, D.W., 2011. A stated preference investigation into the Chinese demand for farmed vs. wild bear bile. PLoS One 6 (7), e21243.

Ehrlich, I., 1973. Participation in illegitimate activities: a theoretical and empirical investigation. J. Polit. Econ. 81 (3), 521–565.

Fischer, C., 2004. The complex interactions of markets for endangered species products. J. Environ. Econ. Manage. 48 (2), 926–953.

Glenny, M., 2008. McMafia: A Journey Through the Global Criminal Underworld, first ed. Alfred A. Knopf, New York.

Gordon, H.S., 1954. The economic theory of a common-property resource: the fishery. J. Polit. Econ. 62 (2), 124–142.

Hardin, G., 1968. The tragedy of the commons. Science 162 (3859), 1243–1248.

Harfoot, M., Glaser, S.A.M., Tittensor, D.P., Britten, G.L., McLardy, C., Malsch, K., et al., 2018. Unveiling the patterns and trends in 40 years of global trade in CITES-listed wildlife. Biol. Conserv. 223, 47–57.

Harris, R.B., 2009. Wildlife Conservation in China. Preserving the Habitat of China's Wild West. Routledge, Oxon.

Heinrich, S., Wittmann, T.A., Prowse, T.A.A., Ross, J.V., Delean, S., Shepherd, C.R., et al., 2016. Where did all the pangolins go? International CITES trade in pangolin species. Glob. Ecol. Conserv. 8, 241–253.

Hinsley, A., Nuno, A., Ridout, M., St. John, F.A.V., Roberts, D.L., 2017. Estimating the extent of CITES noncompliance among traders and end-consumers; lessons from the global orchid trade. Conserv. Lett. 10 (5), 602–609.

Hinsley, A., Roberts, D.L., 2018. The wild origin dilemma. Biol. Conserv. 217, 203–206.

Hotelling, H., 1931. The economics of exhaustible resources. J. Polit. Econ. 39 (2), 137–175.

Hua, L., Gong, S., Wang, F., Li, W., Ge, Y., Li, X., et al., 2015. Captive breeding of pangolins: current status, problems and future prospects. ZooKeys 507, 99–114.

Hutton, J., Webb, G., 2003. Crocodiles: legal trade snaps back. In: Oldfield, S. (Ed.), The Trade in Wildlife, Regulation for Conservation. Earthscan, London, pp. 108–120.

Ingram, D.J., Coad, L., Abernethy, K.A., Maisels, F., Stokes, E.J., Bobo, K.S., et al., 2018. Assessing Africa-wide pangolin exploitation by scaling local data. Conserv. Lett. 11 (2), e12389.

Kremer, M., Morcom, C., 2000. Elephants. Am. Econ. Rev. 90 (1), 212–234.

Liu, Z., Jiang, Z., Fang, H., Li, C., Mi, A., Chen, J., et al., 2016. Perception, price and preference: consumption and protection of wild animals used in traditional medicine. PLoS One 11 (3), e0145901.

Marshall, A., 2009. Principles of Economics: Unabridged, eighth ed. Cosimo Classics, New York.

Mason, C.F., Bulte, E.H., Horan, R.D., 2012. Banking on extinction: endangered species and speculation. Oxford Rev. Econ. Policy 28 (1), 180–192.

Milner-Gulland, E.J., Leader-Williams, N., 1992. A model of incentives for the illegal exploitation of black rhinos and elephants: poaching pays in Luangwa Valley, Zambia. J. Appl. Ecol. 29 (20), 388–401.

Mockrin, M.H., Bennett, E.L., LaBruna, D.T., 2005. Wildlife Farming: A Viable Alternative to Hunting in Tropical Forests? WCS Working Paper No. 23. Wildlife Conservation Society, New York.

North, D.C., 1991. Institutions. J. Econ. Perspect. 5, 97–112.

Ostrom, E., 1990. Governing the Commons: The Evolution of Institutions for Collective Action. Cambridge University Press, New York.

Phelps, J., Carrasco, L.R., Webb, E.L., 2014. A framework for assessing supply-side wildlife conservation. Conserv. Biol. 28 (1), 244–257.

Shairp, R., Veríssimo, D., Fraser, I., Challender, D.W.S., MacMillan, D., 2016. Understanding urban demand for wild meat in Vietnam: implications for conservation actions. PLoS One 11 (1), e0134787.

Swanson, T.M., 1994. The Economics of Extinction Revisited and Revised: A Generalised Framework for the Analysis of the Problems of Endangered Species and Biodiversity Losses. Oxford Economic Papers, 46, 800–821.

Tensen, L., 2016. Under what circumstances can wildlife farming benefit species conservation? Glob. Ecol. Conserv. 6, 286–298.

't Sas-Rolfes, M., 2017. African wildlife conservation and the evolution of hunting institutions. Environ. Res. Lett. 12, 115007.

't Sas-Rolfes, M., Moyle, B., Stiles, D., 2014. The complex policy issue of elephant ivory stockpile management. Pachyderm 55, 62–77.

Veríssimo, D., Wan, A.K.Y., 2019. Characterizing efforts to reduce consumer demand for wildlife products. Conserv. Biol. 33 (3), 623–633.

Williams, S.J., Jones, J.P.G., Annewandter, R., Gibbons, J. M., 2014. Cultivation can increase harvesting pressure on overexploited plant populations. Ecol. Appl. 24 (8), 2050–2062.

Williams, V.L., Loveridge, A.J., Newton, D.J., Macdonald, D.W., 2017. A roaring trade? The legal trade in *Panthera leo* bones from Africa to East-Southeast Asia. PLoS One 12 (10), e0185996.

Yang, C.W., Chen, S., Chang, C.-Y., Lin, M.F., Block, E., Lorentsen, R., et al., 2007. History and dietary husbandry of pangolins in captivity. Zoo. Biol. 26 (3), 223–230.

Zhang, Y., 2009. Conservation and trade control of pangolins in China. In: Pantel, S., Chin, S.-Y. (Eds.), Proceedings of the Workshop on Trade and Conservation of Pangolins Native to South and Southeast Asia. 30 June – 2 July 2008, Singapore Zoo, Singapore. TRAFFIC Southeast Asia, Petaling Jaya, Selangor, Malaysia. pp. 66–74.

PART 5

Conservation Planning, Research and Finance

Conservation strategies and priority actions for pangolins

Rachel Hoffmann[1] *and Daniel W.S. Challender*[2,3]

[1]IUCN Species Survival Commission, IUCN, Cambridge, United Kingdom [2]Department of Zoology and Oxford Martin School, University of Oxford, Oxford, United Kingdom [3]IUCN SSC Pangolin Specialist Group, ℅ Zoological Society of London, Regent's Park, London, United Kingdom

O U T L I N E

Introduction

To effectively conserve threatened species, i.e., those with a higher risk of extinction, there is a critical need to ensure that conservation strategies or action plans guide effective conservation action. To facilitate improved outcomes, this is best achieved through participatory processes, which helps to encourage active engagement by different stakeholders and critically, implementation by all once strategies have been developed. Depending on the context these strategies may be at the local, national, regional and/or global level, or a combination thereof, and ideally mutually reinforcing.

Unlike better studied and iconic taxa (e.g., rhinoceroses [Rhinocerotidae] and lions [*Panthera leo*]), pangolins have received little conservation attention historically. However, since around 2010, the profile of pangolins has increased significantly (see Chapter 21), and much greater attention is being given to determining which conservation actions are needed for each species. As a result, varied stakeholders including governments, non-governmental organizations (NGOs), and practitioners, have initiated conservation programs and projects for pangolins in range countries in both Asia and Africa. However, since there are often insufficient resources available to many practitioners who

Pangolins
DOI: https://doi.org/10.1016/B978-0-12-815507-3.00033-2

are working to conserve species, and with an approach needed that is often complex and multi-faceted, it is vital that conservation actions maximize the likelihood of achieving conservation success for the species involved. This is especially the case where species face multiple threats and/or a high level of extinction risk. This chapter discusses the rationale for developing conservation strategies for pangolins, in order to guide the successful conservation of wild pangolin populations.

Conservation strategies and action plans

In the last three decades most species conservation strategies or action plans (henceforth used interchangeably), have been developed with, or by, IUCN. The IUCN SSC (Species Survival Commission) Action Plan series is considered to be one of the world's most authoritative sources of information to address the most pressing issues relating to conservation of a species (or species group) and to guide effective action. The plans have been published since 1985 and are compiled by Specialist Groups (SGs), the main working units of the SSC. Historically these plans (many in their familiar black-covered jackets), were the culmination of many hours of desk-based research and correspondence, and led the way in providing a summary of the most up to date and comprehensive information (e.g., population status and threats) for the range of species for which they were written. The plans identified conservation priorities to be implemented though the coordinated efforts of relevant stakeholders. Generally, most of these plans had a life span of up to 10 years before significant review and revision. In 2008, the publication of the *IUCN SSC Strategic Planning for Species Conservation: A Handbook* (IUCN/SSC, 2008) signaled a shift away from this review and revision process, by providing recommended methods for developing species conservation strategies. Rather than simply publishing information on species, the handbook emphasized the need for sound conservation science combined with more consultative and participatory processes. This allows for the joint development of management strategies and conservation by all parties with responsibility, interest and concern, or a stake in the survival of particular species. In addition, IUCN has other tools that can be integrated into the action planning process if deemed necessary or appropriate. They include Population and Habitat Viability Assessments (PHVAs), which build upon Population Viability Assessments (PVAs) to set targets for species recovery, and the Integrated Collection Assessment Planning (ICAP) process, through which practitioners working on in situ and ex situ species conservation convene to apply the decision-making process of IUCN's ex situ guidelines (see IUCN SSC, 2014) to regional or global collection planning.

In IUCN's contemporary conservation planning efforts, there is a strong emphasis on the need for an inclusive, participatory approach. This typically entails gathering all relevant (and pre-identified) stakeholders and experts (government, non-government, scientists and practitioners) to share wisdom, familiarity and expertise, identify gaps in knowledge, and determine and assess threats to species, i.e., what threatens the viability of species populations. This may also be accompanied by the assessment or re-assessment of species for The IUCN Red List of Threatened Species. During this process participants agree to leave individual and institutional agendas to one side and, typically over the course of a few days, develop and agree on a shared vision, goals, SMART objectives, and actions and activities to mitigate the threats facing particular species. Cooperation and collaboration are essential, connecting those with the will and capacity to determine responsibilities for action through an evaluation of benefits, costs, the feasibility of potential options, and an appraisal of likely challenges to be encountered at the implementation stage. Similarly, an assessment is made of how the activities should be measured and evaluated to determine their effectiveness in meeting agreed objectives. Often of paramount importance to the successful delivery of conservation action is engagement with local communities (see Chapters 23–27), to ensure that indigenous knowledge and cultural values are incorporated into planning and implementation.

The benefits of going through this process are both direct and indirect. As well as developing a vision and specific goals, objectives and activities, the creation of conservation action plans provide a blueprint to guide the conservation of species, which can be used by funders and donors in identifying and evaluating funding proposals and investment opportunities. Moreover, a key benefit of convening diverse stakeholders over a period of a few days in developing an action plan is that relationships can be built and networks strengthened, which helps to maintain an active workflow from the action plan concept through to delivery, publication and dissemination, and subsequently implementation and evaluation.

Although conservation planning is a priority for many species covered by the hugely diverse network of IUCN SSC SGs, the work of the groups must also encompass a range of other conservation activities. Since 2016, these fall loosely under three essential functions: Assess (assessing the status of species through Red List assessments), Plan (facilitate the identification of the best strategies for conserving biodiversity), and Act (catalyzing action to improve the status of biodiversity). This framework has been defined as the "Species Conservation Cycle" and SGs are being encouraged to plan their objectives around the three components, and to move more strategically (and quickly where possible and necessary) from assessing, through to planning and the effective delivery of conservation action. The Pangolin Specialist Group (PSG) is using this framework to set its objectives annually, and in planning expected outcomes for the IUCN quadrennium (the 4-yearly cycle around which IUCN is operationalized).

Conservation strategies for pangolins

The complex conservation predicament facing pangolins means there is a strong rationale for the development of regional and national strategies to guide conservation of the species. Pangolins face a high or very high risk of extinction, hold near universally high financial value, are widely distributed and trafficked in high numbers along diverse trafficking routes by organized criminal groups, while little is known about particular species (e.g., giant [*S. gigantea*] and black-bellied pangolins [*P. tetradactyla*]), which impedes conservation management. Moreover, in light of the threats the species face, the profile of the species has grown (Chapter 21) and there is increased interest in investing in pangolin conservation, and as discussed, a multitude of stakeholders are initiating conservation projects that focus on pangolins.

As at 2019 such strategies are limited. The first such plan was a PHVA for the Chinese pangolin in Taiwan in 2005 (see Chapter 36 and Chao et al., 2005). Although not an action plan as such, this was followed by recommendations from the TRAFFIC workshop, held in 2008, on "Trade and Conservation of Pangolins Native to South and Southeast Asia" (see Pantel and Chin, 2009). In 2014, the PSG published a fundraiser friendly global conservation action plan, "Scaling up Pangolin Conservation" (see Challender et al., 2014), and more recently, in collaboration with partners including the IUCN SSC Conservation Planning Specialist Group (CPSG), IUCN SSC Asian Species Action Partnership (ASAP) and Wildlife Reserves Singapore, published a Regional Conservation Strategy for the Sunda Pangolin (*Manis javanica*) (see IUCN SSC Pangolin Specialist Group et al., 2018). The PSG also collaborated with the Palawan Council for Sustainable Development Staff (PCSDS), the

Katala Foundation Incorporated, and the Zoological Society of London, to develop a national strategy for the Philippine pangolin (*M. culionensis;* see IUCN SSC Pangolin Specialist Group, 2018), and with multiple partners to update the conservation strategy and PHVA for the Chinese pangolin in Taiwan in 2017 (see Chapter 36). These efforts have been complemented by actions from the First Pangolin Range State Meeting held in Vietnam in 2015 (see Anon, 2015), and CITES (Convention on International Trade in Endangered Species of Wild Fauna and Flora) Resolution Conf. 17.10 on the *Conservation of and trade in pangolins* (see Chapter 19).

However, as most pangolin species lack a conservation strategy at any level, there is a pressing need to hold further workshops using IUCN's best practice guidance to convene all relevant stakeholders to ensure that detailed, regional conservation strategies are in place for all eight pangolin species. This should result in a cohesive and integrative approach to conservation of the species, avoid duplication of effort, ensure lessons learnt are shared, and that existing research and methodologies, allow for the standardization of approaches taken. Of crucial importance, is the engagement in a process that will foster collaboration and build networks and partnerships for implementation of the plans which are developed (which can also facilitate the subsequent development of national actions). This process will enable the various interventions, approaches and tools discussed in this volume (Chapters 17–32, 34–38) to be evaluated and priority actions for pangolins to be set. These efforts could be complemented with other forms of evaluation as appropriate, including PHVAs and ICATs (see also Chapter 31). Critically, there is a profound need to ensure that action plans that are developed are not the end point, but that the actions devised are subsequently implemented, monitored, and evaluated (Fuller et al., 2003).

Conclusion

Conservation strategies are essential for guiding species conservation activities, and IUCN has set the benchmark for developing these strategies in the last three decades. The conservation predicament facing pangolins necessitates the development of conservation strategies that cover all eight species, at least at the regional level, and which can facilitate the development of national action plans. Use of the IUCN approach would ensure that a participatory process is followed, and that action plans are agreed by all key stakeholders, thus avoiding situations where additional and competing strategies are produced by different organizations. This also helps to prevent wasted resources and duplication of effort. As an IUCN SSC SG, the PSG is well placed to convene diverse stakeholders in order to do so and to ensure that conservation funding that is spent on pangolins is done so wisely.

References

Anon, 2015. First Pangolin Range State Meeting Report. 24–26 June, Da Nang, Vietnam.

Challender, D.W.S., Waterman, C., Baillie, J.E.M., 2014. Scaling up Pangolin Conservation. IUCN SSC Pangolin Specialist Group, Zoological Society of London, London, UK.

Chao, J.-T., Tsao, E.H., Traylor-Holzer, K., Reed, D., Leus, K. (Eds.), 2005. Formosan Pangolin Population and Habitat Viability Assessment: Final Report. IUCN/SSC Conservation Breeding Specialist Group, Apple Valley, Minnesota.

Fuller, R.A., McGowan, P.J.K., Carroll, J.P., Dekker, R.W.R. J., Garson, P.J., 2003. What does IUCN species action planning contribute to the conservation process? Biol. Conserv. 112 (3), 343–349.

IUCN SSC (IUCN Species Survival Commission), 2008. Strategic Planning for Species Conservation: A Handbook. Version 1.0. Gland, Switzerland.

IUCN SSC, 2014. Guidelines on the Use of Ex Situ Management for Species Conservation. Version 2.0. IUCN Species Survival Commission, Gland, Switzerland.

IUCN SSC Pangolin Specialist Group, 2018. Scaling up Palawan Pangolin Conservation – Developing the First National Conservation Strategy for the Species. Available from: <https://www.pangolinsg.org/2018/04/23/scaling-up-palawan-pangolin-conservation-developing-the-first-national-conservation-strategy-for-the-species/>. [September 21, 2018].

IUCN SSC Pangolin Specialist Group, IUCN SSC Asian Species Action Partnership, Wildlife Reserves Singapore, IUCN SSC Conservation Planning Specialist Group, 2018. Regional Sunda Pangolin (*Manis javanica*) Conservation Strategy 2018-2028. IUCN SSC Pangolin Specialist Group, Zoological Society of London, London, UK.

Pantel, S., Chin, S.-Y. (Eds.), 2009. Proceedings of the Workshop on Trade and Conservation of Pangolins Native to South and Southeast Asia, 30 June – 2 July 2008, Singapore Zoo, Singapore. TRAFFIC Southeast Asia, Petaling Jaya, Selangor, Malaysia.

CHAPTER

34

Research needs for pangolins

Darren W. Pietersen[1,2] and Daniel W.S. Challender[2,3]

[1]Mammal Research Institute, Department of Zoology and Entomology, University of Pretoria, Hatfield, South Africa [2]IUCN SSC Pangolin Specialist Group, % Zoological Society of London, Regent's Park, London, United Kingdom [3]Department of Zoology and Oxford Martin School, University of Oxford, Oxford, United Kingdom

OUTLINE

Introduction

Pangolins are among the most poorly known mammalian orders. This is largely because they have historically received little conservation and research attention (Challender et al., 2012), in part because of their predominantly nocturnal habits (except the black-bellied pangolin [*Phataginus tetradactyla*]), typically low population densities, and elusive habits (Heath and Coulson, 1997; Willcox et al., 2019). This changed in the 2010s as the profile of pangolins grew (see Chapter 21), but many knowledge gaps remain that are important to the species' conservation. This chapter discusses key research needs for pangolins and their conservation. The focus of the chapter is on conservation research, and research needs are proposed that, if met, should result in the generation of knowledge on pangolins and their threats, and potential means of mitigation, to inform conservation management and action. Future inquiries into pangolin evolutionary history are discussed in Chapters 1 and 2. The research directions proposed have been identified through (1) a review of existing literature, (2) feedback from 65 respondents to a questionnaire on research needs for pangolins completed by pangolin specialists

Pangolins
DOI: https://doi.org/10.1016/B978-0-12-815507-3.00034-4

in 2018 (see IUCN SSC Pangolin Specialist Group, 2018), and (3) chapters in this volume. The research needs discussed are not presented in order of priority, nor are they exhaustive.

Trade, trafficking and policy

There is an increasing volume of literature on legal and illegal trade in pangolins, including trade volumes and routes (e.g., Heinrich et al., 2016, 2017; Ingram et al., 2018, 2019; Nijman et al., 2016). This research is useful to understanding trade and trafficking dynamics, to inform national and international policymaking (e.g., Convention on International Trade in Endangered Species of Wild Fauna and Flora [CITES]; see Chapter 19), and to inform assessments of the impact of exploitation on the species (see *Biology and ecology*). Heinrich et al. (2017) estimated a mean of 29 new pangolin trafficking routes a year between 2010 and 2015, highlighting the need for continued research in this area in order to generate up to date knowledge to inform policy-makers and other key stakeholders. These include those on the ground (e.g., front-line law enforcement personnel). Future research should attempt to elucidate the major contemporary sources of trafficked pangolins (both countries and specific sites) and associated illegal trade routes, transport methods, and networks involved. Inherent to such research is the need to develop more accurate conversion parameters with which to estimate the number of pangolins in illegal trade, for example, from quantities of seized scales. In particular, this applies to the African species (*Phataginus* and *Smutsia* spp.) and the Indian pangolin (*Manis crassicaudata*) for which reliable estimates do not exist (see Chapter 16). Similarly, there is a need to better understand the extent of local and domestic (i.e., national) consumptive use of the species' and the extent to which it has, and is continuing to, contribute to the (over)exploitation of populations, in order to inform local and national management measures (e.g., Ingram et al., 2018).

Reducing consumer demand for pangolin products in Asia and Africa is recognized as a key conservation intervention for pangolins. However, there remains a lack of rigorous, peer-reviewed research on pangolin consumers with which to inform behavior change interventions (see Chapter 22). This is necessary both in order to inform any interventions to change consumer behavior and, critically, to evaluate the impact of any interventions implemented. While existing conservation strategies highlight the need for such research in key markets in Asia (e.g., China and Vietnam) there is a need for research of this ilk in key consumer states in Africa as well, especially where indications are that exploitation is unsustainable. This research should not be limited to traditional concepts of understanding consumers (e.g., through demographic profiles), but would benefit from innovative approaches, e.g., understanding psychographic profiles of consumers (see also Chapter 22). Although pangolin trade and consumption is associated with illegality in most range countries, a suite of research techniques exist which can be used to understand consumers and consumption in these circumstances (see Nuno and St. John, 2015).

All pangolins were included in CITES Appendix I at CoP17 (see Chapter 19). Although intuitively positive to many actors, including some policymakers, the impact that this measure could have on pangolin populations is unknown and warrants research attention. Despite assumptions by some actors that the impact will be positive, there are potential adverse consequences from this change in policy. The increasing illegality of trading in pangolins brought about by changes to national legislation in some range countries could well send trade deeper underground, meaning it is more difficult to monitor. Moreover, it could result in higher prices for pangolin derivatives (e.g., scales), which could increase harvesting incentives in source areas, and thereby hunting and poaching rates, hastening the exploitation of populations. When the black rhinoceros (*Diceros bicornis*) was included in Appendix I in 1977, price increases for rhino horn followed, resulting in the local extinction of the species in at least 18 countries (Leader-Williams, 2003). Research is needed to determine the impact of the Appendix I listing on pangolins and whether it has been positive for the species, or not, by investigating legal and illegal trade dynamics, harvesting incentives, and markets for pangolin products.

As discussed in Chapter 32, attempts are being made to commercially captive breed, or farm, pangolins, and the impact this could have on wild pangolin populations deserves research attention. Challender et al. (2019) concluded that pangolin farming is not feasible in the foreseeable future principally because of difficulties in breeding pangolins on a large scale. However, they also note that there is uncertainty to this assessment, and further research on the potential impacts of farming is required. This includes economic incentives for harvesting, consumer demand, farming costs, and interactions between legal and illegal markets (see Chapter 32).

Forensics

Wildlife forensics is an emerging conservation tool for combatting illegal trade in wildlife. du Toit et al. (2017) investigated methods to extract useable DNA from pangolin scales, and Gaubert et al. (2016, 2018) found distinct geographical structuring within the white-bellied pangolin (*Phataginus tricuspis*), suggesting that this method can be used to determine the geographic origin of trafficked specimens. Future research should focus on expanding accurately georeferenced DNA databases with a combination of mtDNA and nucleic markers (see Chapter 20) to inform understanding of trafficking dynamics and source sites, in order to guide targeted conservation action and policy. Similarly, the development of efficient techniques to detect processed pangolin derivatives (e.g., crushed or powdered scales), and commercial products containing pangolin derivatives (see also Chapter 20) would bolster front-line law enforcement efforts, and commands further research.

Biology and ecology

There is a critical need to accurately and reliably estimate pangolin populations, in order to determine the impact of exploitation for local and international use, and to inform local, national and international management, policy, and conservation action. However, estimating populations is currently impeded in most cases by a lack of established monitoring methods for pangolins (see Chapter 35), and the development of such methods is hindered by a lack of knowledge on the biology and ecology of the species. For example, there are no home range estimates for the Indian, giant (*S. gigantea*), white-bellied or black-bellied pangolins (though see Chapter 8) with which to inform sampling designs.

Existing knowledge of each species is documented in Chapters 4–11. There is a particular need for detailed ecological research on species for which there is an acute lack of knowledge, most notably the giant and black-bellied pangolins. However, there are significant knowledge gaps even for those species that have been comparatively well studied; Temminck's pangolin, for example, has been the subject of a number of ecological studies but these have all been confined to Southern Africa. Further research is thus needed on all species, and should focus on elucidating the factors determining the distribution of the species at the macro- and micro-scale, as well as home range sizes, habitat requirements and potential habitat preferences, and habitat use and variation by season. For the development of indirect monitoring methods, research is needed on burrow creation and occupancy (and other structures used for shelter), including the rate of creation, frequency of use, and potential variation (e.g., by season and lunar phase) and factors determining use of burrows versus other resting structures. As well as being essential for the development of population monitoring methods, the knowledge generated from such ecological research will help to inform management and conservation planning, including the identification of priority areas (i.e., sites) for pangolin conservation.

Conservation planning for pangolins can be further refined through knowledge generated from research into comparative ecologies between natural and artificial and degraded landscapes, and the ability of different species to persist in isolated blocks of monoculture habitat (e.g., oil palm plantations) in the long term, and to breed. Equally, there is a need for research to determine accurate circadian patterns, social and foraging behavior, diet and prey preferences, and if and how they change by season, as well as ecological differences between sympatric species.

There is a lack of knowledge about how pangolin populations behave. The breeding biology and social structure of most species is poorly known, though polygyny and monogamy have been suggested (see Chapters 4 and 11). Research in this area will enable breeding strategies to be understood and will provide insights into gene transfer and genetic diversity. Although there is some understanding of breeding (e.g., seasonal vs. aseasonal), there is little knowledge of female reproductive biology (e.g., estrus cycles), including uncertainty on gestation period. This includes the potential for delayed embryo implantation; see Chapter 9). There is also little or limited knowledge of maternal care and weaning age, growth rates, dispersal behavior, and age at sexual maturity and first breeding. This also applies to generation length (average age of parents in the population), maximum age, and means of ageing pangolins. Knowledge of all these factors is essential for understanding the demography of pangolin populations and recruitment and growth rates, and would inform more robust assessments of conservation status.

Finally, pangolins are thought to play an important ecological role by regulating populations of ants and termites (see Chapter 3). Research to quantify the ecosystem services provided by the species' and better understand the potential impact of their loss may help to inform decision-making regarding investment in their conservation.

Genetics

There is an increasing body of research on pangolin genetics, much of which focuses on genetics for forensic purposes (see Chapter 20). Non-forensic genetic research includes the development of complete nuclear genomes for two pangolin species (Chinese [*M. pentadactyla*] and Sunda pangolin [*M. javanica*]) with the results used to make inferences about pangolin evolution (see Chapters 1 and 2). Transcriptomic analyses have also been performed on the Sunda pangolin and resulted in the identification and mapping of a number of genes. This has provided insights into the biochemistry of pangolins and has yielded insights into the species' reactions to various stimuli, and adaptions to their myrmecophagous lifestyle (Ma et al., 2019; Yusoff et al., 2016).

Gaubert et al. (2016) studied intra-specific variability in the white-bellied pangolin and delimited six discrete lineages, which did not overlap geographically (with one exception), and identified them as Evolutionarily Significant Units (ESUs) that may warrant species or subspecies status (see Chapter 2). Hassanin et al. (2015) found evidence for a cryptic species of white-bellied pangolin in

Gabon based on the sequence of a single individual without supporting data. Further research using nuclear genomics and comparative morphology is necessary to elucidate potential species or subspecies status within the white-bellied pangolin, which would have important conservation management implications.

Nash et al. (2018) used genome-wide markers to reveal highly divergent subpopulations of the Sunda pangolin. This research found distinct lineages in Borneo, Java and Singapore/Sumatra, and the species' wide distribution suggests there may be yet unreported cryptic diversity (Nash et al., 2018). Further investigation is warranted and if additional cryptic diversity were to be confirmed it would have significant implications for conservation management, in particular, policies on releasing Sunda pangolins confiscated from illegal trade across the species' range. Zhang et al. (2015) found evidence of a cryptic lineage of Asian pangolin most closely related to the Sunda pangolin. This purportedly novel lineage may correspond to the Philippine pangolin (*M. culionensis*), a species not included in their analyses, or the results may reflect pseudogene amplification (Gaubert et al., 2015). It is possible that pangolins display greater intraspecific molecular variation than other mammals and this requires assessment within a total evidence framework. There remains a need to clarify the status of the Chinese pangolin sub-species proposed on morphological evidence (e.g., Allen, 1938) but not genetic evidence, including the nominal Formosan pangolin in Taiwan.

There has been little population-level research on pangolins. du Toit (2014) suggested limited geographic structuring within the Southern African population of Temminck's pangolin (*S. temminckii*), specifically a genetic separation between Namibian and South African populations, and populations in Mozambique and Zimbabwe. du Toit (2014) suggests that this may reflect an ancient or more recent human-induced separation, the latter being plausible because there are no apparent natural barriers to gene flow between these populations. These results are based on a small sample size and may also reflect mutation rates of the gene regions used; further research with greater sampling effort from across the species range is ideally needed for clarification. No population-level research has been conducted on the giant, black-bellied or Asian pangolins, with the exception of the Chinese pangolin (see Chapter 4) and those studies discussed. Such research could aid pangolin conservation by providing insights into social structure (see *Biology and ecology*), genetic diversity of populations and species, and provide a means with which to estimate effective population sizes and inbreeding depression, and could be used to inform the release location for trade-confiscated individuals.

Husbandry and veterinary health

There is a need for further research into pangolin husbandry, including nutrition and veterinary health. Large numbers of pangolins are confiscated from illegal trade each year that require a period of captive care for rehabilitation. There are established conservation centers in Asia with appropriate expertise (see Chapters 28–30), but by contrast, African pangolins have largely fared poorly in captivity (an exception is San Diego zoo which maintained a white-bellied pangolin for a number of years). This has made rescue and rehabilitation of pangolins in Africa more challenging. For Temminck's and white- and black-bellied pangolins, the greatest success has involved maintaining animals in a semi-captive environment with individuals being allowed to forage naturally under supervision until they are healthy enough to be released (Tikki Hywood Foundation, unpubl. data). A better understanding of the housing and husbandry requirements for African pangolins and the main reasons for mortality while under care is essential to improving rehabilitation success. However, if ex situ conservation is to contribute to the conservation of African or Asian pangolins in the future (see Chapter 31), there is a need to ensure that adequate husbandry can be provided and long-term care requirements understood, which would necessitate further husbandry research. This includes the improvement of captive diets, or their development where they do not exist for specific species. Research should focus on analyzing wild diets, and the contribution that the gastric physiology and microbiome of pangolins makes to digestion.

Regarding veterinary health, existing research has reported reference intervals for various blood biochemistry and hematology values (e.g., Chin et al., 2015; Khatri-Chhetri et al., 2015). However, some references require validation, which would benefit future clinical care, and should be based on appropriate sample sizes, as well as investigate the effect of seasonality and age (see Chapter 29).

Many parasitic organisms are reported for pangolins (see Chapter 29), and while the zoonotic potential of many of them is known, their significance for the health of individual hosts is not, and requires further research. There is also a need for pathological studies to understand in-depth the underlying causes and progression of disease in pangolins, and possible treatment options, especially for individuals confiscated from illegal trade.

Climate change

Climate change is a potential threat to pangolins. With predicted increases in global temperatures, the geographic range of all pangolin species may contract, shift or be affected in other ways. This applies in particular to the Indian and Temminck's pangolins which inhabit arid and desiccated areas (see Chapters 5 and 11). Research to understand the potential impacts of climate change on pangolins would help to mitigate the potential threat it poses.

Conclusion

Although poorly known mammals, pangolins have started to receive concerted research and conservation attention. However, many knowledge gaps remain, which impedes effective conservation and management of the species. As such, there is a critical need for the generation of knowledge in multiple areas, in particular trade and trafficking dynamics and the impact of policy decisions, pangolin life history, biology and ecology, and populations and how they behave, including the development of monitoring methods, in order to inform and guide future conservation efforts.

Acknowledgments

The authors thank the individuals that completed the questionnaire on research needs for pangolins in summer 2018 which informed this chapter.

References

Allen, G.M., 1938. The Mammals of China and Mongolia. Natural History of Central Asia, vol. XI. Part I. The American Museum of Natural History, New York.

Challender, D.W.S., Baillie, J.E.M., Waterman, C., IUCN SSC Pangolin Specialist Group, 2012. Catalysing conservation action and raising the profile of pangolins – the IUCN SSC Pangolin Specialist Group (Pangolin SG). Asian J. Conserv. Biol. 2, 139–140.

Challender, D.W.S., 't Sas-Rolfes, M., Ades, G., Chin, J.S.C., Sun, N.C.-M., Chong, J.L., et al., 2019. Evaluating the feasibility of pangolin farming and its potential conservation impact. Glob. Ecol. Conserv. 20, e00714.

Chin, S.-C., Lien, C.-Y., Chan, Y., Chen, C.-L., Yang, Y.-C., Yeh, L.-S., 2015. Hematologic and serum biochemical parameters of apparently healthy rescued Formosan pangolins (*Manis pentadactyla pentadactyla*). J. Zoo Wildl. Med. 46 (1), 68–76.

du Toit, Z., 2014. Population genetic structure of the ground pangolin based on mitochondrial genomes.

M.Sc. Thesis, University of the Free State, Bloemfontein, South Africa.

du Toit, Z., Grobler, J.P., Kotze, A., Jansen, R., Dalton, D.L., 2017. Scale samples from Temminck's ground pangolin (*Smutsia temminckii*): a non-invasive source of DNA. Conserv. Genet. Resour. 9 (1), 1–4.

Gaubert, P., Njiokou, F., Olayemi, A., Pagani, P., Dufour, S., Danquah, E., et al., 2015. Bushmeat genetics: setting up a reference framework for the DNA-typing of African forest bushmeat. Mol. Ecol. Resour. 15 (3), 633–651.

Gaubert, P., Njiokou, F., Ngua, G., Afiademanyo, K., Dufour, S., Malekani, J., et al., 2016. Phylogeography of the heavily poached African common pangolin (Pholidota, *Manis tricuspis*) reveals six cryptic lineages as traceable signatures of Pleistocene diversification. Mol. Ecol. 25 (23), 5975–5993.

Gaubert, P., Antunes, A., Meng, H., Miao, L., Peigné, S., Justy, F., et al., 2018. The complete phylogeny of pangolins: scaling up resources for the molecular tracing of the most trafficked mammals on Earth. J. Hered. 109 (4), 347–359.

Hassanin, A., Hugot, J.-P., van Vuuren, B.J., 2015. Comparison of mitochondrial genome sequences of pangolins (Mammalia, Pholidota). C. R. Biol. 338 (4), 260–265.

Heath, M.E., Coulson, I.M., 1997. Home range size and distribution in a wild population of Cape pangolins, *Manis temminckii*, in north-west Zimbabwe. Afr. J. Ecol. 35 (2), 94–109.

Heinrich, S., Wittmann, T.A., Prowse, T.A.A., Ross, J. V., Delean, S., Shepherd, C.R., et al., 2016. Where did all the pangolins go? International CITES trade in pangolin species. Glob. Ecol. Conserv. 8, 241–253.

Heinrich, S., Wittman, T.A., Rosse, J.V., Shepherd, C.R., Challender, D.W.S., Cassey, P., 2017. The Global Trafficking of Pangolins: A Comprehensive Summary of Seizures and Trafficking Routes From 2010–2015. TRAFFIC, Southeast Asia Regional Office, Petaling Jaya, Selangor, Malaysia.

Ingram, D.J., Coad, L., Abernethy, K.A., Maisels, F., Stokes, E.J., Bobo, K.S., et al., 2018. Assessing Africa-wide pangolin exploitation by scaling local data. Conserv. Lett. 11 (2), e12389.

Ingram, D.J., Cronin, D.T., Challender, D.W.S., Venditti, D. M., Gonder, M.K., 2019. Characterizing trafficking and trade of pangolins in the Gulf of Guinea. Glob. Ecol. Conserv. 17, e00576.

IUCN SSC Pangolin Specialist Group, 2018. Methods for monitoring populations of pangolins (Pholidota: Manidae). IUCN SSC Pangolin Specialist Group, Zoological Society of London, London, UK.

Khatri-Chhetri, R., Sun, C.-M., Wu, H.-Y., Pei, K.J.-C., 2015. Reference intervals for hematology, serum biochemistry, and basic clinical findings in free-ranging Chinese Pangolin (*Manis pentadactyla*) from Taiwan. Vet. Clin. Pathol. 44 (3), 380–390.

Leader-Williams, N., 2003. Regulation and protection: successes and failures in rhinoceros conservation. In: Oldfield, S. (Ed.), The Trade in Wildlife, Regulation for Conservation. Earthscan, London, pp. 89–99.

Ma, J.-E., Jiang, H.-Y., Li, L.-M., Zhang, X.-J., Li, H.-M., Li, G.-Y., et al., 2019. SMRT sequencing of the full-length transcriptome of the Sunda pangolin (*Manis javanica*). Gene 692 (15), 208–216.

Nash, H.C., Wirdateti, Low, G., Choo, S.W., Chong, J.L., Semiadi, G., et al., 2018. Conservation genomics reveals possible illegal trade routes and admixture across pangolin lineages in Southeast Asia. Conserv. Genet. 19 (5), 1083–1095.

Nijman, V., Zhang, M.X., Shepherd, C.R., 2016. Pangolin trade in the Mong La wildlife market and the role of Myanmar in the smuggling of pangolins into China. Glob. Ecol. Conserv. 5, 118–126.

Nuno, A., St. John, F.A.V., 2015. How to ask sensitive questions in conservation: a review of specialized questioning techniques. Biol. Conserv. 189, 5–15.

Willcox, D., Nash, H.C., Trageser, S., Kim, H-J., Hywood, L., Connelly, E., et al., 2019. Evaluating methods for detecting and monitoring pangolin (Pholidota: Manidae) populations. Glob. Ecol. Conserv. 17, e00539.

Yusoff, A.M., Tan, T.K., Hari, R., Koepfli, K.-P., Wee, W.Y., Antunes, A., et al., 2016. *De novo* sequencing, assembly and analysis of eight different transcriptomes from the Malayan pangolin. Sci. Rep. 6, 28199.

Zhang, H., Miller, M.P., Yang, F., Chan, H.K., Gaubert, P., Ades, G., Fischer, G.A., 2015. Molecular tracing of confiscated pangolin scales for conservation and illegal trade monitoring in Southeast Asia. Glob. Ecol. Conserv. 4, 414–422.

CHAPTER

35

Developing robust ecological monitoring methodologies for pangolin conservation

Dana J. Morin[1,2,3], *Daniel W.S. Challender*[3,4], *Ichu Godwill Ichu*[3,5], *Daniel J. Ingram*[6], *Helen C. Nash*[3,7], *Wendy Panaino*[8], *Elisa Panjang*[3,9,10], *Nick Ching-Min Sun*[11] *and Daniel Willcox*[3,12]

[1]Department of Wildlife, Fisheries and Aquaculture, Mississippi State University, Starkville, MS, United States [2]Cooperative Wildlife Research Laboratory, Southern Illinois University, Carbondale, IL, United States [3]IUCN SSC Pangolin Specialist Group, ℅ Zoological Society of London, Regent's Park, London, United Kingdom [4]Department of Zoology and Oxford Martin School, University of Oxford, Oxford, United Kingdom [5]Pangolin Conservation Network, ℅ Central Africa Bushmeat Action Group (CABAG), Yaoundé, Cameroon [6]African Forest Ecology Group, Biological and Environmental Sciences, University of Stirling, Stirling, United Kingdom [7]Department of Biological Sciences, National University of Singapore, Singapore [8]Brain Function Research Group, School of Physiology and Centre for African Ecology, School of Animal, Plant and Environmental Sciences, University of the Witwatersrand, Johannesburg, South Africa [9]Organisms and Environment Division, Cardiff School of Biosciences, Cardiff University, Cardiff, United Kingdom [10]Danau Girang Field Centre, Sabah Wildlife Department, Kota Kinabalu, Malaysia [11]Graduate Institute of Bioresources, National Pingtung University of Science and Technology, Pingtung, Taiwan [12]Save Vietnam's Wildlife, Cuc Phuong National Park, Ninh Binh Province, Vietnam

CONTENTS

Pangolins
DOI: https://doi.org/10.1016/B978-0-12-815507-3.00035-6

Introduction

Pangolins have generally received little research attention. Research in the 19th and 20th centuries focused primarily on taxonomic classification and morphology (e.g., Mohr, 1961). However, focus began to shift to ecology and behavior in the mid-late 20th century and primarily concerned African pangolin species (e.g., Heath and Coulson, 1997; Pagès, 1975; Swart et al., 1999). This has continued in the 21st century (e.g., Akpona et al., 2008) and developed to include Asian pangolins (Lim and Ng, 2008; Wu et al., 2002, 2003).

Pangolins are particularly challenging to study due to their elusive behaviors, principally nocturnal activity patterns, typically low population densities and increasing rarity (Challender and Waterman, 2017; Nash et al., 2016; Pietersen et al., 2014). Furthermore, species differ in size, activity patterns, locomotion, and local and regional habitat, requiring different means of investigation. Most previous research has focused on species that are more observable, including Temminck's pangolin (*Smutsia temminckii*; Pietersen et al., 2014; Swart et al., 1999), Indian pangolin (*Manis crassicaudata*; Irshad et al., 2015), and the Chinese pangolin (*M. pentadactyla*; Wu et al., 2003). Other species, most notably the black-bellied pangolin (*Phataginus tetradactyla*), remain virtually unstudied.

The need for robust ecological monitoring methods to inform pangolin conservation is well-recognized and urgent for all species (Anon, 2015; Challender et al., 2014; Lee et al., 2018). Baseline information, including species' distributions, local population densities and population growth rates are required to estimate the status of populations at the site, sub-national, national and international level, and assess the impacts of local use and poaching for international trafficking. Reliable estimates of population status and underlying drivers of population distribution and dynamics are critical to forming an evidence base on which to determine appropriate local management actions, and inform broader conservation strategies and interventions (Chapter 33) and national and international policy (e.g., CITES, the Convention on International Trade in Endangered Species of Wild Fauna and Flora).

Framework for effective monitoring for conservation

Monitoring is not an isolated objective, but should instead be viewed as a component of broader conservation objectives. Effective monitoring efforts should be efficient and structured to discover as much as possible to inform future conservation actions. Framing monitoring efforts in alignment with the principles of targeted and adaptive monitoring (described later) allows for the greatest impact and advancement of knowledge, while repeated review of the success of methods and updated state of knowledge will ensure that conservation efforts produce maximum results (Lindenmayer and Likens, 2009; Nichols and Williams, 2006). Additionally, monitoring is an integral part of adaptive management when properly planned and implemented based on clearly defined conservation objectives (Gibbs, 2008).

Models of system (e.g., species) responses to management actions are one of five essential elements of informed decision-making in conservation (Kendall, 2001; Nichols et al., 1995; Williams et al., 2002). While population status is a primary objective in monitoring, estimating population parameters without gaining information about the factors that produce the observed patterns results in numbers without explanation, or any information on potential recourse (Krebs, 1991). Thus, one of the most important features of a monitoring program may be how monitoring activities are structured to both estimate parameters of interest and provide information about species ecology and effects of conservation efforts (Nichols and Williams, 2006; Yoccoz et al., 2001). This critical objective is defined here as asking "status+" questions (see Box 35.1). In other words,

BOX 35.1

Current understanding of pangolin distribution and density, gaps in knowledge, and research needs

Pangolin distribution and density is dependent on opposing factors: (1) ecological factors that allow for population persistence, and (2) anthropogenic factors that result in population declines and range contraction.

Ecological factors: Most pangolin species are habitat generalists and historical range-wide distribution was limited only by physiological constraints and geographical features. Predation is not thought to be a substantial cause of mortality due to the defense behavior of rolling into a tight ball exposing only the protective scales. However, little is known about the effects of competition for food resources with other myrmecophagous animals or for burrows or other resting sites, and facultative or obligatory use of other species burrows may also influence distribution and density of fossorial pangolin species.

Anthropogenic factors: Human-related factors can be indirect, including land use conversion and landscape fragmentation, or direct, including harvest for local use or for international trafficking. Harvest for wildlife trafficking may be the dominant determinant in population density and distribution for most pangolin species and local populations, especially in Southeast Asia, and perhaps increasingly in Africa (see Chapter 16). However, total extent of loss to all anthropogenic causes and proportional contributions to overall declines for each species and region is still largely unknown.

Status+ questions for pangolin conservation

Pangolin specialists working in 16 pangolin range states identified three overarching gaps in knowledge at an intensive workshop in 2018 organized to improve and standardize monitoring approaches for pangolin conservation (see IUCN SSC Pangolin Specialist Group, 2018). Identification of knowledge gaps allowed for development of existing relevant research questions to improve understanding of factors positively and negatively influencing distribution and densities of pangolin populations (status+ questions). Addressing these research needs in an adaptive monitoring framework will allow for updating of the current state of knowledge, as hypotheses are refined based on the strength of supporting evidence provided by future research (Fig. 35.1). Supplementary gaps in knowledge that would facilitate the study of pangolin populations were also identified.

Research needs, associated parameters, and example status+ questions

Workshop participants split research needs into four categories and defined associated parameters and initial status+ questions for each:

(1) Factors limiting global distribution and local densities (ecological and anthropogenic).

BOX 35.1 (*cont'd*)

- landscape context and severity of direct anthropogenic threats (removal for local use and international trafficking):
  - Example questions: Is white-bellied pangolin (*P. tricuspis*) density greater in areas with active ranger patrols? Are Chinese pangolin densities higher in protected areas? How does Sunda pangolin (*M. javanica*) occupancy change with distance to villages? What human societal factors contribute to the harvest rates for local use?
- habitat associations, physiological limitations, prey preferences and competitors for resources:
  - Example questions: How does the density of Indian pangolin change with soil texture and prey densities? Does giant pangolin (*S. gigantea*) occupancy change dependent on aardvark (*Orycteropus afer*) occupancy?

(2) Thresholds for exploitation compared to overexploitation to support decisions on local use, management and regulation, and theoretically commercial international trade.
- Current densities, survival and reproductive rates, net recruitment and population growth rates, and typical dispersal behavior:
  - Example questions: How many offspring do female Philippine pangolins (*M. culionensis*) have, how often do they reproduce, and what is the survival rate of offspring? What is the current abundance of black-bellied pangolins in Cameroon and how does it change with local human densities?

(3) Tolerance of pangolin populations to indirect anthropogenic threats (land use conversion and fragmentation).
- Comparative ecologies between natural habitat and artificial and degraded landscapes:
  - Example questions: Are Sunda pangolin densities lower in palm oil plantations in Borneo, and do pangolins reproduce at the same rate in areas fragmented by palm oil plantations? Does land use conversion and fragmentation increase vulnerability of local pangolin populations to harvest?

(4) Additional gaps in knowledge to improve study and inference about pangolin species.
- Ideal sampling designs and timing:
  - Individual and seasonal space use in different habitats, burrow use and architecture, factors influencing the use of other resting structures, and circadian patterns and potential seasonal variation.

monitoring questions should be posed that will provide information about what results in changes in species status (targeted monitoring; Nichols and Williams, 2006) and how populations respond to specific management actions (active adaptive management; McCarthy and Possingham, 2007; McDonald-Madden et al., 2010), as opposed to asking questions about status alone (surveillance monitoring).

Initiating monitoring plans within an adaptive monitoring framework will allow for incorporation of new or refined questions and hypotheses concerning factors affecting the state of the system into long-term monitoring programs, without affecting the integrity of key indicators (i.e., status) measured over time (see Lindenmayer and Likens, 2009). Central to the adaptive monitoring framework is a conceptual model of the system being studied (Fig. 35.1) and how components of the system (e.g., species populations) might function and respond to hypothesized factors or conservation actions (Gitzen et al., 2012; Lindenmayer and Likens, 2009). Based on current

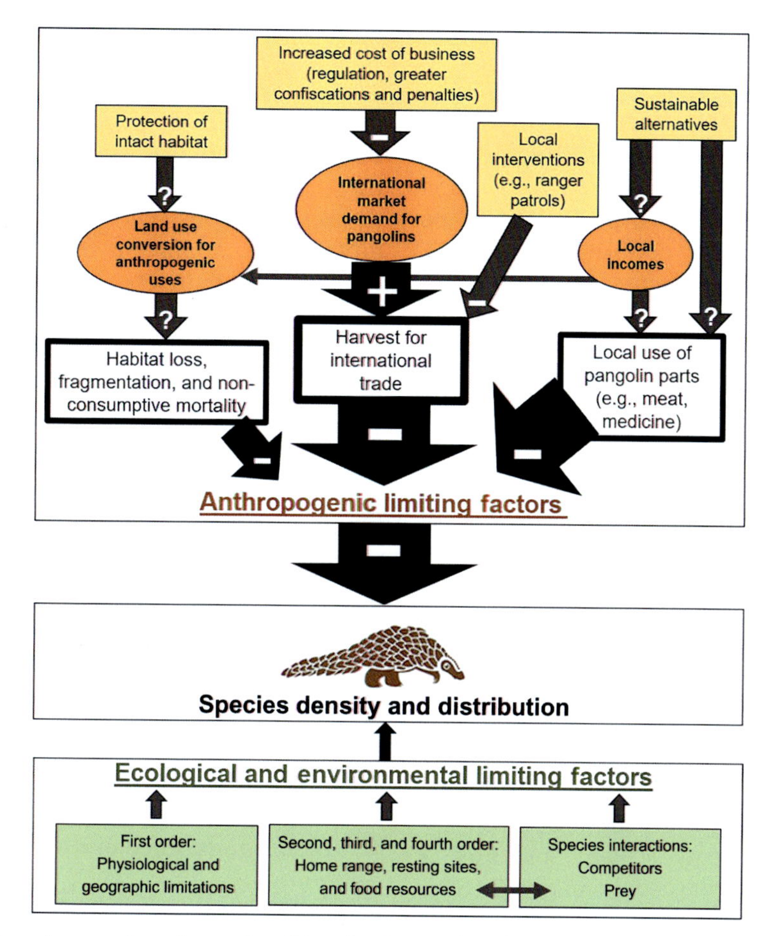

FIGURE 35.1 A systems ecology figure describing the factors that shape the distribution and densities of pangolin populations for use in an adaptive monitoring framework. The conceptual model is generic for application to all pangolin species range-wide, but provides a framework for more detailed model formulation specific to each species, region, and locality. Elements at the top of the figure show threats and potential conservation actions hypothesized to influence pangolin populations range-wide. Threats are in bold boxes, potential factors influencing threats are in orange ovals, current conservation actions that may reduce threats or population declines and extirpations are in gold boxes. Size of arrows suggests hypothesized relative impact and effect: positive (+), negative (–), or unknown (?). Identification of possible conservation actions will guide formulation of status+ questions for targeted monitoring. Elements in green at the bottom of the figure show ecological and environmental factors that limit species distributions and population densities. Current knowledge strongly suggests anthropogenic sources of mortality are substantially more influential on population dynamics and status compared to ecological and environmental factors for most pangolin species in most regions. However, knowledge of ecological and environmental limiting factors informs expectation of population persistence and recovery if conservation actions provide positive impacts, and improves existing knowledge for potential plans for population restoration.

understanding, Fig. 35.1 depicts a generic pangolin population density and distribution as dependent on both anthropogenic, and ecological or environmental limiting factors. Three primary anthropogenic factors are hypothesized to be driving population declines and range contractions: (1) habitat loss, fragmentation and non-consumptive human-caused mortality, (2) harvest for local use, and (3) poaching for international trafficking. All of these factors can affect each species and region to different degrees and the model should be refined specific to the population of study. Conception and use of such a model for pangolin conservation allows questions to be framed specifically to collect data to answer questions about species' response to changes in the environment, exploitation rates, and efforts to conserve populations (Lindenmayer and Likens, 2009). Thus, the adaptive monitoring framework facilitates the targeted monitoring of a population, its status, and factors related to change over time, improving understanding of specific contextual threats and the degree of success of conservation efforts (Fig. 35.1 and Box 35.1).

This does not imply that adaptive monitoring necessarily requires continuous monitoring at the same sites, which can be cumbersome and hindered by unforeseen logistical challenges and costs. However, use of the conceptual model of systems ecology provides a common thread for studies on pangolin ecology and conservation, as information gained can be used to update or support the structure of the conceptual model and underlying hypotheses. Differences found over time or across studies and species can further refine understanding of the specific needs and threats in different regions and for each species.

Finally, as noted earlier, there is little knowledge of some pangolin species and populations and many knowledge gaps to be addressed (see Chapter 34). It may be tempting for scientists to select research questions based on ability to provide answers, or restricted resources. However, researchers are strongly encouraged to prioritize study questions that will provide the most immediate benefit to understanding the threats to pangolins and potential regulatory and management actions. Conservation biology is a "crisis science" (Soule 1985, 1991), which pangolins typify. Loss of some pangolin populations is occurring at alarming rates. Defining and answering status+ questions will identify populations at greatest risk and population strongholds, but also provide valuable information on actions that may stem current population declines.

Designing studies to monitor pangolins

Once suitable status+ questions have been defined within an adaptive monitoring framework, appropriate sampling methods and study designs need to be determined. Choice of both will depend on the parameter to be measured (the parameter of interest) as determined by the conservation context and available resources.

Distribution and densities

Confirming presence and estimating occupancy

Trade-offs exist between spatial scales of studies and intensity of sampling effort to estimate more detailed parameters. For some pangolin species, even basic knowledge of distribution is unknown or uncertain, preventing the possibility of protection in some regions and the inclusion of those areas in planning for conservation actions (Challender and Waterman, 2017). In these circumstances, initial studies may be aimed at confirming presence or site occupancy and relating these parameters of interest to relevant ecological and/or anthropogenic factors. However, it should be noted that while confirming presence can be useful as a first step in monitoring populations, counts or inference based solely on confirming presence of pangolins (i.e., relative abundance indices) are unreliable and subject to severe bias (Sollmann et al., 2013) and should not be used.

Instead, study designs at larger scales should acknowledge imperfect detection of pangolins at sampling sites and incorporate histories of detections *and non-detections* of the target species over temporal or spatial replicates to estimate the probability that the pangolin species occurs at a site (site occupancy), even if presence was not confirmed (MacKenzie et al., 2002, 2017). True occupancy is binary (either a pangolin species occurs at a site or it does not). Hence, while occupancy allows for estimating distribution at large spatial scales and does not require individual identification (see *Estimating density*), the information the parameter provides is coarse; occupancy will only detect changes in population status that result in the extirpation of the species from a site which, in the case of informing future conservation efforts, will be too late for action.

Estimating density

Population densities (abundance/unit area) provide greater information on population status, the effects of ecological and anthropogenic factors, and population responses to conservation measures. Depending on the precision of the density estimates, small changes in local abundance can be detected and used to trigger intervention or management actions prior to loss of the species at a site. However, density is more difficult to estimate and requires more intensive sampling frequency, limiting spatial scales of inference. As density is a highly desirable parameter of interest, many statistical models have been devised to estimate it. These methods can be divided into those that require some form of individual identification and those that do not. The former include capture mark-recapture (CMR; Otis et al., 1978; Williams et al., 2002) and spatial capture-recapture (SCR; Borchers and Efford, 2008; Royle and Young, 2008; Royle et al., 2014). The latter include distance sampling and hierarchical models which estimate density based on counts of animals across space given a latent model of hypothesized individual movement (e.g., random encounter models and unmarked models).

From a study design and estimation stand point, SCR models provide unbiased and often precise density estimates while offering the greatest flexibility in study design and sampling effort, as estimates are typically robust to violations in model assumptions (Royle et al., 2014, 2016). Density estimation with SCR is preferred to traditional CMR methods which do not include additional information on the location of detected individuals. This is because models of SCR estimate density as a parameter in the model, whereas CMR methods transform abundance estimates to densities *ad hoc* using a potentially biased or arbitrary effective sampling area (Borchers and Efford, 2008). It is not possible to identify individual pangolins based on appearance (though see IUCN SSC Pangolin Specialist Group, 2018). Thus, sampling methods for SCR studies must include either the capture and distinct marking of individuals for identification upon re-encounter, or non-invasive sampling and genetics (gNIS) to identify individuals.

Density estimation methods that do not require individual identification instead require strict adherence to model assumptions, or accuracy and precision of estimates may be compromised. Of these methods, distance sampling (Buckland et al., 2012), is the most well-tested and provides reliable estimates as long as model assumptions are met. But as individual pangolins can be difficult to detect due to inactivity (e.g., resting in burrows) and low densities, distance sampling for burrows combined with methods to estimate burrow occupancy are the most promising for estimating density of fossorial species (i.e., Temminck's, giant, Indian and Chinese pangolins) dependent on terrain and associated visibility of burrows. Both random encounter models (Rowcliffe et al., 2008) and "unmarked," or spatially correlated count models (Chandler and Royle, 2013) require strong assumptions or information about individual movement (Burgar et al., 2018; Cusack et al., 2015). However, limited knowledge of space use and home range for most pangolin species means such methods do not yet have application (Challender et al., in prep; IUCN SSC Pangolin Specialist Group, 2018; Willcox et al., 2019).

Active and adaptive sampling

Whether a natural state or due to overexploitation, most pangolin species appear to occur in low densities across their ranges (though see Chapter 9; Challender et al., in prep; Irshad et al., 2015; Pietersen et al., 2014; Willcox et al., 2019). Low density populations are difficult to monitor at the spatial scales and intensities required to achieve population parameter estimates that are useful for assessing change over time or success of conservation actions. In addition, several pangolin species demonstrate cryptic behavior or will be difficult to detect due to limited diel activity (Lim, 2007). Thus, careful consideration is needed to select appropriate sampling methods and in designing sampling schemes. Standard stratified sampling designs using passive detector arrays (e.g., camera trap grids, acoustic arrays, randomized established transects) will require extensive survey effort to result in sufficient detections of pangolins to estimate even coarse population parameters (i.e., occupancy) with reasonable precision (see Khwaja et al., 2019; Willcox et al., 2019). Instead, active sampling methods including searches for individuals or signs with detection dogs, or study designs placing transects based on information gained through recce surveys will likely be most effective. Furthermore, adaptive sampling holds great promise for estimating low density populations, improving both efficiency and precision in parameter estimates including occupancy and density (Conroy et al., 2008; Wong et al., 2018). Adaptive sampling formally incorporates two sampling intensities in the model, utilizing low effort monitoring over large spatial extents with more intensive sampling implemented in areas where detections exceed a threshold indicating increased probability of detection (more return for a given sampling effort and cost). Similarly, creative solutions refining and combining sampling methods will improve effectiveness for population parameter estimation specific for pangolin species (see *Sampling methods for monitoring pangolins*).

Estimating population inputs and outputs

Population size and changes over time (population growth rate) is the result of contrasting population inputs (births and immigrations into the population) and outputs (deaths and emigrations from the population) since the last population census (demographic rates). This is commonly referred to as the BIDE model (Cassell, 2001) and provides a convenient structure for forming hypotheses about population dynamics, assessing specific impacts of harmful or beneficial actions, and predicting how populations will respond to changes in demographic rates. Understanding the demographic mechanisms that produce change in population densities and distribution is equally important to estimating population status, as this information can elucidate the conservation actions most immediately critical to slow population declines. There is no doubt that mortality (removal from the population), principally resulting from human harvest for local use (Ingram et al., 2018) and/or international trafficking (e.g., Heinrich et al., 2017), is the primary driver of population change for most pangolin species and should be a central focus of research to best address urgent conservation needs (see Chapter 34). Additionally, information on life history for each species, including reproductive rates, and dispersal ecology, can shape expectations of how different populations will respond and potentially recover following interventions and protection of populations, aiding in prioritization for direction of resources. In particular, predictions of population response to mortality will be critical to providing science-based justification for regulatory decisions on local use and international trafficking. Methods to estimate demographic rates are diverse (Williams et al., 2002) and a subject of ongoing research in the field of population ecology (Kéry and Royle, 2016; Kéry and Schaub, 2012) and selection of an appropriate model or estimator will depend on the sampling methods available and the detail of data collected.

Sampling methods for monitoring pangolins

Ultimately, selection of a sampling method or combination of methods will depend on the parameters of interest and study objectives. Many commonly used sampling methods have so far yielded poor results for monitoring pangolins (Willcox et al., 2019) and careful thought is required to ensure adequate sampling specific to the target pangolin species. For example, passive use of camera trap arrays, a common monitoring technique for many mammal species (Burton et al., 2015; O'Connell et al., 2010), has not resulted in encounter rates of pangolins sufficient to produce reliable occupancy estimates, except possibly in unexploited populations (Willcox et al., 2019). Furthermore, lack of visual individual identification prevents robust estimates of density using camera trap arrays without strong assumptions about individual space use that is not currently known or confirmable (Burgar et al., 2018; Cusack et al., 2015), even if encounter rates were high. Four pangolin species are arboreal or semi-arboreal so standard camera trap placement along game trails would not be expected to detect these species. The species that do travel along the ground do not appear to use game trails but instead move randomly or through thick brush and microhabitats unsuitable for camera trap placement (E. Panjang, D. Willcox, pers. comm.), and baiting camera traps have not improved results (Marler, 2016). Additionally, the sparse distribution of many pangolin populations would require saturation of prohibitively large spatial extents to detect enough individuals to estimate occupancy or density using a passive sampling array, resulting in extensive effort in establishment and maintenance of the sampling grid and cost in equipment for little return in monitoring outcome. However, camera traps do have application for the study of pangolins when combined with other sampling methods that can inform placement to generate data for specific parameters of interest, as has been successfully used for other ecologically similar species such as armadillos (Cingulata) and gopher tortoises (*Gopherus polyphemus*; Ingram et al., 2019). For example, placing cameras at burrow entrances or complexes can confirm pangolin use or monitor telemetered females with offspring, and placing camera traps in trees could be used to confirm presence of arboreal and semi-arboreal species.

Despite the challenges, several sampling methods have proven application for estimating parameters of interest for certain pangolin species and others may have potential for similar species (Fig. 35.2). Sampling methods involving active searches for pangolins and their signs are the most promising. Searches

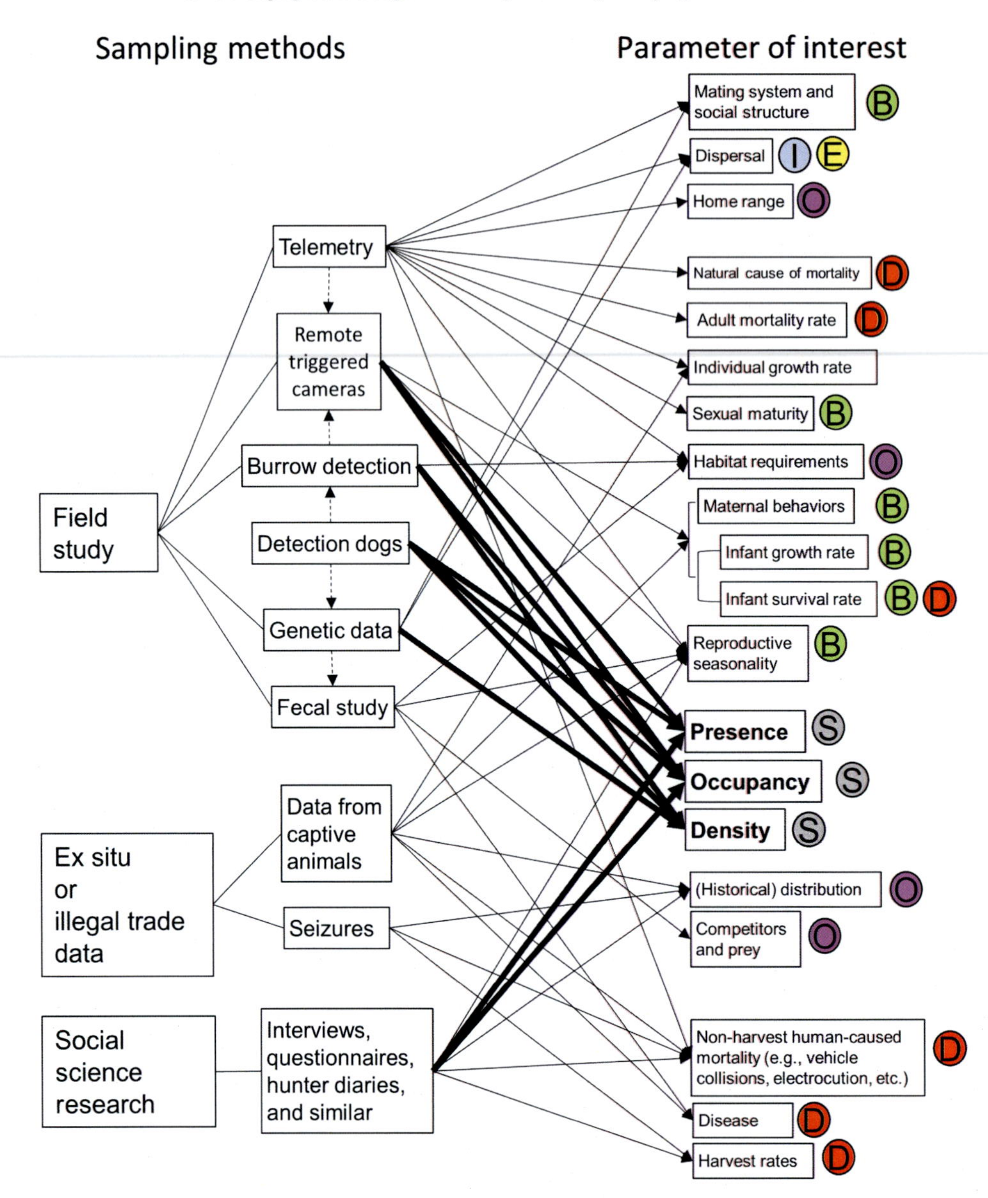

FIGURE 35.2 Sampling methods (left and center columns) with proven or high potential application to pangolin species connected with solid lined arrows to specific parameters of interest (right column). Parameters of interest are categorized based on demographic information provided including status (gray "S"), reproductive recruitment or births (green "B"), immigration (blue "I"), mortality or deaths (red "D"), and emigration (yellow "E"). Parameters noted with "O" (purple) provide other information that will contribute to overall understanding of pangolin distribution and density. Bold arrows connect sampling methods for estimating primary measures of population status (presence, occupancy, and density) with appropriate study design and statistical methods. Vertical dashed arrows in the second column indicate complimentary methods for estimating parameters of interest. For example, setting remote-trigger cameras at burrows can confirm presence or allow for study of maternal behaviors and reproductive seasonality.

for burrows within plots or along transects can be used to estimate density of fossorial species. As pangolins use more than one burrow or multiple pangolins or other species might use the same burrow (Willcox et al., 2019), use of an endoscope or other means is required to confirm pangolin presence. Density should be estimated by combining distance sampling (Buckland et al., 2012) with simultaneous estimation of burrow occupancy rates (see Stober et al., 2017 for an example with gopher tortoises).

Capture and marking of pangolins for SCR may be more fruitful than remote-detector arrays for estimating density. Researchers have drilled holes into a single scale, or series of scales of Temminck's pangolins, using a standardized numbering system for identification on recapture (W. Panaino, pers. comm.). Alternatively, uniquely identifiable microchip implants have worked well for Sunda pangolins, but an electric scanner is required to read the microchip when the pangolin is recaptured (Nash et al., 2018). Temminck's pangolins in the Kalahari leave very distinctive tracks in the soft, sandy soils. Regularly-driven sand roads in the Kalahari are often useful to detect pangolin tracks, which tend to cross directly instead of running along the road, and have been followed to burrows for capture and marking of pangolins. However, finding other pangolin species based on track surveys is more difficult. Field signs of other pangolins can be easily confused with other species and are not generally considered a reliable indicator of pangolin presence. Evidence of digging, claw marks, scratches on trees, smell, and burrows, can be created by a wide variety of other species and finding pangolins in low density populations based on observing signs may require extensive searches.

A trained conservation dog team successfully detected ground-dwelling Chinese pangolins and their feces in Nepal (H.J. Kim, J. Hartmann, pers. comm.). However, extensive effort was required due to the low existing population density, and adaptive sampling strategies will likely be necessary to improve efficiency and allow for adequate detections to estimate population parameters (beyond confirming presence). The detection dog team was unable to detect the feces of semi-arboreal Sunda pangolins, but were able to detect a single individual at a site in southern Vietnam. Handlers speculate the trained dog was ineffective because Sunda pangolins sometimes defecate in water or in trees diffusing the direction of the scent, in addition to the very low population density in the study area due to poaching, but further research is needed to refine the use of detection dogs to locate pangolins and their signs under different conditions. Genetic non-invasive sampling (gNIS) allows for detection of pangolins from their signs, confirmation of species, and potentially, identification of individuals for use in CMR and SCR models (Waits and Paetkau, 2005). Thus, combining gNIS with sampling methods to detect pangolin signs could allow for robust estimates of population status if encounter rates and study design are appropriate. A wide variety of additional topics can also be explored using genetic data collected, including relatedness and mating systems, dispersal ecology, and genetic population viability analysis.

Social science research methods such as collecting local ecological knowledge (LEK) and other forms of human-collected information (e.g., confiscation records) can be highly useful and provide preliminary information on pangolin distribution to direct efficient field sampling and to assess threats to pangolin populations (Nash et al., 2016; Newton et al., 2008). Methods include questionnaires, unstructured and semi-structured interviews and participatory mapping, while novel methods can be used to ask sensitive questions of local community members, indigenous peoples, and other stakeholders (Newing, 2011; Nuno and St John,

2015). Further, if sampling is appropriately structured, social science research methods can potentially be used to estimate pangolin occupancy over large regions (as has been done for forest elephants [*Loxodonta cyclotis*] e.g., Brittain et al., 2018) and inform more intensive field sampling efforts.

Tracking pangolins with VHF radio tags has proven useful for understanding home range sizes, dispersal movement, habitat use, activity patterns, and other aspects of ecology for some species of pangolin (e.g., Temminck's; see Pietersen et al., 2014). These technologies have generally worked better for ground-dwelling pangolins and the semi-arboreal Philippine pangolin, although issues remain in locating pangolins when they spend large parts of the day in burrows hindering signal transmission. For other semi-arboreal pangolins, problems with tag attachment have limited research. In previous studies, the scales of Sunda pangolins have broken after only two to three weeks causing the tag to be dislodged (Lim, 2007; Nash et al., 2018). However, two Sunda pangolins were tracked for approximately 8 weeks with VHF tags in Borneo (E. Panjang, pers. comm.). Satellite tags which can gather finer scale location and temporal data are generally too large to be attached to smaller pangolin species and problems have been reported in forest environments and when fossorial species are in burrows. However, developing technology reducing battery size and new satellite networks for animal tracking hold promise for future pangolin tracking.

Available sampling methods continue to improve, particularly with advancement in new technologies and increased focus on pangolin monitoring (Challender et al., in prep). Some potential methods require only field testing and refinement in pilot studies for application to specific pangolin species. For example, arboreal camera trapping could be used to elucidate distribution and activity patterns of semi-arboreal and arboreal species such as the black-bellied pangolin. Other proposed methods will require proof of concept prior to serious investment and use in conservation monitoring programs. Theoretically, acoustic devices could be used to detect pangolins based on sounds made when breaking apart ant nests and termite mounds. However, as a passive monitoring array, extensive sampling would still be required to achieve adequate detections for parameter estimation. An emerging technology, genetic identification of species from parasitic invertebrates including ticks or Tetse flies (iDNA) could have application for pangolins in the future (Abrams et al., 2018). However, development and potential application will require comprehensive field testing and specialized models before this method will be applicable. Regardless, development of inventive methods for improving sampling, and creative solutions to estimation of parameters are needed to further study of pangolins to best inform existing and future conservation efforts.

References

Abrams, J.F., Hoerig, L., Brozovic, R., Axtner, J., Crampton-Platt, A., Mohamed, A., et al., 2018. Shifting up a gear with iDNA: from mammal detection events to standardized surveys. bioRxiv, 449165.

Akpona, H.A., Djagoun, C.A.M.S., Sinsin, B., 2008. Ecology and ethnozoology of the three-cusped pangolin *Manis tricuspis* (Mammalia, Pholidota) in the Lama forest reserve, Benin. Mammalia 72 (3), 198–202.

Anon, 2015. First Pangolin Range State Meeting Report. June 24–26, 2015, Da Nang, Vietnam.

Borchers, D.L., Efford, M.G., 2008. Spatially explicit maximum likelihood methods for capture–recapture studies. Biometrics 64 (2), 377–385.

Brittain, S., Bata, M.N., De Ornellas, P., Milner-Gulland, E. J., Rowcliffe, M., 2018. Combining local knowledge and occupancy analysis for a rapid assessment of the forest elephant *Loxodonta cyclotis* in Cameroon's timber production forests. Oryx, 1–11.

Buckland, S.T., Anderson, D.R., Burnham, K.P., Laake, J.L., 2012. Distance Sampling: Estimating Abundance of Biological Populations. Springer Science & Business Media, Berlin.

Burgar, J.M., Stewart, F.E., Volpe, J.P., Fisher, J.T., Burton, A.C., 2018. Estimating density for species conservation: comparing camera trap spatial count models to genetic spatial capture-recapture models. Glob. Ecol. Conserv. 15, e00411.

Burton, A.C., Neilson, E., Moreira, D., Ladle, A., Steenweg, R., Fisher, J.T., et al., 2015. Wildlife camera trapping: a review and recommendations for linking surveys to ecological processes. J. Appl. Ecol. 52 (3), 675–685.

Cassell, H., 2001. Matrix Population Models: Construction, Analysis and Interpretation, second ed. Sinauer Associates, Sunderland, Massachusetts.

Challender, D.W.S., Waterman, C., Baillie, J.E.M. (Eds.), 2014. Scaling Up Pangolin Conservation. IUCN SSC Pangolin Specialist Group Conservation Action Plan. Zoological Society of London, London, UK.

Challender, D., Waterman, C., 2017. Implementation of CITES Decisions 17.239 b) and 17.240 on Pangolins (*Manis* spp.), CITES SC69 Doc. 57 Annex. Available from: <https://cites.org/sites/default/files/eng/com/sc/69/E-SC69-57-A.pdf>. [August 2, 2018].

Challender, DWS., Alvarado, D., Archer, L., Brittain, S., Chong, J.L., Copsey, J., et al. (In prep.). Developing ecological monitoring methods for pangolins (Pholidota: Manidae).

Chandler, R.B., Royle, J.A., 2013. Spatially explicit models for inference about density in unmarked or partially marked populations. Ann. Appl. Stat. 7 (2), 936–954.

Conroy, M.J., Runge, J.P., Barker, R.J., Schofield, M.R., Fonnesbeck, C.J., 2008. Efficient estimation of abundance for patchily distributed populations via two-phase, adaptive sampling. Ecology 89 (12), 3362–3370.

Cusack, J.J., Swanson, A., Coulson, T., Packer, C., Carbone, C., Dickman, A.,J., et al., 2015. Applying a random encounter model to estimate lion density from camera traps in Serengeti National Park, Tanzania. Wildlife Manage. 79 (6), 1014–1021.

Gibbs, J.P., 2008. Monitoring for Adaptive Management in Conservation Biology. Network of Conservation Educators and Practitioners, Center for Biodiversity and Conservation, American Museum of Natural History, New York.

Gitzen, R.A., Millspaugh, J.J., Cooper, A.B., Licht, D.S., 2012. Design and Analysis of Long-Term Ecological Monitoring Studies. Cambridge University Press, Cambridge.

Heath, M.E., Coulson, I.M., 1997. Home range size and distribution in a wild population of Cape pangolins, *Manis temminckii*, in north-west Zimbabwe. Afr. J. Ecol. 35 (2), 94–109.

Heinrich, S., Wittman, T.A., Ross, J.V., Shepherd, C.R., Challender, D.W.S. Cassey, P., 2017. The Global Trafficking of Pangolins: A Comprehensive Summary of Seizures and Trafficking Routes From 2010–2015. TRAFFIC, Southeast Asia Regional Office, Petaling Jaya, Selangor, Malaysia.

Ingram, D.J., Coad, L., Abernethy, K.A., Maisels, F., Stokes, E.J., Bobo, K.S., et al., 2018. Assessing Africa-Wide pangolin exploitation by scaling local data. Conserv. Lett. 11 (2), e12389.

Ingram, D.J., Willcox, D, Challender, D.W.S., 2019. Evaluation of the application of methods used to detect and monitor selected mammalian taxa to pangolin monitoring. Glob. Ecol. Conserv. 18, e00632.

Irshad, N., Mahmood, T., Hussain, R., Nadeem, M.S., 2015. Distribution, abundance and diet of the Indian pangolin (*Manis crassicaudata*). Anim. Biol. 65 (1), 57–71.

IUCN SSC Pangolin Specialist Group, 2018. Methods for monitoring populations of pangolins (Pholidota: Manidae). IUCN SSC Pangolin Specialist Group, Zoological Society of London, London, UK.

Kendall, W.L., 2001. Using models to facilitate complex decisions. In: Shenk, T.M., Franklin, A.B. (Eds.), Modeling in Natural Resource Management. Island Press, Washington D.C., pp. 147–170.

Kéry, M., Schaub, M., 2012. Bayesian Population Analysis Using WinBUGS: a Hierarchical Perspective. Academic Press, San Diego.

Kéry, M., Royle, J.A., 2016. Applied Hierarchical Modeling in Ecology: Analysis of Distribution, Abundance and Species Richness in R and BUGS, vol. 1: Prelude and Static Models. Academic Press, San Diego.

Khwaja, H., Buchan, C., Wearn, O.R., Bahaa-el-din, L., Bantlin, D., Bernard, H., et al., 2019. Pangolins in global camera trap data: implications for ecological monitoring. Glob. Ecol. Conserv. 20, e00769.

Krebs, C.J., 1991. The experimental paradigm and long-term population studies. Ibis 133 (s1), 3–8.

Lee, P.B., Chung, Y.F., Nash, H.C., Lim, N.T.L., Chan, S.K. L., Luz, S., et al., 2018. Sunda Pangolin (*Manis javanica*) National Conservation Strategy and Action Plan: Scaling Up Pangolin Conservation in Singapore. Singapore Pangolin Working Group, Singapore.

Lim, N.T.L., 2007. Autecology of the Sunda Pangolin (*Manis Javanica*) in Singapore. M.Sc. Thesis, National University of Singapore, Singapore.

Lim, N.T.L., Ng, P.K.L., 2008. Home range, activity cycle and natal den usage of a female Sunda pangolin *Manis javanica* (Mammalia: Pholidota) in Singapore. Endanger. Sp. Res. 4, 233–240.

Lindenmayer, D.B., Likens, G.E., 2009. Adaptive monitoring: a new paradigm for long-term research and monitoring. Trends Ecol. Evol. 29 (9), 482–486.

Mackenzie, D.I., Nichols, J.D., Lachman, G.B., Droege, S., Royle, J.A., Langtimm, C.A., 2002. Estimating site occupancy rates when detection probabilities are less than one. Ecology 83 (8), 2248–2255.

Mackenzie, D.I., Nichols, J.D., Royle, J.A., Pollock, K.H., Bailey, L., Hines, J.E., 2017. Occupancy Estimation and Modeling: Inferring Patterns and Dynamics of Species Occurrence. Elsevier, New York.

Marler, P.N., 2016. Camera trapping the Palawan Pangolin *Manis culionensis* (Mammalia: Pholidota: Manidae) in the wild. J. Threat. Taxa 8 (12), 9443–9448.

McCarthy, M.A., Possingham, H.P., 2007. Active adaptive management for conservation. Conserv. Biol. 21 (4), 956–963.

Mcdonald-Madden, E., Probert, W.J., Hauser, C.E., Runge, M.C., Possingham, H.P., Jones, M.E., et al., 2010. Active adaptive conservation of threatened species in the face of uncertainty. Ecol. Appl. 20 (5), 1476–1489.

Mohr, E., 1961. Schuppentiere. Neue Brehm-Bucherei. A. Ziemsen Verlag, Wittenberg Lutherstadt.

Nash, H.C., Wong, M.H.G., Turvey, S.T., 2016. Using local ecological knowledge to determine status and threats of the Critically Endangered Chinese pangolin (*Manis pentadactyla*) in Hainan, China. Biol. Conserv. 196, 189–195.

Nash, H.C., Lee, P., Low, M.R., 2018. Rescue, rehabilitation and release of Sunda pangolins (*Manis javanica*) in Singapore. In: Soorae, P.S. (Ed.), Global Re-Introduction Perspectives. Case-Studies From Around the Globe. IUCN/SSC Re-introduction Specialist Group, Abu Dhabi, UAE.

Newing, H., 2011. Conducting Research in Conservation: A Social Science Perspective. Routledge, Oxon.

Newton, P., Nguyen, T.V., Roberton, S., Bell, D., 2008. Pangolins in peril: using local hunters' knowledge to conserve elusive species in Vietnam. Endanger. Sp. Res. 6, 41–53.

Nichols, J.D., Johnson, F.A., Williams, B.K., 1995. Managing North American waterfowl in the face of uncertainty. Annu. Rev. Ecol. Syst. 26, 177–199.

Nichols, J.D., Williams, B.K., 2006. Monitoring for conservation. Trends Ecol. Evol. 21 (12), 668–673.

Nuno, A., St John, F.A.V., 2015. How to ask sensitive questions in conservation: a review of specialised questioning techniques. Biol. Conserv. 189, 5–15.

O'Connell, A.F., Nichols, J.D., Karanth, K.U., 2010. Camera Traps in Animal Ecology: Methods and Analyses. Springer Science & Business Media, Berlin.

Otis, D.L., Burnham, K.P., White, G.C., Anderson, D.R., 1978. Statistical inference from capture data on closed animal populations. Wildlife Monogr. 62, 3–135.

Pagès, E., 1975. Étude éco-éthologique de *Manis tricuspis* par radio-tracking. Mammalia 39, 613–641.

Pietersen, D.W., Mckechnie, A.E., Jansen, R., 2014. Home range, habitat selection and activity patterns of an arid-zone population of Temminck's ground pangolins, *Smutsia temminckii*. Afr. Zool. 49 (2), 265–276.

Rowcliffe, J.M., Field, J., Turvey, S.T., Carbone, C., 2008. Estimating animal density using camera traps without the need for individual recognition. J. Appl. Ecol. 45 (4), 1228–1236.

Royle, J.A., Young, K.V., 2008. A hierarchical model for spatial capture–recapture data. Ecology 89 (8), 2281–2289.

Royle, J.A., Chandler, R.B., Sollmann, R., Gardner, B., 2014. Spatial Capture-Recapture. Academic Press, San Diego.

Royle, J.A., Fuller, A.K., Sutherland, C., 2016. Spatial capture–recapture models allowing Markovian transience or dispersal. Popul. Ecol. 58 (1), 53–62.

Sollmann, R., Mohamed, A., Samejima, H., Wilting, A., 2013. Risky business or simple solution–relative abundance indices from camera-trapping. Biol. Conserv. 159, 405–412.

Soulé, M.E., 1985. What is conservation biology? Bioscience 35 (11), 727–734.

Soulé, M.E., 1991. Conservation: tactics for a constant crisis. Science 253 (5021), 744–750.

Stober, J.M., Prieto-Gonzalez, R., Smith, L.L., Marques, T. A., Thomas, L., 2017. Techniques for estimating the size of low-density gopher tortoise populations. J. Fish Wildlife Manage. 8 (2), 377–386.

Swart, J.M., Richardson, P.R.K., Ferguson, J.W.H., 1999. Ecological factors affecting the feeding behaviour of pangolins (*Manis temminckii*). J. Zool. 247 (3), 281–292.

Waits, L.P., Paetkau, D., 2005. Noninvasive genetic sampling tools for wildlife biologists: a review of applications and recommendations for accurate data collection. J. Wildlife Manage. 69 (4), 1419–1433.

Willcox, D., Nash, H.C., Trageser, S., Kim, H-J., Hywood, L., Connelly, E., et al., 2019. Evaluating methods for detecting and monitoring pangolin (Pholidata: Manidae) populations. Glob. Ecol. Conserv. 17, e00539.

Williams, B.K., Nichols, J.D., Conroy, M.J., 2002. Analysis and Management of Animal Populations. Academic Press, San Diego.

Wong, A., Fuller, A.K., Royle, J.A., 2018. Adaptive Sampling for spatial capture-recapture: an efficient sampling scheme for rare or patchily distributed species. bioRxiv, 357459.

Wu, S.B., Liu, N.F., Ma, G.Z., Xu, Z.R., Chen, H., 2003. Habitat selection by Chinese pangolin (*Manis pentadactyla*) in winter in Dawuling Natural Reserve. Mammalia 67 (4), 493–501.

Yoccoz, N.G., Nichols, J.D., Boulinier, T., 2001. Monitoring of biological diversity in space and time. Trends Ecol. Evol. 16 (8), 446–453.

CHAPTER

36

Conservation planning and PHVAs in Taiwan

Jim Kao[1], Jung-Tai Chao[2], Jason Shih-Chien Chin[1], Nian-Hong Jang-Liaw[1], Jocy Yu-Wen Li[1], Caroline Lees[3], Kathy Traylor-Holzer[3], Tina Ting-Yu Chen[1] and Flora Hsuan-Yi Lo[1]

[1]Taipei Zoo, Taipei, Taiwan [2]Taiwan Forestry Research Institute, Taipei, Taiwan [3]IUCN SSC Conservation Planning Specialist Group, Apple Valley, MN, United States

OUTLINE

Introduction

The Chinese pangolin (*Manis pentadactyla*) is native to Taiwan occurring in lowland regions (Sun et al., 2019). A number of authors (e.g., Allen, 1938; Chao, 1989) consider it an endemic sub-species, the "Formosan pangolin" (*M. p. pentadactyla*), based on morphological characteristics, but further research is needed to clarify its taxonomic status (see Chapter 4; Kao et al., 2019). Overexploitation for local use and international trade and trafficking has caused declines in Asian pangolin populations (see Chapters 14 and 16), but the Chinese pangolin in Taiwan represents one of the few potentially

Pangolins
DOI: https://doi.org/10.1016/B978-0-12-815507-3.00036-8

stable populations existing in close to natural conditions. Conservation and research activities have contributed to this success and offer unique opportunities to expand knowledge and understanding of pangolins and further their conservation.

In the 1950s and 1960s, at least 60,000 Chinese pangolins were harvested annually in Taiwan to meet demand for pangolin leather (Chao, 1989). This resulted in population declines in the late 1970s, and Taiwan resorted to sourcing pangolin skins from Southeast Asia (see Chapter 16). However, hunting for local use of meat and scales (for medicinal purposes) also took place and continued despite population declines, but with declining returns. A survey of local hunters between 1988 and 1989 revealed that the monthly harvest of pangolins had declined compared to previous years, and it was rare for hunters to find wild pangolins (Chao, 1989).

Despite this historic harvest and use, little was known about wild Chinese pangolin populations in Taiwan until the late 1980s when Jung-Tai Chao and the Taiwan Forestry Research Institute (TFRI) initiated research on the biology, reproduction and conservation of the species. Lack of knowledge also constrained efforts to maintain pangolins in captivity. In the 1990s, Taipei Zoo initiated the rescue program "Operation Project for Wildlife Sanctuary and Rescue Center" and began accumulating information on pangolin veterinary care and husbandry. Rescued pangolins experienced high mortality initially due to digestive problems brought about by inadequate diets (Chin et al., 2012). Veterinarians struggled to monitor pangolin health, or misjudged the condition of animals, due to a lack of knowledge on physiology and biology. Pangolin keepers also had little understanding of how microhabitats may influence pangolin behavior and may have provided animals with sub-optimal captive environments.

However, ex situ management provides easy access for observing animals and environmental factors, which can be controlled and monitored. Over time, valuable data and information was collected on pangolin reproductive physiology, husbandry and veterinary care, and a suitable diet was eventually developed (Chin, 2006, 2007, 2008; Taipei Zoo, unpubl. data; Yang et al., 2001; Yang, 2006). Between 1995 and 2004, 67% of mortalities of rescued pangolins at Taipei Zoo were the result of digestive problems (Chin et al., 2012). However, this declined to 13% between 2008 and 2017 (Taipei Zoo, unpubl. data).

Advances in captive care are also reflected in breeding success. The first successful breeding (i.e., conception and parturition in captivity) of a Chinese pangolin at Taipei Zoo was the birth of a male, "Chuan Pan" in December 1997; its parents were rescued in 1995 and 1996. This animal is still alive and more than 20 years old. It has also bred successfully producing a female offspring, which itself has bred producing two second-generation captive-bred Chinese pangolins.

Despite decreasing mortality rates of rescued animals by the early 2000s, wild pangolins were still under threat from poaching, trapping and attacks from feral dogs (Wang et al., 2011). To address the paucity of information on wild pangolins, inform targeted research, and promote pangolin conservation, Taipei Zoo, TFRI, the IUCN Species Survival Commission (SSC) Pangolin Specialist Group (PSG), and the Taiwan Council of Agriculture (COA) collaborated to conduct a Population Viability Analysis (PVA) and develop a conservation strategy for the Chinese pangolin in Taiwan. In 2004, with financial support from the Council of Agriculture, Taipei Zoo invited the IUCN SSC Conservation Planning Specialist Group (CPSG), formerly known as the Conservation Breeding Specialist Group (CBSG), to conduct the first Pangolin Population and Habitat Viability Assessment (PHVA) workshop for the Chinese pangolin in Taiwan.

The 2004 PHVA

A PHVA workshop is a multi-stakeholder, science-based species conservation planning process developed by the CPSG in the 1980s and is consistent with the general planning process recommended by the IUCN SSC (IUCN SSC, 2017). Facilitated plenary and small group discussions are used to analyze problems, set goals, evaluate potential solutions, and recommend conservation actions. This often includes a PVA model component that provides a quantitative tool to help biologists, managers and other stakeholders to understand the primary threats to species populations and to evaluate effective actions for their conservation.

In the 2004 PHVA workshop (Fig. 36.1), 42 participants from universities, government and zoos discussed Chinese pangolin conservation in two threat-based working groups, one centered on habitat issues and one on anthropogenic threats, and a third team worked on population biology and PVA modeling (Chao et al., 2005).

Discussion of habitat-related issues led to the identification of three primary land use activities of concern: road construction, housing, and poor agricultural practices. Specific actions were recommended to address each of these threats to achieve the goals of reducing habitat loss, fragmentation, and degradation (Table 36.1).

The most serious threats posed to pangolin populations by anthropogenic activities were hypothesized to be predation by feral dogs and direct poaching of pangolins, with development and related activities also identified as having a potential negative impact on populations. Although pesticide use and the possible presence of exotic (escaped) pangolins, which could present a disease risk, were discussed, their impact was unknown. Goals identified to mitigate these threats were: to control feral dogs, stop poaching, and reduce demand for

FIGURE 36.1 The 2004 pangolin PHVA workshop at Taipei Zoo, Taiwan.

TABLE 36.1 Habitat-related threats identified at the PHVAs, 2004 PHVA recommendations, progress made up to 2017, and 2017 PHVA recommendations.

Threat	2004 PHVA recommendations	Progress (2004–17)	2017 PHVA recommendations
Road construction	Promote critical review for new road construction Consider corridors to connect habitat Restrict road use in pangolin habitat	Since 2004, the Taiwan Area National Freeway Bureau allocated an annual budget to mitigate the negative impacts of road construction on wildlife, including the creation of habitat corridors, controlling invasive species, and preventing unnecessary road construction In 2017, the Public Construction Commission Executive Yuan developed the "Ecological Checklist Mechanism for Public Construction" for all public construction plans	Initiate research to fill knowledge gaps regarding how road construction affects populations and pangolin conservation Determine the frequency of pangolin roadkill and where it occurs (hotspots) Review existing information on pangolin roadkill and evaluate its impact on populations Create a platform for local educators near roadkill hotspots and pangolin habitat through which to disseminate information
Housing	Reduce lowland hill development Establish protected areas Enforce construction laws Establish stringent Environmental Impact Assessments (EIA) Encourage ecological (eco-friendly) engineering Initiate community awareness programs	Since 2004, nine new protected areas, including National Parks, Nature Reserves and Wildlife Protected Areas have been established Seven of these protected areas are located mainly in lowland areas	Initiate research on the interaction between land use conversion for human use and pangolin presence and density
Agricultural abuse	Improve law enforcement against land abuse (by Taiwan Forestry Bureau) Establish guidelines and hold educational events to improve public awareness of pangolins Encourage habitat restoration Reduce pesticide and herbicide use Encourage eco-farming (organic)	In 2009, the COA established a Rural Regeneration Policy that was endorsed by the Executive Yuan to replace agricultural subsidies, engage in carbon "locking" and promote environmentally friendly agricultural production In 2017, the COA announced the expansion of organic agriculture to double its current area (~ 0.8%) within three years. The Organic Agriculture Promotion Act was decreed on 30 May 2018 The *Satoyama* Initiative, recognized by the Conference of the Parties (CoP10) of the Convention on Biological Diversity (CBD) was supported by the government, private sectors, and NGOs in Taiwan	Initiate research on the interaction between land use conversion for human use and pangolin presence and density

pangolins and their parts in Taiwan. This was to be achieved through better enforcement of the Wildlife Conservation Act, which came into effect in 1989, and the promotion of alternatives to pangolin scales in Traditional Chinese Medicine (TCM) (Table 36.2).

Human activities sometimes lead to injured or displaced pangolins. As mentioned previously, the rescue and rehabilitation of such animals was hampered by the lack of sufficient knowledge of pangolin husbandry and veterinary care. This led to a recommendation to disseminate the expertise developed by Taipei Zoo and to encourage further research on rescue and rehabilitation (Table 36.3).

The PVA development proved challenging due to the paucity of information on, and knowledge of, pangolin life history, populations and threat-related data. A pangolin population model was developed using the *VORTEX* software program. *VORTEX* is a computer program that simulates the complex dynamics of wildlife population growth and how those dynamics may change through time in the presence of threats, including human activities (Lacy, 1993). The 2004 baseline model projected a slow decline in the pangolin population in Taiwan but incorporated some uncertainty. Sensitivity testing was used to identify those parameters most critical to the model results. This enabled the development of a prioritized list of research needs to address important data gaps for a more complete assessment of wild Chinese pangolin population viability (Table 36.4). Data were insufficient in 2004 to use the PVA model to make reliable population projections or to evaluate potential management strategies.

Although there was a limited amount of data available on wild pangolin populations in the lead-up to the 2004 PHVA, the workshop was beneficial for pangolin conservation. It provided the necessary incentive to improve knowledge about pangolins by increasing communication and cooperation among the participants and stakeholders and, importantly, by identifying research needs. With the knowledge and expertise present, the workshop participants were able to develop priority goals and recommend actions that could serve as a framework for a pangolin conservation management strategy.

Activities following the 2004 PHVA

Many conservation actions undertaken in Taiwan between 2004 and 2016 have benefited pangolins. Some emanated from the 2004 PHVA directly while others were the result of broader conservation action, including by the government at the national level.

The 2004 PHVA led to increased research attention on the Chinese pangolin in Taiwan, which generated additional knowledge of the species. Between 2004 and 2016, 11 masters and one Ph.D. student focused on the species, with research topics varying from life history and population genetics, to best practice rescue protocols, ecology and disease (e.g., Khatri-Chhetri et al., 2016). In addition, a wildlife rescue and reporting system was designed and implemented in 2006 (see Wang, 2007). This led to the development of the Taiwan Roadkill Observation Network, which has been actively keeping records of roadkills in Taiwan since August 2011. The Chinese pangolin is one of 31 focal species tracked by this network and there are more than 14,000 members, demonstrating concerted public interest concerning the negative impacts of roads on wildlife.

The Observer Ecological Consultant Co., Ltd. (OECC), established in 2004, helped transform the Taiwan Area National Freeway Bureau into an ecologically oriented construction organization. The Freeway Bureau allocates an annual budget to mitigate the negative impacts of road construction on wildlife. This has funded the creation of corridors

TABLE 36.2 Anthropogenic threats identified at the PHVAs, 2004 PHVA recommendations, progress made up to 2017, and 2017 PHVA recommendations.

Threat	2004 PHVA recommendations	Progress (2004–2017)	2017 PHVA recommendations
Feral/stray domestic dogs	Remove stray dogs from pangolin habitat Microchip pet dogs Effectively control garbage in order to avoid providing stray dogs with food	Increasing numbers of stray dogs is becoming a serious threat to wildlife, including pangolins	Establish a platform to collect scientific evidence about the impact of stray dogs (on pangolins) Promote educational information about the impact of stray dogs and cats on pangolin populations Hold a meeting with government agencies to discuss how to improve the management of stray dogs Improve government operated wildlife hot-line (for reporting injured pangolins) through better standard operating procedures
Poaching	Improve enforcement of wildlife conservation laws Remove traps set to catch wildlife Organize volunteer watch teams to prevent poaching	Public awareness of biodiversity and conservation in general has improved, while wildlife exploitation has decreased	Design a training course for the judiciary including topics on environment, biodiversity, poaching and trapping wildlife Convene relevant stakeholders to review problems with applicable laws and their enforcement[a] Discuss how to improve management of wildlife use, trapping and habitat Conduct research to generate basic information (e.g., on scale of poaching) for the authorities for strategic planning and law enforcement purposes
Demand for pangolin products	Improve enforcement of wildlife conservation laws Reduce demand for pangolins and pangolin products Promote pig hooves as a substitute for pangolin scales in Chinese medicine Organize watch teams to prevent poaching and trafficking	Public awareness of biodiversity and wildlife conservation, in general, has improved The use of wildlife in Taiwan has reduced, in part, because of the implementation of the Animal Protection Act and Wildlife Conservation Act	
Anthropogenic disturbance	Prepare "Rules of Conduct" for visitors in national parks, nature parks and forest recreation areas Enforce park laws Increase public awareness of native fauna and flora	Public awareness of biodiversity and conservation in general has improved, while wildlife exploitation has decreased	Establish suitable education programs to enhance public awareness about protecting the environment, and implement targeted behavior change programs in order to catalyze pro-environmental behaviors

[a] *There is a conflict between two laws, The Indigenous Peoples Basic Law (2005, amended 2018) and the Wildlife Conservation Act (2013). The former allows indigenous people to harvest wildlife, but is contradicted by the Wildlife Conservation Act.*

TABLE 36.3 Ex situ conservation related issues identified at the PHVAs, 2004 PHVA recommendations, progress made to 2017, and 2017 PHVA recommendations.

Issue	2004 PHVA recommendations	Progress (2004–17)	2017 PHVA recommendations
Poor knowledge of veterinary care for rescued pangolins	Establish protocols for quarantine and physical examination Conduct research on anatomy, physiology, nutrition, and medical care of pangolins	Accumulation of experience and research on pangolin veterinary care, husbandry, nutrition, and breeding, through Taipei Zoo's rescue and rehabilitation system	Establish a pangolin rescue integration working group Develop long-term communication among rescue centers
Dietary issues in captivity	Conduct nutritional research Determine appropriate ex situ diets	Progress has been made with specific diets for pangolins including the development of a protocol for juveniles	Conduct further husbandry nutrition, physiology, and reproduction research
Lack of standardized procedures	Establish diets and protocols for quarantine and basic physical examination procedures	Accumulation of experience and research on pangolin veterinary care, husbandry, nutrition, and breeding, through Taipei Zoo's rescue and rehabilitation system	Hold a workshop to bring together the three existing rescue centers to share standard operating procedures and agree on post-release monitoring guidelines Standardize research protocols and methods to make sure the required information is collected from rescued individuals
Insufficient capacity for injured pangolins			Explore the feasibility of an additional rescue center in eastern Taiwan. Establish training programs to increase the number of individuals with pangolin husbandry expertise
Lack of a managed ex situ insurance population		Taipei Zoo has established a breeding population for research purposes	Develop and follow guidelines for retaining non-releasable rescued pangolins Allocate available space for pangolins in the three main rescue centers in Taiwan to maximize rescue capacity

that connect habitats isolated by roads and their construction, thereby helping to reduce the impact of habitat fragmentation and the prevention of unnecessary road construction through intervention during the planning stage.

Due to the efforts of the OECC and like-minded environmental non-governmental organizations (NGOs), the Public Construction Commission Executive Yuan, made an official request in 2017 to 21 central government agencies to adopt and further develop the "Ecological Checklist Mechanism for Public Construction" for all public construction plans in the future. Although this mechanism is at an early stage, its adoption in all public construction should prevent the destruction of wildlife habitats and corridors thereby benefitting Taiwan's biodiversity, including the Chinese pangolin.

TABLE 36.4 Priority research needs identified at the PHVAs, 2004 PHVA recommendations, progress made up to 2017, and 2017 PHVA recommendations.

Issue	2004 PHVA recommendations	Progress (2004–17)	2017 PHVA recommendations
Research management	Establish a database to integrate the information of pangolin research Promote basic research on pangolin biology and populations	Between 2004 and 2016 there were eleven masters theses and one Ph.D. dissertation completed on the Chinese pangolin in Taiwan	Form a research working group and an ethics committee in order to: (1) set research priorities and standardize protocols, (2) coordinate data-sharing and sample exchange between researchers and rescue institutions, (3) ensure funding for research and conservation activities, and (4) develop training course in population monitoring
Taxonomy	Clarify the taxonomic status of the Chinese pangolin in Taiwan		Clarify the taxonomic status of the Chinese pangolin in Taiwan
Species biology	Research on anatomy and physiology	A master's thesis on diet digestibility (Chang, 2004) was completed	Initiate research to fill the knowledge gaps identified (see Kao et al., 2019)
Life history data	Conduct radio-telemetry studies to track reproduction and survival rate of pangolins Develop appropriate capture and restraint methods for research and rescue purposes Develop standard methods to identify the age of pangolins	A master's thesis on breeding and parental care (Chan, 2008) was completed Research programs and ex situ breeding population established at Taipei Zoo	Address important knowledge gaps in key demographic parameters (see Kao et al., 2019)
Population data (size and structure)	Validate burrow counting method for estimating pangolin populations Explore alternative census methods for population research Develop genetic markers for taxonomy research purposes Clarify the genetic structure of the population	A master's thesis on the kinship and social structure of Chinese pangolin populations in Taiwan was conducted based on microsatellite variation (Chang, 2014)	Determine meta-population structure and characteristics
Habitat data	Determine pangolin density across habitats Determine habitat types and available habitat for pangolins	Three masters theses were conducted on home range size, burrow habitat, and habitat distribution	Determine extent of pangolin habitat and habitat quality
Threat data	Determine mortality rates and causes	A Ph.D. thesis on health monitoring and disease surveillance was completed Four masters theses were completed on intestinal parasites, rescue and rehabilitation, and identifying pangolin products	Determine general health condition Quantify and understand how identified threats affect populations and their conservation Investigate the spatial distribution of threats in relation to the known distribution of pangolins in Taiwan

Human impacts on pangolins have increased in the last decade with the development of many lowland areas in Taiwan, including pangolin habitat. However, two main actions led by the government should benefit pangolins. First, three national parks, three nature reserves, four wildlife refuges, and seven major wildlife habitats were established between 2004 and 2018. Five of these 17 new protected areas (Xuhai-Guanyinbu Nature Reserve, Hokutolite Stone Nature Reserve, Taoyuan Gaorong Wildlife Refuge, Feitsui Reservoir Snake-eating Turtle Refuge, and Yunlin Huben Fairy Pitta Major Habitat) fall mainly in lowland areas, where pangolins occur (Fig. 36.2). As protected areas are subject to more stringent development rules, it is hoped these new sites will have a positive impact on pangolin populations.

Second, in 2009 the COA established a Rural Regeneration Policy that was endorsed by the Executive Yuan, Taiwan's highest level government agency. The key points of the policy were to: (1) replace agricultural subsidies with investment in rural communities, (2) engage in carbon "locking" (sequestration) in rural communities, and (3) promote environmentally friendly agricultural production. In 2017, the COA announced the expansion of organic agriculture to double its current area, to ~0.8% of the country's land within three years. Following the announcement, the "Organic Agriculture Promotion Act" was decreed on May 30, 2018 and was executed in 2019.

In addition, the *Satoyama* Initiative, which was recognized at the 10th meeting of the Conference of the Parties (CoP10) to the Convention on Biological Diversity (CBD) as a potentially useful tool to better understand and support human-influenced natural environments and benefit biodiversity and human well-being, was supported by the government, private sector, and NGOs in Taiwan. The *Satoyama* Initiative attempts to restore or rebuild Socio-ecological Production Landscapes (SEPL) through eco-friendly farming practices. This involves significantly reducing, or in some cases totally banning, the use of agrochemicals such as pesticides, herbicides, and chemical fertilizers.

Khatri-Chhetri et al. (2017) found a high prevalence of hepatic and respiratory lesions in dead pangolins from southeastern Taiwan between 2003 and 2014. Sun et al. (2019) proposed this might be a result of long-term exposure to toxins in the environment, due to pesticides and herbicides commonly used on Taiwanese farms. Although the relationship between pangolin mortality and agrochemicals is still not clear and in need of further research, the Rural Regeneration Policy and *Satoyama* Initiative movements should result in cleaner environments, and restoration of natural habitats, which should benefit wildlife, including pangolins.

However, numerous threats to pangolins remain. Increasing numbers of feral dogs is becoming a serious threat to local wildlife, including pangolins. It is estimated that feral dogs in Taiwan numbered ~85,000 in 2009 and increased to ~128,000 by 2015 (Council of Agriculture, Executive Yuan, 2009, 2015). Moreover, pangolin trafficking takes place in East Asia including in Taiwan. Although there is little evidence to suggest that pangolins from Taiwan are trafficked internationally, between the late 1990s and 2018, there have been seizures of pangolins in Taiwan, typically en route to China (Heinrich et al., 2017). For example, in February 2018, customs officials at Kaohsiung Harbor discovered a shipment of 4000 pangolins in an abandoned shipping container that was destined for China (Marex, 2018). Although there isn't evidence of Taiwanese pangolins being trafficked (the pangolins in the aforementioned seizure were sourced from elsewhere, most likely Southeast Asia; see Chapter 16), other species are captured in Taiwan and trafficked (e.g., yellow-margin box turtles [*Cuora flavomarginata*]; see

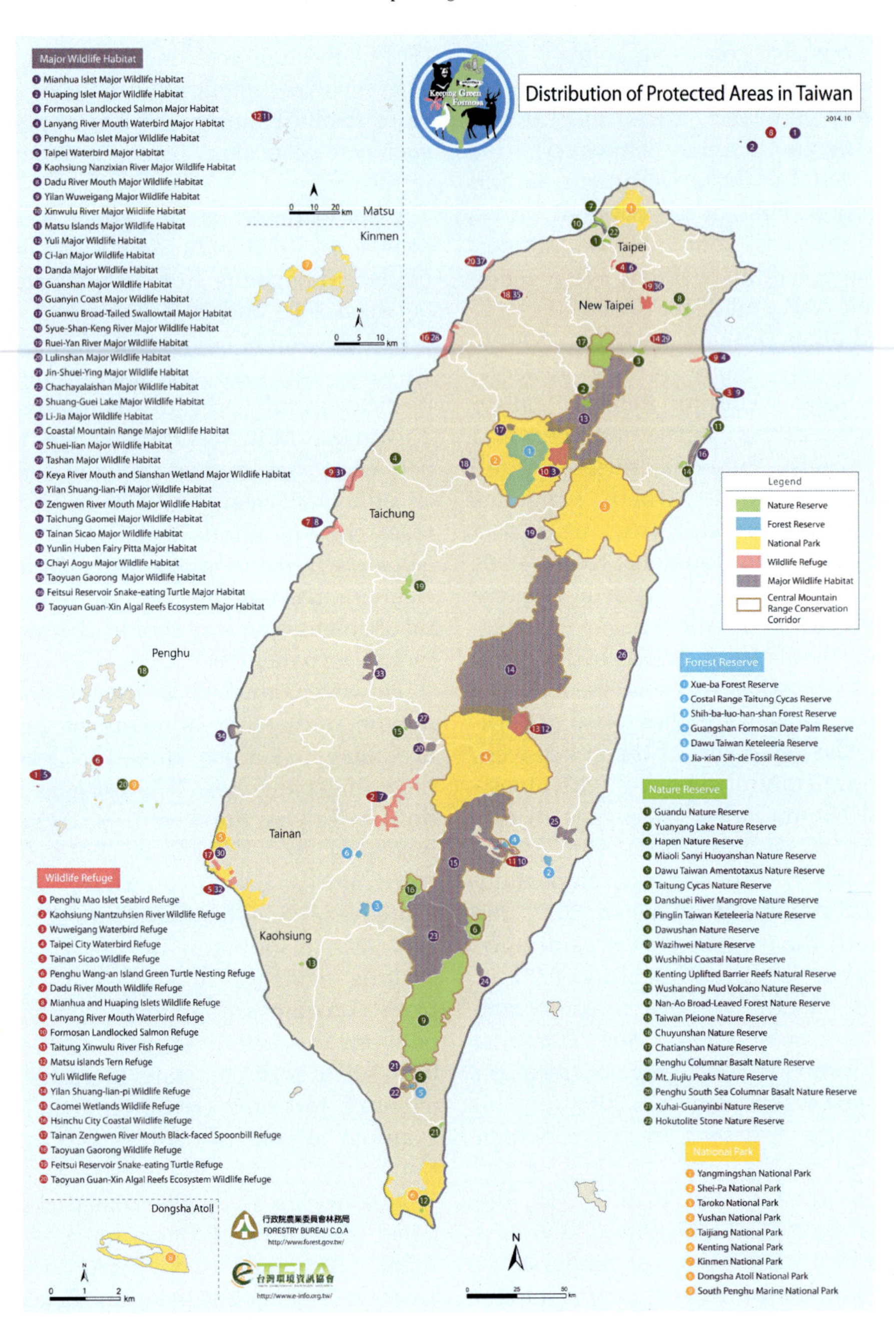

FIGURE 36.2 Distribution of protected areas in Taiwan. Five protected areas were established between 2004 and 2016 and encompass pangolin habitat.

Chen and Wu, 2016) and there is a need for law enforcement agencies in Taiwan to remain vigilant about potential pangolin poaching and trafficking in the future.

Recognizing positive action taken for pangolins between 2004 and 2016, but also ongoing and potential threats, Taipei Zoo and other major stakeholders re-convened experts in 2017 to review the status of the Chinese pangolin in Taiwan and re-evaluate the conservation action plan and associated strategies.

The 2017 PHVA

Thirteen years after the first PHVA for the Chinese pangolin in Taiwan, Taipei Zoo collaborated with the COA, ESRI (Endemic Species Research Institute) and the PSG and CPSG to host the second international PHVA workshop in December 2017 (see Kao et al., 2019). The PHVA was preceded by two one-day preparatory meetings: first, a PVA model development meeting and second, a One Plan Approach conference outlining the species conservation planning process, PVA as a quantitative planning tool, and tools for the integration of ex situ activities into species conservation. A One Plan Approach to conservation planning comprises the development of plans with input from all responsible parties, whether involved in in situ or ex situ conservation, with the ultimate goal of supporting the conservation of species in the wild (see Chapter 31). A combined 130 participants, including over 70 pangolin experts from 13 countries, attended the PHVA workshop and preparatory meetings (Fig. 36.3).

Population modeling

Updated population modeling based on the best available data for Chinese pangolin population size, biology and threats in Taiwan suggested that the wild population is not in decline and is projected to have good viability in the absence of significant undetected threats. Sensitivity analysis identified adult female survival and reproductive rate as the primary drivers of future population growth and viability. Important knowledge gaps in reproduction and survival, and especially information on the degree and impact of various threats to pangolins (e.g., dogs, roads, poaching), require investigation to better assess the future of the Chinese pangolin and to develop effective management strategies for its conservation across Taiwan. While some uncertainty still exists regarding pangolin biology, the resulting model represents the best population model available for pangolins and may serve as a basis for future PVAs for other pangolin species (Box 36.1).

Until better information is available on demographic rates, and especially on population estimates and threats, management actions that promote successful reproduction and minimize the loss of adult female pangolins from the wild are recommended to promote population viability.

Conservation action plan

A plenary discussion involving all workshop participants led to consensus on a vision for pangolins in Taiwan and identification of threats and obstacles to ensuring a viable wild pangolin population. Three concurrent working groups were developed to address: (1) habitat improvement issues and direct threats from humans (combining two group topics from the 2004 PHVA); (2) population biology, distribution and wild status (including data gaps and research); and (3) ex situ management activities to support pangolin conservation.

Participants discussed an extensive list of threats, known and potential, that are caused directly by humans. These included predation

FIGURE 36.3 Participants of the 2017 pangolin PHVA workshop at Taipei Zoo, Taiwan.

by both feral and hunting dogs, roadkill, human development (conversion of land for human use), trapping, illegal trade (potential threat), forest fires (potential threat), and logging (potential threat). Four objectives were proposed, along with actions to address them: (1) to understand and quantify how each of the threats affects pangolin populations and their conservation in Taiwan; (2) to ensure that laws governing the welfare of individual animals and those governing the conservation of wildlife are sufficiently comprehensive and well-enforced, and that related policies work in support of each other; (3) to enhance public awareness, direct targeted behavior change campaigns at priority audiences, and ensure that public education is carried out in the right way, by the right people; and (4) to identify and implement measures to protect pangolins and encourage co-existence with humans, both in natural and in human modified habitats (see Tables 36.1 and 36.2).

Group discussion of population biology, distribution and wild status identified the remaining information gaps most relevant to pangolin conservation, which fell broadly into four main areas: (1) wild Chinese pangolin meta-population structure and characteristics; (2) wild pangolin numbers and habitat quality; (3) wild pangolin health; and (4) taxonomic classification of the Chinese pangolin in Taiwan. This led to the recommendation to form a working group to coordinate research and conservation, the "Formosan Pangolin Core Group" (FPCG; see Table 36.4).

Participants representing the pangolin rescue centers in Taiwan, as well as rescue centers in Hong Kong Special Administrative Region (hereafter "Hong Kong"), Sabah (Malaysia), Thailand and Vietnam, and individuals rescuing pangolins in Nepal, met with Taipei Zoo and field conservation representatives to discuss the potential options for ex situ contributions to pangolin conservation. This was

BOX 36.1

Population viability analysis for the Chinese pangolin in Taiwan

Complex and interacting factors, both natural and anthropogenic, influence the fate of wildlife populations. A population viability analysis (PVA) uses computer modeling to evaluate this complexity and assess a population's status under current and varying conditions. In concert with the 2017 PHVA workshop, a PVA was conducted for the wild Chinese pangolin population in Taiwan using the *VORTEX* (v10.2.14) software program (Lacy and Pollak, 2017). Detailed demographic and population data were compiled by Taipei Zoo and ESRI, and revised with input from pangolin experts prior to the PHVA. While there is incomplete knowledge of Chinese pangolin biology, population size and threats, comprehensive assessment of both published and unpublished data and expert opinion led to valuable analyses. For full PVA details, see Kao et al. (2019).

While there are no reliable population estimates of wild pangolins on Taiwan, studies suggest four sub-populations with restricted connectivity. Habitat suitability modeling produced a spatial distribution model of potential pangolin habitat, and estimates of pangolin density and percent occupancy were applied to develop rough population estimates for modeling purposes.

Field and captive data were examined to develop reproductive inputs for the model. These included long-term polygynous mating, with an average age of first reproduction of 2–3 yrs, maximum age of 15 yrs, and an average inter-birth interval of 1.5 yrs. Mortality rates were generated based on general mammal life history characteristics such as generation time and expected intrinsic growth, zoo data, and information on wild population trends and mortality such as rates for roadkill and rescue.

Given the scarcity of data for wild pangolins, sensitivity testing explored primary model inputs to determine which parameters most affect population viability. The most important knowledge gaps for assessing the viability of wild Chinese pangolin populations in Taiwan and to guide effective management strategies for their conservation are: wild population size and degree of fragmentation; population trend; reproductive rate; causes and rates of loss of pangolins (through death or capture); and regional differences in threats. Understanding pangolin mortality rates, both natural and anthropogenic, is key to understanding the probable future and viability of these populations. In the absence of this specific knowledge, expert opinion and a proactive approach may help to identify probable threats and potential actions to reduce those potential threats. High priority is to understand and minimize threats that affect adult breeding female survival and reproduction.

Without accurate knowledge both of pangolin demographic rates and threats, it is difficult to provide precise estimates of future population viability and the level of losses that the Chinese pangolin population can withstand. An exploration of the impact of such losses on pangolins suggests that effects become noticeable with population losses of about 4% annually, and lead to population decline at about 6–7% annual loss (Fig. 36.4). This includes the loss of pangolins due to poaching as well as other non-natural threats, such as roadkill or predation by feral dogs. The actual number of losses that wild populations can withstand is dependent on actual population size and other sources of mortality as well as other demographic rates.

BOX 36.1 (*cont'd*)

Despite the uncertainty in demographic rates, population size and distribution, and anthropogenic threats of wild Chinese pangolins, there was sufficient information available for PVA methods to provide useful information to help guide future research and potential management. There is no direct evidence either from the field or from modeling that suggests Chinese pangolin populations are in decline in Taiwan, although these conclusions should be interpreted with caution. As knowledge improves of pangolin biology and the sources of mortality and their rates, more reliable viability projections will be possible to guide effective management actions to conserve the Chinese pangolin in Taiwan.

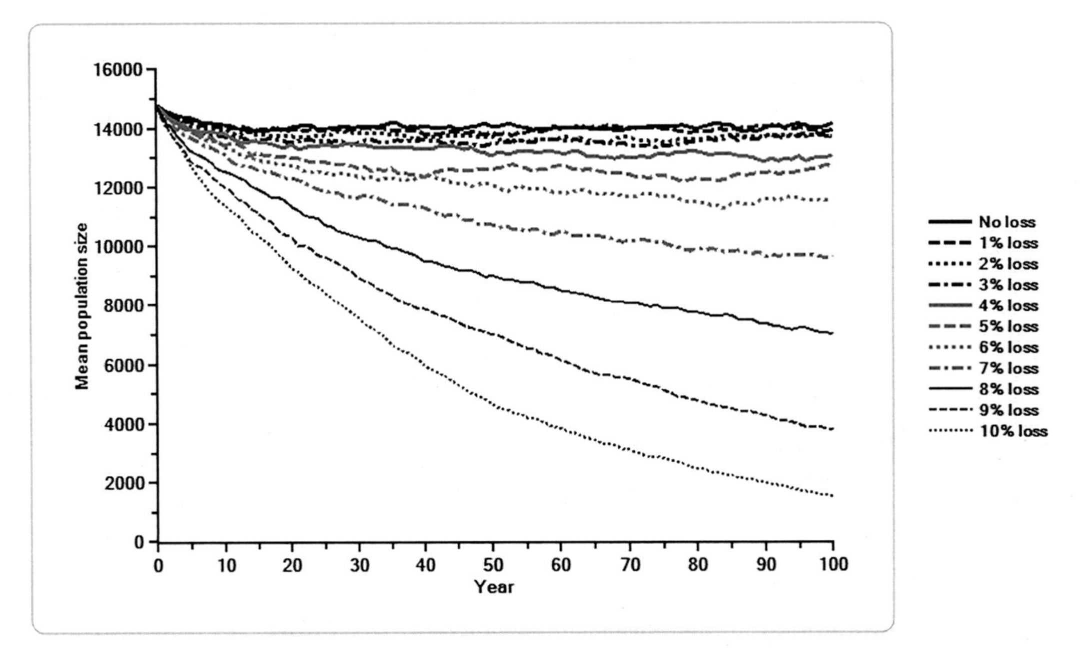

FIGURE 36.4 Projected mean size of Chinese pangolin population in Taiwan using base model values, under varying degrees of annual loss (0–10%) of pangolins across all age and sex classes.

accomplished by applying the *IUCN Guidelines for the Use of Ex Situ Management for Species Conservation* (IUCN SSC, 2014; see also Chapter 31). Three ex situ conservation roles were identified for the Chinese pangolin in Taiwan: (1) rescue, rehabilitation and release; (2) research; and (3) insurance against extinction. Group discussion explored the relative value, risks and feasibility of developing programs to fulfill these roles and to maximize the contribution of the ex situ population to pangolin conservation. This led to recommendations: (1) to conduct best practices for pangolin rescue, rehabilitation and release in Taiwan; (2) to conduct studies on the ex situ population to better understand the species and its management; and (3) to establish a viable ex situ Chinese pangolin breeding population, maintained for the purpose of insurance and founded using retained individuals from rescue operations (Table 38.3).

There is no evidence that pangolin populations in Taiwan have declined in the last decade, and anecdotal evidence that they may even be increasing in some areas. Compared to the global situation for pangolins, Taiwan is perhaps one of the last remaining strongholds where Chinese pangolins can still be studied under relatively natural conditions. However, to ensure long-term population viability, a number of challenges remain. A National Conservation Strategy Planning Meeting was held immediately following the 2017 PHVA workshop. This led to the creation of the FPCG, the purpose of which is to coordinate information sharing and implementation of the National Conservation Strategy and Action Plan. The following sub-groups were also established: the Pangolin Research Group (led by ESRI); the Pangolin Conservation Strategy Group (led by the Forest Bureau of the COA); the Pangolin Education Group (led by Taipei Zoo); and the integrated Pangolin Rescue Group (led by Taipei Zoo; see Fig. 36.5). The workshop process encouraged leadership from a broad range of stakeholders to maximize the likelihood of the recommended actions being implemented.

Comparison of two PHVAs and lessons learned

In comparing the 2004 and 2017 PHVA conservation planning workshops, differences are evident in the types of participants, integration of their roles and activities, and the issues discussed (Table 36.5).

Interdisciplinary research has increased; for example, animal science and entomology scholars from both zoos and universities have worked together to analyze the pangolin's natural diet in order to improve artificial diets, resulting in useful rescue and husbandry techniques. Ecological advisors have also contributed to pangolin conservation by supporting the habitat area analysis with Geographical Information Systems (GIS).

Zoos and rescue centers in Taiwan played a more important role in the 2017 PHVA than in the 2004 workshop. Ten years of valuable records were compiled, and the analyzed results may assist in the improvement of pangolin husbandry, reproduction, veterinary care, growth rates, juvenile development, nutrition needs, artificial diets, and disease prevention in the future. The PHVA inspired rescue centers to collaborate on the development of standard procedures for the rescue and release of pangolins and to create research programs.

Participants of the 2017 PHVA also included five conservation NGOs and rescue organizations from other Asian countries, including the Department of National Parks, Wildlife, and Plant Conservation (Thailand), Kadoorie Farm and Botanic Gardens (Hong Kong), Sabah Wildlife rescue unit (Malaysia), Save Vietnam's Wildlife (Vietnam) and the Small Mammal Conservation and Research Foundation (Nepal). This was the first time that these eight organizations (including three in Taiwan) had convened to share experiences from multiple case studies and to discuss international conservation action for pangolins.

Certain issues raised at the 2004 PHVA are no longer considered of conservation concern, such as the use of pesticides and herbicides and the potential for hybridization with exotic pangolin species (such as the Sunda pangolin). However, there are threats that remain and could become more severe in the future, that are in need of immediate action, including

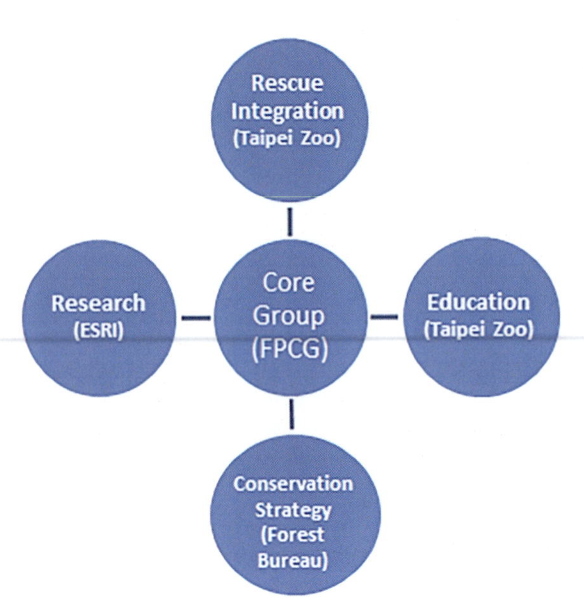

FIGURE 36.5 Implementation framework for the Formosan Pangolin National Strategy and Action Plan.

habitat destruction, hunting and poaching, illegal trapping, illegal trade, roadkill and feral dog attacks.

The 2017 PHVA resulted in a more systematic clarification of the issues first raised in 2004 and greater clarity about what needs to be done to conserve pangolins in Taiwan. Further research is required to fill gaps in knowledge about wild pangolin life history and demographic rates, population size and distribution, genetic characteristics, density and habitat utilization, and veterinary care. The pangolin conservation community in Taiwan is optimistic that recommended actions will be implemented now there is a designated group committed to taking action on agreed priorities.

By conducting two PHVA workshops for the Chinese pangolin in Taiwan, several lessons have been learnt:

- The PHVA process can help as a tool for identifying major threats and for prioritizing actions that should be taken to mitigate these threats.
- These workshops can provide a platform for local scientists and other stakeholders to share and integrate current information, to examine the current status of threatened species, and to collaborate on research and conservation activities.
- The extended period of discussion involved in the PHVA process (3–4 days) can improve collaboration among stakeholders and support consensus building between specialists and stakeholders to establish a multi-disciplinary team.
- The PHVA provides opportunities for media exposure, enhancing public awareness of the threatened species and the threats they face.
- The second PHVA in 2017 provided an opportunity to measure conservation progress since the first PHVA in 2004 and to modify conservation plans and strategies to reflect the current socio-economic and ecological context.
- It is important to integrate multidisciplinary groups of stakeholders in all workshop

TABLE 36.5 Comparison between PHVAs in 2004 and 2017.

PHVA	2004	2017
Date	**October 23–26**	**December 3–8**
Participants	Pangolin researchers, academics and graduate students, zoo staff and other biologists involved in pangolin conservation	Pangolin researchers, academics and graduate students, zoo staff, rescue centers, ecological NGOs, wildlife conservation NGOs, an ecological consultant company, and other biologists involved in pangolin conservation
PHVA Discussion Working Group	Habitat improvement Human-caused threats Population biology/modeling	Population biology Human threats Ex situ management for conservation
Issues	Habitat, environment: land use, loss, fragmentation, degradation, lack of knowledge regarding pangolin habitat requirements Human activities: road construction, housing, agriculture Feral dogs Poaching Hybridization between Chinese pangolin and Sunda pangolin (*Manis javanica*) Husbandry and veterinary care Pesticide use	Status of information about Chinese pangolins Habitat, environment: land use, heavy rain, logging, forest fire, food resource Feral dogs Roadkill Traps and hunting dogs Illegal trade Rescue, rehabilitation and release
Goals	Reduce habitat loss, fragmentation, degradation, minimize human disturbance to habitat, increase understanding of pangolins habitat use Feral dog control Stop poaching and reduce demand for pangolins Increase public awareness Stop or reduce the use of highly toxic pesticides Clarify the taxonomic status of the Chinese pangolin in Taiwan Improve veterinary care for pangolins and knowledge of their biology and medicine To clarify biological knowledge gaps in order to identify conservation research priorities	To gather key information about pangolin biology, distribution, habitat and threats To ensure adequate protection for pangolins and their habitat To maximize the contribution to conservation of ex situ pangolins
Actions	Legislation: critical review of new road construction, improve law enforcement for land use abuse Establish habitat connection: e.g., highway corridors and restrict road construction in suitable habitat Decrease development in lowland hill regions and establish protected areas Encourage habitat restoration and reduce use of pesticides and herbicides Encourage researchers to focus on pangolins	Working groups formed to coordinate conservation actions and review progress every five years: Formosan Pangolin Core Group, Research Group, Conservation Group, Education Group, Integrated Pangolin Rescue Group (see Kao et al., 2019)

discussions to ensure an integrated One Plan Approach to species conservation.
- Establishing a working group to coordinate agreed activities is one of the most important strategies in conservation action implementation.

The future of pangolins in Taiwan

Pangolins are trafficked in high numbers globally. Populations of the Chinese pangolin in China declined dramatically in the late 20th century and are in decline in other range states. Taiwan is seemingly the only place that has stable Chinese pangolin populations. Through two PHVA meetings and over more than 13 years of continuous conservation actions, much knowledge of Chinese pangolins has been accumulated. However, many actual and potential threats remain. Future conservation work will be coordinated and implemented by the FPCG and its sub-groups, sharing experiences and knowledge of this unique species towards the achievement of a collective vision for the future of pangolins in Taiwan: "By 2042, everyone is aware of and values the "Formosan pangolin" and is willing to work together to properly protect its habitat and maintain population viability, based on adequate knowledge, so that pangolins can live in harmony with human beings" (Kao et al., 2019).

Acknowledgments

The authors extend their deep appreciation to Johnnie Junior Cheng and Adriana Santacruz-Castro for their valuable assistance in the production of this chapter.

References

Allen, G.M., 1938. The Mammals of China and Mongolia. Natural History of Central Asia vol. XI. Part 1. The American Museum of Natural History, New York.

Chan, Y.T., 2008. The Breeding Behavior Study and Mother-Young Relationship of Captive Formosan Pangolins (*Manis pentadactyla pentadactyla*). M.Sc. Thesis, National Pingtung University of Science and Technology, Pingtung, Taiwan. [In Chinese].

Chang, C.Y., 2004. Study on the Apparent Digestibility of Diet on Formosan Pangolin. M.Sc. Thesis, National Taiwan University, Taipei, Taiwan. [In Chinese].

Chang, S.P., 2014. The Kinship and Social Structure of the Formosan Pangolin (*Manis pentadactyla pentadactyla*) in Luanshan, Taitung, based on Microsatellite Variations. M.Sc. Thesis, National Pingtung University of Science and Technology, Pingtung, Taiwan. [In Chinese].

Chao, J.T., 1989. Studies on the Conservation of the Formosan Pangolin (*Manis pentadactyla pentadactyla*). General Biology and Current Status. Division of Forest Biology, Taiwan Forestry Research Institute. Council of Agriculture, Executive Yuan, Taiwan. [In Chinese].

Chao, J.-T., Tsao, E.H., Traylor-Holzer, K., Reed, D., Leus, K. (Eds.), 2005. Formosan Pangolin Population and Habitat Viability Assessment: Final Report. IUCN/SSC Conservation Breeding Specialist Group, Apple Valley, Minnesota.

Chen C.F., Wu, L., 2016. Protected Turtles Seized From Taiwanese Vessel on Route to China. Available from: <http://focustaiwan.tw/news/asoc/201611220021.aspx>. [November 22, 2016].

Chin, S.C., 2006. Conservation Medicine Research on Chinese Pangolin (*Manis pentadactyla*). Research Programs of Taipei Zoo, Taipei, Taiwan. [In Chinese].

Chin, S.C., 2007. Physiology, Ecology, Pathology, and Husbandry Research on Chinese Pangolin (*Manis pentadactyla*). Research Programs of Taipei Zoo, Taipei, Taiwan. [In Chinese].

Chin, S.C., 2008. Physiology, Ecology, Pathology, and Husbandry Research on Chinese Pangolin (*Manis pentadactyla*). Research Programs of Taipei Zoo, Taipei, Taiwan. [In Chinese].

Chin, S.C., Yu, P.H., Chan, Y.T., Chen, C.Y., Guo, J.C., Yeh, L.S., 2012. Retrospective Investigation of The Death of Rescued Formosan Pangolin (*Manis pentadactyla pentadactyla*) During 1995 and 2004. Taiwan Vet. J. 38 (4), 243–250. [In Chinese].

Council of Agriculture, Executive Yuan, 2009. The Number of Stray Dogs in Each County in Taiwan. Available from: <https://animal.coa.gov.tw/html/index_06_1_4.html>. [April 2, 2019]. [In Chinese].

Council of Agriculture, Executive Yuan, 2015. The Number of Stray Dogs in Each County in Taiwan. <https://animal.coa.gov.tw/html/index_06_0621_dog.html>. [April 2, 2019]. [In Chinese].

Heinrich, S., Wittman, T.A., Ross, J.V., Shepherd, C.R., Challender, D.W.S., Cassey, P., 2017. The Global

Trafficking of Pangolins: A Comprehensive Summary of Seizures and Trafficking Routes From 2010–2015. TRAFFIC, Southeast Asia Regional Office, Petaling Jaya, Selangor, Malaysia.

IUCN SSC (IUCN Species Survival Commission), 2014. Guidelines on the Use of Ex Situ Management for Species Conservation. Version 2.0. IUCN Species Survival Commission, Gland, Switzerland.

IUCN SSC Species Conservation Planning Sub-Committee, 2017. Guidelines for Species Conservation Planning. Version 1.0. IUCN, Gland, Switzerland.

Kao, J., Li, J.Y.W., Lees, C., Traylor-Holzer, K., Jang-Liaw, N.H., Chen, T.T.Y., et al., (Eds.), 2019. Population and Habitat Viability Assessment and Conservation Action Plan for the Formosan Pangolin, *Manis p. pentadactyla*. IUCN SSC Conservation Planning Specialist Group, Apple Valley, Minnesota.

Khatri-Chhetri, R., Wang, H.-C., Chen, C.-C., Shih, H.-C., Liao, H.-C., Sun, C.-M., et al., 2016. Surveillance of ticks and associated pathogens in free-ranging Formosan pangolin (*Manis pentadactyla pentadactyla*). Ticks Tick-Borne Dis. 7 (6), 1238–1244.

Khatri-Chhetri, R., Chang, T.-C., Khatri-Chhetri, N., Huang, Y.L., Pei, K.J.C., Wu, H.Y., 2017. A retrospective study of pathological findings in endangered Taiwanese pangolins (*Manis pentadactyla pentadactyla*) from Southeastern Taiwan. Taiwan Vet. J. 43 (1), 55–64.

Lacy, R.C., 1993. VORTEX: A computer simulation model for population viability analysis. Wildl. Res. 20, 45–65.

Lacy, R.C., Pollak, J.P., 2017. VORTEX: A stochastic simulation of the extinction process. Version 10.2.14. Chicago Zoological Society, Brookfield, IL, USA.

Marex, 2018. Four Thousand Disemboweled Pangolins Found. Available from: <https://www.maritime-executive.com/article/four-thousand-disemboweled-pangolins-found>. [February 3, 2018].

Sun, N.C.-M., Arora, B., Lin, J.-S., Lin, W.-C., Chi, M.-J., Chen, C.-C., et al., 2019. Mortality and morbidity in wild Taiwanese pangolin (*Manis pentadactyla pentadactyla*). PLoS One 14 (2), e0212960.

Wang, P.J., 2007. Application of Wildlife Rescue System in Conservation of the Formosan Pangolins (*Manis pentadactyla pentadactyla*). M.Sc. Thesis, National Taiwan University, Taipei, Taiwan.

Wang, L.M., Lin, Y.J., Chan, F.T., 2011. Retrospective analysis of the causes of morbidity of wild Formosan pangolins (*Manis pentadactyla pentadactyla*). Taiwan J. Biodivers. 13 (3), 245–255. [In Chinese].

Yang, C.W., 2006. The Research on Gastrointestinal Tract Microbiota and Diet Requirements of Chinese Pangolin (*Manis pentadactyla*). Research Programs of Taipei Zoo, Taipei, Taiwan. [In Chinese].

Yang, C.W., Guo, J.C., Li, C.W., Yuan, H.W., Ttai, Y.L., Fan, C.Y., 2001. The Research on Chinese Pangolin (*Manis pentadactyla*) in Taiwan. Research Programs of Taipei Zoo, Taipei, Taiwan. [In Chinese].

CHAPTER

37

Leveraging support for pangolin conservation and the potential of innovative finance

Oliver Withers[1] *and Tenke Zoltani*[2]

[1]Conservation and Policy, Zoological Society of London, Regent's Park, London, United Kingdom

[2]Better Finance, Geneva, Switzerland

OUTLINE

Introduction

As pressure on wildlife, habitats and natural resources increases, protecting the integrity of the global commons is becoming more urgent and challenging. A transformational shift is needed in global food production and consumption patterns (Poore and Nemecek, 2018), resource use cycles, and energy generation. This entails a rethinking of production-consumption patterns, and the economic drivers thereof, which in turn requires multi-stakeholder

Pangolins
DOI: https://doi.org/10.1016/B978-0-12-815507-3.00037-X

partnerships built across various industries and sectors of society. To change these patterns, mechanisms that offer appropriate rewards to those involved in improving efficiency and increasing circularity are needed.

Exploring innovative pathways to change embedded patterns of use and consumption in a financially sustainable way, while recognizing the needs of the environment and opportunities afforded by natural capital, is a requirement for a growing number of investors and donors. There is also growing demand from investors and donors for approaches committed to producing positive environmental impacts (Reisman et al., 2018). This chapter endeavors to link these themes, identifying innovation pathways with potential for conservation. It uses case studies to demonstrate the potential for building new instruments to achieve positive outcomes for pangolin conservation. The main case study is the use of Rhino Impact Investments for rhinoceros (Rhinocerotidae) conservation. Other natural capital-based solutions are also highlighted.

To increase investment in conservation, and to foster needed solutions, elicits many questions. How can conservation organizations mobilize additional private capital for positive species outcomes? What does an "investable" or "bankable" project look like? And, what are the right structures to engage disparate partners? This chapter proposes a reframing of conservation from a challenge (communities and markets threaten the environment) to an opportunity (innovation is used to deploy capital to yield positive environmental impacts and potential financial returns).

Pangolins and challenges of the traditional conservation funding model

Overexploitation is pushing a wide range of species, including pangolins, towards extinction (Maxwell et al., 2016; Chapters 4–11). For pangolins, most offtake globally is technically illegal and high numbers of the animals and volumes of their derivatives are trafficked internationally, mainly to Asian markets (Chapter 16). Trafficking is complex and affects pangolin populations but it also impacts local communities and indigenous peoples in various ways and can foster corruption, violence and insecurity (see Chapter 23). Preventing trafficking and conserving pangolins, while ensuring sustainability of local livelihoods where the species occur, requires large-scale and long-term political and financial commitments. Success ultimately depends on changing the way in which pangolin conservation is funded.

Despite an increased focus on, and funding for, pangolin conservation since around 2013 (see Chapter 21), there remains insufficient funding from traditional sources to effectively mitigate the threats pangolins face in the long-term. Addressing this challenge requires a targeted and scalable approach, and one that utilizes innovative financing mechanisms to draw on new sources of capital. For example, capital blended from philanthropic and impact investor sources to drive improvements in protected area management effectiveness, which ties funding to outcomes.

For pangolins, the question most pertinent is "how do we enable species recovery and population growth in a sustainable and financially transparent way, while supporting the diverse needs and requirements of donors and funders?" This is immediately frustrated by the absence of standardized population monitoring methods for pangolins and appropriate indicators of pangolin status need to be developed (see Chapter 35). To answer this question, the contemporary challenges of the traditional conservation funding model, which are driven by short-term contracts, limited funds, and a lack of capacity for adaptive management, need to be examined.

Due to increasing pressure on natural resources, including from illegal wildlife trade (IWT), and limited—and in some cases—diminishing budgets (O. Withers, unpubl. data), protected area managers face significant constraints. These include financial resources being diverted away from ex ante resilience- and sustainability-building to suppressing problems like poaching or dealing with fires or other disaster response (Emerton et al., 2006). This can create a vicious cycle in which protected area and park managers are forced to pay for the challenges of today out of funds intended to prevent the challenges of tomorrow.

Funding for conservation is often only secured for short periods (e.g., 2–5 years) and may involve irregular or delayed transfer of funds. This can restrict conservation managers from swiftly adapting interventions in order to respond to rapid changes in threats, and from implementing interventions that may take longer than the specified time to reach the intended outcome. Short-term funding may also not allow sufficient time for the interventions to be institutionalized before the funding stops; as a result, many interventions may achieve their objectives, but the impacts do not last longer than the duration of the funding. This short-term approach prevents the long-term implementation of effective interventions.

The conservation sector is also facing a large funding shortfall. The projected annual cost of global biodiversity protection has been estimated at USD 300–400 billion (Credit Suisse and McKinsey, 2016). Comparably, it is estimated that only USD 52 billion was invested into conservation projects globally in 2016 (Credit Suisse and McKinsey, 2016), and the World Bank (2016) estimated that international donor funding invested in combatting IWT in the period 2010–2016 was only USD 1.3 billion. Insufficient funding for conservation restricts conservation planning and implementation, making it difficult to adequately mitigate the threats species face.

A lack of capacity for adaptive management is also inherent to the sector. Whilst adaptive management has been widely acknowledged to be key to successful conservation (e.g., Stankey et al., 2005) the elements needed, for example, for protected area managers to manage their sites adaptively are often missing from traditional funding models as discussed. In the traditional funding model, many protected areas also remain hampered by a failure to take a holistic and evidence-based approach to management and planning, lacking access to a financial mechanism to make management decisions based on (1) identified gaps in capacity for site management, and (2) a clear evidence-based theory of change with which to design and implement interventions and monitor impact. This applies to pangolin conservation, which has been limited to traditional donor or sovereign funded models, but which would benefit from results-based financing mechanisms in order to incentivize impact in more efficient and transparent models. Alternatives to the traditional funding model are discussed throughout this chapter.

The sector in context: the rise of natural capital approaches and conservation finance

In conservation finance, a direct or indirect financial investment is made to conserve the composition and values of an ecosystem for the long term. This is done by activating one or more revenue streams generated by the sustainable management of an ecosystem to generate cash flows, which in part remain within the ecosystem to enable its conservation, and in part are returned to investors or donors if financial return is sought. Examples follow later in the chapter to illustrate the concept. Within conservation most finance has been sourced from traditional funding sources, as is the case with pangolins, i.e., governments, foundations, corporations and/or private individuals donate money to projects, rather than explicitly seeking cash flows to sustain financing and longevity (Credit Suisse and McKinsey, 2016).

It is crucial that the traditional sovereign or donor-driven model is expanded towards a financially sustainable approach motivated by impact. This is because sovereign funding sources are increasingly stretched by myriad other demands put on governments (e.g., ageing populations, healthcare) and there are limitations to traditional donor (philanthropic) funding (Borgerhoff Mulder and Coppolillo, 2005), while the scale of the need and number of potential investees in conservation increases (Emerton et al., 2006). Long-term, sustainable financing would ensure a more efficient use of funds, which could be tied specifically to goals set ex-ante, and would offer the potential to recycle proceeds from financial returns for reinvestment, creating a virtuous cycle without putting greater demand on funders. One way of doing so is by identifying existing and future potential cash flows from projects, based on the natural capital underlying them, and harnessing that funding for reinvestment, repayment, or building a viable financial model. The source of this financial value, natural capital, can be defined as "the stock of renewable and non-renewable natural resources (e.g., plants, animals, air, water, soil, minerals) that combine to yield a flow of benefits to people" (adapted from Atkinson and Pearce, 1995; Daly, 1994). The need for such an approach is further emphasized recognizing macroeconomic and demographic factors. These include an increasing human population globally (estimated to be 9.77 billion by 2050), food insecurity, increasing numbers of emerging consumers with greater spending power, and a changing climate with visible and increasing damage (United Nations, 2017).

The international development community—including through pursuit of the U.N. Sustainable Development Goals (SDGs)—is catalyzing new results-focused financing instruments, some of which seek to mobilize untapped private sector capital and knowledge, while repositioning global economic and social challenges as investible opportunities. There are also several unifying factors catalyzing financing for conservation. These include the Natural Capital Declaration, which is a global statement demonstrating the commitment of the financial sector to work towards integrating natural capital criteria into financial products and services, which has been signed by many financial institutions. Adherence to environmental, social and governance (ESG) factors that can be material to financial institutions (e.g., investments on behalf of employees and clients) and business supply chain standards and commodity certifications, such as the Roundtable on Sustainable Palm Oil (RSPO) and Better Cotton Initiative (BCI), further contribute to the development of these initiatives. Such initiatives are relevant because pangolins are part of integral and fragile ecosystems that are at risk due to supply chains and business decisions that don't account for the environment and otherwise may negatively impact populations by fragmenting and destroying habitat and opening up previously inaccessible areas to exploitation. Highlighting pangolin conservation through appropriate initiatives will enable investors to intentionally seek positive impacts for pangolins. Essentially, the challenge is to transition from "do no harm" to intentional positive impact, with pangolins as the focus.

From responsible investing to conservation finance

There is enormous opportunity in responsible investments, wherein ESG aspects are considered in addition to financial returns, and which have become mainstream. Globally, assets of almost USD 23 trillion have been allocated to responsible investment strategies (up to 2016; GSIA, 2017). This constitutes an increase of 25% from 2014 and is equivalent to 26% of total global financial assets. Under the responsible investment umbrella, conservation investments can specifically target a measurable positive impact on the environment in addition to financial returns.

Revenue streams derived from natural resources like forests, fish and freshwater are independent from macroeconomic developments like inflation or public equity volatility. This is one reason that financial flows into natural resource protection and sustainable exploitation are expected to grow. Funds, direct investments, real asset investments, fixed income, and pay for performance instruments, all exist to channel capital into underlying ecosystems and establish and maintain ecosystem infrastructure. Such natural capital opportunities afford investment and philanthropy portfolios uncorrelated returns, portfolio diversification, the potential for risk-adjusted rates of return (or return of capital), clear and measurable impacts, and seek to preserve wealth and create sustainable returns. Hence, there is a business case for investment in these areas; for example, approaches that incentivize pangolin protection and conservation. However, the appropriate mechanisms and structures must be found to meet both investor and the species' needs.

Importantly, capital for investment does exist. In 2016, USD 52 billion was invested in conservation projects globally, the majority from public and philanthropic funds (Credit Suisse and McKinsey, 2016). NatureVest and EKO (2014) found that private investors in the United States intended to deploy USD 5.6 billion in conservation impact investments up to 2019, equivalent to USD 1.12bn/year, mostly in the United States, and expected returns (average target IRR [internal rate of return]) were in the range of 5%–9.9%. The same survey found USD 23.4 billion of global conservation impact investments between 2009 and 2013. Investments by development finance institutions (DFIs), such as the International Finance Corporation (IFC), totaled USD 21.5 billion, and private investments accounted for USD 1.9 billion between 2014 and 2019 (NatureVest and EKO, 2014). The conclusion was that USD 300–400 billion is needed to preserve healthy ecosystems (land, air and water), emphasizing the aforementioned funding gap. However, this funding gap creates opportunities for investors and donors, and the overall estimated cost of conservation annually (USD 300–400 billion) is tiny in the realm of total bankable assets globally of USD 175 trillion (that is, financial assets among retail, ultra-high net worth individuals, and institutional investors)—meaning "the ask" for conservation is just 0.1–0.2% of investable capital (Deloitte, 2018).

Motivated by return, yield, or other financial benefits, alongside environmental impact and a home-bias (the proclivity to invest or donate to causes in your own home or geographic region), investors looking for conservation finance solutions tend to be high net worth individuals, private foundations and, to a lesser extent, corporations. There is a common (mis)perception that returns must be sacrificed for impact. Financial innovators are thus trying to address the risk/return criteria of potential investors while maximizing the potential non-financial benefits (i.e., conservation of high-importance lands, resources, or species).

So, why isn't funding flowing more abundantly into the conservation sector? Identified hurdles include high search costs (the finding of suited investors and projects), lack of a track record of project developers, difficulty in securing collateral, weak regulation around land rights, difficulties in monetizing, scaling or replicating opportunities, expensive monitoring and/or auditing, and lack of price predictability and limited capacity, i.e., few specialists that can link conservation and finance and advise on investments (Davies et al., 2016). Conservation investment models for pangolins (as well as other species) must be able to overcome these hurdles.

The diversity of investment needs, risk and return, ecosystems, and funding sources has meant a plethora of financing mechanisms have been developed. Selected mechanisms with potential application to pangolins—conserving the species and their habitats, or the factors influencing their protection—are presented in Table 37.1. Examples of the application of such initiatives to species and habitat conservation are presented in Box 37.1 and 37.2 respectively. These mechanisms are growing in number and importance. One mechanism highlighted below, social and development impact bonds, is expanded on in the discussion of the Rhino Impact Investment (RII) project.

Outcomes-based financing: Rhino Impact Investment Project

Outcomes-based financing instruments are innovative financing mechanisms that have been gaining traction as a way to entice private capital to help address challenging social problems traditionally funded by the public sector. These instruments can potentially unlock additional funding for critical conservation areas and improve management effectiveness. This section illustrates the potential of these instruments for pangolin conservation, using the RII project as a case study.

The objective of the RII project is to transform conservation financing by demonstrating a scalable outcomes-based financing mechanism that directs additional private and public sector funds to improve the management effectiveness of priority rhino populations. It focuses on both black (*Diceros bicornis*) and white (*Ceratotherium simum*) rhinos, but primarily the former, following the technical site selection process.

The RII financing model directs impact investment funds toward selected sites to finance management interventions for rhino conservation. The investors assume the risk of the investment based on an understanding or measurement of the risk and uncertainty associated with the interventions. A Key Performance Indicator (KPI), net rhino growth rate, and pre-agreed outcome targets are used to inform the conservation outcome and measure the success of the investment. Based on the conservation outcome (measured by the KPI), the outcome-payers pay the investors back the original investment plus or minus a percentage which is linked to the outcome achieved relative to the outcome targets set (Fig. 37.1). The hypothesis behind outcome-payers paying investors a financial return is to compensate the investors for taking on the risk of not achieving the targeted outcomes and not realizing their potential financial return. The higher the risk exposure, the higher the financial return expectations become to compensate for that risk. The outcome-payers only pay for results and are guaranteed success because investors have assumed the risk of outcomes not being achieved, and the financial return is a premium outcome-payers offer to pay investors in order to guarantee pre-agreed value-for-money on rhino outcomes and impact. This allows traditional donors to transition into being outcome-payers, and to use the savings accrued from not underwriting unsuccessful interventions and transfer this outcome risk to investors.

The RII pools key rhino sites into a diversified portfolio to offer a single conservation financing instrument at scale (Fig. 37.2). The portfolio approach enables investors' risk, linked to achieving rhino outcomes, to be diversified across multiple sites and countries. This is beneficial for de-risking the product to catalyze investment and for reducing the cost

TABLE 37.1 Selected innovative finance approaches and instruments with potential for pangolins.

Innovative finance mechanism	Description
Biodiversity offsets	Measurable conservation outcomes resulting from actions that compensate for significant residual adverse biodiversity impacts arising from development projects.
Bioprospecting	Systematic search for biochemical and genetic information in nature in order to develop commercially-valuable products and applications.
Debt for nature swaps	Agreement that reduces a developing country's debt stock or service in exchange for a commitment to protect nature.
Ecological fiscal transfers	Integrating ecological services means making conservation indices (e.g., size/quality of protected areas) part of the fiscal allocation formula to reward investments in conservation.
Enterprise challenge funds	Funding instrument that distributes grants (or concessional finance) to profit-seeking projects on a competitive basis.
Green bonds	Green bonds are those where proceeds are invested exclusively in projects that generate climate or other environmental benefits. Green bonds are fixed-income securities that raise capital for a company or project with specific environmental benefits in mind. They provide a means to ring-fence funds for investments in green themes, i.e., sustainable agriculture, energy-efficient buildings, clean energy, industry and transportation, water and waste, biodiversity conservation, or urban farming.
Lotteries	Governments and civil society groups use lotteries as a means of raising funds for benevolent purposes such as education, health, preservation of historic sites and nature conservation.
Payments for ecosystem services	Payments for ecosystem services (PES) occur when a beneficiary or user of an ecosystem service makes a direct or indirect payment to the provider of that service.
Social and development impact bonds	A financial instrument that allows private (impact) investors to upfront capital for public projects that deliver social and environmental outcomes in exchange for a financial interest.
Taxes on fuel, pesticides, or on renewable natural capital	The sale tax any individual or firm who purchases fuel for his/her automobile or home heating pays; or for pesticides; or for water/timber or other natural capital usage. These taxes can reduce the consumption of fossil fuels (and pesticides) and forest cutting, while reducing greenhouse gas emissions and generate public revenues that can be channeled into conservation.
Sustainable agriculture	Sustainable agriculture notes are debt securities that invest in grower cooperatives and agricultural enterprises promoting agricultural practices that improve environmental performance and build food systems while benefiting small-to-mid-sized farmers. They offer semi-annual contingent interest payments as well as targeting social and environmental returns aimed at alleviating poverty, increasing food production and environmental conservation.

BOX 37.1

Pythons and the luxury leather industry

Pythons (Pythonidae) have been the subject of trade from Asia for the European fashion and leather industries for more than 50 yrs. The python skin has become "a classic" and demand is growing. Starting out as a skin sold in an Indonesian village for USD 30, a python skin handbag from famous Italian and French fashion houses can fetch up to USD 15,000. However, experts have raised concerns about the conservation of these species. Animal welfare groups have also campaigned against cruelty in the transport and slaughter of snakes. The "Python Conservation Partnership" was launched in 2013, a collaboration between Kering, the International Trade Centre (ITC) and the International Union for Conservation of Nature, through the IUCN Species Survival Commission (SSC) Boa and Python Specialist Group, to improve the sustainability of the python trade and facilitate industry-wide change (IUCN, 2013).

Source: Kasterine, A., Arbeid, R., Caillabet, O., Natusch, D., 2012. The Trade in South-East Asian Python Skins. International Trade Centre (ITC), Geneva, Switzerland.

BOX 37.2

Using State Revolving Funds for land conservation

State Revolving Funds (SRFs) are used in the United States as a source of low-cost financing for infrastructure projects, and are starting to be used in conservation. For example, if a local public water system needs new storage tanks or a project is devised to recapture storm water, borrowers can apply for loans through the state's Clean Water State Revolving Fund (CWSRF; Martinez, 2018).

The SRF is a state-owned infrastructure bank and uses federal funds to provide low-cost loans. The US government's Environmental Protection Agency provides states with capitalization grants to clean water and drinking water SRFs, which are matched with 20% additional funds from state sources. The grants are leveraged double or triple in value when their impact on communities is measured dollar-for-dollar, for example, for each $1 of federal capitalization, $3 of assistance is provided to communities through the CWSRF (Martinez, 2018). In 2017, the CWSRF nationally received a USD 996 million federal capitalization grant that states matched resulting in nearly USD 1.3 billion dollars (Martinez, 2018). States added to this total as well and made loans for the amount of USD 7.6 billion (Martinez, 2018).

In 2006, the CWSRF in California approved a USD 25 million loan to support the purchase of ~40,000 acres of redwood forest in Mendocino. The Conservation Fund now manages the working forest for low-level, sustainable harvests, and its goal is to restore water quality in, and the productivity of streams (Martinez, 2018).

Martinez, M., 2018. Using State Revolving Funds for Land Conservation. Available from: <https://www.conservationfinancenetwork.org/2018/05/21/using-state-revolving-funds-for-land-conservation>. [May 31, 2019].

of rhino outcomes risk, i.e., the financial return expected or required by investors to compensate for their risk exposure and impact return.

This case study focuses in detail on the technical conservation elements of the RII product development process, which demonstrates quantifiable outcome targets, measurable risk and evidence-based best practice.

There are a number of key differences between the traditional funding model and the RII approach:

(1) **Outcomes-based approach:** Unlike traditional fundraising models, the design of the RII model demands an outcomes-based approach to address conservation challenges. It shifts the focus from inputs and outputs to long-term outcomes and impact, since private investors will be concerned about recovering their investment and ensuring that it is put to most efficient and productive use by service providers who are delivering the interventions. Anchored on metrics and evidence, this model drives stakeholders to critically analyze and understand conservation issues, define progress and focus on results.

(2) **Outcome-payers only pay for results:** The value proposition for outcome-payers is very clear; they only pay if results are achieved and RII delivers its targeted net rhino growth rate. In other words, donors only reward conservation programs that are successful, according to the agreed-upon metrics. This is different from traditional conservation funding in which donors underwrite a program at the beginning, based upon predicted, but not guaranteed outcomes or value for money.

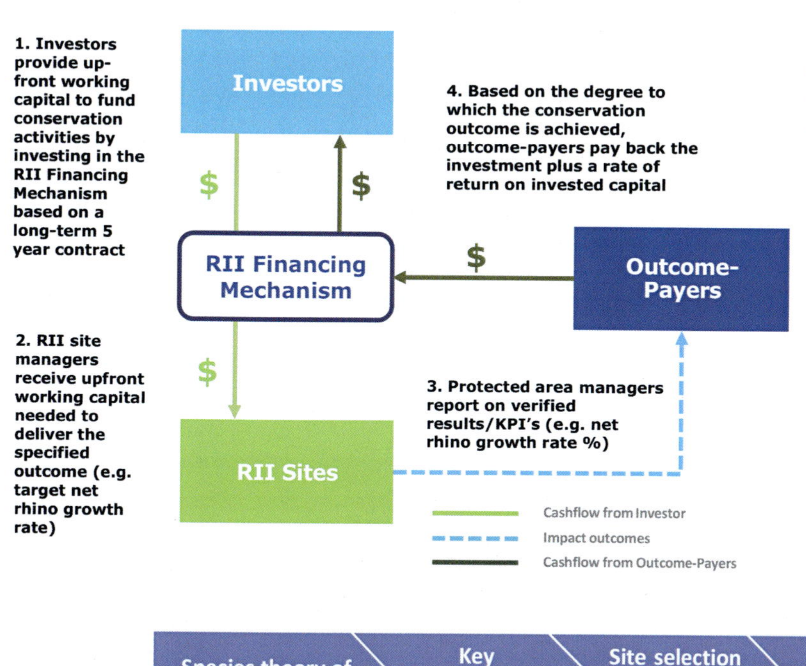

FIGURE 37.1 The RII outcomes-based financing model. *Source: Rhino Impact Investment Project.*

Species theory of change
Key performance indicators
Site selection & portfolio construction
Investment readiness

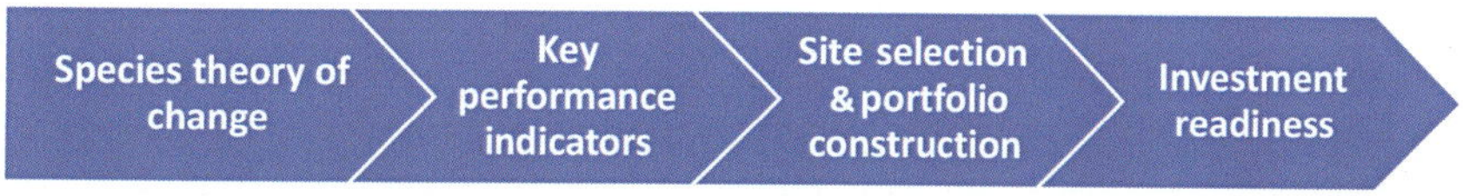

FIGURE 37.2 RII technical conservation product development process. *Source: Rhino Impact Investment Project.*

(3) **Investment capital crowded-in:** This approach transforms a development challenge into an investible opportunity rather than a problem, i.e., it creates an opportunity for impact investment. Impact investors provide up-front working capital to fund rhino conservation activities and interventions, easing current funding shortages and growing the funding pot for financing the conservation sector and reducing donor fatigue.

(4) **Incentivizes adaptive management:** Outcome-payers, investors and protected area managers are incentivized to work together to take a flexible approach to service implementation. This flexibility in management enables greater achievement of impact and ensures performance targets (i.e., rhino population growth) are met.

Theory of change

A Theory of Change (ToC) is "a comprehensive description and illustration of how and why a desired change is expected to happen in a particular context" (Theory of Change Community, 2019). It focuses on "filling in" the missing middle between what a program or particular initiative does (i.e., its activities) and how these lead to the end goals being achieved. This is accomplished by identifying the desired long-term goals and then working back from these to identify the necessary conditions that must be in place and the causal links between them for the goals to be achieved (Theory of Change Community, 2019). Within the species-centric ToC in the RII (Fig. 37.3), four actions (themes) show the causal link, that when optimized, can lead to the outcomes that enable rhino population growth.

Underpinning the rhino ToC framework for turning inputs into impact, a set of enabling conditions were identified by the project developers in consultation with protected area and rhino management experts. These factors are typically organizational and structural in nature and hence should be addressed before species specific interventions are implemented in order to maximize the efficacy of the intervention(s). From the perspective of an investor, these are viewed as clear contributors and prerequisites to the potential success of an investment. The RII ToC was endorsed by a meeting

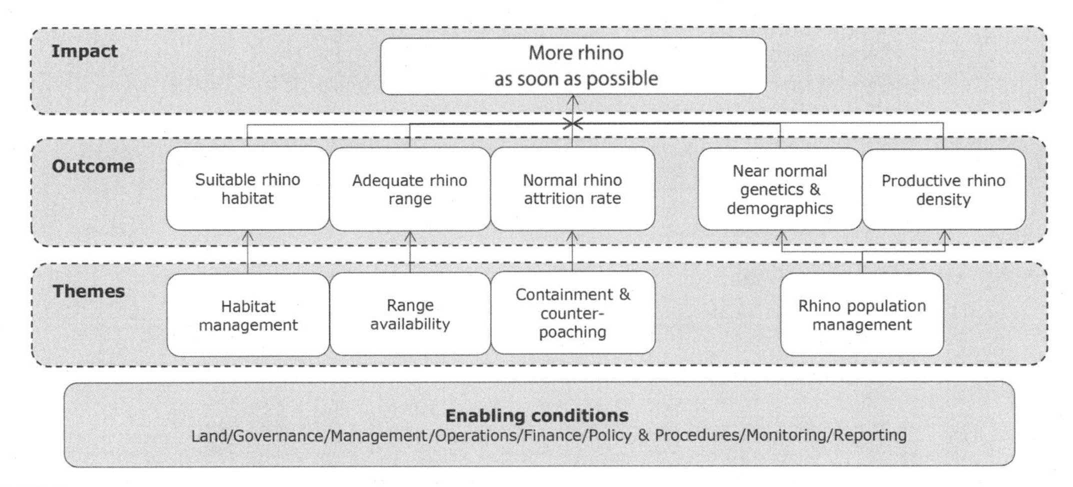

FIGURE 37.3 High level theory of change for wild rhino. *Source: Balfour, D., Barichievy, C., Gordon, C., Bret, R., 2019. A Theory of Change to grow numbers of African rhino at a conservation site. Conserv. Sci. Pract. 1 (6), e40.*

of rhino specialists (managers, researchers and consultants) at the second Rhino Science and Management Meeting (The Dinokeng Workshop) in March 2018 in South Africa.

Outcome metrics and key performance indicators

The RII KPIs for measuring success were identified and informed by the rhino conservation ToC. They were defined in consultation with and endorsed by the Chair and Scientific Officer of the IUCN SSC African Rhino Specialist Group (AfRSG). "More rhino as soon as possible" is measured by net rhino growth rate (Figs. 37.3–37.4). The net growth rate is graphically represented in Fig. 37.5. Management to achieve more rhinos as soon as possible centers on maximizing the natural biological growth rate and minimizing the unnatural mortality rate (Fig. 37.6).

The RII measure of success is the percentage of target net rhino growth rate achieved:

Primary % of KPI Target Achieved=(Actual net growth rate−Baseline net growth rate)/(Target net growth rate−Baseline net growth rate). The target growth rates and baselines are based on historical data at a site and continental level.

Site selection and portfolio construction process

The AfRSG's rhino population classification system was used to identify a portfolio of priority rhino sites. Under this rating system, the populations deemed to be of greatest continental significance—whereby their survival is considered critical for the survival of the species and subspecies—are classified as Key 1, Key 2 or Key 3 populations – Key 1 being the most significant and important.

The logic is that resources available for conservation are limited and if nothing else can be achieved, then the priority is to at least try to conserve the *Key* rated populations of continental significance and especially the highest rated *Key 1* populations.

This system of categorization has shown that by selecting only a subset of populations it is possible to conserve the majority of both black and white rhino. Of the 133 rhino populations rated as either Key or Important, 34 priority rhino sites—25 in Africa and 9 in Asia—comprise 76% of all wild rhinos.

Several of the sites were not considered an appropriate fit for RII for practical reasons. These include technical genetic queries, unconfirmed population sizes, trans-border management

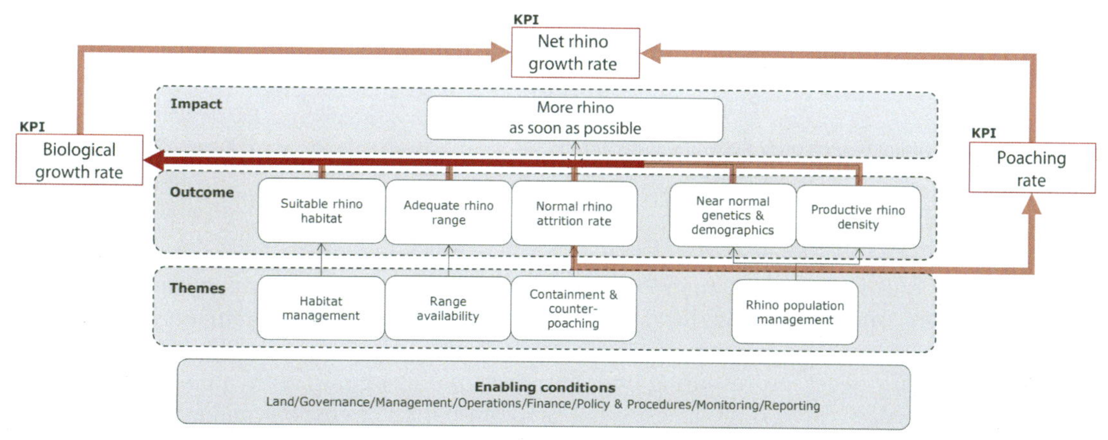

FIGURE 37.4 RII impact pathway and KPI. *Source: Rhino Impact Investment Project.*

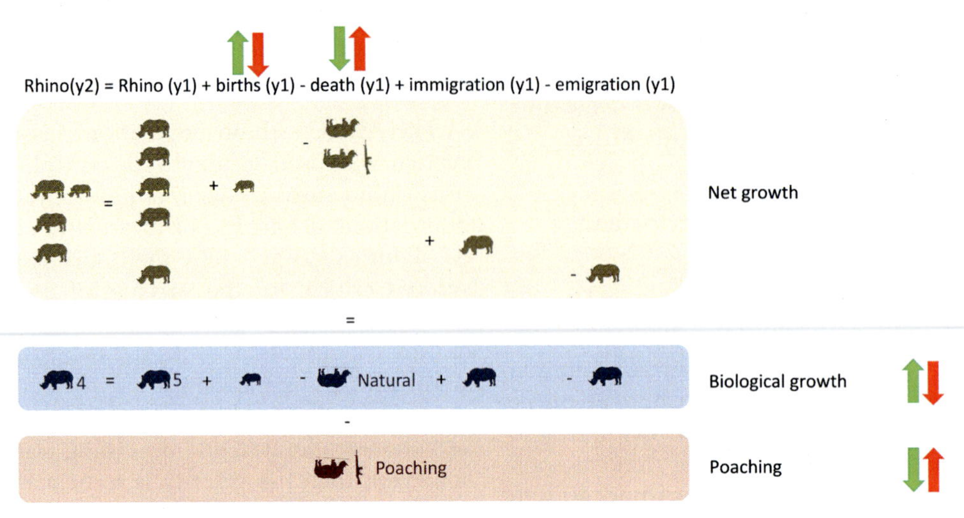

FIGURE 37.5 Net rhino growth rate calculation.

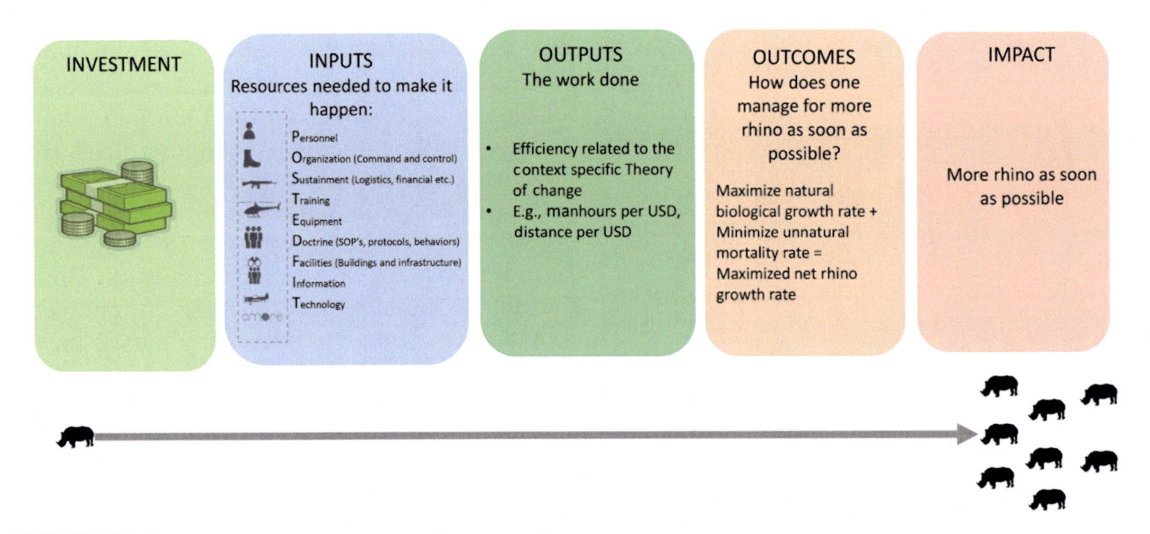

FIGURE 37.6 Managing to maximize net rhino growth rate.

challenges, political sensitivities and strategic and management uncertainty. This filtered the 25 African priority sites down to 18, 15 of which agreed to participate in gap assessments conducted by the RII project, representing over 50% of Africa's rhinos.

The RII developed a bespoke rhino-focused gap assessment tool to assess and score the selected sites' ecological, managerial and financial capacity to achieve impact. The site assessments are an aggregation type approach with a quantitative focus on management defined gaps, and an estimated cost of the interventions. The site assessments allow for both intra-site comparisons, and comparisons between sites to ascertain a state of rhino conservation relative to management defined goals.

The quantitative assessment is based on six primary indicators of rhino conservation area management effectiveness.

(1) **Security**: Can a Protected Area (PA) effectively protect its rhino population?
 a. Overt Security
 b. Covert Security
 c. Investigations
(2) **Monitoring**: Can the PA effectively monitor and manage its rhino population?
(3) **Management**: Is the PA managed effectively?
(4) **Biological Management**: Does the PA have the habitats/conditions/expertise for rhino management?
(5) **Socio-political**: Does the PA involve local stakeholders in rhino conservation?
(6) **Financing**: Does the PA have the necessary operating budget year on year?

Each Category is subdivided into practical aspects of rhino management:

(1) Strategy
(2) Operations
(3) Monitoring and Reporting
(4) Human Resources
(5) Equipment
(6) Infrastructure

For each of these subcategories, a series of questions is asked:

(1) What does the site have?
(2) What does the site need?
(3) How much will it cost to get there?

The output is a measure of the proportion of the target that the management agency has reached, the type of interventions needed and an estimated cost to achieve these targets, and an estimated potential growth rate that could be achieved under these conditions. This informs the context specific ToC developed for each site that is assessed.

Following the gap assessment, the 15 sites were evaluated by a panel according to a set of site-selection criteria drawing on the assessment results. The evaluation panel comprised recognized experts in rhino management, including members of the AfRSG, protected area management, and security personnel with experience of conservation in Africa. The panel evaluated sites based on five criteria:

(1) **The importance of the site/proposal for rhino conservation:** What is the relative importance of the population for global rhino conservation?
(2) **Intervention strategy:** Is there a logical, well thought out ToC to produce the desired results, and how large will the impact of these interventions likely be?
(3) **Agency and site manager track record:** Does the track record and status of the management agency and individuals involved engender confidence that the intervention strategy will be delivered?
(4) **Cost effectiveness:** Can the proposed intervention strategy be implemented more efficiently and/or is there any obvious aspirational spend which does not have a causal link to achieving rhino outcomes?
(5) **Risk profile:** An assessment of the risk of failure of the plan, and therefore the investment.

The criteria allow for subjectivity in appraising sites, and each site was discussed to generate consensus, but the ranking was done on an individual basis per site. All of the sites selected were black rhino populations. This was not intentional or by design, however it is important to note in hindsight that the single species focus allowed for a simpler and more transparent financing mechanism.

Site investment readiness

To be able to deliver on the impact capability identified by the evaluation panel, shortlisted sites received grant funding and technical assistance to meet a minimum level of investment readiness.

The goal of the investment readiness phase was to ensure that sites could absorb the planned investment and could deploy the resources efficiently and accountably to deliver targeted rhino outcomes. This included ensuring sites have mitigation strategies to manage risks in order to achieve targeted rhino outcomes and address any investor concerns.

Each site required the following to achieve investment readiness:

(1) **Enabling Conditions** approved following a due diligence exercise.
(2) **A Theory of Change**, a five year rhino conservation and intervention strategy and budgeted work plan.
(3) **Monitoring** to meet the level required in the context of the RII, including meeting the requirements of RII reporting and auditing:
  a. Sites must achieve an agreed confidence interval on their rhino population estimate.
  b. Sites have to provide evidence of all rhinos at the end of the investment readiness period and once per annum during the investment phase.
  c. The baseline number of rhinos in each site must be audited and judged to be accurate within an agreed confidence interval.

The specific scope of work that was conducted in the investment readiness phase ultimately depended on the context, needs and status of each of the sites selected. These were identified through an investment readiness scoping process. Following the implementation of investment readiness activities, each site was evaluated to assess whether it was investment ready. Sites that had achieved investment readiness status were then considered for inclusion in the live RII financing mechanism, completing the technical conservation elements of the site selection and portfolio construction process. Demonstrating to investors that targeted conservation outcomes can be achieved and measured using evidence-based, best practice is critical to the success of outcomes-based financing approaches.

Opportunity along the impact spectrum

Identifying where and how to make a positive impact for species conservation depends in part on what opportunities are available. These can be based on a range of variables, for example, the geography, knowledge of the target species and its conservation needs and, most importantly, the stakeholders involved. Stakeholders in conservation finance—whether working as implementers, investors, or community members—are diverse, with varying degrees of impact based on their roles. The contributions they can make to protecting a species, and the timing they can accommodate to deploy investment or implement projects, depend on their ability and capacity, as well as their resource availability. Working on projects or making investments in protected areas, or in communities in general, comes with certain geographical risks; land tenure rights create further complexity. Communities, governments, non-governmental organizations (NGOs), investors and donors all come with their own requirements and demands, which somehow must be aligned in a cost-efficient way. This variability among stakeholders presents a challenge for engagement and explains why a diversity of capital sources is required, and why it is challenging to create "bankable" projects, especially at scale.

Small, one-off ecosystem-based protection projects do not address the extensive number of species at risk, nor do they warrant the time and costs of due diligence and execution for institutional investors. As there is a multi-billion dollar need for conservation, and illegal markets for species including both rhinos and pangolins, are worth hundreds of millions of dollars, the potential impacts of successfully scaling species-based conservation are numerous and widespread. Replication will be essential, but demonstrating that models such as the RII can be effective is a crucial first step.

Impact is not just at the species level. In rhino conservation, avoiding landscape degradation does not just protect rhinos. It prevents wildfires, thereby contributing to preserving organic matter in the soil and avoids the release of stored carbon, and does not damage water quality. Protecting the landscape limits erosion, landslides, and flooding. All of this can be valued in terms of cost savings and monetized through various structures.

Proactively restoring landscapes and protected area forest health, for example, helps preserve recreation, working landscapes (such as ranching), and resources with historical and cultural significance, which can improve tourism and generate revenues, and can protect lives, livelihoods, and community health. Natural infrastructure improves water supplies—both for communities and wildlife. It can ensure adequate drinking water supplies in times of drought and improve soil composition (and vice versa, healthy soil better holds moisture). Healthy ecosystems reduce rates of erosion and flooding. Many of these interventions are potentially revenue-enhancing. Translating the measurement of benefits to economic value can help stakeholders understand the economic opportunity of conservation finance and widen stakeholder participation.

Recommendations

Pangolin conservation is indicative of a global challenge that requires an integrated sustainable development approach. To achieve positive outcomes for pangolins and sustained impact it is crucial to harmonize the three core elements of sustainable development: economic growth, social inclusion and environmental protection. This requires a holistic portfolio approach to interventions and financing such that positive impact can be realized directly and indirectly.

Using the case studies discussed, namely the RII, and informed by other innovative financing mechanisms (see Table 37.1) the following recommendations are made to inform future strategic funding opportunities for pangolin conservation:

(1) **Theory of Change:** The conservation sector needs to develop and endorse a Theory of Change which can be used as a framework for identifying impact pathways, designing interventions and developing financial instruments for conserving pangolins.
(2) **Data and metrics:** Metrics and data are key to effective programs and devising successful interventions. As has been shown in the RII, monitoring is paramount to the success of such an instrument, and creating a robust monitoring and measurement framework can lead to clear payout triggers for impact bonds. The conservation finance industry lacks clear guidance on what constitutes the most practical and effective indicators for measuring pangolin outcomes, which is linked to the lack of standardized population monitoring methods (see Chapter 35). Appropriate metrics need to be devised and they should ideally use and promote standardized data and analysis approaches, and where possible should

leverage existing infrastructure that is scientifically based, cost-effective and scalable.

(3) **Stakeholder engagement:** Sites should be chosen selectively considering which PAs, communities and beneficiaries to work with. While a diversity of stakeholders and sites can create a diversity of cash flow possibilities and increase the direct and indirect impact, too many stakeholders and sites can increase transaction costs, complexity and execution risk.

(4) **Revenue and value generation:** In creating for-profit instruments, it is essential to identify the ecosystem services for which beneficiaries are most likely to pay—for protected areas this is often most obvious through ecotourism, park fees, lodges and other safari-related opportunities but can also include quantifying fire breaks and flood barriers and the cost savings accrued from this type of natural infrastructure.

(5) **Blended capital:** Building appropriate financing mechanisms and matching risk-return criteria to investor appetite and stakeholder needs (or PA needs) is crucial. Patient capital (usually in the form of grants) is crucial in the early stages of project development, and once projects are investment-ready, concessionary capital may be used to pilot before financing is raised from institutional and other market-rate-asking capital sources.

(6) **Scale:** Due diligence, transaction fees, contracting, and measurement are predominantly fixed costs that could be prohibitive for smaller projects. To eventually attract institutional investors, investment size must justify associated costs. Aggregating, or pooling, of impact investment opportunities can deliver scale but requires effective co-ordination and collaboration across stakeholder groups.

(7) **Political relevance:** The conservation sector needs to endeavor to promote the political relevance of pangolins. This requires positioning pangolins as providing economic and social benefits in addition to their environmental benefits (see Chapter 3), and alignment with the SDGs, which should boost the relevance of the challenge and opportunities for funding.

(8) **Catalyze private investment:** In practical terms, increasing private investment will require the public sector—both governments and multilateral institutions—to provide a range of financial and policy incentives and disincentives to secure change. The disincentives might include tweaking the International Finance Corporation (IFC) Performance Standards—which is a best practice guide for business operations cognizant of ESG aspects—to better integrate biodiversity aspects, or consider more regulatory changes to promote divestment in companies facilitating IWT (as with e.g., arms, tobacco). The incentives could include tax relief (as in the UK with social impact tax incentives) and the introduction of specific impact investment wildlife bonds. The pangolin conservation community needs to work with the broader conservation community to "mainstream" IWT and conservation into global capital markets.

Conclusion

The field of conservation finance offers significant opportunities for achieving positive outcomes for pangolin conservation. Models such as the social impact bond, applied to rhinos, could be applied to pangolins in the future, assuming suitable indicators and means of monitoring pangolin populations are developed. Measuring and communicating the impact and financial returns attached to ecosystem services and natural capital will hopefully stimulate greater valuing of natural resources, habitats and species, including pangolins, and create a virtuous cycle for ongoing growth in conservation finance and increase funding for pangolin conservation. Thus, going forward, a key message is to encourage the adoption of some of these results-based and outcome-oriented financing tools to mobilize capital for pangolins by improving management and cost effectiveness. The trend in impact investing is positive in this vein, with conservation as the next focus for innovative finance, and is garnering the interest of the public and private sector, donors and investors, to test new models for achieving positive impacts for species. Working hand in hand with pangolin experts, financiers and conservationists can overcome previous hurdles to investing in pangolin conservation, and begin to make a tangible difference in the recovery and population growth of the species through innovative finance.

References

Atkinson, G., Pearce, D., 1995. Measuring sustainable development. In: Bromley, D.W. (Ed.), Handbook of Environmental Economics. Blackwell, Oxford, pp. 166–182.

Balfour, D., Barichievy, C., Gordon, C., Bret, R., 2019. A Theory of Change to grow numbers of African rhino at a conservation site. Conserv. Sci. Pract. 1 (6), e40.

Borgerhoff Mulder, M., Coppolillo, P., 2005. Conservation: Linking Ecology, Economics and Culture. Princeton University Press, Princeton.

Credit Suisse Group AG and McKinsey Center for Business and Environment, 2016. Conservation Finance. From Niche to Mainstream: The Building of an Institutional Asset Class. Credit Suisse Group AG and McKinsey Center for Business and Environment, pp. 1–28.

Daly, H., 1994. Operationalizing sustainable development by investing in natural capital. In: Jansson, A.-M., Hammer, M., Folke, C., Costanza, R. (Eds.), Investing in Natural Capital: The Ecological Economics Approach to Sustainability. Island Press, Washington D.C., pp. 22–37.

Davies, R., Engel, H., Käppeli, J., Wintner, T., 2016. Taking conservation finance to scale. Available from: <https://www.mckinsey.com/business-functions/sustainability/our-insights/taking-conservation-finance-to-scale>. [May 31, 2019].

Deloitte, 2018. The Deloitte International Wealth Management Centre Ranking 2018: The Winding Road to Future Value Creation, third ed. Deloitte Consulting AG.

Emerton, L., Bishop, J., Thomas, L., 2006. Sustainable Financing of Protected Areas: A Global Review of Challenges and Options. IUCN, Gland, Switzerland and Cambridge, UK.

GSIA (Global Sustainable Investment Alliance), 2017. Global Sustainable Investment Review 2016. Global Sustainable Investment Alliance.

IUCN. 2013. Kering, IUCN and the International Trade Centre form partnership to improve python trade. Available from: <https://www.iucn.org/content/kering-iucn-and-international-trade-centre-form-partnership-improve-python-trade>. [May 15, 2019].

Martinez, M., 2018. Using State Revolving Funds for Land Conservation. Available from: <https://www.conservationfinancenetwork.org/2018/05/21/using-state-revolving-funds-for-land-conservation>. [May 31, 2019].

Maxwell, S.L., Fuller, R.A., Brooks, T.M., Watson, J.E.M., 2016. Biodiversity: the ravages of guns, nets and bulldozers. Nature 536 (7615), 143–145.

NatureVest, EKO Asset Management Partners, 2014. Investing in Conservation: A Landscape Assessment of an Emerging Market. NatureVest, EKO Asset Management Partners.

Poore, J., Nemecek, T., 2018. Reducing food's environmental impacts through producers and consumers. Science 360 (6392), 987–992.

Reisman, J., Olazabal, V., Hoffman, S., 2018. Putting the "Impact" in impact investing: the rising demand for data and evidence of social outcomes. Am. J. Eval. 39 (3), 389–395.

Stankey, G.H., Clark, R.N., Bormann, B.T., 2005. Adaptive Management of Natural Resources: Theory, Concepts, and Management Institutions. Gen. Tech. Rep. PNW-GTR-654. U.S. Department of Agriculture, Forest Service, Pacific Northwest Research Station, Portland, Oregon, United States.

Theory of Change Community, 2019. What is Theory of Change? Theory of Change Community. Available from: <https://www.theoryofchange.org/what-is-theory-of-change/>. [May 14, 2019].

World Bank, 2016. Analysis of International Funding to Tackle Illegal Wildlife Trade. The World Bank, Washington D.C.

United Nations, 2017. World Population Prospects: The 2017 Revision, Key Findings and Advance Tables. Working Paper No. ESA/P/WP/248. United Nations, New York.

Supporting pangolin conservation through tourism

Enrico Di Minin[1,2,3] *and Anna Hausmann*[1,2]

[1]Department of Geosciences and Geography, University of Helsinki, Helsinki, Finland
[2]Helsinki Institute of Sustainability Science (HELSUS), University of Helsinki, Helsinki, Finland
[3]School of Life Sciences, University of KwaZulu-Natal, Westville, South Africa

OUTLINE

Introduction

Pangolins have been considered the most trafficked wild mammals in the world because of high volumes of illegal trade in the animals and their meat and scales (see Chapters 16 and 21; Heinrich et al., 2016; Whiting et al., 2013). Thus far, conservation efforts, for pangolins as well as for other species affected by the illegal trade, have focused on disrupting the market, e.g., by improving legislation and law enforcement and monitoring the illegal trade along supply-chains primarily. More recently, some efforts have focused on reducing the consumer demand for pangolin products (see Chapter 22; Challender et al., 2015; Cheng et al., 2017; Nijman et al., 2016; Whiting et al., 2013). However, socio-economic disparities in pangolin range states, where poaching occurs, remain vastly unaddressed, which combined with factors including ineffective law enforcement and corruption, means pangolin hunting and poaching continues seemingly unabated.

Pangolins
DOI: https://doi.org/10.1016/B978-0-12-815507-3.00038-1

Resources for biodiversity conservation are woefully inadequate (McCarthy et al., 2012). Consumptive and non-consumptive use of biodiversity have therefore been promoted as effective means of creating much needed funding to support biodiversity conservation (Di Minin et al., 2013a; Naidoo et al., 2011). Ecotourism is among the fastest growing industries in the world and conservation areas are the cornerstone of the ecotourism industry (Balmford et al., 2015). Through non-consumptive ecotourism, mutual benefits for both biodiversity and humans with whom biodiversity co-exists can be created. Ecotourism can potentially provide an important source of income for human communities in developing countries (Krüger, 2005). However, many concerns about ecotourism, including carbon emissions related to long-distance traveling, destruction of habitat, displacement of local people and disturbance to animals have also been identified (Gössling, 1999). When carried out sustainably though, ecotourism provides important opportunities to maintain land under conservation land use (Di Minin et al., 2013a), and create incentives for enhancing the survival of threatened species (Buckley et al., 2016).

Biodiversity is among the main attractors of ecotourists to conservation areas (Di Minin et al., 2013b; Naidoo and Adamowicz, 2005; Siikamäki et al., 2015). Charismatic species, mostly large-bodied mammals (e.g., tiger [*Panthera tigris*] and elephants [Elephantidae]), are some of the most sought-after species by ecotourists (Di Minin et al., 2013b; Leader-Williams and Dublin, 2000). For instance, the whole ecotourism industry in a country such as South Africa is branded around "the Big 5" charismatic species (Di Minin et al., 2013b). A focus on charismatic species only, however, limits incentives for the conservation of areas that support less charismatic species. Ecotourism markets alternative to charismatic species only, however, do exist (Di Minin et al., 2013b; Hausmann et al., 2017a). In particular, rare and elusive species, such as pangolins, are particularly attractive to more experienced tourists (Di Minin et al., 2013b). Assessing such alternative ecotourism markets would help understand the potential of ecotourism to support the conservation of less charismatic species on-the-ground (Buckley, 2013). For example, revenue from a specialized ecotourism market may help raise crucial incentives for the conservation of areas where pangolins occur, generate support for management and protection (e.g., law enforcement), and help support socio-economic development of communities. However, there has been no previous research that has assessed the potential of ecotourism to contribute to pangolin conservation in the wild.

Assessing preferences of ecotourists for biodiversity has traditionally been carried out by using revealed and stated preference methods (Adamowicz et al., 1994). In revealed preference methods, preferences are inferred from the behavior of tourists in real markets, such as estimating the cost of traveling to natural areas (e.g., Ezebilo, 2016) and the hedonic prices in relation to biodiversity features (e.g., Gibbons et al., 2014). In stated preference methods, instead, surveys are used to assess preferences for biodiversity and ecosystem services. Most widely used stated preference techniques are choice experiment and contingent valuation methods where preferences are elicited from individuals in constructed, hypothetical markets (Adamowicz et al., 1998). Changes in preferences are used to calculate willingness to pay (WTP) of ecotourists for several actions and policies. Choice experiments are increasingly used to evaluate preferences for biodiversity (e.g., Di Minin et al., 2013b; Hausmann et al., 2017a; Veríssimo et al., 2009). In contingent valuation (CV) techniques, respondents are asked to directly report their WTP to obtain a specified biodiversity experience (e.g., Ressurreição et al., 2012).

Novel approaches to understand tourists' preferences for biodiversity include mining data from social media (e.g., Hausmann et al., 2018). Social media platforms are becoming popular means of sharing information and experiences in relation to nature, and a wealth of user-generated data can be used to inform conservation science and practice (Di Minin et al., 2015). Social media data are increasingly being used in a number of fields in conservation science, ranging from assessing visitation rates in conservation areas (Tenkanen et al., 2017) to understanding which socio-economic and biological characteristics affect social media postings in conservation areas (Hausmann et al., 2017b). By comparing social media data to traditional surveys, Hausmann et al. (2018) found that social media can be used as an alternative, time and cost-effective means for assessing biodiversity preferences of ecotourists visiting conservation areas. Social media can also potentially be used to monitor species where ecotourists voluntarily or involuntarily act as citizen scientists by reporting sightings, especially of rare or elusive species (Hausmann et al., 2018).

In this chapter, a contingent valuation method is used, implemented through an online survey, and data mined from two social media platforms (Flickr and Twitter), to examine how ecotourism can contribute to pangolin conservation and to scientific research through citizen science. Specifically, the objectives were to (1) assess willingness to pay to see pangolins or support their conservation, and (2) assess how social media data can be used to infer information on the distribution of pangolins inside protected areas.

Methods

Stated preferences

Ecotourists' preferences for pangolin species were evaluated by using a CV technique, administered through an online survey, in order to assess ecotourists' WTP to see pangolins in the wild and support pangolin conservation. CV is a stated preference method, which has been extensively used in environmental economics to assess the utility value of non-marketed goods and services (Boxall et al., 1996). Here, CV was used to assess both use (i.e., non-consumptive, recreational, value in the wild), and non-use (i.e., the utility value unrelated to current or future use) values of pangolins for current and future generations. Non-use values of pangolins were assessed by asking respondents to express their agreement (on a Likert scale from 0 = not at all important, to 4 = very important) to statements related to (1) the importance of being able to see pangolins in the wild (option value), (2) the importance of knowing that other people are or may be able to see pangolins in the wild (altruistic value), (3) the importance of knowing that future generations are or may be able to see pangolins in the wild (bequest value), and (4) the importance of knowing that pangolins exist in the wild unrelated to the fact that people may be able to see them (existence value).

The use values of pangolins for ecotourism were assessed, by asking respondents whether they intended to travel in the near future (5 years' time) to areas where pangolins naturally occur. This was done by asking respondents to state their WTP for visiting a natural area with different probabilities of seeing pangolins in the wild, specifically either an area where pangolins occur and can be potentially seen, or an area where pangolins occur and can be seen with some certainty (but without touching or disturbing the animal). For those respondents who expressed no intention of traveling, their WTP was assessed for a one-off donation to conservation programs that aim to develop a sustainable ecotourism project related to pangolins. In addition, in order to understand whether previous knowledge of pangolins would affect respondents' WTP, their awareness of the existence and conservation status of pangolins, and whether respondents had ever seen a pangolin in the wild before, were assessed.

Finally, in order to identify a potential market for pangolin-related ecotourism, information about respondents' socio-demographic background, as well as preferences for other biodiversity, was collected. In order to assess preferences for broader biodiversity than pangolins, respondents were asked to indicate which type of biodiversity group (i.e., large-bodied mammals above 5 kg of weight, small-bodied mammals below 5 kg of weight, birds, reptiles, amphibians, arthropods, plants, landscapes, etc.) they were mostly seeking when taking part in wildlife watching activities. People were also asked to rank their top five favorite species, and the top five species they had not yet seen, but would particularly like to see.

Survey implementation

In order to reach a broad audience of potential national and international ecotourists/donors interested in tourism initiatives related to pangolins, an online survey, which was anonymous and voluntary, was implemented. Carrying out face-to-face interviews is expensive and potential respondents are limited to those visiting specific locations in space and time. Therefore, the survey was posted online and shared via the most popular social media groups of ecotourists visiting Southern Africa or interested in its wildlife. As social media groups are quite popular among protected area visitors, and people who are interested in wildlife watching and conservation, they provide an efficient means of capturing a representative sample of all tourists who may be willing to engage in tourism initiatives related to pangolins. In this way, a sample size was determined by using a random draw from the known population of previous, current and potentially future protected area visitors. The online survey was shared via specific social media platforms (e.g., the Africa Geographic Facebook group and on Twitter) and news blogs, and respondents were also identified by using the snowball sampling technique (Newing, 2011), where other respondents were recruited from respondents' social networks.

Social media data

Social media posts containing the word (either as a hashtag or in the description of the picture) "pangolin" were collected from the Application Programming Interface of both Flickr (https://www.flickr.com/services/api/) and Twitter (https://developer.twitter.com/en/docs/tweets/search/api-reference). A total of 105,664 posts collected from January 1970 to April 2018 (most of the posts were made later than the year 2000) for Flickr, and 2100 geotagged posts for Twitter (between 2011 and 2017) were collected. A manual classification of the content was implemented in order to maintain only geotagged posts pertaining to pangolin species. Spatial overlay analyses were then carried out to identify only posts from within the IUCN ranges of all extant pangolin species, and from within protected areas in the World Database on Protected Areas (WDPA) (https://www.protectedplanet.net/).

Results

Descriptive statistics

A total of 395 respondents, from 53 different countries, participated in the survey. Most respondents were from the United States (22%), the United Kingdom (17%), and South Africa (17%). The majority of respondents were women (65%), and with high education level (76% having either a Bachelor, Master or PhD degree). On average, respondents were 43 years old, and were mostly included in age classes 18–30 (26%) and 31–40 (23%) year-olds. Overall, 70% of respondents described themselves as interested in conservation or as working in a conservation-related field. On average, gross annual income of respondents was USD 32,500, even if extreme income classes, i.e., highest (over USD 50,000) or lowest (up to USD 5000) income, were the most frequent (33% and 19% respectively). On average, respondents spent USD 1120 in wildlife-watching travel expenses per year. Similar to the gross annual income, classes of extreme travel expense per year, i.e., highest (more than USD 3000) and lowest (less than USD 50) were the most frequent (27% and 33% respectively). Almost all respondents (97%) were aware of the existence of pangolins, and had read or seen information about the species as well as their conservation status (84%). Only 17% of respondents had seen a pangolin in the wild, although almost half of them (40%) had visited, in the past 5 years, natural areas where pangolins occur. Among all respondents, 80% expressed their intention of traveling to areas where pangolins occur and engaging in wildlife watching activities in the future.

Non-use and use values of pangolins

Overall, non-use values received high to very high scores (between 3 and 4; Fig. 38.1), regardless of whether respondents had seen pangolins in the wild or not. Existence and bequest values scored the highest and were perceived as very important. On average, non-use value was significantly higher among respondents who were specifically interested in small-bodied mammals (ANOVA - F_{1387}=9.31, $P<0.01$, Cohen's d=0.513) and birds (ANOVA - F_{1387}=8.459, $P<0.01$, Cohen's d=0.371).

Among respondents who intended to travel in the future, average WTP was significantly higher (t= − 4.44, df=573.55, $P<0.0001$) for visiting an area where pangolins can be seen with some certainty (236.7 ± 17.4 USD) compared to WTP for visiting an area where pangolins occur but are unlikely to be seen (183.11 ± 16 USD). WTP to see pangolins for sure increased with higher income (Pearson's r=0.313, $P<0.0001$) and travel expenses of respondents (Pearson's r=0.309, $P<0.0001$), and it was significantly higher among respondents who were not interested in seeing large-bodied mammals (ANOVA - F_{1288}=4.7, $P<0.05$, Cohen's d=0.741) (Fig. 38.2). Among respondents who did not intend to travel, average values of WTP for a donation was USD 27.72 ± 6.5 and it was significantly higher (ANOVA - $F_{1.78}$=10.32, $P<0.001$, Cohen's d=0.726) among people who are not interested

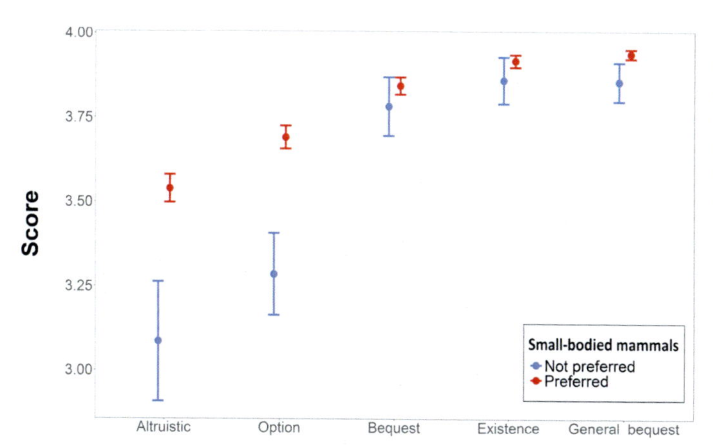

FIGURE 38.1 Average scores for non-use (altruistic, option, bequest, existence) values of pangolins, and for broader biodiversity (general bequest value), for respondents preferring, or not, small-bodied mammals. Bars indicate confidence intervals at 0.95 significance value.

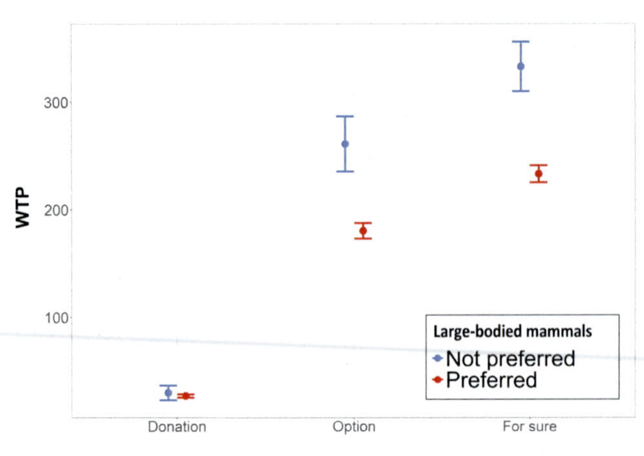

FIGURE 38.2 Average willingness to pay (WTP) in USD for respondents preferring, or not, large-bodied mammals and (a) for donating to an ecotourism pangolin conservation project (WTP donation), (b) visiting an area where pangolins occur but may not be seen (WTP option), and (c) visiting an area where pangolins can be seen with high certainty (WTP for sure). Bars indicate confidence intervals at 0.95 significance value.

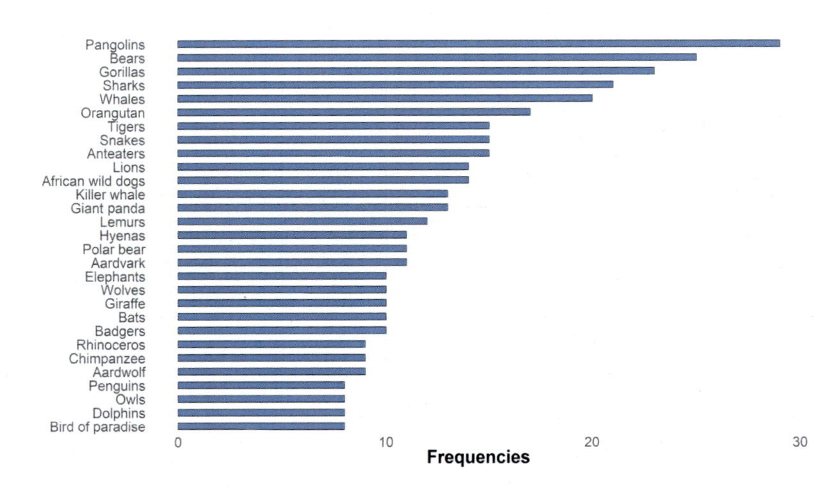

FIGURE 38.3 Favorite species ranked by respondents as top five species, which they have not yet seen but would particularly like to see. Frequencies are weighted according to the position in the rank (i.e., if top 1 = 5 points, top 2 = 4 points, top 3 = 3 points, top 4 = 2 points, top 5 = 1 point).

specifically in wildlife watching, but preferred to experience broader biodiversity when visiting natural areas.

Finally, compared to other species groups, 18% of respondents indicated that pangolins were among their top five favorite species, regardless of whether they had seen the species or not. Moreover, 58% of respondents indicated that pangolins were the top ranked species among the top five species that respondents have not yet seen but would particularly like to see (Fig. 38.3).

Social media and pangolin distribution

A total of 1000 geotagged posts (632 on Flickr and 368 on Twitter) related to pangolins were found to occur within the IUCN geographic ranges of all extant pangolin species (Fig. 38.4). These posts occurred in 41 out of 54 countries in Africa and Asia where pangolins occur (Fig. 38.4). The location with the highest number of geotagged posts on Flickr was Taiwan while the country with the highest number of geotagged posts on Twitter was

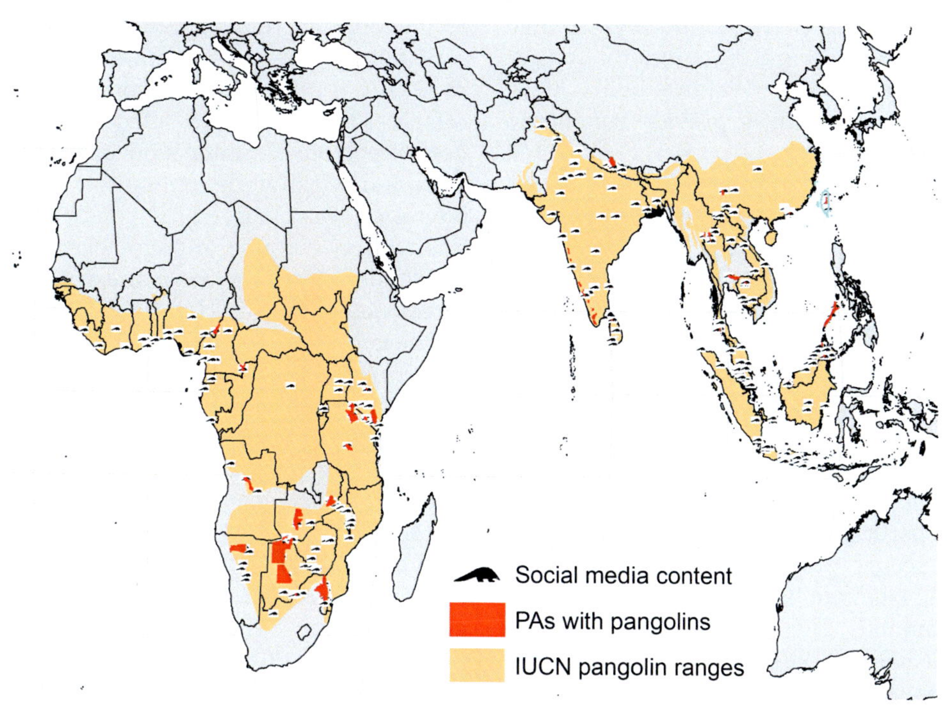

FIGURE 38.4 Geotagged social media posts from Twitter and Flickr on pangolins from within the range of all extant pangolin species. In red are protected areas (PAs) from the World Database on Protected Areas (https://www.protectedplanet.net/) where geotagged social media posts occurred. Note the geographic range maps are from The IUCN Red List of Threatened Species (version 2014-3).

Indonesia. Overall, the geotagged posts confirmed presence of pangolins from within 58 protected areas across all range countries.

Discussion

Our results highlight that, at least in Southern Africa, pangolins can potentially play an important ecotourism role, as most respondents ranked them as their favorite species among those they have not yet seen. Related to this, our results also highlight how the willingness to pay of respondents was higher under the scenario when it was guaranteed the chance to see pangolins compared to the scenario when it was only possible to see them. As our sample of respondents included ecotourists who had also visited other conservation areas in sub-Saharan Africa, our results have wider continental implications beyond Southern Africa. Interest in supporting pangolin conservation was also found among those respondents who were not planning to visit areas where pangolins occur. These results have potential global implications in relation to promoting ecotourism and raising funds to support conservation throughout pangolin range countries. Finally, social media posts were found to provide accurate spatiotemporal information on the global distribution of pangolins.

Pangolins face a high risk of extinction because of pressure from the illegal wildlife trade, largely in meat and scales (Heinrich et al., 2016; Whiting et al., 2013). Pangolins occur in areas, especially in developing countries, where socio-economic and human needs are the highest and resources for conservation often the scarcest (Waldron et al., 2013). Ecotourism can potentially provide benefits to local communities when properly managed (Isaac, 2000). Our results highlight that there is potentially high interest among ecotourists to see pangolins and willingness to pay is higher when the chance to see pangolins is guaranteed. While there are a number of animal welfare (carry out ecotourism activities that do not harm animals) and financial (capital needed to start ecotourism ventures) aspects to consider, it is important to highlight the potential that exists for this reclusive species to be used as an ecotourism flagship, at least in Southern Africa. In particular, our results highlight that WTP for seeing pangolins is higher among ecotourists who are less interested in large-bodied mammals. This suggests that areas that lack charismatic species may be attractive to alternative markets of ecotourists who are interested in pangolins. While these results should be considered carefully, especially in relation to market saturation for ecotourism, they nonetheless indicate the potential to develop region-specific ecotourism projects based on pangolin conservation. Priority should be potentially given to more accessible and stable regions, which are preferred by tourists (Hausmann et al., 2017b), where poaching levels are high. In these areas, ecotourism could generate a number of potential benefits for local communities, ranging from direct employment in ecotourism programs (e.g., as guides), or as rangers, to providing services (e.g., food for lodges), which, in turn, could help support enhanced pangolin protection and conservation.

Our sample of national and international ecotourists, who were interested in visiting conservation areas in sub-Saharan Africa and enthusiastic about conservation, are representative of tourists who choose to travel to Africa for biodiversity experiences (Di Minin et al., 2013b; Hausmann et al., 2017a). The respondents were also well aware of the conservation status and challenges facing pangolins. Our results can therefore be considered robust with regards to adequate representation of ecotourists interested in pangolin conservation. Meanwhile, ecotourists who are less interested in biodiversity conservation, and less willing to travel to areas where pangolins occur, also cared for the conservation of these threatened species and were willing to donate for their conservation. Our results confirm, as in recent studies (Di Minin et al., 2013b; Hausmann et al., 2017a), that there is a market segment that does not necessarily support the notion of ecotourism marketing focusing on charismatic species only, but that ecotourism marketing can also focus on other species, such as pangolins.

Moreover, social media data can potentially provide new insights about the distribution of species (Barve, 2014) and preferences for them (Hausmann et al., 2018). In this chapter, the role of social media data to infer the occurrence of pangolins throughout their range was assessed. It was confirmed that, even in the case of elusive species such as pangolins, social media data can provide additional information to ecological surveys, such as camera trap surveys, that can give new insights on the presence of the species both within and outside protected areas. As this information is also temporal, monitoring of voluntary and involuntary citizen-science data could provide real-time insights about how often pangolins are seen in protected areas, enabling inferences about status and whether local extinction might have happened. While only geotagged social media posts were considered for this study, automatic content classification provides an efficient means of extracting information about pangolins sightings (e.g., by inferring locations from text content) on a larger sample, including non-geotagged posts (Di Minin et al., 2018; Di Minin et al., 2019). However, this information should be used carefully, in order not to disclose exact locations of pangolins, which may be potentially targeted for poaching (Lowe et al., 2017).

Conclusion

This chapter has shown that there is interest and willingness to pay to see pangolins in the wild among ecotourists, which can potentially create incentives for the conservation of the species. However, where this is feasible, there is a need to develop financially sustainable ecotourism projects that can support both pangolin conservation and local livelihoods, and explicitly consider animal welfare concerns. Contributions to local livelihoods might, in turn, create incentives to conserve pangolins rather than poach them. Finally, new ecotourism projects should be developed strategically, in accordance with conservation strategies and priorities, so that they can help achieve conservation and sustainable development objectives (e.g., in areas where poaching is high, or pangolin populations are high).

References

Adamowicz, W., Louviere, J., Williams, M., 1994. Combining revealed and stated preference methods for valuing environmental amenities. J. Environ. Econ. Manage. 26 (3), 271–292.

Adamowicz, W., Boxall, P., Williams, M., Louviere, J., 1998. Stated preference approcaches for measuring passive use values: chioice experiments and contingent valuation. Am. J. Agric. Econ. 80 (1), 64–75.

Balmford, A., Green, J.M.H., Anderson, M., Beresford, J., Huang, C., Naidoo, R., et al., 2015. Walk on the wild side: estimating the global magnitude of visits to protected areas. PLoS Biol. 13 (2), e1002074.

Barve, V., 2014. Discovering and developing primary biodiversity data from social networking sites: a novel approach. Ecol. Inf. 24, 194–199.

Boxall, P.C., Adamowicz, W.L., Swait, J., Williams, M., Louviere, J., 1996. A comparison of stated preference methods for environmental valuation. Ecol. Econ. 18 (3), 243–253.

Buckley, R., 2013. To use tourism as a conservation tool, first study tourists. Anim. Conserv. 16 (3), 259–260.

Buckley, R.C., Morrison, C., Castley, J.G., 2016. Net effects of ecotourism on threatened species survival. PLoS One 11 (2), 23–25.

Challender, D.W.S., Harrop, S.R., MacMillan, D.C., 2015. Understanding markets to conserve trade-threatened species in CITES. Biol. Conserv. 187, 249–259.

Cheng, W., Xing, S., Bonebrake, T.C., 2017. Recent pangolin seizures in China reveal priority areas for intervention. Conserv. Lett. 10 (6), 757–764.

Di Minin, E., MacMillan, D.C., Goodman, P.S., Escott, B., Slotow, R., Moilanen, A., 2013a. Conservation businesses and conservation planning in a biological diversity hotspot. Conserv. Biol. 27 (4), 808–820.

Di Minin, E., Fraser, I., Slotow, R., MacMillan, D.C., 2013b. Understanding heterogeneous preference of tourists for big game species: implications for conservation and management. Anim. Conserv. 16 (3), 249–258.

Di Minin, E., Tenkanen, H., Toivonen, T., 2015. Prospects and challenges for social media data in conservation science. Front. Environ. Sci. 3, 63.

Di Minin, E., Fink, C., Tenkanen, H., Hiippala, T., 2018. Machine learning for tracking illegal wildlife trade on social media. Nat. Ecol. Evol. 2, 406–407.

Di Minin, E., Fink, C., Hiippala, T., Tenkanen, H., 2019. A framework for investigating illegal wildlife trade on social media with machine learning. Conserv. Biol. 33 (1), 210–213.

Ezebilo, E.E., 2016. Economic value of a non-market ecosystem service: an application of the travel cost method to nature recreation in Sweden. Int. J. Biodivers. Sci., Ecosyst. Serv. Manage. 12 (4), 314–327.

Gibbons, S., Mourato, S., Resende, G.M., 2014. The amenity value of English nature: a hedonic price approach. Environ. Resour. Econ. 57 (2), 175–196.

Gössling, S., 1999. Ecotourism: a means to safeguard biodiversity and ecosystem functions? Ecol. Econ. 29 (2), 303–320.

Hausmann, A., Slotow, R., Fraser, I., Di Minin, E., 2017a. Ecotourism marketing alternative to charismatic megafauna can also support biodiversity conservation. Anim. Conserv. 20 (1), 91–100.

Hausmann, A., Toivonen, T., Heikinheimo, V., Tenkanen, H., Slotow, R., Di Minin, E., 2017b. Social media reveal that charismatic species are not the main attractor of ecotourists to sub-Saharan protected areas. Sci. Rep. 7 (1), 763.

Hausmann, A., Toivonen, T., Slotow, R., Tenkanen, H., Moilanen, A., Heikinheimo, V., et al., 2018. Social media data can be used to understand tourists' preferences for nature-based experiences in protected areas. Conserv. Lett. 11 (1), 1–10.

Heinrich, S., Wittmann, T.A., Prowse, T.A.A., Ross, J.V., Delean, S., Shepherd, C.R., et al., 2016. Where did all the pangolins go? International CITES trade in pangolin species. Glob. Ecol. Conserv. 8, 241–253.

Isaac, J., 2000. The limited potential of ecotourism to contribute to wildlife conservation. Wildlife Soc. Bull. 28 (1), 61–69.

Krüger, O., 2005. The role of ecotourism in conservation: panacea or Pandora's box ? Biodivers. Conserv. 14 (3), 579–600.

Leader-Williams, N., Dublin, H., 2000. Charismatic megafauna as "flagship species." In: Entwistle, A., Dunstone, N. (Eds.), Priorities for the Conservation of Mammalian Diversity: Has the Panda Had Its Day. Cambridge University Press, Cambridge, pp. 53–81.

Lowe, A.J., Smyth, A.K., Atkins, K., Avery, R., Belbin, L., Brown, N., et al., 2017. Publish openly but responsibly. Science 357 (6347), 141.

McCarthy, D.P., Donald, P.F., Scharlemann, J.P.W., Buchanan, G.M., Balmford, A., Green, J.M.H., et al., 2012. Financial costs of meeting global biodiversity conservation targets: current spending and unmet needs. Science 338 (6109), 946–949.

Naidoo, R., Adamowicz, W.L., 2005. Biodiversity and nature-based tourism at forest reserves in Uganda. Environ. Dev. Econ. 10 (2), 159–178.

Naidoo, R., Weaver, L.C., Stuart-Hill, G., Tagg, J., 2011. Effect of biodiversity on economic benefits from communal lands in Namibia. J. Appl. Ecol. 48 (2), 310–316.

Newing, H.N., 2011. Conducting Research in Conservtion. A Social Science Perspective. Routledge, Oxon.

Nijman, V., Zhang, M.X., Shepherd, C.R., 2016. Pangolin trade in the Mong La wildlife market and the role of Myanmar in the smuggling of pangolins into China. Glob. Ecol. Conserv. 5, 118–126.

Ressurreição, A., Gibbons, J., Kaiser, M., Dentinho, T.P., Zarzycki, T., Bentley, C., et al., 2012. Different cultures, different values: the role of cultural variation in public's WTP for marine species conservation. Biol. Conserv. 145 (1), 148–159.

Siikamäki, P., Kangas, K., Paasivaara, A., Schroderus, S., 2015. Biodiversity attracts visitors to national parks. Biodivers. Conserv. 24 (10), 2521–2534.

Tenkanen, H., Di Minin, E., Heikinheimo, V., Hausmann, A., Herbst, M., Kajala, L., et al., 2017. Instagram, Flickr, or Twitter: assessing the usability of social media data for visitor monitoring in protected areas. Sci. Rep. 7, 17615.

Veríssimo, D., Fraser, I., Groombirdge, J., Bristol, R., MacMillan, D.C., 2009. Birds as tourism flagship species: a case study of tropical islands. Anim. Conserv. 12 (6), 549–558.

Waldron, A., Mooers, A.O., Miller, D.C., Nibbelink, N., Redding, D., Kuhn, T.S., et al., 2013. Targeting global conservation funding to limit immediate biodiversity declines. Proc. Natl. Acad. Sci. U.S.A. 110 (29), 12144–12148.

Whiting, M.J., Williams, V.L., Hibbitts, T.J., 2013. Animals traded for traditional medicine at the Faraday market in South Africa: species diversity and conservation implications. In: Alves, N., Romeu, R., Lucena, I. (Eds.), Animals in Traditional Folk Medicine. Springer-Verlag Berlin Heidelberg, Berlin, pp. 421–473.

SECTION FOUR

The Future

CHAPTER

39

Taking pangolin conservation to scale

Daniel W.S. Challender[1,2], *Helen C. Nash*[2,3], *Carly Waterman*[2,4] *and Rachel Hoffmann*[5]

[1]Department of Zoology and Oxford Martin School, University of Oxford, Oxford, United Kingdom [2]IUCN SSC Pangolin Specialist Group, ℅ Zoological Society of London, Regent's Park, London, United Kingdom [3]Department of Biological Sciences, National University of Singapore, Singapore [4]Conservation and Policy, Zoological Society of London, Regent's Park, London, United Kingdom [5]IUCN Species Survival Commission, IUCN, Cambridge, United Kingdom

OUTLINE

Introduction

Pangolins are extraordinary species. They are the world's only truly scaly mammals and are evolutionarily distinct, having evolved unique morphological and ecological adaptations over tens of millions of years (Chapter 1). However, they are one of the few mammalian orders where every member species is threatened with extinction (Chapters 4–11). Despite this, they have received scant research and conservation attention historically and remain poorly known. Encouragingly, this has begun to change, following the "rise to fame" of pangolins since the early 2010s (Chapter 21; Harrington et al., 2018). Within a decade, they have risen from virtual obscurity to become high profile ambassadors, alongside more traditionally iconic species (e.g., African elephant [*Loxodonta africana*]), for global campaigns to combat the illegal wildlife trade. This has resulted in a huge leap in recognition of the species by the general public in various parts of the world, and the plight of pangolins has captivated many people who a decade ago would not have known that they exist.

Similarly, within the international conservation community there is growing recognition

Pangolins
DOI: https://doi.org/10.1016/B978-0-12-815507-3.00039-3

of the threats pangolins face and the urgency needed to address them. This is largely thanks to the efforts of a handful of researchers, conservation practitioners and activists who have painstakingly studied, advocated for, and championed pangolins over the past few decades (Chapter 21). The eight species of pangolin are now receiving more conservation attention from governments, non-governmental organizations (NGOs) and practitioners than ever before.

This volume presents the most up to date knowledge of pangolins, their evolution and systematics, natural history, and threats, and discusses a wide range of approaches and interventions as conservation solutions for the species. This chapter considers what the next 20 years may look like for pangolins and reflects on how the international conservation community, through adopting a collaborative and multidisciplinary approach, can take pangolin conservation to scale. There is still time to secure representative populations of each species across a diversity of sites in Asia and Africa and, now that pangolins have the world's attention, the time is ripe to ensure that they receive sustained, strategic conservation focus and support.

Foundations for success

There is already a solid foundation on which to build long-term, sustainable conservation programs for pangolins. Within the global pangolin conservation community the knowledge and expertise on the species, their threats, and what needs to be done to address them, has accumulated in recent decades (e.g., in China, Vietnam and Zimbabwe). This community continues to grow as pangolins attract a new wave of scientists and supporters dedicated to their conservation. As new projects develop, it is imperative that efficient mechanisms are in place for communication, collaboration and knowledge-sharing, to increase the efficiency and effectiveness of conservation actions in the current climate of limited resources and funding.

The Pangolin Specialist Group (PSG), as part of the IUCN Species Survival Commission (SSC), can play a major role in making this a reality. The IUCN SSC is a network of approximately 9000 individuals from around the world constituted into almost 150 Specialist Groups, which are dedicated to the safeguarding and protection of global biodiversity. At the beginning of 2012, pangolins were the only group of mammals not represented by a Specialist Group in the SSC, despite previous incarnations of the group existing between 1992 and 1997 (chaired by Dr. Kevin Lazarus) and 1998–2004 (chaired by Dr. Jung-Tai Chao). In recognition of this gap, the PSG was reformed (see Chapter 21; Challender et al., 2012).

There are a number of benefits for species conservation of having a dedicated SSC Specialist Group, which align closely with taking pangolin conservation to scale. They include the ability to convene stakeholders and use IUCN's neutral position as an intergovernmental organization (IGO) to facilitate dialogue (e.g., amongst scientists, NGOs, conservation practitioners, local communities, civil society and businesses), and leverage political will with governments to help achieve successful conservation outcomes. They also include the ability to initiate and/or contribute to conservation planning for pangolins, to assist in the prioritization and facilitation of focused conservation efforts through stakeholder engagement (see Chapter 33) and to identify geographic areas (e.g., sites) of importance to pangolins. Furthermore, IUCN and its SSC groups are purveyors of independent and credible scientific advice and information. For example, the PSG inputs technical expertise into international policy such as CITES (the Convention on International Trade in Endangered Species of Wild Fauna and Flora), and conducts global assessments of pangolin conservation status and extinction risk for the IUCN Red List. The SSC groups also have the ability to develop

IUCN guidelines, standards, and position statements around issues directly relating to pangolin conservation (e.g., on the role of ex situ conservation). Finally, having pangolins represented in the IUCN SSC provides many opportunities for interdisciplinary collaboration, for example, on thematic issues including sustainable use and livelihoods, conservation planning, wildlife health, and climate change, in order to develop innovative solutions to conservation challenges, and cohesive approaches to research, sharing of knowledge, and increased surveillance of conservation needs and priorities.

This should not imply that all the benefits of a Specialist Group are conferred at an international level. Nationally the PSG is working with government and NGO partners to affect positive outcomes for pangolins within individual range states. As an example, it has collaborated with partners in the Philippines to develop a national conservation strategy to guide investment in priority actions for pangolins in the country.

Conservation priorities for pangolins are also being pursued at a national level, through the establishment of national working groups comprising key stakeholders for pangolin conservation. These groups are proving effective at enhancing collaboration between pangolin researchers and conservation practitioners, and engaging with policy makers and industry, to further pangolin conservation. While not coordinated through IUCN, there is significant overlap in membership between national working groups and the PSG, with active collaborations between the two taking place in Singapore (Chapter 26) and Taiwan (Chapter 36) to develop conservation strategies.

Challenges

While this all offers promise, pangolins are facing ever increasing threats and available evidence suggests that populations are in decline. Of serious concern are the challenges facing pangolin conservation in the forthcoming decades. Pangolins and their derivatives have high financial value in range states and regions of the world where demand exists, and awareness of the species appears to be growing, which presents a key challenge for controlling exploitation levels, especially in rural and austere environments (Chapter 18). This is compounded by those hunting, poaching and trafficking pangolins often perceiving it to be low risk and high reward (Chapter 17); and intercontinental trafficking of African pangolins to Asia, virtually exclusively involving scales, emerged between 2008 and 2019, and appears to be placing an additive pressure on populations (Chapters 8–11, 16). Moreover, international trafficking networks are using increasingly sophisticated smuggling methods and modifying routes to avoid detection (Chapter 16; Heinrich et al., 2017).

A number of key macro-economic factors compound this predicament further. High human population growth rates are forecast in a number of pangolin range states (e.g., Democratic Republic of the Congo [DRC], India, and Nigeria; United Nations, 2019) which will likely increase pressure on natural resources in these countries, including pangolins and their habitats. Global levels of trade are expected to increase (Lloyd's Register et al., 2013), and there are high levels of East Asian investment in Africa (e.g., Zhang et al., 2015), two factors that may well facilitate the trafficking of African pangolins and their scales to Asian markets. The cultural revival of Traditional Chinese Medicine (TCM) and the emergence of China's Belt and Road Initiative (BRI) could also have major implications for pangolins and the sustainability of harvest, use, and trade (see Ascensão et al., 2018).

There is also little evidence that efforts to reduce consumer demand, a recognized conservation priority for pangolins, are efficacious (Veríssimo and Wan, 2019). Yet, even if they are, the question remains whether consumer behavior can be changed on a large enough scale, and within the required timeframe, to contribute to measurably reducing pangolin offtake and trafficking to levels that no longer threaten the species.

Opportunities

Despite these challenges, there is cause for optimism. There is greater knowledge of pangolins and their threats with which to inform short- and long-term conservation interventions and management (Chapters 4–11). More stakeholders comprising governments, NGOs, practitioners, and academics, among others, are investing in the conservation of pangolins locally, nationally, and internationally. The pangolin conservation community is also growing and conservation strategies are starting to be developed (Chapters 33 and 36), with which to guide strategic conservation action and investment. At the country level, national working groups are being formed and providing a key link between national and international conservation efforts (Chapter 26). At the site level, there is better protected area management being brought about through investment, capacity-building and the introduction of tools such as SMART (Chapter 27), and many conservation projects are being implemented in areas where pangolins occur. This offers opportunities for synergy by integrating pangolin conservation priorities into such projects, including developing partnerships with local communities and indigenous peoples (Chapters 23–27). There is also scope to use citizen science to better inform our understanding of the species and contribute to pangolin conservation (e.g., within protected areas; Chapter 38). Donors are also showing greater interest in pangolins, though the challenge will be maintaining this into the future, and overcoming the rising competition for limited conservation funding. As such, innovative approaches will likely be needed to ensure there is adequate support for conservation action (Chapter 37).

A bright future for pangolins?

Taking pangolin conservation to scale means that sufficient knowledge is generated on each species, and population monitoring methods are developed, and they are being used to determine that, at a defined sample of sites representing the species and their habitats, populations are stable or increasing because the threats have been mitigated. Additionally, research on pangolins is ongoing and the species remain high on the conservation agenda, and regional and national conservation strategies are being used to guide conservation. This can be achieved if there is effective site-based protection, effective community engagement, and partnerships are formed at key sites for each species; front-line enforcement agents and judiciaries must also have greater awareness of wildlife crime involving pangolins and take it more seriously, which serves as a deterrent to poaching and trafficking. Consumer demand in key markets must no longer pose a threat to the species. To achieve such a feat, and overcome recognized challenges, will require significant amounts of political will and courage, and concerted and collaborative efforts from the global conservation and scientific communities, including governments and their delegated agencies, academics, and NGOs, and critically, local communities and indigenous peoples all working together at the local, national and international level. There is a strong foundation on which to build to ensure this vision becomes a reality and herein lies the challenge for the next 20 years and beyond.

References

Ascensão, F., Fahrig, L., Clevenger, A.P., Corlett, R.T., Jaeger, J.A.G., Laurance, W.F., et al., 2018. Environmental challenges for the belt and road initiative. Nat. Sustain. 1, 206–209.

Challender, D.W.S., Baillie, J.E.M., Waterman, C., IUCN SSC Pangolin Specialist Group, 2012. Catalysing conservation action and raising the profile of pangolins – the IUCN SSC Pangolin Specialist Group (PangolinSG). Asian J. Conserv. Biol. 1 (2), 140–141.

Harrington, L.A., D'Cruze, N., Macdonald, D.W., 2018. Rise to fame: events, media activity and public interest in pangolins and pangolin trade, 2005-2016. Nat. Conserv. 30, 107–133.

Heinrich, S., Wittman, T.A., Ross, J.V., Shepherd, C.R., Challender, D.W.S., Cassey, P., 2017. The Global Trafficking of Pangolins: A Comprehensive Summary of Seizures and Trafficking Routes From 2010–2015. TRAFFIC, Southeast Asia Regional Office, Petaling Jaya, Selangor, Malaysia.

Lloyd's Register, QinetiQ, University of Strathclyde, 2013. Global Marine Trends 2030. Available from: <https://www.lr.org/en-gb/insights/global-marine-trends-2030/>. [September 27 2019].

United Nations, 2019. World Population Prospects 2019. Highlights. United Nations Department of Economic and Social Affairs. United Nations, Geneva.

Veríssimo, D., Wan, A.K.Y., 2019. Characterizing efforts to reduce consumer demand for wildlife products. Conserv. Biol. 33 (3), 623–633.

Zhang, S., Blanchard, B., Rajagopalan, M., 2015. China just made a $2 billion move in an oil-rich west African nation. Available from: <https://www.businessinsider.com/r-china-agrees-2-billion-infrastructure-deal-with-equatorial-guinea-2015-4?r = UK>. [September 27 2019].

Index

Note: Page numbers followed by "*f*," "*t*," and "*b*" refer to figures, tables, and boxes, respectively.

D

T

X

Y

Z

Printed in the United States
By Bookmasters